# FUNDAMENTALS OF PHYSICS

## WITH SI UNITS

### M. NELKON
M.Sc., F.Inst.P., A.K.C.
*Formerly Head of the Science Department,*
*William Ellis School, London*

## HART-DAVIS EDUCATIONAL

Granada Publishing Limited
Hart-Davis Educational Limited
First published 1967 by Chatto & Windus Educational Limited
Reprinted 1969 – SI edition
Reprinted 1970
Second Edition 1971
Reprinted 1972, 1974, 1975
Third edition published 1977 by Hart-Davis Educational Limited
Frogmore, St. Albans, Hertfordshire
Copyright © 1967, 1971 and 1977
by Michael Nelkon

ISBN 0 247 12721 3

Printed in Great Britain by
Fletcher & Son Ltd, Norwich

# PREFACE TO THIRD EDITION

In this edition the opportunity has been taken to update parts of the text. In particular, the 'newton' has been used extensively as the unit of force in mechanics and fluids, in accordance with recent O-level examinations.

# PREFACE TO SECOND EDITION

In this edition the text has been further modernized in accordance with the revised syllabus of the examining boards. SI units and nomenclature have now been extended in mechanics and heat, and multiple choice questions in all branches of the subject have been added. A new section, Luminous Energy, has been added. The author is grateful to M. V. Detheridge, William Ellis School, London, and Dr. R. P. T. Hills, formerly St John's College, Cambridge, for assistance with this edition. He is also indebted to many teachers for helpful comments, and particularly to A. R. Long, the Henry Box School, Witney, Oxford.

# PREFACE TO FIRST EDITION

THIS book is intended for G.C.E. Ordinary level students in the final two years of their course at independent, grammar and comprehensive schools. It covers the syllabus of the General Certificate of Education in Physics to this standard, which has been modernized recently by the Examining Boards. It is hoped that parts of the book may also be helpful for the Certificate of Secondary Education.

In writing the book, two general aspects have been kept in mind. Firstly, the new approach to the subject sponsored by the Association of Science Education and the Nuffield Foundation – here all teachers gratefully acknowledge the work done by Professor Eric Rogers, Mr John Lewis of Malvern College and their talented collaborators – and secondly, the many sound features of the well-tried and traditional teaching of the subject. I have tried to combine the best of both techniques of teaching. Thus in addition to demonstrations and experiments, I have also aimed at developing an understanding of the subject by logic and discipline and by using the necessary formulae and numerical calculations. To what extent one or the other is adopted for a particular group of students must be left to the judgement of the individual teacher. Each is complementary to the other, and both play an important part in the full scientific education of the pupil.

The book begins with Matter and Molecules. An early introduction to the microscopic view of matter is invaluable for a deep understanding of solids, liquids and gases and their properties. Similarly in Electricity, which plays a key role in modern physics, the carriers of current – electrons, ions and holes – have been introduced early and the part they play in different phenomena has been emphasized. Waves and wave-effects occur in many natural phenomena, and this has been given prominence in a separate chapter, where water waves, sound and light waves and microwaves have all been compared and contrasted.

After Matter and Molecules the book deals with Dynamics, as motion and forces are vital topics in physics. Vectors and scalars, linear momentum, circular motion and the conservation of energy have been discussed. More modern apparatus, such as the ticker-timer for linear motion and the centisecond clock in free fall, have been introduced. Many numerical examples have been given in this section to help the change to the metric system and the Système International, SI, units, which are sponsored internationally. Statics, Pressure in Fluids and Forces due to Fluids, with an introduction to Bernoulli's principle, follow Dynamics. In Heat many measurements have been treated concisely or omitted in accordance with changes in the syllabus, and as the joule is now the unit of heat, the 'mechanical equivalent of heat' has been recast as a thermal equivalent of mechanical energy.

The ripple tank and microwave apparatus, in addition to the ray-box, have been introduced in Optics, so that a wider appreciation of reflection and refraction can be gained. Numerical examples on mirrors and lenses have been given in both Real is Positive and New Cartesian sign conventions, for the convenience of the student who uses one or the other. Sound contains a further treatment of waves and simple harmonic motion applied to the vibrations in pipes and strings.

In Electricity, prominence has been given to the link between potential difference and energy, the use of ferrites, the part played by magnetic fields in the motion of current-carrying conductors in the fields, and the effects on solids and gaseous conductors situated in magnetic fields.

The section on Electrostatics, which discusses the movement of charges in electric fields, precedes the final part of the book on Atomic Physics. The latter contains an introduction to the emission of electrons by cold and hot cathodes and in the photoelectric effect, together with basic applications; a discussion of the principles of Semiconductors and Transistors; and Radioactivity, together with an introduction to basic principles of atomic structure and nuclear energy.

In the presentation, I have tried where possible to begin with an experiment and then followed it with a discussion of the results. The book therefore contains a course of basic experiments, and as an aid to the teacher these have been listed at the front together with a list of manufacturers from whom apparatus may be obtained. Many worked examples have been given in the text in illustration of the subject-matter, and there are numerous exercises to test comprehension. It is hoped that the book will provide a useful introduction to the principles of the subject.

I am very much indebted to Dr David Abbott for his most valuable co-operation in compiling the work; to S. S. Alexander, head of the science department, Woodhouse Grammar School, London, for his advice on many points and his generous assistance at all stages; and to W. G. Sale, head of the science department, Kynaston School, London, for his help on aspects of the new syllabus. I am also grateful to L. J. Beckett, senior physics master, William Ellis School, London, and to his colleagues, C. A. Boyle and C. J. Mackie, for reading parts of the proofs and checking answers; to my former pupil, R. P. T. Hills, St John's College, Cambridge, for his valuable assistance at all stages of preparing the book; to I. D. Taylor, Bedford School, and to G. Severn, for their excellent photographs; to T. E. Walton and G. Pumphrey, for their material help; and to G. Hartfield, for his excellent illustrations of the subject-matter. Finally, it is a pleasure to record with appreciation the very valuable advice received from F. C. Brown, head of the physical sciences department, Institute of Education, London University, and the technical assistance at all stages of the book from G. W. Trevelyan.

M. N.

# ACKNOWLEDGEMENTS

THE author and publishers are very grateful for permission to reprint questions set in past examinations by the following Examining Boards: London University School Examinations Council (*L.*); Northern Universities Joint Matriculation Board (*N.*); Oxford and Cambridge Joint Board (*O.* and *C.*) Cambridge Local Examinations Syndicate (*C.*); Oxford Local Examinations Syndicate (*O.*).

The author is indebted to the examining boards listed for their kind permission to translate numerical values in past questions into SI units; the translation is the sole responsibility of the author.

They are also indebted to the following manufacturing firms and organizations, whose names are given alphabetically for convenience, for supplying photographs to illustrate the work:

*Physical Apparatus.* Griffin & George Ltd, 8A, B, C, D, E, 21D, 22A, B; Morris Laboratory Instruments Ltd, 21A, 22C; W. B. Nicholson (Glasgow) Ltd, 21C, 24C; Panax Equipment Ltd, 24A; Rainbow Radio (Blackburn) Ltd, 22D; Scientific Teaching Apparatus Ltd, 1A, B, 15A, 24B; Teltron Ltd, 23A, B, C, D; Venner Ltd, 21B; White Electrical Co. Ltd, 14A, B, C.

*Applications.* Associated Press Ltd, 5A; Barr and Stroud Ltd, 9A; Bristol University, 19B; John Compton Organ Co. Ltd, 12A; Connoisseur Films, 16C; Dunlop Ltd, 3A, B, C, D; Electrical and Musical Industries Ltd, 11B; English Electric Ltd, 16B; Grubb Parsons Ltd, 10B; Kodak Ltd, 7A, B; Mullard Ltd, 15B, C, D, 17A, B, C, D; 3M Company, 11A; National Physical Laboratory, 5B, C; W. B. Nicolson Ltd, 18A; Parkes Observatory, New South Wales, 10A; Plessey Ltd, 11C, 12B, 13A, B; Science Museum (P. M. S. Blackett), 18C; G. Severn, Esq, 4A, B; I. D. Taylor, Esq, 2A, B, C, D; U.K. Atomic Energy Authority, 9B, 13C, 19A, 20A; Unilever Ltd, 1C; U.S. Information Services, 4C, 6A, 7C, 16A; University of California, 18B; Vauxhall Ltd, 20B.

# CONTENTS

Scientific method. Units. Measurements. Order of Accuracy. Conservation of Energy.

## MOLECULAR THEORY

Solids. Density measurement. Specific gravity. Liquids. Density, relative density, insoluble particles. Gases, Density of air. Molecules in solids, liquids, gases. Size of molecule – oil-film, gold-leaf. Osmosis. Molecules in motion–Brownian motion in gases. Molecular velocity. Molecules in liquids – Brownian motion, effect of temperature. Molecules in solids – motion, forces. Elasticity–Hooke's law, elastic limit, yield point. Friction – static and kinetic (sliding). Rolling friction. Ball-bearings. Surface tension. Detergents. Capillary rise. Cohesion, adhesion. Angle of contact. Viscosity. Lubrication. Streamline, turbulent flow. Forces in gases. Bonding in solids. Forces between molecules. Evaporation. Boiling. Melting. Solidification. Summary.

## MECHANICS

Motion. Ticker-tape. Speed. Velocity. Displacement–time graphs. Velocity–time graphs. Acceleration. Retardation. Formulae. Acceleration due to gravity, $g$. Direct measurement of $g$ by centisecond clock. Pendulum method. Formula for distance with uniform acceleration. Equations of linear motion. Scalars and vectors. Addition and subtraction. Horizontal and vertical components of $g$. Summary.

Force and acceleration. Types of forces. Opposing forces of gravity and friction. Inertial mass. Investigations with ticker-tape. Force due to gravity. Weight. Units. Earth and moon. Practical units. Force on astronaut. Friction and motion. Force and uniform velocity. Terminal velocity. Circular motion and acceleration. Force and Momentum. Momentum changes. Gas pressure. Conservation of linear momentum.

# PLATES

# MECHANICS

| QUANTITY | DEFINITION | FORMULA | UNITS |
|---|---|---|---|
| Velocity ($v$) | $\dfrac{\text{displacement}}{\text{time}}$ | $\dfrac{s}{t}$ | metre per second (m/s) |
| Acceleration ($a$) | $\dfrac{\text{velocity change}}{\text{time}}$ | $\dfrac{v-u}{t}$ | m/s$^2$ $g = 9 \cdot 8$ m/s$^2$ or 10 m/s$^2$ approx. |
| Force ($F$) | mass $\times$ acceleration | $ma$ | newton (N) 1 kgf = 10 N approx. |
| Momentum | mass $\times$ velocity | $mu$ | newton second |
| Work ($W$) | force $\times$ distance | $F \times s$ | joule (J) 1 newton $\times$ 1 metre = 1 J |
| Kinetic energy | energy due to motion | $\frac{1}{2}mu^2$ | joule (J) |
| Potential energy | energy due to level or position | $mgh$ | joule (J) |
| Power | energy (work) per second | $\dfrac{work}{time}$ | Watt (W) 1 h.p. = $\frac{3}{4}$ kW approx. |
| Resolved component of force | force $\times$ cosine of angle concerned | $F \cos \theta$ | (same as force) |
| Moment of force | force $\times$ perpendicular distance | — | newton metre (N m) |
| Mechanical Advantage (M.A.) | $\dfrac{\text{Load}}{\text{Effort}}$ | $\dfrac{W}{P}$ | — |
| Velocity Ratio (V.R.) | $\dfrac{\text{distance/sec. moved by effort}}{\text{distance/sec. moved by load}}$ | — | — |
| Efficiency | $\dfrac{\text{work (energy) obtained}}{\text{work (energy) supplied}} \times 100\%$ | $\dfrac{\text{M.A.}}{\text{V.R.}} \times 100\%$ | — |
| Density ($\rho$) | $\dfrac{\text{mass}}{\text{volume}}$ | $\dfrac{M}{V}$ | g/cm$^3$ or kg/m$^3$ |
| Relative density | $\dfrac{\text{mass of substance}}{\text{mass of equal vol. of water}}$ | — | — |
| Pressure ($p$) | force per unit area | $h\rho g$ | newton per metre$^2$ (N/m$^2$) |

Summaries of Optics, Waves, Sound, Electricity and Atomic Physics are on pages 783 and 784.

# HEAT

| QUANTITY | DEFINITION | FORMULA | UNITS |
|---|---|---|---|
| Joule (J) | work done when 1 newton moves through 1 metre | — | 4·2 J = 1 cal (approx.) |
| Heat capacity (C) | heat to raise temperature of substance by 1°C | $mc$ | J/K or J/°C* |
| Specific heat capacity (c) | heat to raise temperature of unit mass of substance by 1°C | $C/m$ | J/kg K or J/g K (for water, $c = 4200$ J/kg K or 4·2 J/g K) |
| Specific latent heat (l) | heat to change unit mass from solid to liquid, or from liquid to vapour, without temperature change | — | J/kg or J/g |
| Linear expansivity (α) | increase in length per unit length per °C temperature rise | $\alpha = \dfrac{l_2 - l_1}{l_1 \times t}$ | per K or per °C |
| Cubic expansivity (γ) | increase in volume per unit volume per °C temperature rise | $\gamma = \dfrac{V_2 - V_1}{V_1 \times t}$ | per K or per °C |
| Volume coefficient of gas | increase in volume per unit volume at 0°C per °C temperature rise, pressure being constant | $\gamma = \dfrac{V_1 - V_0}{V_0 \times t}$ | per K or per °C ($\gamma = 1/273$ approx. for most gases) |
| Pressure coefficient of gas | increase in pressure per unit pressure at 0°C per °C temperature rise, volume being constant | $\gamma = \dfrac{p_t - p_0}{p_0 \times t}$ | per K or per °C ($\gamma = 1/273$ approx. for most gases) |
| Absolute temperature (T) | temperature measured from absolute zero | $T = 273 + t$ (°C) | K (0 K = absolute zero 273 K = 0°C) |

*1 K = 1°C

## Note to the Student

1. *Summaries.* A list of the main quantities in all branches of the subject, with definitions, formulae, and units, will be found in the front and back of the book.
2. *Multiple Choice Questions.* Papers similar in style to those used in G.C.E. examinations are provided at the end of the book.
3. *SI Units.* This uses the kilogramme (kg) as the unit of mass, the metre (m) as the unit of length, and the second (s) as the unit of time. For Ordinary level, the gramme (g) and centimetre (cm) will also be used where these are considered more convenient units.

# I

# INTRODUCTION

## SCIENTIFIC METHOD · UNITS · MEASUREMENTS · ENERGY

Let the imagination go, guiding it by judgement and principle,
but holding it in and directing it by experiment. FARADAY.

## SCIENTIFIC METHOD

Man has always tried to find some rational explanation of the phenomena of nature in such different aspects as physics, chemistry, biology and astronomy. He therefore carries out experiments and observes what happens. The results of an experiment suggest a theory; the development of the theory suggests other experiments; and so theory and practice help to fertilize each other, and our understanding of nature increases. This combination of hand and head, of practical experiment and theoretical reasoning, is still perhaps the outstanding characteristic of the scientific method.

An essential feature of scientific method is a willingness to abandon previously held theories in the light of newly discovered facts. Very often, in the past, there has been a tendency to believe something merely because an eminent person stated it: and this is the very reverse of the scientific spirit. A good example is to be found in the physics of falling bodies. The famous Greek philosopher and scientist ARISTOTLE had stated, more than three centuries before Christ, that heavy objects always fell to the ground faster than light ones. Aristotle was held in such reverence that no one thought to question this belief for nearly two thousand years. In the early part of the 17th century, however, someone performed a simple experiment to test the theory. He dropped a heavy and a light object simultaneously from a high building and observed that they reached the ground almost together. The theory was thus in conflict with the observed fact; and it was the theory which had to be altered, for facts cannot be altered.

Great advances in science have been made by many famous men. Among them were Sir Isaac Newton, who in the 17th century first stated the main laws of mechanics in a form which has lasted until our own time; Michael Faraday, who in the early 19th century discovered how to produce electricity using magnets; and Lord Rutherford, who during the present century made clear the basic structure of the atom. They were all great seekers after truth. They did not know what material benefits might follow from their discoveries (Faraday can certainly have had no conception of the vast electrical industry of today), but they pushed ahead with their fundamental work.

K. T. Compton, an American scientist, once said: 'When someone is ill, the doctor is called by telephone, visits his patient by automobile, measures his temperature with a thermometer and his pulse with a watch, examines his heart and lungs with a stethoscope and his throat with a light reflector. Every one of these operations uses a tool and technique supplied by a physicist.' This illustrates very clearly how we depend today on scientific invention. Science has not produced merely a few gadgets and mechanical devices. Science has changed, and continues to change, our way of life, bringing many benefits to mankind.

## UNITS OF MEASUREMENT

### Metric System

As science has developed through the centuries, the importance of accurate measurement has become increasingly clear. In order that scientists working in different parts of the world may agree in and compare their measurements, it is necessary for certain basic units of measurement to be generally accepted and defined. The most important of these internationally accepted units are as follows:

The *metre* (*m*) was for many years the distance between two fixed marks on a particular platinum–iridium rod kept under specified conditions in a vault in Paris. It is now defined as the length of a certain number of wavelengths in a vacuum of a particular orange radiation of the krypton-86 atom. Unlike the distance between the marks on the rod, the wavelengths are due to atomic vibrations and remain constant. Thus

$$1 \text{ m} = 1\ 650\ 763 \cdot 73 \text{ wavelengths of the above radiation.}$$

The *kilogramme* (kg) is the mass of a particular solid cylinder made of platinum–iridium alloy kept in Paris, known as the International Prototype Kilogramme.

The *mean solar day* is the average period, taken over twelve months, between successive transits of the sun across the meridian at any part of the earth's surface.

In practice, the following smaller units may also be used:

The *millimetre* (*mm*), which is $\frac{1}{1000}$ part of a metre.
The *centimetre* (*cm*), which is $\frac{1}{100}$ part of a metre.
The *gramme* (*g*), which is $\frac{1}{1000}$ part of a kilogramme.

The *second* was formerly 1/86 400th part of a mean solar day. This unit, used by astronomers, has now been replaced by a unit based on the period of vibration of a particular caesium atom. Dr L. Essen of the National Physical Laboratory designed a caesium clock which measured the period of rotation of the earth to an exceptionally high order of accuracy and this showed clearly the irregularity in the rate of rotation of the earth. Caesium or atomic clocks now used as standard clocks in this country and America. The second is defined as the time for 9 192 631 770 cycles of vibration of a particular radiation from the caesium-133 atom.

The metre-kilogramme-second system of units is commonly abbreviated to m.k.s. units and is used not only by scientists but also by most peoples of the world.

## Volume

Measurements of volume were originally based on the *litre*; this is defined as the volume occupied by a mass of 1 kg of pure water at its temperature of maximum density and under standard atmospheric pressure. The *millilitre (ml)* is $\frac{1}{1000}$ of a litre. Volumes are also based on the cube whose side is 1 cm, or cubic centimetre (c.c.). This is written as 'cm³' to emphasize the cm unit. Thus the density (mass/volume) of mercury, for example, may be given as '13·6 g/cm³'. The cm³ is very slightly larger than the ml.

In the SI system (see below), the cubic metre (m³) is the unit of volume. Water has a density of 1000 kg/m³ in SI units (equivalent to 1 g/cm³); mercury has a density of 13 600 kg/m³ in SI units (equivalent to 13·6 g/cm³).

## British System

In Great Britain, however, different units of length and mass were used:

A *foot* was one-third of the distance (a yard) between two fixed marks on a certain bronze rod kept under specific conditions in London.

A *pound* was the mass of a certain piece of platinum kept in London. In 1963, however, the Weights and Measures Act defined the yard in terms of the metre–1 yard = 0·9144 metre–and the pound in terms of the kilogramme–1 pound = 0·453 592 37 kilogramme.

The foot-pound-second system of units (abbreviated to f.p.s. units) has the disadvantage of inconvenience in calculation, as the following contrast shows:

| | |
|---|---|
| 100 cm = 1 metre | 36 in = 1 yard |
| 1000 m = 1 kilometre | 1760 yd = 1 mile |
| 1000 g = 1 kilogramme | 16 oz = 1 pound |
| 1000 kg = 1 tonne | 2240 lb = 1 ton |

## Système International (S I) Units

A new system of units, known as the Système International (SI) units, has been adopted for all branches of physics. It is based on the metre as the unit of length, the kilogramme as the unit of mass, the second as the unit of time, the ampere as the unit of electric current, degrees Kelvin as units of temperature and the candela as the unit of luminous intensity (p. 446). The unit of force in this system is the newton and the unit of energy is the joule (p. 116).

It may be noted that heat energy is now expressed in *joules* in scientific work, not in calories. A conversion value from calories to joules is defined as follows: The 15°C calorie, defined as the heat required to raise the temperature of 1 gramme of water by 1 deg C at 15°C, is the specific heat capacity of water at 15°C and equal to 4·1868 joules. Similarly, heat energy per second is expressed in joules per second, or *watts* (p. 126).

The interested reader is referred to the publications *Changing to the Metric System* (National Physical Laboratory) and *Système Internationale Units*, obtainable from the Stationery Office, London.

## Gravitational Units

The gravitational system of units, used by engineers, is a system based on the action of gravity. The units of length and time are the same as in the previous systems, namely, the centimetre or metre and second, but a unit of gravitational *force* is chosen as the third unit in place of a unit of mass.

In the gravitational m.k.s. system the unit of force is that due to gravity acting on a mass of 1 kilogramme when the acceleration of gravity is taken as 9·80665 m/s² (see p. 79). This unit is given the symbol of *kgf*. The letter 'f' is used in place of the 'wt' in older notation, so that 1 kilogramme force (1 kgf) = 1 kilogramme weight (1 kg wt) for this particular value of gravity. At other places 1 kg wt differs very slightly from 1 kilogramme force, and we therefore assume 1 kg wt = 1 kgf for practical purposes.

In the SI system, however, the unit of force is the *newton*, symbol N. Approximately (see p. 100),

$$1 \text{ kgf} = 9\cdot8 \text{ N},$$

or, in round figures,

$$1 \text{ kgf} = 10 \text{ N}.$$

## Accurate Measurement of Length · The Vernier

One of the marks of a good scientist is the care he takes to measure things really accurately. Even such a simple operation as the measurement of the length of a straight bar requires careful attention.

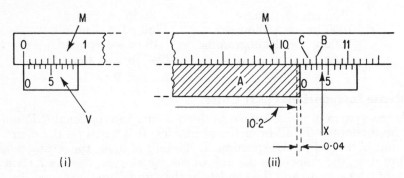

Fig. 1.1 Principle of Vernier

When it is necessary to measure small distances accurately a special type of scale constructed by PIERRE VERNIER is often used. In Fig. 1.1 (i), suppose that the main scale M is graduated in centimetres and millimetres. The vernier scale V is 0·9 cm long and is graduated in ten equal parts. Thus the vernier graduations are 0·09 cm apart, which is 0·01 cm less than the distance between graduations on the main scale.

Now suppose that, in order to measure the length of an object A, the vernier V is moved from the position shown in Fig. 1.1 (i) to that shown in Fig. 1.1 (ii). The length of A clearly lies between 10·2 and 10·3 cm, but to obtain the second decimal place it is necessary to examine the vernier scale and note which division on it coincides with a division on the main scale.

In this case, the coincident vernier division is the fourth, marked X. The third division on the vernier is then 0·01 cm to the right of the gradua-

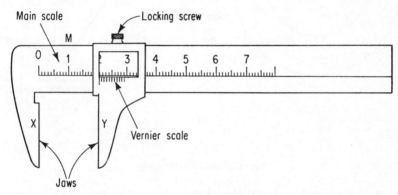

Fig. 1.2 Vernier Calipers

tion B on the main scale, the second division is 0·02 cm to the right of the graduation C and the zero of the vernier is 0·04 cm to the right of the 0·2cm graduation. Hence the length of A is 10·24 cm.

The argument would be precisely the same wherever on the two scales the coincidence had been: the second decimal place in the measurement is given by the number of the vernier scale marking which coincides with a main scale marking.

The same principle is applied in the vernier slide calipers (Fig. 1.2). The object whose length is required is placed between a fixed jaw X and a movable jaw Y. For measuring angles accurately, circular vernier scales are made, graduated in dégrees or half-degrees.

## Micrometer Screw-gauge

The micrometer screw gauge is an instrument for measuring accurately the diameters of wires or thin rods. It utilizes an accurate screw of known pitch such as 0·5 mm, shown inset in Fig. 1.3. This means that for one revolution of the screw the spindle X moves forward or back 0·5 mm (see also p. 157). When measuring the diameter of a wire or rod A one side is rested on the end Y and the thimble is rotated until the end of the spindle X comes into contact with A. Further rotation of the thimble does not make the spindle move forward owing to a ratchet, a device for avoiding excessive pressure on the wire or rod. A locking device is sometimes included so that the reading of the diameter can be retained on removal of A.

An engraved linear millimetre scale L shows the forward movements of the spindle. A circular scale C on the thimble enables fractions of milli-

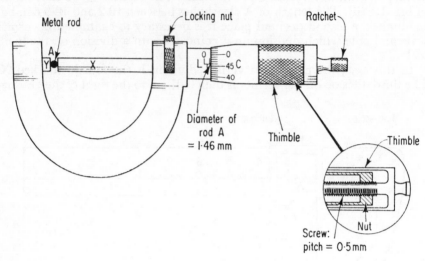

Fig. 1.3 Micrometer gauge

metres to be measured. In the sketch shown in Fig. 1.3, 50 circular divisions = 0·5 mm, or 1 circular division = 0·01 mm. Thus from Fig. 3,

Diameter of rod = 1·0 mm (linear scale L) + 46 × 0·01 mm

(circular scale C)

= 1·46 mm

## The Chemical Balance · Measurement of Mass

The *mass* of an object can be measured by comparing it with 'standard' masses, which are derived from official copies of the 'kilogramme' mass, as stated below. The chemical balance is used to compare masses in this way. Extremely sensitive balances can be designed. The National Physical Laboratory at Teddington in Great Britain has a balance which can measure to one-millionth of a gramme. At this laboratory, which has custody of the official copy of the International Kilogramme, very precise weighings are carried out to assist science and technology.

School balances do not require such high precision. The design of chemical balances has altered so much in recent years that the types used will generally differ from school to school. The main points in the design of a typical general-purpose balance are illustrated in Fig. 1.4.

The beam has knife-edges K of a very hard material (usually agate) built into the centre and the ends. When not in use the balance beam rests on supports A, built into two arms attached to the central pillar B. This pillar is tubular, and a metal rod M, with an agate platform G at its upper end, passes inside the tube. The rod can be raised by turning the handle C and, when raised, the balance swings freely about the central agate knife edge supported on the agate platform. Each scale pan E hangs from a support U which contains a V-shaped groove in contact with the knife edge. When the balance is in the raised position the grooves, and hence the scale pans, are supported on the knife-edges at either end of the beam. When in the lowered position the scale pans rest on the wooden base of the balance, and the supports rest on the balance beam, but not on the agate surfaces. This protects such surfaces from damage.

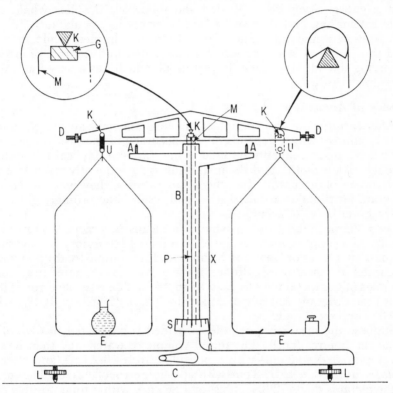

Fig. 1.4 Chemical (common) balance

The balance has a pointer P connected to the centre of the beam, and ends in a tip which traverses a scale S. The wooden base has levelling screws L on its underside which are used to level the balance before use; the level is indicated by a plumbline X and pointer below. Screw-thread extensions D are present on both ends of the beam, and small circular nuts can be screwed along each extension; movement of these nuts enables the balance to be adjusted before use.

The whole balance is kept in a glass-walled case with a sliding front.

Many other devices have been invented to measure accurately not only length and mass but also temperature, time and all the other quantities with which the scientist deals. No matter how delicate or ingenious the device may be, however, it cannot produce accurate measurements without the utmost care on the part of the person using it.

## Errors

There are always sources of errors in experiments. A careful scientist will minimize them in planning his or her experiment. One illustration occurred in the accurate measurement of the charge on an electron. Here one needs to find the speed of fall through air of tiny droplets (p. 699). Early scientists used a cloud of water droplets. Millikan, a famous American scientist, recognised: (i) that the numerous droplets would have different masses, and this would lead to errors in calculations; (ii) that evaporation would take place from the water droplets while they fell, and this would alter their mass. He eliminated both sources of error by observing the fall of a single droplet of oil, which has a very low vapour pressure.

## Order of Accuracy

After carrying out an experiment, the final answer is always given to an *order of accuracy* which depends on the accuracy of the instruments used. Very accurate instruments are used at the National Physical Laboratories throughout the world, and the final result is consequently given to a very high order of accuracy. In school laboratories, however, where less accurate instruments are used, the final result is accordingly given to a much lower order of accuracy.

As a simple illustration of what this means in practice, suppose the density of a lump of metal is found in a school laboratory by measuring its mass on a lever or Butchart balance and its volume by dropping it inside water in a measuring cylinder. If the balance is graduated in grammes, the mass of the metal may be read as 36·2 g, with a possible error of 0·2 g. This is an order of accuracy of 2 parts in 362, or $(2/362) \times 100\%$, which is 0·6% approximately.

Suppose the measuring cylinder provided is graduated in divisions of cubic centimetres (cm³). The initial volume of water may then be read as an estimated 40·2 cm³ with a possible error of 0·2 cm³, and after the metal is dropped in, as 52·8 cm³ with a possible error of 0·2 cm³. The volume of the metal is then (52·8 — 40·2) or 12·6 cm³, with a total possible error of 0·4 cm³. The order of accuracy in measuring the volume is thus 4 in 126 or $(4/126) \times 100\%$, which is 3·2%. The *total* order of accuracy in measuring both mass and volume is hence about 4%.

The density of the metal is calculated by

$$\frac{\text{Mass}}{\text{Volume}} = \frac{36 \cdot 2}{12 \cdot 6} = 2 \cdot 873 \text{ g/cm}^3,$$

to 3 decimal places, by long division or by logarithms. If the final result is given as 2·873 g/cm³, the experimenter claims that the last figure is a '3', not a '2' or a '4'. This is a claim of an accuracy of 1 in 2873 or about $\frac{1}{3000}$, which is 0·03%. For a volume of 12·6 cm³ alone, the observed volume change of water, this would mean an accuracy in reading of about $\frac{1}{3000}$ of 12·6 or about $\frac{1}{300}$ cm³, the volume of a small drop of water. Can you see the fall in level of water in a measuring cylinder caused by one drop? Clearly 2·873 is an exaggerated result. If the result is given to two decimal places as 2·87 this is a claim to an accuracy of 1 in 287 or about 0·4%. The accuracy is still exaggerated.

The result is correctly given as 2·9, an order of accuracy of 1 in 29 or about 3%, near to the estimated order of accuracy. Thus within the estimated order of accuracy of the instruments used, the result for the density of the metal is 2·9 g/cm³.

## Examples of Order of Accuracy

Order of accuracy is an important point to observe in experiments. Details about it must be obtained from more advanced books. The greatest percentage order of accuracy in the individual measurements in an experiment is often a rough guide to the final order of accuracy, but all measurements should be considered. Generally, as we have illustrated, the final result must be given to the correct order of accuracy after the calculation is carried out. Using normal school laboratory instruments, the following results were obtained:

1. *Focal length of lens* (p. 408). Calculation may give $f = 15 \cdot 92$. Result: $f = 15 \cdot 9$ cm (2% error).
2. *Specific latent heat of ice* (p. 286). Calculation may give $l = 314 \cdot 7$. Result: $l = 315$ J/g (5% error).
3. *Electrochemical equivalent of copper* (p. 548). Calculation may give $z = 0 \cdot 000\ 316\ 8$. Result: $z = 0 \cdot 000\ 32$ g/coulomb (3% error).
4. *Resistivity of wire* (p. 504). Calculation may give $\rho = 0 \cdot 000\ 109\ 7$. Result: $\rho = 0 \cdot 000\ 110$ ohm cm (4% error).

## Graphical Methods

Very often, the physicist finds that it is convenient and helpful to examine the results of his experiments with the aid of a graph. For example, suppose that he wishes to find out how the stretching of a spiral spring varies as different weights are suspended from it. He measures the extension for each weight, and enters his results in a table thus:

| Weight (N) | 0·5 | 1·0 | 2·0 | 3·0 | 4·0 | 5·0 | 6·0 |
|---|---|---|---|---|---|---|---|
| Extension (cm) | 0·9 | 1·8 | 3·6 | 5·4 | 7·2 | 9·0 | 10·8 |

He now plots a graph between the weight $W$ and the extension $l$, as in Fig. 1.5. The seven points of the graph lie on a straight line passing through the origin. This shows that, up to weights of 6·0 N, $l$ is directly proportional to $W$; or, in mathematical notation, $l \propto W$. The same result could have been obtained by calculation, but the graphical method is much simpler and quicker and shows the directly proportional relationship immediately.

In other experiments, the results obtained are different. For example, suppose that the physicist wishes to find out how the volume of a gas varies as its pressure is altered, the temperature remaining constant. He measures the volume for each pressure, and enters his results in a table (see over).

In this case, it is clear that, as the pressure $p$ is increased, the volume $V$ is reduced, so he calculates a third line in his table, the inverse or reciprocal of the volume, $1/V$.

| Pressure $p$ | 60 | 30 | 40 | 20 | 50 | 100 |
|---|---|---|---|---|---|---|
| Volume $V$ | 40 | 80 | 60 | 120 | 48 | 24 |
| $\dfrac{1}{V}$ | 0·025 | 0·0125 | 0·0167 | 0·0084 | 0·0208 | 0·0417 |

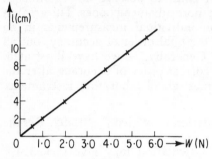

Fig. 1.5 Extension directly proportional to load

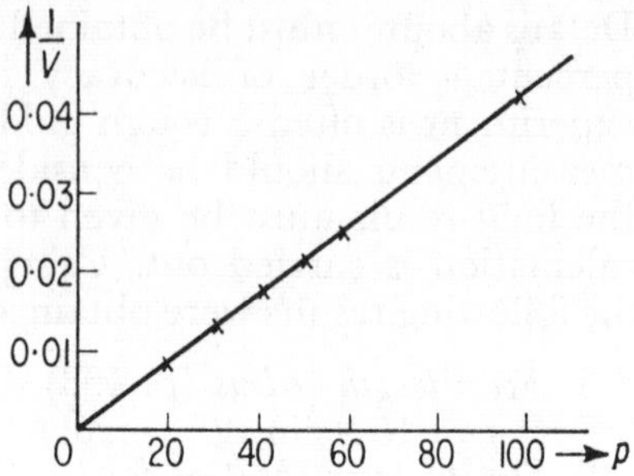

Fig. 1.6 $p$ directly proportional to $1/V$

He now plots a graph between $p$ and $\dfrac{1}{V}$, as in Fig. 1.6. The six points lie on a straight line passing through the origin, showing that

$$p \propto \frac{1}{V},$$

i.e. pressure is inversely proportional to volume.

## The Conservation of Energy

Energy, which may be defined as the capacity for doing work, exists in many different forms. Thus a moving train possesses mechanical energy, and does work in overcoming the frictional resistance to its movements. A fire possesses heat energy, and does work in heating a room or perhaps in converting water to steam. An accumulator possesses chemical energy, and does work in sending an electric current round a circuit. This electrical energy may then do work in ringing a bell or starting up a motor-car engine. Similarly, sound energy does work in setting the air in vibration and eventually causing the ear to hear; while light energy causes the eye to see.

As a result of careful work by scientists over the centuries it has gradually become clear that energy can be transformed from one form to another. Thus when petrol is burned in the engine of a motor-car, chemical energy is converted into heat energy. The heat is used in pushing a piston which is linked to the shafts on the wheels so that the car moves. Thus heat energy is converted into mechanical energy. In electric trains, electrical energy is converted into mechanical energy; dynamos at power stations convert mechanical energy into electrical energy. Similarly, light energy from the sun is changed by the green chlorophyll of plants into chemical energy; television receivers change electrical energy into light energy and

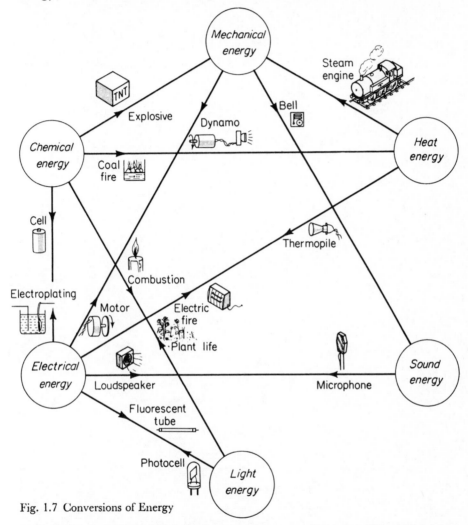

Fig. 1.7 Conversions of Energy

radio receivers change electrical energy into sound energy. Fig. 1.7 shows some transformations of energy.

In all transformations of energy, energy is never destroyed. The total amount of energy remains the same, although it may be changed from one form to another (see also p. 125). The Principle of the Conservation of Energy is one of the most important known to science. In this book we shall study the different forms of energy and the ways in which they have been harnessed to man's service.

# Matter and Molecules

# MATTER AND MOLECULES

## SOLIDS, LIQUIDS, GASES

Water is a common substance on our planet, the Earth. As we travel all over the world it is found in three *states* or *phases*, solid, liquid and gas. As a solid, water is found in the form of ice, which is crystalline, on the peaks of mountains and other very cold places such as polar regions. As a liquid, rain-water is the most common form, and this fills the rivers and oceans. As a gas, water is often called 'water vapour' because it is a liquid under normal conditions, whereas oxygen, for example, is called a 'gas'. Water vapour exists in the air. On cold nights it may separate and be deposited as a liquid in the form of dew and as a solid in the form of hoar-frost, which consists of ice particles.

Generally, all matter, that is, anything which can be weighed, can exist in any of the three states or phases, solid, liquid or gas (vapour).

In this chapter we first consider solids, liquids and gases in bulk and their difference in density. Later we shall discuss the minute particles or *molecules* in the three states of matter and the forces between them, which are closely linked to the properties shown by solids, liquids and gases.

### Solids

Iron, stone, wood and other solids are hard to the touch. Even when they are ground into very tiny particles, such as grains of sand or common salt or iron filings, for example, they are still hard. If tiny specimens of a solid such as wood are examined through a high-powered magnifying glass they all look different in shape. On the other hand, the smallest pieces of copper sulphate crystal have a similar geometrical form. Tiny pieces of washing-soda crystals have similar geometrical form, although this is different from that of copper sulphate crystals. This type of solid is called a 'crystal'. Metals, used in making cars, aeroplanes and many other machines, are crystalline (p. 48), so that researches into the crystal structure of metals are important. Some typical crystals are shown in Fig. 2.1.

### Density

If cubes of equal volumes of different solids, such as wood, aluminium, lead, glass and copper, are lifted, you notice immediately that lead is heavy but aluminium and wood are light. The weight or *mass* of each material in equal volumes is thus different (Fig. 2.2).

*The mass per unit volume* of a substance is called its *density*. Thus a block

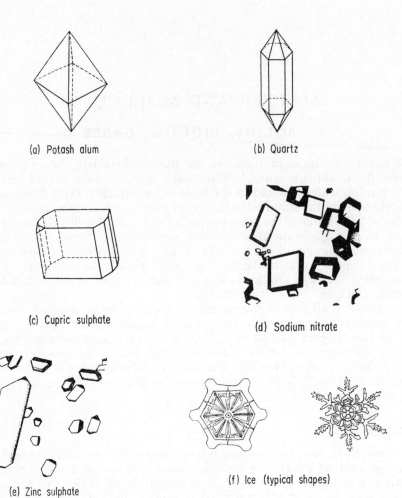

(a) Potash alum

(b) Quartz

(c) Cupric sulphate

(d) Sodium nitrate

(e) Zinc sulphate

(f) Ice (typical shapes)

Fig. 2.1 Crystal shapes

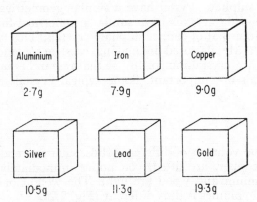

| Aluminium | Iron | Copper |
| 2·7g | 7·9g | 9·0g |

| Silver | Lead | Gold |
| 10·5g | 11·3g | 19·3g |

Fig. 2.2 Masses of 1 cm³ of some metals

of lead of mass 69 g and a volume of 6 cm³ has a density of 69g/6 cm³ or 11·5 g/cm³. Hence:

$$\text{Density} = \frac{\text{Mass}}{\text{Volume}}.$$

The units of density are 'gramme (g)/cm³' or 'kilogramme (kg)/metre³'.

The kilogramme was originally defined as the mass of 1 litre (1/1000 m³) of water at a temperature of 4°C. It should therefore be noted that, for practical purposes,

$$\text{density of water} = 1 \text{ g/cm}^3 = 1000 \text{ kg/m}^3.$$

A mass of water of 50 g thus has a volume of 50 cm³. A mass of water of 2000 kg has a volume of 2 metre.³

*Measurement of Density*

The density of a regular solid can be found by first weighing a rectangular block of it on a chemical balance and then measuring the dimensions and calculating the volume. Then, by calculation,

$$\text{Density} = \frac{\text{Mass}}{\text{Volume}}.$$

The density of an irregularly shaped solid can be found as follows:

EXPERIMENT. *Density of Irregular Solid*

*Method.* Weigh a sample of a metal M (or other convenient solid such as a glass stopper) on a balance. Partly fill a measuring cylinder with

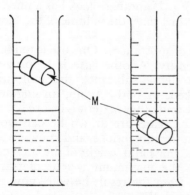

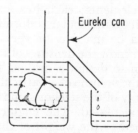

Fig. 2.3 Volume measurement of solid          Fig. 2.4 Volume by Eureka can

water and record the level. Lower the metal gently by thread so that it is completely immersed in the water. Observe the new level (Fig. 2.3). The difference in volume is then the volume of M.

A Eureka can is useful if the sample is too big for the measuring cylinder. This can be filled with water up to the level of a short spout (Fig. 2.4). Lower the sample gently and catch the water displaced in a measuring cylinder. This is the volume of the sample (see Table, p. 18).

| Measurements | Substance | Mass (g) | Measuring cylinder reading | | Volume (cm³) | Density = Mass/Volume (g/cm³) |
|---|---|---|---|---|---|---|
| | | | 1 | 2 | | |
| | | | | | | |

*Calculation.* Calculate the density from density = mass/volume.
Repeat the experiment using a variety of irregular solids.

### Density Values

Some density values are shown below:

| Solid | Density (g/cm³) |
|---|---|
| Gold | 19·3 |
| Lead | 11·4 |
| Copper | 8·9 |
| Iron | 7·9 |
| Aluminium | 2·7 |
| Glass | c. 2·6 |
| Ice | 0·9 |
| Wood (maple) | c. 0·7 |

Since aluminium has a density of 2·7 g (grammes)/cm³, a block of the metal of volume 10 cm³ has a mass of 27 g. If another block has a mass of 54 g then, since 2·7 g occupies a volume of 1 cm³, the volume of the block = 54/2·7 = 20 cm³.

Why is it useful to know the densities of substances? One use is to identify elements; each has a different density. Another use is to provide possible molecular information; in a dense substance the molecules could be massive, or very closely packed, or both. Again, an engineer must know the densities of steel, concrete and other materials before he can begin to design a bridge. Thus after the plans are drawn he can work out the volume of a particular span from the dimensions, and knowing the density of the steel or stone required, the mass and weight of the span can be calculated. The engineer will add the maximum weight of traffic carried to the weight of the span, and from the result he will know the required strength of concrete for the foundations.

### Relative Density (Specific Gravity)

Water is the most common liquid, and it has been found convenient to compare the mass of a substance with the mass of an equal volume of water. This is called its *relative density*. Thus:

$$\text{Relative density} = \frac{\text{Mass of substance}}{\text{Mass of equal volume of water}}$$

If unit volume is chosen as the volumes of the substance and water, the masses are then numerically equal to the densities. Thus also:

$$\text{Relative density} = \frac{\text{Density of substance}}{\text{Density of water}}$$

Note that *relative density* has no units. Density, however, has units such as $g/cm^3$ or $kg/m^3$.

The relative density of lead is 11·4. Thus 1 $cm^3$ of lead is 11·4 times as heavy as 1 $cm^3$ of water, which has a mass of 1 g. The density of lead is thus 11·4 $g/cm^3$. The density of a material in $g/cm^3$ is thus numerically equal to its relative density. This is not the case when the density is in '$kg/m^3$', because the density of water is 1000 $kg/m^3$.

## Liquids

Liquids such as water can be poured to fill the volume of a glass or bottle of any shape. Therefore, unlike solids, liquids have a fixed volume but no geometrical form or shape. Also, an object such as a pencil can be moved about freely inside a liquid, but this cannot be done with a solid.

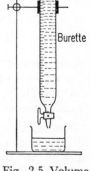

Fig. 2.5 Volume by burette

If a can of oil and a lump of metal of about the same volume are lifted we can tell that the density of the liquid is much less than that of the solid. The liquid mercury, however, is much denser than most metals, as one realizes on lifting a small jar of mercury.

From the definition of density on page 15,

$$\text{Density of a liquid} = \frac{\text{Mass of liquid}}{\text{Volume of liquid}}$$

We cannot measure the dimensions of a liquid as we can with a solid, as it has no geometrical shape. It is therefore necessary to have some vessel for measuring volume. A measuring cylinder or burette can be used. Thus we can fill a burette with alcohol or other suitable liquid in an experiment, weigh a beaker and then run 50 $cm^3$ of liquid into it (Fig. 2.5). On reweighing the beaker the mass of liquid can be found. Then:

$$\text{Density} = \frac{\text{Mass of liquid}}{50\ cm^3}$$

and the density can be calculated.

The results of some liquid density values are given below:

| | | |
|---|---|---|
| Mercury | 13·6 $g/cm^3$ | or 13 600 $kg/m^3$ |
| Turpentine | 0·87 $g/cm^3$ or | 870 $kg/m^3$ |
| Alcohol | 0·79 $g/cm^3$ or | 790 $kg/m^3$ |
| Paraffin | 0·79 $g/cm^3$ or | 790 $kg/m^3$ |

Mercury has a very high density. Light oils have densities less than 1 $g/cm^3$, the density of water. The densities quoted are all measured at room temperature. When a liquid is heated its mass does not alter, but its volume usually increases. Thus the density decreases. Water has an unusual density change with temperature, as we discuss later (p. 250).

### Measurement of Relative Density

As was defined on page 18, the relative density of a liquid is the mass of the liquid compared with the mass of an equal volume of water. The density (specific gravity) bottle is one specially designed to measure the

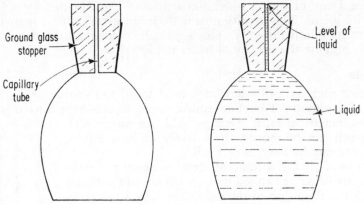

Fig. 2.6 Relative density measurement

relative density of liquids. It has a ground stopper with a fine hole through it, so that the volume of liquid filling it is always constant (Fig. 2.6). When the bottle is filled with a liquid and the stopper is fitted firmly the excess liquid passes through the fine hole and can be wiped off.

*Method.* Weigh a clean and dry density bottle on a chemical balance with its stopper. Fill it completely with water, wiping the excess with a cloth. Reweigh the bottle. Empty the bottle and dry the inside. Then fill it with a suitable liquid such as paraffin or oil and reweigh.

*Measurements.*   Mass of bottle empty = .........
       Mass of bottle + water = .........
       Mass of bottle + liquid = .........

*Calculation.* Relative density of liquid $= \dfrac{\text{Mass of liquid}}{\text{Mass of same volume of water}}$

$$= \text{.........  (no units)}$$

If the relative density of an oil is found to be 0·79, then as explained on p. 19, the density of the oil is 0·79 g/cm³. The density bottle can thus be used to measure the density of liquids.

### Relative Density of Insoluble Particles

The relative density bottle can be used to measure the relative density or density of particles or powders. As an illustration of the method, suppose the relative density of lead shot is required and the following measurements are made:

1. Mass of bottle empty = 15·0 g

2. Mass of bottle *partly* filled with lead shot = 95·2 g

3. Mass of bottle filled with water = 70·4 g

4. Mass of filled bottle when water is added to the lead shot = 143·1 g

*Calculation*

The mass of lead shot used = 95·2 − 15·0 = 80·2 g

Mass of water filling bottle = 70·4 − 15·0 = 55·4 g

Mass of water filling remainder of bottle with lead shot present = 143·1 − 95·2 = 47·9 g

Hence mass of water having same volume as lead shot
= 55·4 − 47·9 = 7·5 g

$$\therefore \text{Relative density of lead} = \frac{\text{mass of lead}}{7·5 \text{ g}} = \frac{80·2}{7·5}$$
$$= 10·7.$$

Draw sketches of the bottle and its contents corresponding to the four weighings above.

## Gases

Solids and liquids are forms of matter which we can see and feel. Gases, however, cannot be felt. Many, such as air, have no colour, but we can observe the effect of air on a windy day, when leaves on trees are moving about. Chlorine gas is greenish-yellow and bromine vapour is reddish-brown. Gases can also be detected by their smell; sulphuretted hydrogen, chlorine and ether vapour are examples.

A small amount of chlorine collected in a test-tube can be transferred to a large flask, which it fills immediately. This illustrates the differences between the three states of matter, solid, liquid and gas.

A *solid* has a definite volume and shape. A *liquid* has a definite volume, but assumes the shape of the vessel into which it is poured. A *gas* or *vapour* has neither of these properties–it will expand to fill any container into which it is passed.

### Density of Gases

Gases are much lighter than liquids and solids of the same volume, and thus have a much smaller density. A simple demonstration which shows that a vapour has mass, although you may not be able to see the vapour, is as follows:

### EXPERIMENT

Place a polythene bag on one pan of a balance and counterbalance this by adding weights to the other side of the balance. Heat up some carbon tetrachloride in a tin fitted with a loose lid for a few moments, until the space above the liquid in the tin contains vapour. Remove the lid and 'pour' some of the vapour into the bag on the balance pan. Notice that the vapour has weight, since the pan on the opposite side to the vapour will rise and the pan on which the polythene bag is placed will sink.

*Measurement of Density of Air*

Air is the most common gas. The density of air can be measured roughly by pumping air into a suitable container, as described in the following experiment:

EXPERIMENT

A 50-litre plastic container A with a tap T is weighed containing air at atmospheric pressure, which is ensured by leaving the tap open (Fig. 2.7). Air is now pumped into the container until it is very hard, using a

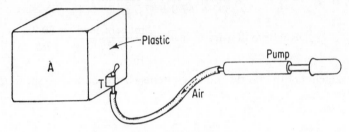

Fig. 2.7 Filling container with air

bicycle or foot pump and a one-way valve in the connecting tube. The tap is then closed and A is reweighed. The increase in weight should be about 7 g – if not, more air is pumped in.

By means of tubing attached to the tap T, the air is now released from the container and collected over water (Fig. 2.8). The collecting vessel is a transparent plastic box B inverted at the open end E, and with internal dimensions 10 cm by 10 cm or a cross-sectional area of 100 cm². A line L is marked on the box at a distance of 10 cm from the closed or upper end F, so that a volume of 1000 cm³ or 1 litre is defined.

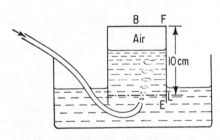

Fig. 2.8 Measurement of air volume

Initially the box is filled with water and inverted as shown. Air is then slowly bubbled through the tube from the plastic container, and the box is moved up or down until the outside and inside water levels are in line with the mark L. One litre of air at atmospheric pressure has now been collected. The tap is now closed, the collecting vessel is refilled with water and the process is repeated. A whole number of litres of air together with an estimated fraction of a litre is then collected. The plastic container A is finally reweighed, and the mass of air collected is thus known.

*Measurements.*     Mass of container A + pumped air   = ........g
                    Mass of A finally                  = ........g
                    Volume of air collected            = ........litre

*Calculation.*    Density of air at atmospheric pressure $= \dfrac{\text{Mass of air (g)}}{\text{Volume of air (litre)}}$

$$= \ldots\ldots\ldots \text{ g/litre}$$

*Results.* The volume, and hence the density, of a gas is very sensitive to changes of temperature and pressure (p. 253). The densities of some gases at standard temperature (0°C) and pressure (760 mm mercury) are:

Air          $0.001\,3$ g/cm³  or $1.3$ kg/metre³
Hydrogen     $0.000\,09$ g/cm³ or $0.09$ kg/metre³
Nitrogen     $0.001\,25$ g/cm³ or $1.25$ kg/metre³
Helium       $0.000\,18$ g/cm³ or $0.18$ kg/metre³

Note that $1.3$ g/litre (air) $= 1.3$ kg/m³; $0.09$ g/litre (hydrogen) $= 0.09$ kg/m³.

It is interesting to note that accurate measurements of the density of air led to the discovery of the rare gases in the atmosphere. One is argon, used for filling electric lamps; two others are helium and neon. Lord Rayleigh and Sir William Ramsay found that nitrogen gas obtained from the atmosphere was very slightly denser than nitrogen gas obtained from a chemical reaction; they came to the conclusion that there was an impurity in the 'atmospheric' nitrogen, and after further investigations they discovered argon and other rare gases.

## Molecular Theory of Matter · Solids

So far we have considered matter in bulk, called a 'macroscopic' view of matter, and found that a gas such as air has a mass of about $1.3$ g/litre, that a liquid such as water has a much larger mass of 1000 g/litre, and that a solid such as copper has a mass of 9600 g/litre.

We now consider how matter is built up, which is a 'microscopic' view of matter.

### EXPERIMENT. *Divisibility of Fluorescein*

Dissolve 1 g of fluorescein in a small volume of water and dilute the solution to 1 litre in a graduated flask. Notice that the solution is green in colour. Fill a burette with the solution and slowly run out 1 cm³, counting the drops (25–30 per cm³). Each drop is coloured deep green and therefore contains dissolved fluorescein. In 1 litre of solution there must be (for 25 drops per cm³), $25 \times 1000$ or 25 000 drops, each one being deeply coloured. If now the solution is successively diluted until the colour is only just discernible, this simple experiment shows that the original 1 g of solid matter can be divided into at least 100 million separate particles.

An alternative experiment can be performed using potassium permanganate instead of fluorescein.

Photographs taken with very high-powered microscopes called electron microscopes and ion microscopes show that solids consist of very tiny particles, invisible to the naked eye. The smallest particles of a substance which have a separate existence are called *molecules*. The molecules of a solid are 'anchored' to a fixed position in the solid, since a solid has a definite shape. Further, as the density of a solid is high, the molecules are packed very tightly together.

## Liquids

Like solids, liquids are made of molecules. To gain an idea of the packing of the molecules in a liquid, consider the following experiment:

EXPERIMENT. *Spaces between the Molecules in Liquids*

Put 100 cm³ of water in a measuring cylinder and stand it on the bench. Weigh out 5 g of potassium nitrate and notice that the salt has bulk, i.e. mass and *volume*. Pour the water into a beaker, and stir in the salt until it has all dissolved. Pour the solution back into the measuring cylinder and notice that the volume is still 100 cm³.

Where has the salt gone? It certainly cannot have disappeared. It is clear that the ions (i.e. charged atoms or groups of atoms) produced in a free state in solution when the salt dissolved in water must have gone into spaces between the molecules in the water.

We therefore conclude there are spaces between the molecules of a liquid. Further, since the liquid has no shape, the molecules must be free to move about. In contrast, the molecules in a solid are anchored to a fixed position (p. 32).

## Gases

Gases can occupy any volume into which they are introduced. Their molecules must therefore be more free than those of a liquid. Further, as we have seen, the density of a gas such as air is 1·3 g/litre, whereas the density of a liquid such as water is 1000 g/litre. Thus the molecules of a gas must be very much farther apart on the average than those of a liquid.

## Size of Molecules

Some idea of the size of molecules can be gained from an experiment to measure the thickness of a very thin film of oil.

EXPERIMENT. *Thickness of Oil Film*

Provided the area of a water surface is sufficiently large and very clean, olive oil spreads on it into a very thin film. Collect a drop of oil on a wire and obtain its volume as described below. Place the drop of oil on the surface of clean water in a large tank, such as a developing tank, which has been lightly dusted with lycopodium powder. Immediately the oil film formed pushes back the powder particles and forms a ring, the diameter of which is measured.

*Example*   Volume of drop on wire (approx) $= 24 \times 10^{-5}$ cm³
Diameter of film (approx) $= 20$ cm
Area of film $= \pi \times 10^2 = 314$ cm²
∴ Thickness of film $=$ Volume/Area
$= \dfrac{24 \times 10^{-5}}{314}$ cm
$= 8 \times 10^{-7}$ cm

Some liquid detergent can also be used. A drop may be obtained by dipping a small piece of wire suspended from the corners of a card in olive oil, so that a small drop collects in the middle of the wire. The drop is then viewed against a transparent $\frac{1}{2}$-mm scale through a hand lens or a microprojector and the diameter estimated. The volume is calculated from $4\pi r^3/3$, the volume of a sphere. Alternatively, a short length of oil can be released from a capillary tube of known diameter.

On extremely clean water, olive oil is considered to spread in a film one molecule thick, a so-called *mono-molecular layer*. Accurate experiments show that the film is about $2 \times 10^{-7}$ cm thick. The oil molecule is long. It has a chain of about 18 carbon atoms standing up at right angles to the water surface and forming the film.

## Molecular Size from Gold-leaf

Another indication of the very small size of a molecule can be obtained from measurements on gold-leaf, which can be made exceptionally thin by skilled craftsmen. After making them it is found that 25 squares of gold-leaf weigh 5 grains or 0·4 g approximately. Now, the density of gold is 19·3 g/cm³. Therefore

$$\text{Volume of gold-leaf} = \frac{0\cdot4}{19\cdot3} = 0\cdot02 \text{ cm}^3 \text{ (approx.)}$$

Each gold-leaf has an area of about 64 cm². Thus all 25 leaves have a total area of about 1600 cm²

$$\therefore \text{Thickness of gold-leaf} = \frac{\text{Volume}}{\text{Area}} = \frac{0\cdot02}{1600}$$

$$= \frac{1}{80\,000} \text{ or } 10^{-5} \text{ cm (approx.)}$$

The thinnest gold-leaf manufactured has been estimated to be $6 \times 10^{-6}$ cm thick.

## Osmosis

The minute size of liquid molecules can also be demonstrated by showing that they are able to pass through a membrane such as parchment or cellophane.

EXPERIMENT. *Demonstration of Osmosis*

Set up the apparatus shown in Fig. 2.9. It consists of a suitable membrane M (e.g. cellophane, pig's bladder, parchment) attached to the end of a thistle funnel T containing a solution of cane sugar (fairly concentrated) and dipping into a beaker of water B. Observe that the level of liquid in the funnel rises, showing that the water molecules pass across the membrane into the sugar solution. When the level of liquid in the tube remains constant the pressure set up is called 'osmotic pressure'. Increase the concentration of the sugar solution and observe the new osmotic pressure.

*Conclusion.* We conclude, therefore, that water molecules can pass across the membrane. The phenomenon whereby a solvent, such as water, passes through a suitable membrane into a solution (or a solvent passes from a dilute to a concentrated solution) is called 'osmosis'. The suitable membrane is said to be 'semi-permeable', since it permits the passage of molecules of the solvent but not those of the solute or dissolved substance, sugar in the above case.

Osmosis is not limited to sugar solutions. It occurs whenever any solution is separated from its solvent by a semi-permeable membrane, or when a dilute solution is separated from a more concentrated one.

Dried prunes and dried peaches have skins which are semi-permeable membranes, and have sugar solution inside. Consequently they swell when placed in water–the molecules of water pass through their skins. A dried bean swells when placed in water for a similar reason.

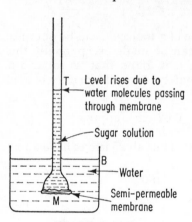

Fig. 2.9 Osmosis

Observe the effect when fresh peaches or plums are placed in water and then in a concentrated solution of sugar. Explain what happens in each case. Many plants obtain moisture by the process of osmosis across the semi-permeable membrane of the root cells. The human body also contains semi-permeable membranes which play an important part in the transfer of liquid from one part to another. Some people have sensitive skins which may be affected by certain detergents. The skin of the hand acts as a semi-permeable membrane, and water then passes from the body into the detergent solution, leading to dry and cracked skin.

A large water purification plant, using the principle of reverse osmosis, is used in America to produce many thousands of gallons of pure water per day. Impure or salt water in a tank is subjected to pressure, and the water molecules are forced through a special membrane and collected as pure water on the other side.

### Ideas of Molecular Size

Pictures of molecules under extremely powerful microscopes, and calculations based on gases, which do not concern us in this book, show that the diameter of a molecule is actually of the order of $10^{-8}$ cm or one-hundred-millionth centimetre. The tiny diameter of a molecule is almost impossible to visualize because we do not live in a microscopic world. Some idea of what it means may perhaps be gained from the following:

1. If a fine hair is magnified until its thickness is that of a wide street a molecule in the hair would then look like a speck of dust in the street.
2. Millions of molecules are on the tip of a needle or pin.
3. In any part of the human body, such as the brow or the hand, there are many millions of pores. When we sweat, only a tiny amount of liquid is obvious. It has come to the surface of the skin by oozing through all the pores, showing that a molecule of liquid must be extremely small.
4. Two grams of hydrogen contain $6 \times 10^{23}$ molecules. If the whole

population of the world were to count such an enormous number of molecules individually at the rate of five per second it would take 100 years to count them all.

5. If a cricket ball was magnified to one kilometre ($\frac{5}{8}$ mile) in diameter a molecule in the ball would then be magnified to about one two-thousandth of a centimetre.

## Molecules in Motion

From consideration of the size of molecules we turn next to consider the movement or motion of molecules. We shall discuss first the case of gases, where the molecules are most free, then liquids and finally solids, where they are least free.

We know from the fragrant smell of roses and other flowers on entering a room that molecules of a gas are free to travel from one place to another. The motion can be investigated by the following laboratory experiment.

EXPERIMENT. *Movement of bromine molecules in air.*

Put a few drops of bromine (TAKE CARE!) into an evaporating basin and place a gas jar over it. Observe the bromine vapour, which rises rapidly into the jar, although bromine vapour is heavier than air. The intermingling of molecules of two gases, here bromine and air, is called *diffusion.* It demonstrates that molecules are in motion.

## Brownian Motion

Another experiment on the movement of gas molecules can be carried out with the apparatus shown in Fig. 2.10 (i). It consists of a small

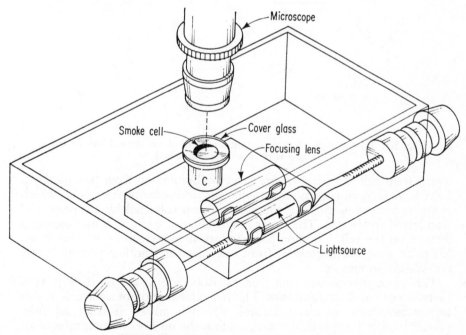

Fig. 2.10(i) Brownian motion demonstration

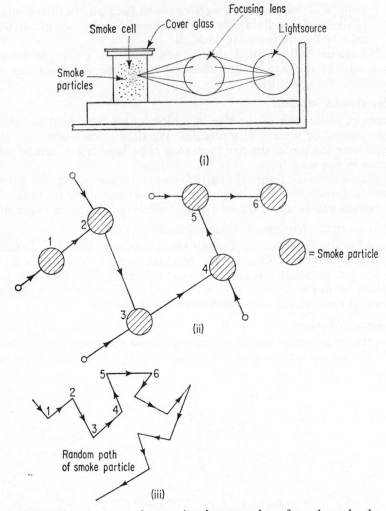

**Fig. 2.10** Brownian or random motion due to resultant force shown by the arrow with small circle attached

transparent cell C with a cover, strongly illuminated from the side by a light L. A piece of cord is set smouldering and some of the smoke, which contains minute particles of the cord, is then collected by a syringe and transferred to the cell. The cover is quickly replaced and a microscope is focused on the cell. The fine particles are then seen to be moving in an irregular way, darting about from one place to another very suddenly and always in motion.

This phenomenon was first seen by Robert Brown (see p. 30). It is therefore called *Brownian motion.* The irregular motion of a particle is due to the movement of air molecules, which bombard it from all sides (Fig. 2.10(ii)). The particle is so small that the number of molecules of air hitting one side is not balanced by an equal number of molecules hitting the opposite side at the same instant. Consequently, the particle moves in the direction of the resultant force. When it moves to another place the same thing happens, as shown. The irregularity of the motion and the incessant movement of the particles shows that the air molecules move rapidly in all different directions and never stop (Fig. 2.10 (iii)).

If the particles in the gas are big they remain fairly still, although they are bombarded on all sides by gas molecules. This time the large number of molecules hitting one side is not *relatively* much greater than the number hitting the other side at the same instant. The difference shows up much more when the number of bombarding molecules is small, as in the case of a very small particle. For this reason, a table-tennis ball (a large 'particle') suspended in air by a thread does not move. You yourself are being bombarded on all sides by molecules of gas in the air; but since you are so large, a few more bombardments on one side than on the other has no appreciable effect.

## Estimate of Molecular Velocity

A rough estimate of the speed of an air molecule can be obtained by considering the height of the atmosphere. Suppose we assume that the atmosphere thins out so much at about 10 000 m that some molecules starting upward with a velocity $u$ at ground level under the opposing pull of gravity, reach a final velocity of zero at the top of the atmosphere. We assume that on the way up the molecules make 'elastic' collisions, and that the velocity will be unchanged.

The retardation due to gravity is about 10 metre per second$^2$ (p. 70). Thus from the equation of motion $v^2 = u^2 + 2as$ (on p. 81), the initial velocity $u$ of the molecule is given by

$$0 = u^2 - 2 \cdot 10 \cdot 10\,000$$
$$\therefore \quad u = \sqrt{2 \cdot 10 \cdot 10\,000} = 450 \text{ m/s (approx.)}$$

This is a speed which is about 1600 km/h. The calculation is very approximate, but it shows that the speed of a molecule in a gas is high. More accurate calculations show that the average speed of an air molecule under normal conditions is about 1800 km/h and that of hydrogen gas is about 6400 km/h.

## Motion of Molecules in Liquids

The movement of molecules in liquids can be investigated by the following experiment:

EXPERIMENT. *Movement of Molecules in Liquids*

Make up a solution by adding some iodine crystals to a beaker containing carbon tetrachloride (or chloroform). Pour a small amount of the purple solution into a boiling tube T (Fig. 2.11(i)). Then very carefully pour some potassium iodide solution, colourless, on top of the solution. Observe that the boundary of the liquids becomes brown in colour. Wait a few days, and note that this colour has spread uniformly into the upper layer (Fig. 2.11(ii)). Thus iodine molecules have moved upwards.

Now repeat the experiment, but using colourless carbon tetrachloride at the bottom of the tube, and above it a layer (brown) of iodine dissolved in potassium iodide solution (Fig. 2.11(iii)). This time observe the pink colour which rapidly spreads through the lower layer. It deepens in colour until the whole carbon tetrachloride layer is purple.

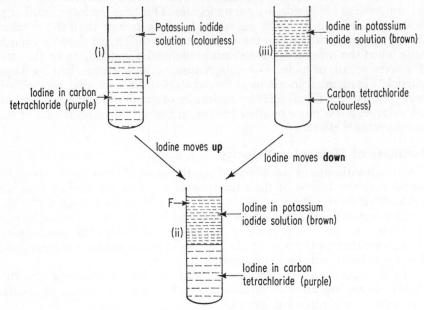

Fig. 2.11 Diffusion (molecular movement) in liquids

The spread of colour indicates a movement of iodine molecules in solution. Thus diffusion occurs in liquids as well as in gases, although it takes place at a slower rate.

## Brownian Motion in Liquids

As with gases (p. 28), we cannot possibly observe the motion of the tiny particles which constitute a liquid, because they are so small. But we can observe what effect the motion of such particles has on finely divided forms of matter, big enough to be seen under a powerful microscope and suspended in a liquid. In 1827 Robert Brown observed this effect when finely ground pollen grains, suspended in water, were viewed under a microscope. He found that the pollen grains were in constant random motion, and this was later called 'Brownian motion'.

EXPERIMENT.

Place a few well-diluted drops of *aquadag* (finely divided graphite particles suspended in water) or *photopake* (a similar suspension used for blacking negatives) on a clean microscope slide and cover the liquid with a cover slip. An image of the slide can then be projected on to a screen by a microprojector so that the particles can be seen, or, alternatively, the apparatus with a microscope described on p. 27 can be used to view the illuminated particles.

When one or two particles are viewed they are seen to be moving in a random and irregular way. This is Brownian motion, which is also obtained with gases (p. 28). It cannot be due to convection currents as there is no general movement of particles in one direction.

As explained on p. 28, Brownian movement is a continual random motion caused by uneven bombardment of the suspended particles by the moving molecules in the liquid, in this case water molecules. The suspended particles are being bombarded continually from all sides, but, since they are so tiny, a few more bombardments on one side than on another causes them to move to one side, where they will again be bombarded unevenly, and so on. Large particles appear unaffected (see p. 29).

**Effect of Temperature**

By raising the temperature, the Brownian movement gets faster, showing that the molecules in the liquid move faster if they are heated. Another simple experiment which shows this is as follows:

EXPERIMENT. *Motion of Molecules and Temperature*

Fill two beakers A and B with hot and cold water respectively. Drop a crystal of potassium permanganate into each beaker, and observe that the colour spreads much more quickly through the water in the hot beaker, A, than in the cold beaker, B. The potassium permanganate dissolves in each case. The more rapid movement of colour in the hot liquid shows that the molecules of the crystal in solution move faster when their temperature increases.

**Motion of Molecules in Solids**

We have seen that the movement of molecules we call diffusion occurs in gases and in liquids. Diffusion also occurs in solids; if a bar of lead is

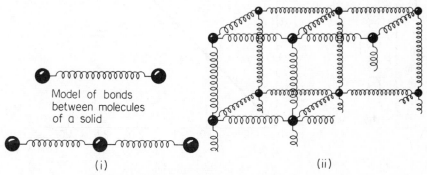

Model of bonds
between molecules
of a solid

(i)                                      (ii)

Fig. 2.12 Model of forces between molecules in solid

placed in contact with one of gold and the bars are left undisturbed for several years, small traces of gold can be detected in the lead bar up to a distance of about 1 mm from the boundary. Thus some gold molecules have moved.

Generally, however, the molecules of a solid remain on the average anchored to one position. Simple observations show that solids become warm when heat energy is given to them, and as they remain a solid we believe, therefore, that their molecules are *vibrating* about their anchored or average position before heat is applied, and that the heat makes them vibrate through greater distances or amplitudes than before.

A rough model which illustrates the vibrations of molecules in solids can be made by using balls joined by springs, as shown in Fig. 2.12. When the springs are pulled and released the balls ('molecules') vibrate about a mean position, but the latter remains constant, so that the structure is unaltered. When the temperature of a solid is raised the vibrations of its molecules continue to have practically the same frequency as before, but their amplitudes of vibration increase. See photographs, Plate 2.

### Forces between Molecules in Solids · Elasticity

We now consider the forces which exist between the molecules of a solid, liquid and gas. We begin first with solids.

From their hardness, the molecules of a solid must be packed tightly together. Some knowledge of the forces between them can be gained by adding weights to a metal, for example, and investigating how it stretches.

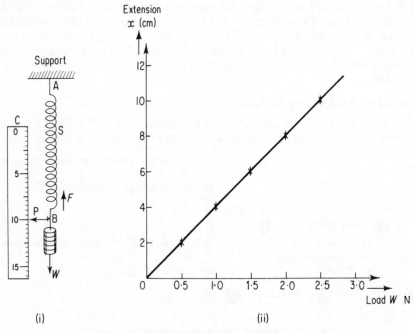

Fig. 2.13 Elasticity of spring

EXPERIMENT. *The Stretching of a Spring*

*Method.* Suspend a spiral spring S vertically from a rigid support at its upper end A (Fig. 2.13(i)). A small pointer P is fixed to the lower end B adjacent to a vertical fixed scale C. First, note the reading opposite P on C. Then add weights W to the lower end of the spring in steps, such as 0·5 N. Note the new position of P for each weight added. Remove the load in steps of 0·5 N and take a second set of readings. Tabulate the results and find the average extension $x$ cm for each load.

*Measurements*

| Load (N) | Readings on C cm | | Extension ($x$ cm) | $\dfrac{Extension}{Load}$ |
|---|---|---|---|---|
| | Loading | Unloading | | |
| | | | | |

*Calculation.* Work out the ratio extension/load for each load.

*Graph.* Plot extension *v.* load, starting from zero on the two axes (Fig. 2.13(ii)).

*Conclusion.* A straight-line graph passing through the origin shows that the extension is directly proportional to the load in the range of loads used. This is known as *Hooke's law* (1679), and it applies up to a load called the 'elastic limit', as seen shortly.

Measure the extension of the spring produced by an object in the same weight range. Then use the graph plotted to find its unknown weight. Repeat with other suitable objects. This is the principle of a *spring-balance* (see p. 99).

When a small weight is attached to the spring the latter alters its shape on extension. When the weight is removed the spring returns to its original shape. The spring is therefore said to be *elastic*. The forces of attraction between the displaced molecules are sufficiently strong to restore the molecules to their original positions. We can investigate further the forces between molecules of a metal by the following experiment:

## Elastic limit · Yield point

EXPERIMENT. *Stretching and Breaking of Copper Wire*

Clamp the end of about 2 m of 26 swg bare copper wire between polythene pads in a suitable clip (see Fig. 2.14) and secure the clip to the bench with a 'G' clamp. Pass the end of the wire over a pulley fixed at the other end of the bench and attach to the end B of the wire a 100 g mass so that it is above the floor. Using glue, attach a length of cotton C to the wire a short distance from the pulley, and pass the thread over a second pulley D placed very close to the first one A. Attach a 10 g mass W to the thread to keep it taut.

Arrange a small length of metal channel M to house a needle as shown, close to the wire and cotton. Put one turn of the cotton round the needle and then push a straw through the needle to act as a pointer.

Add successive 100 g masses to the wire at the load end and, as stretching occurs, the cotton thread will cause the pointer to move. Load up the wire until an additional 500 g has been added and then remove the load, noticing that the pointer returns to its original position. Plot deflection *v.* load. In this range OA of load, therefore, the wire is elastic (Fig. 2.15). Now reload the wire until 1500 g or more has been added. Remove the load, noticing that a permanent extension in the range OP has occurred.

By careful reloading of the wire, find the load for which it just breaks.

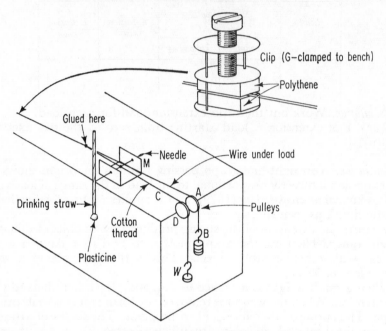

Fig. 2.14 Investigation of elasticity

EXPERIMENT. *Stretching fine Copper Wire*

Take about 2 m of 32 swg bare copper wire and attach each end of the wire to a pencil or wooden dowel. One end of the wire is attached to a fixed point such as a bench by means of a 'G' clamp and the other end is then pulled. Observe: (i) the *elastic property* of the wire on gentle pulling, that is, the wire stretches when pulled, but returns to its original length when released: (ii) the 'plastic' state of the wire just before breaking (strong pull needed) as well as the elongation and reduction in diameter where the wire breaks. Examine the ends of the wire after breaking under a microscope and describe what you see.

*Conclusion.* Beyond a certain load L the molecules do not return to their original position when the load is removed (Fig. 2.15). This load is there-fore called the *elastic limit* of the wire. With loads greater than the elastic limit the molecules are unable to keep their fixed positions in the metal. Elasticity of metals is explained by the distortion of their metal structure; if there is only a small distortion, and the distorting force is removed, the crystals return to their original alignment. When a metal is strained beyond its 'elastic limit' (see Fig. 2.15) a new alignment of crystals is produced. Further discussion is outside the scope of this book.

At a load called the *yield point*, which is reached shortly after the elastic limit, the molecules slide about. This is called the 'plastic stage' of behaviour, and the metal now acts like a liquid. We see, therefore, that the rigid structure of a solid can be broken down, thus freeing the molecules from the strong attractive forces of their neighbours.

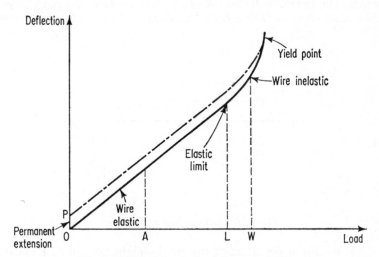

Fig. 2.15 Elasticity, inelasticity and yielding

At this stage it should be noted that a solid also resists compression. It therefore follows that *repulsive forces* exist between the molecules of a solid when they are very close together. When the molecules are farther apart than normal, as we have just seen for the case of a stretched wire, the forces between them are attractive (see Fig. 2.32, p. 50). Forces of attraction and repulsion also exist between atoms.

## Friction

Forces between large groups of molecules occur when one surface is in very close contact with another. This takes place when we are walking, for example. The molecules in the material of the sole of the shoe are then extremely close to those in the material of the ground or floor. The well-known force of *friction* is then produced. Metal surfaces rubbing one against the other produce a large frictional force, and this is not desirable in machines, because energy is then wasted (p. 153).

Friction always opposes motion. The force of friction between two given surfaces can be investigated by the following experiment:

EXPERIMENT

Place a block of wood A on a table and connect it by a light string over a pulley to a scale-pan S hanging vertically (Fig. 2.16). Add increasing weights to S. Observe that the block A does not move at first. The frictional force, which balances the applied force when the block does not move, thus increases. Continue adding weights until A just starts to slip. At this point the frictional force has reached a maximum value. It is called the *limiting frictional force*. Observe the weight in S. Now turn over the block A so that a smaller area rests on the table. Add loads until the limiting frictional force is again reached. Observe that it is practically the same as before.

Increase the normal reaction $R$ of the table surface on A by placing a suitable mass such as 500 g on it as shown. Add masses to S until A

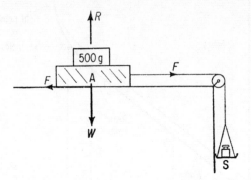

Fig. 2.16 Static friction experiment

just begins to slip again. Record the total weight now in S. Repeat with two or more loads on A to increase the normal reaction $R$. Tabulate the results for $R$ (weight of A + load on A = $R$) and for $F$, the maximum or limiting frictional force on A (weight of S + weight on S). Then work out the ratio $F/R$.

*Measurements*

$$\text{Weight of S} \ = \ .........$$
$$\text{Weight of A} = \ .........$$

| Limiting frictional force $F$ (weight of S + weight on S) | Normal reaction $R$ (weight of A + load) | $F/R$ |
|---|---|---|
| | | |

*Conclusion.*

1. Friction opposes motion.
2. The limiting frictional force is independent of the area in contact.
3. The ratio $F/R$ is practically constant, where $F$ is the limiting frictional force and $R$ is the normal reaction between the surfaces.

These conclusions are called the *laws of solid friction*. They were first stated over 300 years ago. The ratio $F/R$ is called the *coefficient of static friction* between the given surfaces and is denoted by $\mu$. Thus

$$\text{Limiting frictional force } F = \mu R$$

For wood on wood, $\mu$ is about 0·4. Thus if a block of wood of weight 2 N is placed on a flat wooden surface the force $F$ which just moves the block would be:

$$F = \mu R = 0\cdot4 \times 2\,\text{N} = 0\cdot8\,\text{N}$$

## Kinetic or Dynamic (Sliding) Friction

A frictional force is encountered not only when an object is about to slip but also when a surface moves over another surface with a constant velocity. This can be investigated by placing a block of wood, A, for example, on a table and again adding weights to a scale-pan S as shown in Fig. 2.17. This time, however, the block is given a slight push each time a weight is added, and at one point the block continues to slide along the table with constant velocity. The frictional force $F$ is now called 'kinetic'

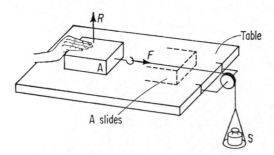

Fig. 2.17 Kinetic or dynamic friction

or 'dynamic' or 'sliding' friction, and its magnitude is equal to the total weight on S plus the weight of the scale-pan.

Experiment shows that the kinetic or sliding frictional force is always less than the limiting static frictional force. The coefficient of kinetic friction $\mu'$ is $F/R$, where $R$ is the normal reaction, and it is therefore less than the coefficient of static friction.

When the brakes of a bicycle are applied the kinetic frictional force between the wheels and the brake blocks slows down the bicycle. If the rider dismounts and slips over, the static frictional force on his or her shoe is insufficient to oppose completely the horizontal force exerted by the shoe on the ground.

## Rolling Friction · Ball-bearings

Friction can be considerably reduced by using rollers, as the following observations show:

EXPERIMENT

Place a book on a flat bench which is somewhat rough. Observe the force required to push the book along the bench. Now place the book on a few round pencils, arranged as rollers. Note the force this time required to move the book with your hand. Next, place some marbles or polystyrene beads beneath the book and observe the force now required to move the book. Finally, place the book on a soft material on the bench and again observe the force required to push the book.

*Conclusion.* The frictional forces on the book are considerably reduced when it moves over round, smooth surfaces such as the marbles or polystyrene beads. This is due to the fact that the marbles or beads roll as the book moves over them, thus reducing the relative motion between the two surfaces in contact. Rolling friction is thus small.

### Ball-bearings

Large blocks of stone are dragged with difficulty over the ground owing to the considerable sliding frictional forces which need to be overcome. It is

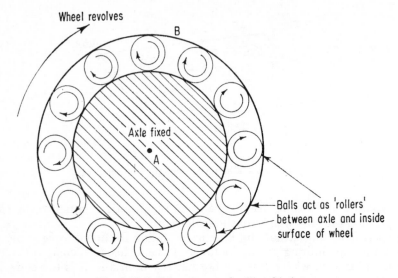

Fig. 2.18 Ball-bearings and rolling friction

easier to pull them by putting circular logs beneath. The logs turn while the stones are dragged, thus reducing the relative motion between the logs and stone surfaces in contact with one another. This reduces considerably the frictional force on the stone.

Frictional forces between sliding metal surfaces would cause wear, and heat would be generated, both of which are undesirable in machines. To overcome friction between the axle A and hub B of a bicycle wheel, *ball-bearings* are placed between them (Fig. 2.18). The ball-bearings tend to roll round the axle as the wheel turns, and, as in the case of the blocks of stones on circular logs, the frictional force is considerably diminished. The bearings are lubricated to reduce friction further. It is important, in practice, that the metal surfaces or ball-bearings which come into contact should be very hard. If not, rolling friction might be more than sliding friction; for this reason an aircraft landing on soft snow may have skis instead of wheels.

## Uses of friction · Motor tyres

Friction helps us in some circumstances, and we try to increase it. Without friction we would be unable to walk, since we would slip. Friction enables a nail to stay tightly in a hole and so hold two pieces of wood together. When brakes are applied to a train the friction between the wheels and brake blocks slows down the wheels. Belt drives on various machines do not slip because of friction.

Similarly, sand may be thrown on to railway lines in bad weather (e.g. ice and snow) in order that the locomotive's wheels can grip the lines better.

The nature of the rubber used in the manufacture of tyres for vehicles, and the pattern of the tyre tread, both play an important part in preventing accidents, particularly on wet roads. Synthetic rubber, specially developed for tyres, is much better than natural rubber. They have a much higher value of coefficient of friction, $\mu$. On a dry day at normal temperature, for example, $\mu$ is about 1·2.

Grip on a road surface is best achieved by a particular tread pattern on the tyre. Basically, this consists of a number of 'ribs' all round the circumference and 'tread bars' at right angles, so that channels and grooves are formed between them (see Plate 3). On wet roads the water is displaced through the channels to the rear of a good tyre, as shown in Plate 3. A fine water film is still left on the road surface beneath the tyre, and this is removed by capillary action, using fine slots called 'microslots' cut in the tyre. Grip is then obtained on the dry road surface.

## Molecular Forces in Friction

When a frictional force opposes the motion of one body sliding over another, atomic and molecular attractive forces at the small *contact areas*

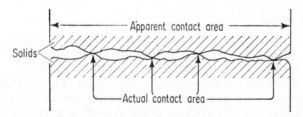

Fig. 2.19 Contact area of touching solids

are largely responsible for the force (Fig. 2.19). No two surfaces are perfectly flat. When they are in contact 'humps' of molecules press against each other and interlock, even though the surfaces appear to be very smooth. Thus only a small part of the geometrical areas is actually in physical contact. The pressures at the points of contact are therefore high, and a type of cold-welding takes place. When one surface is pulled over the other the bonds are continuously broken and re-form at other points.

The frictional force is the force required to break the bonds between large groups of molecules. With heavier objects of the same geometrical area in contact a greater force is applied normal to the surfaces, so that the contact areas increase in size and number. This explains why the force of sliding friction is proportional to the normal force (p. 36).

## Forces between Molecules in Liquids

We now investigate the forces between molecules in liquids. An experiment which demonstrates the existence of such forces can be carried out as follows:

### EXPERIMENT

Fill a beaker of water B completely to the brim (Fig. 2.20(i)). Place a dry needle N on a small piece of blotting paper and gently push the paper

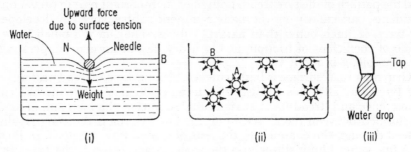

Fig. 2.20 Molecular theory of surface tension

along the rim of the glass towards the water surface. The paper becomes wet and sinks, and the needle remains floating on the water surface. A razor blade can be floated in the same way.

*Conclusion.* The surface acts like a 'skin', covering the liquid, which is able to support the weight of the needle.

## Surface Tension

The property of a liquid surface which enables it to support the needle is called its *surface tension*. It is due to the attractive forces which exist between molecules of the liquid. Inside the liquid a molecule A is attracted equally on all sides by neighbouring molecules (Fig. 2.20(ii)). On the surface, however, a molecule such as B has many more molecules below it than above it in the vapour. Consequently, there is a pull on B tending to move it towards the interior of the liquid. This makes the liquid have the minimum surface area for a given volume. The shape of the liquid which conforms to this condition is a *sphere*. Thus, under surface-tension forces only, a liquid forms a sphere.

This explains why water dripping very slowly from a tap is observed to take a spherical shape while falling (Fig. 2.20(iii)). Lead shot is manufactured by pouring molten lead on to a fine wire mesh at the top of a high tower. Spherical droplets of lead are formed as they fall and solidify as they pass into cold water in a tank, from which they are recovered.

## Spherical Shape of Drops

EXPERIMENT

(a) Pour water into a glass beaker to half-fill it and then gently pour alcohol down the side to form a distinct layer on top of the water. With the end of a pipette, drop a small amount of olive oil into the beaker. Olive oil is insoluble in both water and alcohol, and its density lies between that of the two liquids. Observe that the oil comes to rest near the boundary where the mixture density is about the same as oil, and forms spherical drops (Fig. 2.21). (b) Place a small quantity of aniline in a supported burette, whose tip is just below the surface of water in a beaker (Fig. 2.22). Open the burette tap carefully to release the aniline slowly. Observe the shape of a small aniline drop formed. This can be conveniently shown by

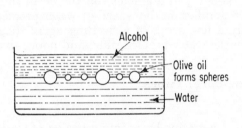

Fig. 2.21 Spheres of oil          Fig. 2.22 Spheres of aniline

using a flat-sided container in place of a beaker and a projector. Aniline has a density similar to that of water at room temperature. The weight of the drop is therefore supported by the upthrust of the water, so that the shape of the drop is due only to surface-tension forces.

## Effect of Impurities · Detergents

If ordinary tap water is allowed to fall carefully on to the surface of a card it forms droplets. If a modern detergent, such as *Teepol*, is mixed with water, however, it enables the water to spread more evenly and fully along the card, i.e. it produces increased surface wetting. This is one of the properties of a modern detergent which makes it so valuable in our homes, since it enables water to spread evenly through the article being cleaned (see also p. 44).

EXPERIMENTS

1. Drop some small camphor shavings into water in a trough and notice that they begin to 'dance' about in rapid motion. A solution of camphor in water has a lower surface tension than pure water. The camphor does not dissolve equally in all directions, and it is drawn away from the side where, instantaneously, it is dissolving most rapidly. When the camphor is in rapid agitation add a drop of oil to the water surface. Describe and explain what happens.

2. A few drops of oil poured on to a water surface soon spread out into a thin film because the forces between oil and water molecules are stronger than the forces between oil molecules themselves. Float a needle or razor blade on a water surface and pour on to the surface a few drops of oil. Explain what happens.

Rescues at sea are sometimes made possible by pouring oil on the water. Boats can ride the swell of a large wave, but are hindered if the wave breaks up. The oil causes a reduction in the surface tension and spreads over the water, thus preventing the formation of ripples which would make the wave break up.

3. Select two clean microscope slides. Coat one with paraffin wax by dipping it into hot wax in a saucepan and allowing the slide to drain and cool.

Make a small pool of water (about $\frac{1}{2}$ in diameter) on the clean slide using an 'eye' dropper, and then another on the waxed or water-proofed one. Using a match-stick, apply a few drops of a 'wetting agent' (e.g. Teepol, Stergene) in the pool on the water-proofed slide, and notice that the water-proofing is spoilt.

## Liquid Rise in Capillary (Fine-bore) Tubes

### EXPERIMENT

Dip a number of capillary tubes with fine bores into a beaker of water so that the inside is wetted, and then support them so that the bottom

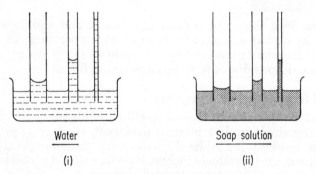

Water                          Soap solution

(i)                              (ii)

Fig. 2.23 Capillary rise

of the tube is the same distance below the surface. Notice that liquid rises up each tube. The height of the water depends on the radius of the tube– the smaller the radius, the greater the height (Fig. 2.23(i)). Repeat the experiment using soap solution instead of water and note the difference (Fig. 2.23(ii)).

## Capillarity

This experiment shows that water rises up a capillary or fine-bore tube. The phenomenon is known as *capillarity* and occurs frequently. For example, blotting paper contains many very fine pores, which act as minute capillary tubes, so that ink is drawn into the paper when pressed down on it. The melted wax of a candle is drawn up the wick by capillary action in the spaces between the fibres of the material.

Bricks are porous—water seeps up into them easily by capillary attraction. To prevent water from the ground saturating the brickwork of a house above ground level, which would damage the interior decorations of a room, a *damp course* is used. This consists of a layer of a non-porous material, such as slate or specially treated felt, which prevents the water in the soil from rising above it.

In summer, hoeing is important to prevent loss of moisture by surface evaporation. It breaks the capillary passages, so that the roots of the plants now absorb the rising water.

## Cohesion and Adhesion

If a narrow glass tube is dipped into mercury one notices that the mercury does not rise as water does, but falls below the surface (Fig. 2.24(i)). The surface, or *meniscus*, curves downwards, whereas a water

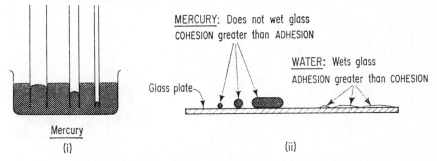

Fig. 2.24 (i) Capillarity; (ii) Cohesion and Adhesion

meniscus curves upwards. Glass molecules attract the water molecules near to it more strongly than does the rest of the water, and consequently the surface curves upwards near the sides of the container (Fig. 2.23(i)). Mercury molecules, however, are attracted more strongly by mercury molecules than glass molecules, and consequently the surface slopes downwards near the sides of the vessel.

*Cohesion* is the name given to the attractive forces between molecules of the same substance. *Adhesion* is the name given to the attractive forces between molecules of unlike substances. Thus the force of attraction between molecules of mercury is a cohesive force; that between a molecule of water and a molecule of glass is an adhesive force.

The adhesion between water and glass is greater than the cohesion of water, whereas the cohesion of mercury is greater than the adhesion between mercury and glass.

When adhesion is greater than cohesion a liquid spreads over a surface and wets it. When cohesion is greater than adhesion the liquid will not wet the surface. On this account water wets a clean glass surface, whereas mercury remains in droplets on the glass (Fig. 2.24(ii)).

*Demonstration of Adhesion and Cohesion*

Suspend a microscope slide horizontally by cotton thread and lower it gently so that its lower surface just touches the surface of water in a beaker. Raise the slide and observe that some water clings to the surface. Repeat with mercury, using a clean slide, and observe any difference. Investigate adhesion and cohesion by using other liquids and other solids than glass. Make a table of your results.

## Angle of Contact

The water meniscus near clean glass curves upwards and runs practically parallel to the glass (Fig. 2.25(i)). We therefore say that the *angle of contact* with the glass is zero. Correspondingly, water spreads along or 'wets' a clean glass surface when some is spilt on it. The meniscus of paraffin oil meets glass, however, at an acute angle of about 26° (Fig. 2.25(ii)). A mercury meniscus curves downwards where it meets glass, as we have seen, and it makes an obtuse angle of contact of about 140° with glass (Fig. 2.25(iii)). Correspondingly, mercury forms droplets when spilt on a flat glass surface, that is, it does not 'wet' glass.

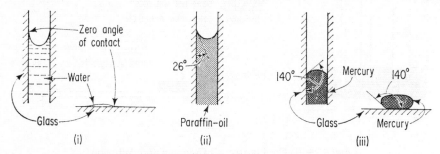

Fig. 2.25 Angles of contact – wetting and non-wetting

The angle of contact changes with impurities in the liquid and with the nature of the solid surface it meets. Thus water forms droplets on unclean glass of windows or when spilt on cardboard. It then resembles mercury in these cases, as the angle of contact has changed from zero to an obtuse angle.

Besides their wetting properties (p. 41), *detergents* produce a change in the angle of contact between dirt of a fatty or grease nature and the fabric material. The detergent solution increases the angle of contact between the grease and the fibre, so that the grease rolls up into a globule (see Plate 1). It is detached from the fibre by agitation in the washing process and passes into the detergent solution.

## Viscosity

Water can be poured from a beaker at a much faster rate than engine oil, which in turn pours faster than treacle. Liquids which pour slowly are termed *viscous* liquids. *Viscosity* is due to the internal friction which exists between layers of liquid or gas in motion.

We can observe the effect of viscosity by filling a large gas jar with water and placing two marks on the outside of the jar with strips of paper, for instance, some distance apart. Release styrocell beads, which are small and light, so that they fall down the centre of the jar, and time their fall between the marks. Repeat the experiment with different distances between the marks.

Heat the beads in boiling water for five minutes. The beads contain an expanding agent, which makes them expand to many times their original volume. Now study the motion of some of these 'expanded' beads in *air*, using two marks on the wall well separated to measure the velocity. Observe the slow fall of the beads, which shows the viscosity of the air.

The bob of a pendulum comes to rest much more quickly with a card attached to it owing to the viscosity of the air. Investigate this effect.

## Comparison of Viscosities

The viscosities of liquids can be compared by the following experiment:

### Experiment

Take a tall, wide cylindrical glass vessel V marked every 10 cm of its length with strips of gummed paper A, B, C, D, E and fill it with glycerine (Fig. 2.26). Allow it to stand until no air bubbles are present. Drop a small ball-bearing P centrally down the vessel and take the time as it passes between successive marks. Repeat with other ball-bearings of equal radius.

Observe that the ball-bearing takes equal times to cover equal distances *after moving down through the liquid.* The velocity thus increases to a maximum. This is called the *terminal velocity* of the falling ball-bearing. We explain later that an object moves with a constant velocity when there is no resultant force on it (p. 104). At first the velocity of the ball-bearing increases because there is a

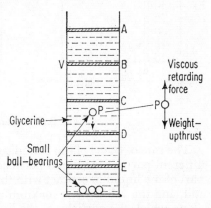

Fig. 2.26 Motion in viscous liquid

downward force on it due to its weight and little opposing frictional force due to viscosity; but as the ball-bearing gathers velocity the frictional force grows, and at one point it neutralizes the effect of the weight. The ball-bearing then moves with constant velocity. The upward force due to liquid thrust has also to be taken into account in a complete discussion, as explained on p. 104.

Repeat the experiment with: (i) water; (ii) engine oil, and observe the difference. Place engine oil, water and glycerine in order of viscosity.

## Lubrication

The subject of viscosity plays an important part in lubrication of engines and other machines, where it is essential to keep metal surfaces from rubbing against each other. If this is not done effectively the metals wear quickly and become hot, and the surfaces seize together. The oil used forms an extremely thin film of molecular dimensions, yet it is sufficiently strong to bear a heavy load. The viscosity of a liquid decreases fairly rapidly with rise of temperature, but engine oils called 'viscostatic' oils can now be used in both summer and winter, without changing them.

### Streamline and Turbulent Flow

The velocities of the different layers or particles of a flowing liquid can be shown by placing some glycerine at one end of a rectangular dish and confining it there by a straight tight-fitting barrier (Fig. 2.27(i)). The edge AB of the liquid is then marked with coloured ink or dye. When the dish is tilted slightly about the edge K and AB is removed the liquid flows down the dish. The coloured boundary is then seen to move along in a wide arc, reaching positions such as ACB and ADB as time goes on (Fig. 2.27(ii)).

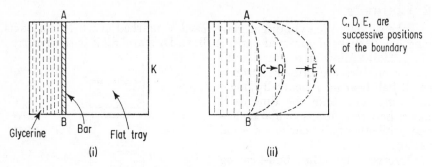

Fig. 2.27 Streamline (uniform) liquid flow

The rate of travel of the boundary is an indication of the velocities of the different parts of the liquid between A and B. It should be noted that the part of the liquid in contact with the sides of the vessel at A and B remains still, and the velocity is a maximum mid-way between A and B. The lines along which the particles of the liquid move are called *flow-lines*. The experiment shows that all the particles at a particular distance from the sides of the vessel flow with the same velocity. This is called *streamline* or *uniform* or *laminar* flow.

When the experiment is repeated with water, made to flow fast by tilting the dish, slow- and fast-moving parts of the liquid now mix. This is called *turbulent* flow.

### Examples of Streamline and Turbulent Flow

Streamline flow is seen when a motor boat travels slowly through the water and turbulence when it travels fast. When a water tap is turned on slightly so that the water velocity is small the jet coming out is seen to be a thin transparent one; this is streamline flow. If the tap is now turned on full the velocity of the water is high and the column from the tap is seen to be turbulent.

metallic crystal, but it has a low strength. If such a crystal is hammered however, it splits up into many smaller crystals differently oriented with respect to each other, and the mechanical strength is then increased. This is called 'cold-working' the metal.

If a metal is heated sufficiently its crystals become more closely aligned and its strength is then decreased. The metal is now softer and is said to be 'annealed'. The strength of a metal can often be increased by forming an alloy with another suitable metal. The atoms of the second metal must be a suitable size, so as to fit in to vacant spaces between the crystals of the first metal and prevent them joining together.

## Forces between Molecules

In general, molecules may contain almost any number of atoms. Those having only one atom are called *monatomic*, those with two are *diatomic*,

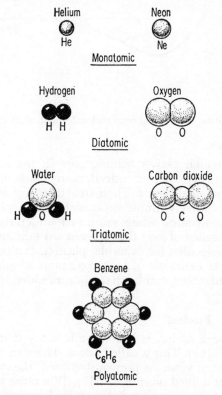

Fig. 2.31 Atoms in molecules

those with three are *triatomic*, etc. Some examples of molecules are given in Fig. 2.31.

Fig. 2.32 shows roughly the force which exists between two molecules. For separations greater than a particular distance, corresponding to a minimum potential energy, the force is attractive and decreases with increasing distance. At closer distances the force changes to a repulsive one, so that it is difficult to compress the molecules together. Thus solids or liquids resist compression. The equilibrium distance of separation, where the attractive and repulsive forces balance, is about 3 or $4 \times 10^{-8}$ cm.

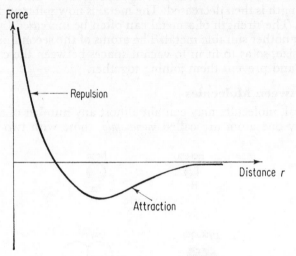

Fig. 2.32 Variation of force between molecules with distance (*exaggerated*)

### Large Molecules

The molecules which play a leading part in the structures of living organisms are often very complex indeed, many consisting of thousands of atoms linked by chemical bonds. Their structures are usually made up of long sequences of relatively simple structural units, arranged in a definite manner. Protein molecules, when examined under the electron microscope, are seen to consist of long thin threads without any apparent inner structure. Large molecules, for example, plastics, can be man-made. The term 'plastic' covers materials of high molecular weight which, in some stage of their manufacture, can be shaped or moulded by pressure or heat or both.

### Change of State · Fusion

When a solid, such as ice, is heated sufficiently it melts and changes completely into a liquid. This is called *fusion*. This can be explained from the molecular point of view. As previously stated, the molecules of a solid are vibrating about a fixed position (p. 47). When the solid is heated the molecules vibrate through greater distances than before. They gain energy because heat is a form of energy (p. 266). By analogy, a light ball, hanging from one end of a piece of elastic and vibrating up and down, will vibrate through a greater distance if it is given more energy by hitting it repeatedly in a downward direction. If the ball is struck very hard so that it gains considerable energy it may break away from the elastic after stretching it. In the same way, if the molecules of a solid are heated sufficiently they obtain enough energy to break away from their 'anchored' position in the solid and become free. The molecules now have translational energy in addition to vibrational energy. Since they are free to move about they no longer form a solid structure.

## Evaporation · Boiling

The molecules in a liquid move with different speeds. Those near the surface with sufficiently high speeds or energies are able to escape through the surface and exist outside as separate particles of vapour or gas. We say the liquid *evaporates*. When the liquid is heated more molecules gain energy to escape, and eventually a continuous stream of molecules escape through the surface. The liquid is now *boiling*–all the liquid changes into a gas. It can be seen that the heat energy supplied is used in overcoming the forces of attraction between the liquid molecules. The gas molecules formed are relatively very much farther apart than the liquid molecules.

When a gas is liquefied a change occurs which is the reverse to that described. On cooling, the thermal energy of the molecules is reduced, the volume occupied by the molecules decreases and at one stage the strong attractive forces between the molecules produce a liquid state. When the liquid is cooled its thermal energy diminishes further and at one stage it solidifies. The molecules then stay at an average anchored position, which gives a definite shape to the solid form.

## SUMMARY

1. Solids have volume and shape; liquids have volume but no shape; gases have no definite volume or shape.

2. The densities of solids and liquids are about 1000 times as great as gases. The packing of the molecules is correspondingly greater.

3. The molecules of gases and liquids are in irregular motion–shown by the phenomenon of Brownian movement and by diffusion.

4. The diameter of a molecule is of the order of several times $10^{-8}$ cm. A rough estimate of the size of a molecule may be obtained by measuring the thickness of an oil-film on water. A higher value may be estimated from the thickness of gold-leaf.

5. The forces between the molecules of a solid are strong, they are less strong in liquids and weak in gases.

6. The molecules of a solid vibrate about a particular position inside it–they have vibrational energy; the molecules of a liquid move freely through the liquid–they have translational energy and some vibrational energy; the molecules of a gas move freely in the space in which they exist–they have translational energy, and vibrational and rotational energy in addition if the molecule is made up of two or more atoms.

7. In solids such as metals, elasticity shows that large forces can change the shape and length of a metal, but it can recover; larger forces produce a yield or plastic stage. A study of crystals is necessary to explain the behaviour of metals. Frictional forces between two surfaces are due to the welding which takes place at small areas of contact between groups of molecules.

8. In liquids the phenomenon of surface tension is due to the attractive force on molecules in the surface by molecules below; drops of liquid under surface-tension forces have a spherical shape. Viscosity or internal friction is due to the movement of one liquid layer over another because the layers have different velocities.

9. The bonds between the molecules in a solid are broken when fusion takes place owing to the additional energy given to the molecules on heating. Some molecules in a liquid near the surface have sufficient energy to escape and exist outside–this is 'evaporation'. At the boiling point all the molecules gain sufficient energy from the heat supplied to escape through the surface.

## EXERCISE 2 · ANSWERS, p. 56

### Density and Relative Density

**1.** A solid cube of iron has each side 10 cm long. Find its mass if the density of iron is 6·5 g/cm³. What volume of iron has a mass of 130 g?

**2.** 'The relative density of brass is 8·5.' What does this statement mean? Find the mass of 500 cm³ of brass.

**3.** A thread of mercury has a mass of 10·2 g and is in a tube of uniform cross-section 0·1 cm². Calculate the length of the thread if the density of mercury is 13·6 g/cm³.

**4.** 100 g of water is mixed with 60 g of a liquid of relative density 1·20. Assume no change in volume, find the average relative density of the mixture.

**5.** Describe an experiment to measure the density of air. Point out any difficulties you would expect in getting an accurate result.

**6.** Distinguish between the density and relative density of a material. Describe how you would use a relative density (specific gravity) bottle to find the density of sand. Point out the precautions to be taken to obtain a good result.

80 cm³ of water are mixed with 140 cm³ of liquid of relative density 0·83. What is the density of the mixture if there is no change in total volume on mixing? (*N.*)

**7.** Define *density* and *relative density*.

Describe how you would determine the relative density of lead, provided in the form of lead shot.

Determine, from the following readings, the value of the relative density of lead:

Mass of empty bottle, 23·05 g.

Mass of bottle and lead shot, 81·55 g.

Mass of bottle and shot, with water added to fill completely, 101·55 g.

Mass of bottle completely filled with water, 48·55 g. (*O.*)

**8.** Distinguish between *density* and *relative density*. How would you use a density bottle to determine the density of: (*a*) turpentine; (*b*) sand? (*L.*)

**9.** Distinguish between *density* and *relative density*. How would you attempt to determine the density of air under standard laboratory conditions?

A cylindrical beaker of mass 50 g, cross-sectional area 25 cm² and height 10 cm is filled with oil of relative density 0·8. What does the whole weigh? A piece of aluminium, of mass 66 g and relative density 2·2, is lowered carefully into the beaker. What volume of oil overflows? What is the final weight of the beaker and its contents after the outside has been wiped to remove overflow liquid? (*O. and C.*)

**10.** Define *density* and *relative density*.

Describe how you would use the density bottle to determine the relative density of sand.

A cube of glass, of 5 cm side and weighing 306 g, has a cavity inside it. If the density of glass is 2·55 g/cm³, what is the volume of the cavity? (*O.*)

**11.** Describe carefully how you would attempt, by a physical method, to discover the proportions in which two known metals are mixed, in order to form a given sample of alloy.

State any assumptions you make and any principle upon which your method is based.

A 1-cm³ sample of tin–lead alloy weighs 8·5 g. The density of tin is 7·3 g/cm³ and that of lead is 11·3 g/cm³. Calculate the percentage by weight of tin in the alloy. You may assume that there is no change of volume when the metals form the alloy. (*O.* and *C.*)

## Molecules

**12.** Describe and explain one experiment in each case to show that the molecules of (i) a liquid, and (ii) a gas have motion. What evidence shows that the molecules of a solid have motion?

**13.** Assuming water has a density of 1 g/cm³ and water vapour a density of 0·8 g/litre under ordinary conditions, estimate the ratio of the volumes occupied by the same number of molecules of water in the gaseous and in the liquid state. If the gas and the liquid are each contained in a vessel shaped in the form of a cube, estimate the ratio of the distances apart of the molecules. Describe an experiment to show there are spaces between the molecules in a liquid.

**14.** Draw a sketch showing roughly the Brownian motion of a particle suspended in a gas or a liquid. What does the motion indicate? State, and explain, in what circumstances a particle shows no Brownian motion.

**15.** Describe an experiment to observe Brownian motion in a gas, and label the diagram drawn. How do you know that the motion observed is not due to convection or to temperature changes?

**16.** A 1 cm³ pipette was filled with oil and 50 drops were counted as it drained completely. One drop of oil was then allowed to fall on a water surface, and spread into a circle of diameter 40 cm. Estimate an upper limit for the diameter of an oil molecule and explain your reasoning.

**17.** 100 pieces of gold foil, each 5 cm square, weigh 0·5 g. Assuming the density of gold is 20 g/cm³, find the thickness of the foil. If each gold molecule is $4 \times 10^{-8}$ cm diameter, how many molecules, approximately, are in this thickness of foil?

**18.** An olive-oil drop was found to have a diameter of 0·05 cm when viewed against a ½-mm scale. When it was dropped on to a water surface it spread into a circle of diameter 10 cm. Estimate the upper limit for the diameter of an oil molecule. Write down a rough value for the actual diameter of an oil molecule.

**19.** Describe one experiment in each case to show that molecules have: (i) motion; (ii) size; (iii) irregular motion.

State and explain the behaviour of (i) a tiny particle, (ii) a large particle suspended in a liquid medium. What effect is observed in each case when the temperature of the liquid is raised?

**20.** Kinetic energy, the energy of motion, can be classified into translational, vibrational and rotational energy. State the kind of kinetic energy which molecules of (i) a solid, (ii) a liquid, (iii) a gas may have and explain your answer.

**21.** Describe fully an experiment to estimate the size of a molecule. List the errors in the experiment and write down an approximate value of the diameter of a molecule.

**22.** Your friend says: *Molecules have no existence.* Explain how you would convince your friend that he or she was wrong.

**23.** What are the chief differences between *solids*, *liquids* and *gases*? State what you know about the relative differences between their molecular forces, giving reasons for your statements.

**24.** What is meant by the term *diffusion*? Describe an experiment you would perform to demonstrate diffusion in a liquid. Apply the kinetic theory to explain the result of the experiment you describe. (*C.*)

**25.** Red blood corpuscles expand rapidly when they are placed in water, but shrivel when placed in salt solution. What phenomenon does this show, and what is the explanation?

**26.** What is *osmosis*? How would you demonstrate the osmotic pressure of a solution of cane sugar? Give two examples of osmosis in everyday life.

## Effects of molecular forces

**27.** Explain the meaning of the term *elasticity*.

What is Hooke's law as applied to: (*a*) the stretching of a spring; (*b*) the stretching of a rod or wire? What kind of change in the material takes place in each case?

**28.** Draw a graph showing the stretching of a sample of steel and indicate on it the *elastic limit* and the *yield point*. If the wire is stretched beyond the elastic limit, what graph is now obtained as the load is slowly removed?

**29.** What change, if any, would occur in the frictional force between a sledge and the ice over which it slides if: (*a*) the sledge runners are lengthened; (*b*) the load is increased; (*c*) the speed is increased? (*N.*)

**30.** Define *coefficient of friction* and describe how it can be determined experimentally.

A body on an inclined plane is on the point of sliding down the plane. Draw a diagram showing the forces acting on it. If the coefficient of friction is $\frac{1}{8}$, what is the inclination of the plane to the horizontal?

Explain the part played by friction between the tyres and road surface in the forward propulsion of a motor-car. (*C.*)

**31.** Show graphically the relation between the extension of a wire and the load attached to it. Label the axes of your graph and point out the main features shown by the curve. (*L.*)

**32.** Describe two simple experiments to illustrate that the surface of a liquid behaves like an elastic skin.

Using a molecular explanation, show why the surface has this property.

**33.** State Hooke's law and describe how you would test its validity for a spiral spring.

Explain briefly how the action of a spring balance depends on this law. A light rod AB, 20 cm long, is supported horizontally by two exactly similar vertical springs attached to its ends. A load of 2 newtons is hung from a point on the rod 15 cm from the end A, and the upper end of the spring at B has to be raised 2 cm in order to make AB horizontal again. Find the force needed to extend each of the springs by 1 cm. (*O.*)

**34.** How is the surface tension of a liquid explained in terms of the molecular theory of matter? Describe an experiment to show the surface tension of a liquid acting in such a way as to cause movement. Explain how the movement is caused. (*C.*)

**35.** What is meant by the viscosity of a liquid?

When a ball-bearing is allowed to fall through glycerine it soon attains a steady velocity. (i) What forces are acting on the ball-bearing? (ii) What relationship exists between these forces? (*O.*)

**36.** Explain the following observations and name the physical effects involved. (i) A drop of water remains for a time on the surface of a dry cloth before being absorbed, but a similar drop on a damp cloth is rapidly absorbed. (ii) A ball-bearing falling in motor oil soon attains a constant velocity. (iii) A few days after a crystal of copper sulphate has been placed at the bottom of a beaker of water a blue colour is observed throughout the water. (*C.*)

**37.** What is the difference between *adhesion* and *cohesion*? Illustrate your answer by reference to the forces between glass, water and mercury molecules when a glass capillary tube is dipped into water and then into mercury.

**38.** Using a molecular explanation, account for the spherical shape of droplets of liquid.

**39.** Draw sketches showing what happens to the level of (i) water and (ii) mercury when three capillary tubes of different diameter are dipped in turn into each liquid.

What is the *angle of contact* between: (i) water and clean glass; (ii) mercury and clean glass? Explain the different effects observed when water and mercury are each spilt in clean glass.

**40.** A toy duck has a piece of camphor at one end, and is observed to move across water on its own. Explain the movement.

**41.** Using a molecular explanation, account for the frictional force between metals when one moves over the other. How is the frictional force reduced?

**42.** What is the *viscosity* of a liquid? To what is it due? Describe an experiment to compare roughly the relative viscosities of two 'thick' liquids such as glycerine and treacle.

# ANSWERS TO NUMERICAL EXERCISES

## EXERCISE 2 (p. 52)

**1.** 20 cm³   **2.** 4250 g   **3.** 7·5 cm   **4.** 1·07   **6.** 0·89 g/cm³   **7.** 10·6
**9.** 250 g, 30 cm³, 292 g   **10.** 5 cm³   **11.** 60%   **13.** 1250/1, 11/1
**16.** $2 \times 10^{-5}$ cm   **17.** $10^{-5}$ cm; 250   **18.** $2 \times 10^{-6}$ cm   **30.** 7°   **33.** 0·5 N/cm

# Mechanics

Mechanics is the foundation of Physics. M. BORN

# 3

# DYNAMICS

*Dynamics* is the study of motion and of the forces which keep an object in motion or oppose its motion. It is a subject which concerns the forces which keep an aeroplane in flight through the air, or a speedboat in motion through water, or a racing car or train in motion along a road or track. All these machines have reached their present development through the skill of scientists and engineers, who have studied and understood the principles of dynamics.

Science, like Man, has evolved slowly through the centuries. As time goes on, new ideas are put forward, and new discoveries are made. The founder of mechanics is generally recognised to be Sir Isaac Newton, who published a work called *Principia* in 1687 in which 'Laws of Motion' were clearly stated for the first time (see p. 93).

## Ticker-timer

The simple motion of a moving object can be studied in a laboratory with the aid of a ticker-tape, on which equal intervals of time can be marked. The apparatus, which we shall call a 'ticker-timer', consists of a

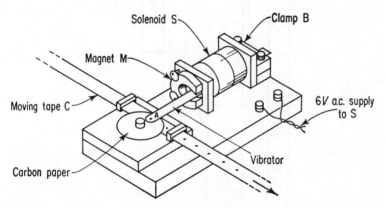

Fig. 3.1 Ticker-timer

flexible strip of soft-iron A clamped at one end B and passing through a solenoid S. Fig. 3.1. The free end of the iron passes between the poles of a strong magnet M, placed at the end of S. When a low alternating voltage from the 50 cycle per second mains is joined to the solenoid terminals the iron strip vibrates at this frequency. A stud beneath the iron at A strikes a carbon paper disc below it, thus marking a moving tape C, running past A, with a series of dots or ticks spaced at equal intervals of $\frac{1}{50}$ second.

**Motion**

A simple illustration of the information the ticker-timer can provide is shown in Fig. 3.2. In this case one end of the tape was held by a pupil, who then moved away through the room and collided with other pupils at intervals.

The tape shows clearly that the pupil started his or her movement at a

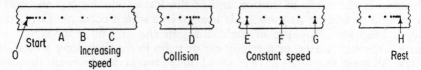

Fig. 3.2 Ticker-tape studies of motion

time corresponding to O, and then began to walk faster, since the distances travelled in equal times increase, as shown by the increasing gaps between A, B and C. A collision with another pupil occurred at D, after which the walk continued. It was a steady walk for a short period as the same distances EF and FG were travelled in successive equal times. Finally, the pupil came to rest at a time corresponding to H.

Fig. 3.3.(i) shows part of the tape after a particular period. To see how the distance $s$ travelled varied with time $t$, successive strips such as 1, 2, 3

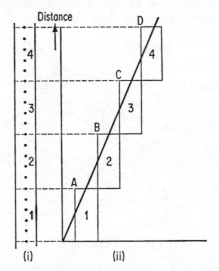

Fig. 3.3 Ticker-tape and distance–time graph

and 4 are cut off the tape. Their respective lengths are the distances travelled in equal intervals of time, 'five-tick' intervals or five-fiftieths ($\frac{1}{10}$) second in this case. We can therefore paste the strips A, B, C, D vertically one above the other on paper as shown in Fig. 3.3(ii), the equal widths of the paper now representing the equal intervals of time, $\frac{1}{10}$ sec. Choosing the middle of the tops of the strips for convenience, we find they join up in a straight line. Thus equal distances are travelled in successive equal times. The motion was therefore regular.

In contrast, the distance–time variation in Fig. 3.4 shows an irregular motion. It is the motion of a trolley X which started off quickly down a slope S at a time A, and then ran up an incline T from B to C. From A to B the distances travelled in equal times increase; from C to D the distances travelled decrease. The trolley's motion is thus irregular.

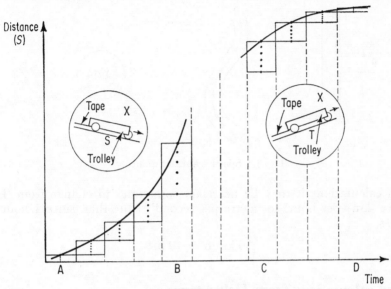

Fig. 3.4 Ticker-tape–irregular motion

## Speed

We can now proceed to the meaning of some common terms used in discussing motion.

If you run round the corner of a street from A to B to see a friend (Fig. 3.5(i)), and travel a distance of 80 m (metre) in 20 seconds, then your average *speed* is

$$\frac{80 \text{ m}}{20 \text{ s}} = 4 \text{ m/s}$$

Similarly, if a car takes 2 hours to travel along a winding road from a town C to another town D 96 km away, Fig. 3.5(ii), the average speed of the car

$$= \frac{\text{Distance}}{\text{Time}} = \frac{96 \text{ km}}{2 \text{ h}} = 48 \text{ km/h}$$

The term 'speed' is always used when the motion of an object changes direction. Thus if a satellite such as *Early Bird*, used for world television communication, takes 24 hours to move round the earth in a circular path 57 600 km long (Fig. 3.5(iii)), then

$$\text{Average speed} = \frac{57\ 600\ \text{km}}{24\ \text{h}} = 2400\ \text{km/h}$$

If the satellite moves through equal distances in equal times, no matter how small the times may be, then the satellite is said to have a constant or *uniform* speed.

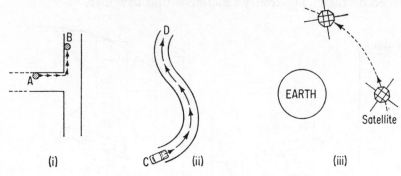

(i)                              (ii)                              (iii)

Fig. 3.5 Speed is a scalar quantity

In calculations it may be necessary sometimes to change from 'kilometre (km) per hour' to 'metre per second'. Note that, since 1 hour = 3600 s,

$$36\ \text{km/h} = 10\ \text{m/s}$$

### Speed calculations from Ticker-tapes

The ticker-tape enables us to find the average speed of objects to which it is attached. As an illustration, Fig. 3.6(i) shows part of the tape A attached to a small trolley moving along a horizontal rough board. In a time of 5 dots or ticks, five-fiftieths or $\frac{1}{10}$ s, a distance of 4·0 cm has been travelled. Hence

$$\text{Average speed} = \frac{4\cdot0\ \text{cm}}{\frac{1}{10}\ \text{s}} = 40\ \text{cm/s}$$

Fig. 3.6(ii) shows that the distances between the dots or ticks at another part B of the tape are now equal, so that the speed is uniform over the time of $\frac{5}{50}$ or $\frac{1}{10}$ s. In this case,

$$\text{speed} = \frac{2\cdot5\ \text{cm}}{\frac{1}{10}\ \text{s}} = 25\ \text{cm/s}$$

We could say that the speed was uniform from one tick to the next, that is, over a very small time-interval of $\frac{1}{50}$ second, if we had a clock which could measure extremely short intervals of time and a mechanism for recording the short distances travelled by the trolley. In practice, the speed of moving vehicles can be recorded directly on an instrument by using radar equipment, as in speed traps, which sends out a radio signal and receives it reflected back from the vehicle. Speedometers in cars record speeds directly.

Speeds can vary from the very slow speed of the tortoise, which may be $\frac{1}{10}$ kilometre per *hour*, to that of radio or light waves, about 300 thousand kilometres per *second*. Astronomers use a unit of distance called a *light-year*

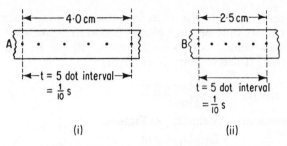

Fig. 3.6 Retardation by ticker-tape

in dealing with the enormous distances in the Universe; it is the distance travelled by light in a year. The nearest visible star, Alpha Centauri, is about 4 light-years from the earth.

### Velocity

Consider a car moving round a circular track ABC with a uniform speed. Fig. 3.7(i). At the instant it reaches A the car points along the direction of the arrow shown, which is a tangent to the circle. At other places, such as B and C, the car points in different directions. If the *direction* as well as the

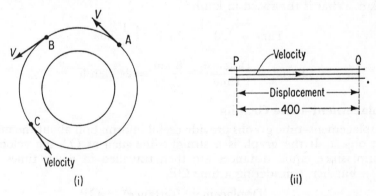

Fig. 3.7 Velocity is a vector quantity

magnitude of the car's motion is considered, we speak of the *velocity* and not the speed. Thus the 'speed' of the car may be the same at A, B or C as it goes round a circle, but the 'velocities' $v$, at the three points are different because the directions are different. Velocity, then, has direction and magnitude. Speed has magnitude only, and we therefore speak of the average speed of a car along a winding road.

If a car moves in a constant direction we use the word 'displacement' rather than 'distance'. Thus along a perfectly straight road PQ 400 m long, Fig. 3.7(ii), which takes a car 20 s to travel,

$$\text{Average velocity} = \frac{\text{Displacement}}{\text{Time}} = \frac{400 \text{ m}}{20 \text{ s}}$$
$$= 20 \text{ m/s}$$

If the displacement along the road in equal times is constant no matter how small the times may be, the velocity is said to be *uniform*.

A train with a velocity of 96 km/h travels 192 km in 2 h. Generally, it can be seen that, if the velocity is uniform,

$$\text{Velocity} = \frac{\text{Displacement}}{\text{Time}}$$

$$\text{Displacement} = \text{Velocity} \times \text{Time}$$

$$\text{Time} = \frac{\text{Displacement}}{\text{Velocity}}$$

*Examples*

1. A car has a velocity of 72 km/h. How far does it travel in $\frac{1}{2}$ minute?

We must *always* use the same kind of unit in calculation. Thus 72 km/h = 20 m/s and

$$\frac{1}{2} \text{ min} = 30 \text{ s}$$
$$\therefore \text{ Displacement} = 20 \times 30 = 600 \text{m}$$

2. A car moving round a circular racing track takes 120 s to do a lap of 8 km. What is the speed in km/h?

$$\text{Time} = 120 \text{ s} = 2 \text{ min} = \frac{1}{30} \text{ hour}$$

$$\text{Speed} = \frac{\text{Distance}}{\text{Time}} = \frac{8 \text{ km}}{\frac{1}{30} \text{ hr}} = 240 \text{ km/h}$$

## Displacement–time Graphs

Displacement–time graphs provide useful information about the motion of an object. If the graph is a straight line such as OA the velocity is uniform, since equal distances are then travelled in equal times (Fig. 3.8(i)). Further, considering a time OP,

$$\text{Velocity} = \frac{\text{Displacement (distance)}}{\text{Time}} = \frac{\text{QP}}{\text{OP}}$$

The ratio QP/OP is called the *gradient* of the line OA. Hence

$$\text{Velocity} = \text{Gradient of OA}$$

The gradient is constant everywhere along OA, since it is a straight line.

Fig. 3.8(ii), however, shows the displacement–time curve OLM for the motion of a ball thrown vertically in the air. It reaches a maximum height at L, and then descends again to the ground at a time M. This is an example of non-uniform velocity. The velocity at any height such as B is now the gradient of the *tangent* to the curve at B, which is BS/RS. At the

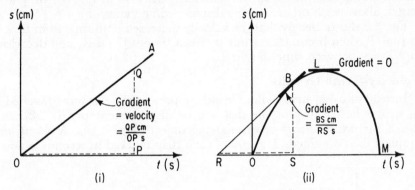

Fig. 3.8 (i) Uniform velocity; (ii) Variable velocity

highest point L the velocity is similarly the gradient of the tangent here, which is zero; momentarily the ball is at rest.

### Velocity–Time Variation with Ticker-tape

Fig. 3.9 shows successive tape strips pasted beside each other after a run by a trolley down an incline AB, then along the horizontal BC and then up an incline CD, as shown inset in Fig. 3.9. Each strip represents the actual

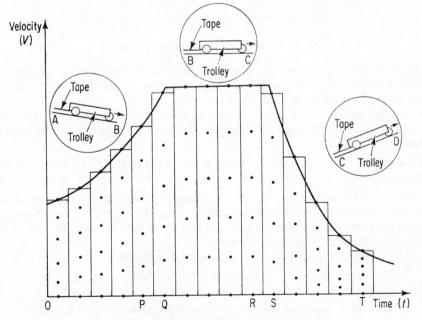

Fig. 3.9 Measurement of velocity changes

distance travelled in successive $\frac{5}{50}$ or $\frac{1}{10}$ s, and thus if the tops are joined the variation in the velocity at the end of equal time-intervals can be seen from the line obtained. Observe carefully that here the strips are pasted beside each other on the time-axis; whereas in Fig. 3.4 they were pasted above each other to show distance–time variation.

Fig. 3.9 shows clearly that the velocity increased with time from time O to time P, then became constant between times Q and R, and decreased in velocity between times S and T.

### Velocity–Time Graphs

Instead of using strips whose lengths represent the distance travelled in $\frac{1}{10}$ s, it is better to plot the distance or displacement per 1 s, because this is the *actual velocity* of the moving object (Fig. 3.10). A *velocity-time* graph is then obtained. The *area* of each strip marked in seconds is now a

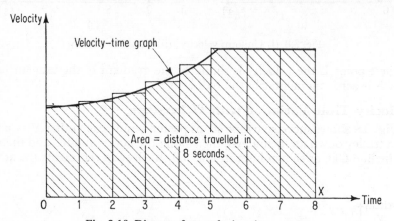

Fig. 3.10 Distance from velocity–time graph

measure of the distance travelled in each second of the motion. Thus the total area beneath the velocity–time graph is a measure of the total distance travelled from time O to time X.

The speedometer of a car measures speed directly. Along a straight road suppose the variation of velocity with time is recorded as follows in a short run:

| Time (s) | 0 | 2 | 4 | 6 | 8 | 10 | 12 | 18 | 24 | 30 | 32 | 34 | 36 | 38 | 40 | 42 |
|---|---|---|---|---|---|---|---|---|---|---|---|---|---|---|---|---|
| Velocity (m/s) | | 0 | 12 | 24 | 36 | 48 | 60 | 60 | 60 | 60 | 60 | 50 | 40 | 30 | 20 | 10 | 0 |

Fig. 3.11 shows the velocity–time graph of the motion. We can see from it that the velocity increased uniformly along the part OA, then became constant over the part AB and came to rest along the part BC when its velocity decreased uniformly to zero.

The respective distances travelled were as follows:

In time OH, Distance = Area of triangle OAH = $\frac{1}{2}$OH . AH
$= \frac{1}{2} \times 10$ (s) $\times 60$ (m/s) $= 300$ m.

In time HK, Distance = Area of rectangle HABK = HK . AH
$= 20$ (s) $\times 60$ (m/s) $= 1200$ m.

In time KC, Distance = Area of triangle BKC = $\frac{1}{2}$ KC . BK
$$= \frac{1}{2} \times 12 \text{ (s)} \times 60 \text{ (m/s)} = 360 \text{ m.}$$
Hence, total distance = 1860 m

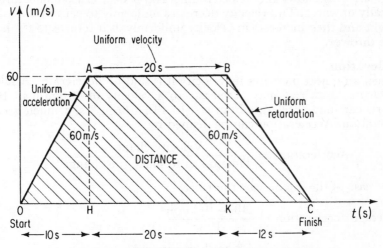

Fig. 3.11 Calculation from velocity–time graph

The total distance travelled would also be given by the area of OABC, which is a trapezium. Hence

total distance = $\frac{1}{2}$ sum of parallel sides × height
$$= \frac{1}{2} \text{ (AB + OC)} \times = \frac{1}{2} \text{ (20 + 42)} \times 60$$
$$= 31 \times 60 = 1860 \text{ m}$$

## Other Velocity–Time Graphs

Fig. 3.12(i) shows the velocity–time graph of a lift whose velocity increases uniformly along OA and then decreases uniformly in velocity to rest along AB. The distance travelled is the area of the triangle OAB.

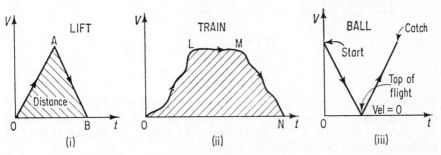

Fig. 3.12 Velocity–time graph

Fig. 3.12(ii) shows the velocity–time graph of a train between two stations. The train's velocity increased along OL as it pulled away from the station, it travelled with fairly uniform velocity for a time along LM, and finally decreased in velocity to rest at N. The distance travelled is the area below the curve OLMN. Similar graphs are obtained in speed tests on railway engines by British Railways.

Fig. 3.13(iii) shows the velocity–time graph of a cricket ball thrown vertically upwards. The velocity decreases uniformly to zero at the top of its flight and then increases in velocity uniformly as it returns to the hand of the thrower.

### Acceleration

When an object increases its velocity it is said to 'accelerate'. A saloon car starting from rest may accelerate to a velocity of 48 km/h in 10 s; a sports car may accelerate from rest to 48 km/h in 3 s, a much greater acceleration. Acceleration can be defined by:

$$\text{Acceleration} = \frac{\text{Velocity change}}{\text{Time taken to make the change}}.$$

In the case of the saloon car, therefore,

$$\text{Average acceleration} = \frac{(48 - 0) \text{ km/h}}{10 \text{ s}}$$

$$= 4{\cdot}8 \text{ km/h per second}$$

In the case of the sports car,

$$\text{Average acceleration} = \frac{(48 - 0) \text{ km/h}}{3 \text{ s}}$$

$$= 16 \text{ km/h per second}$$

The 'zero' in the numerator of the acceleration value has been deliberately inserted. It emphasises that acceleration always refers to a velocity *change*, not to the actual velocity. Thus a car travelling with a uniform velocity of 48 km/h has no acceleration. A car accelerating from 48 to 80 km/h in 4 s has an average acceleration given by

$$\frac{\text{Velocity change}}{\text{Time}} = \frac{(80 - 48) \text{ km/h}}{4 \text{ s}}$$

$$= \frac{32 \text{ km/h}}{4 \text{ s}}$$

$$= 8 \text{ km/h per second}$$

This means that the velocity increases by 8 km/h every second. Thus if the acceleration remains constant after 80 km/h is reached the velocity increases 2 s later by 2 × 8 or 16 km/h. The velocity reached is thus 16 km/h more than 80 km/h, or 96 km/h.

### Non-uniform and Uniform Acceleration

Fig. 3.13(i) shows lengths of successive tape strips when a heavy block of wood first moves down a rough inclined plane. Successive lengths are the distances travelled in equal times of 5 ticks or dots, and hence it can be seen from joining the tops of the strips that the velocity increases in an irregular way with time. This is an example of *non-uniform* acceleration. By contrast, if tape is attached to a trolley on a smooth inclined board, and the trolley is released so that it runs down the board, the tape has the appearance of Fig. 3.13 (ii) after the run. On cutting the tape after successive equal time intervals and pasting the strips beside each other, the

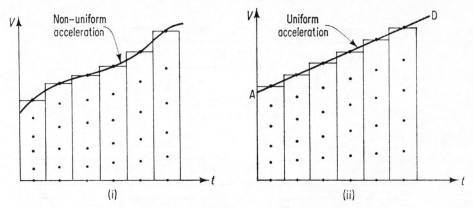

Fig. 3.13 Ticker-tape and acceleration

tapes now lie on a fairly straight line AD. The acceleration of the object is therefore *uniform*.

*Uniform acceleration* is defined as *the motion of an object whose velocity increases by equal amounts in equal times, no matter how small the time intervals may be.*

### Calculation of Acceleration from Ticker-tape

The ticker-tape enables the acceleration of a moving object to be found. Suppose A and B are two parts of the tape attached to a trolley accelerating non-uniformly down an incline with a time-interval corresponding to 100 ticks or 2 sec between them (Fig. 3.14(i)).

During the period of $\frac{5}{50}$ or $\frac{1}{10}$ s on A,

Average velocity $= 2 \cdot 0$ cm$/\frac{1}{10}$ s $= 20$ cm/s

During the period of $\frac{1}{10}$ s on B,

Average velocity $= \dfrac{3 \cdot 0 \text{ cm}}{\frac{1}{10} \text{ s}} = 30$ cm/s

Hence average acceleration $= \dfrac{\text{Velocity change}}{\text{Time}}$

$= \dfrac{30 - 20 \text{ (cm/s)}}{2 \text{ (s)}}$

$= 5$ cm/s²

Uniform acceleration can be calculated in the same way. The first strip at C in Fig. 3.14(ii) has a length of 4·5 cm, and this is the distance travelled during $\frac{5}{50}$ or $\frac{1}{10}$ s. The velocity here is thus

$$4\cdot5 \text{ cm}/\tfrac{1}{10} \text{ s or } 45 \text{ cm/s}$$

At the later time corresponding to the strip D, the velocity is

$$6\cdot5 \text{ cm}/\tfrac{1}{10} \text{ s or } 65 \text{ cm/s}$$

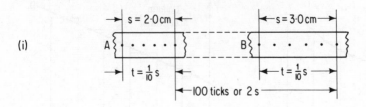

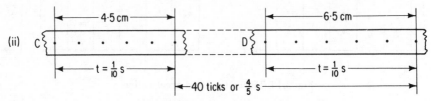

Fig. 3.14 Measurement of acceleration

The time from C to D is $\frac{4}{5}$ s. Hence

$$\text{Acceleration} = \frac{\text{Velocity change}}{\text{Time}}$$

$$= \frac{(65 - 45) \text{ cm/s}}{\frac{4}{5} \text{ s}}$$

$$= 25 \text{ cm/s}^2$$

## Acceleration from Velocity–Time Graphs

We can now see how acceleration can be measured generally from a velocity–time graph. From Fig. 3.15, for example,

$$\text{Uniform acceleration along OA} = \frac{\text{Velocity change}}{\text{Time}}$$

$$= \frac{\text{LA}}{\text{OL}}, \text{ if the time is OL.}$$

The ratio LA/OL is the *gradient* of the line OA. Hence

*Acceleration = Gradient of velocity–time graph.*

Along OA the gradient is constant. In calculating the acceleration from the gradient, we read LA from the velocity-axis and OL from the time-axis.

In Fig. 3.15, AB represents uniform or constant velocity and the gradient or acceleration is zero. The velocity–time graph represented by BC is a line with a steeper gradient than OA, and hence the acceleration here is greater. When the graph shows an irregular velocity variation as along CD, the acceleration at a time corresponding to X, for example, is the gradient of the *tangent* to the graph at this point, as shown (compare p. 65).

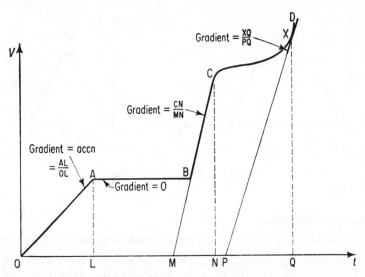

Fig. 3.15 Acceleration from velocity–time graph

## Retardation or Deceleration

If the brakes of a bicycle or car are applied the velocity decreases. The bicycle or car is then said to undergo 'retardation' or 'deceleration'. 'Uniform retardation' means that the velocity decreases by equal amounts in equal times, no matter how small the times may be. Like acceleration,

$$\text{Retardation} = \frac{\text{Velocity change}}{\text{Time taken}},$$

but as the velocity decreases with time it can be considered as 'negative acceleration' (see p. 73).

Suppose a train slows uniformly from 98 to 48 km/h in 10 s. Then 2s after

$$\text{Retardation} = \frac{(98 - 48) \text{ km/h}}{10 \text{ s}}$$
$$= 5 \text{ km/h per second}$$

the velocity of 48 km/h is reached, therefore, the velocity decreases by 5 × 2 or 10 km/h. The velocity is then 48 − 10 km/h or 38 km/h. It can be seen that 9·6 s after 48 km/h is reached the train comes to rest.

Fig. 3.16 illustrates the velocity–time curve OABR of a train which departs from a station with uniform acceleration at a time O to a time P, then moves with a uniform velocity from P to Q, and is uniformly retarded from Q to R at the next station. The retardation is given by the gradient of the line BR, which is BQ cm/s ÷ QR s. The uniform retardation of a lift whose velocity–time graph is OSB, Fig. 3.16(ii), is the gradient SA/AB of the line SB.

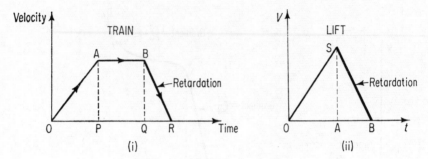

Fig. 3.16  Velocity–time graphs

## Summary of Graphs

We can summarize the information obtainable from graphs as follows:

1. *Displacement* (*s*)–*time* (*t*). The *gradient* at a point represents velocity.
2. *Velocity* (*v*)–*time* (*t*). (i) The *area* between the graph and the time-axis represents displacement or distance.
   (ii) The *gradient* at a point represents acceleration or retardation.

## Formulae with Uniform Acceleration

Suppose a car is moving with a velocity of 24 km/h along a straight road and accelerates uniformly to a velocity of 72 km/h in 10 s. Then

$$\text{Acceleration} = \frac{\text{Velocity change}}{\text{Time taken}}$$

$$= \frac{(72 - 24)\ \text{km/h}}{10\ \text{s}}$$

$$= 4 \cdot 8\ \text{km/h per second}$$

If the units of velocity are expressed in m/s, then, since 24 km/h = 6·7 m/s and 72 km/h = 20 m/s,

$$\text{Acceleration} = \frac{(20 - 6 \cdot 7)\ \text{m/s}}{10\ \text{s}}$$

$$= 1 \cdot 33\ \text{m/s}^2$$

Since the unit 'second' is repeated twice, the acceleration is usually expressed as 1·33 m/s². Note that 'cm/s' or 'm/s' is a unit of velocity or speed; 'cm/s²' or 'm/s²' is a unit of acceleration.

Formulae are very useful tools of science because they express concisely the general relationship between different quantities. Suppose an object is accelerating uniformly, and that

$$u = \text{initial velocity}, \ v = \text{final velocity},$$

at the end of a time $t$. Then the acceleration, which we shall denote by $a$, is given by:

$$a = \frac{\text{Velocity change}}{\text{Time}}$$

$$\therefore \ a = \frac{v - u}{t} \quad . \quad . \quad . \quad . \quad . \quad . \quad . \quad (1)$$

Consider a train accelerating uniformly at the rate of 2 km/h per second for 10 s after it had reached a velocity of 30 km/h. The increase in velocity is $10 \times 2$ km/h or 20 km/h. The final velocity after 10 s is thus (30 km/h + 20 km/h) or 50 km/h. Thus if $a$ represents the magnitude of the uniform acceleration of an object, in a time $t$ the increase in velocity is $a \times t$ or $at$. Assuming the initial velocity, at the beginning of the time, is $u$, the final velocity $v$ is given by:

$$v = u + at \quad . \quad . \quad . \quad . \quad . \quad . \quad . \quad (2)$$

This general formula, which should be memorized, can also be obtained directly from (1). Here we have

$$at = v - u, \text{ so } v = u + at.$$

'Retardation' or 'deceleration' has the same units as acceleration, and in formulae it can be regarded as 'negative acceleration'. This is illustrated by the following examples:

## Examples

### 1. *Acceleration*

A car travelling at 36 km/h accelerates uniformly at 2 m/s². Calculate its velocity in km/h in 5 s time.

We must use the *same units* in problems. The velocity of 36 km/h is therefore changed to metre per second. Since 36 km/h = 10 m/s,

$$\therefore \text{ Initial velocity } u = 10 \text{ m/s}$$

Now
$$a = 2 \text{ m/s}^2, \ t = 5 \text{ s}$$

$$\therefore \text{ Final velocity } \ v = u + at = 10 + 2 \times 5 = 20 \text{ m/s}$$

$$\therefore v = 72 \text{ km/h}$$

### 2. *Retardation*

A train slows from 72 km/h with a uniform retardation of 2m/s². How long will it take to reach 18 km/h?

$$\text{Initial velocity } u = 72 \text{ km/h} = 20 \text{ m/s}$$
$$\text{Final velocity } v = 18 \text{ km/h} = 5 \text{ m/s}$$
$$\text{Acceleration } a = -2 \text{ m/s}^2$$
$$\text{From } v = u + at$$
$$\therefore 5 = 20 - 2t$$
$$\therefore t = 7 \cdot 5 \text{ s}$$

## Acceleration Due to Gravity

A cricket or tennis ball, hit straight up into the air, slows down as it reaches the top of its flight. This is due to the attraction of the earth on the ball, or *gravitational attraction*. At its maximum height the ball momentarily stops. It then falls with increasing velocity to the ground owing to gravitational attraction. This is discussed more fully later. Here we may note that the ball undergoes a retardation as it moves up, and an acceleration while it falls.

The acceleration due to gravity can be roughly studied by attaching a heavy ball A to the end of the tape of a ticker-timer T placed above the

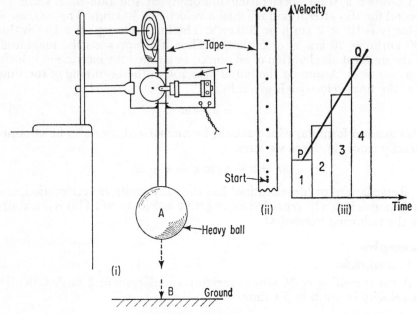

Fig. 3.17 Falling sphere

ground (Fig. 3.17(i)). After the ball strikes the ground at B the tape has an appearance similar to that shown in Fig. 3.17(ii). The average velocity after equal time intervals may be obtained by cutting successive strips corresponding to equal times (see p. 65). The result is shown in Fig. 3.17 (iii). The tops of the strips lie on a straight line PQ. This shows that the acceleration due to gravity of a falling object is uniform or constant. The magnitude of the acceleration can be measured from the strips as explained on p. 70, but the results are affected by the friction between the tape and the guides.

## Direct Measurement of *g* by Free Fall

A better method of measuring *g*, the acceleration due to gravity, is to time a free fall with an electric clock capable of measuring one-hundredths of a second or centiseconds.

The apparatus consists of an electromagnet M which is energized by a switch incorporated with the electrical supply to the clock (not shown) (Fig. 3.18). A steel ball, with a piece of thin paper between itself and the iron core, is held by M. When the switch is pressed, the current in M is cut off and the ball begins to fall. Simultaneously, the clock is switched on. After falling a height $h$ the ball strikes a hinged plate X at B. X then breaks contact with C, and the clock automatically stops.

The time $t$ of fall can be read to $\frac{1}{100}$ s. The height $h$ of fall is the distance from the bottom of the ball when held at M to the plate X. On repeating the experiment with magnetic materials of different mass in place of the ball, practically the same time of fall is obtained. Thus all

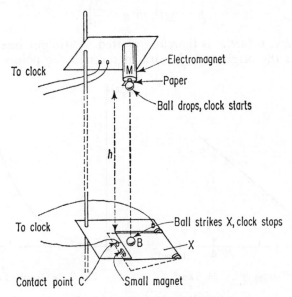

To clock — Electromagnet

M

Paper

Ball drops, clock starts

$h$

To clock

Ball strikes X, clock stops

B

X

Contact point C

Small magnet

Fig. 3.18 Free fall method for $g$

objects, no matter what their mass may be, fall under gravity with the same acceleration.

*Calculation.* In one measurement the height $h$ was 1·20 metres and the average time of fall was 0·50 s. Since the ball falls from rest, the initial velocity $u$ is zero.

From $\qquad s = ut + \frac{1}{2}at^2$ (see p. 80), we then have, since $a = g$,

$$h = \frac{1}{2}gt^2$$

$$\therefore g = \frac{2h}{t^2} = \frac{2 \times 1 \cdot 2}{0 \cdot 50^2} = 9 \cdot 6 \text{ m/s}^2$$

## Graphical Method for $g$

In physics it is always best not to rely upon a single measurement of a quantity. A more accurate and reliable value can often be found by taking many measurements and then plotting a *straight-line* graph (see p. 9).

In the free-fall experiment the height of fall $h$ can be varied by moving the electromagnet M up or down. The average time $t$ of fall on each occasion is measured with the electric clock. One set of results, obtained by a pupil, is shown below:

| $h$ (cm) | 0 | 109·5 | 96·0 | 84·1 | 67·5 | 54·7 |
|---|---|---|---|---|---|---|
| $t$ (s) | 0 | 0·469 | 0·441 | 0·412 | 0·369 | 0·333 |
| $t^2$ (s²) | 0 | 0·22 | 0·19 | 0·17 | 0·14 | 0·11 |

Now if an object drops from rest through a distance $h$, then (see p. 80)

$$h = \tfrac{1}{2}gt^2, \text{ or } g = \frac{2h}{t^2}.$$

A graph of $h$ v. $t^2$ (*not* $t$) is therefore plotted. A straight line OP, which goes through the origin, can be drawn through the points (Fig. 3.19).

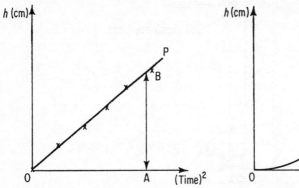

Fig. 3.19 Graph of $h$ v. $t^2$, *Time²*

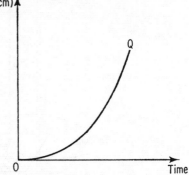

Fig. 3.20 Graph of $h$ v. $t$, *Time*

This result means that $h$ is directly proportional to $t^2$, or that $g$ is a constant. Further,

$$\frac{h}{t^2} = \text{the gradient (slope) of the line OP}$$

$$= \frac{AB}{OA}$$

$$\therefore \quad g = \frac{2h}{t^2} = 2 \times \frac{AB}{OA} = 2 \times \frac{49 \cdot 5 \text{ cm}}{0 \cdot 1 \text{ s}^2}$$

$$\therefore \quad g = 990 \text{ cm/s}^2 = 9 \cdot 9 \text{ m/s}^2.$$

It should be noted that a graph of $h$ v. $t$ is a parabolic curve OQ and not a straight line (Fig. 3.20). This is because $h = \frac{1}{2} gt^2$ is a *square*-law relation between $h$ and $t$. In contrast, the relation between $h$ and $t^2$ is a directly proportional or follows a *linear law*. A straight-line graph passing through the origin is therefore obtained in this case.

Note carefully that if a straight-line graph is obtained from measurements of any two quantities $X$ and $Y$, then $X$ and $Y$ have a linear relationship to each other. If, in addition, the graph passes through the origin, that is, $X = 0$ when $Y = 0$, then $X$ and $Y$ are 'directly-proportional' to each other.

## Oscillations of Simple Pendulum

There are many scientific phenomena in which *oscillations*, or *vibrations*, occur. As we shall see later, the effects produced by radio waves, light or sound, for example, are all due to oscillations, so this is an important type of motion.

The *simple pendulum* was an early example of oscillation in mechanics. It

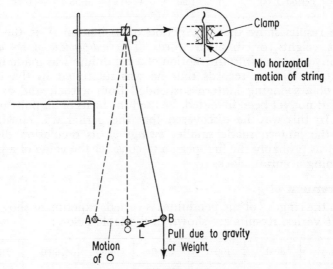

Fig. 3.21 $g$ by simple pendulum

consists of a small heavy object such as a lead bob or brass weight O suspended by a long thread PO from a fixed point P (Fig. 3.21). When O is pushed slightly and released it oscillates or vibrates about its original undisturbed position, moving along the arc AOB of a circle.

When the bob O reaches the end of its oscillation it comes momentarily to rest. This effect is due to the retarding pull of gravity on the bob. When it reaches B, for example, the downward pull or *weight* $W$ has a sideways effect in the direction BL, where BL is the tangent at B to the arc AOB. The bob thus moves back to O, gathering speed as it moves, and again slows up as it rises to A.

## Pendulum period

The *period* $T$ of an oscillation or vibration is defined as the time for a complete or to-and-fro oscillation. In Fig. 3.21, for example, the period is the time taken for the bob to move from B to A and back to B. Or the time taken to move from O to B, then to A and back to O. You should always remember that the oscillation begins all over again at the end of a period. It is incorrect to say that the period is the time from O to B and back to O, because the bob is moving in *different* directions at the beginning and end of the time.

An interesting result for the period $T$ of a simple pendulum is obtained when different small weights are suspended from the same long thread PO. Here are some measurements, made with different *small* amplitudes:

Pendulum length PO = 98·5 cm

| Mass | 100 g | 50 g | 20 g | lead bob |
|---|---|---|---|---|
| No. of oscillations | 20 | 20 | 20 | 20 |
| Time taken (s) | 39·6 | 39·7 | 39·6 | 39·8 |
| Period $T$ (s) | 1·98 | 1·99 | 1·98 | 1·99 |

These results show that the period of oscillation $T$ is the same for different weights, or, in other words, *T is independent of the mass*. The earliest investigation into the motion of a pendulum was made by Galileo about 1600. History records that he was interested in the to-and-fro motion of a swinging lantern suspended from a roof, and as accurate clocks had not yet been invented, he used the beats of his pulse to time the swings. In this way he discovered that the period was constant, even though the lantern made smaller swings as its oscillation died away. Galileo was probably the first person to consider the value of a pendulum in designing accurate clocks.

### Measurement of g

When the length $l$ of the pendulum is varied, experiment shows that the period $T$ varies. Results are shown in the table below:

| $l$ (cm) | 80·0 | 100·0 | 120·0 | 140·0 | 160·0 | 180·0 |
|---|---|---|---|---|---|---|
| $T$ (s) | 1·80 | 2·02 | 2·21 | 2·39 | 2·55 | 2·71 |

Now calculation beyond the scope of this work shows that the period $T$ is related to $l$ and to $g$, the acceleration at the place where the pendulum is used, by

$$T = 2\pi \sqrt{\frac{l}{g}}$$

Thus by squaring,

$$g = \frac{4\pi^2 l}{T^2} \quad \cdot \quad \cdot \quad \cdot \quad \cdot \quad \cdot \quad \cdot \quad (1)$$

$g$ can hence be calculated from the various values of $l$ and $T$. For example, if $l = 180$ cm $= 1·8$ m, $T = 2·71$ s, then

$$g = \frac{4\pi^2 l}{T^2} = \frac{4\pi^2 \times 1·8 \text{ m}}{(2·71)^2 \text{ s}^2}$$

$$= 9·7 \text{ m/s}^2$$

We can make another calculation of $g$ by means of a graphical method. To obtain a straight-line graph, we must plot $l$ against $T^2$, from (1), and then draw the best line passing through the origin and all the plotted points (Fig. 3.22). The gradient of the line is QP/OP and is 73·5/3 from measurements. Now from (1),

$$g = 4\pi^2 \times \frac{l}{T^2} = 4\pi^2 \times \frac{QP}{OP}$$

$$\therefore g = 4\pi^2 \times \frac{73·5}{3} = 970 \text{ cm/s}^2 = 9·7 \text{ m/s}^2$$

In doing a pendulum experiment, note that:

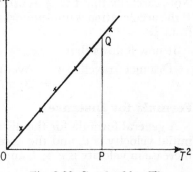

Fig. 3.22 Graph of $l$ v. $T^2$

(i) the formula for $T$ is true only if the pendulum bob swings through very small angles, so that the bob must be pushed *gently* to start the oscillations;

(ii) the count of twenty oscillations should begin and end as the bob swings through the *centre* O of the oscillation, passing a fixed pointer or chalk mark at O for reference. The end of a swing, such as A or B, is not as easy to observe as the middle of a swing.

## Motion under Gravity

Accurate experiments show that the acceleration due to gravity $g$ is about 980 cm/s² or 9·8 m/s². Thus if a ball is released from a height above the ground its velocity after 2 s is given by

$$v = u + at \text{ (p. 80)}$$
$$= 0 + 9·8 \times 2 = 19·6 \text{ m/s}$$
$$\text{Roughly, } v = 0 + 10 \times 2 = 20 \text{ m/s}$$

An object thrown vertically upwards has a retardation of 9·8 m/s². Thus if a ball is thrown up with an initial velocity of 30 m/s, then in 2 s time it will have a velocity $v$, since $a = -10$ m/s² (approx), given by

$$v = u + at = 30 - 10 \times 2 = 10 \text{ m/s}.$$

The ball will come to rest at a time $t$ given by

$$v = 0 = u + at = 30 - 10t$$
$$\therefore \quad 10t = 30, \text{ or } t = 3 \text{ s}$$

## Distance Travelled with Uniform Acceleration

We now have to find the distance travelled by an object moving with uniform acceleration $a$ for a time $t$ from an initial velocity $u$.

First, consider an actual calculation. Suppose a train is moving with a velocity of 10 m/s and accelerates uniformly at 2 m/s² for 10 s. At the end of the 10 s

Velocity reached = initial velocity + increase in velocity
= 10 + 10 × 2 = 30 m/s.

Now if the acceleration is constant, the velocity increases uniformly. In this case,

Average velocity = $\frac{1}{2}$ (initial + final velocity)
= $\frac{1}{2}$ (10 + 30) = 20 m/s.

Note carefully that the '$\frac{1}{2}$' is used only because the acceleration is uniform. If the acceleration is non-uniform the average velocity is a different fraction from $\frac{1}{2}$.

It now follows that

Distance travelled $s$ = Average velocity × Time
= 20 m/s × 10 s = 200 m.

## Formula for Distance

A general formula for the distance $s$ can now be obtained. Suppose the initial velocity is $u$, and the acceleration is $a$ for a time $t$. In a time $t$ the increase in velocity is $a \times t$, and the final velocity $v$ is thus

$v$ = initial velocity + increase in velocity
= $u + at$.

Since the acceleration is uniform,

average velocity = $\frac{1}{2} (u + v) = \frac{1}{2}(u + u + at)$
= $\frac{1}{2} (2u + at) = u + \frac{1}{2} at.$
∴   distance travelled $s$ = average velocity × time
= $(u + \frac{1}{2} at) \times t$
∴  $s = ut + \frac{1}{2} at^2$ . . . . . . . . (1)

Note again that the numerical factor '$\frac{1}{2}$' is due to a *uniform* acceleration. If the acceleration were non-uniform a different numerical factor would be obtained.

*Examples.* As an illustration of the use of the formula, suppose a bicycle accelerates at 1 m/s² from an initial velocity of 4 m/s for 10 s. Then

$u = 4$ m/s, $a = 1$ m/s², $t = 10$ s.
∴ $s = ut + \frac{1}{2} at^2 = 4 \times 10 + \frac{1}{2} \times 1 \times 10^2$
= 40 + 50 = 90 m.

If a train travels at 54 km/h and then accelerates at 2 m/s² for $\frac{1}{4}$ min we must change all the quantities to the *same units* before substituting in a formula. Hence

$u = 54$ km/h = 15 m/s, $a = 2$ m/s², $t = \frac{1}{4}$ min = 15 s
∴ $s = ut + \frac{1}{2} at^2 = 15 \times 15 + \frac{1}{2} \times 2 \times 15^2$
= 225 + 225 = 450 m.

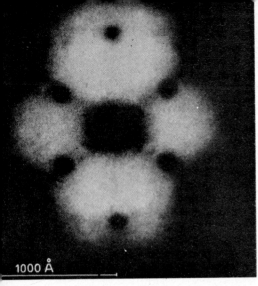

**PLATE 1**
**Atoms and Molecules**

(a) Atoms. Image of a tungsten point, radius about $10^{-5}$ cm, which is a single tungsten crystal. Photograph taken by a field emission ion microscope. The dark regions correspond to the electrons inside the crystal which have high binding energy and the bright regions to those having low binding energy.

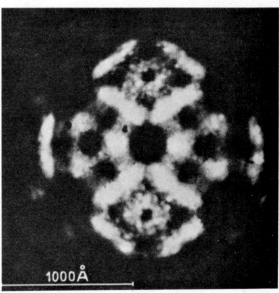

1(b) Visible atoms. Barium atoms, deposited on the tungsten point, are recognizable as coarse grains. On heating the point the thermal vibrations of the atoms can be seen.

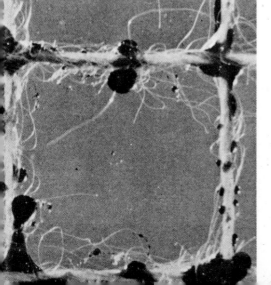

◄ 1(c) Forces between liquid molecules–detergent action. A detergent in solution removing oil from fibres. Observe the spherical shapes of the drops, especially those which are small (p. 41).

**PLATE 2**
**Kinetic Theory Models.**

2(**a**) Layers of molecules (p. 47).

*Kinetic theory models. Photographs of small phosphor-bronze balls, simulating molecules of a solid, taken by Ian Taylor, Esq., Bedford School.*

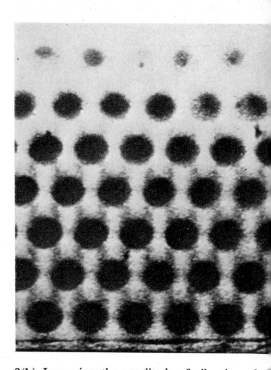

2(**b**) Increasing the amplitude of vibration of the molecules, simulating temperature rise, leads thermal expansion.

2(**d**) Sudden 'cooling' produces a solid of irregular crystal structure, as in quenching metals.

2(**c**) At higher temperatures the molecules begin to move about more freely and the solid approaches a 'liquid' state (p. 50).

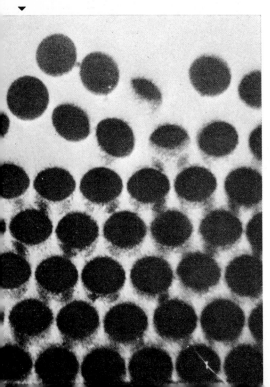

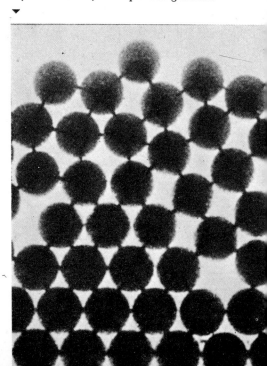

## Relations between *v*, *u*, *a*, *s* · Equations of Linear Motion

So far we have obtained two useful formulae for motion in a straight line, which is called *rectilinear motion*. They are:

$$v = u + at \quad . \quad . \quad . \quad . \quad . \quad . \quad (1)$$
$$s = ut + \tfrac{1}{2} at^2 \quad . \quad . \quad . \quad . \quad . \quad (2)$$

To eliminate the time *t* and obtain another useful relation, we can return to our relation for calculating distance (p. 80). Thus

$$\text{distance} = \text{average velocity} \times \text{time}$$

Using

$$\text{Average velocity} = \frac{v + u}{2},$$

and

$$\text{time } t = \frac{v - u}{a}, \text{ from (1),}$$

then

$$\text{Distance } s = \frac{v + u}{2} \times \frac{v - u}{a} = \frac{v^2 - u^2}{2a}.$$

Cross-multiplying $\therefore 2as = v^2 - u^2$
$$\therefore v^2 = u^2 + 2as \quad . \quad . \quad . \quad . \quad . \quad . \quad (3)$$

The formulae (1), (2), (3) are standard relations for rectilinear (straight-line) motion with uniform acceleration. They should be memorized by the reader.

## Examples in Linear Motion

**1.** A ball is thrown vertically upwards from the ground with a velocity of 20 m/s. Calculate: (i) the maximum height reached; (ii) the time to reach the maximum height; (iii) the time to reach the ground again after the ball is thrown up; (iv) the velocity reached half-way to the maximum height. (Assume $g = 10$ m/s².)

We have $u = 20$ m/s, $a = -10$ m/s², since the ball has a retardation as it rises.

(i) At the maximum height the ball comes momentarily to rest, so that the final velocity $v = 0$.

Using
$$v^2 = u^2 + 2as$$
$$\therefore 0 = 20^2 - 2 \times 10 \times s$$

$$\therefore s = \frac{20^2}{20} = 20 \text{ m} = \text{maximum height.}$$

(ii) To find the time *t* to reach the maximum height, we use

$$v = u + at.$$

Then
$$0 = 20 - 10t$$
$$\therefore t = 2 \text{ s}$$

(iii) The time to reach the ground again is twice the time to reach the maximum height, or 4 s. We can see this by finding the time taken to reach the ground again from the maximum height. In this case

$$\text{Initial velocity } u = 0, a = 10 \text{ m/s}^2, s = 20 \text{ m.}$$

Using
$$s = ut + \tfrac{1}{2}at^2$$
$$\therefore \ 20 = \tfrac{1}{2} \times 10 \times t^2$$
$$\therefore \ t^2 = \frac{20}{5} = 4$$
$$\therefore \ \ t = 2 \text{ s.}$$

The time to reach the maximum height is also 2 s from (ii), thus giving a total time of 4 s.

(iv) Half-way to the maximum height,

$$\text{the distance travelled} = \tfrac{1}{2} \times 20 \text{ m} = 10 \text{ m.}$$

Using
$$u = 20 \text{ m/s}, a = -10 \text{ m/s}^2, s = 10 \text{ m,}$$
then
$$v^2 = u^2 + 2as$$
$$= 20^2 - 2 \times 10 \times 10 = 200$$
$$\therefore v = \sqrt{200} = 14 \text{ m/s (approx.).}$$

**2.** A car travels with a velocity of 18 km/h. It then accelerates uniformly and travels a distance of 50 m. If the velocity reached is 54 km/h find the acceleration and the time to travel this distance.

Here $u = 18$ km/h $= 5$ m/s, $v = 54$ km/h $= 15$ m/s, $s = 50$ m.

From
$$v^2 = u^2 + 2as,$$
$$\therefore \ 15^2 = 5^2 + 2 \times a \times 50$$
$$\therefore \ 225 = 25 + 100a$$
$$\therefore \ \ a = 2 \text{ m/s}^2$$

Also
$$\text{Time} = \frac{\text{Distance}}{\text{Average velocity}}$$
$$= \frac{50 \text{ m}}{10 \text{ m/s}} = 5 \text{ s}$$

(or $\quad v = u + at$
$\therefore \ \ 15 = 5 + 2t$, or $t = 5$ s).

## Scalars and Vectors

In science there are many quantities which have only numerical values. For example, 'temperature' is a number on some scale – it has no directional property. These quantities are called *scalars*. 'Speed' is a scalar quantity. A 'speed of 30 km/h' can have any direction; it can be measured along a winding road or a circular track, for example.

On the other hand, many quantities in physics have direction as well as numerical value or magnitude. These quantities are called *vectors*. 'Velocity' and 'acceleration' are examples of vectors. A car travelling round a circular track may have a constant *speed* of 90 km/h (Fig. 3.23(i)). The *velocity* at the instant the car reaches a point A, however, is 90 *km/h in the direction AP*, where AP is the tangent to the circle. When the car reaches B the velocity at this instant is 90 km/h in the direction BQ. At C the velocity is 90 km/h in the direction CR. The velocity of the car at A, B and C is therefore different, but the speed, the numerical value only, is the same at each point.

'Acceleration' is a vector quantity, because we associate a particular

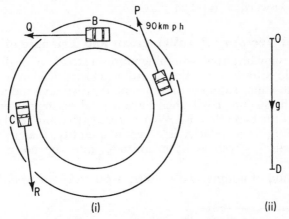

Fig. 3.23 Velocity and acceleration are vectors

direction with it. A falling object, for example, has an acceleration towards the centre of the earth, which is practically a vertical direction. Thus the acceleration due to gravity g can be represented by a straight vertical line OD whose length represents to scale the magnitude of g (Fig. 3.23(ii)). We shall meet other vector quantities later, e.g. 'force' and 'momentum'.

## Representation of Vectors

Suppose a car is moving with a uniform velocity of 48 km/h along a road OA at an angle of 50° N of E of a road OB (Fig. 3.24(i)). Its velocity v

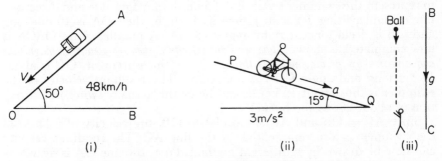

Fig. 3.24 Components of velocity and acceleration

is then completely represented by a line OA 6 cm long at an angle of 50° to OB, where 1 cm represents a velocity of 8 km/h.

Similarly, an acceleration of 3 m/s² of a bicycle down an incline at 15° to the horizontal is represented by PQ in Fig. 3.24(ii), where PQ is drawn 3 cm long and 1 cm represents an acceleration of 1 m/s². Again, a cricket ball thrown into the air undergoes a vertical retardation *g* of 9·8 m/s² as it moves upward. As it moves down, it undergoes a vertical acceleration of 9·8 m/s². In *either* case this effect of gravity is represented by a vertical line BC drawn to represent 9·8 m/s², with an arrow on it pointing downwards (Fig. 3.24(iii)).

### Addition of Vectors · Parallelogram and Triangle of Vectors

Suppose a boy or girl is swimming across a stream of negligible velocity from one side X to the other side Y with a velocity of 5 km/h (Fig. 3.25(i)). At the same time, suppose the current downstream suddenly becomes 10 km/h. The swimmer will then have a total velocity which is the sum or *resultant* of the two velocities. We can *not* say that the resultant is 5 + 10 or 15 km/h, because velocity is a vector quantity, that is, the direction of the velocity, as well as its magnitude, must be taken into account (p. 82).

One method of adding two vectors, a *parallelogram method*, is shown in

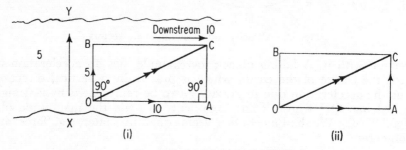

Fig. 3.25 Addition of velocities

Fig. 3.25(i). A line OA is drawn to represent the current velocity of 10 km/h in magnitude and direction. On the same scale, a line OB is drawn to represent the swimmer's velocity of 5 km/h in magnitude and direction. For example, using 1 cm to represent 1 km/h, then OA is 10 cm long and OB is 5 cm long at right angles to OA. A parallelogram OBCA is now completed as shown. *The resultant velocity is then represented in magnitude and direction by the diagonal OC through O.* On measurement, OC is about 12 cm long and angle COA is about 26°. The resultant velocity of the swimmer is thus about 12 km/h, and he or she actually swims in a direction of about 26° to the bank X.

Since AC = OB and AC is parallel to OB, the velocity of 5 km/h of the swimmer is also represented by the line AC. The resultant velocity can now be drawn by a different method. First, the line OA is drawn to represent the velocity of 10 km/h of the stream. Secondly, the line AC is drawn from A to represent the velocity of 5 km/h (Fig. 3.25(ii)). The line OC is then joined. OC, the third side of the triangle, represents the sum or resultant of the two velocities. This may be called a *triangle method* of adding two vectors. It should be noted that the arrows on OA and AC follow each other round.

Another example of adding vectors is shown in Fig. 3.26. An aeroplane flying with a velocity of 300 km/h in the direction OP will go off course in the direction OR if a strong wind of 80 km/h acts in a direction OQ inclined at 60° to OP (Fig. 3.26). OR is the resultant velocity. The pilot

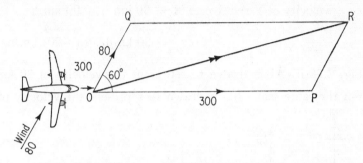

Fig. 3.26 Addition of velocities—parallelogram method

has thus to make a correction by instruments for the wind velocity. Further, the resultant velocity OR is greater than 300 km/h, and hence the plane travels with greater velocity owing to the wind.

## Subtraction of Vectors · Relative Velocity

Consider two ships X and Y travelling parallel to each other during a race with velocities of 30 and 20 km/h respectively (Fig. 3.27(i)). The velocity of X relative to Y is the *vector* difference between the two velocities, or

$$\text{Relative velocity} = \overrightarrow{30 \text{ km/h}} - \overrightarrow{20 \text{ km/h}}$$

The arrows above 30 and 20 km/h respectively indicate that we are dealing with vectors which have direction, and not scalars. Now

$$\overrightarrow{30 \text{ km/h}} - \overrightarrow{20 \text{ km/h}} = \overrightarrow{30 \text{ km/h}} + \overrightarrow{(-20) \text{ km/h}}$$

A velocity of $\overrightarrow{-20}$ km/h is one drawn in the opposite direction to the

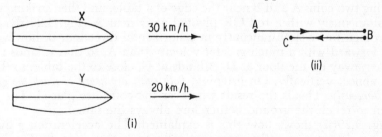

Fig. 3.27 Subtraction of velocities

velocity of 20 km/h, and is therefore represented by the line BC in magnitude and direction (Fig. 3.27(ii)). The velocity of 30 km/h is represented by the line AB, and AB – BC is AC, or 10 km/h.

We have spent some time over a simple example of relative velocity to illustrate how vectors must be subtracted. Now consider two ships X and Y travelling respectively with a velocity of 30 km/h due east and with a velocity of 20 km/h due north (Fig. 3.28(i)). Then

$$\text{velocity of Y relative to X} = \overrightarrow{30} \text{ km/h} - \overrightarrow{20} \text{ km/h}$$

$$= \overrightarrow{30} \text{ km/h} + \overrightarrow{(-20)} \text{ km/h}$$

In Fig. 3.28(ii) AB is drawn to represent the velocity of $\overrightarrow{30}$ km/h to scale. On the same scale, BC is drawn to represent the velocity of $\overrightarrow{-20}$

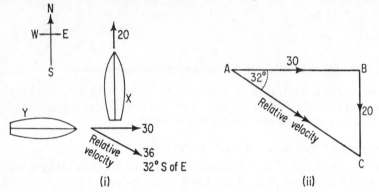

Fig. 3.28 Relative velocity

km/h. *The line AC thus represents the velocity of Y relative to X.* From triangle ABC, AC is 36 km/h (approx.) and angle BAC is 32° (approx.). Thus, as represented in Fig. 3.28(i), the ship Y appears to be moving relative to X with a velocity of 36 km/h at an angle 32° south of east.

### Horizontal and Vertical Motion under Gravity

A simple but informative experiment on motion can be carried out by placing two coins A and B near the edge of a table, and then striking them simultaneously with a rod OR pivoted at O near A (Fig. 3.29(i)). Both coins then move simultaneously in a horizontal direction. B, however, is shot forward with a much greater velocity than A, so that it lands much farther away on the floor at D. A lands at C, close to the table, so that it falls almost vertically. On listening, *both coins are heard to reach the ground simultaneously.* This is the result wherever the coins are placed. The time taken to reach the ground is therefore always the same.

Fig. 3.29(ii) shows how this is explained. The acceleration g due to gravity, a vector quantity, acts vertically on both A and B. The initial forward velocity v of the coin B is a vector quantity acting horizontally. The velocity v thus acts perpendicularly to the acceleration g. Since experiment shows that A and B both reach the floor simultaneously, we deduce that the *vertical* motion of B, due to g, is unaffected by its horizontal motion, due to v. Thus a vector quantity, such as v, which acts horizontally, has no effect in a perpendicular direction, such as the direction of $g$.

The horizontal and vertical motion of an object are hence completely independent. This is illustrated in Fig. 3.29(ii). The vertical distance

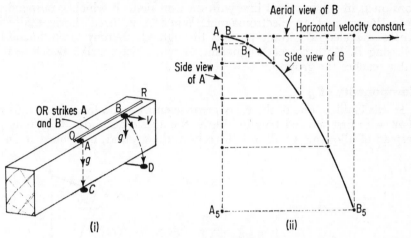

Fig. 3.29 Horizontal and vertical motion under gravity

dropped by the coin B, shown by $B_1$, ... $B_5$, is at all times equal to that of the coin A, shown by $A_1$, ... $A_5$. On the other hand, B has also an initial horizontal velocity $v$. It therefore moves forward, covering a horizontal distance $s$ given by $s$ = velocity $\times$ time = $vt$ since the velocity $v$ is constant.

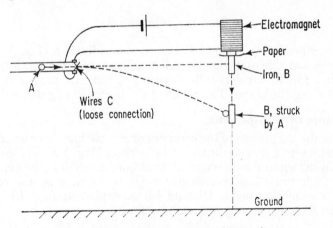

Fig. 3.30 Independent motion under gravity

Thus equal *horizontal* distances are covered in equal times. Vertical distances $s$, however, are covered in a time $t$ given by $s = \frac{1}{2}gt^2$, as this is motion under uniform acceleration $g$, the initial velocity being zero in a vertical direction. Fig. 3.29(ii) shows the vertical distances travelled by both A and B–they fall the same distance $s$ in equal times $t$, according to the square-law relation $s = \frac{1}{2}gt^2$. The horizontal distance of B follows a linear relation with $t$, however, since $s = vt$ in this case.

Fig. 3.30 shows the principle of a 'monkey and hunter' demonstration of the horizontal and vertical motion just discussed. A ball-bearing A is positioned in a horizontal line with an iron plate B which is suspended from the core of an electromagnet. When A is 'fired' horizontally it breaks the electric circuit passing through C, thereby simultaneously releasing B. With correct alignment, A will always strike B before the latter reaches the ground.

## Components of g

It was Galileo, one of the early great scientists, who lived from 1564 to 1642, who first showed how to 'dilute' the acceleration due to gravity $g$. Instead of allowing a ball to fall freely to the ground, he placed it on a

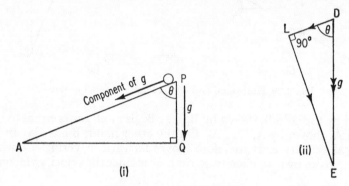

Fig. 3.31 Components of $g$

smooth inclined plane PAQ and allowed it to roll down (Fig. 3.31(i)). In this way he was able to use not the full magnitude of $g$, the acceleration due to gravity, which acts vertically downward, but that part of it which is effective along the plane. This is called the *component* of $g$ down the plane.

## Magnitude of Component

To find the magnitude of the component of $g$, we first draw a vertical line DE to represent $g$ in magnitude and direction (Fig. 3.31(ii)). From D, a line DL is drawn in the direction of the sloping side PA of the plane PAQ, and from E a line EL perpendicular to DL is drawn. Now the vector DE $g$ is the sum of the vectors DL and LE, as explained on p. 84, or, using arrows to show we are dealing with vectors,

$$\overrightarrow{DL} + \overrightarrow{LE} = \overrightarrow{DE}$$

The acceleration $g$ is therefore considered to be equivalent to an acceleration of magnitude DL in the direction of the sloping side PA of the plane PAQ, together with an acceleration of magnitude LE in a perpendicular direction. Now we have just seen from the case of motion under gravity that a vector has no effect in a direction perpendicular to itself. Consequently, the effective part or component of $g$ along the direction PA is the acceleration whose magnitude is DL. Thus if DL makes an angle $\theta$ with DE, which is the angle APQ of the plane,

$$Component\ of\ g = DL = DE\cos\theta = g\cos\theta.$$

The other component of $g$, which is LE, acts perpendicular to DL and affects motion in a direction perpendicular to the plane.

Suppose the angle of inclination of the plane PAQ to the vertical is $60°$. Then an object moving under gravity down the plane has an acceleration given by

$$g\cos 60°, \text{ or about } 10 \times 0.5 \text{ m/s}^2 = 5 \text{ m/s}^2$$

If the plane is made less steep, say $20°$ to the horizontal, then $\theta$, the angle to the vertical, becomes $70°$. The acceleration down the plane is then

$$g\cos 70°, \text{ or about } 10 \times 0.34 \text{ m/s}^2 = 3.4 \text{ m/s}^2.$$

## SUMMARY

1. Speed = distance/time or distance per unit time. It is a scalar quantity – it has no direction. 36 km/h = 10 m/s.

2. Velocity = displacement/time or 'displacement per unit time'. It is a vector quantity – it has direction as well as magnitude.

Uniform velocity is the motion of an object which has equal displacements or travels equal distances in a constant direction in equal times, no matter how small the times may be.

The velocity of a car moving round a circular track is tangential to the track at any instant.

3. Acceleration = velocity change/time. It is a vector and has units such as metre per second$^2$ (m/s$^2$).

Uniform acceleration is the acceleration when equal changes in velocity take place in equal times, no matter how small the times may be.

4. The acceleration due to gravity $g$ is about 9.8 metre per second$^2$ or 10 m/s$^2$ (approx.) It can be measured by free-fall using a centisecond clock or by a pendulum method ($T = 2\pi\sqrt{l/g}$).

5. Equations of motion for uniform acceleration are:

$$v = u + at, \ s = ut + \tfrac{1}{2}at^2, \ v^2 = u^2 + 2as.$$

6. *Graphs · Distance–Time.* With uniform velocity the graph is a straight line inclined to the time-axis. With uniform acceleration a parabola is obtained; but when a graph of $s$ v. $t^2$ is plotted for an object moving from rest at $t = 0$, a straight line is obtained whose gradient is $\tfrac{1}{2}a$, where $a$ is the acceleration.

*Velocity–Time.* With uniform acceleration the graph is a straight line whose gradient is equal to the acceleration. The distance travelled is represented by the area between the velocity–time curve and the time-axis.

7. *Velocity Addition.* Either use the parallelogram method or the triangle method. In relative velocity calculations, subtract the two velocities by a vector method.

**8.** *Components.* (i) Since the acceleration due to gravity *g* acts vertically downwards, it has no component in a horizontal direction. A ball thrown forward has thus a constant horizontal velocity and moves vertically downwards with acceleration *g*. (ii) The acceleration of an object due to gravity down a smooth inclined plane is $g \cos \theta$ if $\theta$ is the angle to the vertical, or $g \sin \theta$, where $\theta$ is the angle to the *horizontal*.

## EXERCISE 3 · ANSWERS, p. 223

(Assume where necessary $g = 10$ m/s².)

**1.** A car travels with a uniform velocity of 36 km/h for 10 s. What is the distance travelled?

**2.** A train moving with a velocity of 10 m/s, starts to accelerate for ¼ minute at the rate of 1 m/s². Calculate the final speed of the train, and the distance travelled, at the end of the time.

**3.** A ball is thrown vertically upwards with a velocity of 20 m/s. What is the retardation of the ball? What distance does it travel before it comes to rest at the top of its motion? (Use $v^2 = u^2 + 2as$.)

**4.** In question 3, find the time taken for the ball to reach its highest point. What is the time taken by the ball to return to the thrower?

**5.** A stone is thrown horizontally with an initial velocity of 15 m/s from a tower 20 m high.

(*a*) How long does it take to reach the ground?

(*b*) At what distance from the base of the tower does the stone reach the ground?

**6.** What is a vector and a scalar quantity? Give ONE example of each.

**7.** A car accelerates from rest at the rate of 2 m/s² for 15 s, travels at a uniform velocity for the next 30 s and then comes to rest after 10 s more. Draw a velocity–time graph of the motion. From the graph, find the total distance travelled and the retardation of the car. Check by formula.

**8.** Give an example of an object moving with (i) uniform velocity, (ii) uniform acceleration, and explain your answer.

**9.** An aeroplane is travelling due north at 200 km/h. A wind of 80 km/h starts to blow from the east. Find the resultant velocity of the aeroplane and its direction of travel: (i) by drawing; (ii) by calculation.

**10.** A ship *X* is travelling due west with a velocity of 20 km/h; another ship *Y* is travelling north-east with a velocity of 15 km/h. Find the relative velocity of *X* with respect to *Y* by drawing.

**11.** The diagram shows part of a tape which was attached to a trolley during a run:

What type of motion is indicated respectively in the regions: (i) AB; (ii) BC; (iii) CD? How could each type of motion have been obtained?

**12.** The tape shown below was attached to a heavy moving ball. (i) Find the average

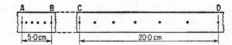

velocity in the regions AB (5-tick interval, each $\frac{1}{50}$ s) and CD (6-tick interval). (ii) Find the average acceleration if there is a 100-tick interval from AB to CD.

**13.** The diagram shows consecutive ten-tick lengths of tape pasted beside each other

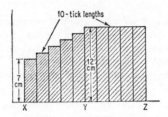

after a run. Calculate (i) the acceleration from X to Y; (ii) the total distance travelled from X to Z. (One tick-interval $= \frac{1}{50}$ s.)

**14.** Explain what is meant by a uniform acceleration of 80 metre per second per minute. What is the value of this acceleration in metre per second²?

Describe an experiment to determine the acceleration due to gravity.

A car starting from rest with a uniform acceleration travels 3 m in the sixth second of its motion. What is the acceleration and how far does the car travel in the first 10 s?

**15.** Explain the meaning of *uniform acceleration* and deduce expressions for (*a*) the velocity acquired in a time *t*, (*b*) the velocity acquired in a distance *s*, by a body which starts from rest and is subject to a uniform acceleration *a*. A train moves from rest with an acceleration of 0·5 m/s². Find the speed, in km/h, which it reaches in moving through its own length, which is 100 m. (*L.*)

**16.** What is meant by *uniform acceleration*? Sketch a velocity–time graph and a distance–(time)² graph for a body starting from rest and moving with uniform acceleration.

A train, starting from rest, accelerates uniformly so that it attains a speed of 48 km/h in 2 min. It travels at this speed for 5 min, and is then brought to rest with uniform retardation in 3 min. Find the total distance travelled by the train.

**17.** Describe an experiment to determine the acceleration due to gravity.

Draw the velocity–time curve of a body which has an initial velocity *u* metre per second and moves for *t* second in a straight line with an acceleration *a* metre per second per second. Use a graph to deduce an expression for the distance travelled by the body in this time. (*N.*)

**18.** Derive, from first principles, an expression for the distance travelled by a uniformly accelerated body moving from rest, in terms of the acceleration and the time.

A ball dropped from rest on to horizontal ground 20 m below rebounds with ¾ of the velocity with which it hits the ground. Find the time that elapses between the first and second impacts of the ball with the ground. (*L.*)

**19.** Define *uniform velocity, uniform acceleration*. Show how, from the velocity–time graph for a body (*a*) the acceleration at any instant, (*b*) the average velocity can be calculated.

A car, starting from rest, is uniformly accelerated at 0·15 m/s² until it reaches a speed of 15 m/s. It travels at this speed for 3 min and is then uniformly retarded so as to come to rest 8 min after starting. Plot the velocity–time graph of the motion and use the graph to determine the average speed of the car. (*O.* and *C.*)

**20.** Describe any experiment which will enable you to verify the formula connecting distance and time for a body starting from rest and moving with uniform acceleration in a straight line.

Draw an accurate graph of velocity against time for a stone thrown vertically upwards with an initial velocity of 20 m/s. Calculate the greatest height reached. (*O.* and *C.*)

**21.** Describe and explain a method of measuring g by free fall. Point out any inaccuracies in your method and show how you would graph your results to obtain the value of g.

**22.** Explain the terms *vector* and *scalar* and give two examples of each. Give a graphical construction for (a) adding, (b) subtracting two vectors of the same kind.

A man is cycling at a constant speed of 10 km/h. At a given instant he is travelling due N; at a later time he has changed his direction to 60° E of N. Make a scale drawing, indicating directions by arrows, to represent these two velocities of the cyclist, and deduce the *change* in velocity. (O. and C.)

**23.** Describe an experiment to illustrate motion with uniform acceleration. Explain how you would establish that the acceleration is uniform.

A body moves from rest with uniform acceleration and travels 90 m in the fifth second. Calculate the velocity of the body 10 s after starting. (N.)

**24.** What is meant by *uniform acceleration, gravitational acceleration*?
Describe a method of measuring gravitational acceleration.

The speed of a train is reduced from 96 to 48 km/h in a distance of 800 m on applying the brakes. How much farther will the train travel before coming to rest assuming the retardation remains constant, and how long will it take to bring the train to rest after the application of the brakes? (O. and C.)

**25.** An aeroplane has an air-speed of 198 km/h at right-angles to a strong wind. The velocity of the machine is 202 km/h relative to the earth. What is the velocity of the wind? (N.)

**26.** Explain the terms: *uniform acceleration, acceleration due to gravity*. Describe a method of determining the acceleration due to gravity at a given place. A tennis ball is hit vertically upwards and returns 6 s later. Calculate: (a) the greatest height reached by the ball; (b) the initial velocity of the ball. (O. and C.)

**27.** Define (a) acceleration, (b) uniform acceleration and explain under what conditions a body is uniformly accelerated.

A freely falling body passes a certain point with a velocity of 30 m/s. Calculate its velocity 3 s later and find how far it will travel in this time. (O. and C.)

# 4

# FORCE · MOMENTUM · ENERGY · POWER

## FORCE AND ACCELERATION

Sir Isaac Newton, 1642–1727, is recognized as one of the greatest scientists who ever lived. He made important discoveries in optics, such as the colours of white light, and he found a universal law of gravitation which explained satisfactorily the motion of the moon round the earth and that of the planets round the sun. In 1687 he published a work called *Principia*, in which, for the first time, the principles and laws of mechanics were clearly stated. Another great scientist, Albert Einstein, developed new ideas on space and time in 1905 in his 'Special Theory of Relativity'. Einstein's work has modified Newton's theories, but the laws of mechanics which Newton stated are basically still true today.

### Force

If you collide with someone while walking, your motion is immediately checked. A *force*, due to collision, thus produces a change in velocity. When a train starts from a station its velocity increases from zero. A force, due to the metal chain or link connecting the train to the engine, again produces a velocity change. If a tennis or cricket ball is hit, or a football is kicked, the force at impact produces a velocity increase. All these examples show that, in general, *a force produces a change in the motion or velocity of an object.*

### Frictional Forces

If a ball is rolled along the ground it will eventually come to rest. This is due to the *frictional force* between the ball and ground which opposes the motion.

For the same reason, if a book on a table is given a gentle push it travels a short distance and then comes to rest (Fig. 4.1(i)). Suppose, however,

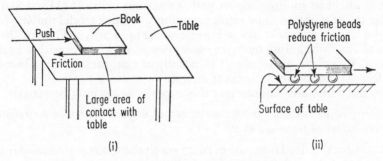

Fig. 4.1 Reduction of friction

that polystyrene beads are sprinkled liberally on the table and the book is pushed gently as before. This time the book slides easily along the table across the beads. The frictional force between the book and the table has thus been considerably reduced. As illustrated in Fig. 4.1(ii), the tiny beads act like millions of small 'rollers', separating the physical contact between the surface of the book and that of the table. The book slides easily along the 'rollers' (p. 38).

In the same way a metal *puck*, which has dry ice (solid carbon dioxide) packed beneath its base, glides smoothly along a plane sheet of glass care-

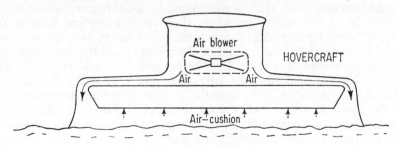

Fig. 4.2 Hovercraft principle

fully levelled if it is given a gentle push. Gaseous carbon dioxide, released between the dry ice and the glass, forms a cushion of gas on which the puck moves. Friction is thus reduced to an extremely small value. The motion of the puck continues for a very long time, rebounding continually from the surrounding sides after collision. The Hovercraft, now crossing the English Channel with passengers, is a vessel which directs a curtain of jets of air downwards below it so that it floats on a cushion of air while moving (Fig. 4.2). It travels at much faster speeds than a steamer ploughing through water because of the reduction in friction.

### Newton's First Law

By using a rotating stroboscope in front of its lens (p. 323), or by using regular flashes of light in a darkened room from an electronic stroboscope, a camera can take successive photographs after equal intervals of time of a puck gliding on its own across glass. The photographs show that the puck with dry ice underneath it, which has practically no frictional force acting on it to affect its motion, moves with a constant velocity (Fig. 4.3(i)). A similar result is obtained by photographing a perspex model moving freely along a linear air-track, like a Hovercraft (Fig. 4.3(ii)). Thus in the absence of any force acting on it, an object continues to move with a uniform velocity. Of course, if the object is initially at rest, then, in the absence of any force, it continues to remain at rest.

Newton's first law summarizes this experience. It may be stated:

*Every object continues in its state of rest or uniform motion in a straight line unless impressed forces act on it.*

Thus a rocket, fired from the earth to reach the moon, eventually reaches outer space, where the gravitational attraction of the earth is very small.

If no other force acts on it, the rocket then continues to move in a straight course with uniform velocity. When it comes within the gravitational field of the moon, however, its path is influenced by the gravitational attraction of the moon, and may therefore change.

Passengers in a fast-moving car which suddenly comes to rest continue

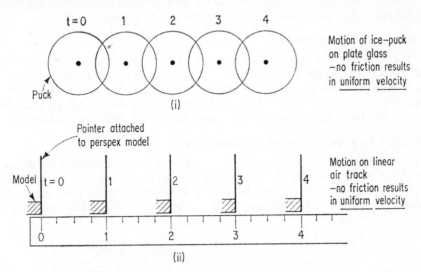

Fig. 4.3 Uniform motion in absence of force

to move forward, since there is little restraining force on them. Safety-straps oppose the forward motion.

## Opposing Forces · Gravity and Friction

If a cricket or tennis ball is thrown straight up into the air its initial velocity is gradually reduced. This is due to an opposing force, the force due to gravity on the ball, which acts vertically downward. The ball thus undergoes a retardation. At the top of its flight it is momentarily stationary.

If the ball is thrown forward on a well-cut lawn the opposing frictional force on it is less than the opposing force due to gravity when it was thrown up. Thus if the initial velocity is the same in each case the ball travels a much greater distance over the lawn. An ice puck (p. 94) travels over considerable distances. The friction is now extremely small owing to the 'cushion' of carbon dioxide gas beneath the base of the puck, and with such small retardation the velocity of the puck is only slightly diminished with time.

## Inertial Mass

Newton's first law recognized that objects have a reluctance to move when they are at rest. They also have a reluctance to stop when they are moving. Objects thus have a certain amount of *inertia*.

A large block of stone can hardly be pushed along the ground. A small wooden block can easily be pushed along the ground. The *mass* of the stone or wooden block is a measure of its inertia. Newton said that the 'mass 'of an object was a measure of the 'quantity of matter' in it, without stating the nature of matter. Masses are measured accurately in terms of a *standard mass*. The standard 'kilogramme' and 'pound' are certain lumps of metal kept in the National Bureau of Weights and Measures of France and England respectively (p. 6). All masses can be measured accurately with a chemical balance as so many kilogrammes or grammes.

## Force and Acceleration

We have seen that a force produces a change of velocity when it acts on an object. It therefore produces an *acceleration* or a *retardation*, a negative acceleration. Racing cars need to accelerate quickly; aeroplanes need a much greater acceleration. We shall now investigate the relationship between the force acting on an object and the acceleration it produces.

## Trolley Investigation of Acceleration

The motion of an object due to a force can be investigated by using a trolley and ticker-tape. A long board B is set up as a runway and a preliminary run is made when the trolley is given a gentle push down the plane (Fig. 4.4). The gradient of B is adjusted until equally spaced dots

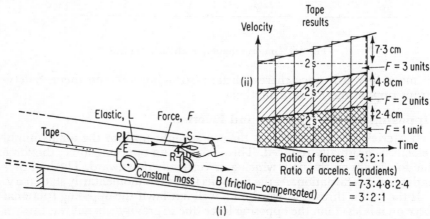

Fig. 4.4 Acceleration and force (constant mass)

are obtained on the tape, showing uniform velocity. The frictional force on the trolley is now compensated by the weight of the trolley down the plane.

A constant force on the trolley can be produced by stretching a piece of elastic L by a constant length (Fig. 4.4(i)). One end, with an eyelet E, is placed on a peg P on the trolley. The other end is pulled between the finger and thumb until it is exactly at the pegs R and S or another fixed reference point, such as the axle of a wheel.

After preliminary trials the stretched length of elastic is kept constant while the trolley is pulled. The trolley then moves down the plane under a constant force, which may be called 'one unit' of force. Forces of two and then three units can be obtained with two other pieces of similar elastic attached to P, each stretched exactly the same length as previously. After each run the tape is removed and a fresh tape is attached to the trolley.

### Force and Acceleration (Constant Mass)

Fig. 4.4(ii) shows the results obtained when each tape is cut into 10-tick units and pasted beside each other. As explained on p. 69, the acceleration is the gradient of the line joining the tops of the strips.

Conclusions are:

1. A constant force produces a constant acceleration.
2. The acceleration is directly proportional to the force.

### Mass and Acceleration (Force Constant)

To investigate the effect of mass on acceleration when the force is constant, one, two and then three trolleys can be mounted one above the

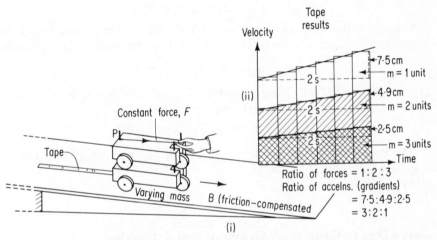

Fig. 4.5 Acceleration and mass (constant force)

other to vary the mass and pulled each time by the same force with one elastic (Fig. 4.5(i)). The results are shown in Fig. 4.5(ii).

Conclusion: The acceleration is *inversely* proportional to the mass. Instead of using other trolleys, the experiment can be carried out by loading one trolley with different masses and measuring the total mass each time.

## Force, Mass and Acceleration · The Newton

All these experimental results can be summarized by one relationship or formula, namely,

$$F \propto ma,$$

where $F$ is the force acting on a mass $m$ and $a$ is the acceleration produced. The magnitude of the acceleration produced depends on the mass and on the force.

The *newton*, symbol N, is the unit of force in the SI system. It is defined as *the force acting on a mass of 1 kilogramme which gives it an acceleration of 1 metre per second²*. From this definition, it follows by proportion that a force on a mass of 10 kg which gives it an acceleration of 2 m/s² has a magnitude of $10 \times 2$ or 20 newtons. Generally, then the relation between force $F$, mass $m$ and the acceleration $a$ produced is

$$F = ma \quad . \quad . \quad . \quad . \quad . \quad . \quad . \quad (1)$$

where $F$ is in newton, $m$ in kilogramme and $a$ in metre per second². Thus if a force of 50 newtons produces an acceleration of 2 m/s² the mass $m$ of the object concerned is given by

$$50 = m \times 2, \text{ or } m = 25 \text{ kg.}$$

Also a force of 12 N acting on a mass of 2 kg produces an acceleration $a$ given by

$$12 = 2 \times a, \text{ or } a = 6 \text{ m/s}^2$$

Suppose a ball of mass 200 g or 0·2 kg is hit so that it moves with an initial acceleration $a = 10$ m/s². The force $F$ on the ball is then given by

$$F = ma = 0{\cdot}2 \times 10 = 2 \text{ N}$$

If a car of mass 2000 kg is acted on by a force of 4000 N, the acceleration $a$ of the car, from $F = ma$, is given by

$$4000 = 2000 \times a, \text{ or } a = 2 \text{ m/s}^2$$

The *dyne* is a c.g.s. unit of force; this is the force which gives a mass of 1 g an acceleration of 1 cm/s².

## Force Due to Gravity or Gravitational Attraction

If we slip and fall over we realize immediately that the earth has an attraction on our mass. A ball thrown into the air comes down again for the same reason. All objects, tiny as well as large, are attracted by the earth. This is called *gravitational attraction*. The molecules of air in the atmosphere are kept round the earth by the force of attraction; a thin 'blanket' of gas stretches more than 11 kilometres high round the earth.

The force of attraction on an object due to gravity is called its *weight*. Since the time of Galileo it has been known that all objects near the earth's surface fall with the same acceleration under gravity, which is about 980 cm/s² or 9·8 m/s². The force on a mass of 1 kg due to gravity is thus

$$F = ma = 1 \times 9{\cdot}8 = 9{\cdot}8 \text{ N (newton)}$$

If the mass is 1000 kg, the force due to gravity acting on it is

$$F = ma = 1000 \times 9{\cdot}8 = 9\,800 \text{ N}$$

Thus the larger the mass of an object, the greater is its weight, and

weight $\propto$ mass.

## Mass and Weight

The *mass* or amount of matter or 'stuff' in a 1 kg box of chocolates carefully sealed is 1 kg all over the world; it would be the same at the equator
or at the poles, or even if it were landed on the moon. On the other hand,
if the force due to gravity is measured all over the world by attaching the

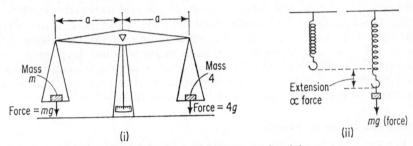

Fig. 4.6 Measurement of mass and weight

box to a spring and measuring the extension, the latter will vary slightly.
It will be greater at the north and south poles of the earth, for example,
than at the equator. The *weight* is therefore greater at the poles. This is
explained by the fact that the earth is not perfectly spherical—it is flattened
at the poles. Objects here are therefore nearer to the centre of the earth,
and hence the force of attraction is greater. Generally, if $m$ is the mass of
an object and $g$ is the acceleration due to gravity at a particular place,

$$\text{weight of object} = \text{mass} \times \text{acceleration} = mg.$$

The Weight $mg$ is in newtons if $m$ is in kilogramme and $g$ in metre per
second$^2$ ($g = 9 \cdot 8$ m/s$^2$, or, in round figures, $g = 10$ m/s$^2$).

We can now see why a common balance compares masses and not
weights. As an illustration, suppose an unknown mass $m$ in one scale-pan
is counterbalanced exactly by a mass of 4 kg on the other scale-pan (Fig.
4.6(i)). The balance is a lever with equal arms $a$. On weighing, therefore,
the moments of the forces about the pivot are equal (p. 138), or

weight on one scale-pan $\times a$ = weight on other scale-pan $\times a$.
$\therefore$  weight on one scale-pan = weight on other scale-pan.

Assuming the scale-pans have equal weight, we then have

$$mg \text{ (weight)} = 4g \text{ (weight)}.$$

*Since g is the same, it can be cancelled* on both sides, so that $m = 4$. The same
result, $m = 4$, would be obtained with a common balance at any part of
the world since it is independent of $g$. Hence a common balance compares
masses. 'Weight' can be measured by means of a spring balance, since its
extension depends on force (Fig. 4.6(ii)) (see p. 33).

### Gravity above Earth · The Moon

When a space rocket is fired the force of attraction of the earth on it diminishes as it gets farther away. The weight of the rocket therefore diminishes, although its mass remains constant. The acceleration due to gravity, from $F = ma$, thus also decreases. At a distance of about 6400 km from the earth, a distance from the centre of twice the earth's radius, the acceleration due to gravity diminishes to one-quarter of its magnitude at the earth's surface, or to about 2·5 m/s².

In July 1969, America succeeded in landing a spacecraft, Apollo-11, on the moon. On this planet the magnitude of the gravitational attraction is about one-sixth that of the earth, since the mass of the moon is much less than that of the earth. Man would therefore tend to 'float' in walking over the moon's surface. Theoretically, high-jumpers could leap very high distances for the same reason. The absence of an atmosphere on the moon is considered to be due to the weak gravitational attraction of matter by the moon owing to the moon's relatively small mass.

### Practical Units of Force

The force due to gravity on a mass of 1 kg at a particular place is taken as the practical unit of force by engineers. It is called the *kilogramme force* and the symbol is *kgf* (see p. 4). Thus 1 kgf = 1 kg wt at this particular place on earth, where the acceleration $g$ due to gravity has the specific value 9·80665 m/s².

Since $g$ varies, the weight or force due to gravity varies all over the world. Thus '1 kilogramme-wt' must not be confused with a 'mass' of 1 kilogramme, which is constant all over the world.

Since force $F = ma$, the force due to gravity on a mass of 1 kg is given by

$$F = 1 \times 9\text{·}8 = 9\text{·}8 \text{ N}$$

assuming $g = 9\text{·}8$ m/s². The differences from this value of $g$ are slight all over the world, and hence, for practical purposes, we may assume that

$$1 \text{ kgf} = 9\text{·}8 \text{ N} \quad . \quad . \quad . \quad . \quad . \quad . \quad (1)$$

### Weights and Masses

In SI units, however, weights are measured in newtons. The weight of a girl of mass 50 kg is about 500 N. The weight of an average size apple of mass 100 g is about 1 N. For future use, note that, in round figures,

$$\text{weight of 1 kg mass} = 10 \text{ N} \quad . \quad . \quad . \quad . \quad (2)$$

and $$\text{weight of 1 g mass} = 0\text{·}01 \text{ N} \quad . \quad . \quad . \quad . \quad (3)$$

### Calculations

. As an illustration, suppose a car of mass 1000 kg is accelerating at 2 m/s². Then the force $F$ acting on it is given by

$$F = ma = 1000 \times 2 = 2000 \text{ N}$$

Suppose that a rope is pulling up a load of mass 80 kg with a constant acceleration $a$. Assuming $g = 10$ m/s², the weight of the load $= 80 \times 10 = 800$ N.

Suppose the tension (force) in the rope is 1000 N. This force acts upward on the load but the weight acts downward. So the net upward force $F$

$$= 100 - 800 = 200 \text{ N}$$

Now $$F = ma,$$

where $m$ is the mass, 80 kg, of the load. Hence

$$200 = 80 \, a$$

So $$a = 2 \cdot 5 \text{ m/s}^2$$

A man in a lift experiences a force of reaction due to the floor of the lift. Suppose the mass of the man is 50 kg and the lift moves *upwards* with a constant acceleration of 2 m/s² at an instant.

The reaction force $R$ at the floor acts upwards on the man; his weight, 500 N, acts downwards. So the net upward force $F = (R - 500)$ in newtons. From $F = ma$, it follows that

$$R - 500 = 50 \times 2 = 100$$
$$\therefore \quad R = 600 \text{ N}$$

So the man feels an upward force due to the floor which is greater than his weight, 500 N.

If the lift moves *downwards* with the same acceleration, similar calculation shows that the force at the floor is now 400 N, which is less than the man's weight.

## Force on Astronaut

When an astronaut is strapped into his seat in the spacecraft the upward force on him due to the reaction of the seat is balanced by his weight $mg$, and no discomfort is experienced. At blast-off the spacecraft and astronaut are lifted with a tremendous acceleration from the firing pad. If $R$ is now the reaction of the seat on him the *resultant* upward force $F$ is $(R - mg)$. Thus if the upward acceleration is $15g$, then, from $F = ma$,

$$R - mg = m \cdot 15g$$

Thus $R = 16mg$. This is an abnormally high force for the human body to withstand. Experiments which achieve this order of acceleration, by whirling him round at very high speeds, are therefore carried out as part of the training of the astronaut (see p. 108 and Plate 4).

## Effects of Friction

Friction is a common force in our everyday life (see p. 39). If we rub our fingers over a table we experience a force of resistance or frictional force. Friction can be very useful. On account of it we can walk across the ground. As you step forward, friction opposes your forward force, so that

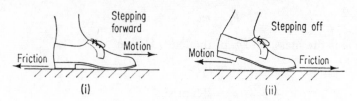

Fig. 4.7 Friction assists walking

your foot does not slip forward (Fig. 4.7(i)). When you raise your heel again to take another forward pace, this time friction acts the other way and prevents the front part of your shoe moving back (Fig. 4.7(ii)). *Friction always opposes motion.* The smoothness or lack of friction between a banana-skin and the ground actually results in an unbalanced forward force when stepping on to the skin, so that we fall over. On account of friction between the ground and the rubber tyre of a bicycle the latter can move forward. To stop the bicycle we use the opposing frictional force between the brake blocks and the wheel rim.

On the other hand, friction can be a considerable handicap in machines. Some part of the force applied to a pulley system used for hauling loads, for example, must also be used to overcome the frictional force at the wheels. This is wasted energy (see p. 153). Ball-bearings in wheels, and oiled or greased axles, are used to diminish the frictional force (p. 38).

## Friction and Motion

We shall now see what effect an opposing force such as friction has on the motion of an object.

Suppose the force on a car of mass 2000 kg due to the engine is 3000 N and the frictional force at the ground and the air resistance together total 1000 N (Fig. 4.8). First, note *all the forces* on the car, which are 3000 and 1000 N respectively. Secondly, note their direction, since they are vectors (p. 83), and draw the corresponding arrows in a diagram. Thirdly, calculate the *resultant force*, which is 3000 − 1000 or 2000 N. This gives the magnitude of $F$, or resultant force, in the formula $F = ma$. Fourthly, note the *mass moved*, $m$; this has no connection whatsoever with 'forces', and is 2000 kg. Remembering that in $F = ma$, the force $F$ must be in newtons, the acceleration is given by

$$2000 = 2000a$$
$$\therefore a = 1 \text{ m/s}^2 \text{ (approx.)}$$

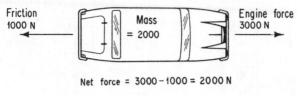

Net force = 3000 − 1000 = 2000 N

Fig. 4.8 Forces on car

Similarly, if the force in a rope pulling a sledge is 100 N, the mass of the sledge is 40 kg, and the frictional force is 20 N, the resultant force $F = 100 - 20 = 80$ N. From $F = ma$,

$$80 = 40a$$
$$\therefore \quad a = 2 \text{ m/s}^2$$

### Uniform Velocity and Force

At first sight it may be puzzling to know why a car can travel along at a uniform or constant velocity if the engine provides a force on the car. The reason is that the force due to the engine counteracts exactly the opposing frictional force, so that the *resultant* force on the car is zero. No acceleration or velocity increase is therefore produced. When the car first started off, the force due to the engine exceeded the frictional force. The resultant force produced an acceleration. As the speed of the car increased, the ground frictional force at the car wheels and the air resistance increased. Eventually, by adjusting the pressure on the car accelerator pedal, and hence the engine force, the car can continue to run at a steady velocity of 30 km per h, for example, as already explained.

If a model trolley is placed on a plane which can be tilted, the frictional force at the wheels can be balanced by the component of the trolley's weight down the plane, so that the trolley moves down with constant velocity. If the plane is tilted farther the component of the weight down the plane increases and the resultant force accelerates the trolley. An ice-puck moves over a smooth flat glass plate for a considerable time. At first, if it is given a gentle push, it accelerates. It would then undergo appreciable retardation if the frictional force at the glass was large and come to rest, but as the latter force is very small (p. 94) the retardation is consequently very small.

### Terminal Velocity

The same general principles apply if a small steel ball-bearing B is dropped gently into water (Fig. 4.9(i)), and then into a thick liquid such as glycerine or syrup, each contained in a tall jar (Fig. 4.9(ii)). In the case of the water, B falls rapidly and hits the bottom of the jar shortly after. In the case of the glycerine, however, experiment shows that B falls slowly and that it takes equal times to fall equal distances AD, DE after reaching the middle of the jar. The velocity of B thus became uniform or constant (see p. 45).

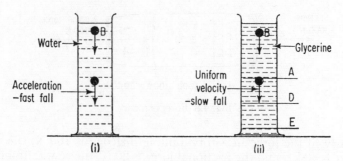

Fig. 4.9 Low and High viscosity

## Explanation

When B is dropped into water the downward force on it (weight less the upthrust of the liquid, p. 206) is much greater than the upward frictional force due to the liquid viscosity. As B gathers speed the frictional force increases, but the downward force is still very much greater as water has a low viscosity, and hence B continues to increase in velocity.

Glycerine, however, has a high viscosity and hence high frictional force on objects moving through it. At first the downward force on B is greater than the opposing frictional force. B thus increases in velocity. The frictional force then increases. At some stage, which is quickly reached, the frictional force counteracts completely the downward force. Consequently, as previously discussed, the resultant force is zero, and at this point B moves with a constant velocity.

The maximum velocity reached by B is called its *terminal velocity*. An object dropped from an aeroplane reaches a terminal velocity when the frictional force due to air resistance completely counteracts its weight.

## Circular Motion · Centripetal Force

If a heavy stone is attached to the end of a long string and whirled steadily round in a horizontal circle by holding the other end of the string we can feel that a force is needed to make the stone move in its circular path. If the stone is whirled faster the force increases. A far greater force can be seen exerted by the muscles of a hammer-thrower as he whirls the heavy hammer round at a fast speed in a circle.

An object moving in a circle thus requires a force towards the centre of the circle to keep it moving, and this is called the *centripetal force* on the object. A force produces an acceleration. Consequently, *the object has an acceleration towards the centre of the circle*, even though it moves round the circle with a constant speed. We shall prove this later (p. 107).

## Experiment on Acceleration in a Circle

The magnitude of the acceleration of an object moving in a circular path depends on its speed $v$ and on the radius $r$ of the circle. A simple apparatus to investigate the relationship is shown in Fig. 4.10(i). It consists of a mass S, such as a rubber stopper, at the end of a long nylon cord C which passes through a glass tube with a rubber grip. The top of the tube is polished, so that there is little friction here on the cord. A variable mass W, consisting of a number of steel washers of equal weight, can be attached to the other end of the cord, as shown. Their total weight provides the centripetal force $F$ acting on $S$ due to the tension in the cord.

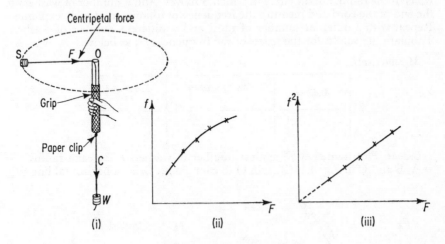

Fig. 4.10 Centripetal force

*Method.* Attach a number of washers to the end of the cord. Keep the length of OS constant at a convenient value such as a metre, and then, holding the tube G vertically, swing the mass S in a horizontal circle at a steady frequency of rotation. Do not move the hand more than necessary in a small circle. Observe the time taken for, say, 30 revolutions of S.

Increase the number of washers, and keeping the length OS the same, observe the new time for 30 revolutions of S as it whirls in a circular path of the same radius. Repeat with more washers, and tabulate five readings of number of washers and time for 30 revolutions.

*Measurements.*

| No. of washers $F$ | Time for 30 rev | No. of rev per s $f$ |
|---|---|---|
|  |  |  |
|  |  |  |

*Graphs.* Work out the number of revolutions per second or frequency of rotation $f$ of S and enter the results in the table. Then plot (i) number of washers $F$ v. $f$ and (ii) number of washers $F$ v. $f^2$.

*Results.* The graph of (i) is a curve (Fig. 4.10(ii)). The graph of (ii), however, is a straight line through the origin (Fig. 4.10(iii)). Hence $F$ *is directly proportional* to $f^2$. Now the force $F$ produces an acceleration on S towards the centre, and $f$ is directly proportional to the *speed* $v$ of S as it moves round the circle, because the greater $v$, the greater is $f$. Thus $F \propto v^2$ for a given radius.

**Acceleration and Radius of Circle.** *Method.* Alter the length of OS, which is the radius $r$ of the circle in which S moves. Add a number of washers to the end of the cord and measure the frequency of rotation as already explained. Repeat with a different number of radii and a different number of washers. Tabulate the results for the *square* of the frequency, $f^2$, as below.

*Measurements.*

| Radius $r$ | No. of washers $F$ | Frequency $^2$ $f^2$ |
| --- | --- | --- |
|  |  |  |

*Graphs.* Plot frequency $f^2$ against number of washers $F$ for each radius $r$ as at A, B and C in Fig. 4.11(i). Join O to each point. Draw a horizontal line PQ,

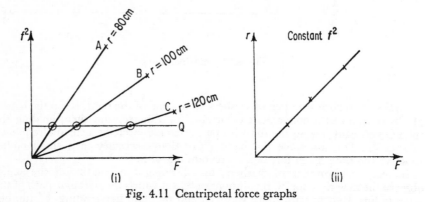

Fig. 4.11 Centripetal force graphs

and from the points of intersection obtain the different forces $F$ required to keep the mass S moving in circles of different radii with the same frequency$^2$. Plot $r$ v. $F$. A straight line passing through the origin is obtained (Fig. 4.11(ii)).

*Conclusion.* For the same frequency of rotation, the acceleration towards the centre of the circle is directly proportional to the radius $r$.

## Summary of Results

We can now combine the results of the two experiments. In the first one the acceleration towards the centre was found to be proportional to $f^2$ for a given radius, where $f$ is the frequency. In the second one the acceleration towards the centre was found to be proportional to $r$, the radius, for a given value of $f^2$. From both experiments, therefore, it follows that

$$\text{Acceleration towards centre } a \propto rf^2 \quad . \quad . \quad . \quad (1)$$

If the constant speed in the circle is $v$ the distance travelled in 1 sec is $v$. The number of revolutions per second or frequency $f$ is then given by

$$f = \frac{v}{2\pi r},$$

because one revolution corresponds to a distance equal to the circumference $2\pi r$ of the circle. Thus

$$rf^2 = r \times v^2/4\pi^2 r^2 = v^2/4\pi^2 r.$$

Hence, as an alternative statement to (1),

$$\text{Acceleration towards centre } a \propto v^2/r \quad \ldots \quad (2)$$

### Acceleration in a Circle

We shall now prove that the acceleration of an object moving with a constant speed $v$ in a circle of radius $r$ is given by

$$\text{Acceleration } a = \frac{v^2}{r}.$$

At a point A in its motion the *velocity* $v$ acts in the direction of the tangent AP (Fig. 4.12(i)). Suppose the object is moving very slowly and moves only a small distance to B in 1 sec. The velocity $v$ is now in a new direction, that of the tangent BQ.

On p. 63 it was stressed that velocity is a vector quantity, that is, it has direction as well as magnitude. If the velocity at A is represented by $\overrightarrow{v_A}$ and that at B by $\overrightarrow{v_B}$, then the two velocities are different, and

$$\text{Change in velocity} = \overrightarrow{v_B} - \overrightarrow{v_A} = \overrightarrow{v_B} + (-\overrightarrow{v_A}).$$

Thus, as explained on p. 86, the change in velocity is obtained by drawing a line OL to represent $\overrightarrow{v_B}$ in direction and magnitude and then adding on to it a line LM which represents $-\overrightarrow{v_A}$, as shown in Fig. 4.12(ii). The resultant is thus

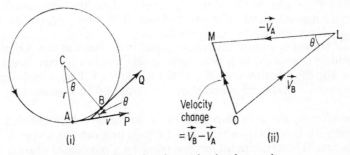

Fig. 4.12 Velocity change in circular motion

the velocity represented by OM. This *change or difference in velocity* points towards the centre C of the circle. It shows that the acceleration acts towards the centre, as already pointed out.

The acceleration $a$ is given by velocity change/time, and we have taken a time of 1 sec as the time to move from A to B. Thus OM is numerically equal to the acceleration $a$. Now triangles OLM and ACB are similar. Hence OM/OL = AB/AC, or OM/$v$ = $v$/$r$, assuming the chord AB is practically equal to the arc AB. Thus OM = $v \times v/r = v^2/r$. Hence

$$\text{Acceleration } a = \frac{v^2}{r}.$$

Although we have taken a movement of 1 sec from A to B, the same result is obtained for any small fraction of a second, as the reader should verify.

We can see in a general way why the acceleration is $v^2/r$ by considering Fig. 4.12(ii). The greater $v$, the longer is OL and hence the greater is the velocity change, which is represented by the length of the line OM. The time to make the change is the time to travel from A to B in Fig. 4.12(i), and this increases with $r$ and decreases with $v$. Hence, in general, the time is proportional to the ratio $r/v$.

Thus the acceleration $\quad\quad = \dfrac{\text{Velocity change}}{\text{Time}}$

$$= \frac{v}{r/v} = \frac{v^2}{r}.$$

## Motion in a Circle

It can now be seen that if an object of mass $m$ is moving in a circle of radius $r$ with a constant speed $v$, a force $F = ma = mv^2/r$ is necessary to keep it in motion, as the acceleration in a circle is $v^2/r$.

Consider a bucket of water which is whirled in a vertical circle of radius $r$. If the speed $v$ is such that the acceleration $v^2/r$ is 12 metre per second² and the mass of the water is 1 kg, then a force $mv^2/r$ of 12 newtons is needed to keep the water in its circular motion. At the top of the vertical circle the weight of the water, about 10 N, provides part of the 12 N centripetal force needed to keep the water moving. The rest of the force is provided by the reaction of the bottom of the bucket on the water, which is therefore 2 N. If the bucket is whirled round more slowly, and $mv^2/r$ is now 8 N, for example, the weight of the water is greater than the magnitude of the centripetal force required. The excess force, 2 N, now acts on the mass of water, which therefore falls out of the bucket.

In their training, astronauts are strapped into chairs at the end of a long beam, which is whirled in a horizontal circle at high speeds. See Plate 4. If $v^2/r = 16g$ their bodies are subjected to a force of $16mg$, which simulates what happens at blast-off when they are strapped in a spacecraft (see p. 102).

In centrifuges, particles suspended in a liquid of a different density are separated very quickly by whirling the liquid in a tube at a very high rate of revolution. The centripetal force needed on a particle of mass $m$ to keep it moving in a circle of radius $r$ is $mv^2/r$. If the particles are less dense than the liquid the mass $m$ of a particle is less than that of an equal volume of liquid. The particle thus needs a smaller centripetal force than this liquid at the same place, and hence the particles move nearer to the axis of rotation. In this way the particles spiral towards the axis and can be skimmed off when the centrifuge is stopped.

When a bicycle or car moves round a circular path, the centripetal force is provided by the frictional force between the ground and the tyres (Fig. 4.13(i)). If the frictional force is insufficient to provide the necessary value of $mv^2/r$ at high speeds the bicycle or car will move outwards or skid at the particular speed.

## Orbits

The centripetal force required to make the planets move continuously in their orbits round the sun, for example, is provided by the gravitational attraction between the sun and the particular planet (Fig. 4.13(ii)). The

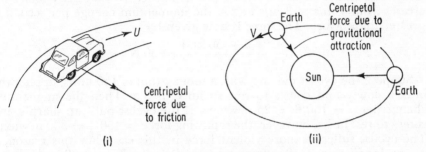

Fig. 4.13 Types of centripetal forces

centripetal force on the moon as it rotates round the earth is provided by the gravitational attraction between the two masses.

A spacecraft in orbit, and an astronaut inside, require a force equal to $mv^2/r$ to keep them moving round the earth, where $r$ is the radius of the orbit. This is provided by the gravitational attraction of the earth. If the weight of the astronaut at this height is just equal to $mv^2/r$ where $m$ is his mass, the gravitational attraction on him is just equal to the force necessary to keep his body moving in the orbit. Likewise for the spacecraft. Since both orbit at the same place, he has no tendency to sink to the floor or rise to the ceiling. Other objects in the spacecraft also have this kind of 'weightlessness', and they will float about if not tied down to the spacecraft.

## MOMENTUM

### Force and Momentum

We now consider an important quantity associated with a moving object, called its linear *momentum*.

The momentum of a moving object was defined by Newton as the product of its mass $m$ and its velocity $u$ or $mu$. Thus a boy of mass 50 kg moving with a velocity of 1 m/s has a momentum of $50 \times 1$ or 50 kg m/s. Now we have already seen that a force produces a velocity change when it acts on an object. Consequently, a momentum change occurs. Newton defined force in his second law by stating:

*Force is directly proportional to the rate of change of momentum produced.*

As an illustration of the definition, suppose a boy or girl on a bicycle, total mass 50 kg, has a velocity of 1 m/s and that, pedalling faster for 5s, the velocity increases to 3 m/s. Then, since 'rate of change' means 'change per second',

$$\text{Average force} = \frac{\text{Momentum change}}{\text{Time}} = \frac{(50 \times 3 - 50 \times 1) \text{ kg m/s}}{5 \text{ s}}$$
$$= 20 \text{ kg m/s}^2.$$

1 kg m/s² is the force acting on a mass of 1 kg which gives it an acceleration of 1 m/s². By definition, this force is the newton (see p. 98). Hence the force is 20 newtons. If the cyclist pedals harder, so that his velocity increases from 1 m/s to 3 m/s in 2 s, the momentum change per second is greater than before. The force is now given by

$$\frac{50 \times 3 - 50 \times 1}{2} = 50 \text{ N}$$

Suppose the cyclist moving with a momentum of 150 units, applies the brakes slowly and comes to rest steadily in 6 s. Then the momentum change per s = 150/6 = 25 newtons. If the cyclist pulls up sharply and comes to rest in ½ s, however, the retarding force = 150/½ = 300 newtons. The cyclist suffers a more violent force in this case. For this reason, a person landing on his feet from a height is well-advised to flex the knees so as to provide a less rapid momentum change on coming to rest.

Since Momentum = Mass × Velocity, the force $F$ acting on a constant mass is given, from Newton's second law, by

$$F = \text{Momentum change per second}$$
$$= \text{Mass} \times \text{Velocity change per second}.$$

But the 'velocity change per second' is the acceleration $a$. Hence

$$F = ma.$$

This result was first given on p. 98.

*Example.* (i) State a unit of momentum in the SI system.

(ii) Sand falls gently at a constant rate of 50 g/s onto a horizontal belt moving steadily at 0·4 m/s. Find the force on the belt in newtons exerted by the sand.

(i) Since force $F$ = momentum change/time, then momentum change = $F \times$ time. A 'momentum' unit is thus *newton second*.

(ii) Force $F$ = momentum change per second = mass of sand per second × velocity change. Neglecting the sand's initial velocity,

$$F = 0·05 \text{ (kg/s)} \times 0·4 \text{ (m/s)}$$
$$= 0·02 \text{ N}$$

## Short-time Forces

We have just dealt with a force acting for a short time. An engine pulls a train with a force which acts for a much longer time. In contrast, a cricket bat or tennis racket exerts a force on a ball for a very small fraction of a second. This is also the case when a football is kicked or a hockey stick strikes a ball or two billiard balls collide. When short-time forces act on an object a very rapid change in momentum may be produced.

A particular case of such a force is that exerted by gas molecules on the walls of a containing chamber. Like a crowd of bees in a closed room, the molecules move about at random in different directions, striking the walls repeatedly for a very short interval of time and rebounding. The effect on the walls is then similar to that produced by an expert boxer hitting a suspended punch-ball rapidly against the protected dial of a machine, which registers the average force. The repeated impacts of molecules on the walls produce an average force or gas pressure, whose magnitude depends on the rate of change of momentum.

## Gas Pressure

As an approximation, we can consider the molecules to be tiny spheres, which rebound from the wall of the containing cylinder like highly elastic balls.

This was an idea used most successfully by Clerk Maxwell in 1870 to

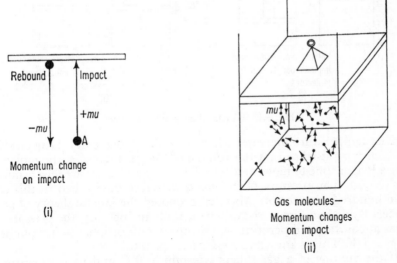

Fig. 4.14 Momentum change and gas pressure

explain gas pressure in terms of molecules. A particular molecule A of mass $m$ and velocity $u$ will have a momentum $mu$ just before striking the wall normally, say, and after rebounding it will have a velocity $u$ in the opposite direction (Fig. 4.14(i)). Since momentum is mass × velocity, and the latter is a vector quantity (p. 83), momentum has direction as well as magnitude. Before striking the wall the momentum is $+mu$; after striking the wall it is $-mu$.

$$\therefore \text{ Momentum change} = mu - (-mu) = 2mu.$$

The total momentum change *per second*, or force, will depend on the speed $u$ of the molecules. If it is fast, molecules will move swiftly towards the wall across the cylinder, and the number of impacts per second will be high (Fig. 4.14(ii)). If the speed is low the number of impacts per second is relatively small. Thus

Momentum change per second $\propto 2\,mu \times u \propto u^2$, omitting the constant value $2m$.

$$\therefore \text{ Force or gas pressure } p \propto u^2.$$

Thus the gas pressure depends on the average value of the *square* of the velocities of all the molecules.

### Volume and Temperature Change

When the volume of a gas at a given temperature is decreased, by pushing a piston down from A to B, for example, Fig. 4.15(i), the molecules take less time to travel between the piston and the wall opposite. Consequently, more impacts per second are made. Thus the pressure of a gas at

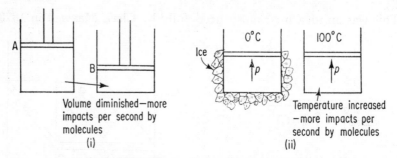

Volume diminished—more impacts per second by molecules
(i)

Temperature increased —more impacts per second by molecules
(ii)

Fig. 4.15 Kinetic theory of gas pressure

a given temperature increases when the volume decreases. Conversely, if the piston is raised so that the volume of the gas increases, the molecules take a longer time to move across to the piston and there are fewer impacts per second. The pressure of the gas is therefore diminished in this case. Two hundred years before Maxwell proposed the kinetic theory of gases, Robert Boyle had discovered experimentally in 1660 that the pressure $p$ of a gas at constant temperature was inversely proportional to its volume $V$, or $p \propto 1/V$. This is known as *Boyle's law* (p. 196).

If the volume of a gas is kept constant at 0°C and its temperature is raised to 100°C, for example, the speed of the molecules is increased (Fig. 4.15(ii)). Both the momentum change and the number of impacts per second are therefore increased. The pressure therefore rises when the temperature rises (see p. 251).

### Action and Reaction

When you lean on a table or desk with your elbow the downward force of your elbow on the table, called the 'action' force, produces an opposite upward force called the 'reaction', which is the force on your elbow due to the table. As you press down harder on the table, thus increasing the

**PLATE 3**
Friction

**3(a)** Friction test, showing the zero frictional force between a completely-worn tyre and a fast-moving drum in contact.

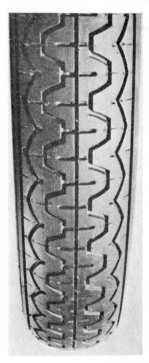

**3(d)** Motor tyre tread (p. 39).

**3(b)** Aquaplaning with poor tyre tread.

**3(c)** Road grip on the same road with good tyre tread.

PLATE 4

**Acceleration Due to Gravity and to Circular Motion**

◀ 4(**a**) Ball falling vertically under gravity, photographed at equal time intervals – observe the increase in velocity (p. 74).

4(**b**) Parabolic path of ball projected horizontally with uniform velocity and rebounding from floor after falling. Observe the acceleration and retardation (deceleration) (p. 87).

*Free fall multiflash photographs, taken by Gordon Severn, Esq.*

4(**c**) Astronauts are subjected on a centrifuge to about 16$g$ acceleration, corresponding to the highest force to be encountered in an actual *Mercury* mission (p. 108).

'action', you experience a greater reaction. Newton, in his third law of mechanics, stated that *action and reaction are equal and opposite.*

Thus if a person strikes a brick wall with his knuckles the force on the wall is equal and opposite to the force on the knuckles at the moment of impact. Although the forces are equal, the effect on the wall is negligible but that on the hand is considerable! The earth attracts you with a force equal to your weight; from the law of action and reaction, you attract the earth with an equal and opposite force.

## Momentum Changes Due to Action and Reaction

We have seen that forces produce a change in momentum. The momentum change produced by forces of action and reaction can be studied by means of two trolleys A and B (Fig. 4.16(i)). One trolley B has a projecting wooden rod R, which can be pushed back to compress a spring and

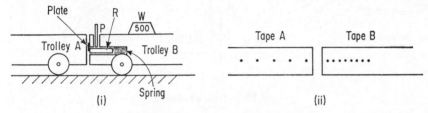

Fig. 4.16 Momentum measurement

then held in this position by a plate. A can then be placed in contact with B. (Alternatively, the spring can be compressed by R and the two trolleys held together by a taut piece of string connected to each.) Ticker-tapes are then attached to A and B.

When a projecting wooden peg P is tapped sharply (or the taut string is burnt) the released spring imparts momentum to A and B, which move off in opposite directions. The tapes are then removed, and the initial velocities are then calculated from the spacing of the dots (p. 62) (see Fig. 4.16(ii)). A mass W of 500 g is placed on B to make its mass appreciably different from that of A.

*Results*

|  | Trolley A | Trolley B |
|---|---|---|
| Mass | 620 g | 1120 g |
| Velocity (initial) | 20 units | 11 units |

*Calculation*

Momentum change of A = mass × velocity change
= 620 × 20 = 12 400 units.

Momentum change of B = 1120 × 11 = 12 320 units.

Similar results were obtained on varying the masses of A and B.

*Conclusion.* Allowing for experimental error,

Momentum change of A = Momentum change of B.

This result can easily be explained. From Newton's law, action and reaction are equal and opposite. The force on A by the rod is therefore equal to the force on B. The change of momentum per second of A is hence equal to that of B. But the forces on A and B act for the same length of time. Hence

*Momentum change of A = Momentum change of B.*

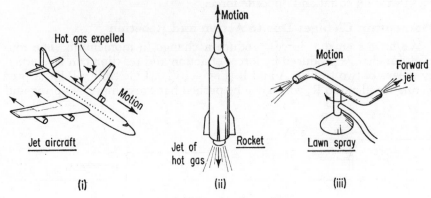

Fig. 4.17 Forces of reaction

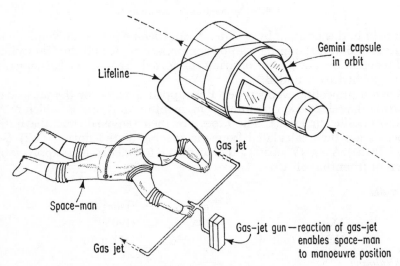

Fig. 4.18 Astronaut gas-jet gun

## Examples of Momentum Changes

There are many examples of momentum change caused by forces of action and reaction. The reaction force was put to a very useful purpose by Whittle when he invented the *jet aeroplane*. Here a jet of very hot gases issues from behind the aeroplane; the speed and mass per second of the gas are so high that considerable momentum is imparted to the stream of gas (Fig. 4.17(i)). An equal and opposite momentum is imparted to the aeroplane, which undergoes a forward thrust. The same principle is used in rockets used for launching satellites. Jets of hot gas issue downwards; an equal opposite momentum is given to the rocket, which then rises (Fig. 4.17(ii)). *Sprinklers* on lawns throw out water in one direction, thus giving it momentum, and themselves rotate in the opposite direction owing to the equal and opposite momentum (Fig. 4.17(iii)). A *gas-jet gun* was used by astronauts to manoeuvre in space (Fig. 4.18).

When a bullet is fired from a rifle the rifle jerks backwards. Suppose a bullet of mass 15 g leaves a rifle of mass 3000 g with a velocity of 100 km/h. Then

$$\text{Momentum change of rifle} = \text{Momentum change of bullet}$$

$$\therefore 3000 \times v = 15 \times 100$$

where $v$ is the backward velocity of the rifle.

$$\therefore v = \frac{1500}{3000} = 0.5 \text{ km/h}$$

### Conservation of Linear Momentum

Further examples of action and reaction occur when objects collide. The momentum changes concerned can again be studied with two trolleys and ticker-tapes.

#### EXPERIMENT

Place some plasticine on the front of trolley B (Fig. 4.19(i)). By using plasticine to keep them in place, fix drawing pins on the front of trolley A

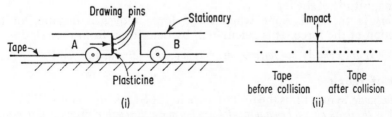

Fig. 4.19 Momentum and collision

with their points towards B. Attach a ticker-tape to A. The trolley A is then given a push so that it travels with uniform velocity and collides with B, which is stationary. Both trolleys then stick together and move off with the same velocity. The tape is then examined. As shown in Fig. 4.19(ii), the spacing of the dots changes abruptly after collision, and the velocity of A before collision, and that of A and B after collision, can be measured from the dot spacings.

*Results.* The following results were obtained by two pupils:

(1) *Before collision.* Mass of trolley A = 615 g
   Velocity of A = 14·5 cm ÷ $\frac{2}{50}$ s = 363 cm/s
(2) *After collision.* Mass of trolleys A + B = 1235 g
   Velocity of A + B = 7·1 cm ÷ $\frac{2}{50}$ s = 178 cm/s.

The momentum before collision $= 615 \times 363 = 220\,000$ (in round numbers)

The momentum after collision $= 1235 \times 178 = 220\,000$ (in round numbers)

Other collision experiments give similar results.

*Conclusion.* When objects collide,

Their total momentum before collision = Their total momentum
after collision.

Provided, then, that no forces other than those due to collisions act on them, the total momentum of colliding objects is constant. This is called the *principle of the conservation of linear momentum*. It should be carefully noted, of course, that momentum is a vector quantity, that is, it has direction as well as magnitude (p. 111). Consequently, the conservation of momentum only applies to that direction along which no external force acts.

## WORK · ENERGY · POWER

### Work

In mechanics we usually associate the term 'work' with movement. Thus an engine which pulls a train along a track is said to do *work*. If the train is pulled 100 m by a force of 5000 N, and then a further 200 m by the same force, more work is done in the latter case. The work done thus depends on the distance moved by the force. Further, if the train is pulled 100 m first by a force of 5000 N and then by a force of 10 000 N, more work is done in the latter case. The work done is thus also dependent on the magnitude of the force.

*Work* is therefore done whenever a force moves a distance in the direction of the force. It is calculated, by definition, from the relation:

$$Work\ done\ W = force \times distance\ moved\ in\ direction\ of\ force \qquad (1)$$

### Unit of Work

The *joule*, symbol J, is the unit of work in the SI system of units. This is defined as *the work done when a force of 1 newton moves through a distance of 1 metre in its own direction*. Hence if a block is pulled steadily 3 m against a frictional force of 4 newtons the work done $= 4 \times 3$ or 12 joules. The work done by a force of 50 N moving through 10 metres in its own direction

$$= 50 \times 10 = 500\,J$$

Memorise that:

$$1\ newton \times 1\ metre = 1\ joule$$

Since the weight of 1 kg mass $= 10$ newtons (approx.),

the work done in raising 1 kilogramme through 10 cm or 0·1 metre is about a joule. Basically, the joule is a unit of mechanical work and energy, but it is also used as the practical unit of electrical energy and heat energy. See pp. 265 and 490.

## Other examples of Work

There are many different examples of work. Consider the case of a watch spring which is wound up. The molecules of the spring are then moved closer together, against opposing or repulsive forces exerted between the molecules (see p. 50). Thus work is done against molecular forces.

There are forces of attraction on tiny particles called electrons inside atoms (p. 733). Normally, the electrons exist in equilibrium inside the atoms. But when light which is sufficiently energetic is incident on a metal plate, some of the electrons in surface atoms are removed completely from their parent atom (see p. 695). Work is thus done against the atomic forces of attraction on the electrons.

When a car engine is working, the volume of the petrol vapour expands and contracts continually (p. 276). As the gas expands, the piston of the engine is forced down. In this case, therefore, mechanical work is done by a gas.

## Calculation of Work

From the relation (1) for work $W$ on p. 116, if an engine force of 5000 newtons pulls a train 100 m,

$$W = 5000 \times 100 = 500\,000\,\text{J}$$

If a sledge is pulled 40 m by a boy with a horizontal force of 200 newtons,

$$W = 200 \times 40 = 8000\,\text{J}$$

It should be carefully noted that in calculating the work done from (1) the 'distance moved' must be that *in the direction of the force*. A horse pulling

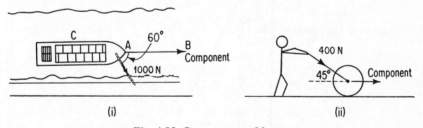

Fig. 4.20 Components of forces

a barge C along a canal by means of a rope at an angle 60° to the bank may produce a tension of 1000 N in the string, but the distance moved along the canal from A to B, for example, is *not* in the same direction as the tension (Fig. 4.20(i)). The force pulling the barge in the direction A to B is actually the *component* of the tension in the direction AB. This is 1000 cos 60° or 500 N (see pp. 89, 166). Hence if AB is 40 m,

Work done $W$ = Force in newton × Distance in metre
$$= 500 \times 40 = 20\,000\,\text{J}$$

Similarly, if a lawn roller is pushed with a force of 400 N at an angle of 45° to the lawn the force acting along the ground is 400 cos 45° or 280 N (Fig. 4.20(ii)). Thus if the roller is pushed along a distance of 20 metres, then

$$\text{Work done } W = 280 \times 20 = 5600 \, \text{J.}$$

## Work Done against Gravity

If a lawn roller is pushed steadily, work is done against the frictional force between the roller and the grass. When a nail is hammered into a piece of wood work is done against the resistance of the wood. Thus work is always done in moving against some opposing force.

A boy or girl climbing a rope does work in raising himself or herself against the force of gravity, which acts downwards towards the centre of the earth. If the weight is 440 newton and the distance climbed steadily is 3 m, then

$$
\begin{aligned}
\text{Work done} &= \text{Force} \times \text{Distance moved in direction of force} \\
&= 440 \, (\text{newton}) \times 3 \, (\text{metre}) \\
&= 1320 \, \text{joule}
\end{aligned}
$$

Climbing stairs, or climbing a mountain, likewise requires work to be done against gravity. If the climb is high a person out of condition will be very conscious that he or she has done work by the end of the climb!

## Energy

A *pile-driver* is a heavy concrete weight used, for example, to drive a thick stake or pile of timber into the sea-bed when constructing foundations for a pier or bridge (Fig. 4.21). By means of an engine, the pile-driver is

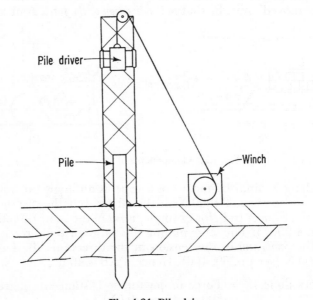

Fig. 4.21  Pile-driver

first raised high above the pile. Here, the poised pile-driver is stationary. But when it falls and crashes into the pile it drives it into the sea-bed. The pile-driver thus does work against the resistance of the sea-bed. Consequently, the pile-driver had *energy* when it was poised stationary above the pile, that is, *it was capable of doing work.* If it was raised higher it was capable of driving the pile farther into the sea-bed; and hence, in its new position, it had more energy than before. *Energy is defined as the capacity for doing work.*

Of course, the energy of the stationary pile-driver high above the pile comes from the energy spent by the connected engine in raising it to this height. We shall discuss the relation between different kinds of energy later (p. 125). Here we must carefully note that *the energy of an object is always equated to the work done in bringing it to its position.* In this case, therefore, if the pile-driver has a weight of 660 newton and is raised 20 m, then

$$\text{Energy of stationary pile} = \text{Work done} = \text{Force} \times \text{Distance}$$
$$= \text{Weight of pile} \times \text{Distance}$$
$$= 660 \ (\text{N}) \times 20 \ (\text{m})$$
$$= 13\ 200 \ \text{J}$$

## Potential Energy · Hydroelectric Energy

We have seen that any object on which work is done has an amount of mechanical energy. If it is stationary, like a poised pile-driver, it is said to

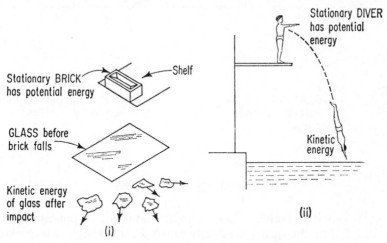

Fig. 4.22 Conversion of potential to kinetic energy

have *potential energy*. Potential energy is defined as energy possessed by reason of level or position.

There are many examples of potential energy. The pile-driver poised above the pile is one. A stationary brick on a shelf above the ground has potential energy; if it is pushed off the shelf and crashes into a sheet of glass near the floor it will splinter the glass into many pieces (Fig. 4.22(i)).

The moving glass splinters, some of which may be hurled through big distances, show that the stationary brick on the shelf had potential or stored energy which it now gives to the glass. A high-diver on a platform, poised to dive, has an amount of potential energy equal to the work done in raising himself (Fig. 4.22(ii)). Thus

*Potential energy = Weight of diver × Height above water.*

When he dives into the water the potential energy is used up in doing work against the resistance of the water.

Hydroelectric power stations utilize the potential energy of water as it falls to a lower level under gravity. Here it moves with considerable kinetic energy and pushes against blades attached to a large wheel, which is therefore set into motion. The wheel is attached to a generator, which then produces a voltage as explained on p. 627.

## Other Potential Energies

These are examples of potential energy due to work done in moving against the force of gravity. A different example of potential energy is that of a wound spring. Unwound, the spring has a certain amount of potential

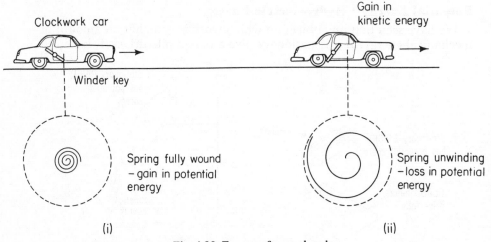

Clockwork car

Gain in kinetic energy

Winder key

Spring fully wound – gain in potential energy

Spring unwinding – loss in potential energy

(i)

(ii)

Fig. 4.23 Energy of wound spring

energy. When it is coiled it has a greater amount of potential energy (Fig. 4.23(i)). As the spring gradually unwinds, the wheels of a clockwork car, for example, move (Fig. 4.23(ii)), so that work is done as the potential energy is used up. Where does the potential energy of the spring come from in this case? It comes from the work done in compressing the spring against the elastic forces in the metal, which are due to forces between molecules (p. 50).

If a box is raised, work is done against the opposing force due to gravity. If a piece of elastic is stretched slightly work is done against the forces of attraction between the molecules of the elastic, which opposes the movement. The stretched elastic thus has potential energy equal to the work done against these forces. If the elastic is released it flies back to its original length under the action of the attractive forces. A stretched catapult first gives a missile potential energy, so that, when released, the latter flies through the air with considerable speed. The bow of an archer gives energy to an arrow in a similar way.

## Kinetic Energy

A fast-moving cricket ball which crashes into a window pane hurls splinters of glass through big distances. A moving train coming to rest at a station may compress powerful springs at the buffers. Thus, since it can do work, these examples show that a moving object has energy. It is called *kinetic energy*. Kinetic energy is energy due to motion; in contrast, potential energy is the energy of a stationary object due to level or position.

In a general way we can see what factors affect the kinetic energy of a moving object. A massive rugby player running with the ball can charge his way through direct opposition more easily than a lighter man moving with the same speed. The magnitude of the kinetic energy thus depends on the *mass* of the moving object; the greater the mass, the greater is the kinetic energy.

The kinetic energy also depends on the speed or *velocity* of the moving object. The faster a rugby player runs, the more easily can he push through opposition directly in his path. Thus the greater the velocity, the greater is the kinetic energy of a moving object.

A bullet is a very light object. But because it moves with very high speed it has a considerable amount of kinetic energy and penetrates deeply into the flesh. A cricket ball is much heavier, and moving with far less speed than a bullet, it can knock a stump out of the ground. Similarly, a heavy mallet moving with slow speed can make a peg penetrate deeply into the ground.

## Formula for Kinetic Energy

The kinetic energy of a moving object thus increases with its mass $m$ and its velocity $u$. We can find the magnitude of this energy by calculating the work it does when it is brought to rest. Suppose that it is brought to rest in a time $t$ by a *constant* opposing force $F$, in a similar manner to the way a moving train is brought to rest by powerful springs in buffers at a station. Then, since the final velocity is zero, the momentum change produced by $F = mu$.

$$\therefore F = \text{Momentum change per second} = \frac{mu}{t}.$$

The distance $s$ moved by the object in a time $t$ as the velocity diminishes uniformly from $u$ to zero is given by

$$s = \text{Average velocity} \times \text{Time} = \frac{u}{2} \times t$$

$$\therefore \text{Work done} = F \times s = \text{Kinetic energy}$$

$$= \frac{mu}{t} \times \frac{ut}{2} = \tfrac{1}{2}mu^2.$$

Thus the kinetic energy K.E. of an object of mass $m$ moving with a velocity $u$ is given by

$$\text{K.E.} = \tfrac{1}{2}mu^2.$$

In the formula for kinetic energy it should be carefully noted that $m$ is the *mass* of the moving object (not its 'weight'). The units of kinetic energy, like those of potential energy, are the same as the units of 'work', that is, they are joules. Here it is important to note that, since $F \times s = \tfrac{1}{2}mu^2$, and $F$ may be expressed in newton, the *units* in the kinetic energy formula are as follows:

When　　　　　　$m$ in kg, $u$ in m/s, then K.E. is in *joule*.

## K.E. Calculations

A boy or girl of mass 50 kg running with a velocity of 2 m/s would have kinetic energy given by

$$\text{K.E.} = \tfrac{1}{2}mu^2 = \tfrac{1}{2} \times 50 \times 2^2 \text{ joules}$$
$$= 100 \text{ J}$$

If the velocity was doubled, the kinetic energy would increase *four* times as much, since this is proportional to the square of the velocity.

A bullet weighing 20 g moving with a velocity of 30 m/s, would have a kinetic energy given by

$$\text{K.E.} = \tfrac{1}{2}mu^2 = \tfrac{1}{2} \times 0{\cdot}02 \text{ (kg)} \times 30^2 \text{ (m/s)}^2$$
$$= 9 \text{ J}$$

## Forms of Kinetic Energy

Kinetic energy, the energy due to motion, can have several forms. A fly or a bee has *translational kinetic energy* when it moves through the air; the word 'translation', in this sense, means that the whole insect moves from one point to another in space (Fig. 4.24(i)). On the other hand, a bowler

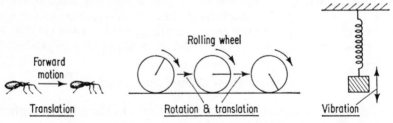

Fig. 4.24 Types of kinetic energy

can make a cricket ball spin through the air, in which case the ball has *rotational kinetic energy* as well as translational kinetic energy (Fig. 4.24(ii)). The earth has rotational kinetic energy; it spins about a north–south axis. Finally, if a small mass is attached to the end of a vertical spring and pulled down and released it will vibrate up and down. The mass has *vibrational kinetic energy* (Fig. 4.24(iii)). Simple gas molecules, those which have one atom, only have translational kinetic energy. More complex gas molecules may have, in addition, rotational and vibrational kinetic energy as they move about.

## Potential and Kinetic Energy under Gravity

Energy can change from potential to kinetic, and vice-versa. As an illustration, consider a box of 2 kg allowed to drop freely from rest from a height 5 m above the ground. At this height its energy is only potential energy P.E. Assuming the weight of 1 kg mass $= 10$ N, then

P.E. $=$ Weight $\times$ Height $= 20$ N $\times$ 5 m $= 100$ J  .          (1)

Just before it hits the ground the energy of the box is only kinetic energy K.E. The velocity $v$ is given by $v^2 = u^2 + 2as$, or, since $u = 0$, and $a = g = 10$ m/s$^2$ (approx.),

$$v^2 = 2 \cdot 10 \cdot 5.$$
$$\therefore v = \sqrt{100} = 10 \text{ m/s}$$
$$\therefore \text{K.E.} = \tfrac{1}{2}mv^2 = \tfrac{1}{2} \times 2 \times 10^2 \text{ J (see p. 122).}$$
$$= 100 \text{ J} \qquad . \quad . \quad . \quad . \quad . \quad . \quad . \quad . \quad (2)$$

From (1) and (2), it follows that the potential energy at the height of 5 m is equal to the kinetic energy near the ground, so that the energy of the object is constant.

When the box has fallen through a distance of 3 m say, from rest, the energy at this instant is partly potential and partly kinetic. The potential energy P.E. is given by

$$\text{P.E.} = \text{Weight} \times \text{Height above ground}$$
$$= 20 \text{ N} \times 2 \text{ m} = 40 \text{ J}$$

The velocity $v$ at this instant is given by

$$v^2 = u^2 + 2as = 2 \cdot 10 \cdot 3 = 60$$
$$\therefore \text{K.E.} = \tfrac{1}{2}mv^2 = \tfrac{1}{2} \cdot 2 \cdot 60 \text{ J}$$
$$= 60 \text{ J} \qquad . \quad . \quad . \quad . \quad . \quad . \quad . \quad (4)$$

From (3) and (4),

Total energy $=$ P.E. $+$ K.E. $= 40 + 60 = 100$ J

It therefore follows that the total mechanical energy of the box is *constant* as it falls to the ground. After hitting the ground and coming to rest the mechanical energy is zero. The mechanical energy has all been changed into *heat energy* as a result of the impact.

## Energy of Pendulum · Galileo's Experiment

Instead of allowing an object to fall freely under gravity, it can be suspended from a rigid support by a long thread and set swinging like a simple pendulum. Consider a small lead bob swinging in this manner (Fig. 4.25(i)). At the end A of its swing the bob is momentarily stationary, so that it has only potential energy. We can consider its lowest point B as a reference 'zero' level, so the potential energy at A is that due to its vertical height above B. As the bob moved down in an arc of a circle the bob increases in velocity until it reaches a maximum kinetic energy at B, where its potential energy is zero. At an intermediate point D the energy is partly kinetic and partly potential. Consequently, the energy gradually changes from potential at A to kinetic at B. The bob then swings to C, which is practically on the same horizontal level as A, where it is momentarily stationary. The kinetic energy of B has thus changed to potential energy equal to the original amount at A. Consequently, the energy of the bob is constant as it swings, omitting from consideration any frictional

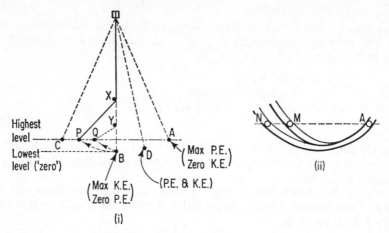

Fig. 4.25 Mechanical energy changes

forces on the bob. At D, for example, the sum of the potential and kinetic energy is equal to the potential energy at A or to the kinetic energy at B.

Galileo demonstrated that the kinetic energy at B was practically equal to the potential energy at A by placing a rigid peg at X well above B. The bob was released from A, and on reaching B the string was stopped by X so that the bob swung upwards. Observation showed that the bob reached a height P, practically on the same horizontal level as A. The kinetic energy at B was therefore equal to the potential energy at A. If the rigid peg was moved to a position Y, and the experiment repeated, the bob again swung up to Q at the same height as P. Thus Galileo's experiment shows that energy can be changed from one form to another.

The same results are obtained by bending metal rails such as curtain rails into a shallow curve (Fig. 4.25(ii)). A marble at A then rolls practically to the same horizontal level at M. If the rails are bent farther into a shallower curve and the experiment is repeated the ball reaches N, again practically on the same horizontal level as A.

### Effect of Friction

Close examination of a bob of a swinging pendulum, or of a marble rolling from rest in a curved metal rail as above, shows that the height reached on the other side of the swing is just below the original height or level. Thus some mechanical energy has been lost. This is due to *friction*, in the one case between the bob and the air, and in the other case between the solid surface of the marble and that of the metal rail. Both the bob and the marble eventually come to rest, so that the energy is completely used up in doing work against the frictional forces. As we shall see shortly, we do not get this energy back–it is changed to *heat* energy and the change cannot be reversed. When a card is attached to the bob of a pendulum so that its plane is perpendicular to the motion the bob comes to rest much more quickly. The amount of air swept is now much more than in the case of the bob alone, so that the frictional force is much greater and the energy is expended more quickly.

## Forms of Energy

The scientific meaning of 'energy' grew slowly over the centuries. Among famous scientists who contributed to a full understanding were Kelvin of Scotland, Joule of England and Clausius and Helmholtz of Germany, all of whom lived in the 19th century.

'Energy' means 'the capacity for doing work'. Today, many different forms of energy are recognized. We have already discussed *mechanical energy*. A steam engine uses *heat energy* to do work. An electric motor uses *electrical energy* to drive an electric train. *Light energy*, falling on a light meter used in photography, causes a pointer to move across a scale. *Sound energy* causes a microphone diaphragm or thin plate to vibrate. *Chemical energy* is the source of energy in our food which makes us grow and also provides us with muscular energy to move objects. *Nuclear energy*, the energy in the nucleus of atoms, produces heat energy, which in turn is used to generate electrical power.

## Energy Conversions

By means of suitable machines or apparatus, energy can be changed from one form to another. This is illustrated on page 11. Thus a steam engine converts heat energy to mechanical energy. Mechanical energy is converted to heat energy when a match is struck. A light meter or photoelectric cell converts light energy to electrical energy. An electric lamp converts electrical energy to light energy. A solar cell converts the heat of the sun to electrical energy. An electric fire converts electrical energy to heat energy. A microphone converts sound energy to electrical energy. A telephone earpiece converts electrical energy to sound energy. A battery converts chemical to electrical energy; a reverse change occurs in electroplating. The energy from the sun produces chemical changes which make plants and trees grow, and the energy is stored underground in coal seams centuries later after the wood is absorbed by the soil and sinks.

An electric plant at a power station illustrates how energy can be changed from one form to another until a desired form of energy is produced. Coal is first burned, so that heat energy is produced from chemical energy. By means of a steam engine or turbine, the heat energy is converted into mechanical energy, which turns the rotors of an electric generator (p. 634). Electrical energy is then produced. Electric lamps and heaters in homes and buildings now convert electrical energy to light and heat energy. Finally, the light energy collected by the eye falls on nerves in the retina, which stimulates the sensation of vision.

The heat energy received by the steam turbine is not all converted into mechanical energy. Some of the energy is used in overcoming the frictional forces in the wheels of the turbine. Likewise, some of the mechanical energy used to turn the rotor of the generator is used to overcome the frictional forces at the bearings. On account of this wastage of energy the Central Electricity Board in England is actively pursuing the generation of electricity by Magnetohydrodynamics MHD, where no such frictional forces occur (see p. 634). Sound energy is also produced by the spinning wheels owing to air disturbance. However, if the whole generating plant receives 100 units of energy, initially in the form of heat from the coal used, the total energy produced, calculated by adding together all the different forms of energy, will still be 100 units.

This leads us to a generalization known as the *Principle of the Conservation of Energy*. It was arrived at after many years of experiment and experience, and it is recognized today as one of the most important principles in science. It states that, *in a given system, the total amount of energy is always constant, although energy may be changed from one form to another.*

## Power

Two boys of the same weight each set out from the bottom of a hill to reach the top walking at a steady speed. One, X, is strong and athletic. The other, Y, is the opposite type, weak and not athletic. X can reach the top in a much shorter time than Y. Walking steadily, each has overcome the same frictional forces at the ground, and each has raised the same weight through the same height. Consequently, each has done the same amount of work. X, however, has done it in a shorter time than Y. We say that X has a greater *power* than Y. In the same way, a large engine can work faster than a small engine of the same kind and is said to have a greater power.

*Power* is defined as the rate of doing work or expending energy, or as the work done or energy expended per second. Thus

$$\text{Power} = \frac{\text{Work done or energy expended}}{\text{Time taken}}.$$

## Units of Power

Suppose an engine raises 100 kg of water through a height of 30 m in 10 s. The power at which it works is calculated, assuming the weight of 1 kg mass $= 9.8$ N, by

$$\text{Power } P = \frac{100 \times 9.8 \text{ (newton)} \times 30 \text{ (metre)}}{10 \text{ second}}$$

$$= 2940 \text{ joule per second.}$$

The *watt W* is a practical unit of power. It is defined as 1 joule per second rate of working. Hence, for the above engine,

$$P = 2940 \text{ watts} = 2 \cdot 940 \text{ kilowatts (kW)},$$

where 1 kilowatt = 1000 watts.

As we shall see later, electrical energy is measured in joules and electrical power in watts or kilowatts. A lamp of 120 watts uses energy at the rate of 120 joule per second. It produces a brighter light than a lamp of 60 watts.

Suppose a car travels with a uniform velocity of 72 km/h or 20 m/s along a horizontal road and overcomes a constant frictional force of 500 newtons while moving. Then

Engine power = Work done per second = Force × Distance per second
= 500 × 20 joule per second
= 10 000 watts = 10 kW

1 *horse-power* (h.p.) is 746 watts, about $\frac{3}{4}$ kW

$$\therefore \text{ Engine power } = \frac{10}{3/4} = 13\tfrac{1}{3} \text{ h.p.}$$

Since 1 h.p. is about 750 watts a small $\frac{1}{6}$-h.p. motor in a vacuum cleaner uses about 125 watts when working.

## Estimation of Power

A rough estimate of a pupil's power can be made by asking him or her to walk steadily from the bottom to the top of a building. Suppose it is 20 m high and the pupil's weight is 600 newton. If the time taken is 60 second, then

$$\text{Power} = \frac{\text{Work done}}{\text{Time}} = \frac{600 \times 20 \text{ joule}}{60 \text{ second}}$$

$$= 200 \text{ watts}$$

$$= \frac{200}{750} \text{ h.p.} = 0 \cdot 3 \text{ h.p. (approx.)}$$

In medical research, the power of an athlete is measured by cycling on rollers fitted with a friction brake, which introduces a known frictional force. The athlete does work against this force and the power can be estimated from the rate or speed with which he pedals.

## Engine Brake-power

An estimation of engine power called brake-power can be made with the apparatus shown in Fig. 4.26. When the machine is switched on it does work against the tension in the band or belt round the wheel, which remains stationary while the wheel rotates. At the end of a measured interval of time $t$ the revolution counter is read and the weight and the spring balance reading, which remains fairly steady, are recorded.

Suppose the number of revolutions made by the wheel is $n$, the wheel radius is $R$ m, the weight is $P$ N and the spring balance reads $S$ N.

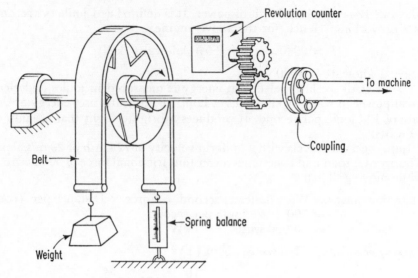

Fig. 4.26 Measurement of Engine Power

Then, since the tension in the belt is $(P - S)$ N and this is the force against which the engine is working,

Power = Work done per second

$$= \frac{(P - S) \text{ (newton)} \times 2\pi Rn \text{ (metre)}}{t \text{ (second)}} \text{ watts}$$

The power of the engine can then be calculated.

### Efficiency of Machine

A machine such as an electric motor or an engine may take in an amount of energy of 200 000 joules, for example, in a given time while working. Some of this energy is always used in doing work in overcoming friction and in other wasteful ways. Suppose 40 000 joules are used up. Then the useful work obtained from the machine is 160 000 joules. The *efficiency* of the machine is defined as

$$\frac{\text{Work or energy obtained from machine}}{\text{Work or energy supplied to machine}} \times 100\%$$

In this case, therefore,

$$\text{Efficiency} = \frac{160\,000}{200\,000} \times 100\%$$

$$= 80\%.$$

Theoretically, a machine with a 100% efficiency would do as much work in a given time as is expended on it. In practice, machines have efficiencies much lower than 100%, as forces due to friction, for example, have to be overcome and some energy is wasted here.

## SUMMARY

1. *Mass* is a measure of inertia. Units: kilogramme (kg), gramme (g). *Force* produces acceleration. Unit: newton (N). The weight of 1 kg mass = 10 N (approx.).

2. $F = ma$, where $m$ is mass, $F$ is force. With friction or other opposing forces, $F$ represents the *resultant* force. Force is a vector.

3. *Terminal velocity*. When an object falls through a viscous medium its velocity increases until the resistance becomes equal to its weight. The maximum velocity is called the 'terminal velocity'.

4. *Circular motion*. An object moving with constant speed $v$ in a circle of radius $r$ has an acceleration towards the centre of magnitude $v^2/r$. This is provided by a force towards the centre of magnitude $mv^2/r$ called the centripetal force, where $m$ is the mass of the object.

5. *Momentum* = mass × velocity. Force is equal to the rate of change of momentum. Momentum is a vector.

6. *Gas pressure* is due to the repeated momentum changes of molecules of gases when they rebound from the walls of vessels. The average pressure depends on the mean of the squares of all the different velocities of the molecules.

7. *Conservation of linear momentum*. When a system of bodies collide or interact, then, if no external forces act, the total momentum in any direction before collision = the total momentum in the same direction after collision.

8. *Work* = force × distance moved in direction of force. Unit: joule (J); 1 joule = 1 newton × 1 metre. Work is a scalar.

*Energy* is the capacity for doing work, and has the same units as work.

9. *Kinetic energy* is energy due to motion and is $\frac{1}{2}mu^2$, where $m$ is the mass and $u$ is the velocity. *Potential energy* is energy due to level or position and is *weight* × *height* above a particular level. Energy is a scalar.

10. *Conservation of energy*. In a given system, energy can be changed from one form to another, but the total amount of energy is constant.

11. *Power* is the rate of doing work or expending energy. Unit: joule per second or watt (1 watt = 1 joule per second). Approx. 1 h.p. = 750 watts.

12. *Efficiency* = work or power obtained/work or power supplied. The efficiency of a machine is never 100% owing to frictional forces.

## EXERCISE 4 · ANSWERS, p. 223

(Assume where necessary $g = 10$ m/s², and weight of 1 kg mass = 10 N.)

**1.** (a) Complete the following: (i) a newton is a unit of . . . ., (ii) a joule is a unit of . . . ., (iii) a watt is a unit of . . . . (b) Which of these is a vector quantity? (*N.*)

**2.** Name a unit of force. Define the unit you have named. (*N.*)

**3.** A car of mass 500 kg is moving with an acceleration of 2 m/s². Find the force acting on it.

**4.** A force of 0·12 newton acts on an object of mass 200 g. What is the acceleration of the object? How far does it travel from rest in $\frac{1}{4}$ min?

**5.** A boy pulls a sledge 6 m with a force of 80 N. Calculate the work done by the boy.

**6.** A man of mass 80 kg is running with a speed of 6 m/s. What is his kinetic energy in joules?

**7.** Niagara Falls are 50 m high. Calculate the potential energy of 5 kg of water at the top, relative to the bottom. What is the kinetic energy of this water just before it reaches the bottom, and what happens to the energy after the water reaches the bottom?

**8.** What is the work done when a force of 0·002 N moves an object 3 m in its direction? Calculate the work done if the force is 0·1 N.

**9.** A boy is free-wheeling on a bicycle down a plane inclined at 20° to the horizontal. The total mass of the boy and bicycle is 50 kg. What is the force down the plane? If the frictional force at the ground is 10 N, calculate: (i) the net force down the plane; (ii) the acceleration of the bicycle.

**10.** A car developing 10 h.p. (7·5 kW) is moving with a uniform velocity of 15 m/s. What energy per second is expended by the engine, and what is the frictional force overcome?

**11.** A bullet of mass 15 g has a speed of 400 m/s. What is its kinetic energy? If the bullet strikes a thick target and is brought to rest in 0·02 m, calculate the average net force acting on the bullet. What happens to the kinetic energy originally in the bullet?

**12.** The diagram shows a tape A which was attached to a moving trolley X of mass 500 g. At C it collided with a trolley Y and both moved together, the tape now being shown by B. Find the mass of Y.

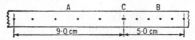

**13.** Two stationary trolleys P and Q, with tapes attached, are pushed apart by a spring plunger inside one of them. P has a mass of 400 g and Q a mass of 600 g.

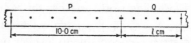

The respective tapes obtained after they are pushed apart are shown above. Calculate the length $l$.

**14.** What is the relation between force and acceleration for a constant mass? Describe in detail a trolley and ticker-tape experiment to demonstrate your answer, listing the important practical details and showing how the result is obtained.

**15.** Define *linear momentum*. What remains constant before and after two objects moving in a straight line collide? Describe an experiment to verify your answer, listing the important practical details and showing how the result is obtained.

**16.** An elastic ball of mass 0·2 kg strikes a wall normally with a velocity of 0·5 m/s and rebounds with the same velocity. Calculate the momentum change. If the ball strikes the wall repeatedly in the same way twice per second, find the average force exerted on the wall.

**17.** A molecule of a gas of mass $m$ strikes a wall of a vessel normally with a velocity $c$ and rebounds with an equal velocity. What is the change of momentum? Explain why the pressure of the gas depends on the average of the *squares* of the velocities of the molecules.

Using the molecular theory, *explain* the change in the gas pressure as the volume of the containing vessel is increased.

**18.** (a) State the principle of the conservation of energy. (b) Name a unit of energy. (N.)

**19.** Define *uniform acceleration*. Find the distance travelled by a body of mass 10 kg while its velocity is reduced from 9 m/s to 4 m/s by a constant force of 70 N.

**20.** A body of mass 0·5 kg, and initially at rest, is subjected to a force of 2 newtons for 1 second. Calculate: (a) the change in momentum of the body during the second; (b) the change in kinetic energy of the body during the second. (N.)

**21.** Define *work, power.*

A car of mass 800 kg starting from rest on a level road travels 60 m in 20 s with uniform acceleration. Assuming there were no frictional forces, what force would be necessary to accelerate the car? If there are constant frictional forces totalling 800 N what is: (i) the work done by the engine during the period considered; (ii) the horse-power developed? (1 h.p. = $\frac{3}{4}$kW) (L.)

**22.** Explain the terms *vector* and *scalar.*

Which of the following are vectors: speed, velocity, momentum, force, energy?

A ball of mass 100 g falls from a height of 3 m on to a horizontal surface and rebounds to a height of 2 m. Calculate the change (a) in momentum, (b) in kinetic energy, of the ball when it strikes the surface. (O. and C.)

**23.** Explain the meaning of the terms: *momentum, kinetic energy, force.*

A motor-car weighing 1000 kg travels at 48 km/h. Calculate (a) the momentum of the car, (b) the kinetic energy of the car, and find the braking force that can bring it to rest in 40 s. (O. and C.)

**24.** What is meant by: (a) *force;* (b) *uniform accleration?*

Sketch a graph showing the relation between time and distance fallen by a body falling freely from rest. How would you use this graph to determine the velocity of the body at any instant?

A body of mass 1 kg falls freely from rest through a height of 75 m and comes to rest having penetrated 7·5 cm of sand. Calculate: (i) the time taken in falling; (ii) the velocity when the body strikes the sand; (iii) the average force exerted by the sand in bringing the body to rest. (N.)

**25.** Describe briefly the changes in the potential and kinetic energies of the bob of a simple pendulum as it goes from one side of its swing to the other. (C.)

**26.** State Newton's laws of motion.

A body of mass 10 kg, resting on a smooth horizontal surface, is acted on by a constant horizontal force of 25 N for 1·5 s. Find the kinetic energy acquired by the body. (N).

**27.** A rocket is held fixed so that it ejects material horizontally with a speed of 160 m/s. If it ejects a mass of 1 kg each second, calculate the force needed to hold it fixed.

If the rocket were to be used to propel a sledge on a frozen lake, find, neglecting friction, the acceleration of the sledge at a time when the total mass was 150 kg. (O. and C.)

**28.** Explain the terms *work, kinetic energy, power,* and define a unit for each in the S I or metric system.

A railway engine pulls a train at a steady speed of 48 km/h along a level track. The tension in the coupling between coaches and engine is 50 000 N. What horse-power is being spent in pulling the coaches?

What additional power would be required to pull the coaches up a slope of 1 in 100 at the same speed, if their mass is 500 000 kg? (O. and C.) [1 h.p. = $\frac{3}{4}$ kW.]

**29.** Give the meaning of the terms *momentum, force, energy* and define a unit in which each can be measured on the metric or S I system.

A man can throw a 0·1 kg ball with a speed of 20 m/s. Calculate the kinetic energy and the magnitude of the momentum that he gives to the ball.

If, while throwing, he exerts a constant force on the ball over a distance of 1 m, calculate: (a) the magnitude of the force in newtons; (b) the time for which the force acts. (O. and C.).

**30.** Define *kinetic energy* and *potential energy*, stating how each is measured and a unit in which each is expressed.

A pile-driver is raised to a height of 6 m and allowed to fall on to a pile to be driven into the ground. Give an account of the changes of energy which occur from the beginning, of the motion until the pile-driver is again at rest. If the mass of the driver is 800 kg calculate the kinetic energy just before it strikes the pile. (*L.*)

**31.** State Newton's laws of motion, and use them to explain rocket propulsion. Assuming that the engines of a rocket eject material vertically downwards at a speed of 300 metre per second, what mass of material must be ejected each second to support a weight of 1000 kilograms? (*O. and C.*)

**32.** Explain the terms *mass, momentum, force* and describe an experiment which illustrates the conservation of momentum.

A firework rocket is burning fuel at the rate of 14 g/s. All the products of combustions are ejected in one direction at a speed of 210 m/s. What is the greatest weight the rocket can have if it is going to move vertically upwards? Give your answer in newtons. (*O. and C.*)

**33.** With the aid of labelled diagrams, describe and explain *one* application in each case of (i) a simple lever, (ii) an inclined plane. Suggest suitable dimensions for each machine, and state the velocity ratio these dimensions would produce.

A lift carrying a 120 kg mass of bricks travels to the top of a building 10 m high in 15 seconds. If the useful output of the engine driving the lift mechanism is 1 kilowatt, calculate the efficiency of the operation of the lift mechanism. ($g = 9 \cdot 81$ m/s².) (*N.*) (See also Chap. 6.)

# 5

# MOMENTS · PARALLEL FORCES · CENTRE OF GRAVITY

## MOMENTS

In dynamics we considered forces which keep an object in motion. In statics, which we now discuss, we consider how forces are used in machines such as levers and how they keep objects such as bridges in equilibrium.

### Types of Forces

A *push* and a *pull* are examples of forces. If a girl pushes a pram she feels the force exerted by her muscles, which become taut (Fig. 5.1(i)). Ropes

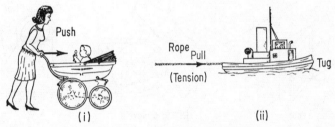

Fig. 5.1 Force may be push or pull

or strings are used to pull or support objects. When a tug uses a strong rope to pull a boat from a sandbank the rope becomes taut, and the force in the rope is called its *tension* (Fig. 5.1(ii)). Fig. 5.2(i) shows the tensions

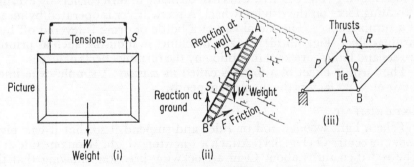

Fig. 5.2 Types of forces

$T$ and $S$ in a string which supports the weight $W$ of a picture. The weight acts vertically downwards; the tensions together act upwards to support the weight.

As we saw on p. 83, forces are vector quantities, that is, they have

magnitude and direction. This is discussed further in a later chapter (p. 164). *In all diagrams in Statics, always draw the directions of the forces acting on the object concerned, showing this with an arrow.* Fig. 5.2(ii), for example, shows the forces on a ladder AB, whose weight $W$ acts at a point G. It rests against a smooth wall; a force of *reaction R* therefore acts at right angles to the wall on the ladder. At B on the rough ground a force of reaction $S$ acts vertically upwards and a force of friction $F$ acts along the ground and prevents the ladder slipping. Fig. 5.2(iii) illustrates the forces on part of a bridge. A joint at A is acted upon by forces $P$, $Q$ and $R$ in the girders meeting there. The forces $P$ and $R$ are called thrusts or struts. The force $Q$ is known as a tension or tie in the girder AB; the same name is used for the force in a rope or a string.

## Moments

Forces can produce turning effects, and this is widely utilized in every-day life and in machines. When a door is opened the force on the handle

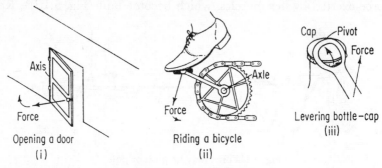

Opening a door
(i)

Riding a bicycle
(ii)

Levering bottle–cap
(iii)

Fig. 5.3 Turning-effects of forces

exerts a turning-effect about the hinges (Fig. 5.3(i)). In the case of a bicycle, the force on the pedals produces a turning-effect which keeps the wheels moving (Fig. 5.3(ii)). Cars can be made to turn corners by exerting a rotating force on the driving wheel. A screwdriver is operated by means of a turning-effect. The metal cap of a bottle of drink is levered off by a turning-effect (Fig. 5.3(iii)). Many machines in industry, such as printing presses and lathes, are kept in motion by the turning effects of forces.

The turning effect of a force is called its *moment*. A simple experiment shows on what factors the moment depends.

### EXPERIMENT

Take a light wooden rod or ruler and suspend it so that it can pivot about its centre O (Fig. 5.4). Attach a 1 newton weight $W$ on one side at B. The rod then turns about O in a clockwise direction, showing that the moment of $W$ about O is clockwise.

When another weight $P$ of 0·5 N is attached to the other side of O, and moved to a position such as A, the rod can be brought back to a horizontal position or 'balanced'. The moment of $P$ about O is anti-clockwise, or opposite to the clockwise moment of $W$ about O. Since the rod is now in equilibrium, the moments must be equal in magnitude.

If $W$ is kept at B the force $P$ can be increased by attaching a 0·1 N weight to it. The rod now tilts in an anti-clockwise direction, showing that the moment of $P$ about O has been increased.

If the 0·1 N weight is removed the road returns to its horizontal position of equilibrium. The distance of $P$ from O can now be increased. When this

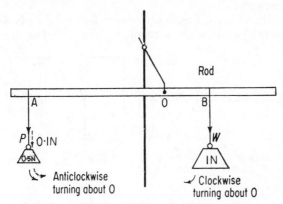

Fig. 5.4 Clockwise and anti-clockwise moments

is done the rod tilts anti-clockwise, showing that the moment of $P$ about O has increased.

*Conclusion.* The moment of a force $P$ about a pivot O depends on the magnitude of $P$ and on its distance in relation to O.

## Moment Definition

This experiment bears out our past experiences on a see-saw. The farther away we were from the turning-point of the see-saw, the greater was the turning-effect about that point; thus the moment increased when

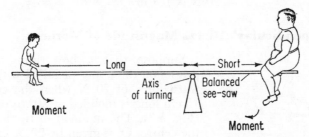

Fig. 5.5 Moment depends on distance and weight

the distance from the point increased. A heavy child placed on the see-saw had a greater moment than a light child at the same place, showing that the moment increased as the magnitude of the weight (or force) increased (Fig. 5.5). The force and the distance are both taken into account in the definition of moment, which is as follows:

*The moment of a force about a point or axis O =*
*Force $\times$ Perpendicular distance from O to line of action of force.*

## Calculation of Moments

In the SI system of units, the force is measured in newtons (N) and the perpendicular distance is measured in metres (m). The moment is therefore expressed in *newton metre* (N m). If the force is due to the weight of a mass, then the moment may also be expressed in N m by expressing the weight in newton.

As a simple illustration, Fig. 5.6(i) shows forces of 40 N and 100 N on a light horizontal beam with a turning-point O. The moment of the

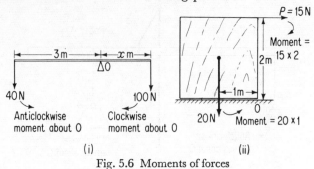

Fig. 5.6 Moments of forces

former force about O is $40 \times 3 = 120$ N m, and the moment of the latter is $100x$ N m. Fig. 5.6(ii) represents a box of 20 N weight pulled by a horizontal force $P$ of 15 N at the top; the moment of the force about O is 30 N m, and the moment of the weight of the box about O is 20 N m. Consequently, the box will tilt about O in the direction of $P$ because it has a greater moment.

Suppose a mass of 2 kg is suspended from a beam which can turn about a point O and the perpendicular distance from O is 40 cm or 0·4 m. Then moment about O, using weight of 1 kg mass = 10 N,

$$= 20 \text{ N} \times 0.4 \text{ m} = 8 \text{ N m}$$

## How 'Perpendicular' Affects Magnitude of Moment

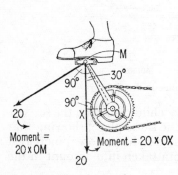

Fig. 5.7 Moment and perpendicular

Suppose that a boy riding a bicycle presses on the upper pedal M with a downward force of 20 N when the crank OM 15 cm long is inclined at 30° to the vertical (Fig. 5.7). The moment of the force about the centre, O, is given by $20 \times OX$, where OX is the *perpendicular from O* to the direction of the force. But $OX = OM \sin 30° = 15 \times 0.5 = 7.5$ cm $= 0.075$ m.

∴ Moment about $O = 20 \times 0.075 = 1.5$ N m.

Suppose, however, that the force of 20 N is directed perpendicular to OM in a downward direction.

The moment of the force about O is now given by $20 \times$ OM, since OM is perpendicular to the direction of the force. As OM $= 15$ cm $= 0.15$ m

$$\text{Moment} = 20 \times 0.15 = 3 \text{ N m}$$

A little thought shows that 3 N m is the greatest possible moment about O of the 20 N force exerted by the rider. It is therefore best to press at right angles to OM in any position of the latter. If the rider presses in the direction MO there is no turning-effect about O; the perpendicular from O to the direction of the force is zero in this case.

It should be noted that it is meaningless to discuss, or calculate, the moment of a force unless the pivot or axis about which the moment is taken is specified or known.

## Sign of Moment

The turning-effect of a force may cause rotation in either a clockwise or an anti-clockwise direction. In Fig. 5.6(i), for example, the moment of the 40 N about O on the beam is anti-clockwise and the moment of the 100 N about O is clockwise. The direction of the moment, as well as its magnitude, must always be taken into account. If the anti-clockwise direction is taken as positive the clockwise direction is negative.

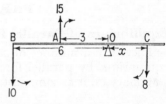

Fig. 5.8 Parallel forces and moments

As an illustration of the sign rule, consider a light beam pivoted at O and acted upon by the forces 15 N, 10 N and 8 N shown (Fig. 5.8). If OB $= 6$ m, OA $= 3$ m, OC $= x$ m, the sum of the moments about O

$$= + 10 \times 6 - 15 \times 3 - 8x = (15 - 8x) \text{ N m}.$$

## Equilibrium and Moments

When a pivoted beam is in *equilibrium* under the action of several forces the relationship between the moments can be investigated with the apparatus shown in Fig. 5.9.

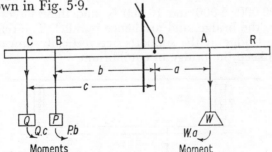

Fig. 5.9 Experiment on principle of moments

### EXPERIMENT

Suspend a light rod or ruler R from its midpoint O. Attach a weight $W$ such as 0.5 N at A and two smaller weights $P$, $Q$ at B and C (Fig. 5.9). Move $W$ along the rod R until it is horizontal or balanced.

*Measurements.* Measure the perpendicular distances, $a$, $b$, $c$ from O to the respective lines of action of the forces $W$, $P$ and $Q$. Record the results:

| W | P | Q | a | b | c |
|---|---|---|---|---|---|
|   |   |   |   |   |   |

*Calculation.* Clockwise moment of $W$ about $O = W \cdot a$.

Total anti-clockwise moments of $P$ and $Q$ about $O = P \cdot b + Q \cdot c$.

Measurements show that $W \cdot a = P \cdot b + Q \cdot c$, to a good approximation.

*Conclusion.* When a pivoted rod is in equilibrium, the total anti-clockwise moments of all the forces about the pivot = the total clockwise moments of all the forces about the pivot.

## PARALLEL FORCES

### Equilibrium of Parallel Forces

There are a number of cases in mechanics where an object is kept in equilibrium by parallel forces. One example is a bridge when traffic is stationary on it at some instant (Fig. 5.10). The weights of the individual

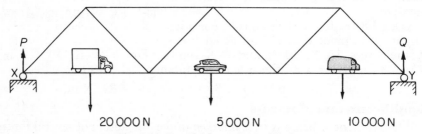

20 000 N      5 000 N      10 000 N

Fig. 5.10 Parallel forces in equilibrium

vehicles are 20 000, 5 000 and 10 000 N, and these forces act vertically downwards on the bridge, and are hence parallel. The reaction forces $P$ and $Q$ on the bridge at the supports act vertically upwards. Here five parallel forces keep the bridge in equilibrium.

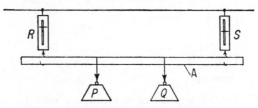

Fig. 5.11 Parallel forces in equilibrium

EXPERIMENT

The equilibrium of parallel forces can be investigated by suspending a light rod A by spring-balances at either end and attaching weights $P$ and $Q$

from any two points on the rod (Fig. 5.11). The readings $R$, $S$ on the balances are observed.

*Measurements.* Typical readings are as follows:

| P | Q | R | S |
|---|---|---|---|
| 0·5 N | 0·5 N | 0·65 N | 0·40 N |

Thus total downward force $= 0{\cdot}5 + 0{\cdot}5 = 1{\cdot}0$ N
Total upward force $\quad= 0{\cdot}65 + 0{\cdot}40 = 1{\cdot}05$ N

*Conclusion.* The total forces in one direction = the total forces in the opposite direction, allowing for experimental error.

## Moment Condition of Equilibrium

Although the total forces acting on an object in one direction may equal the total forces in the opposite direction, it does not necessarily follow that

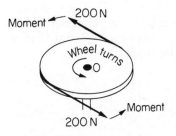

Fig. 5.12 Parallel forces not in equilibrium

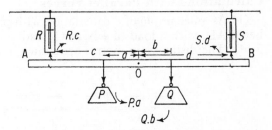

Fig. 5.13 Experiment on equilibrium of parallel forces

the object is in equilibrium. This can be seen from Fig. 5.12, which shows a wheel W, capable of turning about an axle at its centre O, under the action of two forces of 200 N acting tangentially to W. The force in one direction is equal to the force in the opposite direction. But the moment of *both* forces about O act anti-clockwise. Hence the wheel turns; it is not in equilibrium.

EXPERIMENT

In Fig. 5.13, however, the light rod AB is in equilibrium. To investigate the magnitudes of the moments of the forces, we take any point O on the rod, and measure the perpendicular distances of the forces from O. These are shown by $a$, $b$, $c$, $d$ respectively, and results are as follows:

| a (m) | b (m) | c (m) | d (m) |
|---|---|---|---|
| 0·10 | 0·15 | 0·34 | 0·62 |

*Calculation.* With $P = 5 \cdot 0$ N $= Q, R = 6 \cdot 5$ N, $S = 4 \cdot 0$ N,

Total clockwise moment about O $= Q \cdot b + R \cdot c$
$$= 5 \times 0 \cdot 15 + 6 \cdot 5 \times 0 \cdot 34$$
$$= 3 \cdot 05 \text{ N m}$$

Total anti-clockwise moment about O $= P \cdot a + S \cdot d$
$$= 5 \times 0 \cdot 10 + 4 \cdot 0 \times 0 \cdot 62$$
$$= 2 \cdot 98 \text{ N m}$$

*Conclusion.* Allowing for experimental error,
The total clockwise moment of the forces about any point =
the total anti-clockwise moment of the forces about the same point.

*Summarizing*

When an object is in equilibrium under parallel forces two conditions are true:

1. Total forces in one direction = total forces in opposite direction.
2. Total clockwise moments of the forces about any point is equal to the total anti-clockwise moments of the forces about the same point.

## Calculations with Parallel Forces

(1) As an example of a calculation with parallel forces, consider a light beam AD, with a load of 4 N at O, resting on supports at B and C (Fig. 5.14). The distances in metres are shown in the diagram.

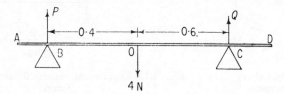

Fig. 5.14 Calculation on parallel forces

To find the forces of reaction $P$ and $Q$ at B and C respectively, we first use the principle that the total upward force = the total downward force.

$$\therefore P + Q = 4 \quad . \quad . \quad . \quad . \quad . \quad (1)$$

Next, use the principle that the total clockwise moment about any point equals the total anti-clockwise moment about the same point. We can take moments about any point on the beam. But to eliminate the force $Q$, for example, take moments about C, where $Q$ acts, *because the moment of Q about a point on its line of action is zero.*

Then    Moment of $P$ about C $= P \times$ BC $= P \times 1 \cdot 0$, clockwise
Moment of 4 N about C $= 4 \times$ OC $= 4 \times 0 \cdot 6$, anti-clockwise
Moment of $Q$ about C $= 0$

$$\therefore P \times 1 \cdot 0 = 4 \times 0 \cdot 6, \text{ or } P = 2 \cdot 4 \text{ N} . \quad . \quad . \quad (2)$$

Hence, from (1),        $Q = 4 - P = 4 - 2 \cdot 4 = 1 \cdot 6$ N

(2) Another example is shown in Fig. 5.15. To find $X$ and $Y$, the re-actions at the respective supports, if the distances shown are in metres,

(i) Total upward force $= X + Y =$ Total downward force $=$
$$40 + 20 + 30 = 90 \text{ N}$$

(ii) Moments about left-hand support,

Total anti-clockwise moment $= 40 \times 4 + Y \times (2 + 4) =$
$$20 \times 2 + 30 \times 9$$
$$= \text{total clockwise moment}$$
$$\therefore 160 + 6Y = 40 + 270 = 310$$
$$\therefore Y = 25 \text{ N},$$

From (i)                           $\therefore X = 90 - Y \; 65 \text{ N}$

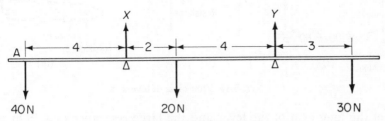

Fig. 5.15  Parallel forces

If increasing weights were placed at A the beam would tilt about the left support at some stage and the reaction $Y$ would then vanish. The magnitude $W$ of the weight then at A can be found by taking moments about the support.

Thus                $W \times 4 = 20 \times 2 + 30 \times 9 = 310$
$$\therefore W = 77\tfrac{1}{2} \text{ N}$$

$$\therefore \text{ Additional weight at A} = 77\tfrac{1}{2} - 40 = 37\tfrac{1}{2} \text{ N}$$

## The Lever . First Class

The turning-effect or moment of a force is used to advantage in the case of the *lever*, a machine which enables a small force or effort to overcome a large weight or resistance. The crowbar, pliers, scissors and nut-crackers are examples of levers.

All levers have a point or axis called the *fulcrum*, about which the lever-age takes place. Levers are divided into three types or classes, depending on the positions of the effort and weight in relation to that of the fulcrum. In the *first class* of levers the effort $P$ and weight $W$ are on opposite sides of the fulcrum O (Fig. 5.16(i)). A crowbar is an example of the first class of levers. The bend O of the crowbar is the point about which leverage takes place, or the fulcrum (Fig. 5.16(ii)). The effort $P$ is applied at the end of the long 'arm' NO of the lever to overcome the weight or resistance at the end of the short 'arm' DO.

Suppose the resistance at D is 200 N and that its perpendicular distance from O is 4 cm. The force $P$ to just overcome 200 N should produce a moment about O slightly greater than the opposing moment of the 200 N. Taking moments, then, if $P$ acts perpendicular to ON and ON is 50 cm, $P$ is practically given by

$$P \times 50 = 200 \times 4, \text{ or } P = 16 \text{ N}$$

Thus a small force, 16 N, can overcome a much larger one, 200 N, by using a lever. Note carefully that the small effort or force is applied at the

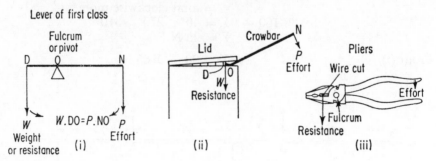

Fig. 5.16  First class of levers

end of the long arm of the lever and the large resistance or weight to be overcome is at the end of the short arm.

Fig. 5.16(iii) shows that *pliers*, used for cutting wire, are an example of the first class of levers.

## Second Class of Levers

A *bottle-top opener* and a *nut-cracker* are examples of levers of the second class (Fig. 5.17(i), (ii), (iii)). Here the resistance $W$ is on the same side of

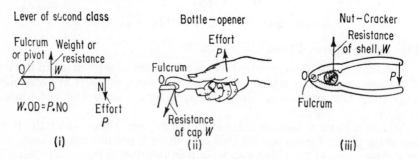

Fig. 5.17  Second class of levers

the fulcrum O as the effort $P$, whereas in the first class of levers they are on opposite sides.

The principle is exactly the same as in the first class of lever. The small effort $P$ is applied at the end of the long arm and overcomes a much larger resistance or weight $W$ at the end of the short arm. Suppose the resistance of the bottle cap in Fig. 5.17 (ii), 30 N, is 1 cm from the fulcrum

O, and the effort $P$ is 15 cm from O. Then if the distances are perpendicular to the forces we have, taking moments about O,

$$15 \times P = 1 \times 30$$
$$\therefore P = 2 \text{ N}$$

## Third Class of Levers

If sugar tongs or coal tongs are used the effort $P$ is *nearer* the fulcrum O than the resistance or weight $W$ (Fig. 5.18(i), (ii). This is the characteristic of the third class of levers.

To find $P$ when a lump of coal has a resistance $W$ of 1 N and the

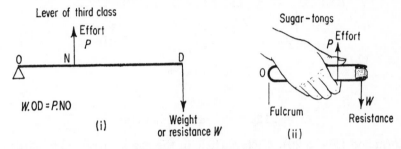

Fig. 5.18 Third class of levers

perpendicular distances to the hinge or fulcrum O are 20 cm and 30 cm respectively, take moments about O. Then

$$P \times 20 = 1 \times 30, \text{ or } P = 1 \cdot 5 \text{ N}$$

An effort larger than the resistance is now required.

The *forearm*, used for picking up objects using the elbow O as a fulcrum, is an example of the third class of levers. The muscular force $P$ of the biceps is nearer O than the weight $W$ (Fig. 5.19(i)). The *lever safety valve*,

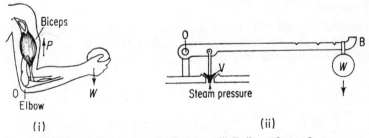

Fig. 5.19 Levers in use: (i) Arm; (ii) Boiler safety-valve

used on boilers, has a valve V fitting into the boiler, and a lever OB carrying a weight $W$ (Fig. 5.19(ii)). When the steam pressure exceeds a dangerous value, fixed by the position of $W$, the valve V rises and allows the steam to escape until the pressure falls. This is an example of a third class of levers.

## CENTRE OF GRAVITY

### Meaning of Centre of Gravity · Centre of Mass

A cricket ball, hit into the air, begins to descend to earth after a short time; a footballer, knocked off his balance, falls downwards. These are effects of gravity, and all objects, no matter how small, or whether gaseous, liquid or solid, are attracted towards the centre of the earth by a force which is proportional to their mass (p. 99).

Consider a rod of wood AB of a constant cross-sectional area, that is, a

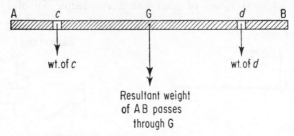

Fig. 5.20 C.G. and resultant weight

uniform rod (Fig. 5.20). Each particle in the rod is attracted towards the centre of the earth by a force which is the particle's weight, and since the centre of the earth is a very long way off, the weights of the various sections of the rod, such as $c$, $d$, act downwards in parallel directions. We thus have a large number of parallel forces. Their resultant, the total weight of AB, acts at some point G. The centre of gravity of a body is the name given to the point through which its total weight appears to act.

The *centre of mass* of a body is the point where its total mass appears to act. For small objects, such as we are concerned with, the centre of mass coincides with the centre of gravity. Thus the centre of mass of the rod in Fig. 5.20 is also at G.

### C.G. Positions

The centre of gravity (C.G.) of a *uniform rod* is at its mid-point. The C.G. of a uniform *circular disc* is at the centre of the circle (Fig. 5.21(i)). The C.G. of a uniform *ring* is at the centre (Fig. 5.21(ii)).

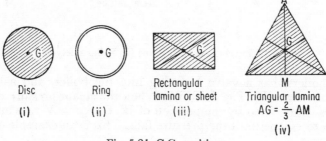

Fig. 5.21 C.G. positions

The C.G. of a uniform *square*, *rectangle* or *parallelogram* thin sheet or lamina has its C.G. at the point of intersection of the diagonals (Fig 5.21(iii)).

The C.G. of a uniform *triangular lamina* has its C.G. at the point of intersection of the medians, which are the lines joining the vertices to the respective mid-points of the opposite sides (Fig. 5.21(iv). By geometry, the C.G. is two-thirds of the distance from the vertex along a median. Thus $AG = \frac{2}{3}AM$ in Fig. 5.21(iv).

## Experiments to Locate C.G. of Lamina

### (1) *Approximate Method*

First, try to balance a lamina such as a thin wooden board on a point. Since the whole weight acts at the centre of gravity G, the position is roughly determined in this way. As this is not precise, a more practical way

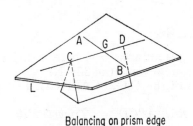

Balancing on prism edge

Fig. 5.22 C.G. by balancing

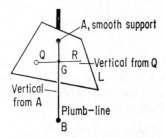

Fig. 5.23 C.G. by plumbline

is to balance the lamina L on a sharp edge, such as the edge of a triangular glass prism, and to draw the line AB of the edge (Fig. 5.22). The C.G. lies somewhere on AB. The lamina is then removed, turned round and balanced again. The new line CD of the edge is drawn on it. Then the centre of gravity G is the point of intersection of AB and CD.

### (2) *Accurate Method*

A better method is to suspend the lamina L from a horizontal axis A such as a pin or nail, making sure that L can swing freely about A (Fig. 5.23). When the lamina is at rest its C.G. will be somewhere vertically below A. A small weight B attached to a long thread, often called a plumb-line, is then hung from A. The line of the thread is vertical, and its direction is drawn on L.

The lamina is then supported from an axis such as Q at a different place, and the procedure is repeated. The new vertical line QR intersects the first one at the centre of gravity G.

The C.G. of an object such as a stool can be found in a similar way by suspending it in turn by its legs, and locating the vertical lines by using thread if necessary tied to the legs.

## C.G. Calculation

If a lamina or sheet of metal is cut into a shape equivalent to a combined rectangle and triangle, for example, the position of its centre of gravity can be calculated by using moments.

As an illustration, consider a lamina cut in the form of a shape ABHCD which can be considered as a rectangle ABCD joined to an isosceles triangle BCH of the same base. Suppose AB = 10 cm, AD = 6 cm and the height HN of the triangle is 12 cm (Fig. 5.24). The C.G. of the rectangle is at $G_1$, which is at a distance $G_1H$ from H. But $G_1N$ = 5 cm, NH = 12 cm, and hence $G_1H$ = 17 cm. The C.G. of the triangle BCH is two-thirds from H along the median, and hence $G_2H$ = 8 cm. Further the area of ABCD is 60 cm², and if 1 cm² of the material of the lamina weighs $w$ N the weight

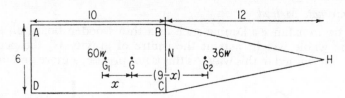

Fig. 5.24 C.G. calculation

of ABCD is 60$w$ N. The triangle BHC has an area = $\frac{1}{2}$ base × height = $\frac{1}{2}$ × 6 × 12 = 36 cm², and has therefore a weight of 36$w$ N.

We are now able to find the position of the C.G. of ABHCD. Let us tabulate our information:

|  | Rect. ABCD | Triangle HBC |
|---|---|---|
| Weight in N | 60$w$ | 36$w$ |
| Distance of C.G. from H | 17 cm | 8 cm |

Now the weight of 60$w$ at $G_1$ and the weight of 36$w$ at $G_2$ are parallel forces. If the whole weight of the lamina can be balanced at G the moment about G of 60$w$ at $G_1$ = the moment about G of 36$w$ at $G_2$. Suppose $G_1G$ = $x$ cm. Then, since $G_1G_2$ = 17 − 8 = 9 cm, $G_2G$ = (9 − $x$) cm.

$$\therefore 60w \times x = 36w \times (9 - x).$$
Solving $\qquad\qquad \therefore x = 3\frac{3}{8}$ cm.

Thus the C.G. is $3\frac{3}{8}$ cm from $G_1$ along the direction $G_1H$.

### Stability and Centre of Gravity

In designing a car or a ship, the engineer must take into account its *stability*, especially at high speeds. A car turning a corner fast or going round a sharp bend in the road at high speed has a tendency to overturn. Ships, which are liable to considerable rolling and lurching in heavy seas, must be designed to be in stable equilibrium even in rough weather. Tests for stability of buses on inclines are carried out by loading the top deck with sandbags representing an equivalent weight of passengers on this part of the bus.

The stability of an object is connected with the position of its centre of gravity and the moment its weight exerts about an axis. As a simple example, consider an oil-drum held on an inclined plane as shown in Fig. 5.25(i). The vertical through its centre of gravity, G, just passes through the edge O of the base OA. When the drum is slightly displaced about O as shown, the weight $W$ now has an anti-clockwise moment about O. This moves the drum farther away from its initial position, and hence it topples over. The drum in Fig. 5.25(i) is hence said to be in *unstable equilibrium*.

Suppose that a cylinder of the same diameter, but a smaller height, is placed on the inclined plane (Fig. 5.25(ii)). The centre of gravity G is now lower than that of the previous cylinder, and the vertical through G may then pass through the base OA, as shown. If the cylinder is slightly tilted about O and then released the weight $W$ now exerts a clockwise moment about O. This restores the cylinder to its original position. Consequently, the cylinder is said to be in *stable equilibrium* in Fig. 5.25(ii).

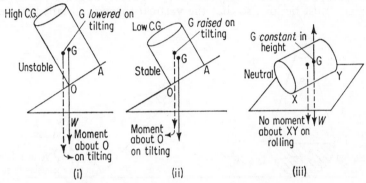

Fig. 5.25 Stable, unstable and neutral equilibrium

If the cylinder is taken and placed with its curved surface XY on a horizontal plane it will be in equilibrium no matter what position it assumes on rolling (Fig. 5.25(iii)). This is due to the fact that in a displaced position, unlike the previous cases the weight $W$ has no moment about the axis XY. The cylinder is therefore said to be in *neutral equilibrium*. Here the reaction of the plane through G is always equal and opposite to $W$.

A criterion can be given for the three types of equilibrium just discussed. If the vertical line through the centre of gravity of an object passes through its base when the object is slightly displaced the object is in stable equilibrium; if it passes outside the base when the object is slightly displaced the object is in unstable equilibrium; and if the vertical always passes through the base, no matter to what position the object is displaced, the equilibrium is neutral.

## Height of C.G.

Another condition for stable, unstable and neutral equilibrium can be obtained from consideration of the potential energy of the object concerned. If a slight displacement produces an increase in potential energy, that is, the height of the centre of gravity rises, then the object will fall back to its original position if released, so that the equilibrium was stable (Fig. 5.25(i)). Similarly, if a slight displacement produces a decrease in potential energy, that is, the height of the C.G. becomes lowered, the object will fall farther away when released, so that the equilibrium was unstable (Fig. 5.25(ii)). If it is in neutral equilibrium there is no change in the height of the C.G. when the object is displaced slightly (Fig. 5.25(iii)).

## Applications

It can be seen from the above that the risk of unstable equilibrium is increased as the height of the centre of gravity of an object is increased; the vertical through the C.G. is then more liable to fall outside the base when the object is slightly tilted. The nearer the centre of gravity is to the ground, the more stable is the equilibrium likely to be.

For this reason, extra passengers are allowed on the lower deck of a crowded bus, but not on the upper deck; the centre of gravity G of bus and passengers must be low, so that the vertical through G falls between the

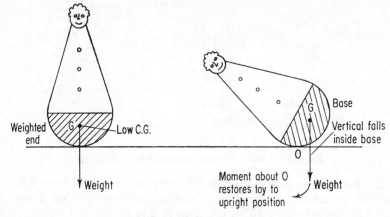

Fig. 5.26 Stability due to low C.G.

wheels even when the bus is rounding a corner. Racing cars are built low for the same reason. The base of the stand of a punchball is a very heavy piece of metal; the C.G. of the whole arrangement is then so low that it is in stable equilibrium, and the ball returns to the boxer however powerfully it is struck. Toys which spring up to the vertical, no matter how they are laid on the table, have a rounded heavy lead base, so that the vertical through the C.G. of the toy always passes through the base (Fig. 5.26). Articles such as bunsen burners and electric lamp-stands are designed with a large and heavy base to make them stable, and oil-lamps have sometimes a lead base for the same reason. Divers wear heavy boots which have lead soles not only to weigh them down but also to help them to maintain stability while moving about the ocean-bed.

## SUMMARY

1. The moment of a force about a point or axis is the product of the force and the perpendicular distance from the point to the line of action of the force. Moments are measured in newton metre or kgf cm or gf cm or dyne cm.

2. The conditions of equilibrium for parallel forces are: (i) the sum of the forces in one direction equals the sum of the forces in the opposite direction; (ii) the sum of the clockwise moments about any point equals the sum of the anti-clockwise moments about that point.

3. In the first class of levers the resistance or weight, and the effort or force, are on opposite sides of the fulcrum. In the second and third classes the resistance and effort are on the same side of the fulcrum. In the third class the effort is nearer the fulcrum than the resistance. The effort has a moment about the fulcrum which just overcomes the moment of the resistance about the fulcrum.

4. The centre of gravity of a rectangle or square is the point of intersection of the diagonals. The centre of gravity of a triangle is two-thirds along the median from any apex.

5. The equilibrium of an object is stable or unstable according as the vertical through the centre of gravity falls inside or outside the base when the object is slightly displaced. The equilibrium is neutral if the vertical always passes through the base, however the object is displaced. A low centre of gravity assists stability.

## EXERCISE 5 · ANSWERS, p. 223

(Assume weight of 1 kg mass = 10 N.)

**1.** A uniform rod 1 m long weighing 0·5 N is supported horizontally on two knife edges placed 10 cm from its ends. What will be the reactions at these supports when a 1·0 N weight is suspended 10 cm from the mid-point of the rod? (*N.*)

**2.** A uniform half-metre scale *AB* is balanced horizontally across a knife edge placed 15 cm from *A*. A mass of 30 g is hung from the end *A*.

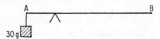

What is: (*a*) the mass of the scale; (*b*) the force exerted on the scale by the knife edge? (*N.*)

**3.** Explain *stable, neutral* and *unstable equilibrium* and give *one* example of each. (*L.*)

**4.** Define the centre of gravity of a body, and explain how you would determine the weight of a closed umbrella if you were given a metre rule, a straight edge and a 0·5 N weight. (*L.*)

**5.** A simple weighing machine is made of a uniform bar 125 cm long, weighing 50 N and pivoted 2·5 cm from one end. Find the weight that must be suspended at the end of the long arm so as to balance a load of 3200 N suspended at the end of the short arm. (*N.*)

**6.** A uniform metre scale is balanced horizontally across a knife edge at the 20-cm mark with a 3·0 N weight hung by cotton from the 11-cm mark. Show in a diagram the three forces acting on the scale and calculate its weight. (*N.*)

**7.** Explain how (i) the moment of a force, (ii) the work done by a force, are each measured.

A trap door of mass 10 kg in a horizontal floor is made of a uniform piece of wood 1 m square, smoothly hinged along one side. What is the least vertical force required to raise the door?

What is the work done when the door is raised through an angle of 45°? (*L.*)

**8.** Define *moment of a force about an axis, centre of gravity.* State the *principle of moments* and describe an experiment to test its validity.

A steel rod 7 cm long consists of two cylindrical parts, 3 cm and 4 cm long respectively, having a common axis. If the area of cross-section of the shorter part is twice that of the longer, find the position of the centre of gravity of the rod. (*L.*)

**9.** Describe an experiment to find the position of the centre of gravity of a piece of sheet metal of irregular shape. Explain why your method gives the required result. (*N.*)

**10.** Explain what is meant by *the moment of a force about an axis*.

State the conditions of equilibrium for a number of coplanar parallel forces. Describe an experiment to verify them.

A uniform rod AB is 180 cm long and weighs 20 N. It balances horizontally on a single support when carrying loads of 10 N and 30 N at distances of 60 cm and 120 cm respectively from A. Find the position of the support and the magnitude of the downward thrust on it. (*L.*)

**11.** Explain what is meant by *centre of gravity, stable equilibrium*.

State and explain, with the aid of diagrams, why the base of a bunsen burner is *heavily* weighted and of *large* area. (*L.*)

**12.** You are supplied with a metre rule, a knife edge, a block of wood, 20-, 50- and 100-g masses, some thread and no other apparatus. How would you use this apparatus to verify the principle of moments?

How does the usual way in which a door is pushed open agree with your knowledge of moments?

A uniform rod *PQ*, 180 cm long and weighing 140 N, is freely pivoted at *P* and has a mass of 7 kg attached to a point 60 cm from *Q*. What vertical force will have to be applied at *Q* to hold the rod horizontally and what is then the reaction at *P*? (*N.*)

**13.** You are supplied with a metre rule, a 100-g mass, a knife edge, thread and an object *M* of mass between 50 and 200 g. Describe the experiment you would carry out, using this apparatus, to find the mass of *M*. State the principle used in the experiment described.

A circular disc of uniform wood has a radius of 5 cm. A circle of radius 2 cm is cut from it with its centre 2 cm from the edge of the disc. Find the centre of gravity of the remaining wood. (*N.*)

**14.** Describe the common beam balance and explain its action. A uniform metal rod *AB*, 60 cm long and of mass 2 kg, is fastened at *A* to the surface of a metal ball 10 cm in diameter and of mass 10·5 kg, the centre of the ball being on the line of the axis of *AB*. The arrangement is suspended by a string attached to the rod at a point *C*, 5 cm from *A*. What mass must be suspended from *B* for the rod to be horizontal? (*L.*)

**15.** What do you understand by the *principle of moments*? Describe how you would verify the principle experimentally for parallel forces. A uniform rod of length 50 cm and mass 100 g is pivoted 8 cm from one end *A*. Loads are weighed by attaching them to *A*, and moving a weight of 5 N along the rod, on the opposite side to the pivot, till the rod balances horizontally. Find the position of the weight when a load of 24 N is attached to *A* and calculate the total upthrust at the pivot. (*N.*)

**16.** What do you understand by *stable, unstable* and *neutral equilibrium*?

Give ONE example of each.

Describe how you would find the position of the centre of mass of a thin plane sheet of material of irregular shape.

Taking the mass of the earth to be 81 times that of the moon, the radius of the earth to be 6560 km and the distance between the centres of the moon and earth to be 60 times the earth's radius, find the distance of the centre of mass of the moon–earth combination from the centre of the earth. (*O.*)

**17.** Explain the term *moment of a force about an axis*.

What is meant by the *principle of moments*?

Show by considering a frictionless wheel and axle how the principle of moments follows from the principle of conservation of energy.

A faulty balance has arms which differ slightly in length. A mass *M* is balanced by 25·2 g when placed in the left-hand pan. When placed in the other pan 23·8 g is required for balance. Calculate the true mass of *M*. (*O. and C.*)

**18.** Define the *moment* of a force about an axis. State the principle of moments and describe how you would test it experimentally.

In the sketch *AB* represents the lowered mast of a sailing boat, which is to be erected by two men. The mast, 10 m long, weighs 600 N, and is freely pivoted at *C*, 1 m from *A*.

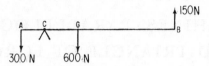

In order just to raise it from the horizontal position, one man exerts a downward force of 300 N at *A*, and the other an upward force of 150 N at *B*. Find the distance from *C* of the centre of gravity, *G*, of the mast. (*O.*)

**19.** Define *centre of gravity* and describe how you would find it experimentally for an irregularly shaped sheet of cardboard. It is proposed to order from a glazier a glass disc 18 cm in diameter with a 2 cm diameter circular hole with its centre 4 cm from the edge of the disc. This disc is to have a small hole bored by the glazier accurately through its centre of gravity. Where should the hole be bored? (*O.* and *C.*)

# 6

# MACHINES · PARALLELOGRAM
# AND TRIANGLE OF FORCES

## MACHINES

A heavy roller or barrel can best be raised up a step by rolling it up an inclined plank of wood with one end resting on the step. The inclined plank, or inclined plane, is an example of a *machine*; it enables a large weight or resistance to be overcome by a small effort. A lever, such as a bottle-top or can opener, is also an example of a machine. The screw is a machine; a small force applied at the end of the screw head is able to move the screw forward against the relatively large resistance of the wood. Pulleys are machines which enable heavy loads to be raised by much smaller forces. We deal shortly with the principles of all these machines.

### Mechanical Advantage · Efficiency

Consider a lever AOB of the first class with a pivot at O, and suppose a small force or effort $P$ is applied at the end B of the long arm (Fig. 6.1).

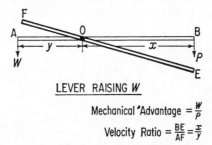

LEVER RAISING $W$

Mechanical Advantage $= \dfrac{W}{P}$

Velocity Ratio $= \dfrac{BE}{AF} = \dfrac{x}{y}$

Fig. 6.1 Lever as machine

If a load or resistance $W$ at the end of A is just raised or overcome, the *mechanical advantage* of the machine is defined as the ratio $W/P$, or

$$\text{Mechanical advantage} = \frac{\text{Load } (W)}{\text{Effort } (P)} \quad \cdot \quad \cdot \quad (1)$$

Suppose the load $W$ is raised steadily through a *small* distance AF when the force $P$ is moved through the large distance BE. Assuming no friction at the pivot, then, from the law of conservation of energy,

$$\text{Work done by } P = \text{Work done in raising } W$$
$$\therefore P \times BE = W \times AF$$

$$\therefore \frac{W}{P} = \frac{BE}{AF} = \frac{x}{y},$$

152

where $x$ and $y$ are the respective lengths of the long and short arms of the lever. A very long arm to apply the effort and a very short arm for the load will hence provide a large mechanical advantage.

In practice, however, some of the work is spent in moving against frictional forces at the hinge or pivot of the lever (Fig. 6.2). Thus:

Work done by $P$ = Work done against $W$ + work done against friction.

If 1 joule of work is done by $P$, and 0·2 joule are used in overcoming friction, then 0·8 joule of work are available for $W$. The *efficiency* of any machine is defined by:

$$\text{Efficiency} = \frac{\text{Work obtained}}{\text{Work supplied}} \times 100\% \qquad \ldots \quad (2)$$

or $\qquad \text{Efficiency} = \dfrac{\text{Work done against } W}{\text{Work done by } P} \times 100\%.$

In the numerical case quoted, the efficiency is hence 0·8 : 1, or 80%.

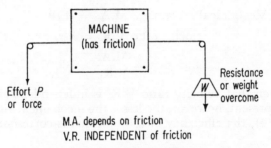

M.A. depends on friction
V.R. INDEPENDENT of friction

Fig. 6.2 Mechanical advantage and velocity ratio

## Velocity Ratio

In the case of the lever, the small effort $P$ moves through a large distance BE, and the large resistance or load $W$ moves through a small distance AF in the same time (Fig. 6.1). The relative distances per second moved steadily by an applied effort and the load overcome is called the *velocity ratio*. Thus:

$$\text{Velocity ratio (V.R.)} = \frac{\text{distance moved by effort}}{\text{distance moved by load in same time}} \qquad (3)$$

Hence in the lever shown in Fig. 6.1,

$$\text{Velocity ratio} = \frac{\text{BE}}{\text{AF}} = \frac{x}{y}.$$

The effort in a machine usually moves through a much greater distance than the heavy load overcome, from the law of conservation of energy. It should therefore be remembered that the velocity ratio is usually much greater than one.

The mechanical advantage $W/P$ of a machine depends on the frictional forces which may be present, since part of the effort has to overcome them.

On the other hand, the relative displacements of the load and effort depend only on the geometry of the moving parts of the machine and do not depend on friction, so that the velocity ratio is independent of friction.

### Relation between Efficiency, M.A. and V.R.

We can now derive a relation between mechanical advantage, velocity ratio and efficiency of a machine.

For any machine, with the usual notation,

$$\text{Efficiency} = \frac{\text{Work obtained}}{\text{Work supplied}} \times 100\%$$

$$= \frac{W \cdot y}{P \cdot x} \times 100\%,$$

where $y$ and $x$ are the respective distances moved by $W$ and $P$ in the same time.

$$\therefore \text{Efficiency} = \left(\frac{W}{P} \div \frac{x}{y}\right) \times 100\%.$$

But          Mechanical advantage M.A. $= W/P$

and          Velocity ratio V.R. $= x/y$.

$$\therefore \text{Efficiency} = \frac{\text{M.A.}}{\text{V.R.}} \times 100\% \quad . \quad . \quad . \quad . \quad (4)$$

We have seen that velocity ratio V.R. is independent of friction. The more friction present, however, the less is the mechanical advantage M.A. Hence, from (4), the efficiency of the machine is correspondingly less.

### Example

A machine with a velocity ratio of 6 requires 800 joules of work to raise a load of 600 N through a vertical distance of 1 m. Find the efficiency and mechanical advantage of the machine.

$$\text{Efficiency} = \frac{\text{Work obtained}}{\text{Work supplied}} \times 100\%$$

$$= \frac{600 \times 1 \text{ joule}}{800 \text{ joule}} \times 100\%$$

$$= 75\% \quad . \quad . \quad . \quad . \quad . \quad . \quad (1)$$

Since          $\text{Efficiency} = \dfrac{\text{M.A.}}{\text{V.R.}} \times 100\%$

$$\therefore 75 = \frac{\text{M.A.}}{6} \times 100,$$

$$\therefore \text{M.A.} = 4\cdot5 \quad . \quad . \quad . \quad . \quad . \quad . \quad . \quad (2)$$

### Wheel and Axle

The *wheel and axle* is a machine which can be used to raise heavy loads such as the anchor of a ship, or loads such as buckets of water from the bottom of a well.

The effort $P$ is applied at one end of a rope tied round a wheel of radius $R$ (Fig. 6.3(i)). The load $W$ is attached to the end of another rope, wound round an axle of radius $r$ in the opposite direction to the rope round the wheel. Suppose the rope round the wheel is pulled down by a constant force $P$ so that the wheel turns exactly through one revolution. The axle also turns through one revolution. Since the effort then moves a distance $2\pi R$ and the load a distance $2\pi r$,

$$\text{Velocity ratio V.R.} = \frac{2\pi R}{2\pi r} = \frac{R}{r} \quad \cdot \quad \cdot \quad \cdot \quad (1)$$

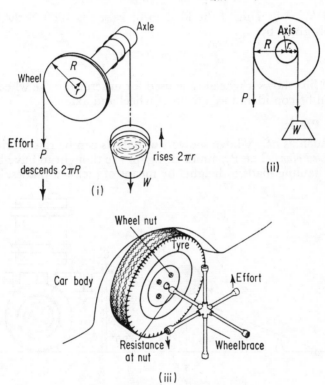

Fig. 6.3 Wheel and axle machine

If the machine was perfect, that is, no friction was present, then work done by effort = work done in raising load.

$$\therefore P \times 2\pi R = W \times 2\pi r$$

$$\therefore \frac{W}{P} = \frac{R}{r}.$$

In practice, the work done by $P$ is greater than $W \times 2\pi r$ owing to friction, and hence the mechanical advantage is less than $R/r$.

From Fig. 6.3(ii), it can be seen that the wheel and axle can be regarded as an example of a lever, with the axis of the wheel and axle as the fulcrum.

Taking moments about the axis, then, in the absence of friction, $P \times R = W \times r$, or $W/P = R/r$.

From (1), a wheel with a radius of 50 cm and an axle of 5 cm has a velocity ratio of 50/5 or 10. If the efficiency is 80% the mechanical advantage M.A. is given by

$$\text{Efficiency} = \frac{\text{M.A.}}{\text{V.R.}} \times 100\%$$

or

$$80 = \frac{\text{M.A.}}{10} \times 100$$

Hence M.A. = 8. Thus if the load to be raised is 1000 N the effort required $P$ is given by

$$\frac{1000}{P} = 8, \text{ or } P = 125 \text{ N}$$

Fig. 6·3(iii) shows a wheelbrace used for unscrewing the wheelnuts of a car. It can be considered as a form of wheel and axle.

## Inclined Plane

Loads such as heavy lawn mowers or barrels can be raised with the aid of an *inclined plane*. The Pyramids in Egypt are thought to have been built by slaves hauling loads to heights by means of inclined planes. The stone

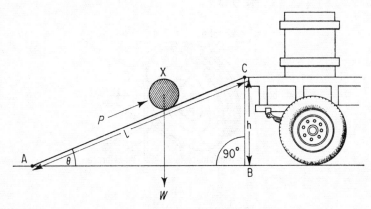

Fig. 6.4 Inclined plane

pillars at Stonehenge, England, are considered to have been raised by the same method, after which the earth was scooped away.

Suppose that a weight $W$ at the bottom A of an incline is raised steadily to the top C by means of an effort $P$ applied at the end of a rope (Fig. 6.4). The distance $l$ moved by the effort is then AC. The load is raised through a *vertical* distance $h$ or BC. Hence

$$\text{Velocity ratio} = \frac{\text{AC}}{\text{BC}} = \frac{1}{\sin \theta},$$

where $\theta$ is the angle of inclination of the plane to the horizontal. Thus if $\theta = 30°$, the velocity ratio = $1/\sin 30° = 2$; if $\theta = 20°$, the velocity ratio

= 1/sin 20° = 3 (approx.). The smaller the inclination of the plane, the greater is the velocity ratio.

If the friction is absent, an ideal case, then the work done by the effort equals the work done on the load.

$$\therefore P \times AC = W \times BC.$$

$$\therefore \frac{W}{P} = \frac{AC}{BC} = \frac{1}{\sin \theta}.$$

In practice, owing to friction, the mechanical advantage is less than the velocity ratio.

### The Screw and Screwdriver · Screw jack

The thread of a *screw* can be regarded as a continuous inclined plane wrapped round a cylinder (Fig. 6.5(i)). When the screw is turned through one revolution by a force applied at the screw head, the point moves forward against wood resistance through a distance equal to its *pitch*, $p$, the name given to the distance between consecutive threads. At the same time the force or effort moves a distance $2\pi a$, where $a$ is the radius of the cylinder or screw. Hence

$$\text{Velocity ratio of screw} = \frac{2\pi a}{p}.$$

A *screwdriver* can be regarded as a wheel and axle (Fig. 6.5(ii)). The handle H can be regarded as a 'wheel' round which the effort $P$ is applied, and the blade of the screwdriver as the 'axle'. The velocity ratio of the screwdriver is thus the ratio $R/r$, where $R$ is the radius of the handle and $r$ is half the width of the blade (p. 155).

Suppose a screwdriver is turned through a complete revolution so that a screw penetrates into wood W a distance equal to its pitch $p$. If the circumference of the handle H of the screwdriver is 4 cm and the force exerted is $F$, the work done is $(F \times 4)$ units. If the resistance of the wood is 2000 N and the pitch of the screw is 0·1 cm, it follows that $F \times 4 = 2000 \times 0·1$, from which $F = 50$ N. Thus the screwdriver is a machine which enables a large resistance to be overcome by the application of a small force; in general, the larger the diameter of the handle H, the greater is the mechanical advantage of the screwdriver.

A *screw jack* consists basically of a long screw rod R passing through a threaded block and a brace B to turn R. Fig. 6.5(iii) shows one form of car jack. Suppose B is turned through one complete revolution in a circle of radius 30 cm. The block is then raised a distance equal to the pitch $p$ of the screw, say 0·32 cm. The velocity ratio is then equal to $2\pi \times 30 \div 0·32$, or about 600. The mechanical advantage is much less owing to friction. If the efficiency of the jack is 25% the mechanical advantage is 25% of 600 or 150. Thus if a load of 15 000 N requires to be raised, the effort needed is only 100 N.

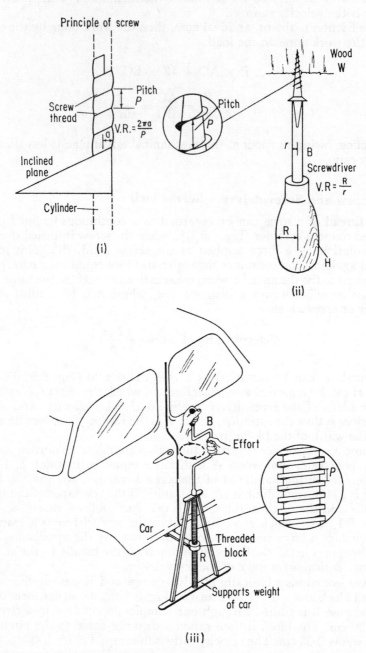

Fig. 6.5 Screw, screwdriver and screw jack

## Pulleys

*Pulleys* are machines used by builders for hauling heavy loads to higher floors. In the form of cranes they are used at docks for lifting heavy cargoes in and out of ships.

A simple pulley is a fixed wheel with a rope passing round its groove. A load $W$ is attached to one end of the rope. It is easier to raise a load by hauling at the other end, rather than raising it directly, because the hauler can use part of his weight, $W_0$, as a pulling force in addition to his muscles (Fig. 6.6).

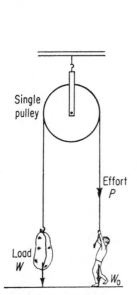

Fig. 6.6 Single pulley

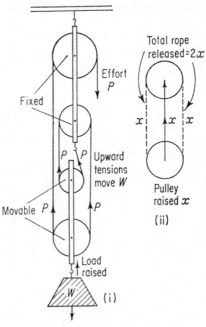

Fig. 6.7 Action of pulleys

## Block and Tackle

A more practical form of pulley system is to use a number of pulleys and to pass a continuous rope round them. Fig. 6.7(i) shows a system of four pulleys. The upper two are fixed, and the lower two are movable. The load $W$ to be raised is attached to the lower pulleys, and the effort $P$ is applied to the end of the rope passing round the top pulley.

Suppose the rope is pulled down with a force $P$ which is equal to 100 N. The tension everywhere in the rope is then 100 N if frictional forces are neglected. Now there are four pieces of rope pulling upward on the load, two round each of the lower two pulleys. Hence the total upward pull is $4 \times 100$ or 400 N. Thus the load $W$ raised is 400 N when the effort $P$ is 100 N.

In practice, owing to the weight of the lower block and the friction at the pulley wheels, the force $P$ needed to raise a load of 400 N is more than 100 N, say 120 N. In this case,

$$\text{Mechanical advantage M.A.} = \frac{\text{Load}}{\text{Effort}} = \frac{400}{120} = 3\cdot3.$$

As the load increases, the weight of the lower block becomes relatively small and thus the mechanical advantage and efficiency rise to a maximum value.

### Velocity Ratio

When the lower two pulleys move up a vertical distance $x$, responding to a movement $x$ of the load $W$, each pulley releases a length $x$ of rope on

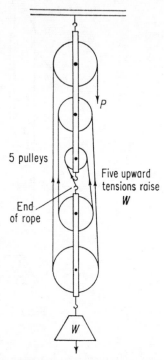

5 pulleys

Five upward tensions raise $W$

End of rope

$P$

$W$

Fig. 6.8 Pulley system – five pulleys

each side or a total length $2x$ (Fig. 6.7(ii)). The effort $P$ then moves down a distance of $2 \times 2x$ or $4x$. Thus

$$\text{Velocity ratio} = \frac{\text{Distance moved by } P}{\text{Distance moved by } W \text{ in same time}} = \frac{4x}{x} = 4.$$

Thus *the velocity ratio is equal to the number of pulleys*, which is a general result for this system of pulleys.

With an odd number of pulleys such as five, the upper set of fixed pulleys is three and the lower movable set is two. The end of the rope is now attached to the lower frame, instead of to the upper frame as shown in Fig. 6.8, giving a velocity ratio of 5.

## Measurement of Mechanical Advantage, Velocity Ratio, Efficiency

### 1. *Mechanical Advantage*

The mechanical advantage of a model of any machine so far described can be found by using a known weight $W$ of, say, 5·0 N as the load, and then attaching a scale-pan to the other end of the string or rope of the machine concerned. Fig. 6.9 shows the case of the pulley system. Weights are then added to the scale-pan until the load begins to rise steadily. If the total weight, plus scale-pan, is 1·6 N, then mechanical advantage $= W/P$ $= 5·0/1·6 = 3·1$.

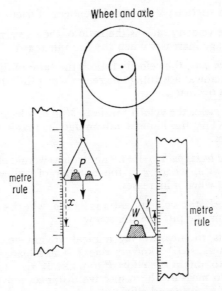

Fig. 6.9 Measurements on machine

### 2. *Velocity Ratio*

Two metre rulers are required, one to measure the distance moved by the effort and the other the corresponding distance moved by the load. The string or rope where the effort is applied is pulled down an observed distance $x$, and the corresponding distance $y$ moved up by the load is measured.

Then Velocity ratio $= x/y$.

### 3. *Efficiency*

Having obtained the effort $P$ which raises the load $W$ steadily, the effort is moved an observed distance $a$, and the corresponding distance $b$ moved by the load is measured. The efficiency can now be calculated from

$$\text{Efficiency} = \frac{W \times b}{P \times a} \times 100\%.$$

## SUMMARY

1. Mechanical advantage M.A. = load $(W)$/effort $(P)$.

Velocity ratio V.R. = (Distance moved by effort)/(Distance moved by load in same time).

$$\text{Efficiency} = \frac{\text{Work obtained (raising } W)}{\text{Work supplied (by } P)} \times 100\%.$$

2.
$$\text{Efficiency} = \frac{\text{M.A.}}{\text{V.R.}} \times 100\%.$$

M.A. depends on friction; V.R. is independent of friction.

3. In the *lever*, the velocity ratio is the ratio of the two arms of the lever. The mechanical advantage increases when the ratio increases.

4. In the *wheel and axle*, the velocity ratio is the ratio of the radii of the wheel and axle. The mechanical advantage increases when the wheel radius increases and the axle radius decreases.

5. In the *inclined plane*, the velocity ratio is $1/\sin \theta$, where $\theta$ is the inclination to the horizontal. The mechanical advantage increases when the angle of inclination decreases.

6. In the *block and tackle pulley system*, one string is wound all round the pulleys. The velocity ratio $= n$, where $n$ is the number of pulleys. The mechanical advantage increases when $n$ increases.

7. In the *screw jack*, the velocity ratio is $2\pi \, a/p$, where $a$ is the radius of the moving arm and $p$ is the pitch of the screw.

8. In experiments to measure mechanical advantage, velocity ratio and efficiency of machines, use (i) a known weight $W$ as a load, (ii) a scale-pan with increasing weights to obtain the effort $P$ to make $W$ rise steadily, and (iii) two fixed metre rules, to measure the respective distances moved by $P$ and $W$ for calculation of velocity ratio and efficiency.

Exercises on *Machines* may be found on p. 172.

# PARALLELOGRAM AND TRIANGLE OF FORCES

In this section we consider forces which are not parallel and which keep objects in equilibrium. Such forces maintain the stability of bridges, for example, and engineers must know how they are added together. The engineer can then apply his knowledge to estimate the relation between the strength of the supports, the forces in the girders and the weight of the bridge.

## Resultant of Two Forces

If two boys pull a sledge in the same direction by ropes, and the tensions $P$ and $Q$ in each rope are 150 and 100 N respectively, the total or *resultant force* is $(150 + 100)$ or 250 N (Fig. 6.10(i)). If the forces act in opposite directions the resultant force is $(150 - 100)$ or 50 N in the direction of the larger or 150 N force (Fig. 6.10(ii)).

Suppose now that the ropes are inclined at 60° to each other and that the tensions are again 150 and 100 N (Fig. 6.10(iii)). Experience shows that the sledge moves forward more in the direction of the 150 N than the 100 N force, as one might expect, and that the total or resultant force on the sledge is now less than (150 + 100) or 250 N.

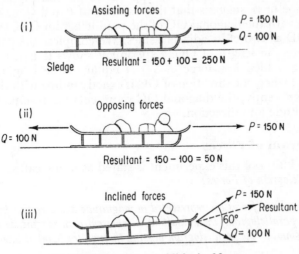

Fig. 6.10 Resultant (addition) of forces

## Experiment on Resultant

An experiment to investigate the resultant of two forces inclined to each other is shown in Fig. 6.11(i). Two strings OX and OY are knotted at O, and weights of 2·0 N P and 1·5 N Q are suspended from the ends of the

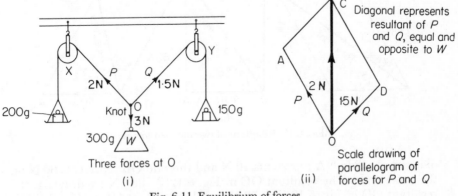

Fig. 6.11 Equilibrium of forces

strings after they are passed round grooved wheels, as shown. A weight of 3·0 N W is suspended from O and hangs vertically.

When the weights have settled in their position and point O is in equilibrium a sheet of paper is placed behind the strings and their three directions are marked on the paper. The two forces $P$ and $Q$, or 2 N and 1·5 N, can now be represented to scale by lines OA and OD respectively, for example, using a scale of 1 cm for 0·2 N, OA is 10 cm long for 2 N and OD is 7·5 cm long for 1·5 N (Fig. 6.11(ii)).

Since the point O is in equilibrium, the resultant of the two forces 2 N $P$ and 1·5 N $Q$ must be equal and opposite to the third force $W$ or 3 N. The direction of $W$ suggests that the resultant of $P$ and $Q$ may have some connection with the diagonal OC of the parallelogram OACD, which has OA and OD as two of its adjacent sides. We complete the parallelogram and measure the length of OC. Experiment shows that it is practically 15 cm long, which is a force of 3·0 N equal to the magnitude of the resultant. Further, the direction of OC is exactly in line with the direction of W. Consequently, the diagonal OC represents the resultant exactly in both magnitude and direction.

## Parallelogram of Forces

The conclusion of this experiment is stated as a law called the *Principle of the Parallelogram of Forces*:

*If two inclined forces are represented in magnitude and direction by the adjacent sides of a parallelogram, their resultant is represented in magnitude and direction by the diagonal of the parallelogram passing through the point of intersection of the two sides.*

Fig. 6.12 shows the resultant $R$ of two forces of 40 and 60 N respectively when they act at angles to each other of: (i) 60°; (ii) 90°; (iii) 120°. In each

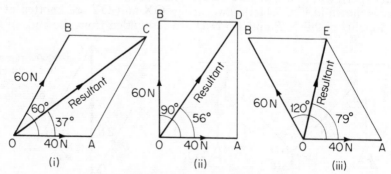

Fig. 6.12 Resultant of two inclined forces

diagram the side OA represents 40 N and the side OB represents 60 N on the same scale. The resultant OC in Fig. 6.12(i) is 87 N by drawing, the resultant OD in Fig. 6.12(ii) is 72 N and the resultant OE in Fig. 6.12(iii) is 53 N, each acting at the angles shown. The magnitude and direction of the resultant thus depends considerably on the angle between the two forces, as shown. In the special case of two *equal* forces inclined at an angle of, say, 70° to each other the resultant acts at an angle of 35° to each or half-way between.

## Example

The resultant of a force of 100 N and one of $P$ N is 150 N acting at an angle of 30° to the force of 100 N (Fig. 6.13). By drawing, find $P$ and the angle it makes with the force of 100 N.

First, draw OC to scale to represent the resultant of 150 N and then draw OA on the same scale to represent the 100 N force acting at an angle of 30° to OC. Join AC, from O draw a line parallel to AC and from C draw a line parallel to AO to intersect the line from O at B. Then OACB is a parallelogram of forces. Measure OB, or $P$; we find $P = 80$ N. Measure the angle AOB; we find this is equal to 69°.

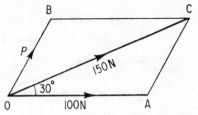

Fig. 6.13 Vector subtraction

## Components of Forces

On occasions, only part of a force is used. A typical case is the use of a lawn-mower. Suppose it is pushed with a force of 200 N as shown at a direction of 30° to the ground (Fig. 6.14). Part of the force then pushes the mower horizontally; this is called the *resolved horizontal component* of the force. The remainder of the force presses the mower vertically into the

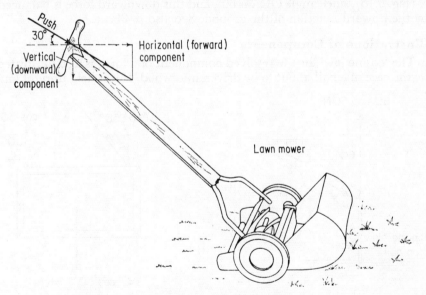

Fig. 6.14 Horizontal and vertical components

ground; this is called the *resolved vertical component* of the force. The useful part of the force pushing a lawn-mower is its horizontal component.

The resolved components of a force can be found by applying the parallelogram of forces. As an illustration, in Fig. 6.15 let CO represent in magnitude and direction the force of 200 N pushing a lawn-mower at an angle of 30° to the ground. From C draw a line CA perpendicular to the horizontal OA, and a line CB perpendicular to the vertical through O. Then, since OACB is a parallelogram (strictly, a rectangle), the forces represented by the sides CB and CA are together equal to the force represented by CO. Conversely, the force represented by CO is equivalent to a force represented by CB acting along the horizontal as shown, together with a force represented by CA acting along the vertical into the ground. CB and CA respectively thus represent the resolved components, or effective parts, of the 200 N force along the ground and the vertical.

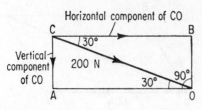

Fig. 6.15 Magnitudes of components

Since CBO is a right-angled triangle, $\dfrac{CB}{CO} = \cos 30°$. Consequently, CB = CO cos 30° = 200 cos 30° = 173 N. The result can be generalised as follows: The component of a force $F$ in a direction inclined at an angle $\theta$ to it is given by

$$F \cos \theta.$$

The component along the vertical direction CA is thus 200 cos 60° = 100 N (Fig. 6.15), since angle OCA is 60°, and this downward force is balanced by the upward reaction of the ground. See also p. 88.

### Illustrations of Components

The 'cosine law' for the resolved component of a force is also illustrated by the case of a nail about to be driven into wood. If it is held upright and

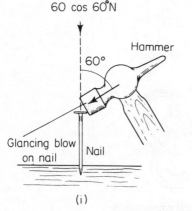

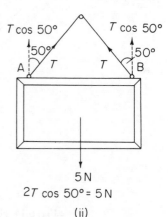

Fig. 6.16 Component forces: (i) in action; (ii) in equilibrium

then hit by a vertical blow of 60 N with a hammer the nail receives the full effect of the blow along its length, which drives it into the wood. If the blow of 60 N is delivered carelessly, so that it strikes the nail at an angle of 60° to it, the nail is driven into the wood by an effective force = 60 cos 60°

$= 60 \times 0.5 = 30$ N (Fig. 6.16(i)). This force is the resolved component of the blow along the nail. If the nail is struck at right angles to its length it never moves into the wood; the resolved component along the length of the nail $= 60 \cos 90° = 60 \times 0 = 0$.

Fig. 6.16(ii) shows a picture of weight 5 N supported by a string on a nail. The tensions $T$ at A and at B together support the weight. Suppose the strings at A and B make angles of 50° with the vertical. Then the vertical component of $T$ at A $= T \cos 50°$; the vertical component of $T$ at B $= T \cos 50°$. Thus the total vertical upward force $= 2T \cos 50°$, and this supports the weight acting vertically downward.

$$\therefore 2\,T\cos 50° = 5 \text{ N}$$

$$\therefore T = \frac{5}{2\cos 50°} = 3.9 \text{ N}$$

## Equilibrium of Three Forces Acting at a Point

There are many cases in engineering practice and in everyday life in which an object is kept in equilibrium by three forces acting at a point. Fig. 6.17(i) illustrates the hinge O of the corner of a bridge resting on a

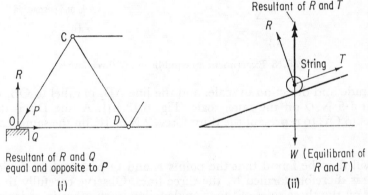

Fig. 6.17 Equilibrium of three forces

support. The equilibrium of O is maintained by the forces $P$ and $Q$ in the girders CO and OD, and the reaction $R$ of the support. In Fig. 6.17(ii) an object is held on a smooth inclined plane by a string along the plane. The forces maintaining the equilibrium are the weight $W$ of the object, the tension $T$ in the string and the reaction $R$ of the plane. *When the plane is smooth the reaction $R$ acts normally to the surface*, a point to be remembered by the student.

Suppose the object resting on the inclined plane in Fig. 6.17(ii) has a weight of $W$ of 100 N. Since it is in equilibrium, the sum (resultant) of the two forces $R$ and $T$ must also be 100 N and must act in an opposite direction to $W$; if either of these two conditions is not obeyed the object cannot be in equilibrium. For the same reason, the resultant of the forces $R$ and $Q$ in Fig. 6.17(i) must equal $P$. The name *equilibrant* is given to the single force which maintains two forces in equilibrium.

## Triangle of Forces

### EXPERIMENT

The relationship between three forces in equilibrium at a point can be investigated by attaching three different weights $P$, $Q$ and $W$ gf to three separate strings knotted at a point O (Fig. 6.18(i)). When the strings have settled down the point O is in equilibrium under tensions equal to $P$, $Q$ and $W$, such as $1\cdot0$ N $P$, $1\cdot5$ N $Q$ and $2\cdot0$ N $W$.

By placing a sheet of paper behind the strings, the directions of $P$, $Q$ and $W$ can be marked with pencil. A line OA is drawn to represent $1\cdot0$ N $P$ in

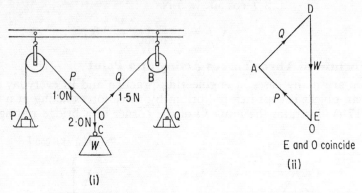

(i)

E and O coincide

(ii)

Fig. 6.18 Experiment on equilibrium of three forces

magnitude and direction to scale, and the line AD, parallel to OB, to represent $1\cdot5$ N $Q$ on the same scale (Fig. 6.18(ii)). A line DE is drawn parallel to OC to represent the third force, $2$ N or $W$, on the same scale.

### Result

It will then be found that the points E and O practically coincide. A *triangle* is therefore formed by the three lines. Observe carefully that the arrows on the three sides of the triangle follow each other in order round the triangle.

## Principle of Triangle of Forces

When any of the three weights $P$, $Q$ and $W$ is varied and they settle in equilibrium, the same result is obtained. This is known generally as the *Principle of the Triangle of Forces*, which states:

*If three forces acting on an object are in equilibrium a triangle can be drawn whose sides, taken in order, represent the forces in magnitude and direction.*

We can understand the reason for the principle by considering three forces $P$, $Q$, $S$ in equilibrium at a point O (Fig. 6.19). The resultant $R$ of $P$ and $Q$ must be equal and opposite to the third force $S$. The diagonal CO of the parallelogram OBCA must hence represent $S$.

Since BC is equal and parallel to OA, the side BC represents the force $Q$. Thus the sides of the triangle OBC represent the respective forces $P$, $Q$ and

$S$, taken in order. In this way it can be seen that a triangle can always be drawn whose sides represent three forces in equilibrium.

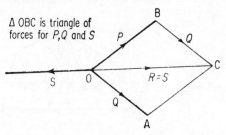

Fig. 6.19 Explanation of triangle of forces

## Application of Triangle of Forces

The triangle-of-forces principle can be used to calculate unknown forces if one force is known.

As a simple illustration, suppose that an electric lamp of 0·5 N is pulled by an attached horizontal string so that the flex makes an angle of 30° to the vertical (Fig. 6.20(i)). To find the tensions $P$ and $T$ in the string and flex, draw a line AO parallel to the known force of 0·5 N to represent it in magnitude on some scale, e.g. 1 cm = 0·1 N (Fig. 6.20(ii)). Then AO is 5 cm long. To obtain the triangle of forces, draw lines from A and from O which are parallel to the other two forces $T$ and $P$, and let them intersect at B (Fig. 6.20(ii)). Then triangle OAB is a triangle of forces for $T$ and $P$ and

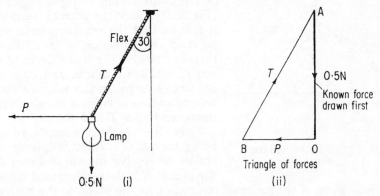

Fig. 6.20 Triangle of forces – small object

the 0·5 N, and OB represents $P$, the force to which it is parallel, while BA represents $T$. Suppose OB is 2·9 cm long and BA is 5·8 cm long by measurement. Then, since 0·1 N is represented by 1 cm, $P = 0·29$ N and $T = 0·58$ N.

## Equilibrium in Bridges

Fig. 6.21 illustrates a Warren bridge, which consists of light girders forming triangles of 60°. If the reaction $R$ at the support $a$ is known, the forces $T$, $S$ in the two girders meeting at $a$ can be found by drawing the triangle of forces. Fig. 6.22 illustrates a part of the Forth bridge, one of the longest in the British Isles, largely designed from drawings by engineering

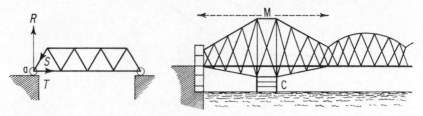

Fig. 6.21 Equilibrium of three forces      Fig. 6.22 Equilibrium in bridge

draughtsmen of the forces required to maintain it in equilibrium. It is known as a 'double cantilever' bridge, as its principal feature is a pair of projecting arms M having a common support C.

## Equilibrium and Resolved Forces

Besides the application of the triangle of forces, unknown forces can sometimes be more conveniently found by resolved components (p. 166).

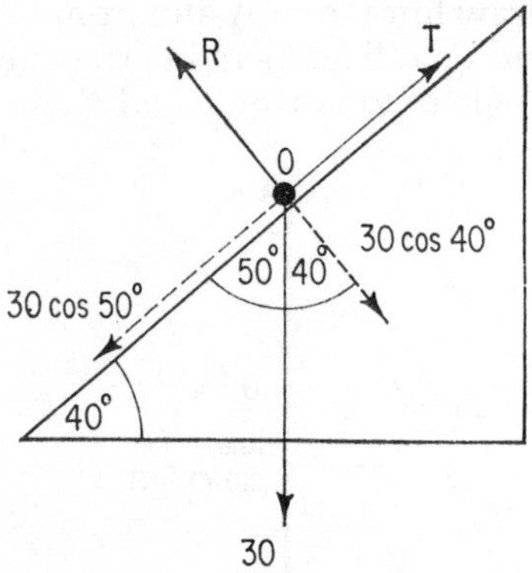

As an illustration, consider an object O of 30 N supported in equilibrium on a smooth plane by a string of unknown tension $T$, and $R$ is the unknown reaction of the plane acting at right angles to it (Fig. 6.23). Since the object is stationary, it follows that the resultant force acting on it in any direction is zero. We may therefore resolve the three forces at O in any direction we desire, and equate the net force to zero. In this way we shall be able to obtain equations to calculate the unknown forces $T$ and $R$.

Since the resolved component of $R$ along the plane is $R \cos 90°$, or zero, we choose to resolve the three forces along the plane; $R$ is then eliminated from the equation. Thus, if the angle of inclination of the plane to the horizontal is 40°,

Fig. 6.23 Equilibrium and components

$$T - 30 \cos 50° = \text{net force upward along plane} = 0$$
$$\therefore T = 30 \cos 50° \text{ N} = 19 \text{ N (approx.).}$$

If we had taken the net force down the plane instead of up we should have had $30 \cos 50° - T = \text{net force} = 0$, from which $T = 30 \cos 50°$ N again.

To find $R$ by resolving the three forces, choose the direction along which the resolved component of $T$ is zero; this is the direction perpendicular to the plane, as $T \cos 90° = 0$.

The net downward force in this direction $= 30 \cos 40° - R = 0$,
as the object is in equilibrium.

$$\therefore R = 30 \cos 40° \text{ N} = 23 \text{ N}$$

If we had chosen to resolve the three forces in a vertical direction, then

$$R \cos 40° + T \cos 50° - 30 = 0.$$

We should need another equation to find $R$ and $T$. Resolving the three forces
horizontally, we should have

$$R \cos 50° - T \cos 40° = 0,$$

as the resolved component of 30 N in this direction is zero. Solving the two
simultaneous equations for $R$ and $T$ would enable these forces to be calculated;
but it should be noted by the reader how much more easily the problem is
solved by choosing to resolve the forces in a direction perpendicular to one of the
unknown forces, since it is then eliminated from the equation.

### Equilibrium of Large Object

So far we have considered the equilibrium of a small object, or point
object, under the action of three forces. In practice, however, the forces on
a large object act at widely different points.

As an illustration, consider a uniform ladder AB with one end B on the

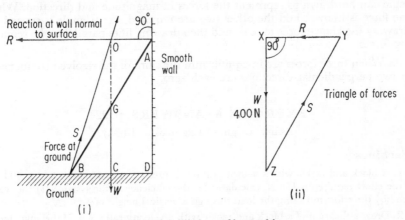

Fig. 6.24 Equilibrium of large object

rough ground and the other end A resting against a smooth wall AD
(Fig. 6.24(i)). The force due to the wall at A, the reaction $R$, acts normally
to the wall here, since it is smooth. The weight $W$ of the ladder acts
vertically downward along GC through its centre of gravity, G, which is
at the middle of the ladder.

The two forces $R$ and $W$ meet at a point O, obtained by producing CG
upwards to meet the line of action of $R$. The resultant of $R$ and $W$ must
pass through O. *Consequently, the third force S at the ground B must pass through
O, otherwise it could not counterbalance exactly the resultant of $R$ and
$W$.* This leads to a general rule: *Three forces acting on any object in equilibrium*

*must* all *pass through one point.* In this case the point is O, where *R* and *W* intersect.

We are now able to determine the magnitudes of *R* and *S*. Suppose the length of the ladder is 20 m, that it is in equilibrium at an angle of 60° to the ground and that its weight is 400 N. The triangle BAD is drawn to scale, with angle ADB = 90°, angle ABD = 60° and AB = 20 m. Next, the line AO is drawn normal to the wall and the line CG is drawn vertically to meet it at O. *Then BO is joined.* This gives the direction of the force *S* at B. The triangle of forces XYZ is now drawn, beginning with a vertical line XZ to represent 400 N *W* and continuing with lines XY and ZY parallel to *R* and *S* respectively (Fig. 6.24(ii)). From measurements of XY and ZY, the forces *S* and *R* can be found.

## SUMMARY

1. The resultant of two forces is a single force which has the same effect in magnitude and direction as the two forces acting together. It is given by the diagonal of the parallelogram of forces. The equilibrant is the single force which maintains the two forces in equilibrium.

2. The resolved component of a force *F* in a direction inclined at an angle $\theta$ is $F \cos \theta$.

3. When three forces are in equilibrium: (*a*) they meet at a point; (*b*) a triangle can be drawn to represent the forces in magnitude and direction. When one force is known, and the other two are required, the triangle is started by drawing the known force to scale and then drawing lines parallel to the remaining forces.

4. When three forces are in equilibrium the sums of the resolved components in two perpendicular directions are each zero.

## EXERCISE 6 · ANSWERS, p. 224
(Assume weight of 1 kg mass = 10 *N*.)

### Machines

1. A block and tackle with a velocity ratio 4 is used to raise a load weighing 400 N. If the effort required is 150 N, calculate: (*a*) the efficiency of the system; (*b*) the work done by the effort in raising the load through a vertical height of 4 m. (*N*.)

2. Draw a diagram of a block and tackle with a velocity ratio of 4. If this machine is used to lift a lump of metal weighing 2000 N and the weight of the lower block and the frictional force are together equivalent to a load of 1000 N, what is the effort required and the efficiency of the arrangement? (*L*.)

3. (*a*) Draw a simple labelled diagram of a lever which has a mechanical advantage greater than 1. (*b*) How is the velocity ratio of an inclined plane calculated? (*N*.)

4. An effort of 100 N applied to a pulley hoist results in a load of 480 N being lifted 1 m vertically upwards. To do this, the effort moves a distance of 6 m. Calculate: (*a*) the mechanical advantage of the hoist; (*b*) the efficiency of the operation. (*C*.)

5. Draw a clearly labelled diagram of *either* a screw-jack *or* a Weston differential pulley, showing where the load and the effort are applied. Derive the formula for calculating the velocity ratio. Explain how the value of the velocity ratio can be found by experiment.

A machine having a velocity ratio of 50 can lift a load of 5000 N by applying an effort of 250 N. Calculate: (*a*) the efficiency; (*b*) the wasted energy when the load is raised 1 m from the ground.

Explain an advantage of having a machine with a low efficiency. (*L*.)

**6.** Explain: (i) how the work done by a force may be measured; (ii) how the efficiency of a machine is defined in terms of work.

A block and tackle with a velocity 5 is used to raise a load of 1000 N weight. If the combined weight of the lower block and the frictional forces is equal to a load of 400 N weight, find: (i) the effort required to lift the load; (ii) the efficiency of the machine. (*N*.)

**7.** An iron girder, weighing 3000 N, is hoisted to the top of a building 27 m high by means of a pulley system with a velocity ratio 4. The effort is applied by means of an electric motor.

Draw a diagram of the pulley system.

Find: (i) the useful work done in shifting the girder; (ii) the work done by the motor if the efficiency of the pulley system is 90%; (iii) the effort exerted on the pulley rope; (iv) the horse-power of the electric motor required to haul up the girder in 2·00 min. [Efficiency of electric motor = 80%; assume 1 horse-power = 750 watt.] (*C*.)

**8.** Draw a labelled diagram of a single rope pulley block and tackle having a velocity ratio of 5. If, using the machine a load of 500 N is raised a distance of 2 m by an effort of 160 N, find: (i) its efficiency; (ii) the work wasted in the machine. (*L*.)

**9.** Define *mechanical advantage, velocity ratio* and *efficiency* as applied to a machine.

Describe a simple screw-jack, and explain its mode of action.

In such a machine the pitch of the screw is 5 mm and the length of the effort arm is 30 cm. If the efficiency is 10%, find the load that can be raised by an effort of 20 N applied at right angles to the effort arm. (*O*.)

**10.** Describe some form of simple machine and explain the meaning of the terms *mechanical advantage* (*M*), *velocity ratio* (*V*) as applied to it.

Why is the ratio *M*/*V* always less than 1 for practical machines?

What is the greatest force that could be applied by tightening a nut on a screw of pitch 1 mm using a spanner 20 cm long and a force of 100 N? (*O. and C*.)

**11.** Illustrate, by referring to a simple pulley system, the meaning of the terms *machine, velocity ratio, mechanical advantage*.

Two men are rolling a 1500 N oil drum up a plane inclined at 30° to the horizontal, by the following method. The men stand at the top of the plane and each pulls a rope fixed to the plane at the top, passing down the plane, under and half-round the drum, then, parallel to the plane, up to the man. What is: (*a*) the least force that each man must exert, and (*b*) the least work each does in rolling the drum a distance of 3 m up the plane? (*O. and C*.)

**12.** Define *mechanical advantage, velocity ratio, efficiency* of a machine.

Give a labelled diagram of a pulley block and tackle of velocity ratio 5 and explain why it has this velocity ratio.

In an efficiency test carried out on this machine the following results were obtained:

| Load in N | 20 | 80 | 140 | 220 | 300 |
|---|---|---|---|---|---|
| Effort in N | 10 | 25 | 40 | 60 | 80 |

Calculate the efficiency in each instance and plot a graph showing how the efficiency varies with the load.

Comment on the variation of the efficiency with the load and give a reason for this variation. (*L*.)

**13.** Describe a wheel and axle and deduce a formula for its velocity ratio.

A certain wheel and axle has a velocity ratio of 6. Its efficiency is 75% when the load is 2000 N. Calculate: (*a*) the least force necessary to lift this load; (*b*) the work done by the effort when the load of 2000 N is raised through a vertical distance of 1 m. (*N.*)

**14.** Define *mechanical advantage* and *velocity ratio*.

Describe how you would determine, by experiment, the efficiency of a machine used for lifting a heavy load.

A load of 21 000 N placed on the top of a screw-jack is lifted by a force of 300 N applied at right angles to the operating arm at a distance of 35 cm from the axis of the screw. If the screw has two threads to 2·5 cm, find its mechanical advantage, velocity ratio and efficiency. (*L.*)

## Parallelogram and Triangle of Forces

**15.** A force of 50 N and a force of 80 N act at the same point and are inclined at 45° to each other. Find the magnitude of the resultant force. (*N.*)

**16.** What is meant by the *parallelogram of forces*? Describe an experiment to illustrate it.

Horizontal forces of 100, 70 and 40 N act on a body on a horizontal plane in directions N, S and E respectively, their lines of action passing through its centre of gravity. Find graphically or by calculation the resultant force in magnitude and direction.

If the mass of the body is 10 kg and there is a frictional force of 20 N, with what acceleration does the body start to move? (*L.*)

**17.** State the conditions of equilibrium when three, non-parallel, co-planar forces act on a body. (*N.*)

**18.** The ends of a rope 3 m long are attached to the lower side of a fixed horizontal beam by two hooks 2·4 m apart. A weight of 160 N hangs from the mid-point of the rope. By drawing or calculation find the tension in the rope. (*N.*)

**19.** State the theorem of the *triangle of forces*.

A slab of concrete weighing 1200 N rests on a plane inclined at 30° to the horizontal. Find, graphically or otherwise, the normal reaction of the plane and the frictional force preventing motion. (*L.*)

**20.** State the parallelogram law for forces and explain how the same law may be applied to velocities.

A man can row at 5 km/h in still water and wants to cross a river to a point exactly opposite. If the river is 150 m wide and is flowing at 3 km/h., find, by means of a scale drawing or otherwise, the direction in which he must set off. How long will it take him to cross? (*L.*)

**21.** State the theorem of the triangle of forces.

A metal sphere, 10 cm in diameter and weighing 4·8 N, is supported by a string 8 cm long attached to a point on the sphere and to a point in a smooth vertical wall. Find: (i) the tension in the string; (ii) the reaction of the wall on the sphere. (*N.*)

**22.** What do you understand by the statement that force is a vector quantity? Explain the parallologram rule for finding the resultant of two forces, and describe how you would test it experimentally.

A steel sphere of weight 4·00 N rests in a smooth symmetrical V-shaped groove, the sides of which are inclined at 45° to the horizontal. Find the forces exerted on the ball by the sides of the groove. (*O.*)

**23.** State the theorem of the triangle of forces. Describe how you would verify this by experiment.

What course must be taken by the pilot of an aeroplane flying through the air at a speed of 300 km/h to enable him to fly due north when a wind is blowing from the west at 50 km/h.? How far would he travel due north in one hour? (*N.*)

**24.** State the conditions under which three non-parallel coplanar forces produce equilibrium. Describe an experiment to illustrate your statement.

A crate is dragged steadily in one direction along a horizontal floor by a rope which makes an angle of 30° with the floor in the vertical plane containing the direction of motion. If the tension of the rope is 450 N, find: (a) the effective force on the crate along the floor; (b) the force tending to lift the crate. (L.)

**25.** What is meant by: (a) the resultant of a number of forces acting at a point; (b) the component of a force in a given direction? How would you find, experimentally, the resultant of two given forces acting in different directions at a given point?

An electric lamp, of weight 9 N, hangs on a flex which is pulled aside by a horizontal string until the flex makes an angle of 30° with the vertical. Find the tension in the string and in the flex. (L.)

**26.** A picture is supported on a wall by two strings attached to a nail. Each string is inclined at 30° to the horizontal, and the weight of the picture is 10 N. Find the tension in each string by resolving the forces vertically, and by one other method.

**27.** State what is meant by the *triangle of forces* and describe, with full explanation, an experiment to illustrate it.

A uniform girder AB, of 450 N, makes an angle of 30° with the ground which is horizontal. The end A rests on the ground and the end B is attached by a rope to a point C vertically above A so that the angle ABC is 60°. Find, graphically or otherwise, the tension in the rope and the force exerted by the ground on the girder. (L.)

**28.** State the conditions for the equilibrium of a body under the action of three non-parallel coplanar forces.

A uniform ladder 16 m long and of weight 180 N rests in equilibrium with one end A against a smooth vertical wall and the other end B on the ground at a horizontal distance of 9·6 m from the wall.

Draw a diagram showing the forces acting on the ladder and find graphically or by calculation the forces acting at A and B. (O. and C.)

# 7

## PRESSURE IN FLUIDS

### LIQUID PRESSURE

**Pressure**

A girl wearing very narrow or 'stiletto' heels can produce an impression of the heels on a wooden floor. We say this effect is due to the *pressure* of the heel on the floor. The same girl wearing wide-heeled shoes exerts a much smaller pressure on the floor. The same force, her weight, is exerted on the floor in each case, but the area of the heel in contact with the floor is much

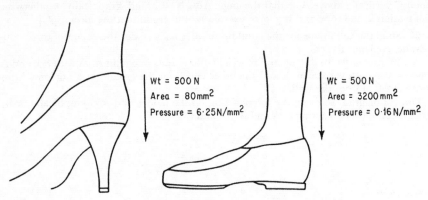

Wt = 500 N
Area = 80mm²
Pressure = 6·25N/mm²

Wt = 500 N
Area = 3200 mm²
Pressure = 0·16 N/mm²

Fig. 7.1  Pressure is greater below narrow heel

less with a stiletto heel. In the latter case, therefore, the 'force per unit area' exerted by the heel is much greater, as calculated in Fig. 7.1.

Scientists define **pressure** as *force per unit area*. Thus the pressure on the head of a diver at work on a sea-bed is the weight of water affecting each unit area of the surface of his head. The units of pressure are *newton per metre²* (N/m²). See also p. 181. If the average pressure on a diver's body of surface area 0·6 m² is 200 000 N/m², the *force* on him is 200 000 × 0·6 = 120 000 N, or the product *pressure × area*. If the total force due to the pressure of water on a plate in the side of a ship is 600 000 N and the area of the plate is 2 metre², the average pressure on the latter is 300 000 N/m². Thus the average pressure $p$ over a surface is given by

$$p = \frac{\text{Total force on surface } (F)}{\text{Area of surface } (A)} \quad . \quad . \quad . \quad . \quad (1)$$

*Hence 'pressure' is not the same as 'force'.* The force $F$ on a surface of area $A$ can be calculated, if the average pressure $p$ is known, by the relation

$$F = p \times A \quad . \quad . \quad . \quad . \quad . \quad (2)$$

## Examples of Pressure

A standing boy exerts a force equal to his weight on the ground below his feet. If the weight is 500 N and the area of his shoes is 100 cm² or 0·01 m², then average pressure $p$ on ground is given by

$$p = \frac{\text{Force}}{\text{Area}} = \frac{500 \text{ N}}{0 \cdot 01 \text{ m}^2}$$

$$= 50\ 000 \text{ N/m}^2$$

If the boy lies flat on the ground his area in contact with the ground increases considerably. Suppose it is now 0·2 m², which is twenty times the original area. The new average pressure $p$ is then 500 N/0·2 m² or 2500 N/m², which is twenty times *less* than before. A person who goes to the rescue of someone who has fallen through ice into a pond would therefore be well advised to crawl over the ice rather than walk.

If a heavy parcel is carried by thin string the small area of string in contact with the flesh produces a large pressure and the string tends to cut into the flesh. Thick string or thick luggage handles produce much less pressure on the hand. A ballet dancer pirouetting on her toes produces a high pressure on the floor. Great pressure is encouraged in the design of ice-skates. The blade makes such a small area of contact with the ice that the pressure below it is very high and ice melts (p. 291). The skate thus moves through a thin film of water.

## Pressure in Liquid

A diver in water experiences a pressure due to the weight of water above him. We can investigate how the pressure in a liquid varies with depth or density by using the apparatus shown in Fig. 7.2.

Thin plastic membranes or covers are tied firmly over the tops of

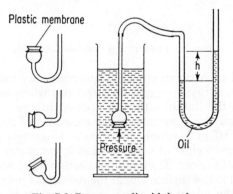

Plastic membrane

h

Pressure   Oil

Fig. 7.2 Pressure at liquid depths

several thistle funnels, each bent at different angles as shown. One of them is connected by rubber tubing to a U-tube containing light oil. If the plastic is pressed lightly with the finger the air inside is compressed and the liquid levels alter. The greater the pressure on the plastic, the greater is the difference in levels. Thus the U-tube acts as a convenient pressure gauge or manometer.

### EXPERIMENT

1. Lower the funnel into the glass vessel, filled with water. Notice that, the deeper it goes, the greater is the difference in levels of the liquid in the U-tube. It follows that the *pressure of water in the vessel increases with depth.*

2. Replace the funnel by others in turn, whose mouths are pointing in different directions. You will observe that the results obtained at a particular depth are the same as those in the first case. It follows that the *pressure in a liquid at one place is the same in all different directions.*

3. Lower a thistle funnel to the same depth in a number of different liquids in turn, such as paraffin and dense copper sulphate solution. Notice that *the greater the density of the liquid, the greater is the pressure at the same depth.*

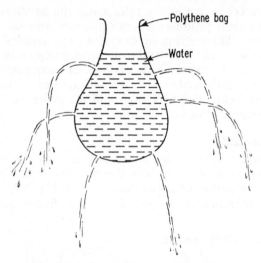

Fig. 7.3   Pressure acts normally to surface

A rubber ball or a polythene can, punctured by many fine holes and filled with water, sprays water in all directions when squeezed (Fig. 7.3). The pressure on the liquid, although applied in one direction outside, is transmitted equally by movement of the molecules of the liquid to all parts of it. The same effect is produced when a watering-can is inclined–the pressure due to the weight of water behind it forces out water through the holes in the rose of the can. It should be noted that the water shoots out in a direction normal to the surface where the particular hole is situated, so that the pressure of the liquid is exerted at right angles to the surface on the part concerned.

Fig. 7.4 (i) shows a simple demonstration that pressure increases with depth. It consists of a vertical tube which was part of a bicycle frame. Holes such as A and B are drilled at equal distances, but the holes should be very slightly staggered to avoid collision of a given jet of water with one from the hole below it. Fill the tube with water as shown and maintain the water level by water from the tap. Notice that the lower holes such as G produce the more powerful jets and that equal jets are produced by holes at the same level.

The hot water from a tap on the ground floor of a building issues with a greater speed than the water from a tap on the third floor, for the same reason, and the bottom of a dam is made much thicker than the top because the pressure of the water increases with the depth (Fig. 7.4(ii)).

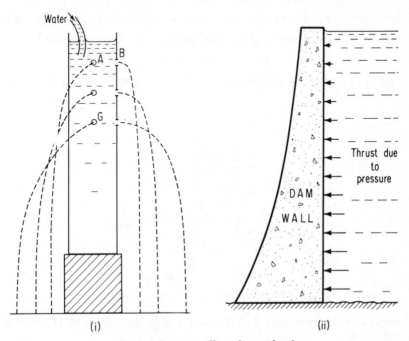

Fig. 7.4 Pressure effects due to depth

## Investigation of Pressure Variation with Depth

So far we have shown that pressure increases with depth, but we do not know the exact relationship between them. An experiment to investigate this relationship may be carried out as follows:

*Method.* A trough A was filled with water (Fig. 7.5). An open glass tube T, with a plastic or polythene cover S held against the lower end, was placed vertically in the water as shown. The pressure of the water on S kept it firmly against the tube T so that no water entered T.

The depth $h$ of S below the water level was measured. A light oil was then slowly poured into T. When the weight of the oil just counteracted the force due to the pressure of the water on S the latter floated away. The height $H$ of the oil in T was then measured.

The experiment was repeated with different depths of water $h$. Each time the height $H$ of oil was measured when the cover S floated away.

*Measurements*

| $h$ (cm) | $H$ (cm) |
|---|---|
|  |  |

*Graph.* Plot a graph of depth $h$ v. length of oil column $H$.

*Deductions.* When S floats away the downward force due to the weight of oil and the atmospheric pressure is then just equal to the upward force due to the water at the depth $h$ and the atmospheric pressure. Thus

upward force due to water pressure = downward force due to weight of oil.

Thus if $A$ is the inside area of the tube and $p$ is the water pressure,

$$pA = \text{weight of oil.}$$

Now the weight of the oil is proportional to its length $H$ and $A$ is constant.

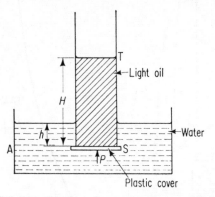

Fig. 7.5 Investigation of pressure change with depth

It follows that the pressure $p$ of the water is proportional to $H$, or $H$ is a measure of $p$.

The graph of pressure or $H$ with depth $h$ is found to be a straight line passing through the origin. Hence the pressure in a liquid is directly proportional to its depth.

### Pressure in a Liquid

We can now obtain a general formula for the pressure at a place below the surface of a liquid of density $\rho$ kg per m³. Suppose a horizontal area B of 1 m² is placed $h$ m below the surface. The pressure on B is due to the weight of liquid acting on it, and the volume of this liquid = $1 \times h = h$ m³ (Fig. 7.6).

Hence the mass of the liquid = volume × density = $h\rho$ kg, and its *weight* (which is a force, p. 98) = $h\rho$ kgf = $h\rho g$ newton where $g = 9.8$.

$$\therefore \text{pressure on B} = \frac{\text{force on B}}{\text{area of B}} = \frac{h\rho g}{1}$$

$$= h\rho g \text{ newton per metre}^2.$$

In general, **Pressure = $h\rho g$** . . . . . . . (3)

At a depth of 50 metres in water, density $= 1000$ kg/m³, we have, using $g = 10$ m/s²,

$$\text{pressure} = h\rho g = 50 \times 1000 \times 10$$
$$= 500\ 000\ \text{N/m}^2$$

From the formulae for the pressure in (3), p. 180, it follows that the pressure in a liquid increases proportionately with the depth $h$ of the place below the surface, as found experimentally, and with the density $d$ of the liquid. The pressure at the bottom of a narrow vessel filled to a height of 10 cm with mercury is thus the same as the pressure at the bottom of a wide vessel filled to the same height of 10 cm. In the latter case the bigger weight of mercury is distributed over a bigger area at the bottom, so the pressure (force per unit area) is the same in each case.

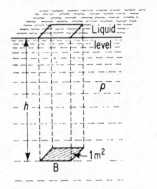

Fig. 7.6 Proof of pressure $= h\rho g$

## Levels of Liquid

When a liquid is poured into a vessel having tubes of different shapes and sizes connected together the liquid is observed to rise to the same height, $h$, in each of the tubes (Fig. 7.7). Since the pressure along the same horizontal line HH in the stationary liquid must be the same (if it were not the liquid would move until the pressure at different points on HH were equalized), this is another demonstration of the fact that the pressure

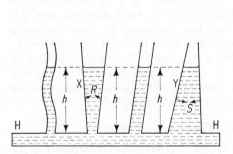

Fig. 7.7 Pressure is independent of area

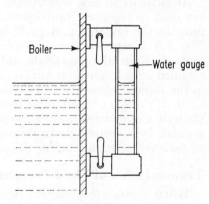

Fig. 7.8 Boiler water-gauge

at a point in a given liquid depends only on the depth of that point below the surface.

Some of the weight of liquid in the vessel X is counterbalanced by upward forces, due to the components of the reactions $R$ of the sloping walls. The net force acting on the narrow base of X is thus less than the weight of liquid. Conversely, the components of the reactions $S$ *add* to the weight of the liquid in the vessel Y when calculating the net force acting on the wide base of Y. The net force per unit area or pressure on the base of X and on the base of Y are equal.

In popular language, it is sometimes said that 'water finds its own level' in different parts of the same vessel. A water gauge in a boiler utilizes the fact that 'water finds its own level', and is used to determine the level of the water in the boiler, which cannot be seen directly (Fig. 7.8).

## Water Supply

Water flows naturally from one place to another at a lower level. Large storage tanks in the roof of a country house may supply water to taps in rooms at a lower level. The reservoirs of water for large towns, containing many millions of gallons, are constructed on very high or hilly regions, and water flows down through pipes to lower levels on release.

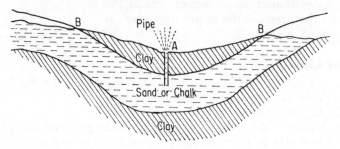

Fig. 7.9 Artesian well

Artesian wells, first constructed in France over a thousand years ago, are used to tap great underground reservoirs of natural water, such as those in the underground chalk basin in counties surrounding London. Fig. 7.9 illustrates the principle. Rain falling on clay does not pass through. Rain falling on chalk and sand, however, seeps through, and it is retained by a lower layer of clay, which forms a basin. To force the water to the surface where it can be drawn off, a tube is bored through to the basin. If there is a 'head' of water above A, as at B, the water is forced up the well. Compressed air has also been used to draw off water from underground. It is pumped down through one pipe to the water basin, and the water is forced to the surface through a concentric pipe.

## Transmission of Fluid Pressure · Hydraulic Press

When a pressure is exerted on a liquid it is transmitted equally in all directions (p. 178). In the last war depth charges exploding at some point below the water gave rise to enormous forces at least 50 yd from the place of the actual explosion. The transmission of pressure through a liquid is utilized in the *hydraulic press*, a machine whose principle of action is illustrated in Fig. 7.10.

By means of a lever L a small force, 20 kgf for example, is applied directly to a piston X of cross-sectional area 30 cm$^2$, say. The *pressure* on the liquid due to the movement of X is then $\frac{20}{30}$, or $\frac{2}{3}$ kgf per cm$^2$, and this is transmitted through the water filling the vessel to a tight-fitting piston Y of much greater area of cross-section than X. If Y has an area of 1200 cm$^2$ the *upward force* $F$ exerted on it is given by $F$ = pressure × area = $\frac{2}{3}$ × 1200 = 800 kgf. Thus a large force is obtained by the application of the comparatively small force of 20 kgf.

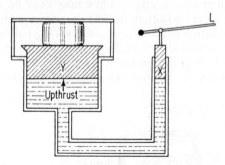

Fig. 7.10 Hydraulic press

If the piston X falls a distance $x$ and the piston Y rises a distance $y$, then, since the volume of water transferred is the same,

$x$ × Area of cross-section ($A_1$) of X = $y$ × area of cross-section ($A_2$) of Y.

$$\therefore \frac{x}{y} = \frac{A_2}{A_1}$$

But $x/y$ is the *velocity ratio* of the machine (see p. 153).

$$\therefore \text{ Velocity ratio of press} = \frac{A_2}{A_1}.$$

In practice, owing to frictional forces, the mechanical advantage of the machine is less than the velocity ratio (p. 154). The larger the ratio of the area of cross-sections, the greater is the mechanical advantage of the press.

## Hydraulic Devices · Car Brakes

The general principle of the hydraulic press, that pressure is transmitted equally to all parts of a liquid, is used in many appliances. Some car jacks, for example, consist of an oil-filled press used for lifting (Fig. 7.11). Most mechanical diggers and bulldozers now use hydraulic principles to power the shovel or blade. Large hydraulic hammers are used to forge red-hot steel into various shapes, and automatic pit props in coal mines are hydraulically operated and moved.

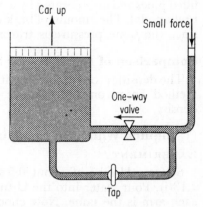

Fig. 7.11 Hydraulic jack

The modern car requires a braking system which retards each wheel equally to minimize the dangers of skidding or locking the wheels when the brakes are applied. The best way is to operate the brakes hydraulically (Fig. 7.12). Each brake consists of two brake shoes, which can be moved apart by hydraulic pressure in a slave cylinder. When the brake pedal is pressed the increased pressure on it is transmitted through oil to the brake shoes, which are then forced apart. They now bear on a revolving brake drum to which the wheel is bolted, so that the wheel is made to stop. When the pressure on the brake pedal is released a spring connected between the two brake shoes contracts and pulls them together again. The drum is then released and the wheel is free to turn.

The slave cylinders on all four wheels are connected together by oil-

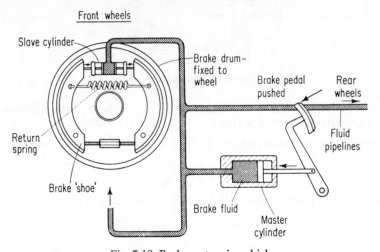

Fig. 7.12 Brake system in vehicle

filled pipes and are operated by a master cylinder connected by levers to the brake pedal. The amount of braking applied to each wheel is exactly equal, since the same pressure is transmitted through the oil or brake fluid.

## Comparison of Densities (or Specific Gravities) of Liquids

The densities (or specific gravities) of two liquids can be compared by a method based on the pressure they exert, since pressure = height × density.

### (1) Liquids Which Do Not Mix

EXPERIMENT

Select a U-tube of about 0·5 cm internal diameter as shown in Fig. 7.13(i). Pour water into the U-tube and notice that the level of water in each arm is the same. Now choose a second liquid which does not mix with water, e.g. paraffin, and pour it into one arm of the U-tube. Notice that the heights of the balancing columns of liquid are different, the extent of the difference depending on the difference in densities of the liquids. Measure the heights $h_1$ and $h_2$ *above the surface of separation*. Repeat for other similar heights by adding more paraffin on one side (Fig. 7.13 (ii)).

*Theory.* Since B and D are on the same horizontal level of the same liquid, water, the pressures at B and D must be equal. Since the atmospheric pressure is the same on each side of the U-tube, it follows that:

pressure at B due to column AB = pressure at D due to column CD.

Thus if $\rho_1$ and $\rho_2$ are the respective densities of the two liquids,

$$\therefore h_1\rho_1 = h_2\rho_2$$

$$\therefore \frac{\rho_1}{\rho_2} = \frac{h_2}{h_1}.$$

Thus the ratio of the densities of the two liquids is *inversely* proportional to their respective heights above the surface of separation. Calculate your

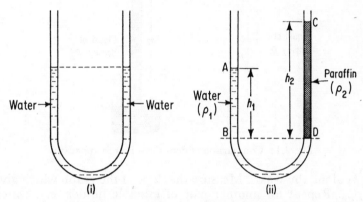

Fig. 7.13 Comparison of densities – U-tube

results for the ratio of the densities of paraffin and water. Repeat the experiment for other pairs of immiscible liquids, e.g. water and olive oil or water and methylated spirit. The density of water can be taken as 1 g per cm³, or 1000 kg per m³, so that actual densities can be found.

## (2) **Liquids Which Mix**

If the liquids mix, an inverted U-tube must be used, commonly in the form of Hare's apparatus (Fig. 7.14). Use water and copper sulphate solution in beakers as the two liquids. By opening the clip and sucking out air, you can make the liquids rise to any desired height, and then remain there when the clip is closed. Measure the two heights $h_1$ and $h_2$ above the levels of the liquid in the respective beakers.

Suppose $\rho_1$, $\rho_2$ are the respective densities of the liquids and $p$ is the pressure of the air above the liquids in the apparatus. Then

$$p + h_1\rho_1 = p + h_2\rho_2 = B \quad . \quad . \quad . \quad . \quad . \quad (1)$$

where $B$ is the atmospheric pressure, which acts on each of the surfaces of the liquids $A_1$, $A_2$ (see Fig. 7.14). Thus $h_1\rho_1 = h_2\rho_2$, from (1), and hence

$$\frac{\rho_2}{\rho_1} = \frac{h_1}{h_2}.$$

The densities are therefore *inversely proportional* to the heights in the two tubes, ignoring surface tension effects.

*Graph.* Vary the heights $h_1$ and $h_2$ and plot a graph of $h_2$ along the y-axis

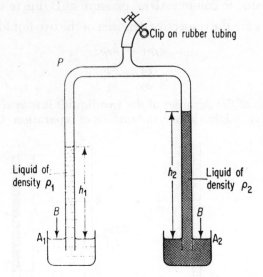

Fig. 7.14 Comparison of densities – Hare's apparatus

against $h_1$ along the x-axis. Measure the slope of the graph which gives the ratio $\rho_1/\rho_2$. Repeat for another pair of miscible liquids, e.g. water and acetone.

## ATMOSPHERIC AND GAS PRESSURE

### Earth's Atmosphere

The earth's surface is surrounded by a thick layer of air called the atmosphere. Air is a mixture of gases, containing about four-fifths by volume of nitrogen, about one-fifth oxygen and a small percentage of carbon dioxide, together with very small quantities of neon, krypton, xenon, helium, hydrogen, ozone and water vapour. At considerable heights traces of ozone are formed by the action on oxygen of the sun's radiation. This ozone layer plays a vital part in shielding the earth from excessive ultra-violet radiation, which in high dosage is harmful to life.

The atmosphere is most dense at ground- or sea-level, estimated at about 1·3 g per litre, and extends to a height of about 80 to several hundred kilometres (50 to several hundred miles). The higher one goes, the less dense is the air, since there is then less weight of air above to compress it. At an altitude of 2000 m (about $1\frac{1}{4}$ miles) the density falls to about 1 g per litre; at 10 000 m (about 6 miles) the density is about 0·4 g per litre. Above the earth's atmosphere in interstellar space there is practically a vacuum—only a very small number of molecules, estimated at about a thousand per litre, exist here, whereas at ground-level the number per litre of air molecules is of the order of $10^{22}$.

### Evidence of Air Pressure

We cannot see the air because it is so 'thin', but the winds are practical evidence of the existence of its molecules. Sailing boats are driven by air pressure, windmills rotate owing to winds and an aeroplane is kept up in flight by air pressure. In the laboratory two convincing demonstrations of air pressure are as follows:

### (1) Collapsing Tin

Obtain a can with a narrow neck, such as a paraffin can. Place a *little* water in the can, and boil for a few minutes (Fig. 7.15(i)). Then remove the

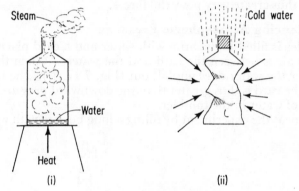

Fig. 7.15 Air pressure effect

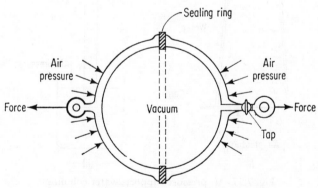

Fig. 7.16 Magdeburg hemispheres experiment

can from the tripod and knock in the rubber bung very tightly. Pour cold water over the can to cool it. Notice that the can collapses (Fig. 7.15(ii)). When the can is removed from the tripod it is filled with steam. After inserting the stopper the can cools down and the steam condenses to give water, of negligible volume compared with the steam. There is therefore practically no air inside the can, and no steam, and the external air pressure on the can now causes the sides to collapse.

Alternatively, the air can be pumped out by connecting the can to a vacuum pump.

### (2) *Magdeburg Hemispheres*

If air is pumped out of two strong metal hemispheres in contact, with grease round their joint to make it air-tight, two boys are unable to pull them apart tugging in opposite directions. The pressure of the air outside has thus exerted a strong force on the hemispheres which kept them together (Fig. 7.16). If air is allowed in, the hemispheres are pulled apart quite easily.

The experiment was first performed in Magdeburg in 1640. Two large hemispheres, about 55 cm in diameter, could not be pulled apart by two teams of eight horses each, pulling in opposite directions! Atmospheric pressure can thus create very powerful forces.

### Counterbalancing Atmospheric Pressure

If a tumbler is filled to the brim with water and a card placed over the top, then no air is between the card and the water.   When the tumbler is turned over the water does not fall out (Fig. 7.17 (i)). The upward air pressure on the card is thus greater than the downward pressure due to the short height of water in the tumbler.

We can repeat the experiment by filling a long burette with water to the

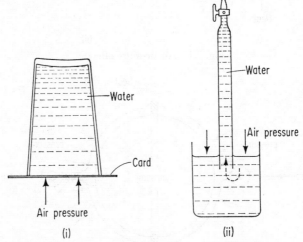

Fig. 7.17  Air pressure supports water columns

brim, then placing a finger over the top so that air is excluded, and inverting the burette in water. The air pressure outside the burette, which is transmitted through the water to the base of the burette and acts upward there, easily supports the water column (Fig. 7.17(ii)). A much denser liquid is therefore needed to counterbalance the atmospheric pressure at its base by the pressure of a column of reasonable height. This was first found over 300 years ago by Torricelli, who used mercury to make the first *barometer* and measure atmospheric pressure.

## Making a Simple Barometer

Take a clean dry thick-walled glass tube about 1 m long. Using clean mercury, fill the tube nearly to the top, then place a finger over the top and invert the tube, so that a large air-bubble runs up and collects all the tiny air-bubbles which may be trapped in the mercury. The tube is turned round again and the finger is removed, and the top is now completely filled with mercury to exclude all air.

Place a finger over the top end of the tube. Invert it, place the end under the surface of mercury in a dish or reservoir and remove the finger. The instant the finger is removed, notice that the mercury drops in the tube to the height $h$ shown in Fig. 7.18. Since air was completely excluded, the space now at the top of the tube is a *vacuum*; it is often called a 'torricellian vacuum'. It can now be seen that the atmospheric pressure is numerically equal to the pressure at the base of a column

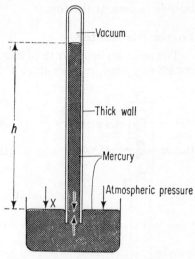

Fig. 7.18 Mercury barometer

of mercury of height $h$, about 76 cm or 0·76 m. Observe carefully that $h$ is the *vertical height above the outside free level* X *of the mercury surface in the trough.* In making a simple barometer, then, a metre rule should be placed (i) vertically, (ii) a little away from the glass tube where surface tension has no effect so that the zero end just touches the free mercury surface.

## Testing the Vacuum

If the vacuum is faulty and contains air or water-vapour the barometer reads less than the true atmospheric pressure. The vacuum can be tested by inclining the tube. The horizontal level of the mercury should then

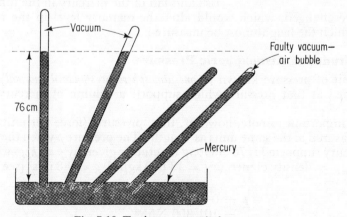

Fig. 7.19 Testing vacuum in barometer

be constant as shown, indicating a constant vertical height above the mercury surface in the trough (Fig. 7.19). When the tube is inclined so that the vertical height of its closed end is less than 70 cm, for example, the whole of the mercury should fill the tube. If the vacuum is faulty an air-bubble is seen at the top.

### Fortin Barometer

The most accurate form of barometer was designed by Fortin and is now commonly used in laboratories. (Fig. 7.20). It has:

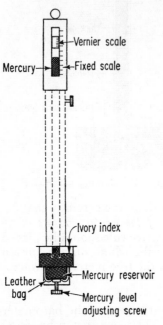

Mercury—

Vernier scale

Fixed scale

Ivory index

Leather bag

Mercury reservoir

Mercury level adjusting screw

Fig. 7.20 Fortin (standard) barometer

(i) a vertical glass tube containing mercury with a vacuum above it;
(ii) a leather bag at the base as a reservoir of mercury;
(iii) a short fixed metal scale, graduated on one side in centimetres and on the other side in inches, and a movable vernier to read the height of the mercury level accurately;
(iv) a fixed ivory index with a sharp point at the bottom, which is the 'zero' of the metal scale.

Before the reading of the mercury height is taken, the level of the mercury surface in the reservoir is adjusted by turning a screw which presses on the leather bag. When the surface of the mercury just touches the point of the ivory index–the mirror-like mercury surface produces an image of the point which helps to make the adjustment very accurately– the height of the mercury is read from the metal scale and vernier. In this way Fortin overcame the difficulty of allowing for the rise and fall of the mercury in the tube as the pressure changed, which would alter the mercury level in the reservoir from which the height must be measured.

### Magnitudes of Atmospheric Pressure

A unit of pressure known as *one atmosphere* or *standard (normal) pressure*, is defined as that pressure which supports a column of mercury 76 cm high.

It is important to note, however, that 'pressure' (force per unit area) is not measured in the same units as 'length'. The pressure $p$ when the column of mercury supported is 76 cm long is actually given by $p = h\rho g$, where $h$ is 0·76 m, $\rho$ = density of mercury = 13 600 kg/m³, $g$ = 9·8 m/s² (see p. 180). Thus

$$p = 0\cdot76 \times 13\ 600 \times 9\cdot8$$
$$= 101\ 300\ \text{N/m}^2$$

If $h$ is the height of a column of *water* supported by one atmosphere of pressure, then, as $p \times h\rho g$, and $\rho = 1000$ kg/m³ for water,

$$h \times 1\ 000 \times 9\cdot8 = 0\cdot76 \times 13\ 600 \times 9\cdot8,$$

Thus cancelling $9\cdot8$

$$h = 0\cdot76 \times 13\cdot6 = 10\cdot34 \text{ m}$$

One atmosphere of pressure therefore supports a height of about 10 m of water, and accordingly it is not surprising that water filling a long inverted glass tube placed in a vessel of water is easily supported by the atmospheric pressure (see Fig. 7.17 (ii) ).

Torricelli's experiment tells us that a column of air 1 m² in cross-section and extending to the top of the atmosphere is equal in weight to a column of mercury of the same cross-section and 0·76 m high. The density of the air decreases as we go up because it is less compressed by the air above, but for a rough estimation of the height $h$ of the atmosphere, suppose the density of air is uniform and about 1·2 kg/m³ as at sea-level. Then, if $h$ is in metres,

$$\text{Atmospheric pressure} = h \times 1\cdot2 \times 9\cdot8 = 0\cdot76 \times 13\ 600 \times 9\cdot8$$

$$\therefore h = \frac{0\cdot76 \times 13\ 600}{1\cdot2} = 8600 \text{ m} = 8\cdot6 \text{ km}$$

Thus a height of about 9 km or 5 miles of air having a uniform density of 1·2 kg/m³ would exert a pressure of 76 cm mercury at its base.

In practice, the height of the atmosphere extends well above these figures, and the total weight of air pressing down on the earth's surface is of the order of a million million million kilogramme force!

Our head and body support a pressure of about 100 000 N/m² due to the atmosphere; for an area of 0·5 m² the total force is hence about 50 000 N. We suffer no discomfort, however, as the pressure of the blood and its dissolved gases counteracts the atmospheric pressure; but airmen ascending to great heights need special apparatus on account of the reduced air-pressure, which results in bleeding through the nose and other parts where the tissues are thin, owing to the greater pressure of the blood. Divers working at great depths in water on salvage work, and men engaged in tunnelling operations in compressed air (situations where the pressure is much greater than atmospheric pressure) have to ascend by degrees to the surface, as a sudden decrease in pressure is harmful to the body.

## Units of Pressure

As we have seen, atmospheric pressure which supports a column of mercury 760 mm or 0·76 m high is calculated by

$$p = h\rho g = 0\cdot76 \times 1000 \times 9\cdot8 \text{ N/m}^2$$
$$= 1\cdot013 \times 10^5 \text{ N/m}^2 = 0\cdot103 \text{ N/mm}^2$$

1 N/m² is also called 1 *pascal*, symbol Pa.

In meteorology a unit of pressure called the *bar* (b) is used. This is defined as $10^5 (100\ 000)$ N/m². Thus

$$1 \text{ millibar (mb)} = \frac{1}{1000} \text{ bar} = 100 \text{ N/m}^2.$$

Atmospheric pressure of 76 cm mercury is thus about 1·013 b or 1013 mb.

### Aneroid Barometer

The aneroid barometer is a form of domestic barometer which contains no liquid. This barometer consists essentially of a steel corrugated cylinder B partially evacuated of air, and the top face C is supported by a strong spring S to prevent B from collapsing under the atmospheric pressure (Fig. 7.21). When the pressure increases the upper face of B is slightly

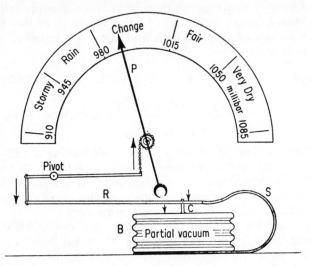

Fig. 7.21 Aneroid barometer

depressed, and the reverse occurs when the atmospheric pressure decreases. The slight movements of B are magnified by means of an attached rod R with a system of small jointed levers, which operate a pointer P controlled by a spring. P moves over a circular dial, as shown, which is graduated in inches of mercury by comparison with a Fortin barometer. The words *stormy, rain, change, fair, very dry,* are printed on the dial, but are only a very approximate guide to the weather.

### Altimeter

As already discussed, the pressure of the air decreases with height. At 650, 1300, 2500 and 5000 m the pressure is respectively about 940, 875, 750 and 550 mb. The dial of an aneroid barometer can thus be calibrated to read altitudes in metre in place of the pressure. The instrument is used in aeroplanes as *altimeters,* and mountaineers, and cars in hilly countries, carry altimeters which indicate the height reached.

## Radio Sonde

Nowadays regular observations are taken of the temperature, pressure and humidity high above the earth. Small balloons, which carry instruments, are released from meteorological stations, and may rise as high as 25–30 km above the earth before they burst owing to the considerably reduced air pressure at this height. The instruments are brought safely to the ground by an automatic parachute. The whole apparatus is known as a 'radio sonde'.

The radio sonde contains a small radio transmitter, which automatically sends out signals at short intervals. These signals are derived from: (i) a small aneroid box, which enables the pressure (and hence the height) to be determined; (ii) a humidity element; and (iii) a thermometer element.

## The Atmosphere

Daily weather maps, which can be seen on television with weather news or in newspapers, carry contour lines of places where the atmospheric pressure is the same. The lines are known as *isobars*. Isobars which are close together indicate a high pressure change in a small distance, and the wind velocity is therefore high in this region. Isobars also give a rough indication of wind direction, as air is forced to flow from a high- to a low-pressure region. However, the rotation of the earth affects the movement of air considerably, so that the air spirals from the high- to the low-pressure region in a direction roughly parallel to the isobars. The winds are also affected by temperature and by land and sea masses over which they pass.

The earth's atmosphere can be divided into several 'layers'. The lowest layer or lower atmosphere, called the *troposphere*, extends from sea-level to about 11 kilometres, where the temperature falls to about −50° C at the top of the layer after decreasing steadily from sea-level. The weather experienced on the earth originates in the winds and clouds in the troposphere. The next layer or upper atmosphere is called the *stratosphere*. Here the temperature is approximately constant and the air is rarefied, so that aircraft cruising in the stratosphere must carry air at atmospheric pressure inside or be 'pressurized'. High above the stratosphere there are layers or belts of ions and free electrons called *Heaviside* and *Appleton* layers. These regions of electrical particles influence radio transmission and reception, as explained on p. 399.

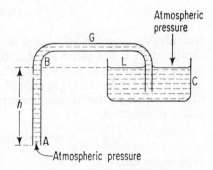

Fig. 7.22 Principle of siphon

## The Siphon

A *siphon* is a simple arrangement for emptying a liquid from a fixed vessel such as a petrol tank C, which is difficult to empty directly (Fig. 7.22). It consists of an open tube G filled with the liquid, which is then placed with one end below the liquid in the vessel, as shown. The liquid in C is then observed to run out of the vessel through A, as long as the other end of G is dipping below the surface of the liquid, and A is kept below the level BL.

The explanation of the siphon action is as follows. Since the pressure at points in the same horizontal level of a liquid is the same (p. 181), the pressure inside the liquid at B is equal to the pressure at L, which is the atmospheric pressure. The pressure of the liquid at A is therefore greater than the atmospheric pressure by the amount $h\rho g$, where $\rho$ is the density of the liquid and $h$ is the height of AB. But the pressure of the air at A is atmospheric, and is therefore unable to support the liquid. It therefore runs out at A. If G is *empty* and is placed with one end below the liquid in C the siphon does not work; there is now no net force on the liquid, as the pressure on it at L, which is atmospheric pressure, is also equal to the air pressure inside the tube G.

### Measuring Gas Pressure · The Manometer

The pressure of the domestic gas supply, which is not much greater than atmospheric pressure, can easily be measured by joining the gas-tap with rubber tubing to one side of a U-tube containing water (Fig. 7.23(i)). When the pressure is steady the water is higher on the open side of

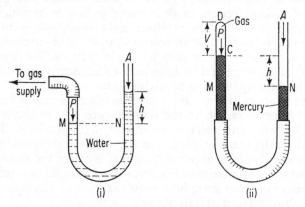

Fig. 7.23 Measurement of pressure

the tube than on the left side, and the difference in levels $h$ is noted. Since the pressure is the same at M and N, points in the same horizontal level of the water, the gas pressure $p$, which is the pressure at M, is equal to the pressure at N. As the right side of the tube is open to the air, the pressure at N is equal to the atmospheric pressure $A$ plus that due to a column of water of height $h$,

$$\therefore p = A + \text{pressure due to height } h \text{ of water.}$$

If $h = 20$ cm,

the height of mercury giving the same pressure $= \dfrac{20}{13 \cdot 6} = 1 \cdot 5$ cm,

as the relative density of mercury $= 13 \cdot 6$.

Thus if A = 76 cm of mercury,

the gas pressure $p = 76 + 1.5 = 77.5$ cm of mercury.

An arrangement for measuring pressure is known as a *manometer*; Fig. 7.23(i) represents a water manometer.

Fig. 7.23(ii) illustrates a gas trapped by mercury in a closed tube, with the mercury level N on the open side $h$ cm *lower* than the mercury level M containing the gas (compare Fig. 7.23(i)). The pressure at $N = A$ (atmospheric pressure) = the pressure at M, a point on the same horizontal level of the mercury. But the pressure at $M = (p + h)$ cm of mercury, where $p$ is the gas pressure in cm of mercury.

$$\therefore p + h = A,$$

if $A$ is in cm of mercury, and hence $p = A - h$. The gas pressure is thus *less* than the atmospheric pressure by the difference in levels of the mercury.

## Boyle's Law

The effect of pressure on the volume of a gas at constant temperature was first investigated experimentally by Robert Boyle, an eminent scientist, about 1660. A simple laboratory apparatus with air as the gas is shown in Fig. 7.24. It consists of dry air in a glass tube of uniform cross-section, attached by tubing to a mercury reservoir.

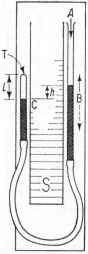

*Method.* Observe the length $l$ of the air column in the tube T from the scale S beside it. This is a measure of the volume of air, since T has a uniform cross-section. Observe the difference in levels $h$ cm of the mercury at C and B. Record the atmospheric pressure $A$ from a Fortin barometer. The pressure $p$ of the air in T is then $(h + A)$ cm mercury.

Increase the pressure by screwing down clips round the tubing or by raising the open tube B of the reservoir. Observe the new difference in level $h$ and the new gas volume $l$. Repeat with several measurements of pressure and volume, taking three readings with B above C (pressure greater than atmospheric pressure) and three readings with B below C (pressure less than atmospheric pressure). Tabulate the results.

Fig. 7.24 Investigation of Boyle's law

Atmospheric pressure $A =$ ...... cm

| | $h$ (cm) | Volume $V$ $l$ (cm) | $p$ | $\dfrac{1}{V}$ | $pV$ |
|---|---|---|---|---|---|
| Above atmospheric pressure $p = A + h$ | | | | | |
| Below atmospheric pressure $p = A - h$ | | | | | |

*Calculations.* Calculate: (i) $1/V$ and tabulate; (ii) the product $p \times V$ and tabulate.

*Graph.* Plot (i) $V$ v. $p$ and (ii) $1/V$ v. $p$.

*Conclusion.* Since the graph of $1/V$ v. $p$ is a straight line through the origin, and also the product $pV$ is a constant, then the pressure and volume of the gas are inversely proportional (see also p.112).

*Boyle's law* states:

*For a fixed mass of gas at a given temperature, the product of the pressure and volume is constant; i.e. $pV$ is constant.*

## Boyle's Law Demonstration with Bicycle Pump

If we close the outlet of a bicycle pump with a finger and then force the plunger inwards we observe that it becomes more and more difficult to compress it as the plunger goes farther and farther inwards.

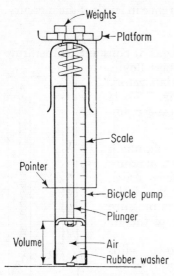

A demonstration of the truth of Boyle's law can be done with a cycle pump, to the top of which is fixed a small metal platform as shown in Fig. 7.25. The plunger is withdrawn to its full extent and a small rubber washer inserted into the end of the pump so that the plunger stays out. The pump is then clamped vertically as shown. A pointer, in the form of a semi-circular piece of wire, moves over a scale painted on the side of the pump. The length of the air column is proportional to the volume of air compressed in the barrel of the pump and any weights added plus the weight of the plunger and platform are together proportional to the pressure on the air. Take several readings of the number $n$ of equal weights added and the corresponding scale reading $r$. Plot a graph of $n$ against the reciprocal of $r$, $1/r$. If Boyle's law is obeyed, then $pV =$ constant. Since $p$ is proportional to $(nw + w_0)$, where $w$ is the weight added each time and $w_0$ is the weight of the plunger and platform, and $V$ is proportional to $r$, then

Fig. 7.25 Boyle's law using bicycle pump

$$(nw + w_0) \, r = \text{Constant.}$$

$$\therefore \quad nw + w_0 \quad = \frac{\text{Constant}}{r} = \text{Constant} \frac{1}{r} \quad . \quad . \quad (1)$$

A graph of $n$ against $1/r$ should thus be a straight line, and it does not pass through the origin on account of the presence of $w_0$ in (1).

## Use of Bourdon Gauge

The Bourdon gauge (see p. 259) provides a useful method of investigating Boyle's law.

The gauge, which was previously calibrated by loading a platform with known weights, so that the pressure was then known, reads pressure directly.

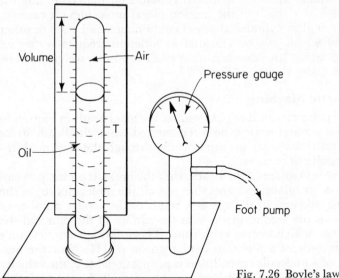

Fig. 7.26 Boyle's law by Bourdon gauge

In the apparatus air is trapped in a wide uniform tube T above oil, and pressure can be applied to it by means of a foot pump connected to the liquid reservoir at the base (Fig. 7.26). The volume of air at any pressure is read directly from a linear scale beside the tube. As before, the pressure values $p$ from the gauge can be plotted against $1/V$, where $V$ is the volume, or the product $pV$ can be calculated.

## Kinetic Theory of Gases

In Chapter 1 we have seen that many forms of matter are composed of small particles called molecules, and a simple experiment was performed to demonstrate the diffusion of bromine molecules into air.

The diffusion shows that gas molecules are in constant motion. Thus a gas in a container can be thought of as an enormous number of molecules, moving in all different directions. Sometimes they collide with each other, and at any given moment many strike the walls of the container. The time of a collision is extremely short. This concept of a gas is called the

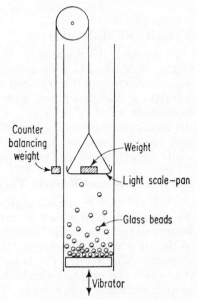

Fig. 7.27 Kinetic machine

*kinetic theory of gases*, and was developed mathematically by Clerk Maxwell over a hundred years ago.

A Leybold or similar kinetic machine is valuable for understanding the behaviour of gases as a result of the motion of gas molecules. It consists, essentially, of a tall glass cylindrical vessel containing small glass or other light spheres and which can be vibrated at different frequencies by an electrical vibrator at the lower end. A scale pan containing weights can be inserted into the vessel shown in Fig. 7.27.

## Use of the Kinetic Machine

With the scale pan in a suitable position, set the glass beads in motion to simulate gas pressure and notice the force exerted when the beads strike the pan. In the same way, gas pressure occurs through the continual impact of gas molecules on a surface.

Notice the random motion of the beads, that the impacts on the pan can be either 'head-on' or 'glancing', and that not all the molecules strike the pan at once. The pressure of the gas is a result of the *average* number of molecular collisions per second made with the surface of the pan and the momentum change which occurs each time. This follows from Newton's law for the magnitude of a force, as explained on p. 111. Now the momentum change of a molecule on collision is proportional to $c$, the velocity of a molecule. The number of collisions per second made by the molecule with the surface of a container is also proportional to $c$ – the faster the molecules move, the greater is the number of collisions per second with the surface. Consequently, as explained also on p. 112,

$$\text{pressure of gas} \propto c \times c \propto c^2$$

## Velocity of Molecules

In practice, $c^2$ is the average of the squares of all the individual velocities of the molecules. It is therefore called the 'mean-square' velocity. The measured pressure of a gas can thus lead to a value for the mean-square velocity of the molecules, and hence to a knowledge of the order of magnitude of that velocity called the *root-mean-square velocity* (not all the molecules necessarily have the same velocity). At room temperature and normal pressure, hydrogen molecules travel at about 6400 km/h while oxygen molecules travel at about 1600 km/h.

## Boyle's Law and Kinetic Theory

With a chosen frequency of vibration, place weights in the scale pan until no movement of the pan occurs. Note the distance measured from the bottom of the tube to the bottom of the pan (since the area of cross-section of the tube remains constant, this length is directly proportional to the volume). Now reduce the volume to half its original value, and notice that twice the original weight has now to be added to the scale pan to keep it stationary once more. The experiment tells us that, by halving the volume, the pressure of the beads has doubled. This is a demonstration of Boyle's law. The velocity of the beads remains the same when their volume is halved, but this time they make twice the number of impacts per second on the surface of the pan because they only have half as far to travel. The pressure is thus doubled.

### Pumps

Air pumps can be constructed either to compress or force air into a given space or to evacuate air from it.

A *bicycle pump* is a simple form of compression pump which has a flexible leather washer at the end of a piston inside the barrel (Fig. 7.28 (i)). It is connected to the tyre, which has a rubber valve in it. When the piston

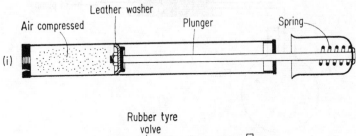

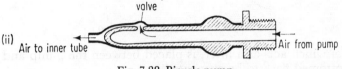

Fig. 7.28 Bicycle pump

is drawn out the air below it expands and the pressure is reduced below atmospheric pressure. Air from outside the pump then flows past the leather washer into the barrel beneath. At the same time the tyre valve remains closed. When the piston is pushed down the air compressed below it forces the leather washer against the sides of the barrel so that no air enters, and is itself forced into the tyre through the valve, which now opens. (Fig. 7.28 (ii)). In this way air from outside is forced into the tyre with each stroke of the piston.

As more air enters the tyre, the pressure inside increases. It then becomes more difficult to force air through the tyre valve, which opens only when the pressure of air in the barrel exceeds that in the tyre.

*Example.* The cylinder of a bicycle pump has a volume of 150 cm³ and is 50 cm long. Find the pressure of air in a tyre of volume 300 cm³ initially at 1 atmosphere, after 2 full strokes of the pump. If a further stroke is made, how far must the handle be pushed in from the top before air enters the tyre?

Each full stroke of the pump forces 150 cm³ of air at 1 atmosphere into the tyre. After 2 strokes, therefore, volume of air entering the tyre is 300 cm³ at 1 atmosphere. Hence total volume of air in tyre at 1 atmosphere = 300 + 300 = 600 cm³.

If $p$ is the actual air pressure in the tyre of volume 300 cm³, then, from Boyle's law,

$$p \times 300 = 1 \times 600$$
$$\therefore p = 2 \text{ atmospheres.}$$

When the handle is pushed down from the top of the cylinder, the air below at 1 atmosphere is compressed. The tyre valve is forced open when the pressure is just greater than 2 atmospheres. Consequently, the handle must be pushed half-way, or 25 cm from the top.

## Evacuation Pump

EXPERIMENT

A demonstration of an air evacuation pump can be made with the simple apparatus in Fig. 7.29. A short length of wide tubing, containing

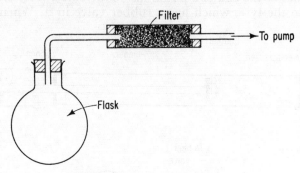

Fig. 7.29  Pump demonstration

glass wool which acts as a filter, is placed between the pump and a large round-bottomed flask to be evacuated.

Blow sufficient cigarette smoke into the flask so that this is visible, seal the flask with the bung, and switch on the pump (Fig. 7.30), which is a

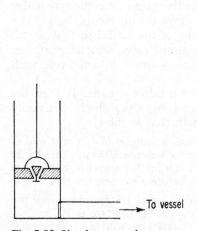

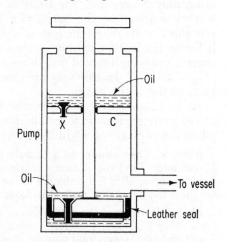

Fig. 7.30  Simple evacuation pump                    Fig. 7.31  Geryck pump

simple one due to von Guericke. Adjust the rate of pumping to be slow and observe how the smoke is cleared from the flask. Repeat the experiment with a modern pump such as a rotary pump, switching it on for a very short time, and observe the difference.

The simple pump used in Fig. 7.30 has two valves, a foot valve and a piston valve, and when the piston is raised the space in the cylinder below it decreases in pressure from Boyle's law. Sometimes it is said that a 'partial vacuum' is formed. The air in the flask then flows through the foot valve, which is forced open, into the cylinder. When the piston is pushed down the foot valve closes and the pressure of the air forces the piston valve open and flows away. In this way the flask is evacuated gradually.

A Geryck pump is a more modern evacuation pump (Fig. 7.31). When the piston reaches the bottom from C the space above the piston is a near-vacuum, since the valve X is closed. Air from the connected vessel therefore expands into the space above the piston. On raising the piston this air is forced out through X, which now opens. The layer of oil on top of the piston acts as a lubricant and as an air-tight seal between the piston and cylinder, and also ensures that no air is left between the top of the piston and C.

## Water Pumps

(a) The garden *syringe*, used for spraying fruit trees, has a piston moving in a cylinder and a narrow nozzle below (Fig. 7.32). When the latter is placed in a liquid and the piston is drawn back, the air below it reduces in pressure, from Boyle's law. Atmospheric pressure acting on the outside surface of the liquid then forces some of it into the cylinder. When the syringe is removed and the piston is pushed forward the liquid is forced out of the nozzle as a fine spray.

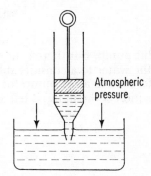

Fig. 7.32  Principle of syringe

(b) The *lift pump* can be used to obtain water from a well. It has a piston moved up and down in a cylinder by a handle outside, and a pipe passes down into the well from the base of the cylinder (Fig. 7.33). The piston has a valve in it which opens upwards, and a foot valve at the bottom of the cylinder also opens upwards.

When the pump handle is worked so that the piston rises in the cylinder the pressure of the air below the piston decreases. At some stage the pressure becomes lower than the air pressure in the pipe. The foot valve $V_1$ then opens, $V_2$ being closed, and the atmospheric pressure forces water into the pipe and the cylinder (Fig. 7.33). When the piston is lowered the valve $V_2$ opens, $V_1$ becomes closed and some water is forced above the piston. As the piston is raised, $V_2$ becomes closed, $V_1$ opens and the water is forced out of the spout. On the downstroke the sequence begins again. Generally, several strokes of the plunger are first necessary to set the pump in full action, because of the residual air in the pipe. It should be noted that the atmospheric pressure does the work of pushing the water from the well into the pipe, and since the atmospheric pressure supports a column of water about 10 m high, the distance from the surface of the water in the well to the valve $V_1$ should not exceed this height theoretically. In practice, however, this distance is less than 10 m, e.g. 7 m, because the valves and piston leak somewhat.

(c) The *force pump* has two valves which also open alternately as in the

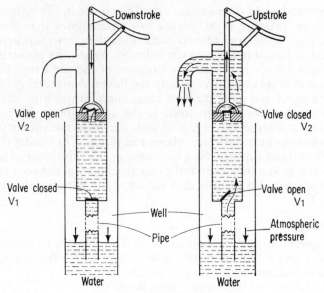

Fig. 7.33 Lift pump

lift pump, but there is no valve in the piston (Fig. 7.34). The piston forces the water to a considerable height, and the pump is therefore used in preference to a lift pump for raising water to heights greater than 8 m. The action of the pump is left as an exercise for the reader.

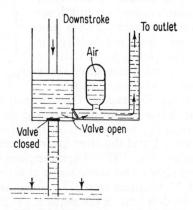

Fig. 7.34 Force pump principle

### SUMMARY

1. Pressure = force per unit area. Units are: newton per metre$^2$ (N/m$^2$).

2. Pressure in liquid = $h\rho g$, where $h$ is depth and $\rho$ is density.

3. Pressure is the same on the same horizontal level of a liquid and acts equally in all directions at a given place.

4. *Hydraulic press.* Velocity ratio = area of cross-section of large piston/area of cross-section of small piston.

5. Densities of liquids which do not mix can be compared with a U-tube – $\rho_1/\rho_2 = h_2/h_1$. If they mix, Hare's apparatus is used – $\rho_1/\rho_2 = h_2/h_1$.

6. Atmospheric pressure supports a column of mercury about 76 cm high. One atmosphere = 76 cm mercury pressure = $0{\cdot}76 \times 13\,600 \times 9{\cdot}8$ N/m² (approx.) = 101 300 N/m². 1 bar = $10^5$ N/m².

7. *Boyle's law* states: For a fixed mass of gas at a constant temperature, the volume is inversely proportional to the pressure, or $pV$ is a constant.

8. *Kinetic theory of gases.* The pressure of a gas is due to the continual bombardment by molecules of the walls of the vessel. The number of collisions per second is proportional to the velocity, and the momentum change at each collision is proportional to the velocity. Hence the average pressure is proportional to the average of the squares of the velocities of all the molecules, which individually may have different velocities.

## EXERCISE 7 · ANSWERS, p. 224

**1.** A rectangular solid block has sides $4 \times 10 \times 20$ cm and a density of 8 g/cm³ or 8000 kg/m³. If it rests on a horizontal flat surface, calculate the minimum and maximum pressure it can exert.

**2.** Describe how you would set up and use a simple barometer to measure the atmospheric pressure.

What are the objections to the use of water as the barometric liquid?

If the reading of a mercury barometer is 75·58 cm at the base of a mountain and 66·37 cm at the summit, what is the height of the mountain?

(Density of mercury = 13 600 kg/m³; average density of air = 1·25 kg/m³.) (*L.*)

**3.** (i) Draw a diagram illustrating the principle of the hydraulic (Bramah) press and explain its action. Calculate the pressure transmitted if a load of 1000 kgf is supported on a circular ram of diameter 1 metre. (Neglect friction.)

(ii) Describe carefully how you would use a siphon to remove the major part of the water contained in a large fixed tank, open at the top. Draw a diagram and use it to explain the action of the siphon. (*L.*)

**4.** Deduce an expression for the pressure due to a liquid of density $\rho$ at a point $h$ below its surface.

Describe the determination of the density of copper sulphate solution by a method which involves balancing liquid columns. Give the theory of the method.

A U-tube of uniform cross-section is partly filled with water. Oil of density 0·75 g/cm³ is poured into one arm until the surface of the water in that arm is depressed a distance of 4 cm. What is the length of the column of oil? (*L.*)

**5.** A uniform vertical tube, 40 cm long, sealed at the upper end, is lowered into mercury until the length of the enclosed air column is 35 cm. Find the depth of immersion of the tube if the atmospheric pressure is 77 cm of mercury. (*N.*)

**6.** With the aid of labelled diagrams explain the action of the following: (*a*) an aneroid barometer; (*b*) a simple exhaust pump.

**7.** 1 cm³ of air at atmospheric pressure, 76 cm of mercury, is introduced into a simple barometer and the level of the mercury in the barometer is lowered by 12 cm. What is:

(*a*) the pressure exerted by the air in the tube; (*b*) the volume of the space above the mercury now occupied by the air? (*N.*)

**8.** Describe an experiment which leads to the conclusion that the pressure exerted by a liquid is proportional to the depth of the liquid. (*N.*)

**9.** Describe a method, based on the principle of fluid pressure, for determining the density of a liquid which does not mix, nor react, with water.

A rectangular tank contains water to a depth of 1 m. If the base of the tank measures 2·4 m by 1·5 m, calculate (i) the pressure, (ii) the thrust, on the base. (1 m³ water has a mass of 1000 kg.) (*N.*)

**10.** Define *pressure*. What do you understand by the pressure at a point in fluid? Obtain a formula for the pressure due to a liquid column in terms of its vertical height and density. Describe the Fortin barometer and the procedure for reading it. (*O.* and *C.*)

**11.** State Boyle's law and describe an experiment to verify it.

Indicate the graphs that should be obtained if (*a*) $p$ were plotted against $1/v$, (*b*) $pv$ were plotted against $p$, where $p$ and $v$ refer to the pressures and volumes obtained from the experiment.

A uniform tube, 96 cm long, sealed at one end, is lowered vertically with its open end downwards into mercury until the length of the enclosed air column is 84 cm. Find the depth of immersion of the tube in the mercury, given that the atmospheric pressure at the time of the experiment is 77 cm of mercury. (*L.*)

**12.** Describe an experiment to determine how the volume of a fixed mass of air varies with its pressure, when the temperature remains constant. How would the results be affected by a rise of temperature in the room during the experiment?

A vertical capillary tube, closed at the lower end, contains a volume of air 21 cm long with a thread of mercury 5 cm long above it. When the tube is inverted the length of the air column increases to 24 cm. Find the atmospheric pressure. (*N.*)

**13.** State Boyle's law and describe an experiment to test it.

In setting up a mercury barometer, a uniform tube was filled only to a level 2·4 cm below the open end when the tube was vertical. It was then sealed, inverted and opened under mercury, in the usual manner, when a 'barometric' height of 66 cm was recorded. Why was this reading too low and what was the correct atmospheric pressure if the length of the tube above the mercury column in the faulty barometer was 20 cm? (*L.*)

**14.** A motor-car weighing 20 000 N rests on a level surface. The four tyres are each inflated to a pressure of $2 \times 10^5$ newtons/metre² above atmospheric pressure, which is $10^5$ newtons/metre². Assuming that the load is spread evenly over the four tyres, find the area of each tyre in *contact* with the ground.

Explain briefly how the molecules in a gas cause a pressure on the walls of the containing vessel.

Describe a simple form of pressure gauge and state how you would use it to measure the pressure in either the mains gas supply or an inflated rubber balloon. (*C.*)

**15.** Define *pressure* and explain what is meant by the pressure at a point in a liquid. How would you show that pressure at a point in a liquid is distributed equally in all directions? Describe some form of hydraulic appliance that makes use of this effect. In a small Bramah press the diameters of the effort plunger and the load plunger are 1 cm to 10 cm respectively. Find the effort needed to raise a load of 1500 N, and the pressure inside the apparatus while this is being done. (*O.* and *C.*)

**16.** Draw a diagram of a hydraulic press showing how an object, such as a bale of cotton, can be compressed. State the hydrostatic principle on which the press functions and describe an experiment which illustrates the principle.

The cylindrical piston of a hydraulic press has a diameter of 30 cm and the plunger a diameter of 5 cm. Calculate: (i) the upward thrust of the piston produced by a thrust of 250 N on the plunger; (ii) the distance moved by the piston when the plunger moves 60 cm. (*O.* and *C.*)

**17.** Define *pressure*. Explain the physical principle involved when a simple liquid manometer is used for the measurement of pressure, and show how the pressure in dynes per cm² may be deduced from the manometer reading.

Two vertical cylinders C and D containing water are joined at their bases by a horizontal tube in which there is a closed tap. The cross-sectional area of C is 5 cm² and the depth of water in it is 10 cm; the area of cross-section of D is 2 cm² and the depth of water in it is 20 cm. Calculate: (*a*) the pressure; (*b*) the total thrust, on the base of each cylinder. What is the final depth in each after the tap in the connecting tube has been opened? (*O.* and *C.*)

**18.** Describe a demonstration of *pressure* due to fast-moving small objects, as in a kinetic machine. Give an explanation of the pressure exerted by a gas on the walls of a containing vessel.

**19.** Using the kinetic theory of gases, show in a general way that the pressure of a gas depends on the mean square of the velocities of the molecules.

**20.** Show how Boyle's law can be explained on the kinetic theory of gases.

# 8

# FORCES DUE TO FLUIDS

## FLUIDS AT REST

### Upthrust of Fluid

If a cork is held below the surface of a liquid and then released it immediately rises to the surface. The cork has thus experienced an upward force or *upthrust* due to the liquid. An upthrust is always exerted on any object immersed in a liquid, whether wholly or partially submerged. Thus ships and submarines would sink if it were not for the upthrust of the sea, and we are able to float on water or swim because of the upthrust on our bodies by the liquid.

Water and other liquids are examples of *fluids*. A gas is a fluid. Liquids and gases transmit pressure and both exert pressure, and consequently both exert an upthrust on objects inside them. A floating balloon has an upthrust on it due to the pressure of air, which counterbalances the weight of the balloon. A liquid exerts a greater pressure than a gas because it is more dense (p. 24), and hence it exerts a much greater upthrust.

### Measurement of Upthrust

The upthrust of a liquid on an object can easily be measured on a small scale. Suppose, for example, that a solid brass object A is attached to a

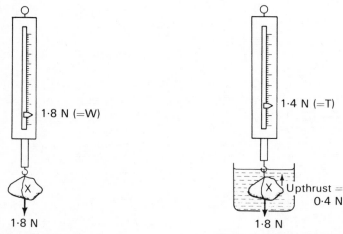

Fig. 8.1 Weighing in air          Fig. 8.2 Upthrust in liquid

spring-balance, which then registers 1·8 N, the natural weight of A (Fig. 8.1). If the latter is completely immersed in a liquid the tension $T$ in the string joining A to the spring-balance is less than 1·8 N by the magnitude of the upthrust of the liquid (Fig. 8.2). If the spring-balance registers 1·4

N, which is the magnitude of the tension $T$, the upthrust of the liquid must be 0·4 N, the difference between the weight and the tension.

The reading of the spring-balance when A is immersed in the liquid is sometimes called the 'apparent weight' of A in the liquid; the difference between the natural and 'apparent' weight of A is known as the 'apparent loss in weight'. The 'apparent weight', it should be noted, is the tension $T$ in the string in Fig. 8.2, and the 'apparent loss in weight' is the upthrust of the liquid.

### Investigation of Upthrust

We can use the spring-balance to investigate the factors which affect the upthrust due to a liquid.

EXPERIMENT

Suspend a metal cube (iron) by means of a piece of thread from a spring-balance and record its weight in air. Now immerse the cube completely in water contained in a beaker and again record the weight. You will notice that the cube apparently weighs less in water than in air, due to the upthrust.

Repeat the experiment using cubes of the same volume but made of different materials. Notice that the apparent loss in weight or upthrust is always the same. Repeat the experiment using an iron cube, but with twice the original volume. Notice that the apparent loss in weight or upthrust is twice the previous value. We conclude that the upthrust is dependent on the volume of the cube but independent of the material of which the cube is made.

Repeat the experiment with the original iron cube using paraffin or other suitable liquid instead of water. Observe that the upthrust on the cube is less than in water, but that once again the upthrust on different cubes of the same volume is constant when they are totally immersed in turn in the paraffin.

*Summarizing*, from all these experiments it follows that the upthrust depends only on: (i) the volume of the solid immersed; (ii) the nature of the liquid in which it is immersed. In a given liquid the upthrust does *not* depend on the nature of the solid immersed.

### Magnitude of Upthrust

We can investigate the magnitude of the upthrust due to a liquid by an extension of the experiment as follows. (Fig. 8.3)

EXPERIMENT

Suspend a large piece of metal A or other convenient substance from a spring-balance S by means of a thread T. When A is hanging freely note the reading on the spring-balance.

Prepare a suitable measuring cylinder C and fill it with water to a convenient graduation such as 50 cm³. Immerse A completely in the water so that it hangs vertically below S and does not touch the sides of C. Note the new reading on S and the new level of water in C.

*Results.* The following results were taken in an experiment:

$$\text{Weight of A in air} \quad = 1 \cdot 2 \text{ N}$$
$$\text{Weight of A in water} = 0 \cdot 9 \text{ N}$$
$$\text{Original level in C} \quad = \quad 50 \text{ cm}^3$$
$$\text{Final level in C} \quad = \quad 80 \text{ cm}^3$$

*Calculations.* Upthrust on A = weight in air − weight in water
$$= 1 \cdot 2 - 0 \cdot 9 = 0 \cdot 3 \text{ N}$$
Volume of water displaced by A = 80 − 50 = 30 cm³
∴ weight of liquid displaced = weight of 30 g = 0·3 N,
since weight of 1 g is 0·01 N in round figures (p. 100).

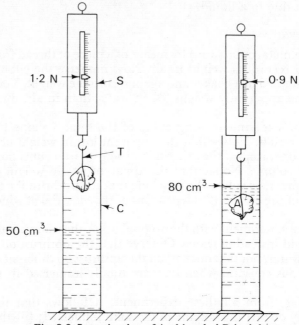

Fig. 8.3 Investigation of Archimedes' Principle

*Conclusion.* The upthrust on an object immersed in water is equal to the weight of water displaced.

Repeat with other liquids and see if this conclusion always holds.

## Archimedes' Principle

More than two thousand years ago Archimedes stated his famous principle or law concerning the magnitude of the upthrust on an object in a fluid. *Archimedes' Principle* states:

> *When a body is totally or partially immersed in a fluid, the upthrust on it is equal to the weight of fluid displaced.* A fluid is a liquid or a gas.

As examples of Archimedes' Principle, consider the following:

1. An iron cube of volume 800 cm³ is totally immersed in: (i) water; (ii) oil of density 0·8 g/cm³; (iii) oxygen gas of density 0·0015 g/cm³. Calculate the upthrust in each case if the weight of 1 g mass = 0·01 N.

Upthrust = Weight of fluid displaced.
(i)   ∴ Upthrust = weight of 800 cm³ of water = 800 × 1 × 0·01 N = 8 N
(ii)  ∴ Upthrust = weight of 800 cm³ of oil = 800 × 0·8 × 0·01 N = 6·4 N
(iii) ∴ Upthrust = weight of 800 cm³ of oxygen =
$$800 \times 0·0015 \times 0·01 \text{ N} = 0·012 \text{ N}$$

2. A metal cube weighs 1·0 N in air and 0·8 N when totally immersed in water. Calculate the volume of the cube and its density.

We have,                 Upthrust = 1·0 − 0·8 = 0·2 N
∴ 0·2 N = weight of water displaced
∴ mass of water displaced = 0·2/0·01 = 20 g
∴ Volume of water displaced = 20 cm³ = Volume of cube
$$\therefore \text{Density} = \frac{\text{Mass}}{\text{Volume}} = \frac{100\text{g}}{20 \text{ cm}^3} = 5 \text{ g/cm}^3$$

3. An iron cube, mass 480 g and density 8 g/cm³, is suspended, half-immersed in an oil of density of 0·9 g/cm³. Find the tension in the suspension.
$$\text{Volume of cube} = \frac{\text{Mass}}{\text{Density}} = \frac{480}{8} = 60 \text{ cm}^3$$
∴ Upthrust = Weight of 30 cm³ of oil = 30 × 0·9 × 0·01 N = 0·27 N
∴ Tension in string (apparent weight) = 4·8 − 0·27 = 4·53 N.

## Density Measurement

*Solid*

Archimedes' Principle enables the relative density or density of solids to be measured. As an illustration, suppose that a brass object A weighs 1·8 N in air, and that 1·6 N is the 'apparent weight' when A is completely immersed in water (see Fig. 8.3). The upthrust of the water is then 0·2 N, and, by Archimedes' Principle, this is the weight of water *displaced by* A.

So the mass of water displaced = 0·2/0·01 = 20 g

Since the density of water is 1 g/cm³, the volume of A is 20 cm³.
$$\therefore \text{Density of brass} = \frac{\text{Mass of A}}{\text{Volume of A}} = \frac{180 \text{ g}}{20 \text{ cm}^3} = 9 \text{ g/cm}^3$$

It should be noted that Archimedes' Principle enables the volume of a solid of *any* shape to be accurately determined. The solid is weighed in air $W$, then suspended by a thread attached to one end of a balance and completely immersed in water, and the tension $T$ in the thread, the 'apparent weight', is determined. The upthrust on the solid $= (W - T)$; and hence, by Archimedes' Principle, the volume of the water displaced by the solid is known as 1 cm³ of water has a mass of 1 g. This is the volume of the solid. The high degree of accuracy in determining volume by this method is due to the fact that weighing is a process which can be performed very accurately.

The *relative density* of a solid is the mass of the solid relative to the mass of an equal volume of water. Since the upthrust on a solid is equal to the weight of water it displaces, the relative density is obviously obtained from the ratio $\dfrac{\text{Weight of solid}}{\text{Upthrust}}$. If we use the above figures for the brass object A, the relative density of brass $= \dfrac{1·8 \text{ N}}{0·2 \text{ N}} = 9.$

*Liquid*

The relative density or density of a liquid can be measured by: (*a*) weighing a solid B of any shape in air, $W$; (*b*) then finding the tension $T$ in the thread supporting it when B is completely immersed in the liquid; (*c*) then determining the tension $T_2$ in the thread supporting it in water. The weight of liquid displaced by $B = W - T_1$, by Archimedes' Principle; and the weight of water displaced by $B = W - T_2$. As the volumes of the liquid and the water are the same, each corresponding to the volume of B,

$$\text{Relative density of the liquid} = \frac{\text{Weight of liquid}}{\text{Weight of equal volume of water}}$$
$$= \frac{W - T_1}{W - T_2}.$$

As an illustration, suppose the object B weighs 8·0 N in air, 6·0 N totally immersed in water and 6·4 N totally immersed in oil. Then

$$\text{Upthrust in water} = 2\cdot0\ \text{N} = \text{Weight of water displaced.}$$
$$\text{and Upthrust in oil} = 1\cdot6\ \text{N} = \text{Weight of oil displaced.}$$

But the volume of the water and the oil are both equal to the volume of B. Hence.

$$\text{Relative density of oil} = \frac{1\cdot6\ \text{N}}{2\cdot0\ \text{N}} = 0\cdot8$$

The mass of water displaced $= 2\cdot0/0\cdot01 = 200$ g. So volume displaced $=$ volume of oil $= 200$ cm³. The mass of oil $= 1\cdot6/0\cdot01 = 160$ g. Thus density $= 160/200 = 0\cdot8$ g/cm³.

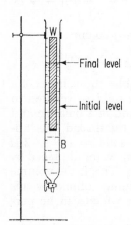

Fig. 8.4 Principle of Flotation

## Floating Objects

So far we have discussed the upthrust on objects totally immersed in liquids; *floating* objects, such as ships, are also subject to an upthrust. From Archimedes' Principle, the upthrust on a submarine floating on the surface is less than when it is totally submerged, since the volume of liquid displaced is then less.

We can investigate the upthrust on a floating object by weighing a narrow wooden rod W slightly weighted at one end. A burette B is now partly filled with water and the level of the liquid taken, after which the rod is placed in the burette so that it floats clear of the sides (Fig. 8.4). The new reading of the water level in B is taken.

The following measurements were obtained in an experiment:

Weight of wood = 0·6 N
Initial reading in burette = 25·0 cm³
Final reading in burette = 85·0 cm³
∴ Volume of water displaced by wood = 60·0 cm³
∴ Weight of water displaced = 60 × 0·01 N = 0·6 N

It is therefore clear that the weight of the wood is equal to the weight of liquid displaced. This experimental result follows from simple reasoning. Any floating object is in equilibrium under two forces: (i) its weight; (ii) the upthrust of the liquid. *Hence the weight of liquid displaced (the upthrust) is equal to the weight of the object.* This is called the *Principle of Flotation.* Some floating objects are shown in Fig. 8.5.

## Calculations on Floating Objects

Suppose the volume of a block of wood floating in water is 50 cm³ and the density of the wood is 0·8 g/cm³; the weight = $50 \times 0·8 \times 0·01 =$ 0·4 N. The upthrust, which is equal to the weight, is hence 0·4 N. But the upthrust is equal to the weight of water displaced, by Archimedes' Principle. So the block displaces $0·4/0·01 = 40$ cm³ of water (Fig. 8.5).

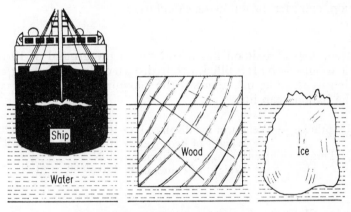

Fig. 8.5 Floating objects

It should be noted that the fraction of the volume of the block immersed is $\dfrac{40 \text{ cm}^3}{50 \text{ cm}^3}$ or $\frac{4}{5}$, which is the relative density of the wood. *Ice* has a density of about 0·9 g/cm³, and, by similar reasoning, it can be seen that ice floats in water with about $\frac{9}{10}$ of its volume submerged (Fig. 8.5).

Consider a block of iron of volume 30 cm³ and density 6 g/cm³; its weight = $180 \times 0·01 = 1·8$ N. If the block is placed in mercury of density 13·6 g/cm³ and floats with a volume $V$ cm³ immersed, the upthrust is $13·6V \times 0·01$ N by Archimedes' Principle.

$$\therefore 13·6V \times 0·01 = \text{weight of iron} = 1·8$$

$$\therefore V = \frac{1·8}{0·136} = 13·2 \text{ cm}^3.$$

Suppose now that the iron block is held below water so that it is completely submerged; the upthrust of the water is then 0·3 N, since the volume of water displaced is 30 cm³. Now this is less than the weight, 1·8 N, of the iron; consequently, the iron *sinks* when it is not supported.

It can thus be seen that, generally, *a solid floats in a liquid if its density is less than that of the liquid.* Mercury is a dense liquid, and hence, as shown above in the case of iron, dense objects float in mercury. The density of the Dead Sea is very high owing to the concentration of salts in it, and swimmers float very easily in it.

## Balloons

Balloons used in weather soundings in the upper atmosphere or in cosmic-ray investigations are filled at ground level with a small amount of a light gas such as helium or hydrogen and so the balloon rises into the air when released. As the balloon rises the gas inside it expands. At some height where the density of the air is much less than at ground level, perhaps as high as 30–40 km, the upthrust is equal to the total weight of the balloon. The latter then ceases to rise and drifts sideways with the wind. As an illustration, consider the following calculation.

*Example*

A meteorological balloon has a volume on the ground of 10 m³ and the fabric weighs 80 N. It is filled with hydrogen of density 0·09 kg/m³, and the air density is 1·2 kg/m³. Estimate the maximum weight of the contents which the balloon can lift.

$$\text{Upthrust} = \text{Weight of air displaced} = 10 \times 1·2 \times 10 = 120 \text{ N}$$
$$\text{Weight of hydrogen} = 10 \times 0·09 \times 10 = 9 \text{ N}$$
$$\therefore \text{Total weight of hydrogen} + \text{fabric} = 89 \text{ N}$$
$$\therefore \text{Weight of contents} = 120 - 89 = 31 \text{ N}$$

## Why Ships Float

At first sight it may not be obvious why ships float in water, because they are made from iron and steel, which are substances denser than water. A ship, however, is not a *solid* block of these metals, in which case it would sink, but is more like a closed *hollow* object, having the exterior surface lined with iron and steel. When such an object is placed in water it sinks to a level such that the weight of water displaced by it is equal to its weight. Thus, if all the iron and steel were melted down its weight would be equal to the weight of water displaced.

The weight of water displaced by a ship is equal to the ship's weight. It is therefore common to refer to the size of a ship as '8000 tonnes displacement', for example, which means that this is the weight of water displaced (1 tonne = 1000 kg). Without a cargo, a ship stands more out of the water, since the weight of water displaced is less.

*Example*

A barge, 40 metre long and 8 metre broad, whose sides are vertical, floats partially loaded in water. If 125 000 N of cargo are added, how many centimetres will it sink?

The increased upthrust of the water $= 125\ 000$ N

$\therefore$ Increase in volume of water displaced $= \dfrac{\text{mass}}{\text{density}} = \dfrac{12\ 500\text{ kg}}{1000\text{ kg/m}^3} = 12.5$ m³

$$\therefore \text{ Distance barge sinks} = \frac{\text{Volume of water}}{\text{Area of water}}$$

$$= \frac{12.5\text{ m}^3}{40 \times 8\text{m}^2}$$

$$= 0.04\text{ m}$$

$$= 4\text{ cm.}$$

## Plimsoll Line

A ship sinks in water until its weight is equal to the upthrust on it, which is the weight of water displaced. Now the density (or specific gravity) of sea-water varies in different parts of the world, according to whether the climate is hot or cold, and hence a ship sinks to different levels if its journey takes it to different regions. When a ship sinks too low in water it is unsafe in heavy seas, and the danger increases if the cargo carried is very heavy. Accordingly, in the 19th century, PLIMSOLL agitated for a safety line to be drawn clearly on the side of a

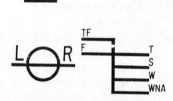

Fig. 8.6 Plimsoll line

ship, and it is now illegal for a ship to be loaded so that the Plimsoll line, as it is called, is below the water level. Fig. 8.6 illustrates markings of the Plimsoll line; the letters LR stand for Lloyd's Register of Shipping, TF for tropical fresh water, F for fresh water, T for tropics, S for summer, W for winter, WNA for winter in the North Atlantic.

## Submarine

The submarine is another application of Archimedes' Principle. In surface trim (Fig. 8.7) the submarine floats with its conning tower and

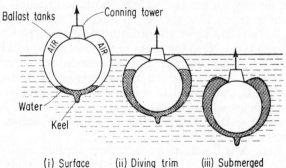

(i) Surface     (ii) Diving trim     (iii) Submerged

Fig. 8.7 Submarine diving

most of the deck clear of the water. The boat is provided with large ballast tanks which may be filled with water, thus increasing the weight so that it sinks lower. Horizontal rudders are tilted to make the submarine dive downwards when the tanks are practically full and the boat is ready to submerge. When the submarine is ready to surface the rudders are moved to drive the boat upwards, and compressed air is forced into the ballast tanks to drive the water out so that the submarine can rise again.

### Floating Dock

A floating dock is merely a very large raft with tall sides capable of supporting a ship. The dock contains many tanks which, when flooded, cause the dock to sink low in the water. In this position the ship is brought into a suitable position on the dock and moored. Then powerful pumps in the dock are employed to empty the tanks, and the whole dock rises and lifts the ship clear of the water so that engineers can perform their work on the hull.

### The Hydrometer

Hydrometers are instruments which measure directly the density or relative density of a liquid. They are used, for instance, by inspectors to

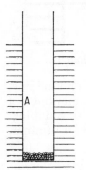

Fig. 8.8 Model hydrometer

test the richness, or otherwise, of milk, which is intimately related to its relative density; the instrument is then given the name of *lactometer*. The strength of spirits must also conform to a certain standard, which is checked by inspectors with the aid of a hydrometer. The instrument is also used to test the relative density of the acid in accumulators, as this gives an excellent guide to the general condition of the accumulator (see p. 514). The relative density of the water in a ship's dock must be entered in the log book by the captain.

A *simple hydrometer* can be made from a test-tube A having its end flattened (Fig. 8.8).

In order to keep it upright when it is placed in a liquid, some sand or lead shot is placed inside A. Suppose the weight of A and its contents is 0·1 N and it is placed upright in water, density 1 g per cm³. Then, from the Principle of Flotation, it sinks to such a depth that its weight is equal to the upthrust of the liquid, which is equal to the weight of liquid displaced. Hence the weight of water displaced is 0·1 N, so the mass = 0·1/0·01 = 10 g. The volume of water displaced is thus 10 cm³. Suppose the uniform cross-sectional area of A is 1 cm². Then the depth to which it is submerged is 10 cm.

Now suppose A is transferred to a liquid of much greater density, say 2 g/cm³. The upthrust when A floats = weight of A = 0·1 N. Hence the volume submerged = 0·1/(0·01 × 2) = 5 cm³. A thus stands more out of the liquid than when in water.

Generally then, A will sink more in lighter liquids and less in denser liquids. A *model hydrometer* can be made, therefore, by placing A in liquids of known density, and marking on a paper scale inside it the values of the density on a level with the liquid surface outside. In this simple hydrometer of mass 10 g the depth $h$ immersed in a liquid of density $\rho$ is given by

10 = Mass of liquid = density × volume immersed = $\rho \times h \times 1$, assuming an area of cross section of 1 cm². Thus

$$h = \frac{10}{\rho},$$

or the depth immersed is inversely proportional to the density. The graduation on the scales between two marked densities on the paper scale are thus not regular, that is, the graduation for a density of 1·5 g/cm³ is not mid-way between the graduations for densities of 1 and 2 g/cm³.

### Practical Hydrometer

One form of practical hydrometer is shown in Fig. 8.9. It consists of: (i) a hollow narrow glass tube or stem T; (ii) a paper scale inside T graduated in densities; (iii) a wide bulb B; (iv) a loaded end S containing lead shot, for example, to keep it upright in liquids (p. 148).

The densities which can be measured depend largely on the volume of the bulb B. For a given hydrometer weight the bulb must sink so that the liquid surface is on the stem. The narrower the stem, the more sensitive is the hydrometer–a small change in volume when the hydrometer is taken from one liquid to another of slightly differing density then produces a large vertical movement along the stem. The density scale on the stem has unequal divisions, and the density values increase downward.

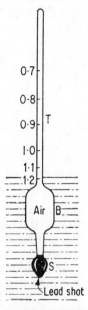

Fig. 8.9 Practical hydrometer

*Example*

A cylindrical tube of uniform section, sealed flat at its lower end and containing lead shot, floats upright in water with 2 cm of its length projecting above the free surface of the liquid. When the water is replaced by a liquid of relative density 0·9 the length projecting above the surface is reduced to 1 cm. Determine the total length of the cylinder. (L)

Suppose h cm is the cylinder length and A cm² is the cross-section area. From the principle of flotation, upthrust = weight of cylinder, W.

Hence for water, $\rho = 1$ g/cm³,       $(h - 2) \times 1 \times A \times 0{\cdot}01$ N = W.
and for liquid, $\rho = 0{\cdot}9$ g/cm³,       $(h - 1) \times 0{\cdot}9 \times A \times 0{\cdot}01$ N = W.
$$\therefore h - 2 = (h - 1) \times 0{\cdot}9$$
$$\therefore 0{\cdot}1h = 1{\cdot}1, \text{ or } h = 11 \text{ cm.}$$

## FLUIDS IN MOTION

In the previous part of the chapter we discussed fluids at rest. In this section fluids in motion are briefly discussed. It is an important subject, because it concerns aeroplane flight, for example, or the motion of a submarine through water.

### Velocity · Streamlines

You will probably have noticed that a stream or river flows slowly when it runs through open country, but flows more rapidly when it comes to narrow openings or constrictions. This is due to the fact that water is practically an incompressible fluid, as now explained.

Fig. 8.10 shows water flowing steadily along a pipe XY, where a part X has a bigger cross-sectional area $A_1$ than the narrower part Y, of cross-sectional area $A_2$. The flow or *streamlines* of the motion represent the direction of the velocities of the particles of the fluid. Since the liquid is incompressible, the volume of liquid moving from P to Q must equal the volume moving from R to S in the same time. Thus $A_1 l_1 = A_2 l_2$, where $l_1$ is PQ and $l_2$ is RS, or $l_2$ is greater than $l_1$. Consequently, the *velocity* of

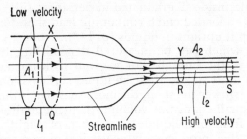

Fig. 8.10 Streamlines and velocity in uniform flow

the liquid at the narrow part or constriction Y of the pipe is greater than that at the wider part X. For the same reason, slow-running water from a tap can be made into a fast jet by placing a finger over the tap to narrow the exit.

It should be carefully noted that the streamlines of the flow along the pipe are closer together in a region such as Y, where the velocity is high, and farther apart in a region such as X, where the velocity is lower.

### Pressure and Velocity

A fluid needs a difference in pressure between its ends to keep it moving steadily. Every part of the fluid, then, has a particular pressure and a particular velocity.

The variation of pressure with velocity of a fluid in motion can be demonstrated with the apparatus shown in Fig. 8.11. It consists of a horizontal tube with a narrower section XY in the middle and six vertical tubes for indicating the pressure at different sections. One end is connected to a 'constant head' apparatus to maintain a steady flow of liquid, water in this case. The different heights of water in the vertical tubes then appear as shown. Observe that if XY had the same diameter as the first part of the horizontal tube, the level at X would have reached P. With the increased velocity of water in XY, however, the level at C is lower than P, showing a *decrease* in pressure.

Similarly, when the water emerges from Y with reduced velocity in the wider tube, the level stands higher at E than it would have been if no change in diameter had occurred. Thus *a reduction in velocity leads to an increase in pressure.*

If the glass apparatus in Fig. 8.11 is inverted and the ends of the vertical tubes are now placed dipping into a trough of water, air can be drawn steadily through the horizontal tube, which is now at the top of the appara-

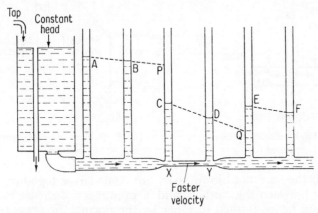

Fig. 8.11 Demonstration of Bernoulli's principle

tus. The water from the trough then rises up the vertical tubes and the variation of the pressure of the air with velocity can now be examined.

## Bernoulli's Principle

Bernoulli, about 1740, was the first person to obtain a relation between the pressure and velocity at different parts of a moving incompressible fluid. If the viscosity of the fluid is negligibly small, which we assume, there are no frictional forces to overcome (p. 44). In this case, considering flow in a horizontal direction, the work done on a given volume of fluid is equal to the gain in its kinetic energy. This is an application, of course, of the principle of the conservation of energy.

Suppose the fluid flows from a region A where the pressure is high to a region C where the pressure is lower. Then, since a given volume gains kinetic energy, the fluid must flow faster at C than at A. Bernoulli stated the relation between the pressure and energy in a general form known as 'Bernoulli's Principle'. It shows, as previously demonstrated, that *at points in a moving fluid where the velocity is high the pressure is low, and where the velocity is low the pressure is high.* The principle has wide applications, as now discussed, and applies only to non-viscous fluids in steady flow.

## Illustrations of Bernoulli's Principle

(*a*) Make a hole in the middle of a piece of cardboard and insert a short hollow tube in it, and then hold a piece of paper XY close to the hole (Fig. 8.12). Blow hard down the tube so that a fast stream of air passes between the card and paper as shown, and release the paper.

Observe that the latter keeps near the card and is not blown away. Explain the phenomenon by using Bernoulli's principle.

(*b*) Suspend two pieces of card X and Y (Fig. 8.13) about 7 cm apart by means of threads. Blow a blast of air from a nozzle N between the cards.

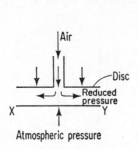

Fig. 8.12 High velocity, pressure reduced

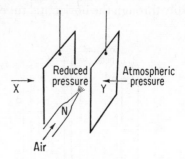

Fig. 8.13 High velocity, pressure reduced

Notice that they are not blown apart but come closer together. The fast-moving air between the cards X and Y produces a decrease in pressure, and they are hence pushed together by the normal air pressure outside. The same 'suction' effect is experienced by a person standing close to the platform at a station when a fast train passes.

## Some Applications of Bernoulli's Principle

### (i) *Filter Pump*

A filter pump is used to draw air through an apparatus (see p. 302). The principle is shown in Fig. 8.14. The pump has a constriction in the centre, so that when it is connected to a water-tap a jet of water flows through a tube of decreasing cross-section. The high velocity of water here sweeps the air away quickly, and thus produces a drop in air pressure. This causes air to flow in from the side tube, and together with the water it is expelled through the lower part of the pump, as shown.

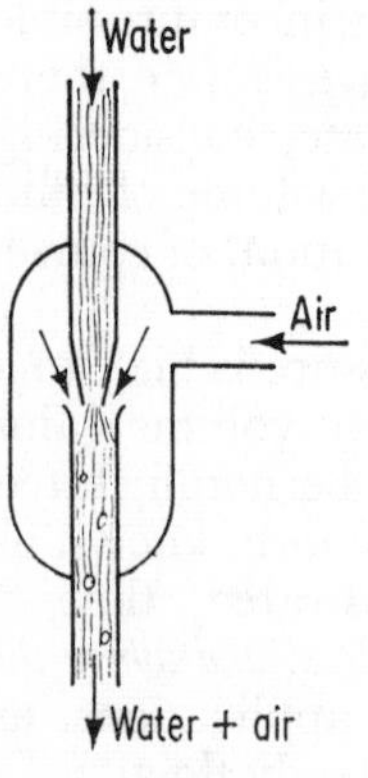

Fig. 8.14 Filter pump action

### (ii) *Carburettor*

The principle of the petrol-engine carburettor used in cars is shown in Fig. 8.15. A float mechanism maintains the petrol level just below the top of a fine jet. The jet is in a constriction known sometimes as a 'venturi', and when the engine is running, air is swept past the jet swiftly during the intake stroke (see p. 276). This causes a drop in pressure at the venturi, from Bernoulli's Principle. The greater atmospheric pressure on the petrol in the reservoir then forces petrol out of the jet in the form of a fine spray, where it mixes with the air and passes into the engine cylinder. This occurs when the accelerator pedal is pressed down. The *choke* control in cars increases the speed of the air flow when pulled out, so that the mixture passing to the engine cylinder becomes richer in petrol, and the car can then be started more easily when cold.

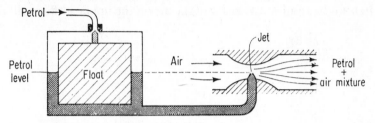

Fig. 8.15 Carburettor action

(iii) *Scent Spray*

A ladies' scent spray works on the same principle as the carburettor. When the bulb is squeezed a stream of air is forced at high speed past the fine nozzle of a tube which passes down into the scent bottle (Fig. 8.16). The air pressure outside the tube is then decreased, from Bernoulli's

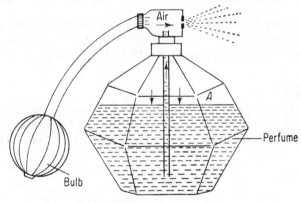

Fig. 8.16 Scent spray

Principle. The atmospheric pressure $A$ on the oil now forces it up, and it emerges as a fine spray mixed with the air.

**Stability and Spin**

A light ball such as a table-tennis ball has considerable stability when supported by an upward blast of air from a vacuum cleaner, for example. If it moves to one side it quickly returns to the air-stream. This is explained by Bernoulli's Principle. Thus suppose the ball moves slightly to one side (Fig. 8.17). The velocity of the air on one side of the ball is then greater than on the other, and the resulting pressure difference causes the ball to return to the air-stream.

A table-tennis player often gives the ball 'top spin', enabling a hard drive to be hit over the net and land on the table. A ball hit with spin swerves considerably in its path. If the ball moves from left to right through still air it is equivalent to keeping the ball stationary and moving the air with an equal velocity to the left. Fig. 8.18 shows the rotating surface of the ball, due to spin, and $V$ the air velocity. The spin of the ball will carry some air with it, and the resultant velocity of the air on the upper surface will there-

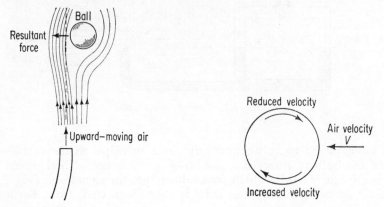

Fig. 8.17 Ball supported on jet          Fig. 8.18 Effect of spin on ball

fore be less than $V$, and on the lower surface greater than $V$. There is therefore a pressure difference between the upper and lower surfaces which deflects the ball, originally hit from left to right, in a downwards direction.

### Aerofoil Lift

The difference in pressure on the upper and lower surfaces of an aerofoil wing can be investigated in a wind tunnel. Fig. 8.19 illustrates the result. A manometer shows that when air flows over the surfaces the pressure

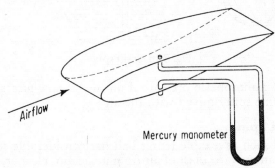

Fig. 8.19 Pressure on aerofoil surface

acting on the lower surface is then greater than the pressure on the upper surface. The streamlines are shown in a photograph on Plate 5, taken at the National Physical Laboratory, Teddington.

As the air travels past the wing it is deflected round the surface. The path of the air is therefore greater than before, and to travel this greater distance the air must increase its speed. By Bernouilli's Principle this means a reduction in pressure. Fig. 8.20 shows the streamlines round a thin wing at an angle of attack to the air stream. It will be seen that the path of the air is longer over the upper surface than the lower, and therefore

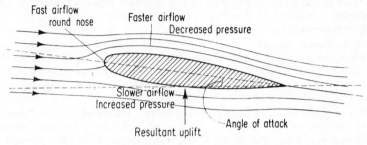

Fig. 8.20 Uplift on aerofoil

the speed is greater on the upper surface. The lift is produced by the difference in pressure between the two surfaces. Increasing the 'angle of attack' or inclination of the aerofoil gives increased lift.

## SUMMARY

1. *Archimedes' Principle states:* The upthrust on an object immersed in a fluid is equal to the weight of fluid displaced.

2. Relative density of a solid = wt of solid/apparent loss in weight in water. Relative density of a liquid = apparent loss in weight of a solid in liquid/apparent loss in weight of same solid in water.

3. Principle of flotation: When an object floats, its weight is equal to the weight of fluid displaced.

4. Practical hydrometers have a narrow stem with a scale graduated in density values which increase down the stem, a wide bulb and a loaded end.

5. *Bernoulli's Principle* shows that the pressure at a point in a non-viscous fluid in steady flow is greatest where the velocity is least, and vice-versa. Owing to the shape of an aerofoil, the streamlines flow faster above it than below, and a lift is hence obtained by the difference in pressure.

## EXERCISE 8 · ANSWERS, p. 224

(Assume the weight of 1 kg mass = 10 N, the weight of 1 g mass = 0·01 N.)

**1.** A cube of wood of side 10 cm floats in water with 4·5 cm of its depth below the surface, and with its sides vertical. What is the density of the wood? (*N.*)

**2.** A solid of weight 6 N is suspended by a string in water. The tension in the string is 4·5 N. Find (i) the volume of the solid; (ii) its density; (iii) its relative density.

**3.** State *Archimedes' Principle* and define *relative density.*
Describe how the relative density of a solid substance may be measured accurately by the application of Archimedes' Principle. Show how you obtain the expression used to calculate the result.
Calculate the volume of a metal cylinder of mass 105 g which has an apparent weight

of 0·925 N when fully immersed in water of density 1·00 g/cm³. When a piece of wax of mass 20 g is attached to the cylinder and both are fully immersed in water the total apparent weight is found to be 0·875 N. What is the density of the wax? (*L.*)

**4.** State Archimedes' Principle and describe an experiment to verify it.

A uniform closed glass tube, weighted internally at one end, has a mass of 50 g. When it floats vertically in water 10 cm of its stem is exposed. When it floats vertically in a liquid of relative density 1·25, 20 cm of its stem is exposed.

Calculate the area of cross-section of the tube. (*N.*)

**5.** State *Archimedes' Principle*.

A weather forecasting balloon of volume 10 m³, contains hydrogen of density 0·090 g per litre and its fabric weighs 65 N. What is the weight of the equipment it carries if it is floating in air of density 1·25 g per litre? Assume that the volume of the equipment is negligible compared with that of the balloon. (*L.*)

**6.** A corked bottle weighs 142 g and has an external volume of 170 cm³. Calculate the volume of a lump of metal of density 8·1 g/cm³ which, put inside the bottle, will cause the bottle just to sink in a liquid of density 1·10 g/cm³. Explain the fact that the same lump of metal attached to the outside of the bottle would not be sufficient to sink the bottle in the liquid. (*C.*)

**7.** Describe, with the help of a labelled diagram, a common hydrometer graduated to record a density range of 0·80 to 1·00 g/cm³. Give reasons for the particular features of its design and explain why it floats at different depths in liquids of different densities. (*L.*)

**8.** What condition must be satisfied for a body to float in a liquid?

A uniform glass tube of cross-sectional area 2·25 cm², closed at its lower end, is suitably weighted to float vertically in a liquid. If the length immersed in water is 15·2 cm, what is the length immersed in a liquid of density 1·16 g/cm³ and what is the weight of the tube and contents? (*L.*)

**9.** How would you use the Principle of Archimedes: (*a*) to explain the upward motion of a hydrogen balloon in air; (*b*) to find the relative density of a cork?

A hollow sphere made of glass floats in water with seven-eighths of its volume submerged, and in dilute acid with three-quarters of its volume submerged. What is the specific gravity of the acid? (*N.*)

**10.** Distinguish between the density and the relative density of a material.

Describe how you would use a density (specific gravity) bottle to determine the density of sand.

A wooden block floats in water with two-thirds of its volume submerged. When placed in oil three-quarters of its volume becomes submerged. Calculate the density of: (i) the wood, and (ii) the oil. (*N.*)

**11.** Give an account of a simple form of hydrometer. What are the advantages and disadvantages of the hydrometer method of finding the relative density of a liquid? Find the mass of cork that must be attached to a solid cube of side 3 cm made of aluminium in order that the combination may be on the point of sinking in water. (Take the relative density of cork to be 0·25 and that of aluminium to be 2·5.) (*O. and C.*)

**12.** State the Principle of Archimedes, and describe the experiment you would perform to verify it. Explain how this principle applies to the ascent of a balloon. A hollow cylindrical tube, with a flat bottom, has an area of cross-section 3 cm² and a mass of 24 g. What mass must be put inside it in order to make it float with its axis vertical and with 16 cm of its length immersed in water? What length is immersed if it is transferred to brine of relative density 1·2? What extra mass would have to be added to make it float with 16 cm of its length immersed in the brine? (*O. and C.*)

**13.** State the Principle of Archimedes, and describe how you would apply it to find the specific gravity of a solid that sinks in water. A closed hydrogen balloon of volume

500 m³ is tethered to the ground by a vertical cable which exerts a tension of 4000 N. Find the mass of the balloon fabric and attachments, given that the density of air is 1·29 kg/m³, and the density of hydrogen 0·09 kg/m³. (*O. and C.*)

**14.** Explain carefully why a piece of wood floats on water with only part of its volume submerged.

Describe the simple hydrometer and show how this instrument is used to find the specific gravity of a liquid.

Calculate the relative density of the material of a piece of metal that weighs 24 g when totally immersed in water and 20 g when totally immersed in a liquid of relative density 1·2. (*O. and C.*)

**15.** What is *Bernoulli's Principle*? To what kinds of fluid does it apply? Describe a demonstration of the principle.

**16.** Why does a ping-pong ball stay on top of a jet of water? Explain how Bernoulli's Principle is involved.

**17.** Draw a sketch of the streamlines round an aerofoil and explain how lift is produced.

**18.** State the principle of Archimedes. Explain how a submarine is made (i) to sink to a point below the surface of the water, (ii) to rise again to the surface. How is the weight related to the upthrust on it in both cases?

A floating crane weighs 206 000 N and floats in sea water of density 1·03 g/cm³ (1030 kg/m³). If the base of the crane is a rectangular block 4 m square and 2·5 m deep, what is the maximum load which can be lifted by the crane when the top of the base is just awash with water? (*N.*)

## ANSWERS TO NUMERICAL EXERCISES

### EXERCISE 3 (p. 90)

**1.** 100 m  **2.** 25 m/s, 262½ m  **3.** 10 m/s², 20 m  **4.** 2, 4 s  **5.** (a) 2 s (b) 30 m
**7.** 1275 m, 3 m/s²  **9.** 215 km/h, N 22°W  **10.** 32·4 km/h, W 19°S
**12.** AB −50 cm/s, CD −167 cm/s; 58 cm/s²  **13.** (i) 25 cm/s² (ii) 105 cm
**14.** 4/3 m/s², 6/11 m/s², 27$\frac{3}{11}$ m  **15.** 10 m/s or 36 km/h  **16.** 6 km  **18.** 3 s
**19.** 1031 cm/s  **20.** 20 m  **22.** 10 km/h 60°E of S  **23.** 200 m/s  **24.** 267 m, 80 s
**25.** 40 km/h  **26.** 45 m, 30 m/s  **27.** 60 m/s, 135 m

### EXERCISE 4 (p. 129)

**1.** (*a*) (i) force (ii) energy (iii) power (*b*) force  **3.** 1000 N
**4.** 0·6 m/s², 67·5 m  **5.** 480 J  **6.** 1440 J  **7.** 2500, 2500 J  **8.** 0·006 J, 0·3 J
**9.** 171 N (i) 161 N (ii) 3·2 m/s²  **10.** 7600 W; 500 N
**11.** 1200 J, 6 × 10⁴N  **12.** 400 g  **13.** 6·7 cm  **16.** 0·2 N s, 0·4 N  **17.** 2*mc*
**19.** 4·6 m  **20.** (*a*) 2 N s (*b*) 4 J  **21.** 240 N. (i) 62 400 J (ii) 4·2 hp  **22.** 1·4 N s, 1 J
**23.** (*a*) 13 300 N s (*b*) 89 000 J, 333 N  **24.** (i) 3·9 s (ii) 38·7 m/s (iii) 10⁴N
**26.** 70·3 J  **27.** 160 N, 1$\frac{1}{15}$ m/s²  **28.** 889,889 hp  **29.** 20 J, 2 N s (*a*) 20 N (*b*) 1/10 s
**30.** 48 000 J  **31.** 33$\frac{1}{3}$ kg/s  **32.** 2·94 N  **33.** 78%

### EXERCISE 5 (p. 149)

**1.** 0·625, 0·875 N  **2.** (*a*) 45 g (*b*) 0·75 N  **5.** 40 N  **6.** 40 N  **7.** 50 N, 35 J
**8.** 2·9 cm from end of shorter rod  **10.** 100 cm from A, 60 N  **12.** 116·7 N, 93·3 N
**13.** $\frac{4}{7}$ cm from centre  **14.** 1 kg  **15.** 43 cm from A; 30 N  **16.** 4800 km  **17.** 24·5 g
**18.** 2·75 m  **19.** 9·1 cm from edge

## EXERCISE 6 (p. 172)

**1.** (a) 67% (b) 2400 J  **2.** 750 N, 67%  **4.** (a) 4·8 (b) 80%  **5.** (a) 40% (b) 7500 J
**6.** (i) 280 N (ii) 71%  **7.** (i) 81 000 J (ii) 90 000 J (iii) 833 N (iv) 1¼ h.p.
**8.** (i) 62·5% (ii) 600 J  **9.** 754 N  **10.** 125 664 N  **11.** (a) 187·5 N (b) 1125 J
**12.** 40, 64, 70, 73, 75%  **13.** (a) 444$\frac{4}{9}$ N (b) 2666$\frac{2}{3}$ J  **14.** 70, 176, 40%  **15.** 120 N
**16.** 50 N, 37° N of E, 3 m/s²  **18.** 133 N  **19.** 1040, 600 N  **20.** 37° to bank, 2 m 15 s
**21.** (i) 5·2 N (ii) 2·0 N  **22.** 2·83 N  **23.** 9·5° W of N, 296 m  **24.** (a) 390 N (b) 225 N
**25.** 5·2 N, 10·4 N  **26.** 10 N  **27.** 225 N, 390 N  **28.** 68 N, 192 N

## EXERCISE 7 (p. 203)

**1.** 0·32, 1·6 N/cm² or 3200, 16 000 N/m²  **2.** 1000 m  **3.** (i) 12 740 N/m²  **4.** 10·7 cm
**5.** 16 cm  **7.** (a) 12 cm mercury (b) 6·3 cm³  **9.** (i) 10 000 N/m² (ii) 36 000 N  **11.** 23 cm
**12.** 75 cm  **13.** 75 cm  **14.** 0·017 m²  **15.** 15 N, 19 N/cm²  **16.** (i) 9000 N (ii) 1$\frac{2}{3}$ cm
**17.** (a) 1000, 2000 N/m² (b) 0·5, 0·4 N 12·9 cm common level

## EXERCISE 8 (p. 221)

**1.** 0·45 g/cm³  **2.** (i) 150 cm³ (ii) 4 g/cm³ (iii) 4  **3.** 12·5 cm³, 0·8 g/cm³  **4.** 1 cm²
**5.** 51 N  **6.** 5·6 cm³  **8.** 13·1 cm, 0·342 N  **9.** 1·17  **10.** (i) 0·67 (ii) 0·8 g/cm³
**11.** 13·5 g  **12.** 24 g, 13·3 cm, 9·6 g  **13.** 200 kg  **14.** 2·2  **18.** 206 000 N

# Heat

# 9

# TEMPERATURE AND THERMOMETERS

## Temperature

If a thimbleful of hot water is poured into a cup of warm water the latter becomes slightly warmer. *Heat* thus passes from the hot to the warm water, although the volume of hot water is considerably smaller than that of the warm water. We say that the *temperature* of the hot water is higher than that of the warm water. 'Temperature' is a property of an object which decides which way heat will flow when it is placed in contact with another object. It must not be confused with heat, which, as we see later, is a form of energy (p. 266). Heat always passes from one object to another at a lower temperature when they are placed in contact.

Temperature measurement is important in everyday life. Thus one of the first actions of the doctor when he is called to a sick patient is to take his temperature. The temperature of a factory or a mill must conform to a value specified by law, otherwise the health of the workers may be impaired. The temperature of the refrigerating plant for storing meat and other food on cargo boats must be a certain value if the food is to remain in good condition over a long period. At the other extreme, furnaces for making glass, cement and different types of steel must be operated at definite high temperatures.

## Mercury-in-glass Thermometer · Fixed Points

The sense of touch is not very reliable for estimating or measuring temperature. Thus warm water feels cool if a finger is transferred to it from hot water. The *thermometer* is a much more reliable instrument for measuring temperature. This uses some physical property of a substance which changes when its temperature is altered. In the *mercury-in-glass* thermometer, for example, which is a common one in everyday use in school laboratories, the volume of mercury changes when its temperature alters.

In making any thermometer, two constant temperatures or *fixed points* must first be marked on it. The 'lower fixed point' is chosen as the *melting point of pure ice*, and the 'upper fixed point' as the *temperature of steam in contact with boiling water at normal atmospheric pressure*, 760 mm mercury.

Over two hundred years ago CELSIUS suggested that the interval of temperature between these two fixed points should be divided into 100 equal parts or *degrees*. The melting point of ice was given the number '0' and the temperature of steam at 760 mm mercury pressure was given the number '100'. The scale of temperature was called the Centigrade scale because the interval was 100 degrees, but the modern name is *Celsius* scale. Thus:

$$\text{Melting point of ice} = 0°C$$

and steam temperature at 760 mm mercury = 100°C

## Making a Mercury Thermometer

The first step in making a mercury thermometer is to choose a capillary glass tube whose bore has the same cross-sectional area all the way along it. One end of the tube is then sealed, and a bulb is blown at this end. The tube is cooled, a funnel is fitted closely over the open end of the tube top and mercury is poured into the funnel, the tube being vertical. When the bulb is warmed gently some air is driven out of the tube through the mercury at the top, and the pressure due to the weight of mercury forces some of the latter inside the bulb when the flame is removed. More mercury is obtained in the bulb by alternate warming and cooling. Finally, the

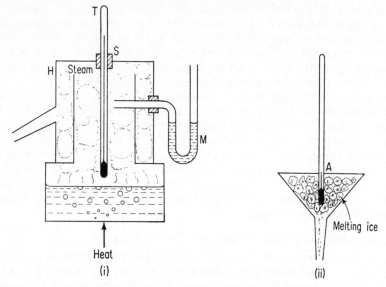

Fig. 9.1 Determination of fixed points

air is driven out of the glass tubing by boiling the mercury; the glass is then maintained at a temperature a little above the highest at which it will be used, and the top is sealed.

To determine the **ice point**, the thermometer is placed with its bulb in *melting* ice, and the level A of the mercury is marked; this is the zero on the Celsius scale (Fig. 9.1(ii)).

The **steam point** has now to be marked on the glass of the thermometer. For this purpose the thermometer T is placed in an apparatus H known as a *hypsometer*, which contains water, a water manometer M for measuring the pressure of the steam obtained when the water is boiled, and an outlet tube to the air (Fig. 9.1(i)). The vessel is double walled so that the inner steam chamber is surrounded by an outer one to maintain its temperature. Now the temperature of boiling water depends on the dissolved impurities, whereas the temperature of the *steam* obtained is independent of any such impurities. The thermometer T must therefore be placed with its bulb *well above* the water, so that the mercury in the tube is all at the temperature of the steam. When the mercury level is steady its position S is marked on the glass tube, and the pressure of the steam, which is read on the manometer M, is noted. The steam point of a thermometer is the temperature of steam at a pressure of 760 mm mercury; the marking S would thus require a minor correction if the manometer and barometer showed that the pressure was not exactly 760 mm.

After marking the upper fixed point on the thermometer as '100°C' on the Celsius scale, the distance between the lower and upper fixed points is divided into 100 equal parts. The thermometer is now ready for measuring temperatures.

## Calibration and Temperature Measurement

*Method.* A simple experiment, which illustrates the principles of calibration and temperature measurement, can be carried out with an *ungraduated* glass capillary tube containing mercury, with a bulb at the

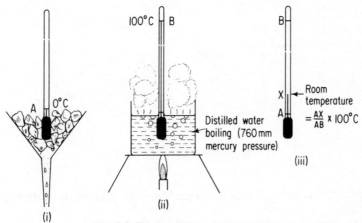

Fig. 9.2 Temperature measurement

bottom to act as a reservoir for the mercury. First, place the bulb in melting ice (Fig. 9.2(i)) and put a mark or rubber band A on the stem corresponding to the level of mercury in the tube.

Next, place the bulb and tube in a beaker of gently boiling water, preferably distilled water, and again put a mark or rubber band B on the stem corresponding to the mercury level (Fig. 9.2(ii)). Remove the tube, and when the mercury level is constant at the room temperature mark the level X on the stem.

*Measurements.* Measure with a millimetre ruler the distance AB and the distance AX between their respective marks.

*Calculation.* The distance AB = 100 degrees C on the Celsius scale

$$\text{Distance AX} = \frac{\text{AX}}{\text{AB}} \times 100°C = 16°C, \text{ for example.}$$

Room temperature is 16°C above the lower fixed point. Since the latter is called '0°C', it follows that the room temperature is 16°C.

Note carefully that this is only an approximate value for the room temperature. The upper fixed point is defined as the temperature of steam at 76 cm mercury pressure and this is *not* the temperature of boiling water.

## Temperature Scales

The *Celsius scale* has been recommended for general use as well as for scientific use. The *Fahrenheit scale* was an earlier scale. The lower fixed point is 32°F and the upper fixed point is 212°F on this scale, so that

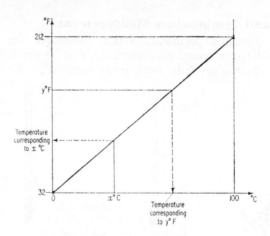

Fig. 9.3 Graph of °C and °F

the interval between the two temperatures is divided into 180 degrees (Fig. 9.3).

Conversion from one scale to another of a particular temperature can be done in several ways.

(i) *Graphically.* On some graph paper, mark Celsius on the horizontal axis and Fahrenheit on the vertical axis. Plot two points, one corresponding to 0°C or 32°F and the other 100°C or 212°F. Draw a straight line between the two points. You can now read off the Fahrenheit temperature corresponding to any value on the scale, and vice versa.

(ii) *Formula.* It can now be seen that

$$\frac{C}{100} = \frac{F - 32}{180}$$

where $C$ represents the actual Celsius temperature and $F$ the actual Fahrenheit temperature. Substitution of one temperature in the formula enables the other temperature to be calculated.

Fig. 9.4 shows temperature scales in common usage. The Celsius and Kelvin scales (see p. 256) are used in all countries for scientific measurements.

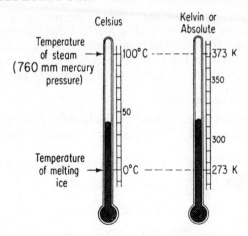

Fig. 9.4 Temperature scales

## Absolute (Thermodynamic) Temperature Scale

The temperature scale used in the SI system of units is the *absolute* or *thermodynamic scale*. On this scale, temperature is measured in 'kelvin', and the symbol is 'K', after Lord Kelvin who suggested the scale. Fig. 9.4.

The lower fixed point is the 'absolute zero', which is given the value 0 K (see p. 255). The upper fixed point is the temperature of the 'triple point of water', the temperature at which ice, water and water-vapour are all in equilibrium. This is given the value 273·16 K. Approximately,

$$- 273 \ °C = 0 \ K, \text{ or } 0 \ °C = 273 \ K$$

It can thus be seen that the absolute temperature $T$ corresponding to a temperature $t$ in °C is given by

$$T = 273 + t$$

A temperature change of 1°C is equal to a temperature change of 1 K. Hence a heat quantity which as a value of '40 joules per °C, or '40 J/°C', can also be said to have a value of '40 J/K' in SI units.

## Alcohol Thermometer

The mercury-in-glass thermometer is widely used in chemical laboratories for temperature measurements in the range 360°C, the boiling point of mercury, to −39°C, its freezing point. In Arctic polar regions, however, where temperatures are extremely low, *alcohol* thermometers are used. Alcohol has a much lower freezing point, about −112°C, than mercury. It has a much lower upper limit, however, because it boils at 78°C.

Unlike mercury, whose silvery surface makes it easily visible even in fine capillary tubes, alcohol must be coloured to make it visible. Further, mercury does not 'wet' glass (see p. 43), and hence responds quickly to a falling temperature, whereas the concave meniscus of alcohol, which 'wets' glass, sticks to the glass as the liquid falls. Mercury has a much greater conductivity than alcohol, and hence reaches its new temperature quickly on contact with a hot body. Alcohol, however, expands about six times as much as mercury for the same temperature rise.

### The Clinical Thermometer

The clinical thermometer is used by doctors for determining the temperature of the human body. It consists of a mercury thermometer with:

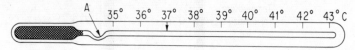

Fig. 9.5 Clinical thermometer

(i) a short range of about 35°–43°C; (ii) a constriction at A in the capillary tubing (Fig. 9.5). When the temperature of a patient is taken with the thermometer the mercury thread easily passes A owing to the large force exerted in expansion. When the thermometer is removed from the patient the mercury thread to the left of A recedes, but the column to the right remains in position owing to the constriction; by this means, the thermometer can still be read. It is afterwards reset by jerking it sharply, when the thread beyond A rejoins the mercury in the bulb.

### Maximum and Minimum Thermometers

It is often necessary to record the maximum and minimum temperatures during each day in studies of the weather.

A *maximum thermometer* utilizes the convex meniscus of mercury, which

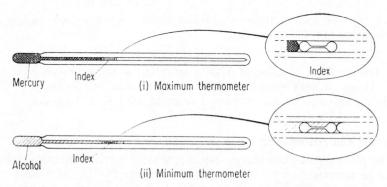

Fig. 9.6 Maximum and minimum thermometers

pushes along a small steel index when the temperature rises and the mercury expands (Fig. 9.6(i)). When the temperature falls, the index stays in position. At the end of the day, therefore, the maximum temperature corresponds to the lower or left position of the index. The index can be reset by tilting the thermometer or by using a small magnet.

A *minimum thermometer* utilizes the concave meniscus of alcohol. This time the index is *below* the meniscus (Fig. 9.6(ii)). Thus when the temperature falls, the index is pulled down by the meniscus; when the temperature rises, the alcohol expands past the index, which stays in position. At the end of the day the minimum temperature corresponds to the upper or right position of the index (compare the case of the maximum thermometer index). The index is reset in the same way as described above.

## Combined Maximum and Minimum Thermometer

In 1782, Six designed a thermometer which registers both the maximum and minimum temperatures over a period of time, as required in weather forecasting (Fig. 9.7). A large bulb is filled with alcohol A and is connected to a bent capillary tube which contains mercury in the lower part. The side D contains alcohol with some air at the top. A steel index X has a light spring attached to keep it in position, a similar index N is above B.

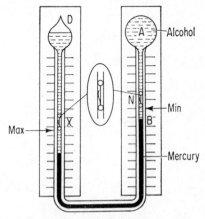

The alcohol in A has the largest volume of the liquids in the thermometer, and alcohol expands relatively much more than mercury for the same temperature change (p. 247). Thus, when the temperature rises the alcohol A expands and flows past N, and the mercury rises and pushes X in front of it. When the temperature falls, the alcohol in A contracts,

Fig. 9.7 Combined maximum and minimum thermometer

and the mercury column at B rises and pushes N in front of it. Hence the lower end of X indicates the maximum temperature on the scale shown, and the lower end of N indicates the minimum temperature. After the temperatures have been read, a magnet is used to reset each index above the mercury columns, and the thermometer is then ready for the next observation.

## Principle of Gas Thermometer

Mercury-in-glass thermometers are not used for very accurate measurement of temperature. Apart from the disadvantage of a relative small range of temperature–the freezing point of mercury is $-39°C$ and its boiling point is $360°C$–the glass also expands and its expansion is irregular.

*Gas thermometers* are used for very accurate temperature measurement at the National Physical Laboratory, for example. A large volume change of gas occurs when its temperature is altered, so that the glass expansion is negligible. A simple experiment, similar to that with an ungraduated capillary tube containing mercury (p. 229), can be performed to illustrate the principle of a gas thermometer.

EXPERIMENT

*Method.* Fit a one-hole rubber bung B containing a short length of narrow glass tubing T tightly into the neck of a conical flask C (Fig. 9.8(i)). Immerse the flask in boiling water contained in a large beaker for about 5–10 min, so that the air inside the flask will be at the temperature of the

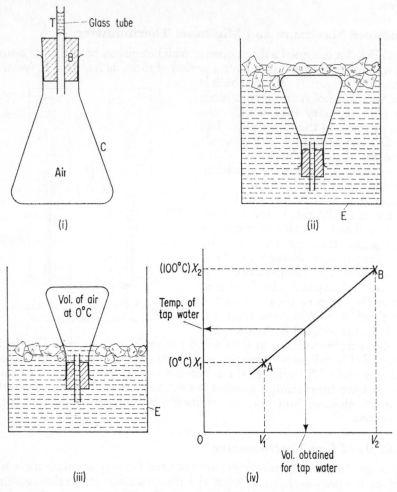

(i)  (ii)

(iii)  (iv)

Fig. 9.8 Simple gas (air) thermometer

water-bath, and at atmospheric pressure. With your fingers firmly over the end of the small glass tube, invert the flask in a beaker E of ice-water, and remove your finger when the neck of the flask is completely below the surface (Fig. 9.8(ii)). Now cover the whole flask with ice-water. In a few minutes the air in the flask will be at the temperature of this mixture.

Before removing the flask from the ice-water, ensure that the pressure of the air inside the flask again equals the atmospheric pressure in the room.

Do this by adjusting the flask until the levels of water inside and outside the flask are the same (see p. 234, (Fig. 9.8(iii)). Place your finger over the end of the glass tube and remove the flask from the ice-water. Then carefully measure: (1) the volume of water inside the flask, and (2) the volume of the whole flask itself (with glass tube and stopper). At the boiling point of water the volume of air is the volume of the whole flask; at the ice-point the volume of air at the same pressure is the volume of the whole flask *less* the volume of water inside. Record the two volumes, $V_1$ and $V_2$.

*Graph.* On the horizontal axis of a piece of graph paper, mark off the two volumes, $V_1$ and $V_2$, you have obtained (Fig. 9.8(iv)). On the vertical axis, which is to represent temperature, mark off two convenient points $X_1$ and $X_2$, which represent 0°C and 100°C respectively. Now mark crosses, A and B, corresponding to the volume at 0°C and 100°C. Join A and B by a straight line.

This line enables, approximately, temperature to be obtained on your simple 'gas thermometer'. Repeat the above experiment but cool the flask this time in tap water. Read off the tap-water temperature from your line AB after you have measured the volume of air in the flask at this temperature. Observe carefully that the temperature is only approximate, as the upper fixed point of a thermometer is the temperature of steam at 760 mm mercury pressure.

## Thermocouple Thermometer

If two *unlike* metals A, B, are connected together and one junction H is warmed while the temperature of the junction C is kept constant (Fig. 9.9(ii)), an electric current flows along the metals. The two joined metals are known as a *thermocouple*, and provide an example of the direct transformation of heat energy to electrical energy. When the temperature of C is kept constant and the temperature of H is increased, the magnitude of the electric current increases. The temperature of H can be deduced from the magnitude of the electric current.

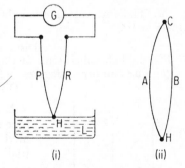

Fig. 9.9 Thermocouple

This is the principle of the *thermoelectric thermometer*, which is used in industry for finding the temperature of molten metals and certain furnaces. Fig. 9.9(i) illustrates the simple arrangement required. The thermocouple may consist of platinum P and platinum–rhodium R wires, with a current-reading instrument G connected in the circuit. G is calibrated to read in deg C. The junction H of the metals is placed in the furnace or liquid L, whose temperature can be read directly from G.

The thermoelectric thermometer is particularly useful for rapidly changing temperatures as there is only a small mass of metal to warm up.

## SUMMARY

1. To change a temperature from °C to °F, or vice versa, the relation $C/100 = (F - 32)/180$ can be used $\left(\dfrac{F - 32}{C} = \dfrac{9}{5}\right)$.

2. The lower fixed point on the Celsius scale is the temperature of melting ice, and the upper fixed point is the temperature of steam above boiling water at 760 mm mercury pressure.

3. The clinical thermometer has: (i) a constriction near the bulb; (ii) a short temperature range (e.g. 35°–43°C).

4. Six's maximum and minimum thermometer utilizes alcohol and mercury, with steel pointers to indicate the maximum and minimum temperatures.

5. The gas thermometer uses the volume change of a gas at constant pressure when its temperature changes.

6. The thermoelectric thermometer uses the current flowing in a circuit when one junction of two metals is heated.

## WORKED EXAMPLES

**1.** The lengths of a degree interval on two mercury thermometers, one having a Celsius scale and the other a Fahrenheit scale, are 2 mm and 1 mm respectively. Calculate: (a) the distance between the upper and lower fixed points of each thermometer; (b) the distance between the lower fixed point and the −8°F mark on the Fahrenheit thermometer.

(a) The interval between the upper and lower fixed points of a Celsius scale = 100°C.

$$\therefore \text{ Distance} \times 100 \times 2 = 200 \text{ mm}$$

For the Fahrenheit scale of 180°F, distance = $180 \times 1 = 180$ mm

(b) The number of degrees = $32 - (-8) = 40$

$$\therefore \text{ Distance} = 40 \times 1 = 40 \text{ mm}$$

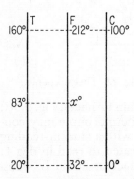

Fig. 9.10 Conversion of temperature

**2.** Describe in detail how the room temperature may be found with the aid of an ungraduated mercury thermometer. The freezing and boiling points of water are marked 20 and 160 on a certain thermometer. Calculate: (a) the temperature on the Fahrenheit scale corresponding to a reading of 83 on the scale of this thermometer; (b) the temperature on the Celsius scale which is represented by the same reading on this thermometer and on the Celsius thermometer. (L).

Fig. 9.10 illustrates the thermometer T, a Fahrenheit thermometer F and a Celsius thermometer C. Suppose $x$ is the temperature on the Fahrenheit thermometer. Then, by proportion (see p. 231),

$$\frac{x - 32}{180} = \frac{83 - 20}{140}$$

$$\therefore \ (x - 32)140 = 180(83 - 20) = 180 \times 63$$

$$\therefore \ x - 32 = \frac{180 \times 63}{140} = 81$$

$$\therefore \ x = 113°\text{F.}$$

Suppose $y$ is the temperature which is the same number on the thermometer T and the Celsius thermometer. Then, by proportion,

$$\frac{y - 20}{140} = \frac{y - 0}{100}$$

$$\therefore \ 100(y - 20) = 140y$$

$$\therefore \ y = -50°\text{C.}$$

## EXERCISE 9 · ANSWERS, p. 320

**1.** State the freezing point of water on: (a) the Fahrenheit scale; (b) the Celsius scale.

What is the Celsius temperature corresponding to 98·4°F? (N.)

**2.** Describe a thermometer for recording the maximum and minimum temperature over a period of 24 h and explain how it acts. (L.)

**3.** The distance between the fixed points of a thermometer is 20 cm. If the mercury level is 4·5 cm above the lower mark, what is the temperature on: (a) the Celsius scale; (b) the Fahrenheit scale? (N.)

**4.** (a) Give three reasons why mercury is a suitable liquid for use in a thermometer. (b) State one way in which the design of a thermometer may be modified so as to increase its sensitivity. (N.)

**5.** Describe experiments to determine the position of the fixed points on an ungraduated mercury thermometer. (L.)

**6.** Draw a labelled diagram of a minimum thermometer. Name the liquid, indicate the temperature range and show where the minimum temperature is read. (N.)

**7.** Define the temperatures 0°C, 100°C. Calculate the rise in temperature, on the Fahrenheit scale, corresponding to a rise of 180° on the Celsius scale. Describe, with a diagram of the apparatus used, how you would test the accuracy of the marking of the upper fixed point of a given mercury thermometer. A mercury thermometer, with only its bulb in an oil-bath, reads 90°C. When lowered deeper into the oil-bath the thermometer reads 91°C. Assuming the oil-bath to be at the same temperature throughout, suggest any one explanation of the difference of the two readings. (C.)

**8.** Explain as far as you can the difference between heat and temperature, and state two properties of matter which would enable it to be used for measuring temperatures.

Give the reasons for the following features of a clinical thermometer: (a) the bulb is small; (b) it is a long cylinder and not spherical; (c) the bulb is made of thin glass; (d) the capillary is of very small cross-sectional area; (e) there is a constriction in the capillary. Convert 95·0°F to °C. (O. and C.)

**9.** What is meant by temperature, thermometer, fixed points, temperature scale? What properties of mercury make it suitable for use in a thermometer, and why? A quartz bulb fitted with a tap is heated in an oven with the tap open to the atmosphere. The tap is closed, the bulb removed from the oven and the tap then opened under water at 0°C. Water enters and fills one-third of the bulb when the air inside is again at atmospheric pressure. What was the temperature of the oven? (*O.* and *C.*)

**10.** Describe, with experimental details, how you would calibrate a mercury thermometer for use between the temperatures of −5°C and 110°C. Describe the construction of a maximum and minimum thermometer. Give reasons why mercury is regarded as a suitable liquid for use in a thermometer. (*O.* and *C.*)

# 10

# EXPANSION OF SOLIDS AND LIQUIDS

If telegraph wires are observed in winter and in summer it will be noted that the wires sag more in summer. This shows that the length of the wires has increased as the temperature has risen. Other observations, as well as experiments mentioned in the previous chapter and below, show that *most materials expand when their temperature increases*, and contract when their temperature decreases.

## EXPANSION OF SOLIDS

### Effects of Expansion

There are many practical consequences of the expansion of solids. The size of the balance-wheel of a watch, which governs the movement of its hands, and the length of a pendulum of a clock are affected by the change

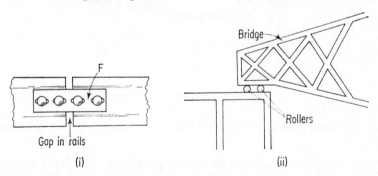

Fig. 10.1 Allowance for expansion

of temperature which occurs from winter to summer. Allowance must be made for the alteration in length of metal bridges and railway lines as the seasons change, since enormous stresses would be set up if the expansion and contraction could not take place freely. Fig. 10.1(i) illustrates the fish plate F which joins one section of a railway line to the next section. The holes in the rail are oval, allowing expansion or contraction to take place. The Forth Bridge, one of the longest in the British Isles, has a total allowance of about 1 m for expansion and contraction, and roller bearings are used at each end of the bridge so that the change in length can take place freely without straining the structure (Fig. 10.1(ii)). For a similar reason, telegraph wires must have sufficient sag in them, as they would otherwise snap in winter when they contracted.

In most central heating systems the joints between the pipes are special non-rigid ones known as *expansion joints*. They are designed to allow one pipe to move relative to the other while expanding or contracting and provide a good seal at the same time. Thus changes of temperature do not affect the seal between the pipes.

## Uses of Thermal Expansion

The large force exerted when solids contract on cooling can be put to practical use:

1. Trains and wagons used on railways have steel tyres fitted round their wheels. The tyre, originally a little smaller than the wheel, is heated in a gas ring until it is slightly greater than the wheel. It is then fitted round the latter (Fig. 10.2). On cooling, the tyre grips the wheel tightly owing

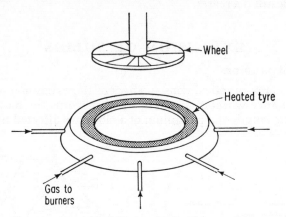

Fig. 10.2 Train wheel and tyre

to the very large force exerted during its contraction. See Example 3, p. 245.

2. Steel plates such as those used in shipbuilding or in large boilers are usually rivetted together using red-hot rivets. Holes are made in the overlapping plates, a red-hot rivet is pushed through and its head held tightly against one plate. The other end of the rivet is hammered tight against the plate. On cooling, the rivet contracts and holds the plates tightly together, this providing a good seal against the sea for ship plates and against steam in large boilers.

## Bimetallic Strips

EXPERIMENT

A bimetallic strip can be made by placing two strips of different metals (e.g. brass and iron) side by side and welding them together along their entire length. One end of the strip is then inserted into a wooden handle (Fig. 10.3).

Heat the bimetallic strip and notice that bending occurs. This shows that one metal expands more than the other when heated through the same temperature. In this case the strip bends with the brass on the outside of the curve, as shown. Hence *brass expands more than iron* for the same temperature rise.

Observe that the bimetallic strip straightens when it cools to room temperature. If it were cooled below room temperature the brass would contract more than the iron. The strip would now bend with the iron on

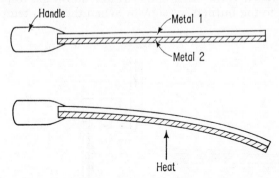

Fig. 10.3 Linear expansion of different metals

the outside. *Invar* is an alloy which has a negligible change in length when its temperature rises, and is therefore very suitable as one of the materials in a bimetallic strip (see p. 242).

### Balance-wheel of Watch

When the air temperature rises, the balance-wheel of a watch would increase in radius if it were made of one metal and the watch would slow down. In addition, the elasticity of the steel hair-spring controlling the movement of the wheel would decrease, and this would also make the watch lose time.

To compensate for temperature changes, the rim of the balance wheel is usually made in two segments of bimetal, for example, brass and iron, with balance or timing screws on them (Fig. 10.4). The more expansible metal, brass, is on the outside. When the temperature rises, the metal spoke carrying the segments increases in length,

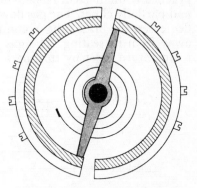

Fig. 10.4 Balance-wheel of watch

but the bimetal segments themselves curve inwards (see Fig. 10.3). The decrease in radius of the segments, and the consequent inward movement of the balance-screws towards the centre would cause the wheel to speed up, and thus the effect of the temperature change has been nullified.

### Thermostats

Bimetal strips are also used in making thermostats for gas cookers and electric blankets. In a modern gas cooker the temperature is controlled using the very low coefficient of expansion of the alloy invar (steel containing 36% nickel). This expands only a millionth of its length per degree C rise in temperature.

Fig. 10.5 illustrates the principle of a gas thermostat. A brass tube Y projecting into the oven, encloses an invar rod X attached to it at one end. A valve is attached at the other end of X and allows gas from the mains to flow through to the burners, as shown. When the oven temperature rises,

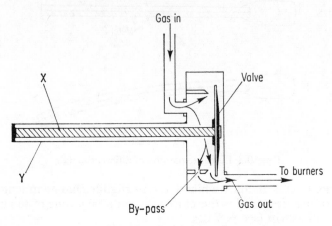

Fig. 10.5  Gas thermostat

the brass tube Y expands and moves to the left, but the invar rod X is practically unaffected. Hence the valve moves to the left nearer its seating and partially closes. The gas flow is then reduced. If the oven temperature decreases, the brass tube Y contracts a little and the valve moves farther away from its seating. The gas flow is then increased. A required oven temperature is set by turning a knob on the regulator. This adjusts the

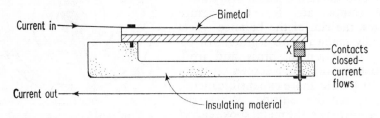

Fig. 10.6  Electric heating control

valve opening so that the gas supply is shut off at the required temperature.

Fig. 10.6 shows the principle of an electric thermostat of the type used in electric blankets for maintaining a steady temperature. The blanket contains a length of insulated resistance wire sewn between two layers of blanket material. When the blanket is first switched on the current flows through a bimetal strip to a fixed metal block *via* contacts between them.

As the temperature rises the bimetal strip curves away from the block, and at one stage the contacts are broken. The current is then cut off. When the bimetal strip cools down contact is again made at one stage and current flows once more.

## Bimetallic Thermometer

This thermometer, used in industry, consists of a thin spiral of bimetal, with one end fixed and the other attached to a pointer which moves over a scale calibrated in degrees (Fig. 10.7). Brass and invar are commonly used.

An intermittent switch circuit is used in trafficators. It incorporates a bimetal contact, as shown in Fig. 10.8. The circuit operates on the same principle as the electric thermostat circuit in Fig. 10.6.

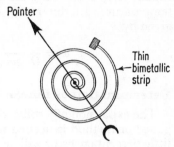

Fig. 10.7 Bimetal thermometer principle

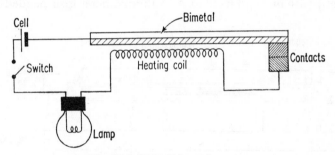

Fig. 10.8 Use of bimetal contact

## Linear Expansivity

For the purposes of industry, as well as research, scientists require to know accurately to what extent metals change in length when their temperature alters; and it is obviously most useful to have a table, or list, showing the increase in length of 1 m (or 1 cm) when the temperature of the metal changes by 1°C. From a knowledge of this quantity, which is known as the *linear expansivity* of the metal, the actual change in length of a metal when its temperature alters can be calculated by simple proportion.

The linear expansivity $\alpha$ of a solid is defined as *the increase in length per unit length of the solid when its temperature changes by one degree.* Thus if 100·00 cm of steel increases by 0·11 cm when its temperature rises by 90°C,

$$\text{Linear expansivity} = \frac{0\cdot11 \text{ cm}}{100\cdot0 \text{ cm} \times 90°C} \quad \cdot \quad \cdot \quad \cdot \quad (1)$$

$$= 0\cdot000\ 012 \text{ per } °C \quad \cdot \quad \cdot \quad \cdot \quad (2)$$

$$= 12 \times 10^{-6}/°C,$$

since $10^{-6} = 1/10^6$. Observe carefully that, since the unit 'cm' appears in both the numerator and denominator in (1), the length unit cancels out. The unit of linear expansivity is thus 'per °C'. Further, the linear expansivity, $\alpha$ is always an average or mean figure for a wide range of temperature. In this range one may state that the mean coefficient is given by

$$\alpha = \frac{Increase\ in\ length}{Original\ length\ \times\ Temperature\ rise}$$

## Determination of Linear Expansivity of Metal

The expansion of a solid when warmed is usually so small that a special gauge or method is needed to measure it accurately. On the other hand, little percentage error will be made if a ruler is used to measure the original length of a long rod.

The apparatus is set up as shown in Fig. 10.9. The length of the rod R of metal used is first measured with a metre rule and it is then placed in the jacket J with one end fixed at A. A thermometer is suspended from the

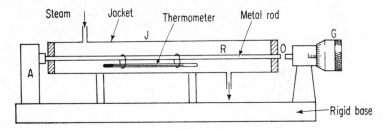

Fig. 10.9 Measurement of linear coefficient

rod by two short wire loops. The micrometer screw gauge G is turned until the end of the screw just touches the end O of the rod, and the gauge reading is noted. The screw is now turned back appreciably to ensure that the end of the rod can expand freely. The reading on the thermometer is noted and steam is now passed through the jacket for several minutes. The micrometer is then screwed up to the rod, the gauge reading is noted and the screw is then turned back again to allow for any further expansion of the rod. This procedure is repeated until the same gauge reading is obtained. The reading is then recorded and the temperature of the rod is observed from the thermometer.

*Measurements.* Suppose the following measurements are obtained:

Original gauge reading = 1·365 cm. Final gauge reading = 1·515 cm.
Original length of rod = 93·3 cm.
Original temperature = 16°C. Final temperature = 99°C.

*Calculation.* Increase in length = 1·515 − 1·365 = 0·150 cm
Increase in temperature = 83°C

$$\text{Linear expansivity} = \frac{\text{Increase in length}}{\text{Original length} \times \text{Temperature rise,}}$$

$$= \frac{0.15 \text{ cm}}{93.3 \text{ cm} \times 83°C}$$

$$= 0.000\ 019/°C.$$

$$= 19 \times 10^{-6}/K \text{ in SI units}$$

The mean linear expansivity for brass $= 18 \times 10^{-6}/K$,
for iron $= 11 \times 10^{-6}/K$
and for Pyrex $= 3 \times 10^{-6}/K$

## Formulae

Since $\alpha$, the linear expansivity, is the increase in length of 1 cm for 1°C temperature rise, it follows that the increase (or decrease) in length of $l$ cm for $t$ °C temperature change is given, by proportion, by

$$\text{change in length} = \alpha\, l\, t$$
$\therefore$ the new or increased length, $l_t = l + \alpha\, l\, t = l(1 + \alpha t).$

*Examples*

**1.** A telegraph wire has a length of 30·00 m and is made of metal of linear expansivity 0·000 02 $(20 \times 10^{-6})/K$. Find the change in length from a hot day, temperature 30°C, to a very cold day, −5°C.

Increase in length of 1 m for 1°C change $= 0.000\ 02$ m.
$\therefore$ Increase of 30 m for 35°C change
$$= 0.000\ 02 \times 30 \times 35 \text{ m} = 0.000\ 02 \times 30 \times 35 \times 100 \text{ cm}$$
$$= 2.1 \text{ cm.}$$

**2.** Part of a steel tape used by a surveyor is 20·00 m at 12°C. What is the overall length measured by using this part of the tape one hundred times on a warmer day corresponding to 22°C?
(Linear expansivity of steel $= 0.000\ 011$ $(11 \times 10^{-6})/K$)

Using the formula, increase in length $= \alpha l t = 0.000\ 011 \times 20 \times (22 - 12)$
$$= 0.0022 \text{ m } (0.22 \text{ cm})$$
$\therefore$ New length of tape $= 20.0022$ m.

Hence, if used 100 times, overall length measured $= 2000.22$ m.

(*Note.* Surveyors' measuring tapes, calibrated accurately by the maker at a particular temperature, are thus subject to error when used at a different temperature.)

**3.** A steel tyre diameter is 150·0 cm at 10°C and is to be fitted on to a train wheel of diameter 151·0 cm. To what temperature must the tyre be heated to just fit the wheel? (Linear expansivity of steel $= 11 \times 10^{-6}/$ K.)

Increase in tyre diameter required $= 151 - 150 = 1$ cm
Using the formula for increase in length, $\alpha\, lt,$

$$\therefore\ 11 \times 10^{-6} \times 150 \times t = 1$$

$$\therefore\ t = \frac{1}{11 \times 10^{-6} \times 150}$$

$$= \frac{10^{6}}{11 \times 150} = 606°\text{C}$$

$$\therefore\ \text{Temperature required} = 616°\text{C}.$$

## Superficial and Cubical Expansivity

When a solid expands, its area alters as well as its length. Thus the metal roof of a shed increases in area when its temperature rises. Area expansion is called *superficial expansion*. The superficial expansivity, $\beta$, is defined as the *increase per unit area per unit temperature rise*. From the definition it follows that the new area $A_2$ of a metal is related to the original area $A_1$ when the temperature changes by $t$ deg C by

$$A_2 = A_1(1 + \beta t).$$

This relation has the same mathematical form as $l_t = l(1 + \alpha t)$ for linear expansivity, and is derived in a similar way.

Fortunately there is not much trouble in calculating the area expansivity if the linear expansivity of the same substance is known. As shown later

*Area expansivity* ($\beta$) = 2 *Linear expansivity* ($\alpha$).

For example, if $\alpha = 0\cdot000\ 018/\text{K}$ for brass the area expansivity $\beta = 0\cdot000\ 036/\text{K}$. Thus 1 cm² of brass increases to $1\cdot000\ 036$ cm² when its temperature is raised 1°C.

A glass stopper which has stuck in the neck of a bottle can be removed by warming the neck, which then expands in volume. Besides an increase in length and area, then, solids such as a water-tank or a metal sphere expand in volume on warming.

The *cubic expansivity* (which concerns volume expansion) of a material is defined as the increase in volume per unit volume when the temperature increases by 1°C. It may also be defined as the fractional increase in volume per degree Celsius rise in temperature.

$$\text{Cubic expansivity } \gamma = \frac{Increase\ in\ volume}{Original\ volume \times rise\ of\ temp.}$$

Thus the new volume $V_2$ of a substance, whether hollow or solid, is related to its original volume $V_1$ when its temperature increases by $t$ deg C by

$$V_2 = V_1(1 + \gamma t).$$

As in the case of the area expansivity, the cubic expansivity $\gamma$ is easily derived from a knowledge of the linear expansivity, since as proved on p. 247,

*cubic expansivity* $\gamma$ = 3 × *linear expansivity* $\alpha$.

Thus if a glass density bottle had a linear expansivity of $9 \times 10^{-6}/\text{K}$, the volume contained by the hollow bottle would expand as if it had a cubic expansivity of $3 \times 9 \times 10^{-6}$ or $27 \times 10^{-6}/\text{K}$.

**Proof that** $\beta = 2\alpha$ **and** $\gamma = 3\alpha$

Suppose a metal square has sides of length 1 cm and $\alpha$ is its linear expansivity. The increase in length of each side of the square when its temperature changes by 1°C is then $(1 + \alpha)$ cm and the new area becomes $(1 + \alpha)^2$ The original area is 1 cm², and hence the increase in area $= (1 + \alpha)^2 - 1 = 2\alpha + \alpha^2$. Now in practice $\alpha$ is of the order of 0·000 01 cm; hence, as the reader can verify for himself, $\alpha^2$ is a negligible quantity compared with $2\alpha$ Consequently the increase in area of 1 cm² when its temperature changes by 1°C is $2\alpha$, which, by definition, is the numerical value of the area expansivity, $\beta$.

To show that the cubic expansivity $\gamma$ is $3\alpha$, consider a cube of side 1 cm and suppose its temperature changes by 1°C. Each side then becomes $(1 + \alpha)$ cm in length, and hence the volume is $(1 + \alpha)^3$, or $(1 + 3\alpha + 3\alpha^2 + \alpha^3)$. Now in practice $3\alpha^2 + \alpha^3$ are very small numbers compared with $3\alpha$. Hence the increase in volume of 1 cm³ when its temperature increases by 1°C $= (1 + 3\alpha) - 1 = 3\alpha$, which is the numerical value of the cubic expansivity. The result is exactly the same whether the cube is solid or hollow.

# EXPANSION OF LIQUIDS

## Liquid Expansion

We now investigate the effect of temperature rise on the expansion of liquids. The measurement of the expansion is always complicated by the expansion of the container itself, which depends on the particular material used for making the container. The following experiment shows the order of magnitude of the expansion of a liquid compared with that of a solid.

EXPERIMENT

Fill a flask F of known volume with water (Fig. 10.10). Push the pipette P down until the liquid level reaches mark M, the 2 cm³ mark of the pipette. Read the thermometer T. Now warm the liquid *gently* and observe its level. Initially, the level falls slightly because the glass expands first and then the level rises slowly and steadily. Note the temperature when the liquid has risen *just* to the tip of the pipette, filling the pipette completely.

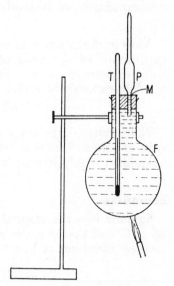

Fig. 10.10 Cubic (volume) expansion of liquid

*Measurements.* The following measurements were taken in one experiment:

| | |
|---|---|
| Volume of pipette | $= 2$ cm³ |
| Volume of liquid in large flask | $= 400$ cm³ |
| Initial temperature of liquid | $= 19$°C |
| Final temperature of liquid | $= 36$°C. |

*Calculation*

Temperature rise $= 36 - 19 = 17°C$.

Hence, 400 cm³ of liquid for a 17°C rise expands by 2 cm³

∴ 1 cm³ for a 1°C rise expands by $2/(400 \times 17)$ cm³ $= 0.0003$ cm³

The liquid expansion is more than this figure, as the vessel expanded slightly.

## Apparent and Absolute (True) Expansivities

If we ignore the expansion of the material of the container, then the 'apparent expansion' is obtained, that is, the expansion relative to that of the vessel. The average *apparent cubic expansivity* of a liquid is defined as the increase in volume, relative to that of the vessel, per unit volume per °C rise, or numerically as the increase in volume relative to the vessel of 1 cm³ for 1°C temperature rise.

The *absolute (true) expansivity* is the *increase in volume per unit volume* per °C temperature rise. Thus

$$\text{Absolute expansivity} = \frac{\text{Increase in volume}}{\text{Original volume} \times \text{Temp. rise}} \qquad . \quad . \quad (1)$$

This gives the average or mean expansivity in the temperature range concerned. Of course, if the vessel expands this must be taken into account. If $\gamma$ is the absolute expansivity, $\gamma_{\text{app}}$ is the apparent coefficient and $c$ is the *cubic* expansivity of the solid container, it can be shown that:

$$\gamma = \gamma_{\text{app}} + c.$$

Thus in the experiment on p. 247, $\gamma_{\text{app}} = 0.0003/K$. Suppose the linear expansivity of the glass of the container is 0.000 01/K, so that its cubic expansivity $= 3 \times$ linear expansivity (p. 247) $= 0.000 03/K$; the absolute expansivity is then

$$\gamma = 0.0003 + 0.000\ 03 = 0.000\ 33/K$$

Note that: (i) the units of 'cubic expansivity' is 'per K' because the units of volume cancel in equation (1) above; (ii) the cubic expansivity of a liquid is roughly 10 times as great as the linear expansivity of a metal.

## Formulae

Suppose $V_1$ is the original volume of a liquid whose temperature *changes* by $t$ °C and that $\gamma$ is the mean cubic expansivity. Then, since $\gamma$ is the increase in volume of 1 cm³ when its temperature changes by 1 °C,

$$\text{the increase in volume} = \gamma V_1 t \quad . \quad . \quad . \quad . \quad . \quad (1)$$

by simple proportion. Consequently the new volume $V_2$ of the liquid $= V_1 + \gamma V_1 t$.

$$\therefore V_2 = V_1 (1 + \gamma t) \quad . \quad . \quad . \quad . \quad . \quad (2)$$

These formulae are similar to those obtained in connection with the expansion of solids (see p. 245).

## Examples

**1.** A mercury-in-glass thermometer has a bulb of internal volume 0·20 cm³ and the capillary tube has an internal cross-section of $2 \times 10^{-4}$ cm². The 0°C level is just above the bulb. Calculate the volume increase of the mercury between 0° and 100°C and the distance between successive divisions on the glass. (Assume apparent cubic expansivity of mercury relative to glass is $16 \times 10^{-5}$/K.)

$$\text{Increase in volume of mercury} = \gamma V t$$
$$= 16 \times 10^{-5} \times 0\cdot20 \times 100$$
$$= 0\cdot0032 \text{ cm}^3$$

since $16 \times 10^{-5}$ cm³ is the increase in volume of 1 cm³ for 1°C rise.

∴ Distance between successive divisions

$$= \frac{\text{Volume between successive divisions}}{\text{Area of cross-section}}$$

$$= \frac{0\cdot0032}{2 \times 10^{-4} \times 100} = \frac{0\cdot0032}{0\cdot02} = 0\cdot16 \text{ cm}$$

**2.** A glass bottle has a volume of 20·00 cm³ at 0°C. It contains 271·34 g of mercury at 0°C when the density of mercury is 13·6 g/cm³. Calculate the temperature when the mercury will just fill the bottle. (Absolute of mercury and glass $= 18 \times 10^{-5}$ and $3 \times 10^{-5}$/K respectively.)

Volume of mercury at 0°C $= \dfrac{\text{Mass}}{\text{Density}} = \dfrac{271\cdot34}{13\cdot6} = 19\cdot95$ cm³

Apparent cubic expansivity $\gamma_{\text{app}}$ $= (18 - 3) \times 10^{-5} = 15 \times 10^{-5}$

Expansion required to fill bottle $= 20\cdot0 - 19\cdot95 = 0\cdot05$ cm³

Since expansion $= \gamma_{\text{app}} V t$, where $t$ is temperature rise,

∴ $15 \times 10^{-5} \times 19\cdot95 \times t = 0\cdot05$

∴ $t = \dfrac{0\cdot05}{15 \times 10^{-5} \times 19\cdot95} = 17$°C (approx.)

## Density of Water · Hope's Experiment

In 1804 HOPE devised a simple experiment to investigate the variation with temperature of the density of water. A tall vessel A containing water

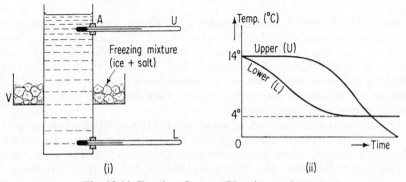

Fig. 10.11 Density of water–Hope's experiment

is surrounded at the middle by a vessel V containing ice and salt (Fig. 10.11(i)). At the top and bottom of A are thermometers, U, L, and readings of their temperatures and the time are taken. The *lower thermometer* L drops from its initial value of 14°C, say, as the water cooled at the middle of A drops to the bottom; but at a temperature of about 4°C the reading on L remains constant and the *upper thermometer* U, which had been constant at 14°C now begins to fall and reaches 0°C (see Fig. 10.11 (ii)).

The water at the bottom of the vessel has the maximum density, and Hope's experiment shows that maximum density is obtained at about 4°C. Water colder than 4°C thus rises from the middle to the top, and the temperature on U continues to fall until ice is formed at the top of the liquid.

## The Anomalous Expansion of Water with Temperature

Ice (a solid) expands slightly when warmed from − 5°C say to 0°C; at 0°C it forms water and contracts (Fig. 10.12(i)) .When the water is warmed it *contracts* from 0° to 4°C, which is exceptional, but from 4° to 100°C it expands, behaving now like most liquids. When it changes into

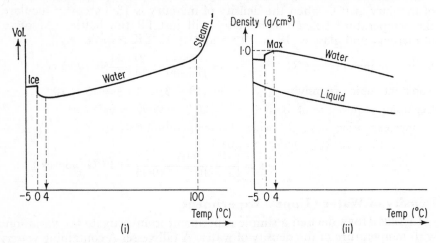

Fig. 10.12 Volume and density changes (*exaggerated*)

steam at 100°C the volume increases 1600 times as much. Water is thus exceptional or 'anomalous' in the range 0° to 4°C. Since 'density' is 'mass/volume', and the mass is unaltered by warming, *water has a maximum density at about* 4°C, as Hope found. Fig. 10.12(ii) shows the density–temperature variation of water compared with most liquids. The latter increase in volume as their temperature rises from 0°C and hence their density diminishes continuously.

This anomalous expansion of water makes it possible for fish to survive in winter-time in Arctic and other cold regions. Consider the air outside a lake, for example, when the temperature begins to fall. The water at the top increases in density, and falls to the bottom of the lake, where the water is less dense. Other water takes the place of the water which sinks to the bottom, and this in turn becomes cooled and sinks. The downward movement of cold water continues until the temperature at the top reaches 4°C. The water at the bottom is then at this temperature, but as the temperature of the top falls to 3°C and lower the *colder water remains at the top* because the density of water below 4°C is *less* than the density at 4°C (see Fig. 10.12(ii)). When the temperature of the air reaches 0°C the water at the top begins to freeze and form ice, and the ice continues to form in a downward direction from the surface. Far below the surface of the frozen lake, then, there is water at 4°C; and the fishes survive.

## SUMMARY

1. The linear expansivity of a solid ($\alpha$) is the increase in length per unit length for 1°C rise in temperature. The area (superficial) expansivity = $2\alpha$. The cubic expansivity = $3\alpha$.

2. **The increase in length of a solid = $\alpha lt$, where $l$ is the original length and $t$ is the temperature rise. The new length, $l_t = l(1 + \alpha t)$.**

3. The **true (absolute) cubic expansivity** of a liquid $\gamma$ is the increase in volume per unit volume for 1°C rise in temperature. The **apparent expansivity** $\gamma$ is the increase in volume relative to that of the container per unit volume for 1°C temperature rise.

4. $\gamma = \gamma_{app} + c$, where $c$ is the cubic expansivity of the material of the container.

5. Increase in volume = $\gamma Vt$. New volume $V_t = V_0(1 + \gamma t)$.

6. Water contracts when warmed from 0° to 4°C and then expands. The maximum density of water is thus at about 4°C.

## EXERCISE 10 · ANSWERS, p. 320

### Solids

1. A metal rod 50·00 cm long at 0°C becomes 50·06 cm long at 100°C. Find the linear expansivity of the metal. (*N.*)

2. In an experiment to determine the linear expansivity of a metal: (*a*) Why is it impracticable to heat the rod directly with a bunsen burner? (*b*) State in words the formula from which this expansivity is calculated. (*N.*)

3. Explain the statement that the linear expansivity of steel is 0·000 012/K.

The new Forth bridge will have a steel span of 1000 m between the main piers. Calculate in centimetres the change in length of the span when the temperature changes from −10° to 30°C.

Describe a method of finding the linear expansivity of a metal. (*N.*)

4. (*a*) Describe one experiment in each instance to show the expansion of a solid, a liquid and a gas on heating.

(*b*) A compound strip of brass and iron is straight at room temperature. Give a labelled diagram to show its appearance when it is heated. Describe one use of such a strip.

A compound strip of brass and iron, 10 cm long at 20°C, is held horizontally with the iron uppermost. When heated from below with a bunsen the temperature of the brass is 820°C and of the iron 770°C. Calculate the difference in lengths of the iron and the brass. (Linear expansivity of brass is 0·000 019 and of iron 0·000 012/K.) (*L.*)

**5.** Define linear expansivity.

Describe how you would measure the linear expansivity of a metal rod.

Show how suitable combinations of two metals of different linear expansivities can be used: (*a*) to operate a thermostat; (*b*) to make a clock pendulum whose period of swing does not alter with temperature. (*O.*)

**6.** Define *linear expansivity* and prove that the cubic expansivity is approximately three times its linear expansivity. Describe an accurate method for measuring the linear expansivity of a metal. The tungsten filament of a wireless valve has a length of 2 cm and a radius of 0·1 mm at 20°C. If the temperature of the filament is raised to 3020°C, find (*a*) the increase in its length and (*b*) the increase in its volume. (Linear expansivity of tungsten = 0·000 007/K. Take $\pi$ as $\frac{22}{7}$.) (*N.*)

## Liquids

**7.** The linear expansivity of glass is 0·000 009/K, and the true cubic expansivity of mercury is 0·000 180/K. What is: (*a*) the cubic expansivity of glass; (*b*) the apparent cubic expansivity of mercury in glass? (*N.*)

**8.** Distinguish between real and apparent expansion of a liquid. Define apparent cubic expansivity and describe an experiment to find this expansivity for water between room temperature and 60°C. What volume of mercury at 0°C must be placed in a flask of volume 100 cm³ at 0°C so that it just fills the flask at 100°C?

(Cubic expansivity of mercury = 0·000 18/K, linear expansivity of glass = 0·000 009/K.) (*L.*)

**9.** What is meant by the apparent cubic expansivity of a liquid in glass? Describe how you would determine its value for a given liquid.

How does the volume of a given mass of water change: (*a*) as the water is cooled from 10° to 0°C; (*b*) when the water is frozen to ice at 0°C? Sketch a rough (not to scale) volume–temperature graph for the range between +10° and −10°C. (*L.*)

**10.** Describe how you would check the accuracy of the fixed points on a mercury thermometer.

What special features in the construction of a thermometer make it sensitive to small temperature changes?

The bulb and stem of a thermometer contain 0·4 ml of mercury up to the zero mark. If the bore has an area of cross-section of 0·03 mm², what is the length of a degree on the scale?

(Apparent cubic expansivity of mercury in glass = 0·000 15/K.) (*N.*)

**11.** Describe a simple experiment which shows that a liquid expands more than glass when its temperature is similarly increased.

A solid floats in a liquid so that it is totally immersed. Describe and explain what happens when the system is heated so that the temperature of the solid and of the liquid are always equal.

An iron rod is 600 cm long at 0°C. It is mounted alongside a copper rod and both are always maintained at the same temperature. When they are heated to 100°C it is found that the difference in their lengths is the same as it was at 0°C. Find the length of the copper rod at 0°C.

(Linear expansivity of iron = 0·000 010 9/K. Linear expansivity of copper = 0·000 016 8/K.) (*C.*)

# PROPERTIES OF GASES · THE GAS LAWS

When a steam engine, a motor-car engine or an aeroplane engine is functioning, gases inside cylinders are expanding and contracting in volume, thus operating pistons which make the machine move. The mechanical engineer must therefore know something of the laws governing the behaviour of gases; he must know, for example, how the volume of a gas changes when its pressure and temperature alter.

The condition or *state* of a given mass of gas depends on its volume, temperature and pressure. Solids and liquids are practically unaffected by small changes of pressure, so that there is no need to take pressure into account in discussing their changes in volume with temperature. A small change in pressure, however, can considerably affect the volume of a gas. Since there are *three* variable quantities for a given mass of gas (volume, pressure, temperature), it is best to keep one of the quantities fixed and investigate the variation of the remaining two quantities.

## Volume Expansion at Constant Pressure

An apparatus to obtain the volume variation of a gas at constant pressure when its temperature changes is shown in Fig. 11.1(i). The dried

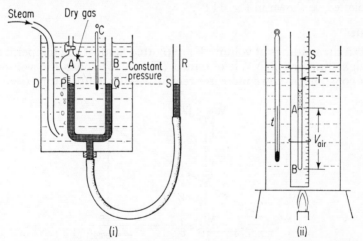

Fig. 11.1 Investigation of Volume–temperature change (constant pressure)

gas is contained in the bulb A above mercury in the U-tube PQ, one side of which, B, is open to the atmosphere. The pressure of the gas is kept constant at atmospheric pressure by moving an open tube R up or down until the levels P, Q and S are the same. The U-tube is placed in a large vessel D containing water, and the gas is warmed by passing steam into the water. The volume of the gas is read from graduations below A.

EXPERIMENT

Equate the levels P, Q and S by moving R, and then read the volume of the gas, dry air for example, in A and its corresponding temperature. Now pass steam into the water until a rise of about 5 deg is obtained. Equate the levels again. Stir the water and read the new temperature and volume. Repeat the procedure for higher temperatures.

*Measurements*

Constant pressure

| Temperature $t°C$ | Volume $V$ | Absolute temperature $T$ |
|---|---|---|
|  |  |  |

*Graph.* Plot the volume $V$ against the temperature $t°C$. Fig. 11.2.

*Capillary Tube Experiment.* A simple apparatus for a class experiment to investigate the expansion of a gas at constant pressure is shown in Fig. 11.2(ii). A column AB of air is trapped by a sulphuric acid pellet, which dries the air, or by a mercury pellet, in a capillary tube T of uniform diameter. This is sealed at the lower end and is attached to a half-metre rule S by rubber bands. The whole arrangement is placed inside a suitable large water-bath which can be heated by a bunsen-burner.

When the temperature is raised the air expands. It does so under the *constant pressure* due to the external atmospheric pressure plus that due to the height of liquid. The volume of the gas is proportional to the length of the column AB, which is read from the attached rule. The temperature is read from the thermometer after the water is stirred. The volume $(V)$–temperature $(t)$ readings are then plotted, as shown in Fig. 11.2.

## Results

When the readings of volume $V$ are plotted against the temperature $t°C$ a straight line graph NY is obtained (Fig. 11.2). Thus the volume of a gas at constant pressure increases regularly with temperature rise.

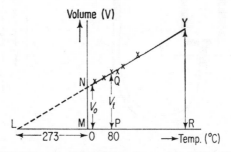

Fig. 11.2 Volume–temperature change (constant pressure)

The *density* of a constant mass of gas is inversely-proportional to its volume. The same experiments on volume can thus be used to investigate the variation of density with temperature by plotting $1/V$ against $t°C$. As the reader may verify, the graph obtained is a *curve* in this case; the density *decreases* with increasing temperature since the volume then rises.

## Volume Coefficient (Constant Pressure)

The *volume coefficient* of a gas at constant pressure is defined in a similar way to that of a liquid (p. 248), but with the important difference that the original or initial volume of the gas is always taken as its volume at 0°C. This is because the volume expansion of a gas is very sensitive to temperature changes. Thus for a fixed mass of gas, the volume coefficient $\alpha$ is the increase in volume per unit volume at 0°C per °C temperature rise, the pressure being constant. Hence

$$\alpha = \frac{\text{Increase in volume from } 0°\text{C}}{\text{Original volume at } 0°\text{C} \times \text{Temperature rise}}.$$

To find the volume coefficient $\alpha$ the line NY is produced to cut the volume-axis at N, thus determining the volume $V_o$ of the gas at 0°C. In the present case, MN $= V_o = 7\cdot1$. The volume $V_t$ at any other temperature, e.g. 80°C, is then taken, and PQ $= V_t = 9\cdot2$. The volume coefficient $\alpha$ is calculated from

$$\alpha = \frac{V_t - V_o}{V_o \times t} = \frac{(9\cdot2 - 7\cdot1)}{7\cdot1 \times 80}$$
$$= 0\cdot0037/°\text{C} = 0\cdot0037/\text{K}$$

Accurate experiment shows that $\alpha$ is $0\cdot003\,66$, or $\frac{1}{273}$/K (approx.) for all gases, a fact which was first discovered by CHARLES in 1780. **Charles' law** states that *a given mass of gas increases in volume by $\frac{1}{273}$ of its volume at 0°C for every degree Celsius rise in temperature, the pressure remaining constant.*

## Absolute Temperature, T

When the temperature of the air in the above experiment is reduced, the pressure being kept constant, the volume diminishes along the straight line YN (Fig. 11.2). Below 0°C, if a freezing mixture were available, the air would diminish in volume along the line YNL, and at a temperature corresponding to L, where the temperature axis is cut, we have the startling deduction that the volume of the air would *theoretically* become zero. The temperature is given the name of *absolute zero*, and in practice a gas liquefies before it reaches the absolute zero.

The temperature of the absolute zero can be found from Charles' law. If 1 cm³ of a gas at constant pressure is at 0°C and its temperature is reduced to $-1$°C the volume reduces to $(1 - \frac{1}{273})$ or $\frac{272}{273}$ cm³.

At $-4$°C its volume becomes $(1 - \frac{4}{273})$ or $\frac{269}{273}$ cm³. In this way it can be seen that at $-273$°C the volume of the gas $= 1 - \frac{273}{273} = 0$. This is true for any volume of gas at constant pressure, because it would shrink from 0°C by $\frac{1}{273}$ of its volume for every degree Celsius temperature fall. Consequently, the absolute zero on the Celsius scale is about $-273$°C. Temperatures very close to the absolute zero have been reached.

## Volume and Absolute Temperature

The absolute zero is called 0 K, after Lord Kelvin, who first proposed the idea of absolute temperature. The temperature of 0°C is 273°C from the absolute zero, or 273 K; the temperature of 100° C is 373 K. (Fig. 11.3.) Generally,

$$t°C = (273 + t) \text{ K.}$$

The absolute temperature is usually given the symbol $T$, and we shall use $t$ for temperature in degrees Celsius. Thus $T = 273 + t$.

Suppose $V_0$ is the volume of a gas at 0°C. Its volume at $t$°C, using the volume coefficient $\frac{1}{273}$ per deg C, is hence given by

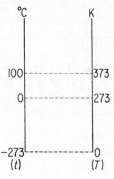

Fig. 11.3 Temperature in °C ($t$) and K ($T$)

$$V_t = V_0 + \tfrac{1}{273} \cdot V_0 \cdot t = V_0\left(1 + \frac{t}{273}\right) \quad . \quad (1)$$

$$\therefore \quad V_t = V_0\left(\frac{273 + t}{273}\right)$$

Now $V_0$ and 273 are constants. Thus $V_0 \propto (273 + t)$. But $(273 + t)$ is the absolute temperature $T$ corresponding to $t$.

$$\therefore \quad V_t \propto T \quad . \quad . \quad . \quad . \quad . \quad . \quad (2)$$

Hence we arrive at the important result that *the volume of a given mass of gas at constant pressure is directly proportional to its absolute temperature,*

i.e. *volume, V* $\propto$ *absolute temperature, T* . . (3)

EXPERIMENT

From your measurements on p. 254, work out in a neighbouring column the absolute temperature $T$ of each temperature recorded. Then plot a graph of Volume v. $T$. See whether it is a straight-line graph which passes throughout the origin. If so, state your conclusion.

Find also the variation of density with $T$ by plotting (1/Volume) against $T$.

## Calculations on Gas Volume at Constant Pressure

As an illustration of this simple relationship, suppose a gas has a volume of 100 cm³ at 15°C and is heated at constant pressure to 30°C. The absolute temperature at 30°C is 273 + 30 or 303 K; at 15°C it is 273 + 15 or 288 K. Since $V \propto T$ at constant pressure, the new volume $V$ at 30° C is given by

$$\frac{V}{100} = \frac{303}{288}$$

$$\therefore \quad V = \frac{303}{288} \times 100 = 105 \text{ cm}^3 \text{ (approx.)}$$

If the temperature falls from 15°C until the volume of the gas at constant pressure decreases to 80 cm³, the new *absolute* temperature $T$ is given by

$$\frac{80}{100} = \frac{T}{288}$$

$$\therefore \quad T = 288 \times \frac{80}{100} = 230 \text{K (approx.)}$$

$$\therefore \text{ temperature} = 230 - 273 = -43°\text{C}$$

## Pressure Changes of Gas at Constant Volume

We next investigate the pressure changes of a fixed mass of gas at constant volume when its temperature varies. Fig. 11.4 illustrates one form of apparatus for this purpose.

B is a large bulb of air, with a narrow (capillary) tube L connected to a mercury manometer for measuring its pressure. B is surrounded by a vessel containing water, which is heated to different temperatures to alter the temperature of the air. The volume of the air will then alter, but by raising or lowering the tube GD, which is open to the air, the mercury on

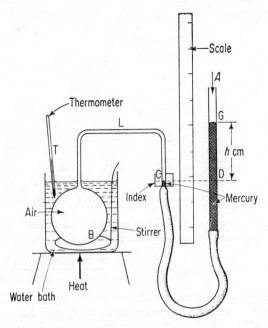

Fig. 11.4 Pressure–temperature change (constant volume)

the other side can always be brought back to a fixed point C. In this way *the volume of the air in B and LC is kept constant.*

The pressure of the air in the bulb B is given by $(A + h)$ cm, where $A$ is the atmospheric pressure in centimetres of mercury and $h$ is the difference in level GD of the mercury levels. If the level in the right-hand tube is *less* than the level at C the pressure of the air in B is given by $(A - h')$ cm, where $h'$ is the difference in levels (see p. 195).

### EXPERIMENT

Vary the temperature of the water in steps of about 10°C. Keeping the volume of the air constant as explained before, record the difference in levels, $h$, of the mercury each time. Then record the atmospheric pressure $A$ and determine the pressure $p$ of the gas as previously explained.

*Measurements*

| Pressure $p$ | Temperature $t°C$ | Absolute temperature $T$ |
|---|---|---|
|  |  |  |
|  |  |  |

*Graph.* (i) Plot $p$ v. $t$, (ii) plot $p$ v. $T$.

*Results.* A straight-line graph is obtained in (i) and (ii).

## Pressure Coefficient

The straight-line graph in (i) is represented by SY in Fig. 11.5. It shows that the pressure varies regularly with temperature change.

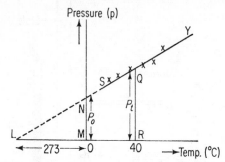

Fig. 11.5 Graph of pressure $v.$ temperature (volume constant)

The *pressure coefficient* is defined as the *fractional increase of the gas pressure at* $0°C$ *for* $1°C$ *temperature rise*, the volume being constant. Thus

$$\text{Pressure coefficient} = \frac{\text{Increase in pressure from } 0°C}{\text{Original pressure at } 0°C \times \text{temperature rise}}$$

To calculate the pressure coefficient from Fig. 11.5 the line YS is produced back to meet the pressure-axis at N, and MN is the pressure $p_0$ of the gas at $0°C$.

Suppose $p_0 = 69.8$ cm of mercury and that the pressure QR at $40°C$ is $80.0$ cm. The increase in pressure from $0°C = 80.0 - 69.8 = 10.2$ cm.

$$\therefore \quad \text{pressure coefficient} = \frac{\text{increase in pressure from } 0°C}{\text{original pressure at } 0°C \times \text{temperature rise}}$$

$$= \frac{10.2}{69.8 \times 40} = \frac{1}{273} /°C = \frac{1}{273}/K$$

Results show that the pressure coefficient is numerically the same as the volume coefficient of the gas, or $\frac{1}{273}$ per deg C. Thus if $p_t$ is the gas pressure at constant volume of a gas at $t°C$ whose pressure was $p_0$ at $0°C$, then, as shown for volume change on p. 256,

$$p_t = p_0 \left(1 + \frac{t}{273}\right).$$

Further,

$$p_t \propto T,$$

where $T$ is the absolute temperature corresponding to $t°C$. The laws are exactly analogous to the volume laws on p. 256. The straight-line graph passing through the origin when $p$ is plotted against $T$ on p. 258 also shows that, experimentally, $p \propto T$.

A car tyre, pumped in the morning at 10°C to a pressure of 0·20 N/mm², increases in pressure as its temperature rises. If the temperature reaches 18°C the new pressure $p$, from $p \propto T$, is given by

$$\frac{p}{0·20} = \frac{273 + 18}{273 + 10} = \frac{291}{283}.$$

$$\therefore p = 0·21 \text{ N/mm}^2 \text{ (approx.)}$$

## Use of Bourdon Gauge

A *Bourdon gauge* is widely used to measure pressure above atmospheric pressure. Basically it consists of a flattened metal tube bent into a circular arc (Fig. 11.6(i)). As the pressure on it increases, the tube tends to straighten

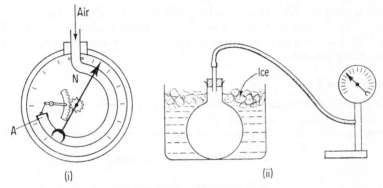

Fig. 11.6 Bourdon gauge investigation of pressure–temperature change

and the closed end therefore moves. The movement is magnified by a lever and gear train and recorded by a pointer moving round a scale previously calibrated.

### EXPERIMENT

Place an open large round-bottomed flask in a large container of ice-cold water. After a while connect it to a Bourdon gauge as shown, and observe the pressure reading and the temperature (Fig. 11.6(ii)). Now warm the ice and water. At suitable rising temperatures until the water boils, observe the corresponding pressure inside the flask and the temperature. *Plot* the pressure against the temperature. Using the pressure value at 0°C, calculate a value for the pressure coefficient of air, as on p. 258. State the sources of error in the experiment.

## Constant-volume Gas Thermometer

For a given alteration of temperature, the pressure change of a gas at constant volume is relatively much greater than the volume change in liquids. For example, the pressure coefficient of a gas is about 0·0037, while the volume coefficient of mercury is only about 0·000 18. Scientists therefore rely on a *gas thermometer* for very accurate work. There is an elaborate gas thermometer at the National Physical Laboratory, but the principle can be shown with the simple apparatus in Fig. 11.6(ii).

## EXPERIMENT

Immerse the bulb of air first in melting ice and then in boiling water; in each case determine the pressure at constant volume, taking the atmospheric pressure into account. The two values of the pressure are then plotted on a piece of graph paper against the corresponding temperatures of 273 K and 373 K. A straight line is drawn through them, since we know that the pressure is directly proportional to the absolute temperature. Now immerse the bulb in a liquid such as tap water at an unknown steady temperature. Determine the pressure $p$ at constant volume. From the straight-line graph, read off the absolute temperature corresponding to $p$. Subtract 273 to obtain the temperature in °C.

This is only an illustration of the principle of a gas thermometer (see also p. 233). In practice, of course, the two fixed points are those defined on p. 227, and more elaborate procedure is necessary to measure an unknown temperature.

## General Gas Law

We can now proceed to a general gas law, in which pressure, volume and absolute temperature may all vary.

From pp. 256, 258, it can be stated that for a given constant mass of gas,

$$p \propto T \text{ (volume kept constant).}$$

and

$$V \propto T \text{ (pressure kept constant).}$$

Combining these two laws into one, in which $p$, $V$ and $T$ may all vary, we write

$$pV \propto T.$$

We can check this is a correct relation by making the volume $V$ constant in it, when we obtain $p \propto T$, and then making the pressure $p$ constant in it, when we obtain $V \propto T$.

It now follows that

$$\frac{pV}{T} = \text{a constant value,}$$

the value depending on the mass of gas used and its nature. Hence *for a fixed or constant mass* of gas, when it has (i) a volume $V_1$ at a pressure $p_1$ and absolute temperature $T_1$, and then (ii) a volume $V_2$ at a different pressure $p_2$ and absolute temperature $T_2$,

$$\frac{p_1 V_1}{T_1} = \frac{p_2 V_2}{T_2}.$$

It may be noted that when the temperatures $T_1$ and $T_2$ are the same, then $p_1 V_1 = p_2 V_2$, which is Boyle's law (p. 196).

## Conversion to S.T.P.

In order to compare the volumes of two gases, it is convenient to convert both volumes to standard temperature and pressure (s.t.p.), which is taken as 0°C and 76 cm mercury pressure.

As an illustration, suppose a volume of (i) 100 cm³ of gas, measured at 10°C and 78 cm mercury pressure, is compared with a volume of (ii) 120 cm³ measured at 50°C and 70 cm mercury pressure. To calculate the respective volumes, $V_2$, at s.t.p., we have:

(i) $V_1 = 100$ cm³, $p_1 = 78$ cm, $T_1 = 273 + 10 = 283\,$K,
$\quad\ V_2 = ?$ $\qquad p_2 = 76$ cm, $T_2 = 273\,$K

From

$$\frac{p_1 V_1}{T_1} = \frac{p_2 V_2}{T_2}$$

$$\therefore \quad \frac{78 \times 100}{283} = \frac{76 \times V_2}{273}$$

$$\therefore \quad V_2 = \frac{78 \times 100 \times 273}{283 \times 76} = 99 \text{ cm}^3$$

(ii) $V_1 = 120$ cm³, $p_1 = 70$ cm, $T_1 = 273 + 50 = 323\,$K,
$\quad\ V_2 = ?$ $\qquad p_2 = 76$ cm, $T_2 = 273\,$K

From

$$\frac{p_1 V_1}{T_1} = \frac{p_2 V_2}{T_2}$$

$$\therefore \quad \frac{70 \times 120}{323} = \frac{76 V_2}{273}$$

$$\therefore \quad V_2 = \frac{70 \times 120 \times 273}{323 \times 76} = 93 \text{ cm}^3$$

It can thus be seen that the measured volume of 120 cm³, which is 93 cm³ at s.t.p., is actually a smaller volume at s.t.p. than the measured volume of 100 cm³, which is 99 cm³ at s.t.p.

## Molecular Explanations of Gas Laws

The effects of a temperature rise on the volume and pressure of a gas can be explained by the kinetic theory of gases, described on p. 111. There it was shown that the pressure of a gas was due to the continual bombardment of gas molecules on the walls of the containing vessel. Heat is a form of energy (p. 226), and when a gas is heated the molecules gain kinetic energy and move about faster. Consequently, for a *fixed volume*, the momentum change produced by the molecules at the walls per impact is greater (p. 112), and the number of impacts per second of each molecule also increases since the speed increases. Thus the pressure increases with temperature rise when the volume is constant. This can be demonstrated with the kinetic model machine described on p. 197, to which the reader should refer. If the vibrational frequency in the machine is increased, corresponding to a temperature rise of the molecules, it is then seen that the pressure exerted by the beads increases when the volume is kept constant.

Conversely, when the temperature of a gas is lowered at constant volume the molecules move more slowly than before. The momentum change at the walls of the vessel per impact, and the number of impacts per second, are then both diminished. The rate of change of momentum or force at the walls is therefore decreased, so that the gas pressure is reduced.

When the gas is heated at *constant pressure*, the molecules again gain kinetic energy and move faster. This time part of the energy is used to do work because the molecules slowly push back the piston which maintains the pressure constant. The rest of the energy remains in the gas, so their temperature is higher and their volume has increased. Although the molecules are now moving faster, the average momentum change per second at the piston is the same as before because the molecules have farther to travel between successive impacts. The pressure is thus constant although the volume has increased.

## Heating by Compression

If a cricketer or tennis player hits a ball hard during play, the ball leaves the bat or racket at a much higher speed than when it approached. This is because the striking surface was moved towards the advancing ball. If, however, the play is defensive, so that the bat or racket is hardly moved, the ball rebounds with approximately the same velocity with which it arrived. This is analogous to what happens when a gas is compressed.

As the piston of the gas container moves down, molecules rebound from it with a higher velocity than when approaching. The average kinetic energy of the molecules is thus increased. This implies that the temperature of the gas is increased. Hence a gas is warmed by compression. Conversely, when the gas is allowed to expand the piston is drawn back, and this time the velocity of molecules leaving the piston is less than on approach. Thus the average kinetic energy, and hence the temperature is reduced. Expansion thus cools a gas.

## Real Gases

The gas laws mentioned in this chapter, such as Boyle's and Charles' laws, are 'ideal gas' laws. They are obeyed closely by actual gases at low pressures if the temperatures are well above their liquefying temperature. In accurate work, however, with actual gases, account must be taken of the volume occupied by their molecules and of the attraction between the molecules. Both become of greater importance as the gas is compressed, until finally, with favourable conditions of temperature and pressure, the attractions between the molecules produces the liquid state (p. 51). Further consideration of real gases is outside the scope of the book.

## SUMMARY

1. The **volume coefficient** of a gas is the increase in volume per unit volume **at 0 °C** for 1°C temperature rise. The **pressure coefficient** is the increase in pressure per unit pressure **at 0 °C** for 1°C temperature rise.

2. Absolute temperature $T$ K $= 273 + t$, where $t$ is the temperature in °C.

3. Charles' law states: For a given mass of gas at constant pressure, the volume increases by $\frac{1}{273}$ of its volume at 0°C for each °C rise in temperature. Thus, **the volume is proportional to the absolute temperature**.

4. For a given mass of gas at constant volume, the pressure is proportional to the absolute temperature.

5. **For a given mass of gas,** $\dfrac{pV}{T} =$ **constant.**

6. In a constant-volume gas thermometer the lower fixed point on the Celsius scale is the pressure of the gas at 0°C and the upper fixed point is the pressure at 100°C.

7. In the kinetic theory of gases, increasing the temperature at constant volume increases the speed of the molecules, and hence increases the number of collisions per second and the momentum change at each collision. The pressure therefore increases. At constant pressure part of the energy when the gas is heated is used to push back the piston, and the gas therefore expands, while the rest of the energy remains in the molecules, which then move faster and have a higher temperature.

## EXERCISE 11 · ANSWERS, p. 320

**1.** A vessel contains 1 litre of a gas at 30°C and at 1 atmospheric pressure. Calculate the temperature at which the volume will, without change in pressure, be 3 litres. (*C.*)

**2.** The volume of a given mass of air at 27°C and 75 cm pressure is 152 cm³. Find its volume at 76 cm pressure, the temperature being −23°C. (*N.*)

**3.** Define *coefficient of expansion of a gas at a constant pressure* and describe an experiment to determine its value.

A capillary tube sealed at the lower end stands vertically and contains a thread of mercury 10 cm long which seals off a column of air 25 cm long at 12°C. What is the pressure on this air if the barometric height is 75 cm of mercury? To what temperature must the tube be heated for the mercury to rise 5 cm?

**4.** State the law which indicates how the pressure of a fixed mass of air varies with its temperature as recorded on a mercury thermometer, when the volume remains constant.

Describe an experiment to verify the law.

80 cm³ of hydrogen are collected at 15°C and 75 cm of mercury pressure. What is its volume at s.t.p.? (*L.*)

**5.** In an experiment to find the relation between the pressure and the temperature of a mass of air at constant volume the following results were obtained:

| Total pressure cm of mercury | 77·8 | 81·4 | 85·9 | 89·5 | 93·2 |
|---|---|---|---|---|---|
| Temp. °C | 20·6 | 34·2 | 51·8 | 65·4 | 79·8 |

Draw a graph to show the relation between pressure and temperature and use the graph to calculate the coefficient of increase of pressure of air at constant volume. (*N.*)

**6.** Describe an experiment to determine the absolute zero of temperature using a mass of air *either* at constant volume, *or* at constant pressure.

In an experiment the pressure of a mass of air at constant volume was found to increase rom 75 cm of mercury at 15°C to 91 cm of mercury at 75°C. From these results what would be: (i) the pressure of the air at 0°C; (ii) the coefficient of increase of pressure of air at constant volume? (*N.*)

**7.** State the laws of Boyle and Charles and show how they can be combined into a single formula describing the effect of temperature and pressure on the volume of a gas.

What is meant by absolute temperature?

A gas thermometer at 27°C contains dry air confined by a horizontal thread of mercury, so that it is at atmospheric pressure, *p*. If it is calibrated for $p = 760$ mm of mercury what temperature will the thermometer indicate if *p* falls by 1%? (*O.* and *C.*)

**8.** Describe an experiment to investigate how the volume of a gas varies with temperature if the pressure is kept constant. Indicate by a free-hand graph how you would expect the density to vary with temperature, showing the result you would expect for a gas whose density is 1 unit at −73°C and which is warmed to 127°C.

Dry air is kept at constant temperature in a cylinder having a well-fitting frictionless piston. To move the piston in by 10 cm requires the total force on the piston (including that due to the external air pressure) to be increased from 200 N to 250 N. How long was the cylinder of air originally? (*O.* and *C.*)

**9.** State Charles' law and describe how you would proceed to verify it experimentally. Express the law in the form of a graph and use this to show what is meant by absolute temperature.

At 27·0°C and 2·00 atmospheres pressure the density of a gas is 0·0400 g per litre. What will its density be at 4·00 atmospheres pressure and 227·0°C? (*O.* and *C.*)

**10.** How would you investigate the way in which the density of dry air varies with pressure at constant temperature? (The actual value of the density is not required.) What result would you expect to find?

At each stroke a pump takes in 500 cm³ of air at atmospheric pressure and after compression forces it into a motor tyre. If the volume of the tyre remains constant at 20 litres and the pressure inside it is originally 152 cm of mercury, what is the new pressure after 10 strokes of the pump?

Take atmospheric pressure to be 76 cm of mercury and assume that the temperature remains constant. (*O.* and *C.*)

**11.** Draw a labelled diagram of the apparatus you would use to investigate the expansion of a sample of air heated at constant pressure. State how you would try to ensure that the example of air was uniformly heated and how you could know whether its pressure remained constant during the experiment.

State the gas equation and state what the symbols in it represent. Calculate the density of air at −100°C and at a pressure of 20·0 atmospheres, given that the density of air at 0°C and at a pressure of 1·00 atmosphere is 1·29 g per litre. (*C.*)

**12.** Using the kinetic theory of gases, explain why the pressure of a gas at constant volume rises when its temperature is increased.

# 12

# QUANTITY OF HEAT · HEAT ENERGY

In the home, heat for cooking is obtained by burning gas or by using electricity. Steam engines burn coal, diesel engines burn oil; and, in general, the greater the amount of fuel, the greater is the quantity of heat obtainable from it.

## Units of Quantity of Heat

The scientist and the engineer have defined *units* of quantity of heat. The unit in scientific work for many years was the *calorie*, which is defined as *the quantity of heat which raises the temperature of* 1 *g of water by* 1°*C*. The **joule** is the unit used nowadays (p. 266). For convenience, 1 large calorie, or 1 kilocalorie, is used to represent 1000 calories.

In this country, however, commercial gas companies use a unit of heat known as the *British Thermal Unit (Btu)*. Approximately, 1 Btu = 1055 joules. The unit on which gas companies base their charge, the *therm*, is about 100 million joules or 100 megajoules (MJ), a value which may be used for the future commercial unit of heat.

## Calorific or Energy Values

The calorific value of one grade of coal is approximately 28 million J/kg. The calorific value of one grade of petrol is 50 million J/kg. The human body is a 'machine' which needs food as fuel to keep working efficiently, and some approximate calorific values, in kilojoules per 100 g, of a variety of foods are given below. Diets which discourage overweight exclude chocolate and sugar, for example, and include lean meat and green vegetables.

| *Proteins* | | *Carbohydrates* | | *Fats* | | *Others* | |
|---|---|---|---|---|---|---|---|
| Cheese | 1680 | Chocolate | 2300 | Butter ⎫ | | Peas | 420 |
| Lean meat | 1200 | Sugar | 1600 | Margarine ⎬ 2900 | | Boiled potatoes | 340 |
| Eggs | 700 | Wholemeal bread | 1000 | Olive oil ⎪ | | Milk | 300 |
| Liver | 600 | | | Fat meat ⎭ | | Fresh fruit | 200 |
| White fish | 300 | | | | | Green vegetables | 150 |

## Heat and Energy · Heat Unit

Early scientists considered that heat was a material substance which they called *caloric*. Thus a body gaining heat increased its caloric content, and one losing heat diminished its caloric content. This conception of heat was generally accepted up to about 1800, when experimental results were obtained which could not be explained by the caloric theory of heat.

One of the first experiments which spelt the doom of the caloric theory was performed by BENJAMIN THOMPSON, or COUNT RUMFORD, as he was later called. Rumford was an American engaged in 1798 in supervising the boring of cannon in a munitions factory in Bavaria. He noted that the borer, the metal bored and the metal chips all became hot, and that heat continued to be produced as long as the boring took place. It seemed highly improbable to Rumford that such a small amount of metal with which he was dealing could contain such an enormous amount of caloric, and he realized that the heat produced was related to the mechanical energy used in boring the cannon. To confirm his view he surrounded a gun-barrel and a borer in a container of water and, to the astonishment of the onlookers, who knew no flame was being used, the water was made to boil. These demonstrations implied that **heat was a form of energy,** and not a material substance. Later, experimental evidence obtained by JAMES PRESCOTT JOULE showed conclusively that this was the true conception of the nature of heat. As we saw in previous chapters, the molecules in solids, liquids and gases gain kinetic energy when heated and then move faster than before.

Since heat is a form of energy, it has been decided internationally to measure quantities of heat in *joules* and heat per second in joules per second or *watts*. About 4·2 J is the heat required to raise the temperature of 1 g of water by 1°C. With a kilogram of water, about 4200 J is needed to raise its temperature by 1°C.

## Heat (Thermal) Capacity

The heat to raise the temperature of a substance by 1°C is defined as its *heat capacity*. A large mass of water such as the sea requires a large amount of heat to raise its temperature 1°C, so that it has a large heat capacity. A small piece of metal, such as the junction of a thermo-couple, has a small heat capacity. Its temperature rises appreciably when it receives a small amount of heat, which is an advantage of the thermocouple in measuring the temperature at a particular part of a hot metal surface (p. 235). Heat capacity, then, is expressed in *J per* °$C(J/$°$C)$ *or in J per* $K(J/K)$.

## Specific Heat Capacity

Experiment shows that 42 joules of heat raises the temperature of 2 g of water 5°C. The same quantity of heat raises the temperature of 2 g of copper 50°C, and the temperature of 2 g of aluminium 20°C. Different substances of equal mass, then, experience different rises of temperature when they absorb equal quantities of heat.

The *specific heat capacity*, $c$, of a substance is defined as *the amount of heat which raises the temperature of unit mass* 1°C. Thus the specific heat capacity of water is about 4·2 J per g per °C, which is written 4·2 J/g K using K in place of °C. With a kilogram mass, the value is 4200 J/kg K. The specific heat capacity of copper is about 0·4 J/g K or 400 J/kg K.

Specific heat capacity, then, is the heat capacity per unit mass of a substance. Thus, for a particular substance,

*Heat capacity = Mass × Specific heat capacity.*

It should be noted that the unit of heat capacity is 'J/K', whereas the unit of specific heat capacity is 'J/g K' or J/kg K.'

## Electrical method for heat capacity, specific heat capacity

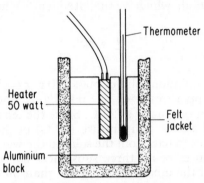

Fig. 12.1 Heat capacity and specific heat capacity by electrical heating

The heat capacity of a metal block, and the specific heat capacity of the metal, can be found by electrical heating. In Fig. 12.1, A is a solid aluminium block of mass 1 kg. An electrical heater H, and a thermometer T, are placed inside holes bored in the block. Both make good thermal contact with the block. A surrounding felt jacket B diminishes heat losses.

In one experiment, a 50-watt heating coil produced a temperature rise from 24°C to 40°C in 5 minutes, or 300 s. Since 50 W = 50 joules per second,

$$\text{heat supplied, } Q = 50 \times 300 = 15\ 000 \text{ J}$$

Since the temperature rise = 40 − 24 = 16°C = 16 K,

$$\therefore \text{ heat capacity of block} = \frac{Q}{t} - \frac{15\ 000 \text{ J}}{16 \text{ K}} = 940 \text{ J/K (approx.)}$$

Also, specific heat capacity of aluminium, $c$,

$$= \frac{Q}{m \times t} = \frac{15\ 000}{1000 \times 16} = 0.9 \text{ J/g K (approx.)}$$

The specific heat capacity of water can also be found by heating water in a polystyrene container. In one experiment, the 50-watt heating coil produced a temperature rise in 400 g of water from 20°C to 34°C in 8 minutes. Then the heat supplied $Q = 50 \times 8 \times 60 = 24\ 000$ J.

$$\therefore \text{ specific heat capacity of water, } c, = \frac{24\ 000}{400 \times 14} = 4.3 \text{ J/g K}$$
$$\text{(approx.)}$$

## Formula for $Q$

Suppose a quantity of heat $Q$ raises the temperature of 200 g of copper, specific heat capacity 0·4 J/g K, by 15°C. Since 0·4 J is the amount of heat which raises the temperature of 1 g of copper by 1°C, the amount of heat $Q$ is 0·4 × 200 × 15 J, by simple proportion, or 1200 J. Similarly, if 50 g of aluminium, specific heat capacity 1·1 J/g K, cools from 100° to 30°C, a temperature change of 70°C, the quantity of heat lost $Q = 1·1 \times 50 \times 70 = 3850$ J. It can now be seen that the quantity of heat $Q$ supplied to a mass $m$ of specific heat capacity $c$ when its temperature *changes* by $t$ °C is given by a basic formula which we shall frequently employ in heat calculations:

$$Q = mct.$$

## Transfer of Heat

Suppose a hot aluminium solid of mass 120 g, $c = 1·1$ J/g K at 90°C, is dropped into a copper vessel of $c = 0·4$ J/g K and mass 100 g, containing 80 g of water originally at 20°C. Since the temperature of the solid (90°C) is greater than the temperature (20°C) of the water and copper vessel, heat continues to pass from the solid to the water and copper vessel until the temperatures of all three substances concerned are the *same*. The temperature of the mixture will be less than 90°C and greater than 20°C.

To calculate the final temperature of the mixture, $x$°C, say, we start with the fact that

*Heat lost by solid = Heat gained by water and vessel* . . (1)

The statement 'heat lost = heat gained' occurs in all problems of heat transfer, as the reader will observe, and follows from the law of conservation of energy. We assume $c$ for water is 4·2 J/g K.

Since the change $t$ in temperature of the solid is $(90 - x)$ °C and of the water and vessel $(x - 20)$ °C,

Heat lost by solid $= mct = 120 \times 1·1 \times (90 - x)$ J
Heat gained by water $= mct = 80 \times 4·2 \times (x - 20)$ J
Heat gained by vessel $= mct = 100 \times 0·4 \times (x - 20)$ J.

Thus, from (1),

$120 \times 1·1 \times (90 - x) = 80 \times 4·2 \times (x - 20) + 100 \times 0·4 \times (x - 20)$
$\therefore 132(90 - x) = 336(x - 20) + 40(x - 20) = 376(x - 20)$
Solving, $\therefore x = 38°C.$

## The Calorimeter

In experiments to measure heat or specific heat capacities, heat exchanges occur between hot and cold objects inside a metal vessel such as the copper vessel mentioned under 'transfer of heat'. It is called a *calorimeter*, and often, though not always, it contains water.

The calorimeter is usually made of copper, a good conductor, so that it attains the same temperature as its contents as soon as possible. It is designed so that negligible heat exchange occurs between the contents and the surroundings. Thus to reduce heat losses by radiation to a minimum the vessel is polished both inside and out, and to eliminate heat losses by convection in the surrounding air and by conduction, the calorimeter is placed in a larger container and surrounded by layers of lagging, such as cotton waste, which is a poor conductor (Fig. 12.2). An insulating lid is sometimes used to cover the calorimeter. This prevents evaporation of the liquid inside.

Cups called Aerocups, made of expanded polystyrene, are very good insulators, and may be used in place of calorimeters in some experiments if desired.

If a copper calorimeter has a mass of 90 g and $c = 0.4$ J/g K, then

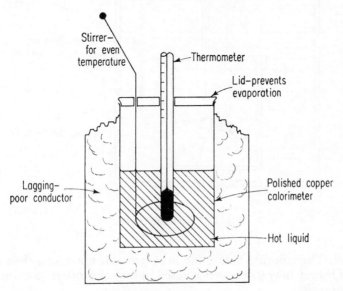

Fig. 12.2 Calorimeter

its Heat capacity = Mass × Specific heat capacity = $90 \times 0.4 = 36$ J/K. The heat gained by the calorimeter if its temperature rises 5°C is then $36 \times 5 = 180$ J.

Suppose that the calorimeter contains 40 g of water, specific heat capacity 4·2 J/g K, and that the water and calorimeter rose 5°C due to a transfer of heat. Then

heat gained by calorimeter = 180 J

and heat gained by water = $mct = 40 \times 4.2 \times 5 = 840$ J

Hence        total heat gained = $180 + 840 = 1020$ J

## Fuel Calorimeter

A fuel calorimeter, used for measuring the quantity of heat obtained by burning fuel or food, is shown in Fig. 12.3. Water in a vessel surrounds a sealed container with the fuel inside, and the combustion is started by an electric heater coil immersed in the fuel. A measured mass of fuel or food can thus be ignited and completely burnt. The heat evolved is found from the temperature rise and thermal capacity of the surrounding water

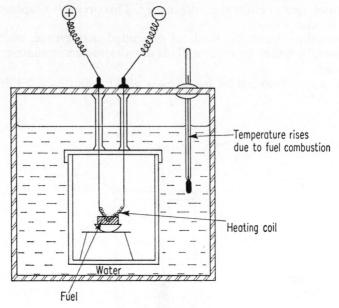

Fig. 12.3 Fuel calorimeter

and vessel. The calorific value, the heat per unit mass, can then be calculated. Oxygen may sometimes be used in the calorimeter to ensure complete combustion.

## Heat Capacity Measurement

In simple laboratory measurements, objects whose thermal capacities are required may be heated to about 100°C by placing them in water which is brought to the boil.

### EXPERIMENT

To measure the heat capacity of a block of metal, first weigh a copper calorimeter, fill it half to two-thirds full of cold water (sufficient to cover the metal) and reweigh. Replace the calorimeter in its lagged jacket. Record the water temperature.

Attach a thread to the metal, place it inside water in a beaker and heat the water to the boiling point. After a few minutes transfer the hot metal *quickly* to the calorimeter and water, shaking the metal to remove any surplus water on it. Stir the water, and note its final steady maximum temperature.

*Measurements.* As an illustration, suppose the measurements are as follows:

> Mass of calorimeter           $= \quad$ 80·0 g $(c = 0·4 \text{ J/g K})$
> Mass of calorimeter + water $=$ 140·5 g
> Initial temperature of water $=$   14·0°C
> Final temperature of water   $=$   19·0°C
> Temperature of hot metal     $=$ 100·0°C

*Calculation.* Heat lost by metal $= Y \times (100 - 19)$ J, where $Y$ is the heat capacity.

> Heat gained by water           $= 60·5 \times 4·2 \times (19 - 14)$ J
> Heat gained by calorimeter $= 80 \quad \times 0·4 \times (19 - 14)$ J

Now heat lost by metal = Heat gained by water and calorimeter, neglecting heat losses.

$$\therefore Y \times (100 - 19) = 60·5 \times 4·2 \times (19 - 14) + 80 \times 0·4 \times (19 - 14)$$

$$\therefore 81\, Y = 1270·5 + 160 = 1430·5$$

$$\therefore Y = \frac{1430·5}{81} = 17·7$$

*Conclusion.* The heat capacity of the metal is 17·7 J/K.

## Specific Heat Capacity of Metal by Method of Mixtures

The specific heat capacity of a solid can be found in the laboratory by warming it to a high temperature, and then quickly transferring it to a lagged calorimeter containing cold water. The water then becomes warmer and the solid cooler, both finally reaching the same temperature. The solid has lost some of its heat to the water and calorimeter.

### Experiment

Weigh a piece of brass or other metal, attach a thread to it and then place it in a beaker of boiling water. Now weigh a clean dry (copper) calorimeter, half-fill it with water and reweigh. Note the water temperature.

After the brass has been in the water at least 5 min transfer it from the beaker to the calorimeter as quickly as possible. Shake off water from the metal during the transfer. Stir the resulting 'mixture' and note the highest water temperature obtained. Make sure that the brass was completely submerged in the water after transfer–if not, serious error occurs (why?) and the experiment should be repeated with sufficient depth of water to cover the brass.

*Measurements.* Suppose measurements are as follows:

> Mass of brass                       $=$ 150·0 g
> Mass of calorimeter              $= \quad$ 60·0 g $(c = 0·4 \text{ J/g K})$
> Mass of calorimeter + water $=$ 150·0 g
> Initial water temperature       $=$   16·0°C $(c = 4·2 \text{ J/g K})$

Final water temperature   $= 26 \cdot 5°C$
Temp. of boiling water   $= 100°C$ (assumed)

*Calculation*

Heat gained by 90 g water $= 90 \times 4 \cdot 2 \times (26 \cdot 5 - 16)$ J
Heat gained by calorimeter $= 60 \times 0 \cdot 4 \times (26 \cdot 5 - 16)$ J
Heat lost by brass $= 150 \times s \times (100 - 26 \cdot 5)$ J

where $c$ is the specific heat capacity of brass. Using

Heat lost by brass = Heat gained by water and calorimeter,

$$\therefore 150 \times c \times (100 - 26 \cdot 5) = 90 \times 4 \cdot 2 \times (26 \cdot 5 - 16) + 60 \times 0 \cdot 4 \times (26 \cdot 5 - 16)$$

$$\therefore 11\ 025\ c = 3969 + 252 = 4221$$

$$\therefore c = 0 \cdot 38$$

*Conclusion.* The specific heat capacity of brass is $0 \cdot 38$ J/g K.

The specific heat capacity of a *liquid* may be measured by heating a metal of known specific heat capacity to $100°C$ and then quickly transferring it to the liquid in a calorimeter. Assuming no chemical action takes place between the metal and liquid, the heat lost by the metal = the heat gained by the liquid and calorimeter; and the unknown specific heat capacity of the liquid can hence be found.

According to the kinetic theory, the molecules of a solid vibrate with increasing amplitude when heat is supplied. Consequently, their kinetic energy increases. Now if the average kinetic energy of the molecules of a solid increases, its temperature increases. The average kinetic energy gained by the molecules per unit mass per degree C temperature rise is the specific heat capacity of the solid.

## Determining the Temperature of the Bunsen-burner Flame

The unknown temperature of a bunsen-burner flame can be determined by the method of mixtures. A metal, such as brass, is suspended in the flame for a period of time, and it is then quickly transferred to a calorimeter containing water. The mixture is stirred, and the final constant temperature is noted.

The temperature $t$ of the bunsen flame is calculated as follows:

Mass of brass $= 65 \cdot 0$ g
Mass of calorimeter $= 42 \cdot 5$ g
Mass of calorimeter + water $= 93 \cdot 6$ g
Initial temp of water $= 14 \cdot 6°C$ ($c = 4 \cdot 2$ J/g K)
Final temp of water $= 68 \cdot 4°C$
Temperature of burner $= t°C.$

Sp. ht. capacity of calorimeter $= 0 \cdot 4$, sp. ht. capacity of brass $= 0 \cdot 36$.
Heat lost by brass = Heat gained by calorimeter + Water

$$\therefore 65 \times 0 \cdot 36 \times (t - 68 \cdot 4) = (42 \cdot 5 \times 0 \cdot 4 + 51 \cdot 1 \times 4 \cdot 2)(68 \cdot 4 - 14 \cdot 6)$$

$$\therefore t - 68 \cdot 4 = \frac{232 \times 53 \cdot 8}{65 \times 0 \cdot 36}$$

$$\therefore t = 600°C.$$

## Electrical Heating

Heat capacities and specific heat capacities can also be found by electrical heating. In this case a 12-volt immersion heater can be placed in a liquid such as paraffin oil, for example, and with an ammeter in series, the heat supplied in a time $t'$ s is given by $IVt'$ J, where $I$ is the constant current in amperes and $V$ is the constant p.d. in volts across the immersion heater measured with a voltmeter (see p. 533). Then if the temperature rise at the end of the time is $t$, the mass of liquid is $m$, its specific heat capacity is $c$ and the heat capacity of the calorimeter is $C$, the heat gained is $(mc + C)t$. Hence the specific heat capacity $c$ can be calculated.

A *joule meter*, which has a dial in units of 100 J marked on it, can also be used to measure directly the number of joules supplied by a 12-volt A.C. supply.

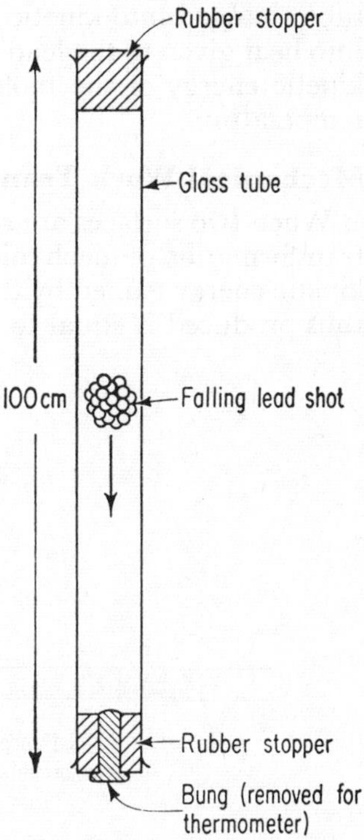

## Heat (Thermal) Equivalent of Mechanical Energy

So far in laboratory experiments heat has been produced by means of burners, which change chemical energy to heat energy, or by the change of electrical energy to heat energy. We now consider two experiments which change mechanical energy directly to heat energy in metals.

*Potential Energy Transformation to Heat*

Take a long wide glass tube about a metre long, having a stopper at one end and a stopper with a rubber bung at the other, so that a thermometer can be introduced here. (Fig. 12.4). Weigh about 250 g of lead shot, note its initial temperature, $t_1°C$, and place it in the glass tube.

Fig. 12.4 Conversion of mechanical to heat energy

Making sure first that both stoppers are firmly sealed at each end, invert the tube smartly about one hundred times. Each time allow the lead shot to fall vertically through the total length of the tube before turning the tube upside down. Count and record the number, $n$, of inversions, and measure quickly the new temperature, $t_2°C$, of the lead shot by removing the bung and inserting the thermometer bulb in the shot.

*Calculation*

(1) *Heat*

The temperature rise $= (t_2 - t_1)$ °C
Let the mass of shot taken $= m$ g, specific heat capacity of lead $= c$
The total heat produced $= Q$
$$= mc(t_2 - t_1) \text{ J} \quad . \quad . \quad . \quad . \quad . \quad . \quad (1)$$

(2) *Mechanical Energy.* Suppose the tube was turned 100 times. Then the potential energy lost $W = 100 \times mgh$ J, where $h$ is the height in m fallen each time by the lead shot, $m$ is the mass in kg of the lead shot and $g = 9.8$.

$$\therefore W = 100 \ mgh \ \text{J} \ . \quad . \quad . \quad . \quad . \quad . \quad . \quad (2)$$

List the errors in the experiment for measuring the heat produced $Q$ and the mechanical energy $W$ transformed, and compare your results in (1) and (2).

In this experiment the potential energy of the lead shot at the top of the tube changed into kinetic energy at the bottom, which, in turn, changed into heat given to the lead shot. Since heat is a form of energy, the average kinetic energy of the molecules of the lead is increased, thus raising its temperature.

## Mechanical Work Transformation to Heat

When two surfaces are rubbed together there is a direct and continuous transformation of mechanical work, done against the frictional forces, into kinetic energy gained by the individual molecules of the surfaces. The heat thus produced is equal to the increase in this kinetic energy.

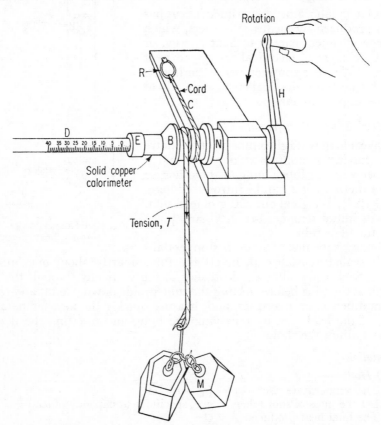

Fig. 12.5 Measurement of mechanical energy and heat produced

A method of measuring the heat produced and the mechanical work done against friction is shown in Fig. 12.5. The apparatus consists of a solid cylinder copper calorimeter B, mounted horizontally on a nylon bush N to provide insulation from the rest of the apparatus. N is connected to a crank H which can be turned. A short-range thermometer D is fixed into a hole bored axially in the calorimeter and held firmly by a plastic collar E. One end of a flexible cord C is attached to a rubber band R and then wound several times round B as shown. Masses M of about 5 kg are attached to the other end of C.

*Method.* The reading on the thermometer is first noted. The crank is then turned at a constant rate of about 1 revolution per second in such a way that the weights are completely supported by the cord and that the rubber band R is fairly slack. After about 300 revolutions are counted the temperature is again noted. The calorimeter is weighed and its diameter is measured by vernier calipers (p. 5).

*Results.* The following results were obtained by a boy:

| | |
|---|---|
| Initial calorimeter temperature | $= 15{\cdot}5\ °\text{C}$ |
| Final temperature | $= 32{\cdot}5\ °\text{C}$ |
| Number of revolutions | $= 330$ |
| Mass of calorimeter ($c = 0{\cdot}4$ J/g K) | $= 184$ g |
| Diameter of calorimeter | $= 2{\cdot}54$ cm |
| Mass on cord | $= 5$ kg. |

*Calculation*

(1) Work done $W$ against tension $T$, using weight of 1 kg mass $= 9{\cdot}8$ N and changing cm to metre,

$$= \text{Tension} \times \text{Distance moved}$$
$$= 5 \times 9{\cdot}8\ (\text{N}) \times \pi \times 2{\cdot}54 \times 330/100\ (\text{m})$$
$$= 1290\ \text{J} \quad . \quad . \quad . \quad . \quad . \quad . \quad . \quad . \quad (1)$$

(2) Heat produced $Q = \text{Mass} \times \text{sp. ht. capacity} \times \text{Temperature rise}$
$$= 184 \times 0{\cdot}4 \times 17\ \text{J}$$
$$= 1250\ \text{J} \quad . \quad . \quad . \quad . \quad . \quad . \quad . \quad (2)$$

Allowing for errors such as heat losses by cooling from the copper calorimeter, it can be seen from (1) and (2) that practically all the mechanical energy expended has been transformed into heat energy.

We can use the apparatus to measure the specific heat capacity $c$ of the metal of the calorimeter. To do this, the experiment is carried out as described above. The heat produced $Q = $ mechanical work done, and can be found from equation (1). Assuming no loss of heat, $Q = mct$, where $m$ is the mass of metal and $t$ its temperature rise. Thus $c = Q/mt$, and knowing $Q$, $m$ and $t$, we can calculate $c$.

## Heat Engines

The experiments performed by Joule on energy changes (p. 266) led scientists to look for other energy transformations and eventually to the idea of a *conservation of energy*. This principle states that energy cannot be created or destroyed, but can only be transformed from one form to another (see p. 126). Hence it is impossible to obtain more energy out of a machine than the energy expended on or supplied to it. 'Perpetual motion machines' designed in the past appeared at first to be able to go on moving for ever once started, without any energy expended on them. Closer examination, however, shows that the design would violate the principle of the conservation of energy.

Heat energy is converted into mechanical energy in a *heat engine*, of which there are many types. The steam engine, the internal-combustion engine, the diesel fuel engine and the jet combustion engine all use different sources of heat. They all use heat energy to provide movement. In the case of the petrol engine, this eventually leads to rotary motion of wheels of cars, for example, and in the case of the jet engine, to linear forward motion of aeroplanes.

It is important to realize that, in practice, machines are less than 100% efficient, that is, the energy obtained is always less than that supplied. In a steam engine some energy is wasted in the exhaust gases; in the jet engine some energy is wasted in the hot gases of the jet as they expand.

## Petrol Engine

The petrol engine, commonly used in motor-cars, obtains its energy from an exploded mixture of air and the vapour of very volatile petrol. It is sometimes called the 'four-stroke cycle' engine because four piston strokes or movements inside the cylinder repeat themselves continuously. The piston strokes are in the order *intake, compression, explosion, exhaust*. The valve positions controlling the cylinder openings, the directions of the piston movements and the state of the gases during each stroke are shown in Fig. 12.6.

*Intake*
As the piston moves down the cylinder the inlet valve opens and a mixture of air and vapour is drawn into the cylinder.
*Compression*
The piston moves up and compresses the mixture to about one-seventh of its volume. At the top of the stroke the mixture is exploded by a spark which passes between the electrodes of the sparking plug.
*Explosion*
The expanding gases force the piston down.
*Exhaust*
As the piston moves up, the exhaust valve is opened and the burnt or exhaust gases are expelled from the cylinder. The cycle is then repeated.

In the four-stroke engine a power stroke is obtained once every four strokes, which corresponds to two revolutions of the crankshaft. A two-stroke engine provides a power stroke once every revolution by combining two strokes of the four-stroke cycle into one. The valves of the four-stroke engine are replaced by openings or ports in the two-stroke engine cylinder, and these are covered and uncovered by the moving piston. The efficiency of the two-stroke engine is low in practice. It is therefore used to drive

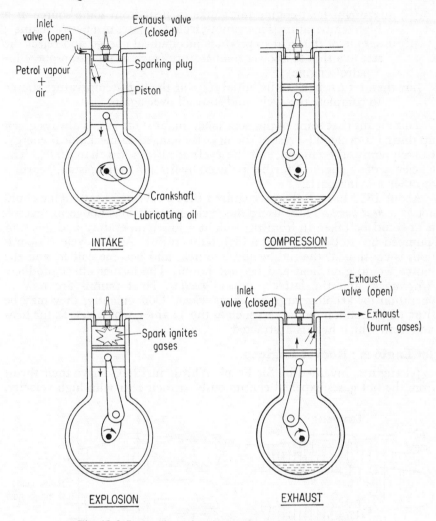

Fig. 12.6 Internal-combustion engine – four-stroke cycle

relatively small machines, such as lawn mowers and mopeds, for example, where the light motor is an advantage and fuel consumption is not the main consideration.

## Cycles and Efficiency · Heat Pump

The four-stroke cycle and other practical cycles were invented after Carnot had shown many years before that the maximum efficiency was obtained when a gas or vapour was made to work in a particular ideal cycle of operations, called later a *Carnot cycle*. It is above the level of this book to go into the details of the Carnot cycle. We may note, however, that:

(i) any useful heat engine must take in heat from some source at a high temperature (for example, from hot gases in a petrol engine);

(ii) use some of this heat to produce mechanical energy and reject the rest to a sink at a lower temperature (such as the atmosphere for a petrol engine);

(iii) then return back to its initial state at the high-temperature source to complete the cycle, and start all over again.

This means that some of the heat taken in, at stage (ii), is always given up during a cycle. The rest of the heat is changed to mechanical energy. Consequently, the efficiency of the cycle is always less than 100%. The efficiency of an engine is further reduced by friction in the piston, bearings or other moving parts.

About 1852 Lord Kelvin recognized that a reversed heat engine could act as a *heat pump*. Thus, using the cycle of operations above in reverse, heat could be taken in from the sink at a low temperature and given or 'pumped up' to the source at a high temperature. As this cycle of operations is continued, the sink would lose heat and become colder, and the source would gain heat and become hotter. The former effect produces *refrigeration*, and the latter produces *heating*. Heat pumps are now in operation for keeping large buildings warm. Conveniently, they may be situated in London near rivers such as the Thames, which acts as the heat sink from which heat is abstracted.

### Jet Engines · Rocket Engines

Jet engines, invented by Sir Frank Whittle in 1940, derive their thrust from the hot gases expelled continuously at their rear with high velocity.

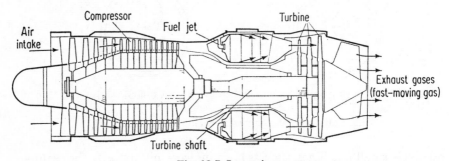

Fig. 12.7 Jet engine

From Newton's law of Action and Reaction, the thrust on the engine is equal and opposite to that on the gas molecules, and this can drive an aeroplane forward at very high speeds.

In a turbo-jet engine, air is drawn in at the front and compressed, and then, after mixing with fuel, it is burnt in a combustion chamber and ejected with very high velocity through an exhaust nozzle at the rear. (Fig. 12.7). The thrust is calculated as given on p. 280. Jet aeroplanes can fly at very high altitudes, where the efficient engine takes in the very large volumes per second of thin air required.

*Rocket engines* are used for launching satellites in orbits round the earth. Unlike jet engines, which need air as fuel, a rocket engine carries its own fuel and oxidizer, and can therefore function at great heights where very little air is present. (Fig. 12.8). Liquid or solid fuel may be used, and very

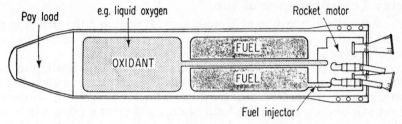

Fig. 12.8 Rocket engine principle

hot gases are expelled from the rear with extremely high velocity. Like the jet engine, the upward thrust on the rocket and load carried is equal and opposite to the downward thrust on the burnt gases. The rocket stops working in outer space as soon as the fuel is used up.

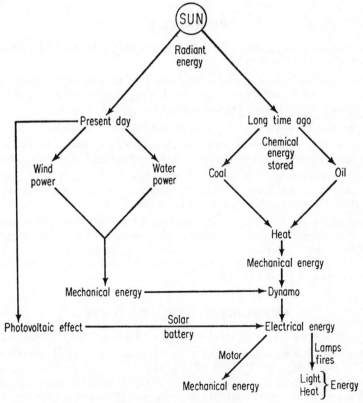

Fig. 12.9 Energy from Sun

The thrust of a rocket or jet engine, then, is due to the rapid change in momentum of the gas ejected from the exhaust. From p. 110 this is equal to the product of the mass of gas issuing per second and its velocity relative to the rocket or engine.

## Energy Transformation of Sun

The sun is the source of most of our energy. It sends out the energy in the form of radiation, and on the earth this is converted into other forms of energy by so-called natural processes. (Fig. 12.9).

Wind power, for example, originates from the unequal heating of land and sea, which produces convection currents. Water power, used in hydro-electric plants, originates from the water which evaporates from the sea and returns to the earth in the form of rain and creates lakes and rivers. Oil has chemical energy derived initially from the sun. It is formed from decayed micro-organisms buried deep in the ground millions of years ago, after using energy from the sun to grow. Coal has chemical energy due to decayed vegetation and wood, buried deep in the ground after absorbing the sun's energy. Today, light and heat energy from the sun are used to generate electrical energy by the photoelectric effect, or in the solar battery, or in the solar furnace.

## SUMMARY

1. Units of heat are: joule, calorie ($4 \cdot 2$ J = 1 cal), kilojoule (kJ), megajoule (MJ). The SI unit of heat is the joule.

2. Heat capacity is the quantity of heat required to raise the temperature of a substance by 1°C. Specific heat capacity is the quantity of heat required to raise the temperature of unit mass of a substance by 1°C.

3. Heat capacity = Mass × Specific heat capacity. Unit J/K.

4. Specific heat capacities of solids and liquids can be found by the method of mixtures–Heat lost = Heat gained – or by the method of electrical heating.

5. Mechanical energy can be transformed into heat energy by doing work against friction. Work done = Force × Distance in direction of force. Heat produced = Heat capacity × Temperature rise.

6. Heat engines convert heat into mechanical energy. The petrol internal-combustion engine explodes a mixture of air and petrol vapour by means of a spark from a sparking plug, and the order for a four-stroke cycle is intake, compression, explosion and exhaust. The jet and rocket engines use reaction forces to produce a forward thrust.

## EXERCISE 12 · ANSWERS, p. 320

(Assume the specific heat capacity of water = $4 \cdot 2$ J/g K or 4200 J/kg K.)

1. How many joules are required to raise the temperature of 500 g of copper from 16° to 116°C? ($c$ of copper = $0 \cdot 4$ J/g K or 400 J/kg K.)

2. A liquid of specific heat capacity 3 J/g K or 3000 J/kg K rises from a temperature of 15° to 65°C and 1 min when an electric heater is used. If the heater generates 63 kilojoules per min, calculate the mass of liquid.

**3.** 4 g of a substance is burned completely, and the heat produced raises the temperature of 1200 g of water from 16° to 26°C. Calculate the calorific value of the substance in megajoules per kilogram.

**4.** Describe and explain the chief features of a *calorimeter* which make it suitable for use in heat experiments.

**5.** Define *three* different units of heat. Which is the SI unit?

**6.** What is meant by the calorific value of a fuel? (*N.*)

**7.** It was found that 0·018 m³ of gas was consumed in raising the temperature of 1·20 litres of water from 20° to 100°C. Find the calorific value of the gas in J/m³. (*C.*)

**8.** Explain the meaning of (*a*) specific heat capacity, (*b*) temperature, (*c*) quantity of heat, and state how they are related.

2000 g of lead at 100°C are dropped into a copper vessel containing 300 g of water at 0°C and rapidly stirred. The final temperature attained by the vessel and its contents is 16°C. Taking the specific heat capacity of lead as 0·13 J/g K or 130 J/kg K, calculate the heat capacity of the copper vessel.

**9.** Explain briefly what is meant by: (*a*) heat; (*b*) temperature.

The calorific value of a good-quality coal is about 35 000 J/g. Describe an experiment to prove this statement, and give a diagram of your apparatus.

**10.** (*a*) Define *heat capacity, specific heat capacity*. State the relation between them.

In order to obtain the temperature of a freezing mixture, a 500-g iron weight is left in it for some time. The iron is then rapidly transferred to a well-lagged copper calorimeter of mass 100 g containing 110 g of water at 40°C. If the final temperature of the mixture is 12·0°C, calculate the temperature of the freezing mixture. (Specific heat capacity of iron is 0·50 and of copper 0·42 J/g K or 500 and 420 J/kg K respectively.)

(*b*) Define the *joule*. A small electric heater immersed in 80 g of water in a calorimeter of heat capacity 33·6 J/K raises the temperature from 9·5° to 20·0°C in 12 min. Using the same apparatus with 60 g of oil instead of water, it is found that the temperature of the oil rises from 6·5° to 23·0°C in 9 min. Find the specific heat capacity of the oil. (*L.*)

**11.** Describe an experiment to determine the specific heat capacity of a liquid. Point out errors which are likely to occur and the precautions you would take to minimize them.

A machine with an input power of 2 kilowatts uses 80% of this power. If all the remaining energy appears as heat, and heats 40 kg of iron, what will be the rise of temperature of this iron in 2 min? (Specific heat capacity of iron = 0·5 J/g K or 500 J/kg K.) (*N.*)

**12.** Define specific heat capacity of a substance, calorific value of a fuel. Describe how you would determine the specific heat of a liquid.

A hollow metal vessel is immersed in 500 g of water contained in a larger vessel. The inner vessel has inlet and outlet tubes through which a supply of oxygen is passed. 1·56 g of coal dust is ignited and completely burnt in the inner chamber. If the temperature of the water increases from 5·6° to 31·8°C during the experiment, and the thermal capacity of the complete calorimeter is 300 J/K, find the calorific value of the fuel. (*C.*)

**13.** A small immersion heater raises the temperature of 250 g of water contained in a vacuum flask from 15°C to 55°C in 5·0 minute. When the experiment is repeated with 500 g of water in the same flask a time of 9·0 minute is required for the same range of temperature.

Determine (*a*) the heat per minute given out by the immersion heater, (*b*) the heat absorbed by the vacuum flask for each °C it rises in temperature. (*L.*)

**Energy Conversions** *(Where necessary, assume g = 10 m/s²)*

**14.** An object of mass 80 kg falls from a height of 100 m. Calculate the heat developed at the ground.

**15.** A water-fall has a height of 50 metre. Considering 1 g of water and the energy it loses, estimate the difference in temperature between the top and bottom of the water-fall. Why is the answer approximate?

**16.** A leather belt is wound round a wheel of diameter 60 cm. The wheel rotates while the belt is kept stationary, and the average frictional force produced is 100 N. What heat is generated when the wheel rotates through 20 complete revolutions?

**17.** State two distinct ways in which mechanical energy may be transformed into heat energy.

Describe and explain an experiment to determine the relation between a unit of heat and a unit of work, stating the units used.

A steam engine develops 195 h.p. and its boiler uses 200 kg of coal per hour. If the calorific value of the coal is $3 \times 10^7$ J/kg, find the efficiency of the plant, (1 h.p. = 750 W) *(L.)*

**18.** Describe an experiment, based on conversion of mechanical or electrical energy, to show that the specific heat capacity of water is about 4·2 J/g K or 4200 J/kg K.

A piece of lead falls 3 m from rest, coming to rest again on the ground. Calculate the rise in its temperature, the specific heat capacity of lead being 0·13 J/g K or 130 J/kg K. State the assumptions you make in the calculation. *(L.)*

**19.** An electric heater of 50 watts is used to heat a metal block of mass 0·5 kg. In 10 minutes, a temperature rise of 12°C is produced.

  (i) How much heat is produced by the heater in 10 minutes?

  (ii) Calculate the specific heat capacity of the metal.

Describe fully how you would measure electrically the specific heat capacity of water.

**20.** State briefly why heat is considered a form of energy. Give an example of a process in which heat energy is translated into mechanical energy. In an experiment a mass of lead shot contained in a vertical cardboard cylinder, falls 60 cm when the cylinder is inverted. Calculate the rise of temperature caused by 70 inversions, taking the specific heat capacity of lead as 0·13 J/g K or 130 J/kg K. What are the probable sources of error in this experiment? *(L.)*

**21.** (*a*) 'Heat is a form of energy.' What form does the energy take in the case of a gas such as helium, whose molecules are single atoms? (*b*) Estimate the difference in temperature between the bottom and top of a waterfall 100 metres high. *(O. and C.)*

**22.** In a rough determination of the specific heat capacity of lead a quantity of lead shot was placed in a cardboard cylinder of length 1 m and was allowed to fall backwards and forwards from one end of the tube to the other 50 times in succession. The temperature of the shot was found to have risen 3·7°C. Calculate the specific heat capacity of lead. *(O. and C.)*

**23.** How do you explain the rise in temperature of a bicycle pump during the action of pumping up a tyre? *(L.)*

**24.** Describe a simple method of measuring the specific heat capacity of water and explain how the result is calculated from the readings. Fifty litres of water are heated by a rotating paddle driven by a ¼-h.p. motor. Assuming that all the work done is used to heat the water, calculate the time required to raise the temperature of the water 5°C neglecting heat losses. (Take 1 h.p. = 746 W) *(O. and C.)*

**25.** What do you understand by the statement that heat is a form of energy? Illustrate this statement with reference to: (*a*) the molecules of a gas in a container at a steady temperature; (b) the energy changes that take place in an internal-combustion engine during the power stroke.

Give a brief account of ONE method for determining the mechanical equivalent of heat.

**26.** Give an account of the principles of: (a) the four-stroke petrol engine; (b) the jet engine.

**27.** A train (of mass $m$ g) has an average specific heat of 0·84 J/g K. It is warmed in winter from $-1\cdot0°$ to $15\cdot0°$C by engine steam. Assuming that the engine could have converted one sixteenth of the heat of the steam which was used to warm the train into useful work through what vertical height would this energy have moved the train and how far would this have taken it up a 1 in 1000 gradient? (O. and C.)

**28.** What is meant by the statement 'heat is a form of energy'? Describe how you would determine the specific heat capacity of a metal block using electrical energy and point out the sources of error in your method.

A lead shot at a temperature of 37°C is fired at a speed of 300 m/s into a resisting target. Assuming that all the heat remains in the shot, what fraction of the shot melts? (Specific heat capacity of lead = 0·13 J/g K or 130 J/kg K; specific latent heat of lead = 25·2 J/g or 25 200 J/kg; melting-point of lead = 327°C. (O. and C.)

**29.** State the meaning of the terms: mechanical energy, joule.

A calorimeter of thermal capacity 42 J/K contains 500 cm³ of water at 20°C. An automatic stirrer dips into the water and is driven by a ¼-h.p. motor. Assuming that all the work done by the stirrer heats the water, calculate the temperature of the calorimeter and its contents after 10 min. (Take 1 h.p. = 746 W.) (O. and C.)

# 13

# CHANGE OF STATE · LATENT HEATS

## Latent Heats

When sufficient heat is given to a solid it melts eventually and forms a liquid. Further heat eventually makes the liquid change into a gas.

As a convenient example of a change of state from solid to liquid and then to a gas, suppose a block of ice is cooled by a freezing mixture to $-10°C$ and then heated steadily. At first the ice, which has a specific heat of about $1 \cdot 7$ J per g per deg C, rises in temperature to $0°C$. It now changes into water, a liquid, but although heat continues to be supplied, the temperature remains constant at $0°C$ until *all* the ice has changed to water. Since the temperature gave no indication of the heat required to change from a solid to a liquid, this was called *latent heat* ('hidden' heat) by Black over two hundred years ago, and the name has persisted. It should be carefully noted that the term is not used when the temperature of a solid rises, because there is no change of state in this case.

When the water formed from the ice is heated further, its temperature rises until it reaches about $100°C$. Here it begins to change into vapour or gas, and the temperature remains constant until *all* the water is changed to vapour. The heat is again here called *latent heat* because a change of state, this time from liquid to gas, occurs. Experiment shows that the latent heat to change a given mass from ice to water is about one-seventh of the latent heat to change the same mass from water to vapour under normal conditions. After a vapour is formed, further heat causes the vapour temperature to rise once more.

## Molecular Explanation

We now know that the molecules of a solid are kept in an average fixed position by strong forces of attraction (p. 33). Thus the latent heat represents energy used in breaking down the forces which hold the molecules of a solid in a regular pattern, and setting them free to slide about, as they do in the liquid state. Similarly, the latent heat to change a liquid to a vapour without temperature change is the energy required to overcome the forces of attraction between the molecules of a liquid so that the molecules are practically independent of each other and exist as a gas. The energy needed in this case is much more than in the former case, and the latent heat is correspondingly greater.

## Fusion

We first consider change of state from solid to liquid, or fusion.

The *specific latent heat of fusion* of a substance is the quantity of heat required to change unit mass from a solid at the melting point to a liquid *at the same temperature*. For example, ice melts at $0°C$, and its specific latent heat of fusion $l$ is the heat required to change 1 g (or 1 kg) of ice at 0°C to water at 0°C. It is about 336 J/g or 336 000 J/kg.

Suppose 5 g of ice 0°C is heated so that all the ice melts and forms water at a final temperature of 10°C. This change occurs in two stages: (1) the ice melts to water *at 0°C*, (2) the water formed rises in temperature from 0° to 10°C (Fig. 13.1). The heat required for stage (1) depends on the value of the latent heat of fusion $l$ for ice, 336 J/g, since it is a change from a solid to a liquid state *at the melting point*. Thus the heat $= 5 \times l = 5 \times 336 = 1680$ J (Fig. 13.1).

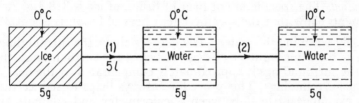

Fig. 13.1  Heat needed to change ice to water at 10° C

The heat for stage (2) concerns only the formula $Q = mct$, since 5 g of water at 0°C rises in temperature to 10°C (no change of state occurs). Thus $Q = 5 \times 4.2 \times 10 = 210$ J. The *total* amount of heat concerned is thus $(1680 + 210) = 1890$ J.

## Specific Latent Heat of Fusion of Ice

The method of mixtures can be used to determine the specific latent heat of fusion of ice.

*Method.* A clean, dry copper calorimeter is weighed empty, together with a stirrer. The calorimeter is then half-filled with water previously heated to about 8 deg above the room temperature, and reweighed. The final water temperature is then noted. Suppose, by this time, it is about 5 deg above room temperature. Now dry small pieces of ice on blotting-paper and add a piece at a time to the water. Stir continuously until the ice is all melted before adding another piece and continue until the final water temperature is 5 deg *below* the room temperature. In this way we have largely compensated for the exchange of heat which always takes place between the calorimeter and contents and the surrounding air when one is at a different temperature. Reweigh the calorimeter to find the mass of ice which has been melted.

*Results.* Suppose the following results were obtained:

| | |
|---|---|
| Mass of calorimeter ($c = 0.4$ J/g K) | $= 56.3$ g |
| Mass of calorimeter + water | $= 139.4$ g |
| Mass of melted ice + calorimeter + water | $= 149.6$ g |
| Initial water temperature ($c = 4.2$ J/g K) | $= 21.0$°C |
| Final water temperature | $= 11.0$°C |

*Calculation.* The ice melts and forms water at 0°C (latent heat of fusion gained), and then the water formed rises to 11°C (no change of state, so heat gained here is given by $Q = mc \times$ temperature change). Thus, since 10.2 g of ice was melted,

Heat gained by ice in forming water at 11°C
$$= 10 \cdot 2l + 10 \cdot 2 \times 4 \cdot 2 \times (11 - 0).$$

Since 83·1 g of water was used in the calorimeter,
Heat lost by water and calorimeter

$$= 83 \cdot 1 \times 4 \cdot 2 \times (21 - 11) + 56 \cdot 3 \times 0 \cdot 4 \times (21 - 11)$$
$$\therefore 10 \cdot 2l + 471 \cdot 2 = 3490 + 225 \cdot 2 = 3715 \cdot 2$$
$$\therefore l = 318$$

*Conclusion.* The specific latent heat of fusion of ice is 318 J/g. Accurate experiments show that the specific latent heat of fusion is about 336 J/g.

*Note.* By warming the water a few degrees above the surroundings, and cooling it the same number of degrees below the surroundings, the heat lost to the surroundings is approximately compensated by the heat gained from the surroundings. This procedure always helps to reduce the error due to exchange of heat between a calorimeter and contents and the surroundings.

### Latent Heat of Vaporization *(Liquid to Vapour Change)*

We now consider the change from a liquid to a vapour state, when 'boiling', or 'vaporization', is said to take place.

The *specific latent heat of vaporization l* of a substance·is *the quantity of heat required to change unit mass of the substance from its liquid to its vapour state at the boiling point.* About 2260 J are needed to change 1 g of water at the boiling point to vapour at the same temperature, so that $l = 2260$ J/g or $2 \cdot 26 \times 10^6$ J/kg for steam.

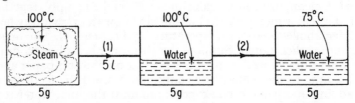

Fig. 13.2 Heat given up when steam condenses to water at 75° C

If 5 g of steam at 100°C, say, condense to water *at* 100°C, $5 \times 2260$ joules of heat are given up (Fig. 13.2). If the steam condenses to water which reaches a temperature of 75°C, however, the heat given up is the sum of two quantities:
(1) The heat given up when 5 g of steam at 100°C condenses to water *at* 100°C, which is $5l = 5 \times 2260$ J. (2) The heat given up when the water decreases in temperature from 100°C to 75°C; from $Q = mct$, this is $5 \times 4 \cdot 2 \times (100 - 75)$, or 525 J. Thus the total heat given up $= 11300 + 525 = 11\ 825$ J.

### Specific Latent Heat of Steam

The specific latent heat of vaporization of steam can be found by passing steam into water in a calorimeter, where it condenses and gives out its latent heat. One form of apparatus is shown in Fig. 13.3. The steam is dried by a water, or so-called 'steam', trap which ensures that only steam enters the calorimeter by the short delivery tube. If no trap were used the small but unknown amount of water entering the calorimeter would cause errors in the calculation, which assumes only dry steam enters. The delivery tube T beyond the water trap can be lagged to prevent condensation of steam.

*Method.* A clean, dry calorimeter C is weighed empty and then half-filled with water and reweighed. The calorimeter is then placed in its lagging material. When the water in the flask F is boiling and steam is coming out freely from the end of T the calorimeter is raised until T is below the water inside it. The steam then passes into the water, where it

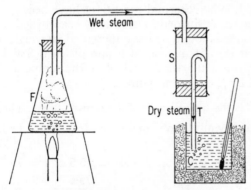

Fig. 13.3 Latent heat of steam

condenses. After the temperature has risen about 25°C above the surroundings the calorimeter is removed and the final water temperature after stirring is noted. The calorimeter is then reweighed to find the mass of steam condensed.

*Results.* Suppose the following results were obtained:

| | |
|---|---|
| Mass of calorimeter ($c = 0.4$ J/g K) | $= 60.0$ g |
| Mass of calorimeter + water | $= 144.0$ g |
| Mass of calorimeter + water + steam | $= 147.8$ g |
| Initial water temperature | $= 17.5$°C |
| Final water temperature | $= 43.0$°C. |

*Calculation.* When the steam passes into the water it first changes from steam at 100°C to water at 100°C (latent heat given up) and then drops in temperature to 43°C (no change of state, heat given up calculated from $Q = mc \times$ Temperature change). Since the mass of steam is 3.8 g from above,

Heat given up by steam $= 3.8l + 3.8 \times 4.2 \times (100 - 43)$
Heat gained by water + calorimeter
$\qquad = 84 \times 4.2 \times (43 - 17.5) + 60 \times 0.4 \times (43 - 17.5)$
$\qquad \therefore 3.8l + 910 = 9607$
$\qquad\qquad l = 2290$

*Conclusion.* The specific latent heat of steam is 2290 J/g or $2.29 \times 10^6$ J/kg. To reduce the error due to loss of heat to the surroundings when the temperature of the water rises as steam is passed in, the water can initially be cooled by melted ice until the temperature is, say, 15°C below the temperature of the room and then steam passed in until it rises about 15°C above room temperature. Compare 'Specific latent heat of fusion of ice', p. 285.

### Latent Heat Examples

**1.** An iron nut of mass 56 g is heated in a bunsen flame until the temperature becomes steady and is then transferred to a deep hole in a large block of ice. The top of the hole is covered immediately with a block of the same ice, to act as a lid. When thermal conditions have become steady the lid is removed and the water formed in the cavity is absorbed by a weighed dry sponge. The sponge is found to have increased in weight by 75·6 g.

Assuming that the latent heat of fusion of ice is 336 J/g and that the specific heat of iron is 0·5 J/g K, calculate the temperature to which the nut was heated by the bunsen flame.

Suppose $t$°C is the temperature of the bunsen flame. Then

Heat given up by nut from $t$° to 0°C = Heat gained by ice changing to water at 0°C

$$\therefore 56 \times 0.5 \times (t - 0) = 75.6 \times l = 75.6 \times 336$$

$$\therefore t = \frac{75.6 \times 336}{56 \times 0.05} = 910°C.$$

**2.** A copper calorimeter of mass 120 g contains 70 g of water and 10 g of ice at 0°C. What mass of steam at 100°C must be passed into the calorimeter to raise the temperature to 40°C?

(Specific latent heats: of ice 336 J/g; of steam 2260 J/g; specific heat capacity of copper 0·4 J/g K.)

Heat given up by steam condensing to water at 40°C
$$= ml + m \times 4.2 (100 - 40) = 2260m + 252m = 2512m \text{ J}$$
Heat gained by water and calorimeter from 0° to 40°C
$$= 70 \times 4.2 \times (40 - 0) + 120 \times 0.4 \times (40 - 0)$$
$$= 13\,680 \text{ J}$$
Heat gained by ice in melting to water at 40°C
$$= 10 \times 336 + 10 \times 4.2 \times (40 - 0)$$
$$= 5040 \text{ J}$$
$$\therefore 2152m = 13\,680 + 5040 = 18\,720$$
$$\therefore m = 8.7 \text{ g.}$$

### Melting Point

In this last section we consider further the effects produced when a solid melts or solidifies, particularly with regard to the change from ice to water and from water to ice. The *melting point* of a substance is defined as the temperature at which the substance changes from the solid to the liquid state.

As an illustration, suppose the melting point of naphthalene, a white, crystalline solid obtained from coal tar and used in moth balls, is required.

### Experiment

Fill a boiling tube with naphthalene and support the tube vertically, by using a clamp and stand. Using a large beaker with water as a water-bath, or using a very small bunsen flame, warm the tube until the naphthalene just melts. Now insert a thermometer into the liquid and continue heating until a temperature of about 90°C is obtained. Turn out the bunsen flame (or remove the water-bath) and record the temperature readings as they fall every half-minute. At a certain stage observe that the temperature remains constant for a time–a solid is now forming–after which the temperature falls once more.

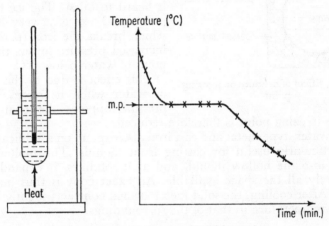

Fig. 13.4 Determination of melting point

*Graph.* Plot a graph of temperature against time (Fig. 13.4). The temperature corresponding to the flat portion is the melting point of naphthalene. At this temperature the heat lost to the surroundings is exactly balanced by the latent heat of fusion, which is given out when the liquid changes to the solid state.

## Expansion and Contraction on Fusion

Because of the expansion of water on freezing, ice is less dense than water, and hence ice floats on water. Icebergs float and drift a long way before they melt. Only about one-tenth of the volume of an iceberg lies above the surface of the sea, the remaining nine-tenths being hidden from view. In January 1966 an iceberg five times as big as outer London ran aground near Russia's south polar base, Molodezhnaya, in Enderby Land. It is probably the biggest one ever seen, measuring 8000 km². The height of the iceberg was 30 m above sea-level.

In very cold weather, when the temperature is below freezing point and the water in household pipes happens to freeze, the ice formed is greater in volume than the mass of water frozen. The pipe is hence subjected to a very great force, and the metal becomes cracked. When the temperature rises above the freezing point the ice in the pipe melts and the water pours through the cracks.

## Expansion of Water on Freezing

### Experiment

Completely fill a small thick cast-iron container with water previously cooled by ice to a temperature near 0°C, and then screw in the tight-fitting iron cap so that air is completely excluded. Now place the container in a freezing mixture of ice and salt (Fig. 13.5). In a little time, when the temperature falls below 0°C, the container is heard to burst. The ice formed expands and exerts a very large force which breaks the iron (p. 239). Thus increased pressure lowers the freezing point of water below 0°C.

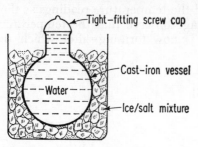

Fig. 13.5 Effect of pressure on freezing point

This effect is due to the expansion of water when it freezes. Increased pressure opposes the change to ice, and the freezing point is therefore reduced.

Like water, type metal and cast iron also expand on solidification. They are particularly useful for casting from moulds. The molten metal is poured into the hollow mould, and as it solidifies, it expands and fills completely all the space available. An exact copy is thus made of the mould. After cooling, the solid metal formed contracts slightly due to the drop in temperature to that of the surroundings.

## Effect of Pressure on Melting Point

*Demonstration*

Pass a thin wire round a large block of ice and attach two heavy weights to the ends of the wire. (Fig. 13.6). Observe that, as time goes on, the wire slowly sinks into the ice and that no 'cut' is visible above it. Eventually, with the block still in one piece, the wire passes through at the bottom.

When ice is subjected to a large pressure its volume tends to decrease. Now when ice changes to water its volume becomes smaller, and hence increased pressure on ice helps it to change to water. It follows that ice can be melted if sufficient pressure is applied to it and, in general, *the melting point of ice is lowered when it is subjected to increased pressure.* The thin wire in the above experiment produces such a high pressure on

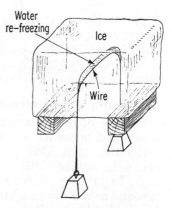

Fig. 13.6 Refreezing (Regelation)

the ice below it that the ice melts. The wire then sinks a little into the ice through the water that is formed, and the latter, being freed from the pressure, immediately freezes again on the other side of the wire. In this way the wire gradually sinks or 'cuts' into the ice-block, and eventually the wire passes completely through the block, leaving it unaltered. This phenomenon is known as *regelation* (refreezing).

## Other Phenomena

Ice-skaters are able to move freely on account of the lowering of the melting-point of ice under increased pressure. The knife-edge runner of the skate has a very small cross-sectional area, and hence the pressure under the knife-edge $\left(\dfrac{\text{Weight of skater}}{\text{Area of knife-edge}}\right)$ is very large. Together with the heat produced by friction, the net effect is the melting of ice under the skate, and the skater moves easily through a thin film of water on top of the ice. If it is too cold, the ice does not melt, and the skate moves with difficulty over the ice.

Two small pieces of ice, pressed one against the other, stick together after releasing them. The increased pressure lowers the melting point, and those parts of the ice under high pressure therefore melt. The water formed flows away to a place where the pressure is less. Here it refreezes, thus binding the two pieces of ice together.

Snowballs are made by compressing the snow very strongly, when some of it melts under the increased pressure. When the pressure is released the water that was formed freezes, thus making a hard ball. If it is too cold the snow does not melt under increased pressure, and it is difficult to make a snowball.

Many substances *expand* when they change from the solid to the liquid state. Tin and paraffin-wax are two examples of these substances. Since increased pressure is unfavourable to a change from the solid to the liquid state in this case, increased pressure *raises* the melting point.

## SUMMARY

1. Specific latent heat is the heat required to change the state of unit mass of a substance without temperature change. Specific latent heat of fusion is the heat to change unit mass of a solid at its melting point to liquid at the same temperature. Specific latent heat of vaporization is the heat to change unit mass of a liquid at its boiling point to vapour at the same temperature.

2. In determining the specific latent heat of ice: (i) the ice must be dried; (ii) the final temperature after it melts must not be too low, otherwise water vapour from the air condenses. In determining the specific latent heat of steam, a water (steam)-trap must be used.

3. Melting points are determined by plotting a cooling curve. As the solid solidifies its temperature remains constant—its latent heat is given out and balances the heat lost to the surroundings.

4. Water expands on freezing. Ice contracts on melting.

5. Increased pressure lowers the melting point of ice.

## EXERCISE 13 · ANSWERS, p. 320

(Assume the specific heat capacity of water $= 4·2$ J/g K or 4200 J/kg K.)

**1.** A beaker contains 500 g of water at 20°C and is heated steadily. If it takes 8 min to reach boiling point of 100°C, how long will it then take for all the water to boil away? ($l = 2268$ J/g or $2·268 \times 10^6$ J/kg.)

**2.** What heat is required to change 6 g of ice at 0°C to water at 12°C? ($l = 336$ J/g or 336 000 J/kg.)

**3.** 20 g of ice melts when a jet of steam at 100°C is passed into a hole drilled in a large block of ice. What mass of steam was used? ($l = 2268$ J/g for steam and 336 J/g for ice.)

**4.** Calculate the quantity of heat required to change 20 g of ice at $-10$°C to steam at 100°C. (Specific heat capacity of ice $= 2·1$ J/g K. Specific latent heat of ice $= 336$ J/g. Specific latent heat of steam $= 2268$ J/g.)

**5.** 4 g of steam at 100°C is passed into 150 g of water at 10°C. Calculate the final temperature of the water. ($l = 2268$ J/g.)

**6.** What mass of ice at 0°C needs to be added to 100 g of water at 22°C to lower the temperature to 5°C? (Specific latent heat of fusion of ice $= 336$ J/g.)

**7.** A piece of copper of mass 230 g at a temperature of 900°C is quickly transferred to a vessel of negligible thermal capacity containing 200 g of water at 20°C. If the final temperature is 100°C what mass of water will boil away? (Specific heat capacity of copper $= 0·42$ J/g K and specific latent heat of steam $= 2268$ J/g.) (*N.*)

**8.** Distinguish between specific heat capacity and specific latent heat. Describe an experiment by means of which you could estimate the temperature of a bunsen flame. A vessel of thermal capacity 42 J/K contains a mixture of ice and water at 0°C of total mass 100 g. When 200 g of copper at 100°C are dropped into the vessel the final temperature is 10°C. What mass of ice was originally present in the mixture? (Specific heat capacity of copper $= 0·42$ J/g K, specific latent heat of fusion of ice $= 336$ J/g.) (*L.*)

**9.** Define melting point and specific latent heat of fusion.

Describe how you would determine the melting point of a solid whose melting point is between 70° and 80°C.

When 120 g of ice at $-6·0$°C is stirred into 930 g of water, initially at 22·5°C, until all the ice has melted, the final temperature is 10·5°C. Calculate the specific latent heat of fusion of ice. (Specific heat capacity of ice $= 2·1$ J/g K.) (*C.*)

**10.** Explain in terms of the molecular theory: (i) the change of a substance from a liquid state to a gaseous state; (ii) latent heat.

Steam is generated in a boiler at the rate of 20·0 g per min under normal atmospheric conditions. A quantity of 1800 g of water, contained in a vessel and initially at 16·0°C, is heated to 100°C by bubbling the steam through it. If it takes 16·1 min to heat the water to 100°C, find the thermal capacity of the vessel. (Specific latent heat of steam $= 2268$ J/g.)

What will happen to the steam if its passage is continued after the temperature of the water has reached 100°C? (*C.*)

**11.** Define specific latent heat of vaporization. How would you determine the specific latent heat of vaporization of water? Explain how the result is calculated. A copper calorimeter of mass 80 g contains 100 g of water at 20°C. A 500-g piece of copper at 400°C is quickly placed into the water. Assuming that heat losses are negligible, calculate the mass of water that is vaporized. (Take $c$ for copper to be $0·42$ J/g K and the specific latent heat of vaporization of water to be 2268 J/g.) (*O.*)

**12.** Describe a method of determining the specific latent heat of fusion of ice.

An electrically heated road has a layer of ice 0·50 mm thick upon it. What power consumption, in watts per square metre of road surface, is needed to melt the ice in

70 min? If the water formed does not run away, at what rate will its temperature begin to rise? Assume that all the heat produced goes into the ice or water. (Relative density of ice = 0·92; specific latent heat of fusion of ice = 336 J/g.) (*O. and C.*)

**13.** One gram of ice at −20°C is heated steadily until its temperature is 120°C. What changes in volume and state would occur? By means of a rough graph, explain how the volume changes with temperature.

**14.** Define specific heat capacity and specific latent heat of fusion.

Describe an experiment to determine the specific heat capacity of a metal, stressing the precautions that should be taken and showing how the result is calculated. Find the rate at which heat must be extracted in a refrigerator to produce 60 g per s of ice at −6°C from water at 20°C. State why, in practice, this rate must be exceeded. (Specific heat capacity of ice = 2·1 J/g K, specific latent heat of fusion of ice = 336 J/g.) (*L.*)

**15.** Define the terms melting point and latent heat of fusion.

State the effect, on the melting point of a pure substance, of an impurity which dissolves in that substance. Describe how you would show the effect in any one particular example.

Some liquid, of mass 200 g and at a temperature of 18·0°C, was put in a refrigerator cabinet which abstracted heat from the liquid at a rate of 1890 J/min. The temperature of the liquid fell at a uniform rate to −7·0°C in 4·0 min, and then remained steady at that temperature for 14·0 min before falling again. Calculate the specific heat capacity and the specific latent heat of the liquid. (*C.*)

## 14

# EVAPORATION · VAPOURS · RELATIVE HUMIDITY

### Evaporation

It is a common observation that small pools of water, formed on roads after rain, soon disappear. This change of water or other liquid from a liquid to a gaseous state is known as *evaporation*.

Liquids vary in the ease with which they change into vapour.

Pour a few drops of methylated spirit, then olive oil and finally water on your hand. Each time observe if evaporation occurs and whether your hand feels cold. Note that only the spirit evaporates easily and that your hand feels cold at the same time.

Liquids, such as methylated spirit and ether, which evaporate very easily have low boiling points. They are called *volatile liquids*. Latent heat is needed to change a liquid to vapour at the same temperature, and this heat is absorbed from the hand when a few drops of a volatile liquid are placed on it. Consequently, one feels colder. We shall see later that volatile liquids are used in refrigerators.

### Demonstration of Cooling by Evaporation

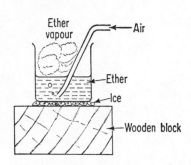

Fig. 14.1 Evaporation produces cooling

Another demonstration of the cooling produced by evaporation is shown in Fig. 14.1. A beaker B is placed on top of a little water W on a wooden block, and some liquid ether is then poured into the beaker. Air is now blown through the ether with the aid of a bellows or foot-pump. The ether evaporates rapidly, and after a time it is found that the beaker is stuck to the wooden block. A thin layer of ice is formed between them, showing that the latent heat of evaporation for the ether was taken from the water, which then turned into ice.

### Factors Affecting Evaporation

For a given liquid, evaporation is affected by the following factors:

(i) *Temperature of the Liquid.* Wet clothes on a clothes line dry more rapidly on a warm day. Thus the higher the temperature, the greater is the rate of evaporation.

(ii) *Area of Exposed Surface.* A wet sheet on a line dries more quickly when opened than when left folded. Thus the evaporation increases as the area increases.

(iii) *Rate of Removal of Vapour.* Wet clothes dry quicker on a windy, cold day than on a calm, cold day. This can also be felt when the finger is moistened and is then moved rapidly about in the air.

## Molecular Explanation

On the kinetic theory, evaporation is explained by the random motion of molecules in a liquid. They all have different velocities, and at any instant they are moving in many different directions. Near the surface, some molecules with high velocities in an upward direction may have

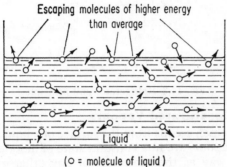

(○ = molecule of liquid)

Fig. 14.2 Kinetic theory explanation of evaporation

sufficient kinetic energy to break through the surface and exist outside as molecules of vapour. (Fig. 14.2). Others which escape may encounter molecules of air or vapour and return through the liquid after collision. A wind, which removes molecules of air and vapour, thus assists evaporation.

It can now be seen that evaporation is due to the escape of molecules of the liquid which have greater energy than the average. The average kinetic energy of the liquid remaining is therefore reduced, and hence its temperature becomes lowered. This explains the cooling which accompanies evaporation.

## Refrigerator

Fig. 14.3 shows the principle of the action of a domestic refrigerator. A volatile liquid is contained in the pipes surrounding the freezing chamber where food is stored. The liquid evaporates, thus cooling the pipes and surrounding air, from which it extracts the necessary latent heat. The vapour formed is removed by a pump P, which compresses it into a condenser Y mounted outside the refrigerator. Here it condenses and gives out latent heat, and this is dissipated by means of metal cooling fins fixed round Y. From Y the liquid is forced into X, where it again evaporates, so that continuous cooling of the refrigerator chamber takes place. A valve (not shown) between X and Y maintains X at a lower pressure than Y to ensure steady evaporation of the volatile liquid.

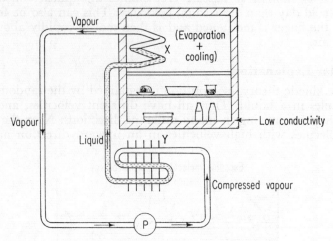

Fig. 14.3 Refrigerator unit principle

## Saturated and Unsaturated Vapours

The molecules which escape from the surface of a liquid, and which then form part of the vapour, are in a constant state of motion, exerting a pressure in the same way as the molecules of any other gas. This pressure is called the *vapour pressure* of the liquid.

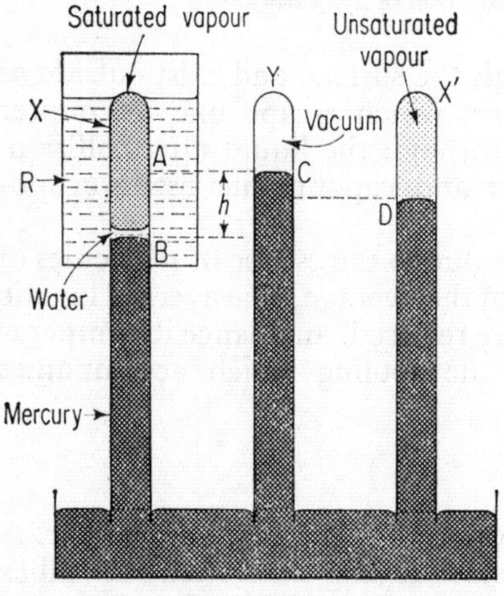

At ordinary temperatures the vapour pressure of water (or any other liquid) can be measured by means of the apparatus shown in Fig. 14.4. Two long tubes, X and Y, closed at one end, are completely filled with mercury and then inverted in a trough of the liquid, a vacuum being formed at the top of each tube (p. 189). By means of a bent pipette a little water is introduced into the bottom of X. Since water is less dense than mercury, the water rises to the top of X, and evaporates. The pressure of the vapour then forces the mercury from level C to level D, as represented in the tube X', and the vapour pressure is equal to the difference in levels of C and D. If a little more water is introduced into the space at the top of X it again evaporates, and the mercury level is depressed further. So long as more water evaporates, the space above the mercury is *unsaturated*.

Fig. 14.4 Saturated and unsaturated vapours

As more water is introduced into X, a point is reached when water remains, as at B, on the top of the mercury column; the level of the mercury now remains practically constant, however much water is introduced into X. The space above the mercury can now take up no more water vapour; it is *saturated* with water vapour.

## S.V.P. and Temperature

The saturation vapour pressure (S.V.P.) of water is found to be constant at a given temperature. If, however, the top of X is enclosed in a jacket R containing water which can be warmed (Fig. 14.4), observation shows that the water at the top of X evaporates as the temperature is increased. If more water is introduced into X a stage is again reached when water is observed at the top of the mercury column, and the space is now saturated at the higher temperature.

The following table shows how the saturation vapour pressure (S.V.P.) of water increases with temperature:

| Temp. in °C | 0 | 1 | 2 | 3 | 4 | 5 | 6 | 7 | 8 | 9 | 10 | 20 | 30 | 50 | 100 | 120 | 140 |
|---|---|---|---|---|---|---|---|---|---|---|---|---|---|---|---|---|---|
| S.V.P. (mm mercury) | 4·6 | 4·9 | 5·3 | 5·7 | 6·1 | 6·5 | 7·0 | 7·5 | 8·0 | 8·6 | 9·2 | 17·5 | 31·8 | 92·5 | 760 | 1450 | 2710 |

The bursting of a boiler is due to the rapid rise of the saturation vapour pressure with temperature. This rise is much more rapid than the rise of pressure with temperature of an unsaturated vapour or gas, which is given by the law on p. 258.

## Molecular Explanation of Saturation Vapour Pressure

We explain vapour pressure and saturation vapour pressure from the movement of molecules of the liquid. As the temperature of the liquid

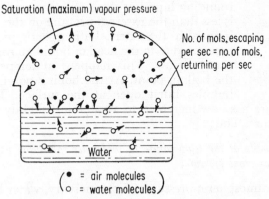

Fig. 14.5 Kinetic theory explanation of S.V.P.

rises, the speed of the molecules increases and hence the number of molecules escaping from the surface increase, that is, the vapour pressure becomes greater.

Suppose a liquid is in a closed space, such as the layer of water at the top of the tube at B (Fig. 14.4). As already explained, molecules continually leave the liquid at the surface and form molecules of vapour. Some of the molecules pass through the surface again to the liquid. Thus at any instant molecules are leaving the liquid and others are returning to it. When the vapour is unsaturated more molecules leave the liquid than return to it per second, and so the liquid evaporates until it all changes into vapour.

When the space above the liquid becomes saturated, that is, some liquid is present, the number of molecules leaving the liquid per second is then equal to the number per second returning to it. We call this 'dynamic equilibrium'. The space above the liquid now has the maximum possible number of molecules at the particular temperature. (Fig. 14.5). It is a *saturated vapour*. An unsaturated vapour, from above, is one not in equilibrium with its liquid. Its pressure is less than the saturated vapour pressure (S.V.P.). At a higher temperature the S.V.P. increases, as more molecules per second then escape from the liquid.

## Boiling and S.V.P.

When water in a beaker is raised to the boiling point small bubbles, containing dissolved air and saturated water vapour, are observed to form at the bottom. As they rise to the surface the bubbles expand owing to the decreased pressure on them, and as soon as they reach the surface the bubbles burst open. Unlike evaporation, which occurs only at the surface of a liquid, note that boiling takes place throughout the whole volume of the liquid—bubbles are formed everywhere. (Fig. 14.6).

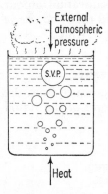

Fig. 14.6 Boiling and S.V.P.

The vapour pressure inside a bubble is always equal to the saturation vapour pressure of the surrounding liquid at the temperature concerned. If this is less than the external pressure on the bubble, which is equal to the external atmospheric pressure plus the pressure due to the head of liquid above it, the bubble cannot grow. This happens at temperatures below the boiling point. At the boiling point, however, the bubbles burst open at the surface. The saturation vapour pressure has now increased to a value just slightly more than the external pressure. Thus:

*A liquid boils at the temperature at which its saturation vapour pressure is equal to the external pressure.*

At normal atmospheric pressure, 76 cm mercury, water boils at 100°C.

## Boiling under Reduced Pressure

A demonstration of boiling water without any flame is easily arranged. A round-bottomed flask is filled with a little water, and a tube in the tight-fitting rubber cork is connected to a vacuum pump (Fig. 14.7). When the pump is switched on bubbles are soon seen rising through the water, which begins to boil at the room temperature.

The pump lowers the external pressure above the water until it becomes equal to the S.V.P. at the temperature of the room. The water then begins to boil, as already explained.

As one goes higher, the atmospheric pressure decreases. At about 3000 m above sea-level water boils at 90°C, and at 4000 m it boils at 85°C. Certain foods, such as beans, cannot be properly cooked in the open in very high regions, as the boiling point of water in these places is too low. The decrease of the boiling point of water with decreasing pressure is put to practical use at stages in the manufacture of sugar, when the latter is boiled in 'vacuum pans'. These are containers in which the pressure is low, so that the sugar is prevented from charring. It is of interest to note that the hypsometer, the apparatus for determining the boiling point of a liquid (see p. 228), was formerly used by mountaineers to determine the height they climbed, as the boiling point of water depends on the height. The hypsometer has been superseded by the altimeter (p. 192).

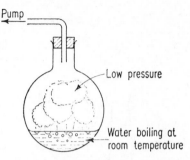

Fig. 14.7 Boiling point depends on external pressure

## Pressure Rise and Boiling Point

Since the S.V.P. of water increases with temperature rise, it follows that water boils at a temperature above 100°C when the external pressure is greater than 76 cm mercury. At a pressure of two atmospheres, for example, the boiling point is raised to about 120°C (see *Table*, p. 257).

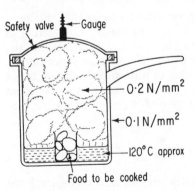

Fig. 14.8 Pressure cooker

In a *pressure cooker*, used by the housewife for rapid cooking and by the bacteriologist for sterilizing culture media, water is boiled under an increased pressure. If the gauge is set for 0·1 N per mm², or about 1 atmosphere above the external atmospheric pressure, the pressure inside the cooker is then increased to about 0·2 N per mm² or two atmospheres. When the water is heated its saturation vapour pressure rises. The water does *not* boil when a temperature of 100°C is reached because the saturation vapour pressure at this temperature is 1 atmosphere, and this is not equal to the external pressure now present above the liquid. At 120°C approximately, however, the saturation vapour pressure of water is 2 atmospheres. Thus the water begins to boil. Food cooked in water boiling at higher temperatures than 100°C takes less time to prepare. Fig. 14.8 illustrates the pressure cooker principle.

Hot springs and geysers exist in New Zealand and the United States of America. To explain how they occur, suppose that water from a stream enters a vertical shaft in the ground, where it is heated from below by heat from a volcano and is eventually raised to its boiling point. (Fig. 14.9). Now suppose that the shaft is so narrow that convection currents (see p. 311) cannot be set up very easily. Water near the top of the shaft will boil at 100°C, since the external pressure is 1 atmosphere. But water near the bottom of the shaft (where the pressure is that due to the atmosphere plus that due to a column of water of considerable length) will require a temperature of, say, 125°–130°C before it can boil. Here, therefore, its temperature rises as it is heated, and at about 130°C it begins to boil. Bubbles now begin to form and rise rapidly to the top, where, as their pressure is so high, they explode with violence. Columns of hot steam and water are shot into the air.

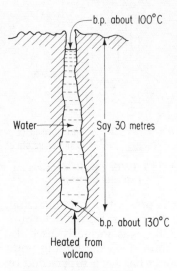

Fig. 14.9 Action of geyser

An impurity in a liquid also causes its boiling point to alter. When salt is added to water, the latter boils at a temperature higher than 100°C. You can cook food more rapidly by boiling it in salt water than by using pure water.

## RELATIVE HUMIDITY OR HYGROMETRY

On a fine day, with a cool breeze blowing, we are conscious that the air is 'dry'; whereas on 'muggy' days the air feels 'wet' or *humid*. This feeling of comparative wetness or dryness is due to the water vapour in the air. Since, however, the actual mass of water in the air may be more on the 'dry' day than on the 'muggy' day, the degree of humidity in the air is not measured by the mass of water alone. A *ratio*, known as the **relative humidity**, is chosen as a measure of the degree of 'wetness' in the air, and is defined by the relation:

$$\text{Relative humidity (R.H.)} = \frac{m}{M} \times 100\% \qquad . \qquad . \qquad (1)$$

where $m$ is the mass of water vapour actually present in a certain volume of the air, and $M$ is the mass of water vapour required to *saturate* the same volume of air at the same temperature.

Suppose that a sealed room had 1·5 g of water vapour present inside it. If a hole were made in a wall, and steam (water vapour) passed gently through the opening into the room, the pressure of water vapour would rise at first. Soon, however, it would be noted that the surfaces of polished metal articles in the room suddenly became misty; at this point the water vapour in the room becomes *saturated*. The introduction of more steam causes condensation to take place, but does not lead to any further increase in the amount of water *vapour* in the room.

If the mass of water vapour in the air $M$ is now 2·5 g the original relative humidity R.H. is given by $\frac{1·5}{2·5} \times 100\%$, or 60%, from (1).

## Formula for Relative Humidity

It is not an easy matter to measure the mass $m$ of water vapour present in the air and the mass $M$ required to saturate the same volume at the same temperature. Scientists have therefore developed another method, which assumes that Boyle's law applies to vapour.

The mass of water vapour in a given volume is proportional to the density of the water vapour present. But, according to Boyle's law,

$$\text{Pressure} \propto 1/\text{Volume} \propto \text{Density}$$

for a given mass at a given temperature: i.e. the density is proportional to the *pressure* of the water vapour. Consequently, at any given temperature the mass $m$ of the water vapour present in air is proportional to its vapour pressure $p$; and the mass $M$ of water vapour required to *saturate* the air is proportional to the saturation vapour pressure $P$. Thus,

$$\text{Relative humidity} = \frac{m}{M} \times 100\% = \frac{p}{P} \times 100\% \quad . \quad (2)$$

On a warm day, at say 19°C, the pressure of the water vapour in the air may be 8 mm mercury, whereas the saturation vapour pressure (S.V.P.) at 19°C is 17·5 mm. Thus the relative humidity is only $8/17·5 \times 100$ or 46%, and so sweat evaporates readily and the air feels dry. On a cold day, at say 10°C, the pressure of water vapour may again be 8 mm mercury, but, since the S.V.P. at 10°C is only 9·2 mm, the relative humidity is as high as $8/9·2 \times 100$ or 87%. We thus perspire slowly that day and the air feels moist. Our feeling of discomfort increases as the relative humidity of the air approaches 100%.

## Dew-point

When ice is added slowly to water or other liquid in a drinking glass, the surface outside is seen to become misty at one stage. The water vapour in the air near the glass, which was hitherto invisible, has now condensed as water droplets on the cold surface. The same effect is produced by breathing on a cold mirror or window surface. The air temperature at which water vapour first condenses when the air is cooled is called the *dew-point* of the air.

Since the amount of vapour in the air has not altered during the cooling of the glass surface, the pressure of the water vapour in the air at the original temperature is equal to the saturation vapour pressure (S.V.P.) at the dew-point. From (2), it follows that

Relative humidity

$$= \frac{\text{Vapour pressure at original temperature } (p)}{\text{S.V.P. at original temperature } (P)} \times 100\%$$

$$= \frac{\text{S.V.P. at dew-point}}{\text{S.V.P. at original temperature}} \times 100\% \quad . \quad . \quad . \quad . \quad (3)$$

This last relation is the basis of accurate methods of measuring relative humidity.

### Regnault (Dew-point) Hygrometer

Regnault designed an apparatus for measuring the dew-point, and hence, from the above relation, the value of the relative humidity. It has two wide tubes with the outside surfaces of the closed ends highly polished silvered metal. (Fig. 14.10.) One tube B contains a little ether, and by con-

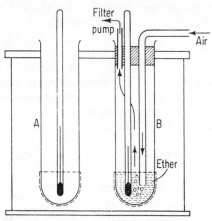

Fig. 14.10 Dew-point (Regnault) hygrometer

necting a filter pump to draw air through a glass tubing dipping into the liquid, air can be bubbled through the ether at a controlled rate. This makes the ether evaporate more rapidly, and the latent heat required is taken from the liquid and container, which therefore becomes cooled.

As the temperature of the outside surface of B drops, a mist forms at one stage. Water has now condensed from the air on to the surface. The polished surface of A is used as a 'comparison' surface—the observer watches A and B, and as A remains unaffected, he or she is able to tell when mist is first formed on B. The temperature on the thermometer in B is then recorded. The pump is now disconnected so that the temperature of B begins to rise, and when the mist first disappears, the temperature is again noted. The dew-point is taken as the average of the two temperatures thus obtained. To ensure that water vapour in the breath does not unintentionally create a mist on B, a glass plate can be placed between the observer and B, or a telescope can be focused on B.

Suppose the dew-point is found to be 4°C and the actual air temperature, observed on the thermometer, is 11°C. We now look up from tables the S.V.P. of water vapour at 4°C and at 11°C. From (3) above,

$$\text{Relative humidity} = \frac{\text{S.V.P. at } 4°C}{\text{S.V.P. at } 11°C} \times 100\%.$$

The S.V.P. at 4°C = 6·1 mm, the S.V.P. at 11°C = 9·8 mm, and hence

$$\text{The relative humidity} = \frac{6·1}{9·8} \times 100\% = 62\%.$$

## Wet-and-dry Bulb Hygrometer

The Regnault hygrometer just described requires an experienced observer to give quick and reliable results of relative humidity. The 'wet-and-dry bulb' hygrometer, however, can be used by an untrained person.

The instrument consists of two thermometers, A and B, standing beside each other. (Fig. 14.11). A piece of muslin is wrapped round the bulb of B, and its lower end dips into water. Thus the bulb of B is always wet and that of A is always dry. The water round the bulb of B evaporates into the air at a rate which depends on the relative humidity, and hence the bulb is cooled and B usually reads lower than A.

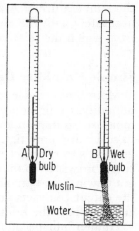

Fig. 14.11 Wet-and-dry bulb hygrometer

To take an extreme case, no water evaporates if the relative humidity is 100%, as the air is then saturated, and the temperature on B is then exactly the same as the temperature on A, which records the temperature of the air. If the air were far from being saturated the water in the cloth surrounding the bulb of B would evaporate quickly, and hence the temperature on B would be much lower than that on A.

If the two thermometers are placed outside in the air a wind produces steady evaporation from B. Otherwise, used as a 'sling psychrometer', the hygrometer can be whirled round steadily by a handle with a wet muslin

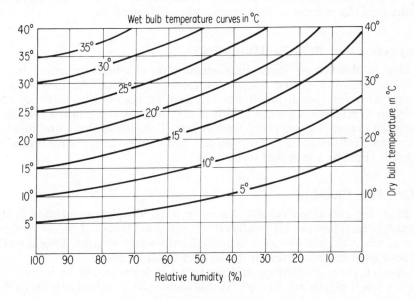

Fig. 14.12  Relative humidity curves

or cloth round the bulb of B, and the steady temperature read after a short time.

Tables have been drawn up which relate the difference in temperature, the temperature of A and the corresponding relative humidity (obtained accurately with the aid of a Regnault hygrometer). Fig. 14.12 is a chart from which the relative humidity can be read directly.

### Simple Humidity Devices

More simple devices than hygrometers can convey a rough measure of humidity. A fir cone, for example, is hygroscopic, that is, it absorbs moisture, so that on a dry day its scales stand out more. On a humid day its scales are more tightly together. Hair and catgut are hygroscopic – catgut tends to contract when it absorbs moisture. In a toy hygrometer the catgut is twisted so that when it is a dry day, and therefore longer, the catgut unwinds and an old lady comes out of a house. When the air is humid the catgut contracts and produces a rotation, and this time an old man emerges from the house and the old lady returns inside.

Human hair becomes more moist as the relative humidity rises, and if freed from grease its length then increases slightly. This is used practically in a form of hygrometer called a *hair hygrometer*.

Basically, the instrument contains a fine hair which is fixed at one end and passes round a grooved wheel to a spring, which keeps it taut. When the hair becomes more moist and increases in length a pointer attached to the wheel then rotates. The scale of the instrument has values of relative humidity on it, previously obtained by calibration with a dew-point hygrometer (p. 302). The hair hygrometer is not accurate but has the advantage that the relative humidity value can be read directly from the position of the pointer so that no experiment is needed.

### Importance of Relative Humidity

One of the factors taken into account in weather-forecasting is the relative humidity of the air, and meteorological stations have special instruments for measuring it quickly and accurately. In factories and mills, as a protection for the health of the workers, the relative humidity must not exceed a value prescribed by law. Tobacco in store, and meat in large-scale refrigerating rooms, will not 'keep' unless the relative humidity is maintained at a specified level.

### Mist, Fog, Cloud

Mist, fog and cloud consist of minute water droplets suspended in the atmosphere. They are all produced when the air is cooled below its dew-point, when the air becomes saturated with water vapour. Excess water vapour then condenses out in the form of small droplets.

When this occurs near ground level mist or fog is said to be formed. At higher levels clouds are formed. Mist is sometimes formed at night. In this case the earth cools down rapidly owing to its low specific heat, and the air

**PLATE 5**
Static and Aerodynamic Forces

◀ 5(a) A new suspension bridge, spanning the Danube at Budapest, is tested before opening by loading with vehicles.

5(b) Streamlines. Flow round an aerofoil at an angle of 9° to the stream. The flow paths are photographed by light reflected by tiny polystyrene spheres in the wind stream.

5(c) Flow when the aerofoil is stalled at an angle of 14° to the stream.

## PLATE 6
### Rain-making

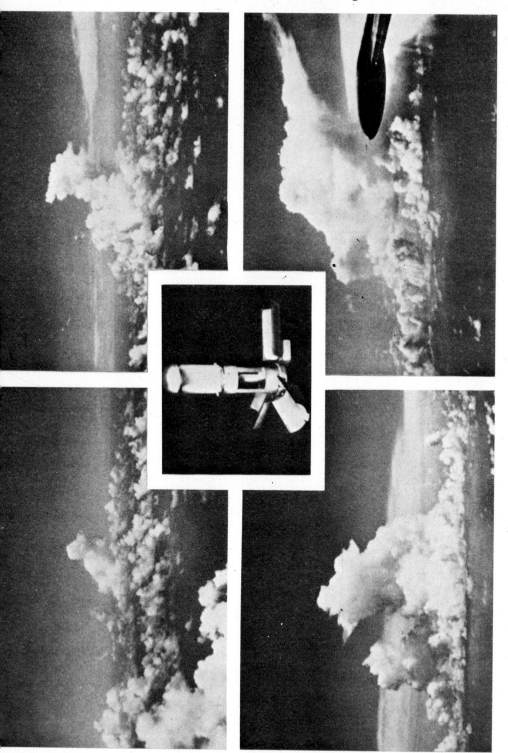

6(a) Sequence of events after a tropical cumulus cloud is 'seeded' with silver iodide vapour, obtained from generators (centre) dropped by an aeroplane through the cloud. *Upper left* – time of seeding, *upper right* – 9 min later, *lower left* – 19 min later, *lower right* – 38 min later.

in contact with it above the ground is cooled below its dew-point and forms a layer of mist.

Fog is formed when two air currents, one warm and almost saturated and the other cold, mix or pass each other. Here the warm air is cooled below its dew-point and fog is formed.

Water droplets of mist, cloud or fog always need minute particles as 'nuclei' or centres to form on, otherwise they cannot grow to an appreciable size. Dust particles, free ions and the larger molecules of gas in the atmosphere may all act as nuclei for drop formation. Near industrial towns, where chimneys produce a considerable amount of smoke particles, dense fog or *smog* may be formed.

The path of a high-flying aircraft is sometimes shown by vapour trails, which are due to the condensation of water vapour in the cooled air behind it. In a dust-free cooled atmosphere, where no nuclei are available, C. T. R. Wilson showed in 1911 that the water vapour will condense on ions formed by alpha-particles streaking through the space concerned (see p. 727). The atmosphere must first be supersaturated for this to happen, that is, the water vapour in it is above its normal saturation pressure value for the particular temperature. The vapour trail produced shows vividly the path of the alpha-particles (see p. 728).

### EXPERIMENT. *Cloud Formation*

For this demonstration you will require a Winchester bottle fitted with a football valve. Place a few ml of water inside the bottle to saturate the air there and attach a bicycle pump to the valve. Pump air into the Winchester and then remove the pump, wait for the air to return to room temperature, and then allow the air to expand. Notice the formation of a cloud inside the Winchester.

Cooling of near-saturated warm air may occur when it rises. It may, for example, make contact with cooler air above it. In addition, the expansion of the air due to decrease of pressure will also produce cooling, as the energy needed to expand against the atmospheric pressure is taken from the thermal energy of the air itself. If the air is cooled below its dew-point water condenses out. A cloud is formed. The cloud consists of water droplets if the temperature in it is above 0°C, but if it is below 0°C it will consist of minute ice particles. *Cirrus clouds*, the tiny wisps seen high in the sky on fine days, consist of ice particles. *Cumulus clouds*, the low, billowy, cauliflower-shaped clouds, and *stratus clouds*, low, flat layers of cloud, may both consist of water droplets or ice particles, depending on the temperature.

### Rain, Snow, Hail

If small drops in a cloud coalesce, and thus become heavier, they fall to the ground as rain. If the raindrops pass through air much below 0°C the water freezes and *hailstones* are formed. A section of a hailstone may show several layers of ice due to continued up-and-down movement in the convection currents existing in large storm clouds.

*Snowflakes* are formed when a falling ice particle unites with others to form crystals of ice. *Dew* is formed at night when the ground cools down and lowers the air temperature directly above it to below the dew-point.

Water vapour then condenses as water droplets on the ground beneath. *Hoarfrost* or ground frost occurs when the dew-point is below 0°C. The water vapour then condenses out as solid ice particles or frost on the ground.

## SUMMARY

1. Evaporation depends on the liquid temperature, its surface area and the wind present. It takes place from the surface of the liquid–since the more energetic molecules escape through the surface, the average kinetic energy of the liquid decreases, and hence its temperature falls.

2. A saturated vapour is one in equilibrium with its liquid. S.V.P. of water can be measured with the aid of two barometer tubes, one of which has water at the top and the other has a vacuum.

3. *Boiling*. A liquid boils when its saturation vapour pressure is equal to the external atmospheric pressure. Increased pressure increases the boiling point. Boiling occurs throughout the whole volume of liquid and occurs at one particular temperature, unlike evaporation, which is a surface phenomenon.

4. The relative humidity R.H. of the air $= (m/M) \times 100\%$, where $m$ is the mass of water-vapour per litre in the air and $M$ is mass required to saturate the volume.

$$\text{Also, R.H.} = \frac{\text{S.V.P. at dew-point}}{\text{S.V.P. at air temperature}} \times 100\%.$$

5. Relative humidity can be measured by means of Regnault's hygrometer or by the wet-and-dry bulb hygrometer.

## EXERCISE 14 · ANSWERS, p. 320

**1.** Define *boiling point* of a liquid.
Describe one experiment in each instance to show how the boiling point of water is affected by: (i) a change in the pressure, (ii) the addition of salt to the water. (*L.*)

**2.** Define *dew-point*. Describe an experiment to determine its value at any particular place. Explain one use for a knowledge of the dew-point. (*L.*)

**3.** What is meant by: (i) *boiling*; (ii) *saturation vapour pressure* of a liquid? Describe an experiment to show that the pressure on the surface of a liquid affects the temperature at which it boils. (*L.*)

**4.** Distinguish between a *saturated* vapour and an *unsaturated* vapour.
Explain why: (i) a film of moisture forms on the outside of a glass of cold water when it is brought into a warm room; (ii) the boiling point of a liquid rises when the pressure on its surface is increased. (*L.*)

**5.** State the effect of an increase of pressure on: (*a*) the boiling point of water; (*b*) the melting point of ice. State the effect of a decrease in volume on the pressure of: (*c*) an unsaturated vapour; (*d*) a saturated vapour. (*N.*)

**6.** (*a*) Describe an experiment to show that the saturated vapour pressure of water at its boiling point is equal to the pressure on its surface.
(*b*) Describe a form of hygrometer and explain how it is used to measure the dew-point. (*N.*)

**7.** Distinguish between evaporation and boiling of a liquid. Describe and explain an experiment to show that evaporation requires heat. (*L.*)

**8.** State and explain the conditions under which a liquid can: (*a*) disappear by evaporation; (*b*) boil. Give the meaning of the terms saturated vapour, saturated vapour pressure.

Describe a method of determining the saturated vapour pressure of water at temperatures between, say, 10° and 100°C. Sketch the apparatus. (*O. and C.*)

**9.** Explain the terms saturated and unsaturated vapour, and define dew-point. The table gives the results of an experiment to determine the saturation vapour pressure of water at various temperatures. Plot a graph of these results and use it to deduce: (*a*) the vapour pressure of water in a room where the dew-point is 8°C; (*b*) the boiling point of water at a pressure of 120 mm of mercury.

| Temperature (°C) | 0 | 10 | 20 | 30 | 40 | 50 | 60 |
|---|---|---|---|---|---|---|---|
| Vapour pressure | 4·6 | 9·2 | 17·5 | 31·8 | 55·3 | 92·5 | 149·4 |
| (mm of mercury) | | | | | | (*O. and C.*) | |

**10.** Explain why water starts to cool when dry air at the same temperature is bubbled through it.

Dry air at room temperature, 15°C, is bubbled through 50 g of water initially at 20°C and is then passed through a drying tube. When the water temperature has fallen to 10°C the mass of the drying tube is found to have increased by 1·0 g. Make an estimate of the specific latent heat of vaporization of water at 15°C given that the thermal capacity of the container is 42 J/K. (*O. and C.*)

**11.** What effect, if any, is produced upon: (*a*) the melting point of ice by an increased pressure; (*b*) the volume when ice changes to water at 0°C; (*c*) the volume of a mass of water when heated from 1° to 2°C; (*d*) the pressure of a saturated vapour by decreasing its volume? (*N.*)

**12.** Describe a form of hygrometer and explain how it is used to determine the dew point.

Describe and explain how the saturated vapour pressure of water at room temperature can be determined. (*N.*)

**13.** Describe simple experiments, *one* in each case, which you would perform in order to carry out the following instructions: (*a*) determine the saturation vapour pressure of alcohol at room temperature; (*b*) show that, at constant temperature, the pressure exerted by a saturated vapour is independent of the volume which it occupies; (*c*) show that the temperature at which water boils depends on the pressure acting on it. (*N.*)

**14.** State the hypotheses of the kinetic theory. Use kinetic theory and inter-molecular attraction to explain briefly *two* of the following: (*a*) gaseous diffusion through a porous plug; (*b*) the pressure of a gas or vapour; (*c*) evaporation and its cooling effect. (*N.*)

# 15

# CONDUCTION · CONVECTION · RADIATION

## Conduction

The end of a spoon dipping into hot tea, and the handle of a poker in a fire, both become hot in a short time. These are examples of *conduction* of heat along metals.

If two rods of similar dimensions, one of iron and the other of glass, are held in a bunsen flame the end of the iron soon feels warm, but the end of the glass does not. Some solids thus conduct heat well; others are poor conductors. This can also be shown by the following experiment.

### Conductivities of Copper and Wood

Take a long wooden rod, wind some copper foil tightly round the middle and nail the foil to the rod (Fig. 15.1). Then wind a piece of paper round the rod so that it overlaps the wood and copper and stick it so that it

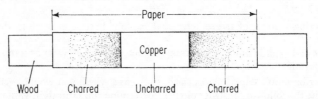

Fig. 15.1 Copper is a good conductor, wood a bad conductor

makes firm contact with them, excluding air. Hold one end of the rod and move it to and fro so that the paper passes repeatedly through the flame of a bunsen burner. Observe that the paper is unaffected where it covers the copper, but is charred where it covers the wood.

*Conclusion.* Since the temperature of the paper round the copper stays below its burning point, the heat of the flame is conducted away quickly by the copper. Conversely, heat is conducted away slowly by the wood because the paper becomes charred. Thus copper is a good conductor, wood is a bad conductor.

### Experiment to Compare Roughly Conductivities of Metals

The apparatus is set up as shown in Fig. 15.2. Essentially it consists of a watertight container filled with some water which can be heated electrically. Rods of metal of equal size, such as copper, lead, aluminium and iron, project vertically from the base, so that the ends are heated by the hot water. Small rings, which just slide on the rods, are coated with paraffin wax and slid to the top. The rods are also coated with wax.

The heater is switched on, and soon the top of the rods reach a high temperature. As heat is conducted along the rods the wax inside the ring melts and the ring now moves farther down. After a while it can be seen that the ring has travelled farthest down the copper rod and least down the lead rod, showing that copper is a much better conductor than any of the four metals and that lead is the worst conductor.

*Molecular Explanation.* In conduction, heat is transferred from particle to particle throughout a body without any visible signs of movement. Our theory of solids on p. 32, however, tells us that their molecules are vibrating, and when one end of a solid is heated the energy supplied makes the molecules move through greater distances than before. Neighbouring molecules are therefore jostled more than before, and they, too, acquire more energy. (Fig. 15.3). This goes on all the way along the solid. Con-

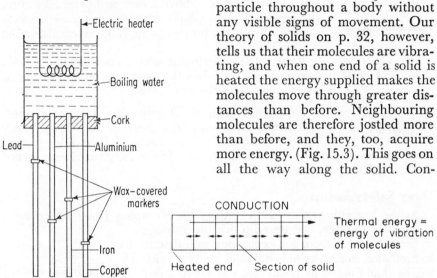

Fig. 15.2 Metals have different conductivities

Fig. 15.3 Molecular explanation of conduction

sequently, energy or heat is transferred along the solid, without any alteration in the *average* position of each molecule.

## Conductivity of Liquids

We can investigate the conductivity of water by the following experiment:

By using a piece of metal gauze above or round it, keep a lump of ice at the bottom of a test-tube filled with water (Fig. 15.4). Hold the tube inclined with a paper grip and bring the top to a bunsen flame as shown. After a time the water at the top is seen to boil, but the ice below remains unaffected. Thus little heat is conducted from the top to the bottom of the water.

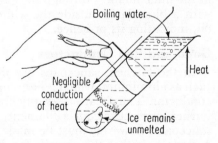

Fig. 15.4 Water is a bad conductor

*Conclusion.* Water is a poor conductor.

Generally, substances which are liquids at ordinary temperatures are poor conductors, with the exception of mercury.

## Ignition Point of Gas

Gases burn when their ignition point or temperature is reached, but if a metal, a good conductor, is placed in the gas it can prevent the ignition of the gas.

EXPERIMENT

Place a metal gauze a short distance above a bunsen burner. Turn on the gas and apply a light *below* the gauze. The gas is seen to burn here, but no flame is obtained above the gauze (Fig. 15.5).

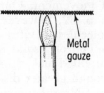

Metal gauze

Repeat the experiment with a cool gauze, but this time light the gas *above* the gauze. A flame is obtained above the gauze but not below.

Fig. 15.5 Ignition point of gas

*Explanation.* The metal gauze is a good conductor. It therefore conducts the heat of the flame away rapidly from the gas on the unlighted side, so that the gas here never reaches its ignition point. When the gauze becomes red hot this gas may be ignited.

## Davy Safety Lamp

At the beginning of the 19th century Sir Humphry Davy, an eminent scientist of the day, was urged to invent a device to protect miners from gas explosions, mainly due to methane, caused by naked flames used for lighting at that time. The result was the Davy safety lamp, in which a gauze metal cylinder surrounds a burning wick 'fed' by oil from the base (Fig. 15.6). Holes at the top of the cover let the air in and the combustion products out. As in the demonstration with the metal gauze in Fig. 15.5, the heat of the flame is conducted away from any dangerous gas outside. It is thus kept below its ignition point. When there is inflammable gas in the mine it enters through the gauze and burns with a blue flame outside the white flame of the oil, indicating its presence.

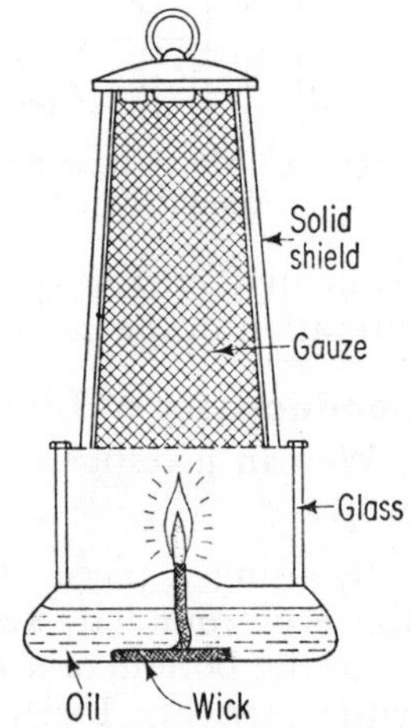

Solid shield

Gauze

Glass

Oil    Wick

Fig. 15.6 Davy safety lamp

## Good and Bad Conductors

In the kitchen, saucepans are made of metal such as aluminium, which is light and a good heat conductor. Thus heat passes quickly from the gas burner, or electric-cooker filament, to the food inside. The handles of the saucepans, however, must be made of insulators, such as plastic or wood materials, so that the utensils can be handled when hot. The handles of kettles and oven doors must likewise be made of insulators.

Air, like all gases, conducts heat extremely slowly, and is therefore a very good insulator. Pockets of air account for the insulation properties of many materials, of which wool is one example. In winter birds look 'fatter' than in summer. They fluff their feathers more to trap pockets of air, which help to keep them warm. If you examine the underside of some car bonnets you will find a felting which keeps the engine warm on account of the many pockets of air in the material.

When large steel boilers used in industry are heated by gas flames a thin layer of gas may be permanently present on the underside of the boiler. As the gas is an insulator, the heat now flows at a much slower rate to the water inside, and hence the thermal efficiency of the boiler is lower. Periodically, therefore, the underside of the boiler is cleaned. The inside of the boiler is also cleaned at intervals because scale, deposited from the minerals in the water, is a bad conductor. To avoid further deposit of scale, the same water should be retained in a closed hot-water system such as that used in the home for domestic heating.

## Heat Insulation in the Home

Any heat which is allowed to escape from a house in winter will add to the bills for fuel. Heat losses may occur through the ceiling, the loft, the chimneys and the walls, doors and windows. Fibre glass or felt can be used to insulate the loft. Expanded polystyrene, a light material full of pockets of air, is an excellent insulator. Floors can be insulated by having a thick layer of felt under the carpet. Double glazing of windows, widely used in the Scandinavian countries and the Soviet Union, provides an insulating layer of air between the outside and inside windows and reduces heat losses appreciably.

## Convection Currents in Liquids

EXPERIMENT

By using a long glass tube, drop a tiny crystal of potassium permanganate through it to the bottom of a flask filled with water (Fig. 15.7). Now gently heat the bottom of the flask just below the crystal. Observe the upward movements of the coloured liquid from the region of the crystal, which dissolves slowly and helps to show the liquid movement.

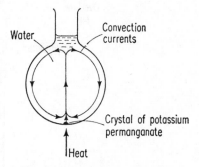

Fig. 15.7 Convection currents in liquid

The upward movement of the coloured liquid is due to *convection*, and the circulation of liquid is called a *convection current*. It is due to the effect of heat at the bottom of the liquid. Heat causes a fluid (liquid or gas) to expand. This makes the warm fluid less dense than the cold fluid above it, and it therefore rises. The cold and denser fluid takes its place by moving down. The phenomenon is called *convection*. Unlike the case of conduction, where the average position of the molecules remain the same, the heat is carried to other parts of the liquid by movement of the warm liquid itself. In boiling water in a kettle, then, convection currents circulate continually, keeping the water stirred up as it is heated.

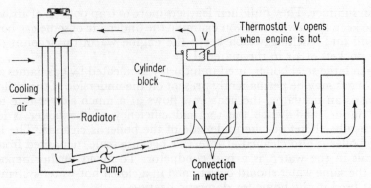

Fig. 15.8 Engine cooling by convection

As shown in Fig. 15.8, water circulates by convection currents round the engine of a car and keeps it cool. Air round the radiator provides the cooling of the water, and a pump helps the circulation. A thermostat, which opens when the engine rises to a suitable high temperature, prevents overheating.

### Domestic Boiler

A hot-water circulation system, used for domestic supplies, is shown in Fig. 15.9. A boiler heats cold water, which then rises up a pipe X to a

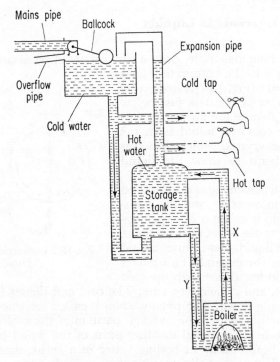

Fig. 15.9 Domestic hot-water convection system

storage tank. Fresh cold water returns to the boiler from a cold-water tank, usually in the roof of the house, through a pipe Y. Here it is heated and rises again to the storage tank. In this way the latter stores hot water for eventual use. When some is drawn off by opening hot taps in the kitchen or bathroom, for example, an equal volume of cold water then enters the cold-water tank from the mains. A ball-cock maintains the level of water constant in the cold-water tank. If the water in the storage tank boils and gives off steam, the expansion pipe at the top discharges it harmlessly into the cold-water tank.

## Convection in Gases

Convection of heat occurs much more readily in gases than in liquids because they expand very considerably when their temperature rises (p. 255).

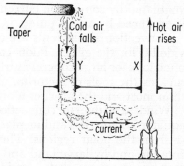

Fig. 15.10 Convection air current

EXPERIMENT

A simple demonstration of convection currents in air can be given with the apparatus in Fig. 15.10. It consists of a closed rectangular box (as a shoe box) with a plane glass front so that the interior is visible. Two wide glass tubes are fixed in holes at the top of the box as 'chimneys'. A candle is now lit below one tube X and a smouldering taper is held above the other tube Y. The smoke movement shows that an air convection current passes down through Y and up through X.

*Explanation.* The air above the candle flame becomes hot and less dense and rises up X. Colder air then flows down Y and takes its place, giving rise to a circulating convection current.

Chimneys help to circulate fresh air in rooms in this way.

## Land and Sea Breezes

Land and sea breezes on the coast are natural convection currents. In summer in daytime the sun warms the land to a higher temperature than the sea, as it has a lower specific heat than sea-water and the sea is in continual motion. The air above the land is heated and rises, and its place

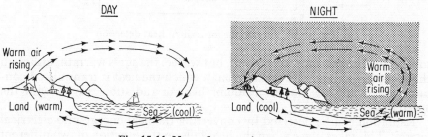

Fig. 15.11 Natural convection currents

is taken by cooler air above the sea moving inland (Fig. 15.11). Air higher in the atmosphere completes the circulation, and hence a *sea-breeze* is obtained.

At night the sea temperature drops only slightly, since it is warmed to a considerable depth during the day. On the other hand, the land temperature drops considerably at night. This time, therefore, a convection current is obtained in the opposite direction to daytime, and this is a *land-breeze* (Fig. 15.11).

### Some Uses of Convection Currents

A more vigorous convection current can be seen at the time of bonfires, when the rapid upward movement of air often carries glowing cinders upwards to considerable heights. The sun often casts a 'shadow' of the hot air rising above a radiator in a room (p. 318). Years ago, a coal mine was ventilated by convection currents produced by fires, but today ventilation is produced by an air pump.

Convection air currents are useful to the glider pilot because they provide lift for the glider. Pilots call the rising air currents 'thermals'. They often look for large concrete surfaces, such as roads, or large metal surfaces, such as the roofs of factories, as the air above these surfaces is warmer than most and the air rises more.

Winds are due to convection currents in the atmosphere. Near the equator hot air rises and its place is taken by cooler and denser air flowing in from the polar regions. The circulating air is affected by the earth's rotation and by large land and sea masses over which it passes, as well as by temperature, and the earth's wind system is therefore complex.

### Radiation

When the sun rises over the horizon in the early morning one can immediately feel its heat. The heat of the sun reaches us by *radiation*. In the transfer of heat by conduction and convection, some material substance, or

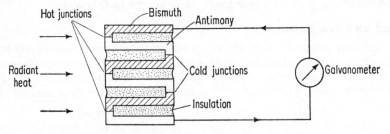

Fig. 15.12 Thermopile for radiant heat detection

solid, liquid or gas, must be present, but we feel the sun's warmth, although the space high above the earth through which the heat is transmitted contains practically no matter. Radiant heat or radiation travels with the speed of light (p. 331).

Radiation can be detected by converting the heat energy into electrical energy. This takes place when the heat falls on the junction of two different metals, the other end of the metals being joined to the terminals of a galvanometer, a sensitive current-measuring instrument, which then registers a current. Bismuth and antimony are two metals used in the *thermopile*, an instrument for detecting radiant heat. In order to magnify the effect, many bismuth–antimony junctions are joined in series. The radiation then falls on a set of junctions at one end, as shown in Fig. 15.12. The metal bars are carefully insulated from each other, and then encased, insulated, in a metal cylinder. The free ends of the antimony and bismuth bars are joined to terminals, and a galvanometer is connected to the terminals when the thermopile is used. A cone, with a high-reflecting surface inside, is fitted over the end.

Place a hot metal ball or bunsen flame or your hand near the thermopile cone and observe the galvanometer deflection.

## Radiation from Different Surfaces

The magnitude of the radiation from different surfaces kept at the same temperature can be investigated with the aid of a hollow cube of metal, called a *Leslie cube* after the inventor, Sir John Leslie (Fig. 15.13). One verti-

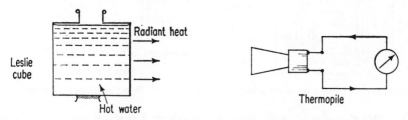

Fig. 15.13 Radiation from different surfaces

cal side is coated dull black, a candle flame can be used for this purpose, another side is highly polished silvery-bright, and the remaining two sides painted with different colours such as grey and white.

### Experiment

Fill the cube with hot water. Place the thermopile close to the cube, so that the cone would intersect the vertical face of the cube if it is imagined produced to the cube. Turn the cube round so that each face in turn is presented to the cone, which collects the radiation. Record the steady deflection each time. Repeat the experiment with hotter water and compare your results.

*Results.* The dull black surface produces the greatest deflection and the polished surface the least. The roughened surface lies second and the white surface third.

We therefore conclude that the radiation from a body depends on the nature of its surface and on its temperature. It also depends on its area—the larger the area, the greater is the radiation. *For a given temperature, a body radiates most heat when its surface is dull black and least when its surface is highly polished.*

## Emission and Absorption

Are good emitters of heat also good absorbers? We can investigate this by using the apparatus shown in Fig. 15.14.

(*a*) A and B are two air-filled bulbs connected to a U-tube C, which is partially filled with a light oil. This is called a *differential-air thermometer*, because a temperature rise of one bulb such as A causes the oil level X on this side to drop owing to the expansion of the air.

A is painted matt black. B is painted gloss white or a colour different from black. A small electric lamp L, placed exactly midway between A and B, is switched on. It is then observed that the oil level falls on the side

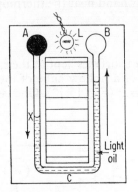

Fig. 15.14 Experiment on absorption          Fig. 15.15 Relative absorbing powers

of A. We therefore conclude that a blackened surface, which is a good emitter as previously seen, is also a good absorber of radiation. Conversely, a white or highly polished surface is a poor absorber as well as a poor emitter of radiation.

(*b*) Investigate the result of (*a*) for yourselves using the apparatus of Fig. 15.15. Use two sheets of tinplate, one blackened and the other polished brightly. Fix a cork on to the reverse side of each sheet using paraffin wax. Place a bunsen burner mid-way between the sheets, fixed vertically a short distance apart. Light the burner and record what happens. Explain your observations.

## Applications of Radiation

Our experiments have shown that *a dull matt surface is a good radiator and a good absorber, especially when black, whereas a light polished or silvery surface is a poor radiator and poor absorber.* You should now be able to explain the disadvantage of a black-painted car in hot weather, the advantage of a tea-pot with a silvery surface and the purpose of a silvered surface at the base of an electric iron.

Factory roofs are sometimes coated with an aluminium (light) paint, which reduces absorption of heat during the day and reduces radiation during the night, so that a fairly steady temperature is obtained in the factory. The cooling of the earth at night is due to radiation; on a cloudy night much heat is reflected back, so that it is not so cold as on a clear night.

Glass appears to be transparent to radiation from bodies, such as the sun, which are hotter than about 500°C, and opaque to radiation from cooler bodies. A greenhouse thus acts as a 'trap' for the heat from the sun. Good absorption of radiation by dark bodies causes ice or snow on a mountain to melt where a stone 'hides' it from the sun, while remaining frozen in full sunlight.

### Infra-red Radiation

Radiant heat, like visible light, consists of *electromagnetic waves*. They have much longer wavelengths than the longest wavelengths in visible light, which is red, and they are therefore called *infra-red rays* (p. 441). Infra-red rays are thus not visible to the eye.

When infra-red rays fall on an opaque object, part of it is absorbed and converted into heat, and part is reflected. A highly polished silvered surface, as we have seen, is an excellent reflector of infra-red rays. Infra-red rays obey the same laws of reflection as light, as we can show.

### Reflection of Infra-red Rays

EXPERIMENT

Place a small electric radiator S, which emits heat rays, at the focus of a large concave reflector $M_1$ (Fig. 15.16). Some distance away, place a match at the focus X of another concave reflector $M_2$, whose reflecting surface

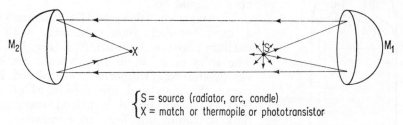

{ S = source (radiator, arc, candle)
{ X = match or thermopile or phototransistor

Fig. 15.16 Reflection of infra-red rays

is turned towards $M_1$. Shield the match from direct heat from the radiator. After a time the match will burst into flame.

Replace the electric radiator by a candle and the match by a shielded thermopile. Observe the change in deflection in the galvanometer used when the candle is lit and when extinguished.

Replace the thermopile by a phototransistor Mullard OCP71, which is sensitive to infra-red rays. Observe the change in current when the candle is lit and when extinguished.

*Conclusion.* The heat or infra-red rays are reflected by $M_1$ so that they are brought to a focus by $M_2$. The rays therefore obey the same laws of reflection as light rays (p. 369).

Infra-red electric heaters, used in bathrooms, for example, have a filament embedded in a rod of silica. When it is hot the silica emits infra-red rays. These are reflected into the room by a concave reflector behind the rod. *Solar furnaces* collect heat rays from the sun by mirrors.

### Radiation and Convection in Heating Rooms

Metal radiators, used in a central-heating system for heating rooms, or electrically heated panels on the walls, warm the air in contact with them. The warm air rises and its place is taken by cooler air, and the process is repeated. In this way *convection currents* circulate in the room. Some heat is radiated directly into the room, of course, from the radiators, but since they warm the room mainly by convection, they are more correctly termed 'convectors'. Thus radiators may be painted in light colours without seriously affecting their heating efficiency.

A coal or log fire warms a room partly by radiation. Some of the heat, however, warms the air above it, part of which then rises up the chimney. Cooler air from open windows, or through openings in doors, then flows into the room and takes its place. Thus fires with an open chimney not only help to heat a room but also to circulate fresh air in it. If a chimney fireplace is sealed, air heated by radiators or electric fires in the room does not escape and the air is then not as fresh as before.

### Vacuum or Dewar Flask

About 1890 Sir James Dewar was engaged in researches on liquid air.

Fig. 15.17 Thermos flask

He designed a vessel to keep the very cold liquid air at its low temperature once it had been made. Equally, as a *thermos flask*, it can keep a liquid hot.

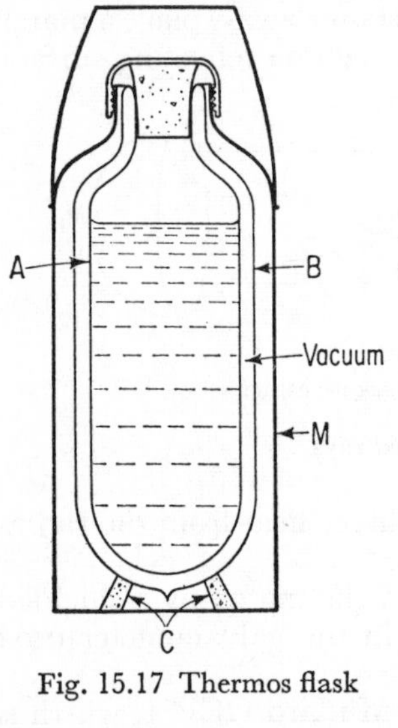

The vacuum (thermos) flask consists of a double-walled glass container, with a vacuum between the walls, A, B (Fig. 15.17). The sides of A, B, in the vacuum are silvered, and cork supports C are placed between the glass and the outside M of the flask. If a hot liquid is placed inside the flask it cannot lose heat by conduction or convection, as the space between the walls A, B is a vacuum. The small amount of heat lost from the liquid by radiation is diminished by the silvering on the wall A, and any radiation striking the wall B is reflected, as the latter is also silvered. A thermos flask should always be handled carefully. If the glass seal is broken by accident the vacuum becomes filled with air and the flask is ruined.

### SUMMARY

1. In conduction, heat is transferred without any change in the average position of the molecules concerned. In convection, heat is carried by the molecules themselves to other places. In radiation, the medium appears to play no part in transfer of heat.

2. Metals are good conductors of heat. Water is a poor conductor.

3. Dull-black surfaces are good emitters of radiation and good absorbers. Highly polished silvered surfaces are the reverse.

4. A miner's lamp is based on the principle that a metal gauze is a good *conductor* of heat. A hot water circulation system transfers heat by *convection*. A thermos flask has a double-walled container, with a vacuum between the walls, and any *radiated* heat is reflected back by the silvering on the outer wall.

## EXERCISE 15

**1.** Explain, using diagrams, how heat is transferred by: (*a*) conduction; (*b*) convection; (*c*) radiation. Which heat transfer, or transfers, could not be obtained in: (i) solids; (ii) liquids?

**2.** Is the specific heat of the land higher or lower than that of water? Describe and explain how sea and land breezes are produced.

**3.** (i) Describe how the air in a room is warmed by a hot radiator in it.
(ii) Explain how you would insulate a house to keep it warm in winter.

**4.** Describe the various types of process involved in the transfer of heat to the occupants of a room from the furnace flames in a central-heating system fitted with a circulating water pump.

What methods of heat transfer could *not* take place: (*a*) inside a sealed 'space station' in orbit around the earth; (*b*) in space itself? (*O*. and *C*.)

**5.** Explain: (*a*) how some of the heat is wasted when a room is heated by means of a coal fire burning in an open grate; (*b*) how an occupant of the room sitting well away from the fire is warmed by it; (*c*) why an eiderdown is very effective in keeping a person warm in bed; (*d*) why it feels warmer to the feet to step out of bed on to a rug rather than on to linoleum. (*N*.)

**6.** Describe with a diagram a Dewar flask. Show how the loss of heat by conduction, convection and radiation is reduced by the way of flask is designed.

Crumpled sheets of aluminium foil are used in the cavities of walls in houses to reduce heat losses from the rooms to the outside air. Why is the foil: (*a*) thin; (*b*) crumpled; (*c*) polished? (*O*. and *C*.)

**7.** Give a brief account of the ways in which heat energy can be transferred from one body to another at a distance.

State, with reasons, which of these ways are involved in the case of: (*a*) the sun and the earth; (*b*) the boiler of a central-heating system and the occupants of a room. (*O*. and *C*.)

**8.** Give a short account of the methods by which a vessel containing a hot liquid loses its heat.

Draw a diagram of a vacuum flask and explain how the rate of loss of heat by a hot liquid placed in a flask is reduced to a minimum. (*L*.)

**9.** How could it be shown that water is a bad conductor of heat?

Show by means of a diagram the essentials of a domestic hot-water system and explain how it works. (*L*.)

**10.** (*a*) Explain why heat losses are reduced but not completely eliminated when a warm object is: (i) wrapped in cotton-wool; (ii) separated from the surrounding air by an evacuated space.

(*b*) Explain and compare the various means by which heat may be transferred from the white-hot filament of a gas-filled electric lamp to a nearby black surface vertically above it. (*N*.)

**11.** What are the methods by which heat is transferred from one place to another? Describe a laboratory experiment to illustrate each of these modes of transference. Describe how, in the hot-water method of heating a house, heat is transferred from the boiler to a person in a room on an upper floor. (*O.* and *C.*)

**12.** Describe the various ways in which heat is lost by a red-hot metal sphere suspended from the ceiling of a room by a metal wire. What would be the effect of using a suspension of quartz instead of a metal? Explain the cause of convection in a gas or liquid and describe an experiment which shows the process in operation. (*O.* and *C.*)

## ANSWERS TO NUMERICAL EXERCISES

### EXERCISE 9 (p. 237)

**1.** 36·9°C   **3.** (*a*) 22·5°C (*b*) 72·5°F   **7.** 324°F   **8.** 35°C   **9.** 136·5°C

### EXERCISE 10 (p. 251)

**1.** $1·2 \times 10^{-5}$/K   **3.** 48 cm   **4.** (*b*) 0·062 cm   **6.** (*a*) 0·042 cm (*b*) $3·96 \times 10^{-5}$ cm³
**7.** (*a*) $2·7 \times 10^{-5}$ (*b*) $15·3 \times 10^{-5}$/K   **8.** 98·5 cm³   **10.** 0·2 cm   **11.** 599·8 cm

### EXERCISE 11 (p. 263)

**1.** 636°C   **2.** 125 cm³   **3.** 85 cm, 69°C   **4.** 75 cm³   **5.** 0·0036/K
**6.** (i) 71 cm (ii) 0·0038/K   **7.** 30°C   **8.** 50 cm   **9.** 0·0480 g/1
**10.** 171 cm mercury   **11.** 40·7 g/1

### EXERCISE 12 (p. 280)

**1.** 20 000 J   **2.** 420 g   **3.** 12·6 MJ/kg   **7.** $22·4 \times 10^{6}$ J/m³   **8.** 105 J/K
**10.** (*a*) −44°C (*b*) 2·4 J   **11.** 2·4 K   **12.** 40 300 J
**13.** (*a*) 10 500 J/min (*b*) 262·5 J/K   **14.** 80 000 J   **15.** 0·1°C   **16.** 3770 J
**17.** 8·8%   **18.** 0·23°C   **20.** 3·2°C   **21.** 0·24 K   **22.** 0·14 J/g K
**24.** 1·56 h   **27.** 84 m, 85 km   **28.** 29%   **29.** 72°C

### EXERCISE 13 (p. 292)

**1.** 54 min   **2.** 2318 J   **3.** 2·5 g   **4.** 60 900 J   **5.** 26·4°C   **6.** 20 g   **7.** 4·4 g   **8.** 8·75 g
**9.** 334 J/g   **10.** 1134 J/K   **11.** 11·8 g   **12.** 36·8 W/m²; 1·14°C/min   **14.** 26 kW
**15.** 1·51 J/g K, 132 J

### EXERCISE 14 (p. 306)

**9.** (*a*) 8·0 mm (*b*) 55°C   **10.** 2520 J

# Waves

# 16

# WAVE PROPAGATION · WAVE EFFECTS

## Sun's Radiation

The sun is the source of most of the energy falling on the earth. It is our natural light source, that is, it sends out a particular kind of energy called *luminous energy* which stimulates the sensation of vision. The sun also radiates energy which does not stimulate vision but produces warmth or heat – we say this is due to *infra-red* radiation from the sun. Another type of invisible radiation from the sun, called *ultra-violet* radiation, produces 'sun-tan' and makes certain minerals fluoresce (see also p. 440).

## Galaxies · Radio-emission

Our world consists of the earth and other solar planets such as Mars and Jupiter moving round the sun. The sun is one of the stars in our *galaxy*, in which astronomers estimate there are a hundred thousand million stars. There are many other galaxies in the universe, also with many millions of stars. *Radio waves* emitted from galaxies at enormously long distances from the earth have been detected by radio-telescopes at observatories in England such as Jodrell Bank and Cambridge. *Quasars* are the brightest emitters so far observed.

## Waves by Ripple Tank

Light energy and radio energy reach us across millions of miles of empty space from the sun and other stars. It is difficult to visualize how energy travels from one place to another in its journey. We can gain an understanding of how this may happen, however, by studying the way energy travels or spreads across water from one place to another.

This can be done with the aid of a *ripple tank* (Fig. 16.1(i)). Basically, the tank consists of: (1) a pool of water W in a shallow rectangular dish with a clear glass base and with the edges lined with sponge; (2) a *dipper* for producing water waves – this may be either a vertical long plane metal strip D or one of the small spheres S beneath a bar B; (3) a *vibrator* in the form of an off-centre weight on the rotating armature of a small electric motor rigidly attached to B; (4) a *lamp* with a small filament which illuminates the water surface from below or above, so that the ripples of the wave can be seen and also projected on a screen.

When the height of the metal strip is adjusted so that it just dips into the water and the electric motor is switched on, the strip or vibrator produces ripples or waves. They spread across the water surface and are absorbed at the edge by the sponge. If the lamp is placed above the water the waves can be seen and viewed from above.

A *stroboscope* is used to make the waves appear stationary. A simple form of hand stroboscope consists of a circular wooden disc with evenly spaced slits round it and an axle in the centre (Fig. 16.1(ii)). It is held by the axle in one hand by an observer and spun round with the other. The waves can be seen moving in quick succession on looking at the water surface through the revolving slits. As the stroboscope speed is increased from zero, the waves appear stationary at one stage. They can then be studied closely.

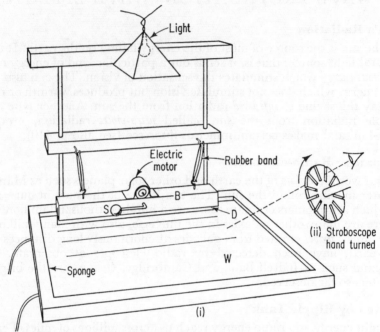

Fig. 16.1 Ripple tank

## Plane and Spherical Waves

With the straight edge of the metal strip M dipping into the water, parallel *plane waves* can be seen spreading along the surface (Fig. 16.2(i)). With one sphere S dipping into the water, *spherical* or *circular waves* spread out having S as their centre (Fig. 16.2(ii)).

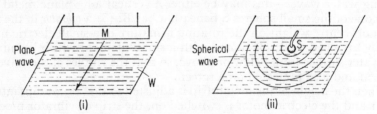

Fig. 16.2 Plane and spherical waves

In both cases small pieces of cork on the water will bob up and down, or vibrate, as the wave passes, but they remain at the same place. This shows that the water itself does not move bodily from place to place along the surface as the wave travels. Only *energy*, the energy of movement, travels along the surface, and water is the material or 'medium' in which the energy travels.

## Transverse and Longitudinal Waves

A similar wave effect is obtained with a long rope tied at one end to a wall. When a number of white ribbons is attached at intervals to the rope and the free end is then jerked up and down, a wave or pulse travels along the rope from one end to the other. The ribbons move up and down, showing the energy of movement of the wave, but they remain at the same place while vibrating. The wave travelling along the rope is now carried by the rope material or medium. It is also instructive to watch waves spreading along a long horizontal flexible coil of wire such as the 'Slinky' type, when one end of the coil is moved rapidly up and down. The waves can be seen travelling along the metal while the average position of one of the coil turns remains the same.

In every wave, then, we distinguish between:

(1) the vibrations, which produce the wave;
(2) the wave itself, which is the energy travelling along the medium.

In both the water wave and the rope wave the vibrations are up and down or vertical, whereas the wave direction is horizontal. Since the vibrations are perpendicular to the wave, these are called *transverse* waves.

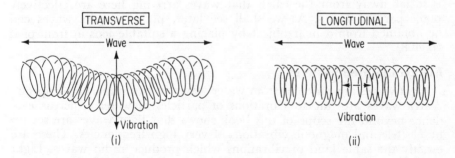

Fig. 16.2A Transverse and longitudinal waves

If the long loose ('Slinky') coil spring is spread along the bench or floor and one end is moved to and fro repeatedly parallel to the axis, a wave can be seen travelling along the wire to the other end. This time the wave is due to vibrations of the coil *parallel* to its direction of travel. It is called a *longitudinal wave*. If the coil is stretched out more, or a stiffer coil is used, the speed of the longitudinal wave is seen to increase.

Both transverse and longitudinal waves occur in nature. Sound waves are longitudinal (see p. 328), whereas light waves are transverse.

## Wave Theory of Light

Having seen how energy can spread from place to place along water by means of waves, it is reasonable to ask whether light energy can travel in a similar way. The 'wave theory' of light was first suggested about 1660 by Huygens, a famous Dutch scientist, and it has proved very fruitful in explaining light phenomena, as we shall see.

For a small visible object S, such as a small electric-lamp filament, we imagine *spherical* or *circular waves* spreading out with the speed of light so that the energy reaches our eyes (Fig. 16.3(i)). The sensation of vision is

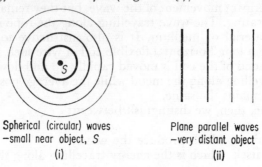

Spherical (circular) waves
—small near object, S
(i)

Plane parallel waves
—very distant object
(ii)

Fig. 16.3 Spherical and plane waves

then produced. The source S continues to be seen as long as it radiates luminous energy. The flame of a lighted match is visible until its luminous energy disappears.

A long way from the lamp the circular waves have a very large radius. They are then effectively *plane parallel waves* (Fig. 16.3(ii)). The sun is so far away from the earth that waves arriving here are effectively plane parallel waves. As we shall see later, plane parallel waves can be obtained from a near object by placing a suitable lens in front of it (p. 408).

## Electromagnetic Waves

Light waves are different from water waves in two important respects. Water waves are due to vibrations of particles of water, whereas evidence beyond the scope of this book shows that light waves are set up by electric and magnetic vibrations of very high frequencies. These are exactly the same kind of vibrations which produce radio waves. Light waves and radio waves are therefore both examples of *electromagnetic waves*.

Further, water waves are carried by water, which is a material medium. Light waves which reach us from the sun travel through large regions of empty space, and hence no material medium is needed. This was not acceptable to scientists of the past, and a medium called 'aether' was therefore invented in which electromagnetic waves travelled. We shall not be concerned with the aether in this book.

Observations show that light waves are transverse waves, that is, the electric and magnetic vibrations are in a direction perpendicular to the direction of travel of the light wave.

## Medium for Sound Waves

When an electric bell is placed in an air-tight vessel and the air is pumped out gradually from the vessel, the sound of the bell also gradually dies away (Fig. 16.4). The clapper of the bell, however, can still be seen striking the gong, although no sound is heard. Thus sound cannot pass through a vacuum. A material medium is necessary to carry the sound, whereas light or electromagnetic waves can pass through a vacuum. We usually hear sounds carried by an air medium, but the sound of horses' hooves can sometimes be detected through the earth by placing an ear to the ground. Tapping at one end of a long pipe circulating a building can be heard in other parts of the building, the sound being carried through the metal. A submarine's propellers moving below the sea give rise to sound waves in the water, which can be detected by a sensitive microphone.

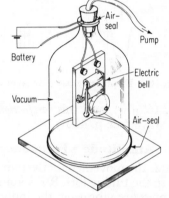

Fig. 16.4 Effect of medium on sound waves

## Sound Waves

Sound travels through the air in waves, but, unlike water waves, these sound waves are not transverse but longitudinal: the to-and-fro movement of the air particles is in the same direction as the movement of the waves (Fig. 16.5).

We can understand how a sound wave travels by supposing a tuning fork is sounding in air, so that the prongs are vibrating. When the prong R first moves to the right it disturbs the layer of air next to it (Fig. 16.5). This layer pushes the layer next to it, and so on. Thus at the first 'out-stroke' of the prongs the layers of air near R are compressed, as shown by $c$ in Fig. 16.5(i). When the prongs return to their normal position a short time later the *compression c*, as the crowding of the layers may be termed, has travelled a short distance from the fork (Fig. 16.5(ii)). The 'disturbance' has thus been passed on from one layer to the next; *the layers themselves have not moved* bodily with the disturbance. When the prongs swing inward a short time later (Fig. 16.5(iii)) the compression $c$ has travelled farther away from R. As the vibrations of the prongs continue new compressions are started, and a number of compressions are present in the air at any instant, as represented in Fig. 16.5(iv). When the diaphragm of the ear of an observer is reached it undergoes a slight displacement due to the movement of the layer of air near to it, and a sound is heard. We shall see later that layers of air at particular places are farther apart than normal at some

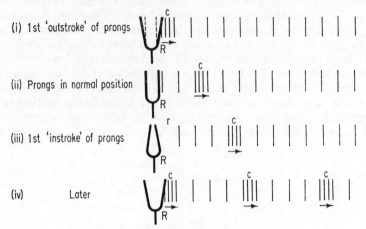

(i) 1st 'outstroke' of prongs

(ii) Prongs in normal position

(iii) 1st 'instroke' of prongs

(iv)        Later

Fig. 16.5 Displacements of air layers

instants, e.g. at *r* in Fig. 16.5(iii), and this *rarefaction* of air also travels in air as part of the sound wave (see p. 456).

### Vibrations of Layers · Amplitude · Frequency

We have now to consider the movement of the individual layers of air in a little more detail. Suppose the tuning fork R is sounding, creating a sound wave in air; since the prongs are vibrating, the individual layers in the air are also each vibrating. The layer C, for example, vibrates continuously through a very small distance on either side of its original (undisturbed) position (Fig. 16.6). The *maximum* distance moved on either side of the original position is known as the *amplitude* of the vibration of C, and in practice is a very small fraction of an inch. The number of complete to-and-fro oscillations of C in 1 sec is its *frequency f* of vibration. This is ob-

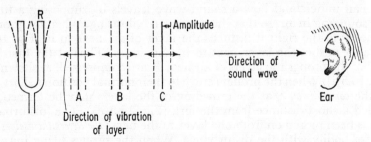

Fig. 16.6 Vibrations of air layers

viously the frequency of the tuning fork's vibration, so that if the latter is 256 per second, C makes 256 complete oscillations or *cycles* in 1 sec while the sound wave passes along. Other layers between the fork and the ear are also vibrating at the same frequency; but the vibrating layer B near C is a little 'out of step' with the latter's movement, and a vibrating layer A near B is a little more 'out of step' with C's motion, since it takes time for the wave to travel from one place to another. We say A, B and C are out of *phase*.

## Displacement Curves · Wavelength

To illustrate the movement of the layers, consider Fig. 16.7(i), in which the original (undisturbed) positions of the layers are shown exaggerated. Fig. 16.7(ii) illustrates the new positions of these layers at some instant when a sound wave is travelling through the air. At this instant some of the layers are displaced to the right of their original positions, others are just passing through their original positions and some are displaced to the

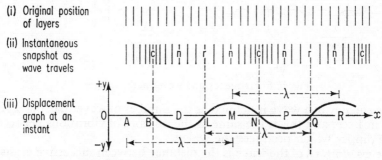

(i) Original position of layers

(ii) Instantaneous snapshot as wave travels

(iii) Displacement graph at an instant

Fig. 16.7  Sound waves and air displacements

left of their original positions. If we represent the displacement $y$ to the right as positive, and to the left as negative, the displacement graph at this instant is illustrated in Fig. 16.7(iii). At A the layer is at the end of its vibration to the right (Fig. 16.7(iii)); at B at this instant the layer is just passing through its original position; at D the layer is at the end of its vibration to the left; and so on at L, M, N, P, Q. Corresponding to A, M or R at this instant, we have *crests* of the wave; corresponding to D or P at the same instant, we have *troughs* of the wave. The *wavelength* $\lambda$ of the sound wave in the air is the distance between successive crests or troughs, just like the wavelength of a water wave. Thus, from Fig. 16.7(iii), the wavelength $\lambda = $ AM, or MR, or LQ. As time goes on, the positions of the layers change, and the crests and troughs appear at another instant at different places from those shown in Fig. 16.7(iii); but the displacement graph at any instant has always the same wave-form, and the wavelength, as defined above, is the same.

## Experiment

Sound waves are converted by a microphone into electric current variations of the same frequency and wave-shape. Connect a microphone to a cathode-ray oscillograph (p. 687) and examine the sound waveforms produced by speech, a whistle and a musical instrument.

Water waves travel in a similar way to sound waves, by slight variations in the positions of the layers or particles of the water which is passed along the water. Light waves travel with variations in electric and magnetic field strengths which cannot be seen, but their effects are felt by the eye and other suitable detectors.

## Wavelength and Velocity

We can understand further wave ideas by studying again the water wave in a ripple tank and making measurements on water waves.

When the electric motor of the ripple tank is speeded up the vibrations of the dipper increase. The plane waves in the water are now generated at a greater rate, and more are seen travelling across the water in the tank (Fig. 16.8(i), (ii)). Thus the *frequency f* of the wave has been increased.

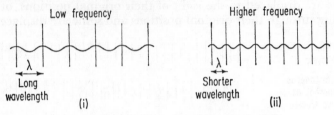

Fig. 16.8 Frequency and wavelength changes

The frequency of a wave is defined as the number of waves passing a given place in 1 sec, and this is the same as the number of vibrations per second of the dipper which produces the waves.

The *wavelength* λ of the wave is the distance between successive crests, or successive troughs, of the wave. At a higher frequency it should be carefully observed that the wavelength becomes shorter (Fig. 16.8(ii)).

The wavelength of a water wave can be measured by placing a transparent glass rule on the base of the dish along the direction of travel of the wave. The water surface is then viewed through a stroboscope. When the speed is increased from zero the waves are seen stationary at one speed.

The following measurements were made by two pupils:

*Wavelength*

Distance for 10 successive crests (9 waves) = 17·1 cm

$$\therefore \text{Wavelength } \lambda = \frac{17 \cdot 1}{9} = 1 \cdot 9 \text{ cm.}$$

*Velocity*

The velocity of the wave, the distance travelled per second, can also be found.

Time for 30 revolutions of stroboscope = 26·0 s

$$\therefore \text{Time for 1 revolution} = \tfrac{26}{30} \text{ s}$$

With 10 slits, time for successive views $= \tfrac{1}{10} \times \tfrac{26}{30} \text{ s} = \tfrac{26}{300}$

= Time for waves to travel one wavelength.

$$\therefore \text{Velocity} = \frac{\text{Distance}}{\text{Time}}$$

$$= \frac{1 \cdot 9}{\tfrac{26}{300}} = \frac{1 \cdot 9 \times 300}{26} = 22 \text{ cm/s.}$$

## Velocity Formula for Wave

We can obtain a general formula for the velocity $V$ in terms of frequency $f$ and wavelength $\lambda$. Suppose a dipper makes 4 vibrations per second; then 4 waves travel out per second. If the wavelength is 5·5 cm, the distance occupied by 4 waves is $4 \times 5·5$ or 22 cm. This is the distance travelled per second by the wave, or its velocity. Consequently, it can always be stated that:

$$Velocity\ (V) = Frequency\ (f) \times Wavelength\ (\lambda)$$

## Sound Waves

This result for velocity, $V = f\lambda$, is a general one for all waves, no matter how they are produced or in which medium they travel. *Sound waves*, for example, have a velocity of about 340 metres per second in air. A person with a deep voice singing a note of frequency 200 cycles per second is sending out sound waves whose wavelength is given, from above, by

$$\text{Wavelength} = \frac{\text{Velocity}}{\text{Frequency}} = \frac{340 \text{ m/s}}{200/s} = 1·7 \text{ m.}$$

If the note rises to 400 cycles per second, which is twice as high, the wavelength decreases to 0·85 m, which is half as long as before.

1 *cycle per second* is now called 1 *hertz* or 1 Hz (Hertz discovered radio waves in 1888). 1000 cycle per second = 1 kHz (kilohertz) and $10^6$ (1 million) cycle per second = 1 MHz (megahertz).

## Electromagnetic Waves

*Radio waves*, which are electromagnetic waves (p. 326), have a velocity of about $3 \times 10^8$ m/s. Broadcasting stations may thus send out long waves of 1500 m wavelength having a frequency of

$$\frac{\text{Velocity}}{\text{Wavelength}} = \frac{3 \times 10^8}{1500} = 2 \times 10^5 \text{ Hz} = 0·2 \text{ MHz.}$$

The British Broadcasting Corporation radio programme sent out on a wavelength of about 330 m uses radio waves of frequency

$$= \text{velocity/wavelength} = \frac{3 \times 10^8}{330} = 10^6 \text{ (one million) Hz (approx.).}$$

The speed of light waves is the same as that of radio waves, about $3 \times 10^8$ metres per second (or $3 \times 10^5$ kilometres per second), as they are also electromagnetic waves. We shall see later that light waves have a very short wavelength of about $6 \times 10^{-5}$ cm (p. 341). Infra-red, ultra-violet, X-rays and gamma-rays are also electromagnetic waves, and hence travel with the same velocity as light. They all have different wavelengths, forming a continuous spectrum ranging from about $10^{-10}$ cm to more than 1000 m, whose order of magnitude is represented below:

| RADIO | INFRA-RED | VISIBLE | ULTRA-VIOLET | X-RAYS | GAMMA-RAYS |
|---|---|---|---|---|---|
| $10^{-1}$ cm | $10^{-4}$ cm | $7 \times 10^{-5}$ | $10^{-6}$ cm | $10^{-8}$ cm | $10^{-10}$ cm |
| 1000 m | | Red | | | |
| | | $4·5 \times 10^{-5}$ cm | | | |
| | | Violet | | | |

Fast aeroplanes may travel at 1000 km per hour, which is about ¼ km/s, a negligible speed compared with that of light. Electrical particles used in nuclear investigations, however, can be speeded up to enormously high velocities approaching that of light, but no particle can ever have a velocity exceeding the velocity of light in a vacuum.

From his Theory of Special Relativity, the great scientist Einstein showed that matter or mass was a form of energy. The two quantities are related by the famous *mass-energy relation*:

$$E = mc^2,$$

where $E$ is the amount of energy produced in *joules* when a change of mass $m$ *kg* occurs, $c$ being the numerical value of the velocity of light in *metres per second* $(3 \times 10^8)$. The speed of light $c$ is thus an important constant in the subject of nuclear energy, where mass changes are produced and energy is released (see p. 738).

## WAVE EFFECTS – DIFFRACTION · INTERFERENCE

### Diffraction of Waves

We now investigate what happens when waves pass through large and small apertures. The results obtained by using water waves in a ripple tank are illustrated in Fig. 16.9(i). In the case of a wide aperture AB between

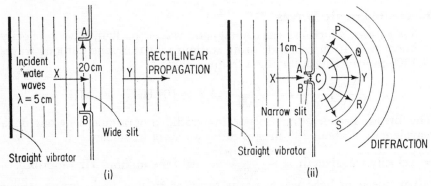

Fig. 16.9  (i) Rectilinear propagation. (ii) Diffraction

two metal strips, plane waves incident from X pass through the aperture as plane waves to Y (Fig. 16.9(i)). The boundaries or edges of the plane waves emerging correspond to the edges A and B of the aperture.

But when the gap C between the two metal strips A and B is made very narrow, and plane waves are incident on it, the waves emerge as *circular* waves (Fig. 16.9(ii)). Thus waves spread round C– they no longer travel in a straight line. Their effect is hence experienced not only in the straight-through direction Y, but also round the edges of the aperture C at P, Q, R and S. The waves are said to be *diffracted* at C. Huygens, who suggested the wave theory of light (p. 326), stated that every point *on a wavefront itself* acts like a source emitting wavelets. We can therefore imagine that when the wavefront reaches the gap C, each point in the gap sends out spherical wavelets. These spread round the gap at the edges, and they are not noticeable here until the gap becomes very small and comparable to the wavelength, as we have seen.

Generally, then, straight-line or *rectilinear propagation* of waves occurs when the opening is very wide compared to the wavelength. Diffraction always occurs when waves are incident on openings, but it becomes noticeable when the opening is comparable to the wavelength.

## Effect of Wavelength

Diffraction is a phenomenon obtained with all waves, whatever their nature.

A person speaking loudly in a room can be heard round a corner without

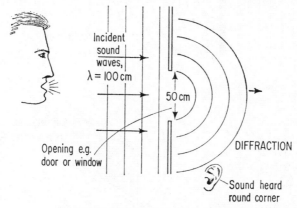

Fig. 16.10 Diffraction of sound waves

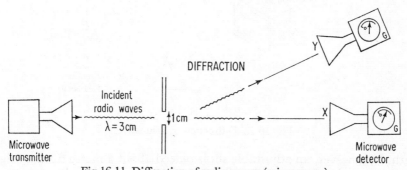

Fig.16.11 Diffraction of radio waves (microwaves)

being seen. *Sound waves* are therefore diffracted when passing through openings such as doors or windows, perhaps a metre wide (Fig. 16.10). This shows that sound waves have wavelengths of this order (see p. 331).

*Microwaves (radio waves)* of 3 cm wavelength, however, are diffracted by openings about 1 cm wide. The microwave receiver, which has a sensitive galvanometer G, collects the waves at X directly from the transmitter (Fig. 16.11). At Y, at an angle to the incident direction, a large deflection is also obtained on G. The waves are hence diffracted.

*Light* passing through a wide opening like a window or letter-box into a darkened place has a sharp boundary where the shadow begins. No diffraction is seen. A narrower opening, such as a slit in a ray-box comb a few millimetres wide, still allows light to pass through in a straight-line

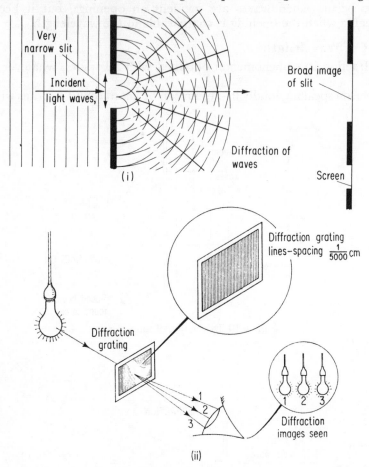

Fig. 16.12 Diffraction of light waves

path. If, however, an adjustable slit is placed about a metre from a bright over-run ray-box, or motor headlamp, bulb, the patch of light on a screen beyond the slit is seen to broaden as the slit is narrowed. Some light has spread round the edges, as shown in the exaggerated sketch in Fig. 16.12(i). The blurred image obtained in a pin-hole camera when the hole is made *very* small is due to diffraction.

An even more impressive demonstration of light diffraction is obtained when a *diffraction grating* is used. This is a piece of transparent film on which there may be as many as 5000 parallel opaque lines, equally spaced, occupying only a distance of 1 cm. The clear spaces between the lines are thus $\frac{1}{5000}$ cm apart. When the grating is placed close to the eye and a bright lamp viewed through it several images are seen (Fig. 16.12(ii)). The light from the lamp has therefore emerged from the clear spaces or openings in several different directions, showing diffraction of the light waves. The wavelength of light is therefore of the order of $\frac{1}{5000}$ cm. Experiments described shortly show it is more nearly $\frac{1}{18000}$ cm. The very short wavelength of light explains why light passes through most openings without diffraction and produces sharp shadows.

## Stationary Transverse Waves

Waves which spread outwards from a vibrator in a ripple tank or from a light source such as a lamp are called *progressive waves*. Their progress is changed abruptly if they strike a wall, for example, so that they are reflected back along the path of the incident or oncoming waves. Using a Slinky coil spring (p. 325) fixed at one end, a transverse wave which travels to this end can be seen to reverse in phase after reflection here, that is, an upward movement of the spring is reflected as a downward movement. We thus have two waves travelling in opposite directions along the spring, which are said to 'interfere'.

The resultant wave, or interference effect, can be demonstrated by

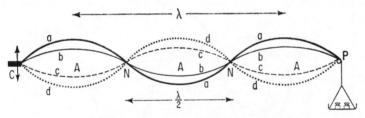

Fig. 16.13  Stationary transverse waves in thread

attaching a piece of thread to a vibrator such as the clapper C of an electric bell from which the gong is removed (Fig. 16.13). The thread passes over a pulley P, and a scale pan is attached to the other end. When weights are placed on the scale pan and adjusted, the thread can be seen to be vibrating with *stationary loops* along its length (Fig. 16.13). The successive positions of the wave are shown by $a$, $b$, $c$ and $d$. This is called a *stationary transverse wave*. Unlike a progressive wave, in which every point in the medium vibrates as the wave spreads out from the source of disturbance, the points P, N, N and C on the thread are permanently at rest as the wave on it moves between P and C. Further, the extent of the vibration increases to a maximum half-way between the points at rest, as shown.

Fig. 16.14(i) and (ii) illustrate briefly how the interference pattern is produced at two different times along the thread. One wave, the forward wave W, travels from C to P. The reflected wave R travels from P to C. At some instant the two waves have the appearance shown in Fig. 16.14(i). The displacements of the thread due to each are then in opposite directions everywhere, and hence the resultant, obtained by adding the displacements, is zero at all points. At a later time, corresponding to a $\frac{1}{4}$ of that for a complete vibration or period $T$, the two waves have moved to positions shown in Fig. 16.14(ii). A point X on the wave W in Fig. 16.14(i) is now at P. So is the point Y on the wave R in Fig. 16.14(i). Since the two displacements in Fig. 16.14(ii) are now everywhere in the same direction at all points of the thread, the resultant wave is that shown.

We have only taken two instants of time, but it is sufficient to show that the interference between the two waves produces a wave in which some points marked N and called *nodes* are always at rest and that other points

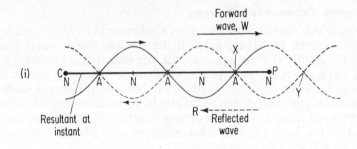

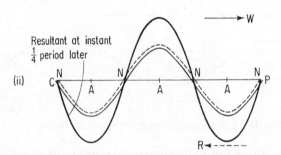

Fig. 16.14 Formation of stationary wave

between them are vibrating with varying amplitudes. The points marked A and called *antinodes* are vibrating with the maximum amplitude. It can be seen that if the wavelength is $\lambda$,

*The distance between successive nodes or antinodes* $= \lambda/2$,
*The distance between a node and the nearest antinode* $= \lambda/4$.

## Stationary Longitudinal Waves

The 'Slinky' coil of wire may also be used to show a stationary longitudinal wave pattern. The coil is placed on the bench or floor or suspended to reduce friction, and extended until it is very long. One end Y is attached to a fixed block so that it cannot move. The other end Z is now moved to and fro repeatedly, so that a longitudinal pulse travels continually from Z to Y and back. By altering the pulse timing at Z a stationary wave can be

**PLATE 7**
Infra-red and Solar Radiation

◄ 7(a) Photograph taken by visible light from ordinary electric lamp.

◄ 7(b) Same photograph taken by infra-red radiation from two hot electric irons, invisible in the dark (p. 442).

◄ 7(c) Solar device producing electrical power in a cylindrical generator by means of a thermoelectric effect (p. 235).

PLATE 8
Waves –
Reflection, Refraction, Interference

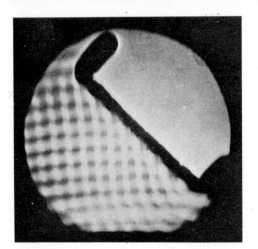

8(a) Reflection of plane waves, coming from below, at plane surface (p. 358).

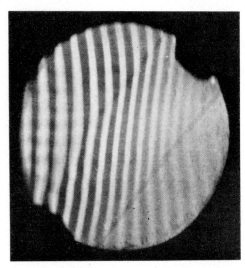

8(c) Refraction of plane waves, coming from left, entering shallower medium sloping upward whose outline is visible, and emerging from medium. Observe the smaller velocity in the shallower medium (p. 381).

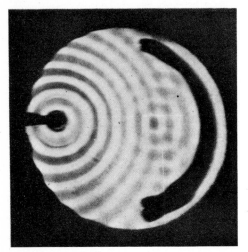

8(b) Reflection of circular waves by concave surface – point source produces a point image near reflector (p. 368).

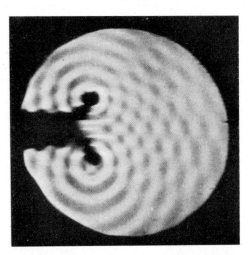

8(d) Interference of waves from two close sources vibrating in phase with the same frequency, showing regions of constructive and destructive interference (p. 338).

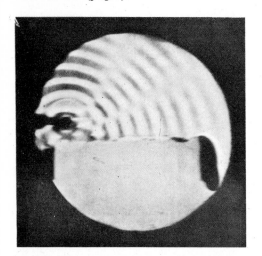

8(e) Interference of waves from one source and a near plane reflector (Lloyd's mirror) (p. 339).

set up with an antinode A at the middle and nodes N, N at the ends Y and Z (compare stationary wave in transverse vibrating string, p. 335). The longitudinal wave now travels from Z to Y and back (one wavelength) in the time between successive pulses at Z (Fig. 16.15(i)).

Other stationary waves may be set up, for example, by increasing the frequency of the pulses at Z three times (Fig. 16.15(ii)).

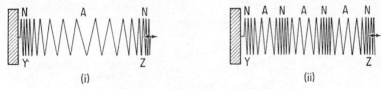

(i)          (ii)

Fig. 16.15 Stationary longitudinal wave in coil

## Stationary Sound, Light and Radio Waves

The stationary wave is an important wave, as it occurs in many different branches of physics. It is produced along the vibrating string of a violin or other stringed instrument when it is played, owing to reflection at the ends. It is the wave set up in the vibrating air inside an organ pipe or wind instrument for the same reason. This is discussed later (p. 466).

Stationary light waves are rarer than sound waves owing to their extremely short wavelength, but photographs showing bright and dark bands, antinodes and nodes, reveal their presence. Stationary waves of current are set up in aerials owing to reflection of current waves at the ends of the metal. Stationary waves are also considered to be set up in those orbits round the nucleus of an atom which the circulating electrons occupy (p. 738).

## Interference of Waves

Waves spreading out from two close vibrating sources can be expected to *interfere* with each other where their paths cross. We can observe the interference of water waves by using two small close vibrating sources A and B just dipping into water in a ripple tank (Fig. 16.16). Although waves are seen travelling outwards from A and B, the water surface is undisturbed along directions such as O in Fig. 16.16(i) and higher than normal along directions such as S. When the two sources A and B are moved closer together the directions of constructive and destructive interference becomes more widely spaced (see Fig. 16.16(v)).

We can explain this effect by adding together the two waves which arrive at any point from the sources A and B. The vibrators A and B are always in step and send out waves which have the same wavelength and the same amplitude. Suppose the point P is equidistant from A and B, or AP = BP (Fig. 16.16(ii)). The two waves arriving at P are then always in step or *in phase* with each other, that is, the crests of the two waves arrive simultaneously, and then, later, the troughs coincide and so on. The combined or *resultant* wave R is obtained by adding the two waves. It is a wave of double the amplitude of that due to A or B. This is therefore known as *constructive interference* of the two waves, as shown in Fig. 16.16(ii).

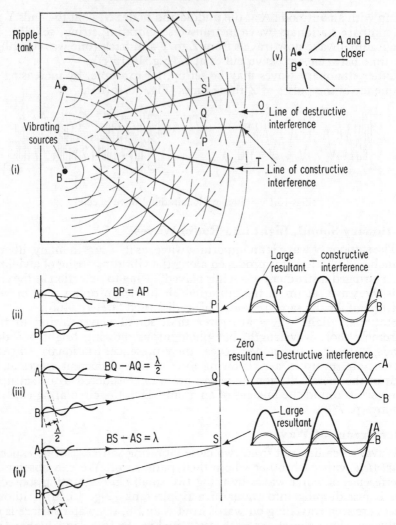

Fig. 16.16 Interference of waves

On the other hand, suppose we move from P in a direction parallel to AB to another point Q such that BQ is longer than AQ by λ/2, half a wavelength (Fig. 16.16(iii)). This time the crests of the wave from A arrive simultaneously with the troughs of the wave from B, and vice-versa. The resultant is therefore always zero in this case. This is called *destructive interference* of the two waves. If we now move to a point S such that BS is longer than AS by λ, one wavelength, then the crests and troughs of the two waves now arrive exactly in step (Fig. 16.16(iv)). Constructive interference is therefore obtained again.

Since BS − AS = λ, the interference pattern of the water waves enables its wavelength to be found. We measure BS and AS, and subtract the two lengths. If this is 2 cm, then the wavelength is 2 cm.

Fig. 16.17 shows how interference occurs between waves from a vibrating source O and those produced after reflection from a metal plate M. The water surface in this case is shown in Plate 8, to which the reader should now refer. This demonstration of interference was first produced in light by Lloyd using a plane mirror and a source of light very close to it.

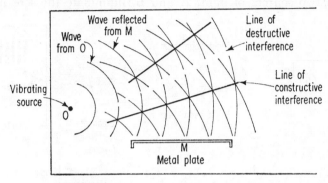

Fig. 16.17 Interference of waves using reflector (Lloyd's 'mirror')

## Interference of Light Waves

An analogous interference phenomenon is shown by two light waves. Light waves have very short wavelengths, such as $\frac{1}{18000}$ cm. The two sources of light must therefore be extremely close to each other for their waves to interfere. We must also ensure that the two light sources are sending out waves always in step with each other or *coherent*, otherwise a permanent interference effect will not be seen. Two lamp filaments close to each other in front of a white screen will not do. The waves from each lamp are due to vibrating atoms in the two metal filaments, and they are always out of step or out of phase with each other or non-coherent. The screen thus shows increased brightness due to the two lamps, but nothing more. If, however, two narrow slits are illuminated by the *same* light source suitably positioned the waves which emerge from the slits are always coherent because they come from the same source. We now have two light 'vibrators' analogous to the two vibrators A and B in Fig. 16.16, which produced an interference effect in water waves because they were always in phase and had the same frequency.

Fig. 16.18(i) shows a simple light interference demonstration. Two narrow slits A and B are cut in a microscope slide originally coated with an opaque solution of aquadag so that they are very close to each other, for example, 0·4 mm apart. They are illuminated by bright light from an over-run ray-box lamp, 12-volt, 24-watt, with a straight filament, which passes through a narrow slit S about 1 mm wide. The two slits A and B act as sources of light waves, and when a white paper screen is placed at W some bright and dark bands are seen.

This is exactly analogous to the case of the water waves from two close sources, as explained on p. 337. The light sources A and B send out waves which interfere in the region shown shaded in Fig. 16.18(i). The dark bands on the screen show destructive interference, that is, troughs and crests arrive simultaneously, and the resultant is zero or absence of light. The light bands between the dark bands are bright and show the constructive interference of the two light waves from A and B, that is, the crests (and troughs) arrive simultaneously. The presence of dark bands on the screen, although two sources of light, A and B, illuminate the screen, can be

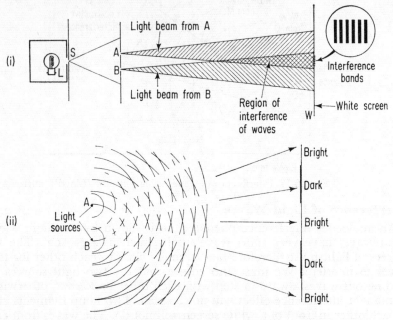

Fig. 16.18 Interference of light waves – Young's experiment

explained only by assuming that A and B send out waves. The constructive and destructive interference of the waves is illustrated in Fig. 16.18(ii).

This demonstration of interference was first carried out about 1800 by a scientist named Sir Thomas Young. It is therefore known as *Young's interference experiment*.

## Measurement of Wavelength

Young's experiment also provides a measurement of the wavelength of light. We can see how this comes about by studying the exaggerated sketch in Fig. 16.19. Here O is the point on the screen W which is equidistant from the two slits, so that NO is normal to AB. If they are in step when they start off the two waves from A and B arrive in step simultaneously at O, since each travels the same distance. Thus the crests and troughs of the waves arrive together, and hence a bright band of light is obtained here.

Now suppose we move from O to another point P near it on the screen. If BP is exactly one wavelength longer than AP the two waves again arrive in step at P. Hence another bright band is obtained here.

The distance PB is longer than AP by $\lambda$, the wavelength of the light waves. Suppose we cut off a length PQ on BP equal to AP. Then, as shown, BQ is equal to $\lambda$. If PN is joined we can see that, as A and B are really very close to each other, the normal NO to AB has veered round to NP through an angle PNO or $\theta$ equal to angle BAQ. Suppose the shift PO is $y$, NO is $D$ and AB is $a$, as shown. From similar triangles PON, BAQ, we have

$$\frac{y}{D} = \frac{\lambda}{a}$$

$$\therefore y = \frac{\lambda D}{a} \quad \text{or} \quad \lambda = \frac{ay}{D}.$$

This result for $y$, the separation between two bright bands, enables us to find out whether blue light has a longer or shorter wavelength than red light. Place

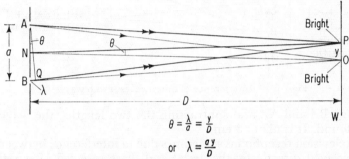

$$\theta = \frac{\lambda}{a} = \frac{y}{D}$$

$$\text{or} \quad \lambda = \frac{ay}{D}$$

Fig. 16.19 Wavelength calculation

a blue filter and then a red filter in quick succession over the light box slit S (Fig. 16.18(i)). Observation shows that the blue bands are closer together than the red bands. Now from above, since the separation $y$ is smaller for blue light, the wavelength of blue light is shorter than that of red light.

To find a rough value of the wavelength of light, measure the average distance between the bright bands on the screen in Fig. 16.18(i) with a millimetre scale, the distance $D$, and the separation of the slits A and B. In one experiment, $y = 1.5$ mm, $D = 100$ cm and $a = 0.4$ mm. Converting to centimetres,

$$\therefore \lambda = \frac{ay}{D} = \frac{0.04 \times 0.15}{100} = 0.000\,06 \text{ cm or } 6 \times 10^{-5} \text{ cm.}$$

This is an approximate value for the wavelength of light. As we have seen, the wavelength varies with the colour of the light. The extreme ends of the visible spectrum are about $4.5 \times 10^{-5}$ cm for violet and $7.0 \times 10^{-5}$ cm for red light. The eye is thus sensitive to a very narrow band of wavelengths.

### Interference of Microwaves

A similar interference demonstration can be carried out with 3-cm radio waves or *microwaves*. A transmitter S of the microwaves is arranged behind three screens P, Q and R which have gaps A and B about 1 cm wide between them, as shown in Fig. 16.20. A and B act as sources each of which send out identical waves. The receiver has a microammeter G which records reception of the waves, and is placed in front of A and B.

Starting from a position corresponding to the perpendicular bisector of AB and then moving it sideways, the receiver collects the two waves reaching it from the two sources A and B. At X the instrument G shows a large deflection–the waves interfere constructively. At Y there is no deflection–the waves interfere destructively. On moving further to Z, G shows a large deflection again–there is now constructive interference. This time, then, $BZ - AZ = \lambda$, one complete wavelength. Thus by

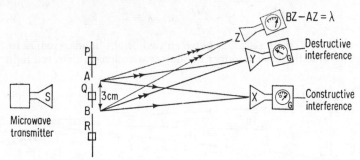

Fig. 16.20  Interference of microwaves (radio waves)

measuring BZ and AZ and subtracting the two lengths, the wavelength can be found. It is about 3 cm.

Poor television reception is sometimes due to interference between radio waves received directly by the aerial and those received after reflection from a neighbouring wall or obstacles such as metal tanks (compare Fig. 16.17, *Lloyd's mirror*).

### Wavelength by Diffraction Grating

The diffraction grating (p. 334) is a very convenient apparatus for measuring the wavelength of light by an interference method.

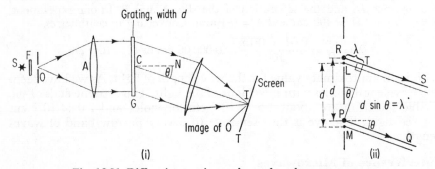

Fig. 16.21  Diffraction grating and wavelength measurement

A bright lamp S, with a red filter F in front, is placed behind a narrow slit O (Fig. 16.21(i)). A converging (convex) lens A is positioned with O at its focus so that a parallel beam illuminates the grating G normally, as shown. The emerging parallel beam is then collected by moving a lens round from the direction CN until a red image I of the slit is first received on the screen T. The angle $\theta$ between: (i) the line IC from the image to the grating passing through the centre of the lens, and (ii) the normal CN to the grating is then measured. The wavelength $\lambda$ is then calculated from $d \sin \theta = \lambda$, where $d$ is the width of the grating element, the distance occupied by one slit and one line. If the grating has 10 000 lines per centimetre, for example, then $d = \frac{1}{10000}$ cm.

*Theory*

The lens collects rays making an angle $\theta$ to the grating normal. Consider two rays from points P and R separated a distance $d$ in neighbouring clear spaces of the grating (Fig. 16.21(ii)). If the path of the ray RS from R is one wavelength longer than that of the ray PQ, then there will be constructive interference when the two rays are brought together by the lens. All the other corresponding rays in the neighbouring spaces, such as those from L and M, which are separated a distance $d$, will also have a path difference of $\lambda$. Consequently, the constructive interference from all the waves produces a *bright image* at the focus of the lens. If PT is the normal from P to RS, then RT is the path difference. Hence $RT = PR \sin \theta$, from the right-angled triangle PRT, since angle RPT is equal to $\theta$. Thus $d \sin \theta = \lambda$. If the red filter is replaced by a blue filter a diffraction first image is obtained at a smaller angle $\theta$ than with the red filter, showing that the wavelength of blue light is shorter than red light. It should be noted that rays diffracted from the clear spaces of the grating in a direction *normal* to the grating can also be collected by a lens pointing directly at the latter. In this case all the rays have no path difference. Consequently, a bright image is seen at the focus of the lens.

From above, for a grating of 5000 lines per cm, or $d = \frac{1}{5000}$ cm, and yellow light of wavelength $\lambda = 6 \cdot 0 \times 10^{-5}$ cm, the *first order* image, corresponding to a path difference $\lambda$, is given by $(\frac{1}{5000}) \sin \theta_1 = 6 \times 10^{-5}$. Thus $\sin \theta_1 = 0 \cdot 3$, or $\theta_1 = 17°$. At a larger angle than $\theta_1$, say $\theta_2$, the path difference between rays from corresponding points on neighbouring clear spaces collected by the lens may be $2\lambda$. Hence another image, the *second order* image, is now obtained. In this case $d \sin \theta_2 = 2\lambda$, from which $\sin \theta_2 = 0 \cdot 6$, or $\theta_2 = 37°$. A third order image, path difference $3\lambda$, is obtained at an angle $\theta_3$ given by $\sin \theta_3 = 0 \cdot 9$, but no more images are possible.

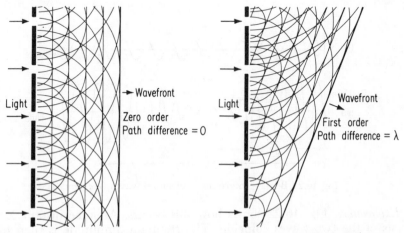

Fig. 16.21A Constructive interference of waves emerging from grating

Summarizing, when a diffraction grating is used, waves are diffracted at each clear space or slit of the grating and interfere with each other in different directions. Constructive interference occurs only in those directions from the normal where $d \sin \theta = \lambda$ or $2\lambda$, etc. Thus a lens, pointed at the grating in any of these directions to collect all the waves, produces a bright image at its focus. See Fig. 16.21A

The spectrometer, described on p. 437, is a compact arrangement of lenses ideally suited for the diffraction grating experiment.

### Interference of Sound Waves · Beats

As we saw previously, sound waves have wavelengths very much longer than microwaves or light waves and very much lower frequencies. A tuning fork of 256 Hz, for example, sends out waves in air about 100 cm long.

. Interference between sound waves can be demonstrated easily by taking two tuning forks A and B each of frequency 256 Hz, and 'loading' the prongs of one of them, B say, with plasticene or wax so that its frequency is slightly lowered to 252 Hz for example. Sounded individually, A and B both emit clear continuous notes of their respective frequencies. Sounded together, however, the note heard by an observer is a wavering or throbbing note, called a *beat* note.

#### EXPERIMENT

Using a microphone connected to a cathode-ray oscillograph operating at a suitable time-base frequency, observe the waveform when the sounding tuning-fork A is brought near the microphone. Repeat with the other fork B of slightly different frequency. Then sound A and B together and observe the resultant waveform, which has a varying amplitude of frequency the same as the beat note.

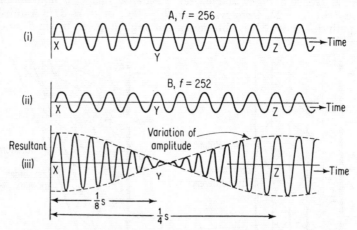

Fig 16.22 Beats – interference of sound waves (*exaggerated*)

*Explanation.* Fig. 16.22 shows how this occurs. At some instant X the crests of the two waves coincide. The resultant amplitude is then high. A loud sound is now heard. After ⅛ second a crest of one wave coincides with a trough of the other. The resultant amplitude is a minimum. A soft sound is then heard. In a further $\frac{1}{8}$ second a loud sound is heard again. Beats are therefore heard at a frequency of 4 Hz.

EXPERIMENT

Suspend independently two pendulums of slightly different length so that they have slightly different frequencies. Arrange one pendulum behind the other with their bobs X and Y on the same horizontal level. Simultaneously start both bobs swinging together. Observe how the faster bob, X say, gradually gains half a cycle on the other, Y, and is then completely out of phase with Y. After an equal interval of time observe that X has now gained one complete cycle on Y, so the two bobs are then swinging in phase again. X then goes out of phase once more and the action is repeated at regular intervals. This illustrates how the vibrations of two tuning forks of close frequency go out of phase and then go in phase, as described.

## Resonance

As we have seen, waves are sent out by different kinds of vibrating systems, mechanical (the vibrator used in the ripple tank), acoustical (the

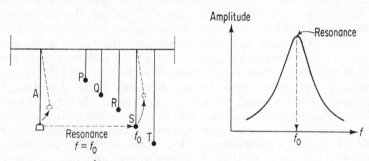

Fig. 16.23 Resonance experiment

tuning fork), electrical (the microwave transmitter) and optical (the lamp filament). The pendulum is a simple mechanical system which can be set into vibration. If the bob is pushed gently and then left, it vibrates with a *natural* frequency which depends on the length of the pendulum (p. 78). All vibrating systems have a natural frequency of vibration, whether mechanical or as above.

EXPERIMENT

Suspend five different pendulums, P, Q, R, S and T, of varying length from a horizontal string, each carrying a weight of about 100 g (Fig. 16.23(i)). Suspend another pendulum A of the same length as S beside them, carrying a heavier weight such as 1 kg. Start A oscillating, keeping the amplitude of swing constant. Observe that all the pendulums have small swings except for S. This has a swing of large amplitude, so that it has responded considerably to the vibrations of A through the string.

*Conclusion.* Since S has the same length as A, and therefore the same natural frequency of swing, a vibrating system such as S can be set into a large amplitude of vibration by a vibrating force of exactly the same frequency such as A.

By using a ruler behind the bob of S, its amplitude of swing can be roughly measured when the length and hence frequency $f$ of A is varied. Fig. 16.23 shows ideally what happens in any vibrating system of natural frequency $f_0$. The amplitude becomes a maximum when the frequency $f$ is exactly equal to $f_0$. The system is then said to be in *resonance* with the applied vibrating force.

## Other Examples of Resonance

Resonance phenomena occur in many branches of physics. In mechanics, a diver jumping repeatedly at one end of a diving-board will set it into resonant vibration and so gain considerable uplift before he dives. In sound, a high-pitched loud note from a singer has been known to shatter a glass near by, as the glass is set into resonant vibrations. In electricity a radio receiver is 'tuned' to a station when an electrical vibrating circuit inside is set into resonance by the incoming radio waves (p. 674).

Optical resonance can also be produced. A natural example occurs in a study of the sun's spectrum. In the heart of the sun the particles of the hot gases are miniature vibrating systems sending out light of all different wavelengths. When the radiation passes into the cooler atmosphere of gases surrounding the sun the atoms here are set into resonance by those radiations which have the same frequency as their natural frequency. The resonating atoms radiate their energy all round them. The energy in a particular direction towards the earth is thus relatively small. The visible sun's spectrum observed on the earth is then found to be crossed by a number of dark lines, which correspond in position to the wavelengths emitted by the resonating atoms. These lines are called *Fraunhofer lines*, after the first discoverer in 1814. They are an example of *absorption spectra*, so-called because the atoms in the atmosphere round the sun absorb incident radiation of the same frequency as their natural frequency. Elements can be identified from their absorption spectra, as the frequencies or wavelengths of the dark lines are characteristic of their vibrating atoms.

## SUMMARY

1. Water and sound waves are due to vibrations of layers or particles in the medium concerned. Light and radio waves are due to electromagnetic vibrations.

2. Sound waves are longitudinal waves—the wave travels in the same direction as the vibrations. Water, light and radio waves are transverse waves—the waves travel at right angles to the vibrations.

3. For all waves, *Velocity = Frequency × Wavelength*. Light waves have a very short wavelength, such as $\frac{1}{16000}$ cm, and travel with a velocity of 300 million

metres per second (186 000 miles per sec). Sound waves have a long wavelength, such as 50 cm, and travel with a velocity in air of about 340 m per sec (1100 ft per sec) at normal temperatures.

4. *Diffraction.* All waves may spread round corners or be diffracted. If the opening is very large compared with the wavelength the incident wave travels straight through without change of direction (rectilinear propagation). If the opening is comparable to the wavelength diffraction occurs. Light is diffracted by very narrow openings and sound by wide openings.

5. *Interference.* A stationary wave is due to interference of two waves travelling in opposite directions. Two close sources, vibrating in step, produce interference effects. In Young's experiment bright and dark bands are produced, from which the wavelength of light can be measured. *Beats* are due to interference in sound between waves of slightly differing frequency.

6. *Resonance.* A vibrating system produces its largest amplitude of vibration when the frequency of an external vibrating force is equal to the natural frequency of the system. Mechanical, acoustical, electrical and optical resonances can all be produced.

## EXERCISE 16 · ANSWERS, p. 348

**1.** Describe a *ripple tank.* Explain how you would use it to demonstrate plane waves and spherical waves. What effect is observed in each case as the frequency of vibration is increased? Illustrate your answer.

**2.** What is a *stroboscope?* Explain its purpose for ripple-tank experiments. Describe how you would measure the velocity of a water wave.

**3.** (i) Explain the difference between a *longitudinal wave* and a *transverse wave.* Name an example of each.

(ii) Describe an experiment which demonstrates the effect of the medium on sound waves. What does the same experiment show about light waves?

**4.** What is the relation between the frequency, velocity and wavelength of a wave? Write down approximate values for the wavelengths of: (i) sound waves; (ii) microwaves; (iii) light waves.

In the British Broadcasting Corporation the Light programme is transmitted on a long wavelength of 1500 m and a frequency of 200 000 Hz. Calculate the velocity of radio waves. Find the frequency for the Home service programme which is transmitted on a wavelength of 330 m.

**5.** Describe how you would demonstrate the diffraction of: (i) water waves; (ii) microwaves; (iii) sound waves. Illustrate your answer by diagrams and give approximate dimensions where necessary.

**6.** Describe and explain how an interference pattern can be demonstrated using water waves. Draw sketches to illustrate the appearance of the pattern when the vibrators are placed farther apart and then closer together. In the latter case, how would you measure the wavelength approximately?

**7.** Describe how you would demonstrate the *interference of light waves.* Draw sketches to illustrate your answer and give approximate dimensions. Explain why two close sodium lamps do not produce an interference pattern on a screen in front of them.

**8.** What is a *diffraction grating?* Describe how you would measure the wavelength of red light and of blue light with a diffraction grating. Which colour has the shorter wavelength?

**9.** Some effects are transmitted from one place to another by 'waves' and some by particles travelling from the source to the receiver. Describe TWO properties of sound which indicate which of these two processes is involved, and give briefly some experimental evidence for any statements you make about sound.

Indicate TWO properties of light which are evidence that it is a wave motion.

Because you can hear, but not see round an obstacle such as a building, does it follow that light and sound must be propagated by different mechanisms? (*O.* and *C.*)

**10.** Sound and light waves possess some similar and some opposite properties. Describe experiments, one in each case, to illustrate ONE similar and ONE opposite property.

**11.** Why are the lines on a diffraction grating regularly spaced? Why are they closely spaced?

A grating of 6000 lines per cm is illuminated normally by yellow light of wavelength $6 \cdot 0 \times 10^{-5}$ cm. (i) Why do 'interference' and 'diffraction' occur, (ii) at what angles to the normal are the first and second order images obtained, (iii) if infra-red rays are incident normally on the grating, what wavelength will coincide with the second order image?

## ANSWERS TO NUMERICAL EXERCISES

### EXERCISE 16 (p. 347)

**4.** $3 \times 10^8$ m/s, 90 091 Hz   **8.** Blue   **11.** (ii) 21° 6′, 46° 3′ (iii) $1 \cdot 2 \times 10^{-4}$ cm

# Optics

# 17

# RECTILINEAR PROPAGATION OF LIGHT

## Luminous and Non-luminous Objects

In the subject of optics or light we are concerned with luminous energy, that is, energy which causes the sensation of vision. The sun is a *self-luminous* object, and so are the stars. Some living creatures such as the glow-worm or fire-fly are self-luminous. Examples of artificial or man-made luminous sources are the electric lamp and the candle.

By contrast, the pages of this book are *non-luminous*. If you are reading it in daylight, then light from the sun falls on the page concerned and some of it is scattered or reflected back to your eyes by the white page. The black print absorbs the light. Consequently, the print stands out on the page. Similarly, light from an electric lamp is scattered back to your eyes if you are reading at night. Likewise, a driver at night can see a 'Halt' sign or the rear reflector of a bicycle or car by light reflected back from his headlamps. Most objects are non-luminous. A person's face, or the pattern on a tie, for example, is seen by light falling on it which is reflected back.

## Rays of Light · Ray-box

The direction or path along which light energy travels is called a light *ray*. The line OABC is a ray from a point source O (Fig. 17.1(i)). ODEF

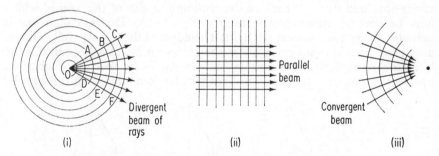

Fig. 17.1 Divergent, parallel, convergent beams and rays

is another ray from O in a different direction. It should be noted that rays are always *perpendicular* to the spherical surfaces of the *wavefronts* (p. 324) which spread out from the source O.

A collection or beam of rays which spreads out from O as shown in Fig. 17.1(i) is called a *divergent beam*. A *parallel beam* is shown in Fig. 17.2(ii). A *convergent beam* is shown in Fig. 17.1(iii). For demonstration purposes, rays of light are conveniently obtained from a ray-box B (Fig. 17.2). This has a small filament lamp L in an enclosed box, with a cylindrical convex lens C and a 'comb' S containing parallel slits in front of it. Rays of light then emerge from C. By moving the lamp L, a diverging or converging or parallel beam of rays can easily be obtained.

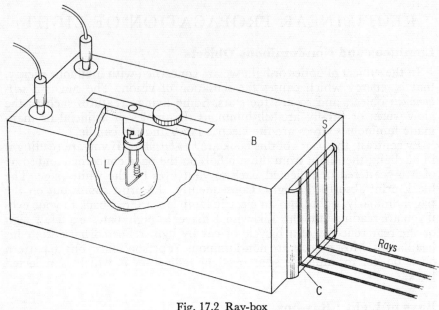

Fig. 17.2 Ray-box

## Light and Large Apertures

If a ray-box is placed on a sheet of white paper and switched on, a thick ray is obtained through each of the rectangular slits of the 'comb' which may be several millimetres wide (see Fig. 17.2). This so-called ray is actually a very narrow beam of light. The edges of the ray are well defined, and it can be said that *light travels in straight lines in this case*. We call this the 'rectilinear propagation' of light. With a much-wider aperture, such as an open letter-box or a window, sunlight can be seen streaming through by

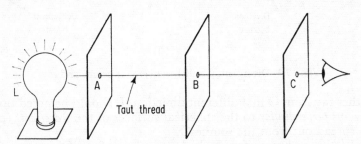

Fig. 17.3 Light usually travels in straight lines

light scattered from dust particles. The edges of the wide beam are again straight, showing that light travels in straight lines through wide openings. The rectilinear propagation of light can be demonstrated by placing an illuminated lamp L in front of a hole A in a cardboard, and then moving two other cardboards, each with holes B and C, until the light through A can be seen (Fig. 17.3). It will then be found that a thread passing through A, B and C and pulled tightly is perfectly straight along ABC.

### Pin-hole Camera

The pin-hole camera was the earliest camera. It was invented about 1550, well before lenses were utilized, and uses the fact that light travels in straight lines.

A pin-hole camera can be made simply by removing the cover from a closed box or tin, and then replacing it by semi-transparent paper so that

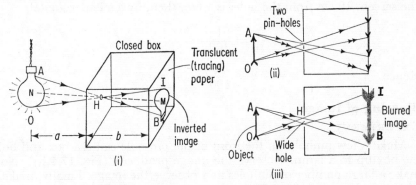

Fig. 17.4 Pin-hole camera and blurring

the box is sealed again from outside light. A small hole H is then made in the middle of the box opposite to the paper (Fig. 17.4(i)). When the hole is held up to a bright lamp in a darkened room an inverted small image of the lamp can be seen on the paper. Move the box towards the lamp and then back, and observe how the size of the image changes.

An image is always formed by rays of light coming from the object. A ray from a point O on the lamp, for example, passes through the hole and strikes the paper screen at I (Fig. 17.4(i)). A ray AH from A produces an image B. If the hole H is small every point on the object produces a corresponding point image. The image is then fairly sharp.

The pin-hole camera has been used by surveyors. It is preferred to the lens camera, as the lens produces distortion. Its disadvantage is the long time required for an image to be developed on the photographic plate, as the amount of light passing through the hole is small.

### Length of Image

From similar triangles AHO, BHI, Fig. 17.4(i), it can be seen that

$$\frac{BI}{AO} = \frac{IH}{HO} = \frac{MH}{HN}.$$

Thus

$$\frac{\text{Image length}}{\text{Object length}} = \frac{\text{Image distance from H}}{\text{Object distance from H}} = \frac{\text{Length of camera}}{\text{Object distance from H}}$$

Suppose, then, that the length of the camera box is 10 cm and an object of length 2 m is photographed when it is 100 cm from the hole. Then

$$\frac{\text{Image length}}{2 \text{ m}} = \frac{10 \text{ cm}}{100 \text{ cm}} = \frac{1}{10}$$

$$\therefore \text{ Image length} = \frac{1}{10} \times 2 \text{ m} = 20 \text{ cm.}$$

A more distant object, such as the sun, will produce a smaller image. Approximately, a disc of diameter 1 cm will just 'cover' the sun when held 100 cm from the eye. If the length of the diameter of the sun's image on the screen, 10 cm from the hole, is $x$ cm, then, by similar triangles,

$$\frac{x}{10} = \frac{1}{100}.$$

Hence
$$x = \frac{10}{100} = 0 \cdot 1 \text{ cm}$$

## Blurring of Image

EXPERIMENT

Make a few pinholes in the front of the pin-hole camera box and hold the box up to a lamp. Observe the image produced (Fig. 17.4(ii)). Now make a large number of pinholes and observe the image. Finally, make a wide hole H and observe the image again (Fig. 17.4(iii)).

*Conclusion.* With many pin-holes or with a wide hole, the image is blurred.

We can explain this result by noting that each pin-hole produces its own image on the screen. With several pin-holes the images overlap, and the image of the lamp is therefore blurred (Fig. 17.4(ii)). A wide hole is equivalent to a very large number of pin-holes. Light from a point such as O now reaches a small *area* of the image round I. (Fig. 17.4(iii)). It will be noted that parts of the area are also reached by rays from other points near O. Consequently, overlapping occurs and the image is blurred.

It is instructive to place a suitable converging (convex) lens in front of the many pin-holes or wide hole H and in contact with the box. A *sharp* image can now be obtained on the screen. The lens thus brings all rays from a point on an object to one unique point on the screen, a property utilized in the lens camera (p. 421).

## Shadows

*Shadows* are due to light travelling in straight lines. When light rays arrive at an opaque obstacle, that is, one which absorbs light, the rays just grazing the edges of the obstacle produce the outline of a shadow. The kind of shadow obtained depends on the size of the luminous object sending out the rays. Thus in Fig. 17.5(i) a ray-box with a small filament, or *point source* of light, produces a sharp shadow of an opaque solid ball B on a screen placed behind B. The shadow is uniformly dark. On the other hand, when the lamp is replaced by a *large source* of light such as a pearl lamp P the shadow is no longer uniformly dark (Fig. 17.5(ii)).

The innermost shadow or *umbra*, the region of full darkness, is not reached by any light. The edge of the umbra corresponds to the rays PL and QM, which just graze B. As we go away from the edge the darkness diminishes in intensity. At a point R on the screen, for example, the grazing ray HLR shows that R receives light from that small part of the lamp corresponding to PH. Likewise, the grazing ray KM shows that the point S receives light from the small part QK. Thus R and S are in

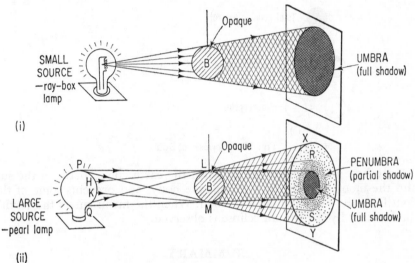

Fig. 17.5 Shadows. (i) Small source, (ii) large source.

'partial shadow', called the *penumbra*. The grazing rays QLX and PMY show that X, Y are on the edge of the penumbra. Above X or Y, points on the screen are illuminated by the whole of the lamp, and hence no shadow is obtained here.

### Eclipses

The moon and earth are both non-luminous. In a clear sky the moon can be seen by light from the sun, which is scattered by the moon's surface. The eclipse of the sun occurs when the moon, an opaque object, comes between the sun and the earth. This is illustrated in Fig. 17.6(i). Parts of the earth lie in the umbra and penumbra of the moon's shadow. The appearance of the sun from various parts of the earth is illustrated in Fig. 17.6(ii). A person at *a* sees a *total eclipse*, another at *b* or *d* sees a *partial eclipse*, and another at *c* or *e* sees no eclipse. On another occasion, different from that shown in Fig. 17.6(i), the earth and moon may be in positions when the extreme rays intersect before reaching the earth. An observer in a position corresponding to *a'* in Fig. 17.6(i) then sees an *annular eclipse*, a ring of light round the shadow of the moon.

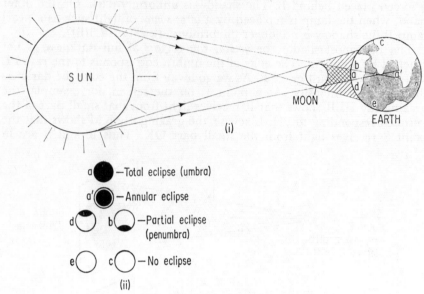

(i)

a ⬤ —Total eclipse (umbra)

a' ⬤ —Annular eclipse

d ◖ b ◗ —Partial eclipse (penumbra)

e ◯ c ◯ —No eclipse

(ii)

Fig. 17.6 Eclipses of Sun

An eclipse of the moon occurs when the earth comes between the sun and the moon. The earth is an opaque body, and prevents some of the light from the sun reaching the moon. When the moon enters the earth's shadow cast by the sun an eclipse is observed.

## SUMMARY

1. Luminous objects radiate light; non-luminous objects are seen by diffusely reflected light.

2. Beams of light may be parallel, convergent and divergent.

3. Shadows and eclipses are due to the rectilinear propagation of light. The *umbra* is the region of full shadow (total eclipse); the *penumbra* is the region of partial shadow (partial eclipse).

4. In the pin-hole camera the image is inverted. The size of the hole must not be too large, otherwise the image becomes blurred—a point on the image is now illuminated by several points on the object.

## EXERCISE 17 · ANSWERS, p. 452

**1.** In the diagram below, *A* represents a spherical source of light and *B* an opaque spherical object, both of which are placed as shown. Draw a diagram showing the shape of *A* as seen from *P*. (*N*.)

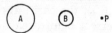

**2.** A fixed source of light is larger than a certain fixed opaque spherical object. With the help of diagrams, describe how the character of the shadow of the object on a screen changes as that screen is moved away from the object. Indicate on your diagrams the regions of umbra and penumbra, and also explain what would be seen in these regions by an eye looking towards the source.

How do eclipses of the sun occur and under what circumstances would you expect the eclipse to be: (a) total; (b) partial? (O.)

**3.** Draw a labelled diagram to show how an annular eclipse of the sun may be observed from the earth. Mark the position of the observer. (N.)

**4.** Describe an experiment to demonstrate the rectilinear propagation of light, and explain the formation of a total eclipse of the sun. (W.)

**5.** What effect, if any, has an increase in the size of the hole of a pin-hole camera on: (a) the size of the image; (b) the brightness of the image; (c) the sharpness of the image? What is the chief disadvantage of the pin-hole camera in use? (N.)

**6.** Describe and explain a method of taking a photograph without a lens, stating the property of light on which your method depends. What is the disadvantage of this method?

Explain with the aid of a diagram in each case how an eclipse of the sun may be: (a) total; (b) partial; (c) annular. (N.)

**7.** Explain with a diagram the conditions necessary for a partial eclipse of the moon to occur. Why do eclipses not occur every lunar month? (N.)

**8.** Describe how a pin-hole camera works.

The distance between the pin-hole and screen of a pin-hole camera is 12·5 cm, and the plate is 20 cm long. At what minimum distance from the pin-hole must a 1·8 m tall man stand if a full-length photo is required?

**9.** Describe, and illustrate by a diagram, the effect on the image in a pin-hole camera of making *two* pin-holes close to each other.

The sun is just covered by a disc of 2 cm diameter placed about 2 metres from the eye. If the length of the diameter of the sun's image formed by a pin-hole camera is 0·5 cm, calculate the distance from the pin-hole to the screen.

# 18

## REFLECTION AT PLANE SURFACES

When one surface of a flat piece of glass is coated with silver almost all the light which falls on the opposite surface is reflected. The mirrors so formed are used as looking-glasses and as car mirrors. They are also employed in the sextant, a navigational instrument, in which angular distances are measured by means of the reflection of the sun's rays at two plane mirrors. The properties of plane mirrors depend on the regular way in which they reflect light, as we shall see later.

### Reflection of Waves

The effect produced when waves meet a plane surface can be observed by means of the ripple tank. A vibrator C generates plane waves, which travel a short distance and are then incident on a plane metal strip AB in

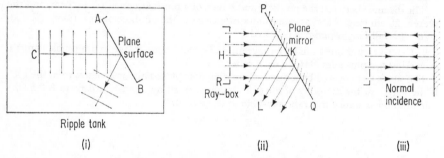

Fig. 18.1 Reflection of water and light waves

the water (Fig. 18.1(i)). The plane waves are seen to be reflected from AB in a definite direction, which varies with the position of AB. In one special case the waves are reflected straight back along their original course. This happens when the waves are incident *normally* on AB, that is, at 90° to AB.

### Reflection of Rays

The reflection of light rays can be investigated with a *ray-box* R and a plane mirror PQ on a sheet of white paper (Fig. 18.1(ii)). A parallel beam of light HK is seen reflected by the mirror in the direction KL. When the mirror is turned, the direction of the reflected rays changes. When the light rays are incident normally on the mirror they are reflected straight back (Fig. 18.1(iii)).

### Diffusion of Light

A plane mirror thus reflects a parallel beam of light in a definite direction. This is due to the high smoothness and polish of the reflecting surface.

But most objects do not reflect light with such regularity. Paper and clothing materials are examples. This is because of the lack of smoothness, or grain, of the surface, which is readily seen under a high-power microscope. Rays of light incident on such surfaces are reflected in different directions, $R_1$, $R_2$ and $R_3$, and are said to be *scattered* or *diffusely reflected* (Fig. 18.2). The pages of a book, or a person's face, are seen by a diffusely-reflected light.

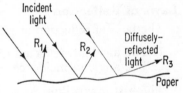

Fig. 18.2 Diffuse reflection

## Experiment on Reflection

Reflection of light can be examined in more detail as follows. A mirror is supported vertically on a sheet of white paper, and ON, normal to the mirror, is drawn from O, a point near the middle of the mirror (Fig. 18.3(i)). Further lines are then drawn on the paper so that they are inclined at angles such as 20°, 45°, 60° and 80° respectively to ON. A ray-box R is now placed so that a *single ray* follows one of the drawn lines, say AO. The position of the reflected ray OB is marked by dots. The ray-box is then moved so that the light is incident in directions corresponding to the other drawn lines, and the positions of the reflected rays are indicated as

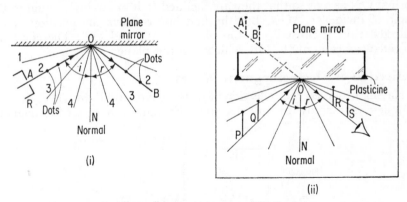

Fig. 18.3 Investigation of laws of reflection

before. The different angles of incidence $i$ and reflection $r$, which are the angles made with the normal, can now be measured.

Alternatively, the direction of an incident ray can be defined by two pins, P and Q, which are placed along one of the drawn lines (Fig. 18.1(ii)). Their images A and B are observed in the mirror, and two more pins R and S are then positioned by eye exactly in line with A and B. The line PQO is an incident ray reflected along ORS, which is then drawn. The experiment is repeated by moving P and Q to other incident lines, and the different angles of incidence $i$ and reflection $r$ are measured.

*Results.* The experiment shows that, in all cases, the angle of incidence $i$

= the angle of reflection *r*, both angles being measured from the normal ON (Fig. 18.4(i)).

## Laws of Reflection

The laws of reflection at plane surfaces are summarized as follows:

1. The incident, reflected ray and the normal all lie in the same plane.
2. The angle of incidence = the angle of reflection.

The first law is illustrated in Fig. 18.4(ii). A ray AO is incident on a plane mirror at O. Many lines, such as OC, can be drawn round the

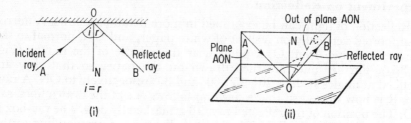

Fig. 18.4 Laws of reflection

normal ON as axis so that they are inclined to it at an angle equal to the angle of incidence AON. But the first law restricts the position of the reflected ray; it must lie in the plane of AO and ON. Therefore the reflected ray can only be OB, which is in this plane.

## Reflection of Electromagnetic Waves

Having observed the reflection of water waves and light waves, we can complete our study by investigating the reflection of electromagnetic

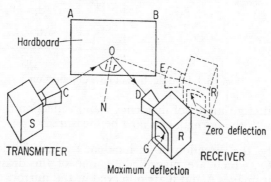

Fig. 18.5 Reflection of microwaves (radio waves)

waves. 3-cm electromagnetic waves or microwaves can be generated by an electronic apparatus S and detected by another R which has a sensitive galvanometer G, as stated on p. 333 (Fig. 18.5). If waves are received by R, the pointer on G is deflected.

### EXPERIMENT

The transmitter S is set up in front of a smooth wall or metal sheet or hardboard AB so that waves are incident in the direction CO (Fig. 18.5). The receiver R is then kept pointed at O on the other side of the normal ON, and moved round from OB. At E no deflection is observed in the galvanometer G. As R is moved round farther the deflection increases, and at some position corresponding to D the deflection is a maximum. The incident waves are reflected along the direction OD. Measurement of the angle of incidence NOC, and the angle of reflection NOD, show they are practically equal. The same result is obtained when S is moved to vary the angle of incidence.

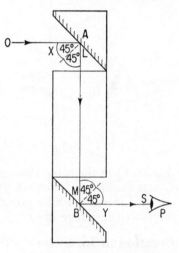

*Conclusion.* Electromagnetic waves obey the same law of reflection at plane surfaces as light waves.

### Simple Periscope

A simple periscope uses plane mirrors to reflect light in suitable directions (Fig. 18.6). Two parallel plane mirrors, A and B are fixed at 45° at the opposite ends of a long tube with two openings X and Y. A ray OL from an object O falls on mirror A at L, and is consequently reflected along LM at 45° to the normal at L, i.e. at right angles to the incident ray. It is then re-

Fig. 18.6 Simple periscope

flected by mirror B, on which it is incident at 45°. The reflected ray BS is therefore at right angles to LM. It passes through the hole Y and is seen by the eye at P. By this device it is possible to view objects obscured by the heads of a crowd. Note that the periscope will not function efficiently if the mirrors are not in their correct positions, that is, they should be parallel and at 45° to the horizontal.

### Formation of Images · Virtual Images

When a small object O is viewed by reflection in a plane mirror, the apparent position of the image I is determined by the direction in which the reflected light enters the eye. An observer at E, for example, looking into the mirror, sees the rays AX, BY reflected as if they come from I, the point of intersection of these rays produced backwards (Fig. 18.7(i)). He or she is not conscious of the fact that the rays started initially from O. Any other ray from O which enters the eye after reflection also appears to come from I. The *image* of O is therefore at I.

It should be noted that: (1) a normal ray ON is reflected back along NO, so that I lies on ON produced; (2) no rays actually pass through I because a *divergent* beam is obtained by reflection. The image I is therefore described as a *virtual image*; it would not be formed on a screen placed in front of the mirror. In contrast, a *real image* is one which could be focused on a screen and is formed by a convergent beam (see p. 409).

Fig. 18.7(ii) shows how the wave theory explains the formation of the point image. The object O sends out spherical waves. When the wavefront CPD reaches the mirror the point P sends back a wavelet towards O. It

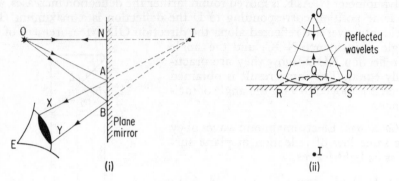

Fig. 18.7 Reflected rays and waves

reaches Q at the same time as R and S are reached by the wavelets from C and D. The reflected wavefront is thus RQS, which appears to spread out as if it came from a point I on the other side of the mirror to O. I is therefore the image of O.

### Location of Images · No Parallax

The image of an object O, such as a pin, in a plane mirror M can be located by placing two other pins, A and B, in line with the image I observed in the mirror (see Fig. 18.8(i)). This is repeated with two more pins in a different position, such as at C and D or at E and F. After re-

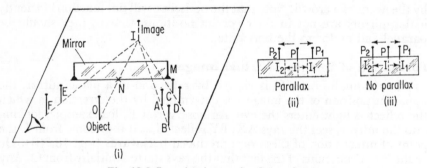

Fig. 18.8 Locating images in plane mirrors

moving the pins, the lines AB and CD (or AB and EF) are produced. Their intersection, which is behind the mirror, gives the position of the image I.

The image of the pin O can also be located directly by a method of 'no parallax'. In this case a tall pin P is placed *behind* the mirror, so that initially the top of P and the image of O are observed in line with each other as shown in Fig. 18.8(ii). If P and I are *not* at the same place, a movement of the observer's head to the right makes the image I move to $I_1$, and P to $P_1$. $I_1$ and $P_1$, which are now displaced from each other, are said to have 'parallax' between them. Likewise, a movement of the observer's head to the left makes the image $I_2$ move relative to $P_2$.

If P and I are exactly at the same place, a movement of the head either way shows *no parallax* between the displaced pin and image, as at $P_2$, $I_2$ in Fig. 18.8(iii). The position of the image I can thus be located by moving the pin P behind the mirror until no parallax is obtained between P and I in the way described.

*Results.* Measurements from experiments show that the perpendicular distance IN of the image I from the plane mirror is always equal to the perpendicular distance ON of the object O from the mirror. Thus the image is as far behind the mirror as the object is in front.

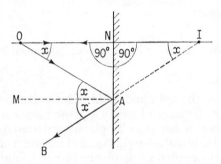

Fig. 18.9 Position of image in plane mirror

### Geometrical Proof

This experimental result, IN = ON, can be proved geometrically as follows:

From Fig. 18.9, angle of incidence OAM($x$) = angle of reflection BAM.

But                angle OAM = angle AON (alternate angles),
                     angle BAM = angle AIO (corresponding angles),
    ∴ angle AON = angle AIN.

It can be seen that triangles OAN, IAN, are congruent since AN is common.
$$\therefore\ ON = IN.$$

### Large Object Experiment

The image of a large object in a plane mirror can be investigated by a class with the aid of a plane glass surface, obtained, for example, by removing the front glass cover of a chemical balance. The glass G is supported vertically, and the fingers of the left hand L are spread horizontally on the bench on one side near G (Fig. 18.10(i)). By moving the *right* hand behind G, and looking through G at the same time, the image I can be just covered by the right hand.

Measurement of the perpendicular distance ON from the top O of the left thumb to G, and of the perpendicular distance IN from the top I of the right thumb to G, shows that IN = ON, or the image is as far behind G as the object is in front.

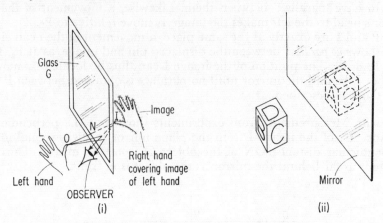

Fig. 18.10  (i) Locating image of hand. (ii) Lateral inversion

### Lateral Inversion

Since the image of the left hand looks exactly like the right hand to an observer, the image of a large object in a plane mirror is said to be *laterally inverted*. The inversion is also seen by placing a lettered cube in front of a plane mirror (Fig. 18.10(ii)). A right-handed batsman or tennis player will therefore appear left-handed if his or her stance is observed in a mirror. Words on a blotting-paper can be seen as they were originally written by holding the blotting-paper to a mirror. The famous artist and inventor Leonardo da Vinci used to write from 'right to left' to avoid ink-smudges, for he was left-handed. His writing can be read by holding it up to a mirror.

*Summarizing*, the image in a plane mirror is: (1) laterally inverted; (2) the same size as the object; (3) virtual; and (4) as far behind the mirror as the object is in front.

### Inclined Mirrors · The Kaleidoscope

When an object is placed between two inclined mirrors, an observer can see several images whose number depends on the angle between the mirrors. Fig. 18.11 shows an object O placed between two mirrors $M_1$ and $M_2$ at 90° to each other. An image $I_1$ is due to reflection at $M_1$, and an image $I_2$ is due to reflection at $M_2$. The image $I_{12}$ is due to reflection at $M_1$ and then at $M_2$, and Fig. 18.11 illustrates how rays starting from O reach the observer E by repeated reflection.

The *kaleidoscope* is a toy in which multiple images are formed by two mirrors, usually inclined at 60° to each other. The mirrors are fixed at one end of the tube, and coloured pieces of tinsel are placed here. The five images of the tinsel, viewed through the other end of the tube, are seen symmetrically in a circle. The tinsel pieces themselves are also seen directly. When the tube is shaken the positions of the tinsel change and other image patterns are seen. Designers are sometimes assisted in their search for new colour patterns by the kaleidoscope.

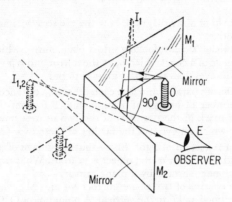

Fig. 18.11 Reflection at two mirrors

## SUMMARY

1. Water waves, light waves and electromagnetic waves are all reflected by plane surfaces in a similar way.

2. *Laws of reflection*: (1) The incident ray, the normal and the reflected ray all lie in the same plane. (2) The angle of incidence = the angle of reflection.

3. Images in a plane mirror can be located by means of pins in line with them or by the method of no parallax.

4. A simple periscope contains two parallel plane mirrors inclined at 45° to the horizontal.

5. The image in a plane mirror is laterally inverted, the same size as the object, virtual and as far behind the mirror as the object is in front.

## EXERCISE 18 · ANSWERS, p. 452

1. The diagram shows a letter F in a horizontal plane in front of a vertical plane mirror AB. Draw the image of the letter seen by reflection and the path of ONE ray of light from the bottom of the letter to the eye E. (*N.*)

2. Describe how you would find experimentally the position of the image of a pin formed by a plane mirror. State the nature and position of the image. (*L.*)

3. State the laws of regular reflection of light. Describe how you would verify them experimentally. (*O.*)

4. Explain: (*a*) why the image of an object formed by a plane mirror is called a

virtual image; (b) why the image of each point on the object is as far behind the mirror as the object point is in front; (c) why the image appears laterally inverted.

Describe, with the help of a ray diagram, the action of a simple form of periscope. (O.)

**5.** State the laws of reflection of light, and describe how you would verify them experimentally.

Explain, with the help of a ray diagram, how the eye sees the image of a bright point formed by a plane mirror.

A man sits in an optician's chair, looking into a plane mirror which is 2 m away from him, and views the image of a chart which faces the mirror and is 50 cm behind his head. How far away from his eyes does the chart appear to be?

**6.** Two plane mirrors are parallel and face each other at a distance of 10 ft. On looking into one mirror a number of images of a lamp bulb 3 ft from this mirror are seen. Draw a diagram and mark in the positions of the two nearest images. Draw the path of a pencil of rays by which an eye sees the farther of these two. (N.)

**7.** State the laws of reflection of light. Show that the image of an object in a plane mirror is as far behind the mirror as the object is in front. What experiment would you perform to locate the image formed by a plane mirror?

A simple periscope consists of two plane mirrors $M_1$ and $M_2$, which are 90 cm apart vertically and each inclined at 45° to the vertical. A point object O, 60 cm away from the upper mirror $M_1$, is viewed by an eye E, 30 cm away from $M_2$, both these distances being taken horizontally. Draw a ray diagram tracing the paths of two rays from O to E, and find the distance from E of the final image seen. (O.)

# 19

# CURVED SPHERICAL MIRRORS

Motor vehicles often have curved mirrors on their wings so that drivers are able to see traffic behind. These mirrors curve outwards towards the observer and are called *convex mirrors* (Fig. 19.1(ii)). Shaving mirrors, however, curve inwards away from the observer (Fig. 19.1(i)). They are called *concave mirrors*. Searchlights, such as those used to illuminate the front of buildings for display purposes, have a concave mirror of a special shape to reflect light strongly a long way. The reflecting concave mirrors used in

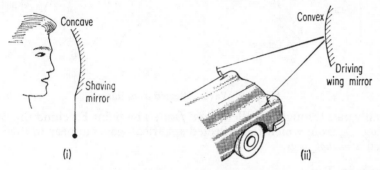

Fig. 19.1 Shaving and driving mirrors

car headlamps have a similar shape. At present the largest concave mirror in the world is at Mount Palomar Observatory in California. It is 200 in in diameter and forms the main part of an enormous telescope used for collecting light energy from distant stars and for investigations in astronomy. Curved mirrors are thus of considerable practical importance.

## Waves and Curved Mirrors

The effect of a curved mirror on *waves* can be studied with the aid of the ripple tank. Plane waves incident from X towards a concave metal strip AB are seen reflected from it as spherical (circular) waves such as P, Q and R, which appear to converge to a point F in front of AB (Fig. 19.2(i)). The concave surface thus 'adds' a curvature similar to its own to the plane wave after reflection. The converging action of the surface can be explained by the wave theory. When a plane wavefront ALB reaches the surface the points A and B reflect back wavelets. They reach P and N at the same time as the wavelet from L reaches M. The reflected wavefront is therefore PMN, which converges to F.

When the metal strip is turned round so that it presents a convex surface Q to the incident waves (Fig. 19.2(ii)), the plane waves are again seen to have a spherical curvature after reflection. This time, however, the reflected waves, such as S, T and V, spread outward as shown. Hence the convex surface also adds a curvature similar to its own to the plane wave incident on it. The reflected waves S, T and V seem to spread outwards from a point F *behind* Q. The wave theory explains the diverging action of the surface. When a plane wavefront AQB reaches the surface, Q sends back a wavelet which reaches P at the same time as the wavelets from A and B reach the mirror. The reflected wavefront is thus PS.

Figs. 19.2(i) and (ii) should be compared with each other. The point F in front of the concave surface AB, where the reflected spherical waves

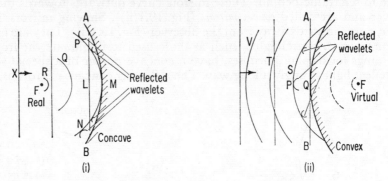

Fig. 19.2 Waves reflected from mirrors

actually pass through, is called a *real focus*. The point F behind the convex surface Q, from which the reflected spherical waves appear to diverge, is called a *virtual focus*.

## Common Terms

Concave and convex mirrors are made by depositing vaporized aluminium on a glass surface which is part of a sphere, like a watch-glass. The width AB of the mirror is called the *aperture*; the centre P of the mirror is called the *pole*; the centre C of the sphere of which the mirror is part is called the *centre of* curvature; the line PC is called the *principal* (chief) *axis* of the mirror (Fig. 19.3). The *radius of curvature r* is the distance PC; the *focal length f* is the distance PF. As we see later, the action of a mirror depends considerably on the magnitude of its radius of curvature or focal length.

## Rays and Curved Mirrors

The effect of concave and convex mirrors on *rays* can be studied with the aid of the ray-box (p. 352). The effect produced on a sheet of white paper by a concave strip of mirror is shown in Fig. 19.3(i). Rays parallel and close to the principal axis CP converge to a point F on the principal axis; this point is therefore called the *principal focus* of the mirror.

By contrast, rays parallel and close to the principal axis diverge from the reflecting surface of a convex mirror (Fig. 19.3(ii)). They appear to diverge from a point F on the principal axis behind the mirror. The principal focus of a concave mirror is a *real* focus, that is, reflected rays actually pass through it. The principal focus of a convex mirror is a *virtual* focus—reflected rays do not actually pass through it.

If the ray-box is made to produce a beam converging to the principal focus F, then, for both the concave and convex surfaces, a *parallel* beam is reflected back (Fig. 19.4). Figs. 19.3 and 19.4 should be compared with

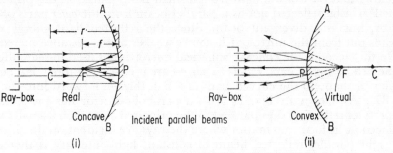

Fig. 19.3 Concave and convex mirrors

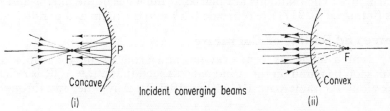

Fig. 19.4 Reflection of parallel beams

each other. They show the paths of the light rays are the same, so that light rays are reversible. This is an illustration of a general law called the *Principle of Reversibility of Light*.

## Parabolic Mirrors

So far we have used a *narrow* parallel beam of light close to the principal axis in studying reflection from mirrors. For a concave mirror, all the

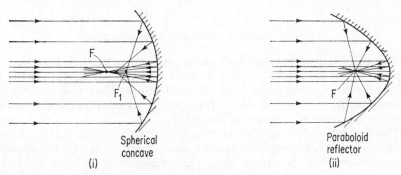

Fig. 19.5 Spherical and paraboloid reflectors

reflected rays then pass through one definite point, the principal focus F (Fig. 19.5(i)). If, however, a *wide* parallel beam covering the whole of the aperture is incident on the mirror, experiment shows that the reflected rays well away from the axis are brought to a focus at different points, such as $F_1$. The parallel beam thus produces a blurred focus, and this is called *spherical aberration*. From the principle of the reversibility of light, it follows that part of the light from a small lamp placed at the principal focus F will be reflected not as a parallel beam from the *outer* parts of the mirror, but as a divergent beam. Since the reflected light energy now spreads out from the mirror, it becomes weaker as the distance increases.

On this account the concave spherical mirror is not used in searchlights or the headlamps of cars. *Parabolic mirrors* are used – a 'parabola' is a curve similar in shape to the curved path of a ball thrown forward into the air (p. 87). As shown in Fig. 19.5(ii), a parabolic mirror has the suitable property of reflecting rays parallel to the principal axis when a small lamp is placed at its focus, no matter where the rays are incident on the mirror. A bright parallel reflected beam of constant light intensity is therefore obtained. Parabolic mirrors are also used behind the straight filaments in electric fires. The filaments are placed at the focus of the mirror and heat rays (infra-red, p. 317) are reflected back into the room as a parallel beam.

## Centre and Radius of Curvature

By using a single ray from a ray-box and a concave mirror strip, a point C can soon be found on the principal axis such that a ray ACR is reflected back along the same path RCA (Fig. 19.6(i)). Any other ray through C,

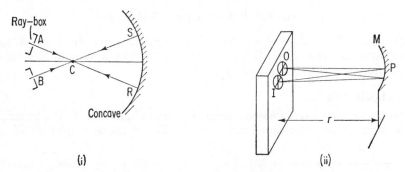

Fig. 19.6 Centre of curvature – measurement of $r$

such as BCS, is also reflected back along the same path SCB. Now this can happen only if the ray CR strikes the mirror *normally* at R and the ray CS strikes the mirror normally at S. It then follows that CR and CS are normals to the mirror at R and S respectively. Consequently, the point C must be the *centre of curvature* of the mirror.

The position of the centre of curvature of a concave spherical mirror can be found by a simple experiment. An illuminated object O, for example, two cross-wires in the centre of a white screen with a lamp behind, is placed in front of the mirror M and in line with its middle or pole P (Fig. 19.6(ii). When the mirror is moved towards or away from M a blurred image is seen near O. But at some point the image of the cross-wires becomes clear, and the sharpest image I is obtained near O, as shown. The object O and image I are now practically coincident in position.

Rays from O are thus now reflected back practically to itself. From our previous discussion, it follows that O must be the centre of curvature of the mirror. The distance OP from O to the mirror is now measured, and this is the radius of curvature $r$. The radius of curvature of any concave surface can be determined by finding where the object and image coincide.

## Focal Length Approximate Value

The *focal length* of a concave mirror can be found quickly, but not accurately, by holding up the mirror M at the back of a room so that the

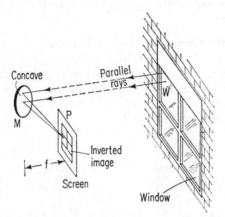

Fig. 19.7 Quick method for $f$

reflecting surface faces the window W (Fig. 19.7). A sheet of paper P is then moved in front of the mirror until the clearest image of the window-frames is seen or, better, the image of any object such as a tree outside the window. It will be noted that the images are upside down. The distance from M to P is then measured. Now the mirror M receives parallel rays from objects a long way from it (see p. 326). Since parallel rays are reflected from M towards its focus, the distance from M to P is practically equal to the focal length $f$ of the mirror.

## Relation between $f$ and $r$

From the two experiments already described, the value of the radius of curvature $r$ can be compared with $f$. It is then found that, practically, *the focal length is half the radius of curvature*, or

$$f = \frac{r}{2}.$$

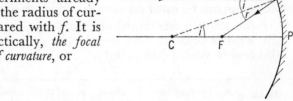

Fig. 19.8 Showing $r = 2f$

This simple relationship always holds for curved spherical mirrors and is explained theoretically by geometry. Suppose a ray AB parallel to the principal axis CP is incident on the mirror at B, and C is the centre of curvature (Fig. 19.8). The ray is then reflected at B to F, the principal focus (p. 368). Now the angle of incidence at B = the angle made by AB with the normal BC at B = the angle of reflection CBF. But, by alternate angles,

$$\text{angle ABC} = \text{angle BCF}.$$
$$\therefore \text{ angle BCF} = \text{angle CBF}$$
$$\therefore \text{ CF} = \text{BF}.$$

Now when B is close to P, that is, when rays are incident on a small area of the mirror round P, BF = PF. Hence CF = PF, or F is midway between P and C. Thus FP $= f = $ CP/2 $= r/2$.

This result, which should be memorized, means that if the radius of curvature is 20 cm the focal length is 10 cm; or if the focal length is 15 cm the radius is 30 cm.

### Investigation into Concave Mirror Images

The nature of the image obtained with a concave mirror depends on how far or how close the object is to the mirror. An investigation can be carried out in the laboratory by using illuminated cross-wires as the object O, a concave mirror M on a stand, a white screen S and a ruler for measuring distances (see Fig. 19.9).

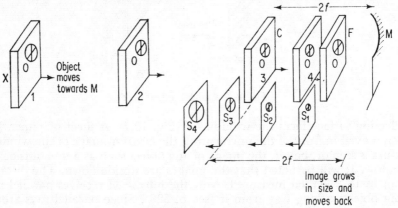

Fig. 19.9 Image changes

After placing the mirror M at the same height as O, and aligning them parallel with S, the distance of O from M can be slowly decreased, starting a long way off at X. The observations at distances similar to those in the table below can be listed on each occasion, stating whether the image is:

    (i) inverted (upside down) or erect (same way up);
    (ii) smaller or larger than the object;
    (iii) real or virtual (see p. 361).

| | 1. | 2. | 3. | 4. | 5. | 6. |
|---|---|---|---|---|---|---|
| Position of O | Long way from M | Nearer, but farther than C | At C, distance 2f | Between C and F | At F | Nearer than F |
| Image I Observation | (i) ... (ii) ... (iii) ... | | | | | |

RESULTS

It will be noted that as the object O moves forward towards the mirror M the image moves back in the opposite direction and grows in size. The results of the experiment can be summarized as follows:

## Real Images

(1) When the object O is a long way from M the image is upside down, smaller and real, $S_1$ in Fig. 19.9. (2) When O is moved nearer the image is still upside down and real and grows bigger, $S_2$ in Fig. 19.9. (3) When O is at C, the centre of curvature, the image is upside down, real and the same size as O, $S_3$ in Fig. 19.9. (4) Between C and F, the image is upside down and larger than O, and it is now behind O, $S_4$ in Fig. 19.9. The image grows larger as the object approaches F, and the screen must move well back from M to receive the clear image.

## Virtual Images

When the object is at F an image cannot be received on the screen. Between F and the mirror, an image again cannot be received on the screen, so the image is *virtual*. If the mirror is now held close to one's face, however, a large image is seen. It is virtual and erect.

Summarizing, when the object is farther from the mirror M than *f* the image is upside down, real and the same size. When the object is nearer than *f* the image is larger, erect and virtual. At the centre of curvature C, at a distance 2*f* from the mirror, the inverted real image has the same size as the object, and C is thus a useful reference point.

## Ray Diagrams · Drawing of Images

The images obtained in mirrors can always be drawn to scale by means of a *ray diagram*. As a small central part of the mirror is used, the reflecting surface is represented by a straight line. The object is drawn as a straight line perpendicular to the principal axis, with an arrow to represent its head and its foot O placed on the axis (Fig. 19.10). The image of A, the top of the object, is found by drawing two rays from it. They may be:

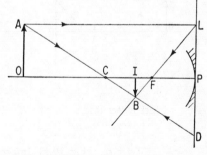

Fig. 19.10 Ray diagram of image

(1) a ray AL parallel to the principal axis, which passes through the principal focus F after reflection;

(2) a ray AC through the centre of curvature C, which is reflected back along the same path ACD after striking the mirror at D (see p. 370); or

(3) a ray AF through F, which is reflected parallel to the principal axis.

The rays (1) and (2) intersect at B below the principal axis after reflection. Thus B is the image of A, or the top point of the image. The image of the foot O of the object lies on the principal axis, since the ray OCP strikes the mirror normally at P and is reflected straight back. Now the mirror does not distort the object. Once the point B is obtained, therefore, the image is completed by drawing a perpendicular BI from B to the principal axis.

### Position of Image · Magnification

The ray diagram in Fig. 19.10 shows that the image IB is inverted and real and smaller than the object OA. If the diagram is drawn accurately to scale, with $FP = f$, $CP = 2f = r$ and the object placed at the given distance OP from the mirror, then the image distance IP can be measured accurately.

The *linear* or *transverse magnification* $m$ produced by the mirror can also be found from the drawing. This is defined as:

$$m = \frac{Height\ of\ image}{Height\ of\ object} = \frac{IB}{OA}$$

Thus by representing the height of the object OA by some convenient length, and measuring the height IB of the image obtained from a ray-diagram, the magnification can be calculated.

### Ray Diagrams of Images

The following ray diagrams of images (Fig. 19.11) should be drawn by the reader and the results compared with those obtained by experiment, described on p. 373.

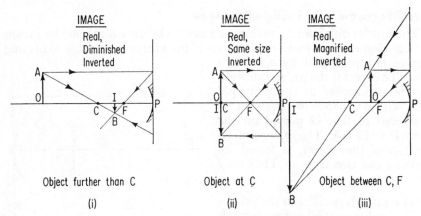

Fig. 19.11 Images in concave mirror

### Virtual Erect Image

Between F and the mirror, a large erect image is seen. In this case a ray AL parallel to the principal axis is reflected along LF to pass through F (Fig. 19.12). A ray AD, drawn in the direction CAD from the centre of curvature C, is reflected back along the same path DA. Since DA and LF intersect at B *behind* the mirror, the image is virtual. It is also magnified and erect. On this account, the concave mirror can be used as a shaving mirror or as a dentist's mirror; in each case, of course, the mirror is placed close to the object so that the latter is inside the focal length(see p. 367).

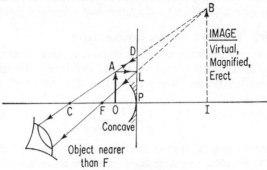

Fig. 19.12  Magnified image

## Convex Mirror

If a convex mirror is held up and objects reflected in it are observed, the images all appear to be the same way up, or *erect*, and *smaller* than the object. This is the case for all object distances from the mirror. The difference between the images produced by a concave and a convex surface can be seen very quickly with the aid of a brightly polished table-spoon. When one's face is observed in the convex surface it appears erect and small; when the spoon is turned over so that a concave surface is obtained, and held away from the face, an inverted image is obtained. Experiments with an illuminated object such as cross-wires show that the light reflected from a convex mirror can never be focused on to a screen. The image seen in this mirror is therefore *virtual*.

Fig. 19.13(i) shows a ray diagram of a typical image produced by a

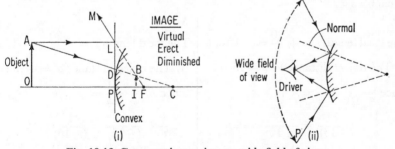

Fig. 19.13  Convex mirror – image, wide field of view

convex mirror. A ray AL parallel to the principal axis is reflected along LM as if it diverged from F, the principal focus, which is behind the mirror (see p. 369). A ray AD, which would pass through the centre of curvature C if produced, is reflected back along the same path DA. Since ML and AD intersect at B behind the mirror, the image IB is virtual. If the object OA is moved nearer the mirror and the image is redrawn it will still be found virtual, erect and smaller than the object.

As the image is always erect, the convex mirror is used as a driving mirror. Objects such as P at a wide angle round the mirror can be observed (Fig. 19.13(ii)). Thus it also has the advantage of a wide field of view and overtaking traffic can be seen. A plane mirror is usually used to observe objects directly behind a car, and has a limited field of view.

## Formulae with Mirrors · Sign Convention

If an object is at a distance $u$ from a curved mirror, concave or convex, of focal length $f$ and radius $r$, and the image is formed at a distance $v$ from the mirror, experiment and theory show that the distances are related by a general formula.

$$\frac{1}{v} + \frac{1}{u} = \frac{1}{f} = \frac{2}{r}.$$

*When this formula is used, it is necessary to add positive (+) or negative (−) signs to each of the distances, according to a sign rule or convention:*

REAL IS POSITIVE (RP). On this convention, real objects and real images are regarded as being at a positive (+) distance from the mirror; virtual images are regarded as being at a negative (−) distance from the mirror.

A concave mirror has a real focus. The focal length is therefore positive. A convex mirror has a virtual focus; its focal length is therefore negative.

NEW CARTESIAN (NC). On this convention distances measured to the right of the mirror are positive; distances to the left are negative. The object is always placed on the left of the mirror.

A concave mirror has a real focus, which is left of the mirror; its focal length is hence negative. A convex mirror has a virtual focus, which is right of the mirror, its focal length is therefore positive.

The following examples illustrate how the sign rule is applied.

1. An object is placed (i) 15 cm, (ii) 5 cm in front of a concave mirror of radius of curvature 20 cm. Calculate the position and nature of the image in each case.

*RP.*

The focal length $f = +\dfrac{20}{2} = +10$ cm.

(i) The object distance $u = +15$ cm.

Since $\dfrac{1}{v} + \dfrac{1}{u} = \dfrac{1}{f}$

$\therefore \dfrac{1}{v} + \dfrac{1}{(+15)} = \dfrac{1}{(+10)}$

$\therefore \dfrac{1}{v} + \dfrac{1}{15} = \dfrac{1}{10}$

$\therefore \dfrac{1}{v} = \dfrac{1}{10} - \dfrac{1}{15} = \dfrac{1}{30}$

$\therefore v = 30$ cm.

The image is hence 30 cm from the mirror and since $v$ is +ve the image is real (see Fig. 19.11(iii)).

*NC.*

The focal length $f = -\dfrac{20}{2} = -10$ cm.

(i) The object distance $u = -15$ cm.

Since $\dfrac{1}{v} + \dfrac{1}{u} = \dfrac{1}{f}$

$\therefore \dfrac{1}{v} + \dfrac{1}{(-15)} = \dfrac{1}{(-10)}$

$\therefore \dfrac{1}{v} - \dfrac{1}{15} = -\dfrac{1}{10}$

$\therefore \dfrac{1}{v} = -\dfrac{1}{10} + \dfrac{1}{15} = -\dfrac{1}{30}$

$\therefore v = -30$ cm.

The image is hence 30 cm on the left of the mirror, on the same side as the object, and is hence real (see Fig. 19.11(iii)).

(ii) The object distance $u = +5$ cm.

$$\therefore \quad \frac{1}{v} + \frac{1}{(+5)} = \frac{1}{+10}$$

$$\therefore \qquad \frac{1}{v} + \frac{1}{5} = \frac{1}{10}$$

$$\therefore \qquad \frac{1}{v} = \frac{1}{10} - \frac{1}{5} = -\frac{1}{10}$$

$$\therefore \qquad v = -10 \text{ cm.}$$

The image is thus 10 cm from the mirror, and since the sign of $v$ is negative the image is a virtual one behind the mirror (see Fig. 19.12).

(ii) The object distance $u = -5$ cm.

$$\therefore \quad \frac{1}{v} + \frac{1}{(-5)} = \frac{1}{(-10)}$$

$$\therefore \qquad \frac{1}{v} - \frac{1}{5} = -\frac{1}{10}$$

$$\therefore \qquad \frac{1}{v} = -\frac{1}{10} + \frac{1}{5} = \frac{1}{10}$$

$$\therefore \qquad v = 10 \text{ cm.}$$

Since $v$ is +ve, the image is 10 cm on the right of the mirror, or virtual (see Fig. 19.12).

2. The image in a convex mirror of radius of curvature 16 cm is 3 cm from the mirror. Calculate the position of the object.

*RP.*

Since the mirror is convex,

$f = -\dfrac{16}{2} = -8$ cm. Now the image in a convex mirror is virtual (p. 375). Hence image distance $v = -3$ cm.

Since $\qquad \dfrac{1}{v} + \dfrac{1}{u} = \dfrac{1}{f}$

$$\therefore \quad \frac{1}{(-3)} + \frac{1}{u} = \frac{1}{(-8)}$$

$$\therefore \quad -\frac{1}{3} + \frac{1}{u} = -\frac{1}{8}$$

$$\therefore \qquad \frac{1}{u} = -\frac{1}{8} + \frac{1}{3} = \frac{5}{24}$$

$$\therefore \qquad u = \frac{24}{5} = 4 \cdot 8.$$

The object is thus 4·8 cm from the mirror. It is a real object since $u$ is positive.

*NC.*

Since the mirror is convex,

$$f = +\frac{16}{2} = +8 \text{ cm.}$$

The image in a convex mirror is virtual (p. 375).

$$\therefore \text{ image distance } v = +3 \text{ cm.}$$

Since $\qquad \dfrac{1}{v} + \dfrac{1}{u} = \dfrac{1}{f}$

$$\therefore \quad \frac{1}{(+3)} + \frac{1}{u} = \frac{1}{(+8)}$$

$$\therefore \qquad \frac{1}{3} + \frac{1}{u} = \frac{1}{8}$$

$$\therefore \qquad \frac{1}{u} = \frac{1}{8} - \frac{1}{3} = -\frac{5}{24}$$

$$\therefore \qquad u = \frac{24}{5} = -4 \cdot 8.$$

The object is thus 4·8 cm on the left of the mirror.

## Linear (Transverse) Magnification

Experiment

As stated on p. 374, the linear or transverse magnification $m$ produced by a mirror is defined as:

$$m = \frac{\text{Height of image}}{\text{Height of object}}$$

An experiment to measure the magnification at different distances of the object from a concave mirror M can be carried out with an illuminated object O, such as a cross-wire or the graduations on a transparent Perspex rule (Fig. 19.14). The clearest image IB is then obtained by moving a screen S. Large images are easier to measure, in which case S is farther from M than the object OA. The length BI of the image is then measured

with dividers or a ruler, or graph paper can be used on the screen.

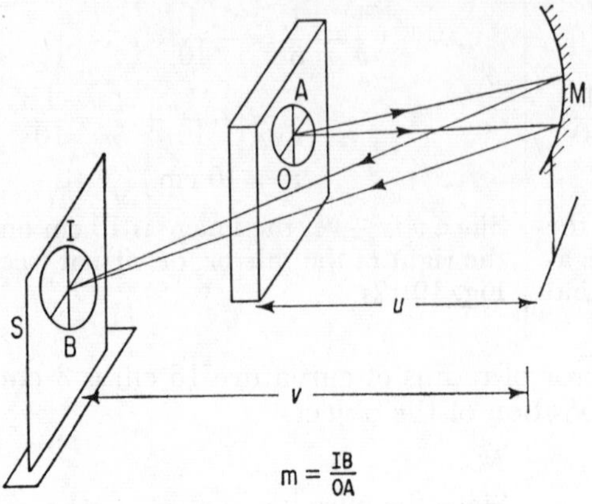

m = $\frac{IB}{OA}$

Fig. 19.14 Magnification experiment

*Result.* The distances $u$ and $v$, and the lengths OA and IB, can be measured for varying object distances. By comparing the magnification or IB/OA, with the corresponding ratio $v/u$, it will be found that

$$m = \frac{v}{u}.$$

This formula for $m$ can be shown to be theoretically true. It is left as an exercise in the geometry of similar triangles to the reader, who should apply this to ray diagrams such as Fig. 19.11.

### Example on Magnification

A concave mirror of focal length 20 cm produces an erect image of magnification 3. Calculate the distance of the object from the mirror.

RP.

Here $m = 3$

$$\therefore \frac{v}{u} = 3, \text{ or } v = 3u.$$

Now object distance (real) $= +x$ cm say

$\therefore$ image distance (virtual) $= -3x$ cm since erect image is virtual (see p. 375).

From

$$\frac{1}{v} + \frac{1}{u} = \frac{1}{f}$$

$$\therefore \frac{1}{(-3x)} + \frac{1}{(+x)} = \frac{1}{(+20)}$$

$$\therefore \quad -\frac{1}{3x} + \frac{1}{x} = \frac{1}{20}$$

$$\therefore \quad \frac{2}{3x} = \frac{1}{20}$$

$$\therefore \quad 3x = 40$$

$$\therefore \quad x = 13\tfrac{1}{3}$$

$\therefore$ Object distance $= 13\tfrac{1}{3}$ cm.

NC.

Here $m = 3$

$$\therefore \frac{v}{u} = 3 \text{ or } v = 3u.$$

Now object distance $= -x$ cm say

$\therefore$ image distance $= +3x$ cm, since erect image is right of the mirror (see p. 375).

From

$$\frac{1}{v} + \frac{1}{u} = \frac{1}{f}$$

$$\therefore \frac{1}{(+3x)} + \frac{1}{(-x)} = \frac{1}{(-20)}$$

$$\therefore \quad \frac{1}{3x} - \frac{1}{x} = -\frac{1}{20}$$

$$\therefore \quad -\frac{2}{3x} = -\frac{1}{20}$$

$$\therefore \quad 3x = 40$$

$$\therefore \quad x = 13\tfrac{1}{3}$$

$\therefore$ Object distance $= 13\tfrac{1}{3}$ cm.

*Graphical Solution.* The object distance in the above example can also be found by a graphical method (Fig. 19.15).

Two lines PQ and RS are drawn parallel to the principal axis CF at heights above the axis in the ratio 1 : 3. QF is then joined and produced to meet RS at B. Then B is the top point of the image, which is therefore BI. Now join B to C, intersecting PQ at A. Then A is the top point of the object, which is therefore AO. The distance from O to the mirror can be determined from the scale used for the focal length in the drawing.

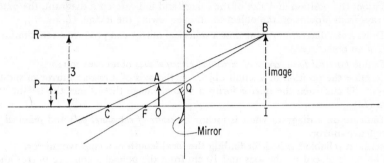

Fig. 19.15 Drawing of image

## SUMMARY

1. A concave mirror produces convergent waves and rays. A convex mirror produces divergent waves and rays.

2. A concave mirror has a real focus; a convex mirror has a virtual focus.

3. The images in a concave mirror are real and inverted until the object is nearer the mirror than its focal length, when the image is erect, magnified and virtual. At C, the centre of curvature, which is $2f$ from the mirror, the image is the same size as the object. The image in a convex mirror is always virtual and diminished.

4. *Mirror formulae.* (i) $\dfrac{1}{v} + \dfrac{1}{u} = \dfrac{1}{f}$. (ii) $m = \dfrac{v}{u}$.

When using the formulae, the sign conventions must be used:

*RP*. Real is +ve, virtual is −ve.

Concave mirror, $f$ is +ve.
Convex mirror, $f$ is −ve.

*NC*. Distances to the right of the mirror are +ve, to the left of the mirror they are −ve.

Concave mirror, $f$ is −ve.
Convex mirror, $f$ is +ve.

## EXERCISE 19 · ANSWERS, p. 452

**1.** In the sketch, the mirror on the left is concave and that on the right is a convex spherical mirror. In each diagram two rays of light are shown incident on the mirror. Draw the approximate path of each ray after reflection. (*N.*)

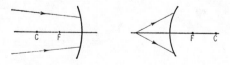

**2.** Define principal focus of a convex mirror. A concave mirror is used to form an image of an object pin. Where must the object be placed to obtain: (a) an upright, enlarged image; (b) an image the same size as the object? (N.)

**3.** Define *principal focus* of a concave mirror.

An object 2 cm high is situated on and perpendicular to the axis of a concave mirror of radius of curvature 30 cm and is 10 cm from the mirror. Find, graphically or by calculation, the position and size of the image and indicate on a diagram, the paths of three rays from a point on the object to an eye viewing the image. (L.)

**4.** Draw a ray diagram to show how a spherical mirror can produce a real diminished image of an object. (N.)

**5.** Define *principal focus, centre of curvature, principal axis* of a *convex* mirror.

Determine the position of a small object on the axis of a *concave* mirror of radius of curvature 30 cm when the mirror forms a virtual image three times the height of the object. (L.)

**6.** Indicate on a diagram what is meant by *centre of curvature, pole* and *principal focus* of a concave mirror.

Describe a reliable method for finding the focal length of a concave mirror.

An object is placed on the axis and 10 cm from the pole of a concave mirror and an erect image is formed three times the size of the object. Find the radius of curvature of the mirror and give a ray diagram showing how the image is formed. (L.)

**7.** Distinguish between *real* and *virtual* images.

A concave mirror of focal length 15 cm gives a virtual image of a small object on the principal axis of the mirror. Determine the position of the object if the image is formed 60 cm from the mirror. What would be the magnification? (N.)

**8.** Describe, with the help of ray diagrams, how a concave mirror can form: (a) a real image; (b) a virtual image. Briefly explain the construction of each diagram.

The diameter of the sun is 1·3 million km, and its distance from the earth 150 million km. What is the diameter of the image of the sun formed on a concave mirror of focal length 20 m situated on the earth's surface? (O.)

**9.** What is the principal focus of a concave mirror?

Describe an experimental method of finding conjugate points (conjugate foci) for a concave mirror. Show how to use your result to find the focal length of the mirror. (N.)

**10.** Draw labelled ray diagrams showing how a concave mirror can form: (a) a real enlarged image; (b) a virtual enlarged image. (L.)

# 20

# REFRACTION AT PLANE SURFACES

It is well known that the bottom of a swimming pool appears to be nearer the surface than it really is. Also, the letters in print seem to be brought nearer by placing a thick block of glass over them. These observations show that changes occur when light travels from water or glass into air. The phenomenon also occurs when light passes through the glass of lenses, and a study of it has assisted the making of efficient microscopes and telescopes. It is called *refraction* of light.

### Refraction in a Ripple Tank

The behaviour of waves moving from one medium to another can be observed with a ripple tank. A transparent plate G is fixed inside the tank by means of plasticene, applied to its corners *a*, *b*, *c* and *d*, so that it is level and just below (about 1 mm) the surface of the water (Fig. 20.1(i)). The

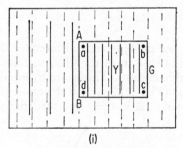
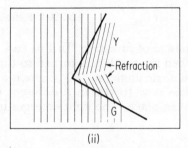

(i)                                    (ii)

Fig. 20.1 Velocity change and refraction of water waves

best effect is produced by waves of low frequency. The region of shallow water above the plate increases the frictional resistance to the oncoming wave, so that it behaves as a different medium. As the incident waves pass AB and enter the 'new medium' at Y over G it is observed that *they travel more slowly*. It can be seen that, effectively, the wavelength is decreased. When the plate G is turned at an angle to the incident waves a similar change in velocity occurs at Y (Fig. 20.1(ii)).

The change in velocity occurring when a wave enters a different medium leads to refraction. This behaviour is characteristic of all waves, including sound and radio waves, as well as light. For example, sound waves in air travel at greater speeds in regions where the air temperature increases, such as near the sand in hot deserts. Light and radio waves move more slowly when they pass from air into a different medium. The speed of light in air is about 300 million metres per second; in glass about 200 million metres per second; and in water about 225 million metres per second.

## Refraction and Ray-box

The refraction of rays of light can be investigated in more detail with a ray-box. A parallel beam of light from the box R is incident on a rectangular glass block ABCD (Fig. 20.2(i)). When the rays are normal, i.e. 90° to the side AB, they pass straight through the glass and emerge without

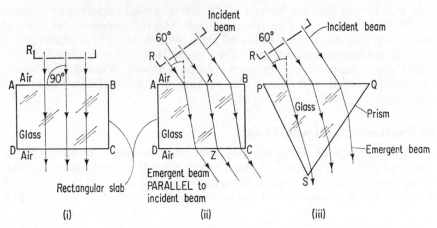

Fig. 20.2 Refraction of light – air to glass

change of direction. In this case we can see no apparent difference. But when the ray-box is turned so that the light rays meet the glass surface at an acute angle, say 60°, there is a noticeable change. The beam inside the glass now travels in a different direction (Fig. 20.2(ii)). Refraction has taken place. The light emerges into the air at Z in the same direction as

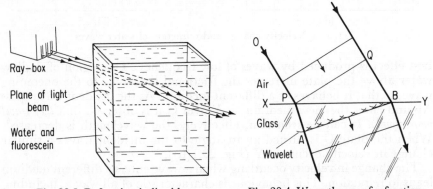

Fig. 20.3 Refraction in liquid

Fig. 20.4 Wave theory of refraction

the incident beam at X, but it is displaced sideways. This refraction, or change in direction, is even more marked when a triangular glass prism PQS is used (Fig. 20.2(iii)). The direction of the emerging beam is now quite different from that of the incident ray. Refraction through water can be studied by placing a little fluorescein in a rectangular tank of water.

The change of direction of the rays when a beam enters the back face of the tank is then clearly visible (Fig. 20.3).

To understand why a change of direction takes place on the wave theory suppose a plane wavefront, travelling in a direction OP, in air, is incident on a plane glass boundary XY (Fig. 20.4). When the wavefront reaches the position PQ the point P sends out a wavelet into the glass which reaches A at the same time as the wavelet from Q reaches B. BA thus represents the new wavefront in the glass, and this travels in the direction PA. As the speed of light in glass is less than in air, PA is less than QB. It can therefore be seen that the wavefront PQ has altered direction on entering the glass. Refraction is thus due to the change in velocity of light when it travels from air to glass.

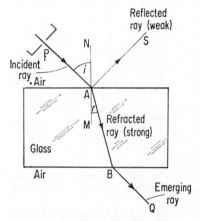

Fig. 20.5 Refraction through rectangular block

If the light travels in a time $t$ from Q to B in air with a velocity $c$, then $QB = ct$. In the same time $t$, the light travels in glass from P to A with a velocity $v$. Thus $PA = vt$. Hence the ratio $QB/PA = ct/vt = c/v =$ constant, since $c$ and $v$ are constants. Now the angle made by OP with the normal at $P = i$, the angle of incidence (see p. 386). By geometry, this is equal to angle QPB, so that $sin\ i = QB/PB$. The angle made by PA with the normal at $P = r$, the angle of refraction, = angle PBA. Hence $sin\ r = PA/PB$. Thus the ratio $sin\ i/sin\ r = QB/PA =$ constant from above. The wave theory hence explains Snell's law of refraction, p. 386.

## Refraction at Air–Glass and Glass–Air Boundaries

By means of a single ray PA from a ray-box incident on a rect-angular block of glass at A, the following refraction effects can be observed, using the normal line NAM to the boundary as a reference line (Fig. 20.5):

(i) The incident ray PA in air is refracted along AB *towards* the normal NAM after it enters the glass.

(ii) The ray AB, incident on the glass at B, is refracted *away* from the normal along BQ on emerging into the air.

Note also that some of the incident light at A is weakly reflected along AS at the air–glass boundary, in accordance with the law of reflection at plane surfaces. Similar results are obtained for air–water and water–air media, as in Fig. 20.3. Thus light travelling from one medium such as air to an optically denser medium such as glass or water is refracted *towards* the normal. From glass or water to air the light is refracted *away* from the normal. In Fig. 20.5, note that BQ emerges *parallel* to PA.

## Apparent Depth

We can now explain why the swimming pool appears shallower than it really is. Consider an object O at the bottom of the pool (Fig. 20.6). An observer at E sees O by rays from the object which enter the eye. Ray ON, normal to the upper surface, passes into the air along NR, its original direction, unaltered. Ray OA, slightly inclined to ON, is refracted into the air along AP, whose direction depends on the refractive index of water. So the rays NR and AP enter the eye (the inclina-

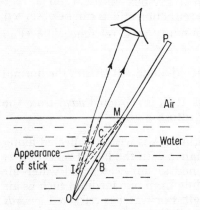

Fig. 20.6 Apparent depth – near-normal incidence (*exaggerated*)

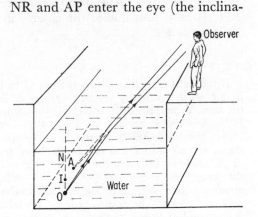

Fig. 20.7 Image in water

tion of OA and AP to the normal, $y$ and $x$, is exaggerated for clarity), and the observer sees the object in the direction from which the rays appear to come. O is therefore seen at I, the point of intersection of PA and RN. But I is nearer to the surface than O. Thus O appears nearer to the water surface.

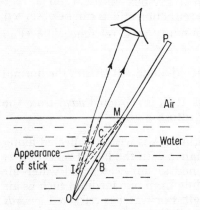

Fig. 20.8 Refraction from water to air

The above discussion dealt with an observer viewing an object directly below him. If instead he or she views the object from the side, for example when standing on the edge of a swimming pool, the apparent position of the object O at the bottom of the pool is now at A (Fig. 20.7). A is higher than I, the apparent position of O when viewed directly overhead and is on the right of the normal ON to the water surface.

### Part-immersion in Water

A stick partly immersed in water seems to be bent at the water surface. This is due to refraction. If PO is the stick and MO that part of it below water, the end O appears to be at I, a point nearer the surface, as explained before (Fig. 20.8). Similarly, another point B on the stick appears to be at C, and so on for all points between O and M. The image of OM is therefore IM, which is not in the same straight line as PM. The stick therefore appears bent, as shown.

## Bringing Objects into View

If an object O is placed in a vessel C so that it is just hidden from an observer at E it can be brought into view by pouring sufficient water into the vessel (Fig. 20.9). In Fig. 20.9(i), rays from O just pass below the edge

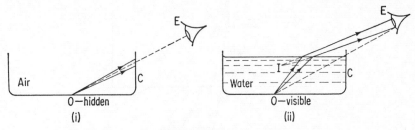

Fig. 20.9 Effect of refraction

of C, so that it cannot be seen by E. When water is poured in to a sufficient depth, refraction occurs as shown (Fig. 20.9(ii)). The rays from O, refracted away from the normal, now enter the observer's eye and the object appears to be at I.

## Laws of Refraction

More precise investigations on the refraction of light from air to glass can be performed with a single ray from a ray-box and a rectangular block of glass (Fig. 20.10).

### Experiment

A line XY is drawn on a sheet of paper. Using a protractor, five lines, such as AO, are drawn radiating from O at convenient angles, say 25°, 35°, 45°, 55°, 65°. The normal ON at O is drawn. The glass block G is then placed on the paper so that its larger boundary or edge coincides with XY.

| $i°$ | $r°$ | $i/r$ | $\sin i/\sin r$ |
|------|------|-------|-----------------|
|      |      |       |                 |

A single ray from R is now shone along the first line AO, incident, say, 25° to the normal ON. Two points C, D are marked on the ray emerging from the block, and they are then joined to intersect the lower glass boundary LM at P. Then OP is the refracted ray in the glass corresponding to the incident ray AO in the air. The angle of refraction $r$ is the angle POQ made by OP with the normal NOQ and is measured.

The ray-box is then moved so that the ray is incident at 35° to ON and the experiment repeated. In this way a table of measurements can be made of the angles of incidence $i$ and the corresponding angles of refraction $r$ as shown.

*Calculations.* The angle $r$ increases as $i$ increases. The ratio $i/r$ is cal-

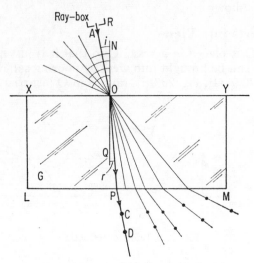

Fig. 20.10 Investigation of refraction

culated for each pair of angles $i$ and $r$, and the results entered. The trigonometrical ratio $\sin i/\sin r$ is also calculated for each pair of angles.

*Result.* The ratio $\sin i/\sin r$ is constant to a good approximation. The ratio $i/r$ is *not* a constant.

### Snell's Law · Laws of Refraction

The refraction of light was known more than two thousand years ago, and many scientists had tried to discover the laws governing refraction. It is recorded, for example, that in A.D. 100 PTOLEMY made hundreds of measurements of angles of incidence and refraction, without being able to discover the relation between them. But in 1621 a Dutch professor called SNELL found that the trigonometrical ratio *sin i/sin r is always constant* for a given pair of media, such as air and glass. This is therefore known as *Snell's Law of refraction.* Thus in one experiment with glass, $i = 60°$, $r = 35°$ by measurement;

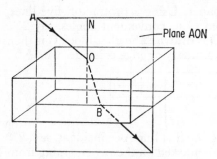

Fig. 20.11 Law of Refraction

then $\sin i/\sin r = \sin 60°/\sin 35° = 0.866/0.574 = 1.51$. In another experiment with the same glass, $i = 70°$, $r = 39°$; then $\sin i/\sin r = \sin 70°/\sin 39° = 0.940/0.629 = 1.52$. (See p. 383 for a proof of Snell's Law.)

A second law of refraction fixes the *plane* in which the refracted ray travels. This states:

*The incident ray, the normal, and the refracted ray all lie in the same plane.*

Thus, in Fig. 20.11, the incident ray AO, the normal ON and the refracted ray OB in the glass all lie in the plane AON shown in faint outline (compare *Law of reflection*, p. 360).

## Refractive Index

The ratio sin $i$/sin $r$ is called the *refractive index* from air to glass. It is a number which gives a measure of the refraction or 'bending' of light when it travels from one medium to another. We shall denote the refractive index by the symbol $n$, so that $n = \dfrac{\sin i}{\sin r}$.

Fig. 20.12 shows refraction from air to crown and flint glass, and from water and glass to air. In each case the angle of incidence $i$ is the angle in

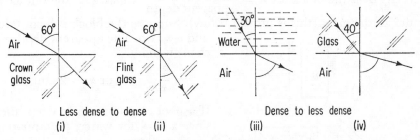

Less dense to dense        Dense to less dense

(i)      (ii)      (iii)      (iv)

Fig. 20.12 Examples of Refraction

the first medium, and the angle of refraction $r$ is that in the second medium. Suppose the respective angles of refraction are 35°, 32°, 42°, 75°. Then

    (i) $n$, air to crown glass,   = sin 60°/sin 35° = 1·51
   (ii) $n$, air to flint glass,     = sin 60°/sin 32° = 1·63
  (iii) $n$, water to air,        = sin 30°/sin 42° = 0·75
  (iv) $n$, glass to air,        = sin 40°/sin 75° = 0·67

It should be realized that it is meaningless to talk about the refractive index of a medium, such as glass, without specifying the medium, such as air or water, in which the light was *originally* travelling before refraction took place. Scientists, however, have compiled tables of refractive indices for many important media on the assumption that the light originally travelled in a *vacuum* before entering the medium. The refractive index of air is only very slightly greater than 1, so that, except when extreme accuracy is needed, the refractive index of glass, for example, can be measured with light incident on it from air.

Lenses for eye-glasses, telescopes and microscopes are usually made from crown glass or flint glass, which have different refractive indices. The construction of the lenses needed for optical instruments depends on an accurate knowledge of the refractive indices of the glass used.

### Glass–Air and Water–Air Refractive Indices

The value of the refractive index from air to glass is sin 60°/sin 35° from (i) previously, which is about 1·5 or $\frac{3}{2}$. When light travels from glass to air its path reverses exactly, so that the angle of incidence is now 35° and the angle of refraction is 60°. Hence the refractive index for glass to air $= \dfrac{\sin 35°}{\sin 60°} = \dfrac{1}{1·5}$ or $\dfrac{2}{3}$. We therefore *invert* the refractive index value from air to glass to find the value from glass to air. Thus, since the refractive index from air to water is 1·33 or $\frac{4}{3}$, the refractive index from water to air is $\frac{3}{4}$ or 0·75.

### Apparent Depth and Refractive Index

We have already explained why an object O at the bottom of a glass block appears to be at I, nearer the surface B, when viewed from above. The *true depth* OB of the block is related to its *apparent depth* IB, viewing directly over the block, by $n = \dfrac{\text{True depth}}{\text{Apparent depth}}$, where $n$ is the refractive index of the glass (this is proved below). Suppose the block is 3 cm thick and $n = 1·5$. The apparent depth $x$ is then given by

$$\frac{3}{x} = 1·5, \text{ or } x = 2 \text{ cm.}$$

If a pool of water is 6 m deep, then since $n = \frac{4}{3}$ for water, the apparent depth $= \dfrac{6\,\text{m}}{\frac{4}{3}} = 4\frac{1}{2}$ m, on viewing the water normally to its surface.

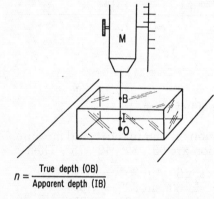

$$n = \frac{\text{True depth (OB)}}{\text{Apparent depth (IB)}}$$

Fig. 20.13 Refractive index by apparent depth

### Experiment

The relationship $n = true\ depth/apparent\ depth$ can be used to measure accurately the refractive index of glass or water. A travelling microscope M is first focused on an ink-mark O on paper (Fig. 20.13). The narrow part of the block is then placed over O, and the microscope is raised to bring the ink-mark back into focus because the mark now appears to be at I. Finally, the microscope is focused on chalk dust or lycopodium powder or paper on top of the block at B. From the three measurements obtained, $n$ can be calculated from the relation $n = \text{OB/IB}$. A similar experiment can be performed to find the refractive index of water, using a beaker with an object such as a pin or a coin at the bottom.

### Proof of Apparent Depth Relation

Suppose a ray OA, incident in glass at A, emerges into air making angles of incidence and refraction $i$ and $r$ respectively (Fig. 20.14). Then

$$n, \text{ glass to air} = \frac{\sin i}{\sin r}$$

$$\therefore n, \textit{ air to glass} = \frac{\sin r}{\sin i}$$

From Fig. 20.14, angle NOA $= i$ and angle NIA $= r$ by geometry.

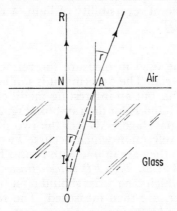

Fig. 20.14 Apparent depth proof

$$\therefore \sin r = \text{NA/IA and } \sin i = \text{NA/OA}.$$

$$\therefore n, \text{ air to glass} = \text{NA/IA} \div \text{NA/OA} = \text{OA/IA}.$$

When OA is very near to ON we can say OA $=$ ON and IA $=$ IN, so that

$$n = \frac{\text{ON}}{\text{IN}} = \frac{\text{Real depth}}{\text{Apparent depth}}.$$

## Refraction with a Glass Prism

*Prisms* are used in optical instruments. They consist of a glass block with inclined faces, so that a section such as ABC is a triangle (Fig. 20.15(i)). Prisms with angles of 45°, 45° and 90°, isosceles right-angled prisms, are

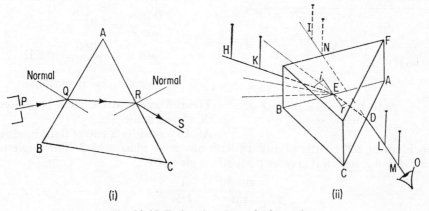

Fig. 20.15 Refraction through glass prism

used in prism binoculars. Equilateral prisms with angles of 60° are also used in optical experiments.

The refraction of a ray of light obtained with a glass prism can be observed with a ray-box. When the ray PQ enters the glass it is refracted towards the normal at Q, along QR (Fig. 20.15(i)). When the ray emerges into the air again from the glass it is refracted away from the normal along RS. From the principle of reversibility of light, a ray incident along SR would emerge along QP.

EXPERIMENT

Prisms of 60° can be used to measure the refractive index of glass or other transparent material. The experiment is similar to that described on p. 385 with rays of light, but an interesting modification can be made by substituting two pins H and K for the ray-box (Fig. 20.15(ii)). An observer at E looking into the glass face FAC sees the two pins at I and N respectively, because of the refraction which occurs. Two other pins L and M are then placed so that they are in line with N and I. The points L and M are joined to meet AC at D. *Then ED is the refracted ray in the glass.* As explained on p. 385, to which the reader should refer, the angles of incidence and refraction, $i$ and $r$, are then measured. The refractive index is calculated by $\sin i / \sin r$, and the average taken for several angles of incidence.

**Direction of Refracted Ray**

We have shown how the direction of a refracted ray can be found experimentally. We will now show how the direction of this ray can be obtained from a knowledge of the angle of incidence and the appropriate refractive index of the medium.

(1) *Calculation.* Suppose the direction of the refracted ray is required for light incident in air at 60° on a plane water surface. $n$ for water is 1·33 (Fig. 20.16). The angle of refraction $x$ is then given by

$$\frac{\sin 60°}{\sin x} = n = 1·33$$

$$\therefore \frac{0·866}{\sin x} = 1·33$$

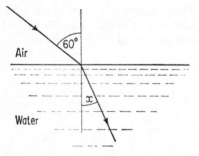

or

$$\sin x = \frac{0·866}{1·33} = 0·651$$

$$\therefore x = 41° \text{ (approx).}$$

Fig. 20.16 Calculation of angle of refraction

The refracted ray can now be drawn with a protractor.

Again, consider a ray of light in glass incident at 30° on the boundary with air. $n$ for glass = 1·5. The angle of refraction $y$, which is greater than 30°, is given by

$$\frac{\sin 30°}{\sin y} = \frac{1}{1·5}.$$

$$\therefore \sin y = 0·75 \text{ and } y = 49°.$$

(2) *Drawing.* The refracted ray can also be obtained by drawing. Suppose the ray AO is incident in air at a *known* angle $i$ on a plane glass surface (Fig. 20.17(i)). If the value of the refractive index $n$ is 1·5, construct two circles of radii 1 and 1·5 units respectively. Produce the incident ray to cut the smaller circle at B. From B draw BM perpendicular to the surface and produce MB to meet the larger circle at C. Join OC. Then OC is the required refracted ray in glass.

This can be proved as follows:

Angle MBO = Angle $i$, the angle of incidence
Angle MCO = Angle $r$, the angle of refraction

$$\therefore n = \frac{\sin i}{\sin r} = \frac{\sin \angle MBO}{\sin \angle MCO} = \frac{OM/OB}{OM/OC} = \frac{OC}{OB}.$$

But

$$\frac{OC}{OB} = \frac{1\cdot5}{1} = n.$$

Therefore OC must be the refracted ray.

To find the refracted ray for light travelling from glass to air, the con-

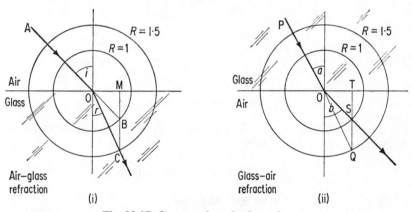

Fig. 20.17  Construction of refracted rays

struction is modified accordingly (Fig. 20.17(ii)). PO is the incident ray making a known angle $a$ with the normal. This time, produce PO to cut the *larger* circle at Q, and draw a perpendicular QT to the plane boundary, cutting the smaller circle at S. Join OS. Then OS is the refracted ray. It is left to the reader to show that $\frac{\sin b}{\sin a} = 1\cdot5$ from Fig. 20.17(ii).

If the perpendicular from Q to the surface does not intersect the smaller circle no refracted ray is possible. This case is important in practice, as we shall see shortly (p. 393).

### Field of View under Water

A fish E under water, or a man swimming under water on his back, will see rays of light such as BC which are refracted from air along CE in the water (Fig. 20.18). The extreme or limiting rays from the outside reaching E correspond to those like LO which just graze the air–water boundary. The angle of incidence in air is then 90°. If the angle of refraction in the water is $r°$ it follows that light from everywhere outside the water is concentrated into a cone whose half-angle is $r°$.

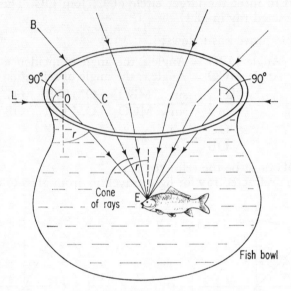

Fig. 20.18 Fishes' view

The angle of the cone can easily be calculated, since $\sin 90°/\sin r = n = \frac{4}{3}$ for water.

$$\therefore \frac{1}{\sin r} = \frac{4}{3}, \text{ or } \sin r = \frac{3}{4} = 0.75$$

$$\therefore r = 49° \text{ (approx)}.$$

Thus a fish can see everything above the water by rays within a cone of half-angle about 49° inside the water.

**Total Internal Reflection**

When a ray of light HK is incident in air on a rectangular glass block most of the light is refracted along KL into the block (Fig. 20.19). A *weak reflected* ray KT is also obtained at the boundary, and from the law of reflection this makes an equal angle $i$ with the normal. When the angle of incidence HKN is increased a strong refracted ray and a weak reflected ray is always obtained. This is the case with XK, incident on the glass nearly at 90°. A strong refracted ray KY and a weak reflected ray KZ result.

A striking change occurs when the situation is reversed optically, so that light passes from *glass to air*. This is best investigated with a semicircular glass block G and a ray-box R with a single ray (Fig. 20.20).

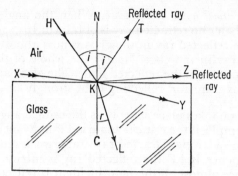

Fig. 20.19 Refraction and reflection

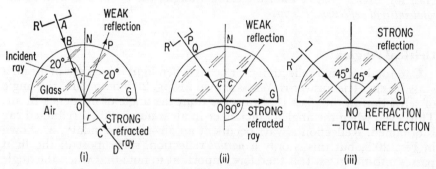

Fig. 20.20 Investigation of total internal reflection

## Experiment

G is placed on a sheet of white paper and the centre O of its plane side is marked. Since any line from O is a radius, and therefore normal to the semicircle, a ray such as AB *directed towards* O will pass into the glass without any change of direction. As we are interested in what happens when light is refracted from glass to air, we must think of BO as the incident ray *in the glass*, angle BON as the angle of incidence and OD as the refracted ray in air. The angle BON is now increased from zero by moving the ray-box round from the normal position, and the effects at the glass–air boundary are noted.

## Results

When the angle of incidence in the glass is increased, the results obtained are as follows:

(1) *Small angle of incidence, e.g.* 20°. There is a strong refracted ray OCD and a weak reflected ray OP in the glass (Fig. 20.20(i)).

(2) *Critical angle of incidence c.* As the angle of incidence in the glass is increased, the angle of refraction in the air increases. At one special angle of incidence, called the *critical angle c*, the refracted ray OG travels along the glass–air boundary. The angle of refraction in the air is then 90° (Fig. 20.20(ii)). A weak reflected ray is obtained, making an angle c with normal ON.

(3) *Beyond the critical angle of incidence.* When the angle of incidence is greater than $c$ no refracted ray is obtained. Instead, a new effect is seen. The reflected ray in the glass is now almost as bright as the incident ray in the glass. Nearly the whole of the incident light energy appears in the reflected light. The glass–air boundary thus behaves as a 'mirror' reflecting light strongly (Fig. 20.20(iii)).

This phenomenon takes place sharply as the critical angle of incidence is exceeded. For example, if the critical angle is 42° a ray incident in the glass at 41° will produce a strong refracted ray and a weak reflected ray. But if the angle of incidence is 43° the reflected ray is intense and there is no refracted ray. Since almost all the light incident is reflected back into the glass for angles greater than the critical angle, the effect is known as the *total internal reflection of light.*

## Critical Angles

Can total internal reflection ever occur when light is incident in air on a glass block? The answer is *No.* As shown in Fig. 20.5 on p. 383, the angle of refraction in the glass is always *less* than the angle of incidence in air. Thus even when the angle of incidence in air is nearly 90°, a refracted ray is obtained. Reflection always occurs at an air–glass boundary, as shown in Fig. 20.5, but this is only a *partial* reflection, since most of the light passes into the glass. It is therefore important to note that when the angle of incidence is increased, *total internal* reflection occurs only when light passes from one medium to an optically *less* dense medium, for example from glass to air or from water to air.

Suppose the angle of incidence in glass is $c$, the critical angle, so that the angle of refraction in air is 90°. Total internal reflection occurs for any angle greater than $c$. If the glass has a refractive index $n$ of 1·5 or $\frac{3}{2}$ for light passing from air to glass, then for light travelling from glass to air the refractive index is $\frac{2}{3}$ (see p. 388).

$$\therefore \frac{\sin i}{\sin r} = \frac{\sin c}{\sin 90°} = \frac{2}{3}$$
$$\therefore \sin c = \tfrac{2}{3} \times \sin 90° = \tfrac{2}{3} \text{ or } 0·6667.$$
$$\therefore c = 42° \text{ (approx).}$$

An angle of incidence of 45° in glass thus produces total internal reflection if the medium on the other side is air.

The refractive index for air to water is $\frac{4}{3}$. Therefore the refractive index for light passing from water to air is $\frac{3}{4}$. The critical angle $c$ for water–air media is therefore given by

$$\frac{\sin c}{\sin 90°} = \frac{3}{4}$$
$$\therefore \sin c = \frac{3}{4} \times \sin 90° = 0·75$$
$$\therefore c = 49° \text{ (approx).}$$

## Multiple Images

When the image of an object in plane mirrors is observed closely, several faint images can be seen besides a prominent one. The presence of multiple images is due to partial reflection and refraction at the non-silvered glass surface of the mirror, as illustrated in Fig. 20.21.

A weak image $I_1$ is formed by reflection from the glass surface of part of the incident light at O. Most of the light passes through the glass and is then reflected at the silvered surface and refracted again at the glass to

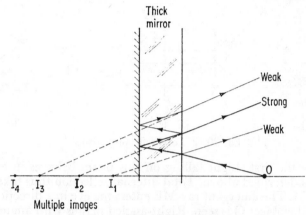

Fig. 20.21 Multiple images

form the main image seen, $I_2$, which is bright. A small part of the light reflected from the silvered surface is reflected at the glass, and further weak images, $I_3$ and $I_4$, due to reflection and refraction, are thus obtained. The images are more widely separated if the glass is thicker.

## Total Reflecting Prisms

On account of multiple images, plane mirrors are not used for high-quality periscopes, such as those employed in submarines where accurate sighting of other ships is required. A special glass prism, producing total internal reflection, is used in this case, which has the great advantage that only one image is produced. As in prism binoculars, which are short telescopes, *right-angled isosceles prisms*, ones with angles of 90°, 45°, 45°, are used as reflectors.

Fig. 20.22(i) illustrates the optical action of a totally reflecting prism. Consider a ray OP from an object O, incident normally on one of the sides containing the right angle C of the prism. The ray passes straight through AC and falls on the hypotenuse face AB at S at an angle of 45°. Since this angle of incidence in glass is greater than the critical angle (42°) for glass of refractive index 1·5, no light passes into the air from S. Instead, total internal reflection takes place at S. A bright reflected ray ST is obtained, making an angle of 45° with the normal at S in accordance with the law of reflection. ST is therefore at right angles to PS, and is almost as bright as PS. ST is consequently incident on BC normally and emerges into the air, perpendicular to its original direction. The emergent light is almost as bright as the incident light from O. Moreover, only one reflected image is produced, unlike a plane mirror.

A periscope is completed by placing a second right-angled isosceles

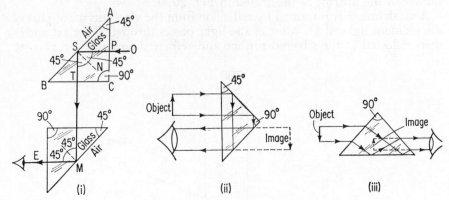

Fig. 20.22 Total reflecting prisms

prism below the first. Light falling on this prism behaves in exactly the same way as described above. Total internal reflection thus takes place in the glass at M. The emergent ray ME enters the eye in a direction parallel to OP, and the object O is seen. Right-angled prisms with angles 90°, 30°, 60° do not act as totally reflecting prisms, as the reader should verify.

Fig. 20.22(ii), (iii) illustrate two other cases of total reflection by the use of right-angled 45° prisms. Fig. 20.22(ii) shows how an object is inverted, as in prism binoculars (p. 428). Total internal reflection accounts for the observed brilliance of diamonds. The refractive index of a diamond is high, so that its critical angle is correspondingly low. Much of the light entering the diamond is therefore totally reflected to the observer.

### Optical Fibres

Nowadays total internal reflection is also applied in a very useful device known as *optical fibres*, following an idea due originally to Baird, the

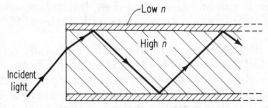

Fig. 20.23 Optical fibre action

founder of television, in 1927. Optical fibres consist of many tens of thousands of long fine strands of high-quality glass coated with glass of lower refractive index. The strands may be $\frac{1}{2000}$ cm in diameter and the refractive indices of the respective glasses about 1·7 and 1·5.

When light is incident on one end of the fibre at an angle less than about 60° it passes inside, where it undergoes repeated total internal reflection at the walls (Fig. 20.23). The angle of incidence here is greater than the critical angle between the high- and low-refractive index glass. The trapped light thus travels along the fibre, no matter how it may be curved, and emerges with high intensity at the other end. A bundle of flexible fibres thus enables an image of an object at one end to be seen at the other end. If the fibres are tapered a magnified image can be produced.

Optical fibres are thus flexible guides which can 'pipe' light to the other end. They are used medically, for example, to examine the inside of the throat, and for many engineering uses.

## Mirages

Total internal reflection can arise in gases such as air, as well as in solid media such as glass. The essential condition again is that the light must be travelling from a denser to a less-dense medium.

In hot weather the layers of air close to the ground are hotter than the

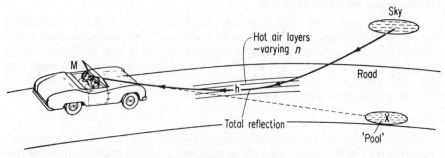

Fig. 20.24 Mirage

layers higher up, because they are heated by the hot ground. A ray from part of the sky, passing from a colder to a warmer air layer (i.e. from a denser to a rarer medium) will therefore bend away gradually from the incident direction until it enters a layer of air $h$, where total internal reflection occurs (Fig. 20.24). The ray is then reflected upwards into the denser air. After undergoing refraction in a gradually increasing upward direction the ray may finally enter the eye of an observer M. To him, the ray seems to come from a place X, which gives the illusion of a reflecting pool of water in the road some distance away. In hot deserts the illusion may be created of an inverted image of a tree in a pool of water, standing below the actual tree position.

## Other Refraction Phenomena

The sun continues to be visible to an observer at A after it has set below the horizon. This is also explained by atmospheric refraction (Fig. 20.25(i)). Such refraction also causes a star S to appear more elevated than it really is (Fig. 20.25(ii)). Astronomers allow for this source of error in their determinations of stellar positions. A correction is also made in navigational measurements of the sun's altitude for the same reason.

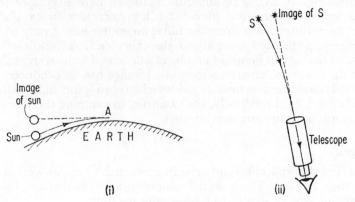

Fig. 20.25 Atmospheric refraction

## Refraction of Microwaves

We have seen that water waves and light waves undergo refraction when travelling from one medium to another. The effect produced when 3 cm radio waves, microwaves, travel from air to a new medium can be studied by making large prisms of solid paraffin-wax, one with angles of 60°, 60°, 60° and the other with angles of 90°, 45°, 45°. Liquid paraffin in hollow prisms may also be used (Plate 22).

EXPERIMENT

With the transmitter S in the position shown in Fig. 20.26(i), waves are incident on the 60° prism P in the direction LM. The galvanometer shows no deflection when the detector R is placed at A on the other side of the

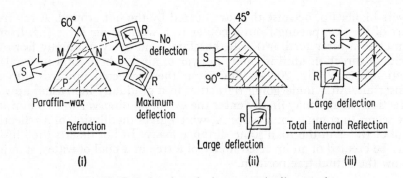

Fig. 20.26 Refraction of microwaves (radio waves)

prism pointing at S, but when R is moved round to a position B a large deflection is obtained. The radio waves, which are incident on the prism in the direction LM, have thus been refracted, and emerge in the direction NB.

With the right-angled 45° prism, the transmitter is placed at S so that the waves pass into the prism and are incident on the hypotenuse face (Fig. 20.26(ii)). A considerable deflection is then obtained in the galvano-meter when the detector R is positioned as shown. The electromagnetic waves are hence totally reflected. Fig. 20.26(iii) shows the positions of the transmitter S and the receiver R when a large deflection is again produced in the galvanometer, due to total internal reflection at two faces.

These experimental results are exactly similar to those obtained with light waves or rays and glass prisms, described earlier in the chapter. They show that radio waves can be refracted when entering a new medium. Light and radio waves are both electromagnetic waves (p. 331).

## Appleton and Heaviside Layers

In 1925 Sir Edward Appleton discovered by experiment the existence of a layer or belt of highly concentrated electrical particles–electrons and ions–situated about 170 miles above the earth. The *Appleton layer*, as it is

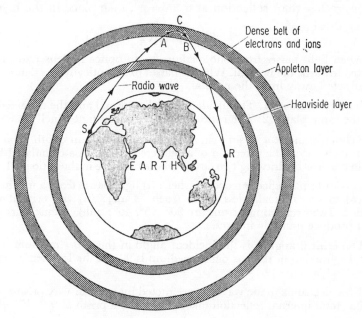

Fig. 20.27 Refraction by Ionosphere

called, reflects radio waves (Fig. 20.27). Lower down, about 70 miles above the earth, Appleton found the existence of another concentrated region of electrons and ions, called the Heaviside layer after Oliver Heaviside, who suggested the existence of the layer about 1905. Both layers are produced by the action of the sun's energetic rays, the ultra-violet rays. At night the Heaviside layer disappears and the Appleton layer remains.

In 1901 Marconi succeeded in sending radio signals from one side of the Atlantic to the other. Radio waves travel in straight lines, and do not follow the curvature of the earth. How they travel such large distances round the earth remained unexplained until Appleton's discovery. The Appleton and Heaviside layers enable the waves to be sent from one side S of the earth, England for example, to the opposite side R such as Australia (Fig. 20.17). At night, radio (electromagnetic) waves travel skywards from S in a straight line to A, the lowest part of the Appleton layer. Here they enter a new medium, one which has a high concentration of electrical particles, and refraction occurs. The layer has a varying concentration or density of particles, and as shown, the waves are refracted in a gradual curve away from the normals until they meet a particular region at C. Total reflection now takes place. The waves then turn back and are refracted along a curve CB. After emerging from the Appleton layer the waves travel to R, the receiving station. (Compare the similar effect in the '*Mirage*', p. 397.)

Without a 'mirror in the sky', as the Appleton and Heaviside layers have been popularly called, radio reception on the other side of the world would not be possible. It should be noted that refraction and total internal reflection, rather than reflection at a 'mirror', take place in the layers.

## SUMMARY

1. When light passes from one medium to an optically denser medium it is refracted towards the normal. When it passes to an optically less dense medium it is refracted away from the normal.

2. *Laws of refraction*: 1. The incident ray, the normal and the refracted ray all lie in the same plane. 2. The ratio $\sin i/\sin r$ is a constant (Snell's law).

3. Refractive index $n = \sin i/\sin r$. Real depth/apparent depth $= n$, at near-normal incidence. The direction of a refracted ray can be calculated by using $\sin i/\sin r = n$ or by drawing two circles whose radii are in the ratio $n : 1$.

4. Total internal reflection occurs when: (i) light passes from a medium (such as glass) to one optically less dense (such as air); (ii) the critical angle is exceeded. Total reflecting prisms are 90°, 45°, 45°; unlike plane mirrors, they do not produce multiple images.

5. The critical angle $c$ is the incident angle in the *denser* medium when the angle of refraction in the less dense medium is 90°. If the less dense medium is air, $\sin c = 1/n$.

6. 3-cm electromagnetic waves are refracted by paraffin-wax prisms or liquid paraffin; total internal reflection takes place with a prism of 90°, 45°, 45°.

7. Radio waves can be refracted and totally reflected at different layers of electrical particles, such as the Appleton and Heaviside layers. Light waves are refracted and totally reflected by different layers of hot air when a mirage is produced.

## EXERCISE 20 · ANSWERS, p. 452

**1.** Explain the statement 'the refractive index of glass is 1·5'. (*N.*)

**2.** Define *refractive index*.
Explain, with the help of a ray diagram, why water in a swimming-pool appears to be shallower than it really is. What is the apparent depth of a swimming-pool if its real depth is 2·4 m and it is viewed from a springboard vertically above? (Refractive index of water is $\frac{4}{3}$.) (*L.*)

**3.** State the *laws of reflection* of light.

If a small illuminated object is placed in front of a thick plate glass mirror, silvered on the back, several images are seen. Explain the formation of these images and indicate which one is usually brightest. (*L.*)

**4.** Explain the statement that the refractive index from glass to air is $\frac{2}{3}$.

A rectangular glass block measures 15 cm by 5 cm. Trace the path of a ray from a point in air which is at a perpendicular distance of 5 cm from the middle of one of the longer sides, for an angle of incidence of 40°. Measure the length of the path of the refracted ray in the glass. (*L.*)

**5.** Define *refractive index*.

A ray of light is incident along a normal to the face AB of a glass prism ABC with angle BAC = 30°. Calculate the angle at which the ray emerges from the face AC. Draw a diagram and clearly show on it the path of the ray of light. (Refractive index of glass = 1·50.) (*C.*)

**6.** Explain why a pool of water appears to be less deep than it really is.

Describe an experiment to determine the refractive index of water by measuring the real depth and the apparent depth. (*L.*)

**7.** State the laws of refraction and illustrate your statement by a ray diagram, marking clearly thereon the angles of incidence and refraction.

Describe two methods of measuring the refractive index of glass, supplied in the form of a rectangular glass block.

Water is poured into a beaker to a depth of 12 cm. To an eye looking vertically down through the water surface, the bottom of the beaker appears to be raised 3 cm. Calculate the refractive index of water and prove any formula you use. (*O. and C.*)

**8.** (*a*) Describe an experiment to find the refractive index of glass. (*b*) A concave mirror produces an upright image, magnified three times, of a small object 10 cm away from the mirror and at right angles to the principal axis. What is the focal length of the mirror? How far must the mirror be moved in order to produce an inverted image magnified three times?

If the answer is obtained by calculation, state the sign convention used. (*N.*)

**9.** Explain the statement 'the critical angle of water is 48° 36″. (*N.*)

**10.** (i) Draw a ray diagram to show why a pool of water appears to be only three-quarters of its real depth when viewed vertically from above.

(ii) Find the critical angle of a medium of refractive index 1·65. (*N.*)

**11.** (*a*) State what is meant by *refraction of light* and describe an experiment to demonstrate it.

(*b*) Give the conditions for total internal reflection to occur and illustrate with a diagram.

Draw a diagram to show how a glass prism can be used to deviate a parallel beam of light through 90° by means of total internal reflection. Explain the action of the prism.

Calculate the minimum value for the refractive index of glass for which a 90° deviation is possible. (*L.*)

**12.** Show, with the help of diagrams, how a ray of light can be turned through 90° using: (i) a glass 45°, 90°, 45° prism; (ii) a plane mirror. State, giving reasons, which of the two methods is preferable in a periscope. (*C.*)

**13.** Explain, with diagrams, the meaning of 'critical angle' and of 'total internal reflection'.

Describe an experiment to find the refractive index of glass.

Find the refractive index of a medium for which the critical angle is 40°. (*N.*)

**14.** What is meant by *total internal reflection* of light? State the conditions under which this occurs. Define *critical angle* and explain how the critical angle for a material in air depends on its refractive index.

Name ONE practical application of total internal reflection.

Find the angle of incidence of a ray of light on one face of a 60° prism if the ray is just totally internally reflected on meeting the next face. (Take the refractive index of glass to be 1·5.) (*O.*)

**15.** State the laws of refraction, and indicate clearly on a diagram the corresponding angles of incidence and refraction.

Describe the circumstances in which total internal reflection occurs and calculate the refractive index of a medium for which the critical angle is 45°. Why would you expect the critical angle of glass for red light be be greater than that for blue light? How would you demonstrate this fact? (*O.* and *C.*)

**16.** How can it be shown by experiment that the reflected ray is turned through twice the angle through which a plane mirror is turned, if the direction of the incident ray remains unchanged? Show how this result follows from the laws of reflection.

Why are triangular prisms often used as reflectors in optical instruments in preference to plane mirrors? (*N.*)

# 21

# REFRACTION THROUGH LENSES

Lenses consist of pieces of glass of varying thickness from the middle to the edges, bounded by spherical surfaces on one or both sides. They are used in spectacles to correct defects of vision, and in optical microscopes for looking at objects too small to be seen by the naked eye. They are also used in telescopes and prism binoculars for seeing distant objects, and in cameras and film projectors.

Lenses have been used to concentrate light since ancient times, and they were then called burning glasses. The first sunshine recorder, made in 1857, utilized a glass sphere or globe to char a paper graduated in fractions of an hour, and they are still used at the Air Ministry and seaside resorts to record the hours of sunshine.

## Types of Lenses

Fig. 21.1 illustrates six common types of lenses. A lens thicker in the middle than at the edges, as in Fig. 21.1(i), is known as a *converging* or *con-*

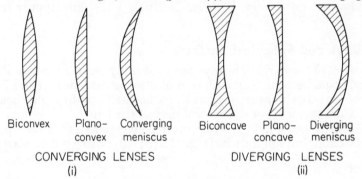

Biconvex  Plano-convex  Converging meniscus  Biconcave  Plano-concave  Diverging meniscus

CONVERGING LENSES (i)     DIVERGING LENSES (ii)

Fig. 21.1 Converging and diverging lenses

*vex* lenses. A lens thinner in the middle than at the edges is known as a *diverging* or *concave* lens (Fig. 21.1 (ii)). Lenses which are biconvex or biconcave are commonly used in the laboratory. Plano-convex and plano-concave lenses have only one curved surface and are used in optical instruments. Converging and diverging meniscus lenses are used as 'contact lenses' to fit the curvature of the eyeball.

The eye has a natural converging lens, and a camera has a glass converging lens. Some spectacles may contain diverging lenses, others may contain converging lenses. One type of telescope has both a converging and a diverging lens inside it.

## Refraction of Waves by Spherical Surfaces

The effect on water waves of refraction at spherical surfaces can be illustrated by immersing an inverted watch glass W in a ripple tank so that its highest point is just below the water (Fig. 21.2). The variation of water depth makes the waves move more slowly as they approach the middle of W, and with adjustment, plane waves of moderate frequency are then seen to slow down on passing over W and emerge with a *converging* curvature

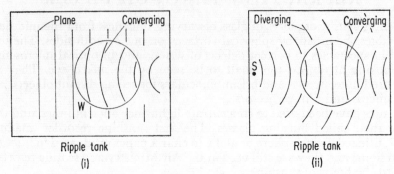

Fig. 21.2  Converging water waves

(Fig. 21.2(i)). When one small sphere S on the ripple tank (see p. 324) is used as a vibrating point source, circular waves spread out and change their curvature on passing W (Fig. 21.2(ii)). They now appear to converge.

### Light Rays and Spherical Surfaces

The effect of lenses on light rays can be seen by using a ray-box R. The lens used could be a model flat glass or plastic convex lens L (Fig. 21.3(i)), or, better, an actual converging lens X, pushed halfway through a board, so that rays are incident on the central part of the lens.

A parallel beam of light is then seen to converge to a focus at a point F after refraction through the lens (Fig. 21.3(i), (ii)). Using a diverging beam

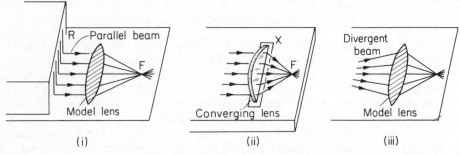

Fig. 21.3 Converging lens action

from R, a converging beam is again obtained after refraction, as shown in Fig. 21.3(iii).

Fig. 21.4 shows the action of a diverging lens on a parallel beam. After refraction through the lens a diverging beam of light is obtained which would pass through a point F behind the lens if produced back. The effect of this type of lens is therefore opposite to that of the converging lens.

Model diverging lens
(i)

Actual diverging lens
(ii)

Fig. 21.4 Diverging lens action

## Wave Theory

Fig. 21.5(i) shows how the wave theory of light explains the converging action of the converging lens. When a plane wavefront APB reaches the lens as shown, the wavelets from A and B reach R and S respectively in the same time as the wavelet from P reaches Q. R is farther away from A than Q is from P, because the velocity of light in air is greater than in glass (p. 381). The curved wavefront RQS emerging from the lens therefore converges to F. In the case of a diverging lens (Fig. 21.5(ii)), a plane wavefront APB becomes a curved wavefront RQS after passing through the lens. This time the wavelet from P travels a greater distance than those from A and B in the same time, because the latter now travel through a

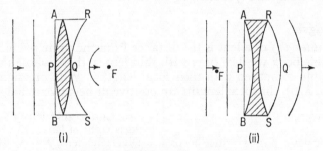

Fig. 21.5 Effect of lenses on plane waves

greater thickness of glass. RQS appears to diverge from a point F on the left of the lens, unlike the case in Fig. 21.5(i).

## Refraction of Single Rays

A ray-box R with a single ray shows very strikingly how different parts of a lens have different effects on rays incident on their surfaces.

As shown in Fig. 21.6(i), when ray 1 strikes the centre part C normally it passes straight through undeviated. The whole ray now lies along the *principal axis* of the lens, which is the line joining the mid-points of the two opposite faces of the lens. When another ray 2 is a small distance above and parallel to the principal axis, it converges to a point F on the principal axis.

Another ray 3, parallel to the principal axis but a little higher than 2, also converges to F after refraction. This ray is deviated more from its original path than 2. The *principal focus* F is the point on the principal axis to which rays parallel and close to the principal axis converge after refraction through the lens. A parallel beam of light, travelling in a direction inclined to the principal axis, is brought to a focus at a point $F_1$, *below F*, called a 'secondary focus' (Fig. 21.6(ii)).

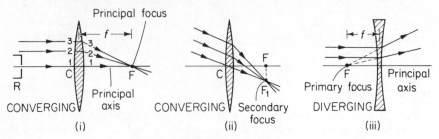

Fig. 21.6 Refraction of single rays

The action of a diverging lens on rays parallel to the principal axis is shown in Fig. 21.6(iii). The rays appear to diverge from a point F on the principal axis, which is called the *principal focus* of the diverging lens. Comparing Fig. 21.6(i) and (iii), it can be seen that:

1. *The principal focus of a converging lens is real.*
2. *The principal focus of a diverging lens is virtual.*

### Focal Length

The *focal length f* of a lens is the distance from the principal focus to the lens. (In this book we deal only with thin lenses, those whose thicknesses are negligible compared with their focal lengths.) With our sign convention (see p. 376), the focal lengths are positive or negative as follows:

*RP.* Converging lens, $f$ + ve since focus is real.

Diverging lens, $f$ − ve since focus is virtual.

*NC.* Converging lens, $f$ + ve since focus is right of lens.

Diverging lens, $f$ − ve since focus is left of lens.

### Lenses as Prisms

It should be noted that a lens can be regarded as built up of a very large number of *prisms* fitted together. The general idea is illustrated roughly in Fig. 21.7. Thus a converging lens consists of prisms with their triangular top parts missing, except at the extreme ends (Fig. 21.7(i)). The centre of the lens is a parallel-sided piece of glass. The sides of the prisms slope more and more as we go higher up from the centre. As shown in the section on prisms, p. 389, (1) rays incident on one side of the prism are deviated towards the bottom or base after refraction, and (2) the amount of the deviation increases as the angle between the sloping sides increases, as shown.

A similar explanation holds for refraction by a diverging lens (Fig.

21.7(ii)). This time, however, the bases of the prisms are upwards. Hence the rays are deviated upwards, giving rise to a diverging beam.

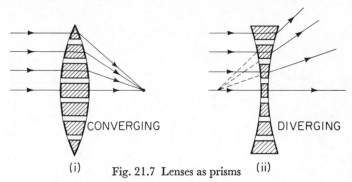

Fig. 21.7  Lenses as prisms

## Refraction of Special Rays

By using a single ray from a ray-box, the following rays can be directed on to a convex (converging) lens and the results of refraction noted:

1. *A ray 1 parallel to the principal axis*—this is refracted through the principal focus F (Fig. 21.8).

2. *A ray 2 incident on the centre C*—this passes straight through. As shown

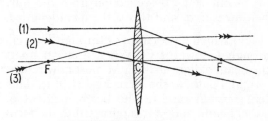

Fig. 21.8  Refraction of rays

in Fig. 21.7(i), the central part of the lens is a parallel-sided piece of glass, and this only slightly displaces the ray but does not change its direction (p. 383).

3. *A ray 3 through the principal focus F*—this is refracted parallel to the principal axis.

## Focal Length of Converging Lens

EXPERIMENT

*Quick method.* In the chapter on curved mirrors, the focal length of a concave mirror was found quickly by using the window frame, or an object outside it, as a distant object, in which case parallel rays arrive at the mirror (p. 371). A similar experiment can be carried out with a converging lens L by holding it as far away from the window W as possible, and moving a paper screen S until a sharp clear image is obtained (Fig. 21.9(i)). Since parallel rays are brought to a focus, the distance LS is equal to the focal length $f$ and is measured. It is an approximate and quick method for $f$, but not accurate because the rays are not perfectly parallel.

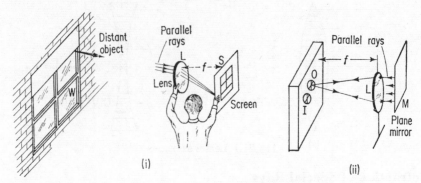

Fig. 21.9 Methods for $f$. (i) Quick. (ii) Accurate

## Accurate Method

The focal length of a converging lens L can be measured with good accuracy by placing a plane strip mirror M behind it and an illuminated object O in front (Fig. 21.9(ii)). When O is moved, a clear image I is obtained near it at one stage by reflection from the mirror M. The distance LO is then measured, and this is the focal length $f$.

*Explanation.* Light which passes from O through L is reflected by M and passes back through L towards O. When a sharp image I is obtained back near O, the beam of light striking M must now return from M practically along its original path, as shown in Fig. 21.9(ii). This means that the beam is incident *normally* on M. The beam incident on L from M is therefore a parallel beam, and so it is refracted to the focus, which is at O.

## Images in Converging Lens

EXPERIMENT

The nature and size of the images produced by a converging lens can be investigated by using illuminated cross-wires as an object and moving it slowly towards the lens from a long way off. Typical distances from the lens at which the image may be observed on a screen are: (1) a long way off; (2) farther than $2f$, where $f$ is the focal length; (3) at $2f$; (4) between $2f$ and $f$.

*Results.* The results are shown in Fig. 21.10. Generally, as the object moves nearer the lens the image increases in size and remains real and inverted. At a distance $2f$ the image is the same size as the object (compare *concave mirror*, p. 373). The image becomes larger as it is moved nearer to the focus. At the focus, however, no image is received on the screen; the rays emerging from the lens are parallel and the image is said to be formed 'at infinity'.

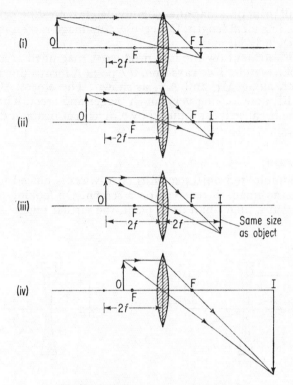

Fig. 21.10 Images in converging lens

## Magnifying Glass or Simple Microscope

If the illuminated object is moved nearer the lens than its focal length no image is received on the screen. But if the lens is placed close to a finger-nail, for example, the eye sees a *magnified and erect image*. A converging lens can thus be used as a magnifying glass or simple microscope. To look

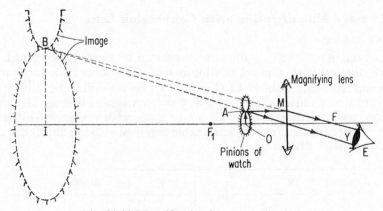

Fig. 21.11 Magnification by converging lens

closely at small print or a map the magnifying glass must be held nearer to the object than its focal length. Otherwise, the image is seen upside down and not particularly large.

Fig. 21.11 illustrates how the lens produces a magnified erect image of the pinions of a watch. Two rays from the point A form a diverging beam after refraction along MF and AY, as shown. The object AO is hence seen by E at BI, now looking very much bigger and erect. The image BI is a virtual one because it is produced by a divergent beam and cannot be seen on a screen.

### Diverging Lens

When an illuminated object such as cross-wires is placed in front of a diverging lens no image is received on a screen. On looking through the lens, however, at a finger-nail or at print, the virtual image is seen to be

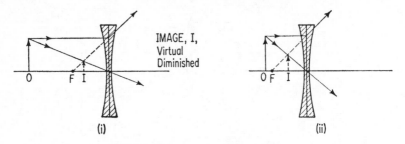

Fig. 21.12 Images in diverging lens

*diminished* and *erect*. This is always the case wherever the object is positioned; whereas the images in a converging lens are sometimes upside down and on other occasions erect, depending on the object position.

Fig. 21.12 illustrates by a ray diagram why the image in a diverging lens is always diminished and erect and virtual.

### Real Image Magnification with Converging Lens

EXPERIMENT

The variation of the magnification produced by a converging lens L can be investigated with the aid of illuminated cross-wires or a transparent graduated rule as an object O (Fig. 21.13). By adjusting the distance of OA from L and moving the screen S, a sharp magnified image IB can be obtained on S. Measurement of IB can be made with a ruler or with graph paper. The results are recorded in a table together with the object distance $u$ and the image distance $v$.

| IB | $u$ | $v$ | $m = BI/AO$ |
|----|-----|-----|-------------|
|    |     |     |             |

*Calculations.* The transverse or linear magnification is defined by $m =$ BI/AO. Calculations of $m$, and of the corresponding ratio $v/u$, are carried out after the experiment.

*Conclusion.* From the results, we find $m = v/u$ to a good approximation.

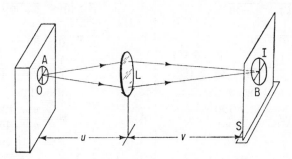

Fig. 21.13 Magnification experiment

*Explanation.* From the ray diagram on p. 409, Fig. 21.10, or p. 410, Fig. 21.12, similar triangles show that, by geometry,

$$\frac{\text{Length of image (BI)}}{\text{Length of object (AO)}} = \frac{\text{Image distance}}{\text{Object distance}}$$

$$\therefore\ m = \frac{v}{u}\,.$$

## Lens Formula

Measurements of the object and image distances, $u$ and $v$, from a converging lens, and of the focal length $f$, show that $u$, $v$ and $f$ are related by a general formula given below. This formula, which can be proved theoretically, applies both to converging and diverging lenses provided a sign convention, stated on p. 376, is applied.

*RP.* On the 'Real is Positive' convention,

$$\frac{1}{v} + \frac{1}{u} = \frac{1}{f}$$

Remember that for a converging lens, $f$ is +ve, and for a diverging lens, $f$ is −ve (p. 406).

*NC.* On the New Cartesian convention,

$$\frac{1}{v} - \frac{1}{u} = \frac{1}{f}$$

Remember that for a converging lens, $f$ is +ve, and for a diverging lens, $f$ is −ve (p. 406).

*Examples*

1. A converging lens has a focal length of 30 cm and an object is placed (i) 40 cm, (ii) 10 cm from the lens. Calculate the image position and its magnification in each case.

*RP.* The lens is converging; hence
$$f = +30.$$

(i) $\qquad u = +40$ cm

From $\qquad \dfrac{1}{v} + \dfrac{1}{u} = \dfrac{1}{f}$

$$\dfrac{1}{v} + \dfrac{1}{(+40)} = \dfrac{1}{(+30)}$$

$\therefore \qquad \dfrac{1}{v} + \dfrac{1}{40} = \dfrac{1}{30}$

$\therefore \qquad \dfrac{1}{v} = \dfrac{1}{30} - \dfrac{1}{40} = \dfrac{1}{120}$

$\therefore \qquad v = 120$

Also, $\qquad m = \dfrac{v}{u} = \dfrac{120}{40} = 3$

(ii) $\qquad u = +10$ cm

$\therefore \dfrac{1}{v} + \dfrac{1}{(+10)} = \dfrac{1}{(+30)}$

$\therefore \qquad \dfrac{1}{v} + \dfrac{1}{10} = \dfrac{1}{30}$

$\therefore \qquad \dfrac{1}{v} = \dfrac{1}{30} - \dfrac{1}{10} = -\dfrac{1}{15}$

$\therefore \qquad v = -15$

Also, $\qquad m = \dfrac{15}{10} = 1\cdot 5.$

Since $v$ is negative, the image is virtual and erect (see p. 409).

---

*NC.* The lens is converging; hence
$$f = +30.$$

(i) $\qquad u = -40$ cm

From $\qquad \dfrac{1}{v} - \dfrac{1}{u} = \dfrac{1}{f}$

$$\dfrac{1}{v} - \dfrac{1}{(-40)} = \dfrac{1}{(+30)}$$

$\therefore \qquad \dfrac{1}{v} + \dfrac{1}{40} = \dfrac{1}{30}$

$\therefore \qquad \dfrac{1}{v} = \dfrac{1}{30} - \dfrac{1}{40} = \dfrac{1}{120}$

$\therefore \qquad v = 120$

Also, $\qquad m = \dfrac{v}{u} = \dfrac{120}{40} = 3$

(ii) $\qquad u = -10$ cm

$\therefore \dfrac{1}{v} - \dfrac{1}{(-10)} = \dfrac{1}{(+30)}$

$\therefore \qquad \dfrac{1}{v} + \dfrac{1}{10} = \dfrac{1}{30}$

$\therefore \qquad \dfrac{1}{v} = \dfrac{1}{30} - \dfrac{1}{10} = -\dfrac{1}{15}$

$\therefore \qquad v = -15$

Also, $\qquad m = \dfrac{15}{10} = 1\cdot 5.$

Since $v$ is negative, the image is virtual and erect (see p. 409).

---

2. A diverging lens has a focal length of 20 cm and an image is 8 cm from the lens. What is the object distance and the magnification?

*RP.* Since the image is virtual, $v = -8$, and as this is a diverging lens, $f = -20$.

From $\qquad \dfrac{1}{v} + \dfrac{1}{u} = \dfrac{1}{f}$

$\therefore \dfrac{1}{(-8)} + \dfrac{1}{u} = \dfrac{1}{(-20)}$

$\therefore \quad -\dfrac{1}{8} + \dfrac{1}{u} = -\dfrac{1}{20}$

$\therefore \qquad \dfrac{1}{u} = -\dfrac{1}{20} + \dfrac{1}{8} = \dfrac{3}{40}$

$\therefore \qquad u = \dfrac{40}{3} = 13\tfrac{1}{3}$

$\therefore \qquad m = \dfrac{v}{u} = \dfrac{8}{13\frac{1}{3}} = 0\cdot 6.$

---

*NC.* The image (virtual) is left of the lens, and hence $v = -8$. Also, $f = -20$ (diverging lens).

From $\qquad \dfrac{1}{v} - \dfrac{1}{u} = \dfrac{1}{f}$

$\therefore \dfrac{1}{(-8)} - \dfrac{1}{u} = \dfrac{1}{(-20)}$

$\therefore \quad -\dfrac{1}{8} - \dfrac{1}{u} = -\dfrac{1}{20}$

$\therefore \qquad \dfrac{1}{u} = -\dfrac{1}{8} + \dfrac{1}{20} = -\dfrac{3}{40}$

$\therefore \qquad u = -\dfrac{40}{3} = -13\tfrac{1}{3}$

$\therefore \qquad m = \dfrac{v}{u} = \dfrac{8}{13\frac{1}{3}} = 0\cdot 6$

3. The image in a converging lens is erect and magnified four times: (i) calculate the object distance if the focal length is 20 cm; (ii) find the object distance by drawing a scale drawing (no calculation permitted).

(i) *RP.*    $m = \dfrac{v}{u} = 4$          |          (i) *NC.*    $m = \dfrac{v}{u} = 4$

$$\therefore\ v = 4u$$

Thus if $x$ cm $= u$, $4x$ cm $= v$.

Now an erect image is a virtual one (p. 409).

$$\therefore\ v = -4x,\ u = +x$$

From          $\dfrac{1}{v} + \dfrac{1}{u} = \dfrac{1}{f}$

$$\therefore\ \frac{1}{(-4x)} + \frac{1}{(+x)} = \frac{1}{(+20)}$$

$$\therefore\ -\frac{1}{4x} + \frac{1}{x} = \frac{1}{20}$$

$$\therefore\ \frac{3}{4x} = \frac{1}{20}$$

$$\therefore\ \ 4x = 60,\ \text{or}\ x = 15$$

$\therefore$ Object distance $= 15$ cm.

---

(i) *NC.*    $m = \dfrac{v}{u} = 4$

$$\therefore\ v = 4u$$

Thus if $x$ cm $= u$, $4x$ cm $= v$.

Now an erect image is left of the lens (p. 409).

$$\therefore\ v = -4x,\ u = -x$$

From          $\dfrac{1}{v} - \dfrac{1}{u} = \dfrac{1}{f}$

$$\therefore\ \frac{1}{(-4x)} - \frac{1}{(-x)} = \frac{1}{(+20)}$$

$$\therefore\ -\frac{1}{4x} + \frac{1}{x} = \frac{1}{20}$$

$$\therefore\ \frac{3}{4x} = \frac{1}{20}$$

$$\therefore\ \ 4x = 60,\ \text{or}\ x = 15$$

$\therefore$ Object distance $= 15$ cm.

(ii) *Drawing.* Draw CQ to represent the position of the lens, and mark the focus F according to some scale. Then draw two lines LP and MQ parallel to

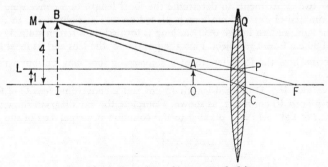

Fig. 21.14 Drawing of image and object

the principal axis at distances of 1 : 4 above the axis, using any suitable scale. Join FP and produce it to meet MQ at B. Then B is the top point of the virtual image. BI is the image. Now join BC, cutting LP at A. Then A is the top point of the object, and AO is the object. The distance OC is the object distance, which can be found from the scale used to represent the focal length, CF.

## SUMMARY

1. *The focal length of a converging lens* is the distance from the lens to the principal focus; the latter is the point on the principal axis to which rays parallel and close to the principal axis converge after refraction through the lens. The focal length is a +ve length.

2. *The focal length of a diverging lens* is the distance from the lens to the principal focus; the latter is the point on the principal axis from which rays parallel and close to the principal axis appear to diverge after refraction through the lens. The focal length is a —ve length.

3. R.P. Convention: $\dfrac{1}{v} + \dfrac{1}{u} = \dfrac{1}{f}$, $m = \dfrac{v}{u}$

   N.C. Convention: $\dfrac{1}{v} - \dfrac{1}{u} = \dfrac{1}{f}$, $m = \dfrac{v}{u}$

4. Images in a converging lens are inverted and real when the object is farther from the lens than the focal length. Nearer the lens than $f$, the image is erect, magnified and virtual (principle of magnifying glass or simple microscope).

5. Images in a diverging lens are always virtual, diminished and erect.

6. Focal length of converging lens – for quick method use window as distant object, for accurate method use a plane mirror or the lens formula.

## EXERCISE 21 · ANSWERS, p. 452

**1.** Describe an approximate method, and an accurate method, of finding the focal length of a converging lens.

An object 4 cm high is at right angles to the principal axis of a diverging lens of focal length 20 cm and 30 cm from it. Determine the position of the image and its size. (*N.*)

**2.** Explain the terms *principal focus* and *focal length* as applied to a converging lens.

Describe two experiments to determine the focal length of a converging lens.

An illuminated object 1·05 cm long is placed on and at right angles to the axis of a converging lens, and an image 0·35 cm long is formed on a screen suitably placed at a distance of 80 cm from the object. Find the position of the lens and its focal length. (*L.*)

**3.** Describe how the focal length of a converging lens may be found using a plane mirror. (*L.*)

**4.** A diverging lens D of focal length 15 cm and a converging lens C of focal length 25 cm are placed 10 cm apart, as shown. Complete the ray diagram showing the path of the beam of light which is parallel to the common principal axis of the lenses. (*N.*)

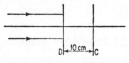

**5.** A converging lens of focal length 12 cm is used to produce a real image of an object enlarged four times. How far must the object be placed from the lens? (*N.*)

**6.** AOB represents the position of a thin diverging lens; F, F are its principal foci. Determine the position of the image of the point object P and complete the ray diagram. (*N.*)

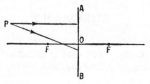

**7.** Define *focal length* of a converging lens and describe a good method of determining its value for a given lens.

A converging lens of focal length 10 cm is used to produce an image three times the size of the object. Find the distance of the object from the lens when the image is: (*a*) virtual; (*b*) real. (If the answers are obtained by calculation, state the sign convention used.) (*L.*)

**8.** Describe briefly an approximate method, and then in detail an accurate method, of finding the focal length of a converging lens.

An object 1 cm high is placed 6 cm away from a converging lens of 12 cm focal length. Find by calculation, or by drawing, the position, height and nature of the image. (*N.*)

**9.** Explain the meaning of *real image*, *virtual image*.

Draw scale diagrams to show the passage of a pencil of rays from a point on a small object, through a converging lens of focal length 10 cm for the following distances: (*a*) 15 cm; (*b*) 5 cm; (*c*) 10 cm. The point must not be on the axis of the lens.

In the case (*b*) deduce the magnification, and state the nature of the image. (*O.* and *C.*)

**10.** Define *principal focus* and *focal length*. Describe how you would determine the focal length of a thin converging lens with the help of a plane mirror.

Draw a diagram showing how a thin converging lens forms an image of a very distant extended object, such as the sun.

The image of the sun formed by a lens of focal length 1·8 m is 1·7 cm in diameter. Taking the distance of the sun to be 150 million km, calculate the sun's diameter. (*O.*)

**11.** Two converging lenses, A and B, have their centres 17·0 cm apart and a common principal axis. On this axis is a point source of light P which is 4·0 cm from A and 21·0 cm from B. The light from P is focused 12·0 cm from A and, after passing through B, it merges as a parallel beam. Draw a ray diagram to illustrate this, and find the focal length of each lens.

Draw a ray diagram to show the effect of moving P 0·8 cm from the principal axis. Explain the fact that the distance moved, as viewed through B, appears magnified. (*C.*)

**12.** Describe how you would find the focal length of a converging lens using an illuminated object and a plane mirror. Mark the focal length on a diagram of the arrangement.

Draw diagrams to show how a converging lens focuses: (*a*) a parallel beam of light parallel to the principal axis; (*b*) a parallel beam of light inclined at a small angle to the principal axis.

A pin-hole camera is used to photograph the sun. Calculate the diameter of the image on the screen placed 100 cm from the pin-hole, if the angle subtended by the sun is 30 min ($\frac{1}{2}$°). What difference would there be in the image if the pin-hole were replaced by a circular hole of about 1 cm diameter? What further effect on the image would result from inserting a converging lens of focal length 100 cm at the circular hole? (*C.*)

# 22

# APPLICATIONS OF LENSES

## The Eye

The eye is one of the most intricate and sensitive instruments devised by nature. Its chief optical features are (Fig. 22.1):

(1) the *eye-lens*, which focuses light entering the eye;
(2) *ciliary muscles*, which are attached to the eye-lens surfaces and alter the focal length;
(3) the *retina*, the light sensitive area of cells at the back of the eye;

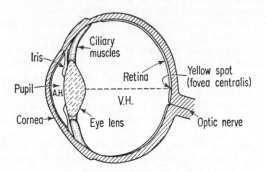

Fig. 22.1 Optical features of Eye

(4) the *yellow spot* (*fovea centralis*), the most light sensitive spot on the retina;
(5) the *iris*, the coloured circle round the eye-lens;
(6) the *pupil*, the circular opening or diaphragm in the iris through which light passes;
(7) the *cornea*, the thick protective transparent covering material in front of the eye-lens which refracts the light most:
(8) the *aqueous humour* (A.H.) and *vitreous humour* (V.H.), which are respectively liquids in front of and behind the eye-lens in which the eye-lens floats.

## Binocular Vision

If you close one eye and attempt to pick up a pen or other object at the corner of the table you will find that a sense of distance is lacking compared with using two eyes. Binocular vision provides two images of the same object which are slightly different in perspective. The brain combines the two images and gives an impression of depth or solidity which is missing if only one eye is used. In Fig. 22.2(i), for example, a view of a pyramid P is seen slightly different by the left eye L and the right eye R, as indicated in (ii). Two retinal images of slightly differing perspective are always necessary to provide depth or a stereoscopic effect. A photograph appears flat because only one image is formed by the camera lens. A red and blue superimposed image, taken with a slightly differing perspective,

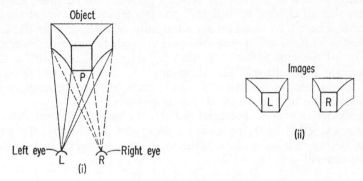

Fig. 22.2 Binocular vision

and each viewed simultaneously with a red filter in front of one eye and a blue filter in front of the other, will look three-dimensional.

## Position of Eyes

Binocular vision is possible only if the fields of view of the two eyes overlap. This clearly depends on the position of the eyes in the head. In small birds, such as the blackbird, the eyes are at the sides of the head. In this case each eye views space independently of the other, that is the visual fields are entirely separate, and the blackbird cannot judge distance by binocular vision. On the other hand, in birds of prey, such as the hawk or owl, the eyes have migrated towards the front of the head, so that this bird possesses stereoscopic vision. The development of binocular vision is most marked in the more advanced creatures. Man possesses this advantage to a greater degree than any other animal.

## Persistence of Vision · The Iris

When a bright light is viewed and then switched off the sensation of vision lasts for a short but definite time. This is called *persistence of vision*. If a light flickers more than 10 times per second, then, owing to persistence of vision, the light is seen constant in intensity. Cinema films are a rapid succession of bright images projected on to a screen, which merge into a continuous record of events by persistence of vision. If the film slows down by accident a flickering effect is then seen.

In a dim light, the iris or diaphragm round the eye opens more, thus increasing the size of the pupil and allowing more light to enter the eye. If a bright light is then viewed, the iris automatically closes more. This reduces the amount of light entering the pupil, thus protecting the eye from excessive brightness and the possibility of damage to the retina.

## Accommodation

The ability of the eye lens to focus points at different distances on to the retina is called its *power of accommodation*. Accommodation is produced by the action of the ciliary muscles. When a person with normal vision looks at a very distant object, at 'infinity', the eye lens focuses parallel rays on the retina. The ciliary muscles are then fully relaxed, so that the accommodation is least. When a near object is viewed, the muscles increase the curvature of the lens surfaces, so that a 'fatter' lens, one with a shorter focal length, is produced. The diverging rays are then focused on the retina. At the nearest point for distinct vision, the ciliary muscles are at maximum strain and accommodation is greatest.

With defective vision, discussed shortly, the normal accommodation of the eye is exceeded in trying to focus an object. In old age, the eye lens tends to become inelastic and unable to accommodate, and spectacles are again necessary.

## Normal Vision

People with normal vision are able to focus near objects such as reading books when they are about 25 cm or 10 in from the eye. This is the closest

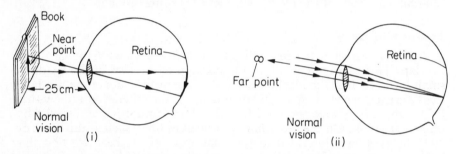

Fig. 22.3 Normal vision. (i) Near, (ii) far point

point at which the object can be seen clearly, and it is called the *near point* of the eye (Fig. 22.3(i)).

People with normal vision are able to see clearly objects which are a long distance away, or at 'infinity'. The farthest point from the eye which can be seen distinctly is called the *far point* of the eye. For normal vision, then, the far point is at infinity (Fig. 23.3(ii)).

## Defects of Vision · Long Sight

If you wear spectacles for reading, your near point N is usually farther than 25 cm from the eye. This defect of vision is therefore called *long sight*. It may be due to an eyeball which is too short. Rays from a point A 25 cm from the eye are then brought to a focus beyond the eyeball, instead of on the retina R (Fig. 22.4(i)).

Suppose the near point N is 30 cm from the eye. Then an object at N can be seen distinctly, as illustrated in Fig. 22.4(i). With the aid of refraction by a suitable *convex (converging) lens* L, light from A can be made to enter the eye *as if it came from N* (Fig. 22.4(ii)). The object at A is then seen distinctly. The converging lens thus 'corrects' the defect of vision.

The focal length $f$ of the lens L can be easily calculated if it is noted that

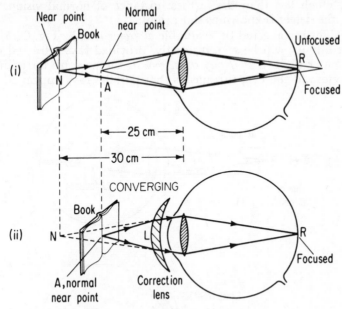

Fig. 22.4 Long sight and correction

an object at A produces a virtual image at N due to refraction through the lens (Fig. 22.4(ii)). We then have:

Object distance = 25 cm, Image distance = 30 cm.

Applying the sign convention to find $f$:

RP. 

$$\frac{1}{v}+\frac{1}{u}=\frac{1}{f}$$

$$\therefore \frac{1}{(-30)}+\frac{1}{(+25)}=\frac{1}{f}$$

$$\therefore -\frac{1}{30}+\frac{1}{25}=\frac{1}{f}$$

$$\therefore \frac{1}{150}=\frac{1}{f}$$

or $\qquad f=150$ cm.

NC. 

$$\frac{1}{v}-\frac{1}{u}=\frac{1}{f}$$

$$\therefore \frac{1}{(-30)}-\frac{1}{(-25)}=\frac{1}{f}$$

$$\therefore -\frac{1}{30}+\frac{1}{25}=\frac{1}{f}$$

$$\therefore \frac{1}{150}=\frac{1}{f}$$

or $\qquad f=150$ cm.

A converging lens of 150 cm focal length is therefore required to correct for the far sight.

### Short Sight

Some spectators are unable to see cricketers or footballers clearly unless they wear spectacles, and people with such a defect of vision must wear spectacles while driving. In these cases the eyeball is too long. Parallel rays from objects a long way off, at infinity, are then brought to a focus in front of the retina (Fig. 22.5(i)). The farthest point A seen clearly may now be 200 cm, much less than the farthest distance of normal vision. Consequently, the defect is known as short sight.

Short sight is corrected by a suitable *diverging* lens L (Fig. 22.5(ii)). An object at infinity will be seen distinctly, provided rays refracted through the lens enter the eye *as if they came from A*, the person's far point. With the lens close to the eye, this means that parallel rays appear to diverge

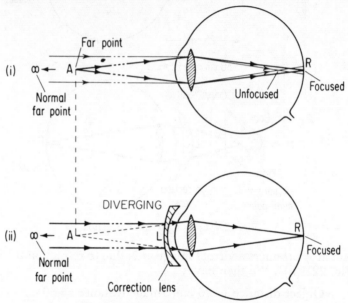

Fig. 22.5 Short sight and correction

from A after refraction, as shown. Thus A is the focus of the lens, and AL is hence the focal length. A diverging lens of focal length 200 cm is therefore required to see distant objects clearly. If another person's far point is 150 cm from his or her eye a diverging lens of focal length 150 cm is required.

With age, when the eye lens becomes inelastic and unable to accommodate, two pairs of spectacles may be necessary. *Bifocal lenses* have an upper portion for long-distance viewing and a lower portion used for reading.

### Demonstration of Defects of Vision

The defects of vision and their correction can be demonstrated with the aid of:

(1) a 5-litre flask filled with fluorescein to act as the eye and a lens of 12·5 cm focal length ($+8D$) attached to its surface by plasticene to act as an eye-lens with normal vision (Fig. 22.6);

(2) a lens of 9 cm focal length $(+11D)$ attached to another part of its surface to act like an eye-lens with short sight;

(3) a lens of 18 cm focal length $(+5·5D)$ at another part to act like an eye-lens with long sight;

(4) a diverging lens of 33 cm focal length $(-3D)$ to correct for short sight, and a converging lens of 40 cm focal length $(+2·5D)$ to correct for long sight.

('$D$' stands for *dioptres*, a convenient optical unit for focal length. The number of dioptres $D$ is calculated from $D = 100/f$, where $f$ is the focal length in centimetres.)

A strong point source of light provides a beam of light which is shone

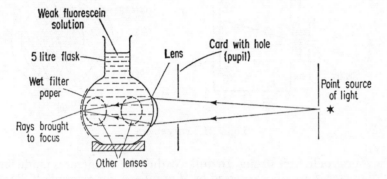

Fig. 22.6 Model eye

into the flask through the 12·5 cm or $+8D$ lens. The passage of the converging rays through the fluorescein can be clearly seen in a darkened room, and the source position is adjusted until they are focused on the far surface of the flask, which acts as the retina for a normal eye. This represents normal vision. The flask is now turned round to demonstrate short sight by the lens concerned and then its correction by adding a lens. Long sight and its correction can be likewise demonstrated.

## Lens Camera

A lens camera is a device similar in action to the eye. It consists basically of a closed box with a converging lens L, a circular opening or *diaphragm* which allows light to fall on a *light-sensitive film* at the back only through the central part of L, and a *shutter*, which allows light to fall on the lens for a short period (Fig. 22.7).

A good-quality lens is built up of several lenses to reduce image distortion. The front surface is also 'bloomed' to prevent interference from reflected light, that is, the surface is coated with a suitable thin film of a fluoride.

The amount of light falling on the film through the lens depends on: (i) the *time*, (ii) the diameter of the opening or *aperture* of the diaphragm (Fig. 22.7(ii)). Generally, the longer the time and the greater the diameter, the larger is the amount of light falling on the film.

The *exposure time* is the time for which light is allowed into the lens, and this is controlled by the shutter speed. The exposure used depends on the film sensitivity to light and on the speed of the object photographed. A fast-moving tennis player or a car must be taken with a short time exposure such as $\frac{1}{500}$ sec, otherwise a blurred image would result owing to the rapid movement. A longer exposure time, such as $\frac{1}{30}$ sec, can be used for a still group of people, such as a cricket or tennis team posing for photographs. At night or in artificial light, long exposure times such as 1 sec or more may be used.

The diameter of the aperture is made smaller in sunlight, for example,

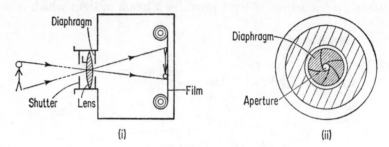

Fig. 22.7 Lens camera

as the source of light is strong. In dull weather the diameter is made larger. The diameter is usually expressed by '*f*-numbers', for example '*f*-8' means a diameter which is $\frac{1}{8}$ of the focal length *f* of the camera lens and *f*-22 is a diameter equal to $f/22$. An aperture with a smaller *f*-number has a larger diameter than one with a higher *f*-number. For good results it is necessary to use the right combination of exposure and aperture when photographing a subject at a given distance.

## Modern Camera

To judge the light intensity, professional photographers always use a *light meter* or *exposure meter*. This is an instrument containing a light-sensitive metal such as selenium which produces a small electric current when light falls on it. A pointer records the light intensity on a previously calibrated dial. Modern cameras have a built-in light-sensitive exposure meter. They may also have a built-in range-finder to allow for the distance of the subject. A rectangular frame, seen through the view-finder, indicates the area which will be photographed.

Before a photograph is taken, the speed rating or sensitivity of the film is first set on a dial in the instrument which is also connected to the exposure meter. The subject or scene is then viewed, the range-finder is correctly adjusted (the lens position is then altered slightly), the required exposure time is set and finally the necessary aperture is automatically obtained by setting two pointers to coincide. One represents the light intensity and the other a combination of exposure time, aperture and film sensitivity.

## Projection Lantern

The lens camera produces a small image on a film. The projection lantern is used for showing audiences large images of slides or other objects on a white screen. The essential features are:

(1) a very powerful small source of light S;
(2) a condenser C of two converging lenses – this collects the light and beams it towards the slide O, which is therefore strongly illuminated;
(3) a projection lens L near the slide, which produces an enlarged inverted image of O on a distant screen M.

Without the condensing lenses C, the slide would be weakly illuminated. The condenser collects light and sends it through O. A strong source of

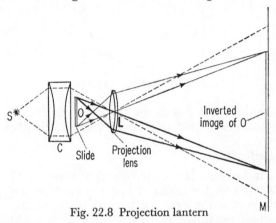

Fig. 22.8 Projection lantern

light is essential as O is non-luminous. The lens L is slightly farther from O than its focal length, so that an enlarged image is obtained on M. The image is inverted, and consequently the slide must be inserted upside down into the projector to see a picture on the screen the right way up. It should be noted that the image of the lamp filament S is produced at L itself, and hence the projection lens does not give rise to an image of the filament on the screen, which would spoil the picture.

## Visual Angle

Unlike the projection lantern, where the image is formed on a screen and then viewed, the image produced by an optical instrument, such as the microscope or telescope, is formed on the retina. It is therefore seen *directly* by the eye.

The apparent height of an object O seen by the unaided eye depends on its distance from the eye. Suppose O is 25 cm from the eye (Fig. 22.9). The image on the retina is then I, say. (A useful guide-line or ray path is one from the top of the object O, which passes through the top of the image I.) Suppose O is now moved farther back to A. A new image B is then formed on the retina. It is seen smaller than before because the layer of light-sensitive cells on the retina covered by B is less than those covered by

I. Consequently, B is smaller than I. Hence the object at O, nearer the eye, is seen smaller when it is moved to A farther away.

The height or length of an image such as I or B is proportional to the *angle* subtended at the eye-lens by the image. By vertically opposite angles, this is also *the angle subtended at the eye by the object outside it.* Thus if O subtends an angle of 2° at the eye and A an angle of 1°, then O is seen twice as tall as A. The angle subtended at the eye by an object viewed is called the

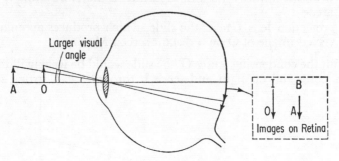

Fig. 22.9 Images on retina

*visual angle.* Its importance lies in the fact that the size of an object seen through an optical instrument is proportional to the visual angle.

### Angular Magnification

Microscopes and telescopes are used for viewing near and distant objects respectively. Viewed by the unaided eye, the visual angle, or angle subtended by the object at the eye, is extremely small. Using the optical instrument, however, the image is usually brought to a distance of 25 cm from the eye. Here it subtends a large visual angle. Consequently, the eye sees a large image. If the visual angle without the instrument is only 0·1°, and with the instrument it is 6°, the *angular magnification* or *magnifying power* is given by the ratio 6°/0·1° or 60.

### Microscope

A microscope is an instrument for magnifying close objects. In the section on lenses we have already seen that a converging lens can act as a

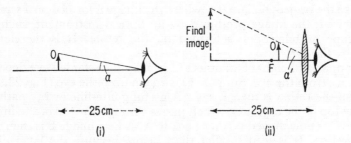

Fig. 22.10 Simple microscope (magnifying glass)

magnifying glass or *simple microscope*. This means that the angle $\alpha'$ subtended by the final image 25 cm from the eye (Fig. 22.10(ii)) is much larger than the angle $\alpha$ subtended by the object O at the unaided eye (Fig. 22.10(i)).

If a box of converging lenses of different focal lengths is taken, and the lenses in it are held close to the eye and used as magnifying glasses, experiment shows that the greatest magnifying power is obtained for the more 'bulging' lenses, or those which have *short* focal lengths. It is left to the reader to show by drawing that the largest visual angles are obtained in these cases.

## Compound Microscope

There is a limit to the angular magnification or magnifying power produced by a single lens, as it is difficult to make lenses of very short focal length. A *compound microscope*, which produces high magnification, is made with two converging lenses. It is used extensively in biological researches and in some researches in physics. Under the microscope, objects as small as bacteria, and other carriers of disease, have been magnified sufficiently for detailed examination.

## Making a Compound Microscope

EXPERIMENT

An elementary form of compound microscope can be made by illuminating a transparent graduated scale A by a lamp S, and moving a converging lens O of short focal length away from it until a clear inverted

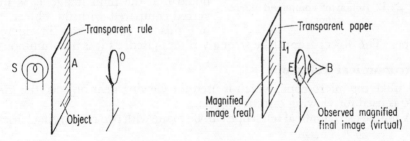

Fig. 22.11 Forming a compound microscope

image $I_1$ is obtained on a semi-transparent paper screen (Fig. 22.11). With the eye close to it a second converging lens E of short focal length is then moved near to $I_1$, until a large clear magnified image is seen through E. This lens is now acting as a magnifying glass. The paper screen can now be removed. On looking at the scale A through *both* lenses, and adjusting E if necessary, a large inverted image of A is seen. This is the final image $I_2$ shown in Fig. 22.12. Experiments with different lenses show that the greatest magnifying power is obtained when both lenses have short focal length.

## Formation of Image

Fig. 22.12 illustrates the basic principle of the compound microscope. The lens O nearer the object A is called the *objective lens*, and it is positioned

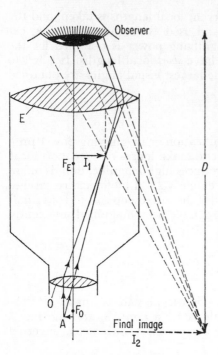

so that A is just *farther* from it than $F_O$, the principal focus. A real, inverted and magnified image $I_1$ is thus obtained. The lens E through which $I_1$ is viewed is called the *eyepiece lens*. This is moved until $I_1$ is nearer to it than its principal focus $F_E$. The lens then acts as a magnifying glass, and a magnified image of $I_1$ is hence obtained at $I_2$ (compare Fig. 22.10(ii)). The observer automatically adjusts the eyepiece so that $I_2$ is at his or her near point, 25 cm from the eye for normal vision, when it is seen most distinctly.

It can be seen from the diagram that the visual angle due to $I_2$ is much larger than the visual angle obtained if the object A were placed at 25 cm from the unaided eye in normal vision. By using converging lenses of short focal length for O and E, high magnifying power such as $\times 500$ can be obtained. It should be noted that the final image $I_2$ is inverted compared with the object A, but this is no handicap in micro-

Fig. 22.12 Action of compound microscope

scopes. The object A must be strongly illuminated, as it is non-luminous.

## Astronomical Telescope

Unlike the microscope, which is used for viewing near objects, the *telescope* is used for viewing distant objects.

A simple astronomical telescope can be made with two converging lenses.

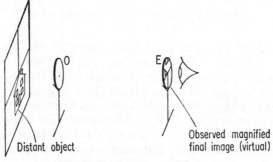

Fig. 22.13 Forming a telescope

They can be chosen from a box containing a number of lenses of long focal length, such as 50–100 cm, and short focal length, such as 5–10 cm. Place one lens O in front of a window through which a distant object can be seen (Fig. 22.13). Using semi-transparent paper, obtain a clear small inverted image on it (see Fig. 21.9(i)). Then move the second lens E away from the image, keeping the eye close to E, until it can clearly be seen magnified and inverted. Vary the focal lengths used for O and E. Experiment shows that highest magnifying powers are obtained when O has a long focal length and E has a short focal length.

### Formation of Telescope Image

The ray diagram of the image formed by an astronomical telescope is shown in Fig. 22.14. The front lens, the objective O, collects light from the distant object. It should be carefully noted: (i) that parallel rays are

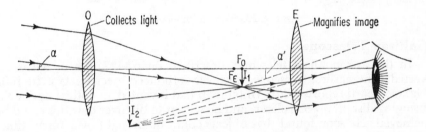

Fig. 22.14 Action of astronomical telescope

incident on this lens; (ii) that the parallel rays are intended to come from the top point or head of the distant object, and are hence inclined at a small angle $\alpha$ to the principal axis. The image $I_1$ is formed at the principal focus of the objective. To draw $I_1$, note that its head lies on the ray through the centre of the objective lens, which passes through undeviated. The eyepiece E is once more used as a magnifying glass, and is moved so that $I_1$ lies inside its focal length, or nearer to the lens than its focus. The final image $I_2$ subtends a large visual angle $\alpha'$ at the eye placed close to E, so that a large image is seen. High magnifying power occurs when the objective has a long focal length and the eyepiece a short focal length, as the reader can verify by drawing.

### Prism Binoculars

Prism binoculars are short telescopes consisting of a converging objective and a converging eyepiece, with two total reflecting prisms of 90°, 45°, 45° (see p. 396). The edges of the prisms are perpendicular to each other, and light which passes through the objective is reflected in turn from one prism to the other and then emerges through the eyepiece. This is shown in Fig. 22.15. In the absence of the prisms the image would be upside down, as shown in Fig. 22.14. One prism turns the image round and corrects for lateral inversion (p. 364), and the second prism inverts the image, as shown in Fig. 20.22, p. 396.

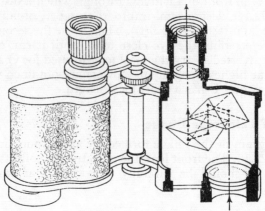

Fig. 22.15  Prism binoculars

## Galilean Telescope

In the astronomical telescope, described on p. 427, the image is inverted. The earliest telescope, however, produced erect images. In 1609 Galileo heard reports that a Dutchman had invented an optical device for bringing distant objects closer, and he set out to discover how this could be achieved. He soon found that objects could be brought over thirty times closer through a combination of a diverging eyepiece and a converging objective. He turned the telescope to the sky, and so discovered the mountains and craters of the moon and the satellites of Jupiter. The *Galilean telescope*, as it is known, is used today as opera glasses.

Fig. 22.16 shows how the erect image is formed by the two lenses. As with the astronomical telescope, the objective collects parallel rays, A.

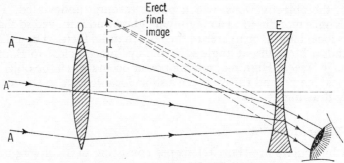

Fig. 22.16  Galilean telescope (opera glasses)

This time, however, the diverging lens intercepts the refracted rays before they can form an image at the focus of the objective. As the diverging lens produces a diverging beam, the top point of the image is above the principal axis, as shown. The eye thus sees an erect final image I.

## Hale (Reflecting) Telescope

The largest telescope in the world for astronomical observations is at Mount Palomar in California, where the atmosphere is exceptionally clear. Unlike the telescopes already described, this one consists of a huge concave parabolic mirror, 200 in in diameter, with a polished coating of aluminium on the front surface which acts as a reflector.

The mirror is made large in order to collect as much light as possible from the stars and planets to be observed, and the absence of lenses

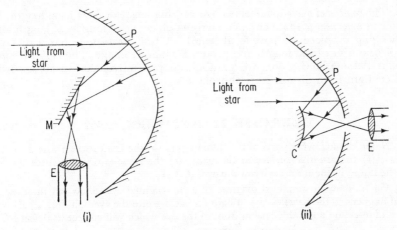

Fig. 22.17 Hale (Reflecting) Telescope

eliminates colour due to dispersion (p. 433) and other defects of images inherent in using lenses, such as the absorption of light by the glass. The idea of using a mirror for an astronomical telescope was first put forward by Newton.

Fig. 22.17(i) illustrates the path of rays from a star, for example, incident on the concave paraboloid mirror P. The light is reflected towards the focus, but is intercepted by a plane mirror M so that the light enters an eyepiece E through which it can be observed. This was suggested by Newton. Fig. 22.17(ii) illustrates the use of a convex mirror C instead of a plane mirror, a method suggested by Cassegrain.

The whole telescope system at Palomar has a special clockwork control for keeping the eyepiece trained on a particular object while the earth rotates on its axis. Photographic plates are also used in place of the eyepiece E to photograph the stars and planets. The Hale telescope continues to make a great contribution to astronomical theories of the origin of stars and to theories of the origin of the Universe.

## SUMMARY

1. The eye has an eye-lens whose focal length can be varied by ciliary muscles for accommodation, a coloured iris with a pupil whose diameter varies with the brightness of the light and a retina with a very light-sensitive spot (yellow spot).

2. Binocular vision gives a sense of perspective or solidity. Persistence of vision enables cinema films to be seen as if action was continuous. Accommodation is the ability of the eye lens to focus points at different distances.

3. The near point for normal vision is about 25 cm from the eye; the far point for normal vision is at infinity. Short sight (a far point nearer than infinity) is corrected by a suitable diverging lens; long sight (a near point farther than 25 cm) is corrected by a suitable converging lens.

4. The length of the image on the retina is proportional to the angle subtended at the eye by the object.

5. The *compound microscope* has two converging lenses of short focal length. The simple *astronomical telescope* has a converging objective of long focal length and a converging eyepiece of short focal length. A *Galilean telescope* has a diverging eyepiece of short focal length and produces an erect image. The *Hale telescope* at Mount Palomar has a concave parabola-shaped mirror 200 in across, which collects light from distant stars and planets.

## EXERCISE 22 · ANSWERS, p. 452

1. Name the following parts of the human eye: (*a*) the front surface where the light enters; (*b*) the sensitive surface at the back; (*c*) the muscles which control the lens; (*d*) the liquid in the front cavity of the eye. (*C*.)

2. Fig (i) shows a simplified diagram of a shortsighted eye. Show, on the diagram, what happens to the parallel rays of light on passing into the eye. On Fig. (ii) show the effect of inserting a suitable lens in front of the eye which will correct its defect. (*C*.)

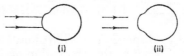

(i)                    (ii)

3. Describe a simple photographic camera and explain how it works.

State TWO similarities and TWO differences between the optical arrangements of the camera and those of a human eye.

A person whose least distance of distinct vision is 20 cm uses a converging lens of focal length of 5 cm as a magnifying glass. What is the magnification when the lens is held close to the eye and the image is formed at the least distance of distinct vision? (*N*.)

4. Give a labelled diagram of the human eye (no description is necessary). How does the eye adjust itself to deal with (*a*) light of varying intensity, (*b*) objects at different distances?

Young children often read comfortably with the book very close to the face. Why is this? Why does the sight of such children often become normal as they grow older? (*L*.)

5. What is meant by the *magnification* produced by a lens?

Deduce the magnification produced by a lens in terms of the distances of an object and its image from the lens.

Draw ray diagrams to illustrate the use of a lens: (*a*) to correct long sight; (*b*) as a magnifying glass.

What spectacles are required by a person, whose minimum distance of distinct vision is 100 cm, to enable him to read a book at a distance of 25 cm? (*O*. and *C*.)

6. State the conditions necessary to obtain a real, enlarged image of an object by a single lens.

Describe, with the aid of a diagram, the action of a projection lantern.

Explain how a converging lens may produce some dispersion of white light.

The focal length of a watchmaker's lens is 4·0 cm and the lens is 3·0 cm from his eye. What magnification will be obtained when he views the inside of a watch held 6·5 cm from his eye? (*N.*)

**7.** You are supplied with converging lenses of focal length 50 cm, 5 cm and 0·5 cm. Which would you use as: (*a*) the objective of a compound microscope; (*b*) the eyepiece of a compound microscope; (*c*) the objective of a telescope? (*N.*)

**8.** Name the type of lens used as: (*a*) the eyepiece of a Galilean telescope; (*b*) a spectacle lens to correct short sight; (*c*) a magnifying glass. (*N.*)

**9.** Explain, with ray diagrams, the use of a lens (*a*) as a magnifying glass, (*b*) in a camera. State the characteristics of the image formed in each instance.

The distance between the film and the lens in a camera is 6·25 cm when it is focused on an object 1·5 m from the lens. How far must the lens be moved, and in what direction, in order to focus the camera on distant objects? (*L.*)

**10.** A converging lens of focal length 5 cm is used to produce on a screen an image of a film with a magnification of 3. What are the distances of the screen and the film from the lens? Draw a ray diagram showing how the image is formed. (If the results are obtained by calculation the sign convention used must be stated.)

Give a labelled diagram showing the structure of a projection lantern. Draw the paths of suitably chosen rays to illustrate the action of the instrument. Give two reasons why the image produced is said to be real and account for any slight coloration which is sometimes seen although white light has been passed through a black and white slide. (*L.*)

**11.** With the aid of a labelled diagram describe the construction and explain the action of a compound microscope.

State one way in which a compound microscope could be modified to give greater magnification.

A magnifying glass gives a five times enlarged image at a distance of 25 cm from the lens. Find, by ray diagram or by calculation, the focal length of the lens. (*N.*)

**12.** A slide 7·5 cm × 7·5 cm is used in a projection lantern to produce a picture 180 cm square on a screen in a lecture hall. The slide is 30 cm from the lens. What kind of lens is used? What can you say about: (i) its focal length; (ii) the length of the hall? (*O. and C.*)

**13.** State the theoretical formula connecting the distances of the image and object from a converging lens with its focal length. How may it be proved experimentally for real images?

If you were desiring to make an astronomical telescope of magnifying power 8, what kind of lens would you purchase for the object glass and the eyepiece respectively? What focal lengths would you select? Draw a typical ray diagram showing the path through this instrument of two parallel rays incident at an angle with the principal axis. (*O. and C.*)

**14.** How would you find experimentally the focal length of a converging lens? Give full details and also explain clearly how you would obtain the result from your readings.

You are provided with two converging lenses of focal lengths 5·0 mm and 50 mm. Show on a diagram how you would place the lenses to make a compound microscope. Show also on the diagram the approximate positions of: (i) the focal points; (ii) the object; (iii) the images. Describe the character of the final image. (Calculations are not required.) (*C.*)

# 23

# COLOURS OF LIGHT · THE SPECTRUM

### Colours in White Light

When a ray-box R, with a single slit O emitting a narrow beam or 'ray' of white light, is placed opposite a white screen S, a white image of the slit

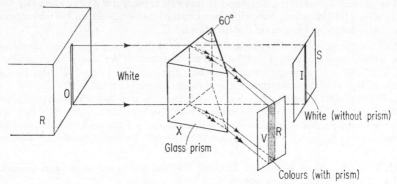

Fig. 23.1 Dispersion by prism

is seen on S (Fig. 23.1). A 60° glass prism X is now placed to intercept the white beam, so that refraction occurs. When the screen S is moved round so that the light emerging from the prism falls on it, a *coloured image* is now observed on S. The edges are red and violet respectively. Further, the coloured image is appreciably wider than the narrow white beam of light seen originally on S.

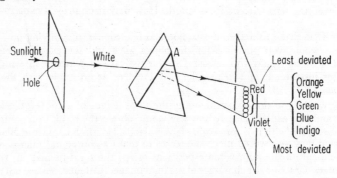

Fig. 23.2 Newton's experiment

The prism experiment outlined here was first performed by Sir Isaac Newton in Cambridge in 1666. He made a small circular opening in a shutter in a darkened room and placed a prism near the hole, so that the light was refracted on to the opposite wall (Fig. 23.1). The spectrum pro-

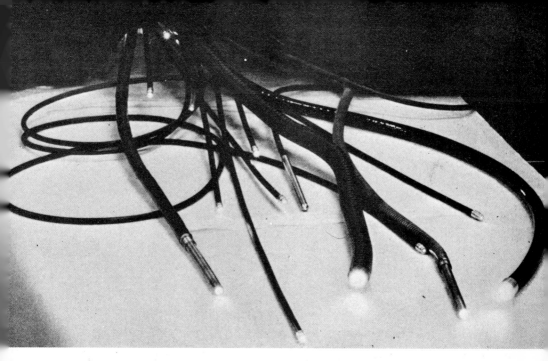

9(a) Flexible light guides of various shapes. A common light source (hidden) is used at one end for all the guides (p. 396).

PLATE 9
Optical Guides –
Refraction of Light

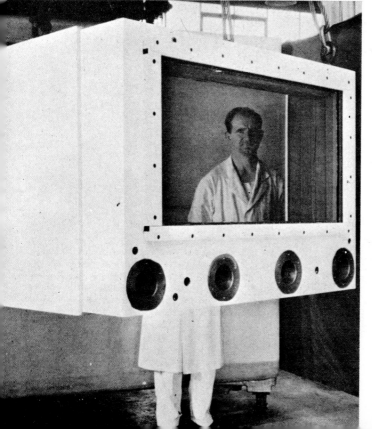

◀ 9(b) Refraction through glass. A technician at Harwell, England, standing behind a very thick block of glass (about one metre thick) of a protective window about to be fitted in front of a dangerously radio-active chamber (p. 383).

PLATE 10
**Radio- and Optical
Telescopes**

**10(a)** Radio-telescope at Parkes Observatory, New South Wales, Australia (p. 323).

◀ **10(b)** Manufacture of Isaac Newton 2·5 metre reflector telescope, for Royal Greenwich Observatory, England. Lowering the Pyrex disc into place on the polishing machine, where it is polished to an accuracy of one-millionth of a cm. The concave front surface will have a depth of 5 cm and a reflecting coating of aluminium (p. 429).

duced in this way, although impure, was composed of the following colours, in order from the apex side A of the prism:

*Red, Orange, Yellow, Green, Blue, Indigo, Violet (ROYGBIV).*

### Dispersion and Recombination with Prism

Newton experimented further to find out whether the colouring was imposed on the white light by the prism, or whether the colouring was already in the white light *before* it passed through the prism. We can follow his experiment by placing another 60° glass prism Y, similar to X, to intercept the colours further (Fig. 23.3(i)). On moving S round, the

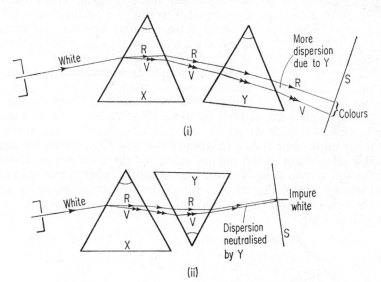

Fig. 23.3 Two prisms. (i) Dispersion, (ii) no dispersion

colours are now seen to be more widely separated than before, but are still respectively red, orange, yellow, green, blue, indigo and violet. The prism Y does not therefore create colours.

The prism Y is now turned round so that the angle enclosing the refracting sides points the other way to that of the prism X and the sides of the prisms are parallel (Fig. 23.3(ii)). When the screen S is moved it is seen that the image on S is *white*. The prism Y has now neutralized the colour effect produced by X.

Newton therefore concluded that the colours were present in white light. The glass prisms simply serve to separate them; they travel in different directions in glass. We say that a prism produces *dispersion* of the colours in white light. A single lens also produces a coloured image of an object.

Dispersion is due to the different speeds of all the colours in glass. Each colour in a ray of white light is then refracted in a slightly different direction on entering a glass prism (see p. 382). Since red is deviated least from its original direction in air, its change of speed on entering glass is least. Hence red light travels in glass with the greatest speed. As violet light is deviated most on entering the glass, it has the slowest speed in glass. In a vacuum, all colours travel with the same speed.

### Newton's Colour Disc · Adding Coloured Lights

In order to see the results of adding colours, Newton constructed a circle

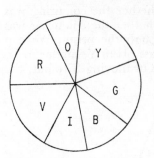

and divided the circumference into seven parts. The seven colours were arranged in the sectors in a similar way to Fig. 23.4. If a disc is painted in this way and spun fast, its colour seems to be grey-white. As the disc slows down the individual colours are seen again. This experiment confirms that 'white' is a colour due to a blend of seven main colours. The sun, for example, is yellowish-white, so that more yellow than any other colour is present in sunlight.

Fig. 23.4 Newton colour disc

Another simple demonstration of colour combination can be made with the aid of plane mirrors and blue and red filters on the sides of a ray-box and a green filter in front (Fig. 23.5(i)). Three coloured images, each of a wide slit in the ray-box, can then be projected beside each other on a white screen in front of the box. On turning the mirrors to deviate the blue and red images more, the three colours overlap. The single image is then seen to be an impure white. These three colours are called *primary colours*. Fig. 23.5(ii) shows the colours obtained by adding primary colours. Red and green lights, for example, produce yellow light (see p. 439).

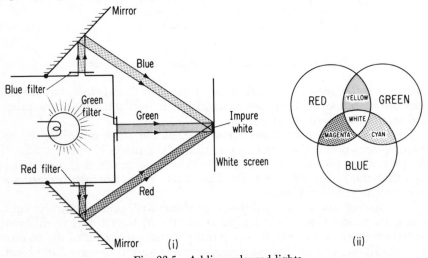

Fig. 23.5   Adding coloured lights

### Rainbow

The colours red, orange, yellow, green, blue, indigo and violet are called the *spectrum* of white light. Our natural white light is due to the sun.

The beautiful colours in the rainbow are due to dispersion of sunlight by water droplets suspended in the air after rain has occurred. A similar colour effect is produced by drops of water from a garden hose when sunlight is present.

Each drop of water is a tiny sphere. When a ray of white light is incident on a drop such as A, dispersion occurs. Fig. 23.6 shows what happens to the red and violet rays in the white light. They are internally reflected inside the drops. The coloured rays then emerge into the air again and

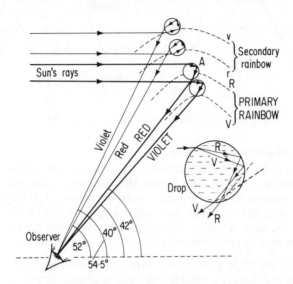

Fig. 23.6  Rainbow – action of water drops

enter the observer's eye. Most of the red rays, for example, emerge from a droplet at an angle of 42° to the line through the observer's eye in the direction of the sun. Different droplets in the sky all round this line at the same angle also produce the same colour, which enters the observe ''s eye. Consequently, an arc or bow of colour is seen in the sky. Other coloured arcs are produced, as shown in Fig. 23.6. The primary rainbow is due to light which undergoes one internal reflection inside the water drop. A secondary rainbow sometimes seen is formed by two internal reflections.

## Dispersion and Deviation

As we have already seen, prisms produce dispersion and deviation. It should be noted carefully that when a ray AO of white light is incident on a prism Y, colours are produced only at the first glass face at O (Fig. 23.7(i)). As the coloured rays, such as OR and OB, travel in a different direction in the glass compared with their original direction AO, deviation also occurs at O. When the rays are incident on the second prism face no further colours are produced, but they are deviated farther.

A rectangular slab X also produces dispersion and deviation of a white ray at its first face (Fig. 23.7(ii)). At the second face, however, the colours emerge travelling in directions parallel to the original direction of the white ray, and the colours are not so noticeable. A prism such as Y, however (Fig. 23.7(i)), produces deviation of colours at both faces. The emerging rays are then more widely dispersed. On this account, *prisms* are used in the laboratory to produce spectra.

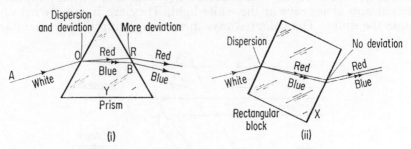

Fig. 23.7 Action of prism and rectangular block

### Formation of Pure Spectrum

Experiments or demonstrations with a prism so far described produce an *impure spectrum*, that is, one in which the different colours overlap. If a study of the colours from a hot gas in an atomic-energy experiment, for example, is undertaken, some method is required which produces a *pure spectrum*, that is, one in which the colours are separated from each other. The experimental requirements for a pure spectrum are:

(1) a *narrow slit* O as a source of light–this produces a series of narrow coloured images, which minimizes the chances of overlapping colours;

(2) a *lens* A placed with O at its focus, so that a beam of parallel light is produced;

(3) a 60° prism for dispersion;

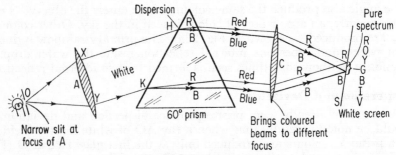

Fig. 23.8 Pure spectrum

(4) a *lens* C for collecting the parallel beams of different colours;

(5) a screen S at the focus of C on which the pure spectrum can be projected, or an eyepiece near the focus through which the pure spectrum can be observed (Fig. 23.8).

If the spectrum of white light is required a bright filament lamp can be used in front of the slit O. In Fig. 23.8 observe that the rays such as XH and YK in the incident white light are parallel, that the red rays inside the prism are hence parallel and that the red rays emerging from the prism are also parallel. The blue rays at H and K travel in different directions to the red rays, but they, too, are parallel to each other inside and outside the

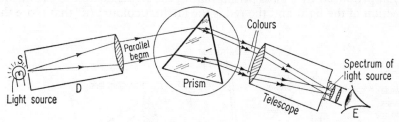

Fig. 23.9 Spectrometer

prism. A parallel red beam and a parallel blue beam, each in a *different* direction, are hence incident on the lens C. Now a lens brings parallel rays to a focus. Since the red and blue beams are in different directions, the lens forms a red image R at a different focus from that of the blue image B (see p. 406). All other colours are likewise brought to a different focus between R and B. A pure spectrum is thus formed.

The slit and lenses, together with an eyepiece, can all be housed in an arrangement called a *spectrometer* (Fig. 23.9). This instrument has a closed tube D with a lens to provide a parallel beam of light from the illuminated slit S, and a telescope which collects the coloured beams as shown and brings them to a focus, where they are observed by E.

### Fairly Pure Spectrum

If only one lens L is available, a fairly pure spectrum can be obtained by first projecting an image I of the slit S on a screen (Fig. 23.10). The beam

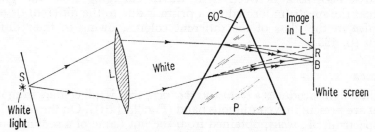

Fig. 23.10 Fairly pure spectrum – single lens

converging to I can then be intercepted by a prism P, which brings the coloured beams to a different focus, as at R and B. The screen is moved round if necessary. The best spectrum is obtained when P is placed symmetrically with respect to the directions of the incident and emerging beams. The spectrum is then formed as near to I as possible.

## Spectrum with Diffraction Grating

Fig. 23.11 shows how a spectrum is formed by a *diffraction grating*. The grating, described on p. 334, takes the place of the prism in the arrangement shown in Fig. 23.9, so that parallel light falls on the grating. A line filament (motor headlamp bulb) may be used as a source. Observe: (i) that the colours in the spectrum of white light are the other way round to that produced by a glass prism, that is, the *blue* is nearest to the incident direction of the light and the red is the farthest colour; (ii) that more than

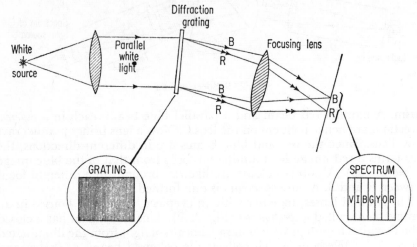

Fig. 23.11 Spectra by diffraction grating

one spectrum may be obtained and may overlap with each other—a grating with say ten thousand lines per centimetre or higher is a convenient grating to use; (iii) that the spectrum of colours is due to constructive *interference* between light waves emerging from the grating, whereas the spectrum formed by a prism is due to the different degree of *refraction* in the glass of the different colours, owing to their different speeds (p. 433).

## Spectra

White light produces a *continuous spectrum*, that is, all colours from red to violet are present in a continuous band (Fig. 23.12(i)). On the other hand, the spectrum of *sodium*, obtained from the hot flame of a sodium burner, consists only of a yellow line (strictly, two lines extremely close together). It is an example of a *line spectrum* (Fig. 23.12(ii)). Mercury vapour has a line spectrum of mainly green and blue lines. Hydrogen gas which glows has a line spectrum consisting of many colours (Fig. 23.12(iii)). Carbon dioxide gas has a *band spectrum*, consisting of many bands of fine lines.

The spectra of hot gases or metals is due basically to energy changes in their atoms or molecules. Like a miniature broadcasting station, which sends out radio waves, the energetic atoms or molecules radiate light as they lose energy.

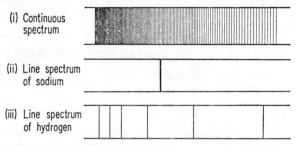

Fig. 23.12 Types of spectra (*diagramatic*)

## Colours of Objects

An object such as paper, which is white in daylight, reflects all the colours of the spectrum, red, orange, yellow, green, blue, indigo, violet. When it is illuminated in the dark room by blue light, therefore, the paper appears blue. If illuminated by red light it appears red. An object which is black in daylight absorbs all the colours. Thus, if it is illuminated by blue light only it appears black. A red rose has this colour because it absorbs all the colours of white light except red. Consequently, the rose appears black if illuminated in a dark room by blue light. Similarly, a blue tie absorbs all the colours in white light except blue, and hence looks black if illuminated by yellow light.

## Additive Colour Mixing

White light can be produced by a mixture of red, green and blue light. This can be demonstrated by means of three lanterns containing red, green and blue filters. When the three beams of suitable intensity fall on a white screen, whiteness is observed where they cross. If one lantern is now removed, say that emitting the red beam, a blue-green colour called *cyan* remains. This is known as the *complementary colour* of red light. In general, any two colours are said to be complementary if they add to produce white. The complementary colours of green and blue are called respectively *magenta* and *yellow* (see p. 434).

In such experiments it is found that all colours can be produced by suitable mixtures of the red, green and blue lights. These three colours are therefore called *primary*. Others such as yellow are called *secondary colours*. 'Additive colours' are those obtained by adding coloured lights.

## Principle of Colour Television

Additive colour mixing is also used in colour television. The light from a particular scene is split up into three coloured components, red, green and blue, by allowing it to fall on special mirrors called dichroic mirrors, built into the television camera (Fig. 23.13). These mirrors have specially prepared layers of transparent material on their surfaces for this purpose.

The three colours obtained are then incident on three separate camera tubes (see p. 697). This produces a corresponding electrical signal which is transmitted to distant colour television receivers, where the colours are reproduced on the screen by three electron beams and add together.

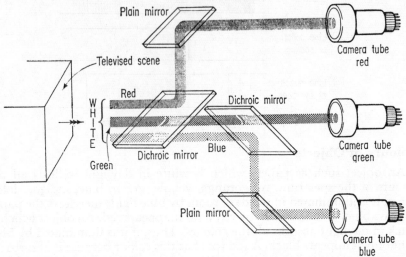

Fig. 23.13  Principle of colour transmission

### Subtractive Colour Mixing

Although the mixing of blue and yellow *lights* creates whiteness, the mixture of blue and yellow *pigments* produces a green. This is because the results of paint mixing are due to a subtractive process. Blue paint absorbs all light except blue and green. Yellow paint absorbs all light except yellow, orange and green. The only light reflected by *both* yellow and blue pigments is green, and this is the resulting colour seen.

Red and blue paints produce purple. Red and yellow paints produce orange. Red, blue and yellow paints produce black. In processing colour film, dyes are used which produce colour by a subtractive process.

### Infra-red and Ultra-violet Rays

The sun also emits rays which are invisible, but which produce the sensation of heat. They are called *infra-red rays*. Soon after, it was found the sun emitted invisible rays beyond the violet end of the spectrum. These cause certain materials to fluoresce and also affect photographic plates. They are known as *ultra-violet rays*.

Radio waves, infra-red rays, the visible rays from red to violet, ultra-violet rays and X-rays are all *electromagnetic waves*. They differ only in having different wavelengths (Fig. 23.14). Radio waves from the British Broadcasting Corporation stations may have a wavelength of about 330 m, and special transmitters can generate wavelengths a few centimetres long, called 'microwaves' (see p. 333). Infra-red rays may have wavelengths of about $\frac{1}{1000}$ cm, yellow light has a shorter wavelength of about

$\frac{1}{16000}$ cm and ultra-violet rays may have very short wavelengths of about $\frac{1}{200000}$ cm. X-rays have wavelengths hundreds of times shorter than ultra-violet rays. Fig. 23.14 shows very roughly the family of electromagnetic waves; there is no sharp boundary between the various waves, and over-lapping occurs.

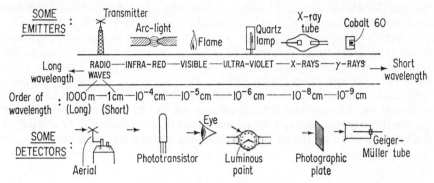

Fig. 23.14 Electromagnetic waves

## Detection of Rays

The presence of infra-red rays can be detected in the laboratory by allowing them to fall on a thermopile (see p. 315). An electric current then flows in a sensitive galvanometer connected to the instrument. Some transistors (p. 714) are very sensitive to infra-red rays and produce an electric current in a galvanometer when exposed to these rays. Ultra-violet rays can be detected by the liquid fluorescein or mineral lubricating oil, which fluoresce when exposed to the rays, or by some photoelectric

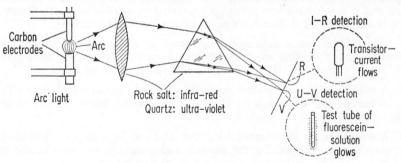

Fig. 23.15 Infra-red and ultra-violet rays

cells sensitive to these rays, which then produce a current in a galvano-meter. See p. 697.

Both infra-red and ultra-violet rays can be produced in the laboratory with the apparatus shown in Fig. 23.15. An arc light is used as the source of light, which emits these rays in addition to those of the visible spectrum. Ordinary glass absorbs infra-red and ultra-violet rays. Consequently, the lens and prism should be made of rock salt for detection of infra-red rays and of quartz for detection of ultra-violet rays.

## Uses of Rays

Infra-red rays are not scattered by fine particles as much as rays in the visible spectrum. Clear photographs of scenes have been taken through haze or mist by using camera lenses with filters to allow only infra-red light to pass through and photographic plates sensitive to infra-red light. In the Second World War aeroplanes were equipped at night with infra-red cameras to see enemy planes, and to photograph enemy territory.

A mercury-vapour lamp with a quartz envelope is a source of ultra-violet rays. Dark spectacles must be worn when the body is exposed to the radiation for health purposes, as the rays are dangerous to the eyes. Costumes on the stage, with fluorescent materials on them, produce spectacular effects when the theatre is darkened and ultra-violet lamps are switched on. Ultra-violet light is often used in the laboratory in place of visible light to photograph an object through a microscope, as a sharper definition and more detail is obtained with shorter wavelengths.

## SUMMARY

1. The spectrum of white light consists of red, orange, yellow, green, blue, indigo and violet colours.

2. A prism produces dispersion and deviation of colours at its first face, and further deviation of colours at its second face.

3. Colours produced by a prism can be recombined by using another prism with its angle pointing in the opposite direction. Newton's colour disc can be used for adding colours.

4. A pure spectrum of white light can be formed by using a narrow slit, two lenses and a prism. One lens produces a parallel beam of white light, the other collects the emerging different coloured beams, which are parallel.

5. White light has a continuous spectrum of all colours. Sodium has a characteristic yellow-line spectrum. Mercury vapour has an orange-, green- and blue-line spectrum. Hydrogen has a line spectrum of all colours.

6. A red flower looks red when exposed to red light because it absorbs all the colours except red. For the same reason, it looks black when exposed to blue light.

7. A white screen illuminated by green and red light appears yellow–the two colours are reflected by the screen, and yellow is the sum of the two colours.

8. Pigments (paints) produce a colour by a subtractive process. Blue paint absorbs all the colours except for blue and green; yellow paint absorbs all the colours except for yellow, orange and green. When the paints are mixed a green colour is hence obtained.

9. Ultra-violet, infra-red and visible rays, together with radio waves and X-rays, are all electromagnetic waves with different wavelengths. Ultra-violet rays have shorter wavelengths than visible rays and produce fluorescence. Infra-red rays have longer wavelengths than visible rays and produce the sensation of heat.

## EXERCISE 23 · ANSWERS, p. 452

**1.** What is meant by *refractive index*? Illustrate your answer with a diagram.

Draw a diagram to show the path of a narrow parallel beam of red light refracted through a triangular glass prism and use the diagram to explain the term *deviation*.

Draw another diagram to indicate what happens if white light is used instead of red, and use this diagram to show that the refractive index of the glass is less for red than for blue light. (*L.*)

**2.** Draw a diagram to illustrate the deviation and dispersion of a ray of white light by a triangular glass prism. (*N.*)

**3.** Explain why a mixture of blue and yellow pigments appears green when illuminated by white light. What is the appearance of the mixture if viewed through a sheet of red glass? (*N.*)

**4.** (*a*) What is a pure spectrum?

(*b*) In the formation of the spectrum of white light by a prism: (i) Which colour is deviated least? (ii) Which colour is deviated most? (*N.*)

**5.** Explain the meaning of the term *deviation, dispersion, pure spectrum.*

Describe how a pure spectrum of white light can be projected on a screen, using two lenses and a prism. Draw a ray diagram to illustrate your answer.

Why does the colour of a body vary with the colour of the light in which it is viewed? (*O. and C.*)

**6.** Distinguish between *deviation* and *dispersion* of a beam of light. Draw a diagram of an optical arrangement suitable for obtaining, on a screen, the spectrum of the light from a slit illuminated by an electric lamp, and show the paths of two rays from a point on the slit to the screen. State the adjustments necessary to produce a pure spectrum.

Describe and explain the appearance of a red tie with blue spots when observed in (*a*) red light, (*b*) green light. (*L.*)

**7.** What is meant by a secondary colour? Describe an experiment to demonstrate the production of a secondary colour. (*N.*)

**8.** Describe the appearance of two strips of cloth when a spectrum of white light is projected on to them, if one of the strips is pure yellow and the other a mixture of pure yellow and pure green.

Two football teams have striped jerseys, the first pure green and white and the second black and pure yellow. They were to play a match on a ground floodlit with mono-chromatic yellow lamps, until it was realized that both team's jerseys would appear to be of the same colours. Describe and explain the appearance of the jerseys in the yellow light. (*C.*)

**9.** Describe how a pure spectrum from a white-light source may be formed on a screen. How would you show that the colours of the spectrum can be recombined to give white light?

State what you would expect to observe if the spectrum were cast on a sheet of red paper. Giver a brief explanation of your statement. (*O.*)

**10.** Shadows of a vertical rod are cast on a vertical white screen by two small coloured lamps, one red and one green, placed at either side of the perpendicular from the screen to the rod. Explain, with a diagram, the positions and colours of the shadows which do not overlap. What will be the colour of the remainder of the screen? (*N.*)

**11.** Draw a labelled ray diagram to show the production of a pure spectrum of white light.

On looking into a thick plate of red glass, two images are seen of a small source of white light which is on the same side of the plate as the eye. Explain how these images are formed, and to illustrate your answer sketch the path of a single ray of light from the source. What would be seen if these images were then viewed through a thin sheet of blue glass? (*N*.)

**12.** (*a*) What are complementary colours? (*b*) Name TWO complementary colours. (*N*.)

**13.** Explain the terms *dispersion* and *pure spectrum* with reference to white light passing through a prism. Explain how you would arrange to produce on a white screen a pure spectrum of the light of an electric filament lamp. Draw a ray diagram showing the passage of red and blue rays from lamp to screen, and mark on your diagram whereabouts on the screen you would expect to find ultra-violet and infra-red rays, assuming that they were present. Name ONE material which allows ultra-violet rays to pass through it. (*O.* and *C.*)

**14.** (*a*) Define *refractive index*.

Two of the sides, AB and AC, of a glass prism meet at an angle of 20°, and the side AC is silvered. A ray of magenta light (a mixture of red and violet) is incident normally on the side AB and after reflection from AC the light emerges from AB. Draw a sketch (which need not be to scale) to show the course of the light, and show clearly where any dispersion occurs. Assuming that, at the point where the light finally leaves the prism the angle of refraction (in the air) for violet light is twice the angle of incidence, deduce the refractive index of the glass for violet light.

(*b*) Name the sequence of colours to be seen on a white screen which is receiving the spectrum of radiation from a white-hot source. Describe and explain the changes in the appearance on the screen if it is coated with a red powder which fluoresces green in ultra-violet radiation. At what place on the screen could infra-red radiation be arriving and how would you verify your answer by experiment? (*C.*)

# 24

# LUMINOUS ENERGY

## Photometry

It has already been mentioned that light is a form of energy which stimulates the sense of vision. The amount of energy reaching the eye is an important factor in everyday life. We can read at night, and often in the daytime in winter, only by light energy emitted by lamps. Motorists and the general public are dependent for safety on proper lighting at night. The illumination engineer plays a very important part in lighting design, both indoor and outdoor. He deals with measurements on light or *luminous energy*, that is, energy producing the sensation of vision, and this is a subject called *photometry*. Factories and operating theatres in hospitals, for example, must have minimum standards of illumination by law in Great Britain.

## Luminous Intensity of Source

The filament of an electric lamp is a common source of luminous energy. The luminous energy emitted per second is given the name of *luminous flux*, the word 'flux' meaning a 'flow' of energy. The luminous flux from the lamp filament spreads all round in space in three dimensions (Fig. 24.1). We therefore say it spreads through a *solid angle*. On the agreed system of measurement, the solid angle in space all round a point is $4\pi$ units; thus one unit of solid angle is $1/4\pi$ or about 8% of the total solid angle all round a point.

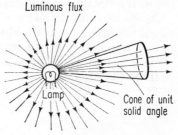

Fig. 24.1 Luminous intensity

The *luminous intensity* of a source of light is defined as: *the luminous flux emitted per unit solid angle*, or *the luminous energy emitted per second per unit solid angle*. The luminous intensity of a lamp is thus a measure of the light energy per second emitted by it. Experiments shows that it varies with the direction relative to that of the luminous lamp filament.

An electric lamp has only a moderate light intensity because all the energetic atoms in the metal filament radiate waves which are not coherent (see p. 339).

It is interesting to note, however, that in the *laser* (*l*ight *a*mplification by *s*timulated *e*mission of *r*adiation) millions of atoms in a ruby (solid state laser) or in a helium–neon gas mixture (gas laser) are stimulated to radiate coherent light waves. The total light intensity of the beam produced is then enormously high.

## Standards and Units of Luminous Intensity

Years ago, before photometry was put on a proper scientific basis, the luminous intensity of a source of light was expressed in terms of a unit defined by reference to the light energy from the flame of a candle of a certain size. The unit was called a *candle-power* (c.p.). There was an appreciable uncertainty in this standard, as the candle could not be reproduced to a high degree of accuracy. The luminous intensity of the flame of an oil lamp operating under specified conditions, named the Vernon Harcourt pentane lamp, was later agreed as a standard having 10 c.p.

As scientific instruments increase in accuracy, so scientists become more demanding in the accuracy of their units. In 1948 it was decided to adopt a unit of luminous intensity called the *candela* (*cd*), a standard based on the luminous intensity of molten platinum under specified conditions. The 'candela' is only slightly different from the 'candle-power'.

At the National Physical Laboratory in Great Britain the photometric laboratory is engaged daily in checking the luminous intensity of electric lamps and other sources of light sent to them by manufacturers. A particular lamp may be classified as 50 candelas, and one which burns more brightly as 150 candelas. The luminous intensity of a powerful searchlight lamp is of the order of many thousands of candelas. As a rough guide, a 60-watt lamp has a mean luminous intensity of 50 candelas.

### Efficiency of Lamps

The *luminous efficiency* of lamps is expressed in 'lumens per watt'. The lumen is the luminous flux emitted by a lamp of 1 cd in a unit solid angle; the watt is the unit of electrical power supplied to the lamp (p. 526).

A 100 W tungsten filament lamp (p. 527) has a luminous efficiency of about 15 lumens per watt. Fluorescent lamps, now widely used for lighting, have a much greater efficiency. A 40 W tubular lamp, for example, can provide as much lighting as a 100 W filament lamp and can be used in the average-size kitchen.

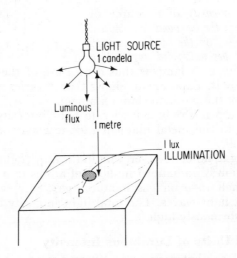

Fig. 24.2 Illumination and SI unit

## Intensity of Illumination or Illumination of Surface

We now turn from the source of light to the place or *area* on which the light falls. For example, the road surface or the walls of a room or the top of a table which are illuminated at night by lamps.

The *intensity of illumination*, or *illumination*, of a surface is defined as the *luminous flux per unit area* incident on it.

The difference between 'luminous intensity' and 'illumination' should be carefully noted. The former refers to a *source* of light, whereas the latter refers to an *area* or surface illuminated. We must therefore expect different units.

The *lux* is the SI unit of illumination. It is defined as the illumination round a point P on a surface when a lamp of 1 candela is 1 metre away from P in a direction perpendicular to the surface (Fig. 24.2).

A minimum illumination of about 150 lux is recommended for libraries, 300 lux for offices and 3000 lux for work in industry involving the handling of minute parts of watches.

## Illumination and Distance

We now see how illumination varies with distance. Suppose a small lamp O sends out a total of 1000 units of light energy per second equally in all directions in space (Fig. 24.3). A sphere A of centre O and radius 2 m has an area equal to $4\pi r^2$, where $r$ is the radius, or $4\pi \times 2^2$ m$^2$. The illumination at any part of the sphere's surface

$$= \text{Light energy per second per unit area}$$

$$= \frac{1000}{4\pi \times 2^2} \text{ units} \qquad \cdots \cdots \cdots \quad (1)$$

Suppose now we consider the surface of a sphere B of centre O and larger radius, say 4 m. The same light energy per second falls on this surface, but it is spread more thinly because the area is larger. The illumination is now

$$\frac{1000}{4\pi \times 4^2} \text{ units} \qquad (2)$$

It can now be seen, from (1) and (2), that

$$\frac{\text{Illumination at 2 m}}{\text{Illumination at 4 m}} = \frac{1/2^2}{1/4^2}$$

or, generally,   illumination $\propto 1/d^2$,

where $d$ is the distance from the small lamp.

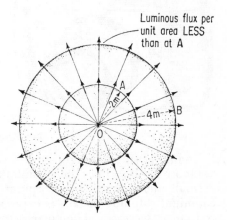

Fig. 24.3 Illumination and distance

### Relation between Illumination *E*, Luminous Intensity *I*, Distance *d*

We can now obtain a formula for the illumination $E$ of a surface at a distance $d$ away from a small lamp of luminous intensity $I$. To simplify conditions, let us assume the light always falls *normally* on the surface.

Then, at a given distance, $E \propto I$, that is, doubling the luminous intensity will double the illumination.

Also, for a given lamp, $E \propto 1/d^2$, as shown.

Combining both relationships it follows that

$$E = \frac{I}{d^2}.$$

When $I$ is in candelas, and $d$ is in metres, then the illumination $E$ is in lux. Thus a small 50 candela lamp will produce an illumination $E$ on a surface 10 m away, and at right angles to the light, given by:

$$E = \frac{I}{d^2} = \frac{50}{10^2} = 0\cdot5 \text{ lux}$$

### Light Meters

As was mentioned on p. 422, light meters are used when taking good-quality photographs. Fig. 24.4 (i) shows the front of a modern camera with a built-in light meter. The latter contains a light-sensitive surface such as selenium metal, on which the luminous energy from the subject falls. The luminous energy is converted to electrical energy and an electric current flows in the electrical meter connected to the selenium. Fig. 24.4 (ii). The magnitude of the current increases as the illumination of the light-sensitive surface increases.

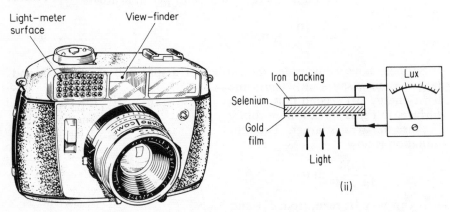

Fig. 24.4 Light meter

Cadmium sulphide light meters are widely used in photography. They are more sensitive than the selenium type. Unlike the selenium light meter, a small mercury cell is incorporated in the electric circuit. When the cadmium sulphide surface is exposed to more illumination, its electrical resistance decreases. The current in the electrical meter hence increases with increased illumination.

The scale of the light meter is calibrated with the aid of a standard lamp, say 10 cd. The lamp is placed 1 metre directly in front of the light-sensitive surface, in which case the illumination $E = I/d^2 = 10/1^2 = 10$ lux. The pointer deflection thus corresponds to an illumination of 10 lux.

## Photographic Film

Film used in cameras is coated with material which converts luminous energy to chemical energy.

Normal, or black and white, film has a thin layer of emulsion consisting of finely divided silver bromide, or other silver halide salt, suspended in gelatine. The composition of the emulsion can be varied to suit different photographic uses, for example, when photographing at night or during the day, and it is coated on to a flexible celluloid base. The silver salt is very sensitive to light, which reduces it to minute particles of metallic silver.

When a film is exposed, the brightest parts of the scene or object photographed affect the film most. The darkest parts affect the film least. Thus an image of the scene or object is 'engraved' on the film.

## Developing a Film

The film can be developed with a chemical solution which reduces the parts of the film affected to minute silver particles which are black. Those

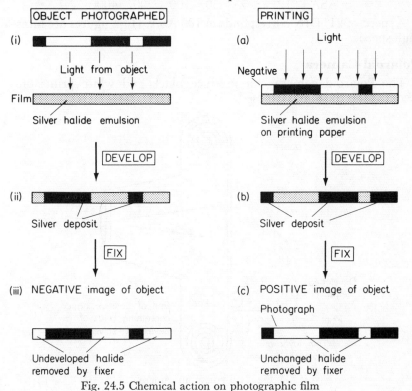

Fig. 24.5 Chemical action on photographic film

parts unaffected by the light remain unchanged. The film now carries a *negative* image of the scene photographed. Fig. 24.5. This image can be *fixed* by treating it with another chemical solution. The solution removes the undeveloped silver halide, so that the film is no longer affected by further exposure to light. A 'positive' image can be obtained by a similar process in printing, as shown in Fig. 24.5.

## Film speed

One of the most important properties of photographic film is its 'speed'. The film speed is a measure of how quickly the film will become correctly exposed when in use, that is how long some of the light-sensitive chemicals on it will take to change suitably. A 'fast' film needs a relatively short exposure time for correct exposure. A 'slow' film needs a much longer exposure time. Fast films are thus used in conditions of poor lighting. Here, the relatively long time needed with a slow film would be inconvenient. Slow films can be used in taking photographs of portraits or still objects, or in very good lighting conditions such as bright sunlight.

Commercially, the speed of a film is measured in units either in the DIN or ASA system. DIN units are continental units and are indicated with a degree (°) sign. For comparison, some values are compared in the table below:

| Din | 12°, | 15°, | 18°, | 21°, | 24°, | 27° |
| ASA | 12, | 25, | 50, | 100, | 200, | 400 |

A speed of 21° DIN corresponds to 100 ASA. High figures are films with high speeds.

## Polaroid Camera

The Polaroid camera can produce black and white prints only 10 seconds after the film is exposed.

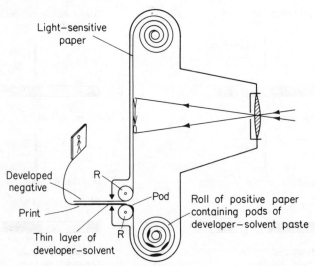

Fig. 24.6 POLAROID camera

Fig. 24.6 shows the principle of this type of camera. As in the normal camera, a sharp image is formed by the lens on the surface of a light-sensitive paper by opening the shutter for a brief period. A positive print is obtained by covering the surface of the paper with a layer of developer combined with a silver halide 'solvent', and then pressing a sheet of blank paper into contact with the sensitive surface of the paper. This is arranged in the Polaroid camera by pulling both the light-sensitive paper and the blank 'positive paper' out of the camera between two pressure rollers RR. The positive paper has small 'pods' of developer solvent paste at intervals along its length. When the paper is pulled from the camera, the pod is first burst by the rollers, which then spread the developer over the light-sensitive surface in contact with it. Simultaneously, the rollers press the blank 'positive' paper into contact with the developed surface. After 10 seconds the sensitized paper is removed, leaving the positive paper with a positive print. This may be permanently 'fixed' by covering with a film of fixer, supplied with the Polaroid film.

## SUMMARY

1. *Luminous intensity* of a source is the light energy emitted per second, or luminous flux, per unit solid angle. The unit is the candela (cd).

2. The *illumination* of a surface is the light energy falling on it per second per unit area. The unit is the *lux*.

3. Illumination $E = I/d^2$, if the light falls normally on the surface.

4. A light meter converts luminous energy to electrical energy. The current flowing increases with the illumination.

5. Photographic film is coated with a silver salt sensitive to light and converts luminous energy to chemical energy. The 'negative' image is converted into a 'positive' image in printing.

## EXERCISE 24 · ANSWERS, p. 452

**1.** How far away from a 200-cd lamp must a screen be placed, normal to the incident of light, in order to receive an illumination (or intensity of illumination) of 8 lux? (*N.*)

**2.** The intensity of illumination due to a point source of light varies inversely as the square of the distance from source to object, while that due to a searchlight is to a large extent independent of the distance. With the help of diagrams explain the reason for this difference. A small 50-cd lamp is surrounded by a globe which cuts off 40% of the light falling on it. What is the intensity of illumination received by a book on which the light falls normally at a distance of 5 m from the lamp? (*N.*)

**3.** Explain carefully why you would expect the intensity of illumination of a screen, illuminated by a small light source, to vary inversely with the square of its distance from the source. If a photographic print can be made with 4 s exposure at a distance of 4 m from a 32-cd lamp, what exposure will be required if the negative is held at 2 m from a 16-cd lamp? (*O.* and *C.*)

**4.** Describe one form of *light meter*. Explain how it works. How is the scale of a light meter calibrated?

**5.** Describe how photographic film produces a 'negative' image and how a 'positive' image is printed.

ANSWERS TO NUMERICAL EXERCISES

**EXERCISE 17** (p. 356)

**8.** 1·125 m    **9.** 0·5 cm

**EXERCISE 18** (p. 365)

**5.** 4·5 m    **7.** 180 cm

**EXERCISE 19** (p. 379)

**3.** 30, 6 cm    **5.** 10 cm    **6.** 30 cm    **7.** 12 cm, 5    **8.** 17·3 cm

**EXERCISE 20** (p. 400)

**2.** 1·8 m    **4.** 5·5 cm    **5.** 48·6°    **7.** 4/3    **8.** (b) 15 cm; 10 cm    **10.** (ii) 37·3°    **11.** 1·41
**13.** 1·56    **14.** 28°

**EXERCISE 21** (p. 414)

**1.** 12 cm from lens, 1·6 cm    **2.** 60 cm from object, 15 cm    **5.** 15 cm    **7.** (a) $6\frac{2}{3}$ (b) $13\frac{1}{3}$ cm
**8.** 12 cm, virtual, 2 cm    **9.** (b) 2    **10.** 1·4 million km    **11.** $f$A = 3, $f$B = 5 cm
**12.** 0·87 cm

**EXERCISE 22** (p. 430)

**3.** 5    **5.** $f = +33\frac{1}{3}$ cm    **6.** lin. mag. = 8    **7.** (a) 0·5 (b) 5 (c) 50 cm
**9.** 0·25 cm nearer film    **10.** 20, 6·7 cm    **11.** $6\frac{1}{4}$ cm
**12.** Converging (i) 28·8 cm (ii) 7·5 m at least

**EXERCISE 23** (p. 443)

**14.** 1·53

**EXERCISE 24** (p. 451)

**1.** 5 m    **2.** 0·8 lux    **3.** 2 s

# Sound

# 25

## BASIC PRINCIPLES

When a bicycle bell is rung and then lightly touched it can be felt to be vibrating, i.e. it is moving to and fro continuously through a small distance. This vibratory movement is common to all objects emitting a sound, such as the violin, the piano, the organ-pipe and the radio loudspeaker.

A sounding object vibrating in this way has a certain amount of energy. This can be demonstrated by touching a suspended piece of cork X with a prong of a sounding tuning fork F. The cork moves through a surprisingly large distance to Y as a result of the impact of the prong (Fig. 25.1(i)).

### Sound Vibrations

Fig. 25.1(ii) illustrates a method of showing the vibrations due to a sounding tuning fork. The fork F has a bristle B attached by wax to one of

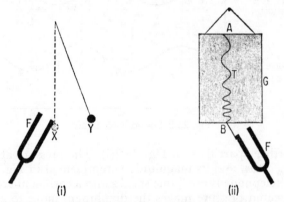

(i)                                     (ii)

Fig. 25.1 Energy and vibration of tuning fork

its prongs, and B presses against the lower part of the face of a vertical glass plate G which has been covered with lamp-black. The fork F is fixed in position, and when it is sounded by means of a bow the prongs vibrate horizontally and the bristle B marks a visible horizontal line on G. The plate G is suspended from a nail by a thread A, and when the fork is next sounded the thread is burnt at the top. The downward movement of the plate results in a visible trace T of the horizontal vibrations of the bristle, and the experiment has resulted in the 'writing of sound' (see also page 477).

### Sound Wave Pressure Variation

As explained on p. 328, to which the reader should refer, a vibrating tuning-fork produces a *sound wave* in air by setting layers of air into vibration and the velocity $V$ of the wave is related to the frequency $f$ of vibration and the wavelength $\lambda$ by $V = f\lambda$. In addition to the changes in the layers' positions (see p. 328), there is a continuous variation of *pressure*. The layers which are crowded together at a place in the air make the pressure there a little above normal at that instant, and this effect is known as a *compression*. In Fig. 25.2(i), (ii), which represents the actual positions of the layers, $c$ represents a compression of the air; compressions are thus obtained at the places B and N (Fig. 25.2(iii)), and the excess pressure above the normal is represented in Fig. 25.2(iv). By following the actual positions of the layers in Fig. 25.2, it can be seen that the pressure at the instant considered is normal ($n$) at D, M and P, and less than normal at L and Q, where the

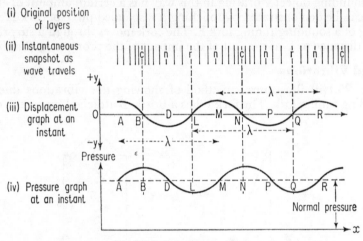

Fig. 25.2 Sound wave in air

layers are farther apart than in Fig. 25.2(i). The pressure at L and Q is termed a *rarefaction*, and its magnitude is represented in Fig. 25.2(iv).

When the telephone is used, one speaks into a carbon microphone. The variation of sound pressure makes the diaphragm move to and fro, and, as explained on p. 591, a varying electric current is produced in the telephone wires. At the other end of the 'line' a person is listening with a telephone earpiece, an apparatus which changes the varying electrical currents back again into variation of air pressure (see p. 592), so that a sound wave is again obtained. Essentially, then, a microphone converts sound energy to electrical energy, while a telephone earpiece does the reverse.

## Characteristics of Sound

A sound note can be completely identified from all other notes by three characteristics: (1) its pitch; (2) its intensity; (3) its quality (or timbre).

## Pitch

The *pitch* of a note is analogous to colour in light. A demonstration of pitch can be made using Seebeck's siren W, which consists of a circular metal plate with different sets of holes equally spaced round it (Fig. 25.3).

EXPERIMENT

Attach a motor to the axle through the centre of W. Connect a piece of glass tubing T with a pointed end to a foot air pump or to the outlet of a heated flask with water which generates steam. Start up the motor so that W is rotating steadily. Hold T so that a jet of air or steam 'puffs' through one set of holes in swift succession as W rotates. Observe the pitch of the note produced. Move the jet to a new set of holes and observe the rise in pitch when the number of holes is increased and decreased. Raise, and then lower, the speed of revolution of the wheel and observe the change of pitch in each case.

Pump the air harder, or generate the steam faster, and observe the effect on the pitch as the number of holes and the speed of the wheel remain constant.

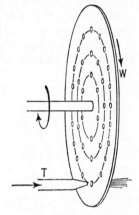

Fig. 25.3 Seebeck's siren

*Conclusions*

1. Since the number of vibrations per second or frequency of the note produced is proportional to the number of holes and to the speed of the wheel, the pitch increases when the frequency of the note increases.

2. The pitch of the note is independent of the air or steam pressure used to produce the note.

From these experiments, it follows that the pitch of a note depends only

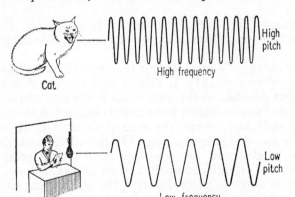

Fig. 25.4 Pitch depends on frequency (amplitude constant)

on its frequency. A note of high pitch has a high frequency, i.e. it is due to vibrations which are very many per second; 1000 cycles per second (1000 Hz) is a high-pitched note, for example. A low note has a low frequency, for example, 100 Hz is a hum heard in a radio receiver prior to the valves becoming warm (Fig. 25.4). Dogs and cats are capable of hearing notes beyond the upper limit of a human being, which is about 20 000 Hz.

## Intensity

The *intensity* of a note is a measure of the sound energy produced, which controls the loudness of the note. As the volume control of a radio receiver is turned up, the note from the loudspeaker becomes louder and louder; the sound energy thus increases. At the same time the amplitude of vibration of the loudspeaker cone can be felt to increase. In general, the intensity, or loudness, of a note is proportional to the square of the ampli-

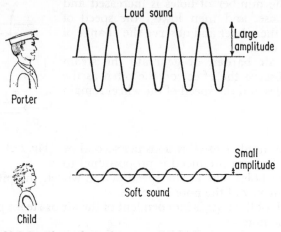

Fig. 25.5 Loudness depends on amplitude (frequency constant)

tude of vibration (Fig. 25.5). The amplitude of the air vibrations decreases the farther we go from a source of sound, so that the loudness also diminishes.

The kinetic energy of a mass $m$ of air moving with an average velocity $v$ is given by $\frac{1}{2}mv^2$ (see p. 121). It follows that the sound energy from a vibrating object depends on the mass of air it sets into vibration, and the larger the mass of air, the louder is the sound obtained. A loudspeaker cone has a large surface area, so that the mass of air set into vibration by it is large, and the sound intensity is large. On the other hand, the vibrating diaphragm in a telephone earpiece has a small surface area, so that the mass of air set into vibration is small, and sound can be heard only when one's ear is close to the earpiece. By itself, the violin string sets into vibration a very small mass of air; but the hollow box to which it is attached has a comparatively large surface area, and a large mass of air vibrates when the box is set into vibration by the sounding strings. A loud note is thus obtained from the violin (see also 'Sonometer', p. 467). Similarly, the note from a sounding tuning fork is a soft note, but when the base of the fork is placed on a table the note becomes loud, as a much larger mass of air, in contact with the table, is then set into vibration.

## Quality (or Timbre) of a Note

If the same note is sounded on a piano, an organ or a violin the source of the sound can immediately be recognized by ear. In technical language, we say that the 'quality' or 'timbre' or 'tone' of the note is different when it is sounded on different instruments.

In practice, it is very difficult to get a pure note, one which contains only one frequency. The note from a tuning fork is a near approach to a pure note. The note from a piano, however, contains a 'background' of other notes, of higher frequency than the one heard. For example, if a note of 256 Hz is sounded, notes of 512, 768 and 1024 Hz are also present; but the latter have a much smaller amplitude than the note of 256 Hz, so that their intensity, or loudness, is much lower. Similarly, an organ pipe

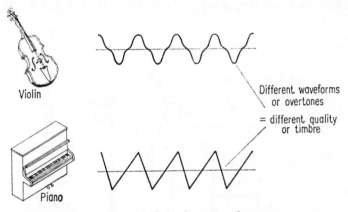

Violin

Different waveforms
or overtones

= different quality
or timbre

Piano

Fig. 25.6 Quality (timbre) depends on waveform or overtones

emitting a note of 256 Hz also produces notes of 768 and 1280 Hz at a much lower intensity. Notes which provide a 'background' to the note heard are called *overtones*, and are registered subconsciously by the mind. The overtones of the same note vary with different instruments, and *the quality of a note is due to the overtones which accompany it.* (Fig. 25.6).

If a note X has a frequency of 300 Hz this is called its *first harmonic*, and a note of frequency 600 Hz is known as the *second harmonic* of X. The third harmonic of X has three times its frequency, i.e. 900 Hz, and so on. The overtones of a note are always harmonics of it, as illustrated by the figures given, but some harmonics may be missing from the overtones (see p. 471).

## Velocity of Sound Waves

In 1864 REGNAULT, a famous French scientist, carried out an experiment to find the speed with which sound waves travel in air. Gunpowder was fired on the top of a mountain, and the flash of light which followed was observed on another mountain some miles away. A short time $t$ afterwards the sound was heard, and since light travels with an enormous speed compared with the speed of sound, $t$ may be taken as the time for the sound to travel the distance $d$ between the mountains. The velocity of sound $V$ was then calculated from $V = d/t$. The experiment was not capable of yielding accurate results; not only were the instruments unable to measure the time $t$ accurately but the time taken for the observer's reactions to the light and to the sound could not be determined and taken into account.

## Measurement of Velocity of Sound

The velocity of sound in air can be measured in a school laboratory with the aid of a cathode-ray oscillograph (p. 687).

### EXPERIMENT

An oscillator is connected through an amplifier to a loudspeaker, which is turned to face a large smooth board or metal reflector about a metre away (Fig. 25.7). A frequency of 2000–3000 Hz is suitable. Stationary

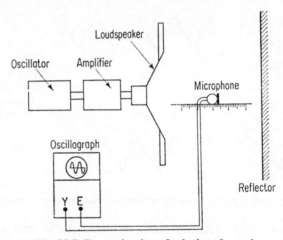

Fig. 25.7 Determination of velocity of sound

waves (p. 336) are then set up between the loudspeaker and the reflector. A small microphone, connected to the Y-plates of a cathode-ray oscillograph, is now moved along a ruler towards the board or away from it. Wave traces are seen which successively diminish to a minimum (node) and increase to a maximum (antinode). The wavelength $\lambda$ is twice the average distance between successive nodes, as explained on p. 336.

*Calculation.* Knowing the frequency $f$ of the sound and $\lambda$, the velocity can be calculated from $V = f\lambda$.

An echo method for the velocity of sound is also described in the booklet *Demonstrations and Experiments in Electronics*, Mullard Educational Service, London. See also p. 473.

By varying the frequency, it can be shown experimentally that the velocity of sound is independent of frequency. This is also confirmed on listening to an orchestra. Although different instruments send out simultaneously sound waves of different frequencies, when the sound reaches us we hear the orchestra playing in time.

## Velocity in Media

It has been shown that the velocity of sound in free air at sea level is about 340 metres per second, about $\frac{1}{3}$ km per second; the velocity of light in air is about 300 000 kilometres per second. The difference in velocity explains why the action of kicking a football, observed some distance away, is heard a little time after the football has left the kicker's foot. Another striking example is the observation that the whistle of a distant engine is heard after the steam is seen. The distance of a storm-centre can be determined by timing how long the thunder takes to reach one's ears after the lightning is seen, and using the fact that the velocity of sound in air is about $\frac{1}{3}$ kilometre per second.

We have seen that the particles of a medium are disturbed when a sound wave passes along it, and that compressions and rarefactions occur. Consequently, stresses are set up in the medium as the wave passes. Now different media, such as wood, iron, water, air, react differently to the same stresses imposed on them; it should therefore occasion no surprise that the velocity of sound varies with the medium concerned. The velocity of sound in water is about 1500 m/s, in iron about 5000 m/s, in hydrogen about 1300 m/s and in carbon dioxide about 250 m/s.

## Velocity of Sound

The velocity of sound in air varies with the temperature of the air. Experiment and theory show that *the velocity V of sound in a given gas is proportional to the square root of its absolute temperature T*. In symbols,

$$V \propto \sqrt{T}.$$

It follows that the speed in air is greater on a warm day than on a cold day, as the absolute temperature $T$ is then greater. The velocity of sound in air increases approximately by 0·6 m/s for each degree Celsius rise in temperature.

On the other hand, experiment and theory show that *the velocity of sound in air is independent of the air pressure*. Thus the speed of sound is the same at the top of a mountain as at the foot of it, if the temperatures are the same.

## Reflection of Sound

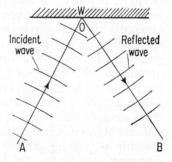

Fig. 25.8 Reflection of sound wave

Sound waves, like light waves, are reflected by surfaces such as walls. Fig. 25.8 illustrates the reflection of sound waves incident in a particular direction AO on a smooth plane W, such as a wall. The wall reflects the waves in a direction OB such that the angle of reflection is equal to the angle of incidence, which is the same as the case of light reflected from a plane mirror.

An *echo* is due to the reflection of sound waves back to the person shouting. The 'whispering gallery' of St Paul's Cathedral is circular in shape, and a person talking quietly on one side of the gallery can be heard on the other side by continuous reflection of sound waves round the wall. The captain of a ship can roughly find the direction and distance of an iceberg by sounding the ship's foghorn and listening for the echo. If the interval between sounding the foghorn and hearing the echo is 2 sec the total distance travelled by the sound = $2 \times 340$ m = 680 m, assuming the velocity of sound in air to be 340 m/s. The distance of the iceberg from the ship is thus 340 m.

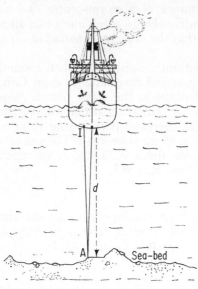

Fig. 25.9 Echo-sounding

By means of a method similar in principle, submarines have been detected when submerged below the water by an apparatus known as *Asdic*. These instruments send out very high-frequency sound waves, called *supersonics* because they have frequencies well above the audio-frequency range, and the waves are reflected back to the apparatus on meeting any object in water. The position of the object can be located from a knowledge of the time-interval elapsing before the return of the echo and the velocity of the sound waves in water.

Many merchant vessels have an apparatus for finding the depth of the water. The *echo-sounder*, as the device is called, sends out a supersonic note from its position at I, which travels through the water to the sea bed at A (Fig. 25.9). Here it is reflected back to the echo-sounder, and if the time of travel is 1 s and the speed of sound in water is 1500 m/s, the distance travelled is 1500 m. The depth of the sea bed $d$ is thus 750 m.

An echo-sounder contains a stylus, or pen, which describes an arc of a circle as it moves to and fro; the arc is graduated in fathoms. A short burst of supersonic waves is sent towards the ocean bed, and simultaneously the pen is released from the zero position and begins to swing across the arc. When the sound-echo is received back at the apparatus the pen is made to record a brown mark on a sheet of recording paper impregnated with a chemical solution of starch and potassium iodide, and the depth is the reading on the scale corresponding to the mark on the paper. By making the paper move slowly downwards past the pen, successive echo marks appear one above the other. A continuous thick line is then obtained, which corresponds to the profile of the sea bed over which the ship is passing. Shoals of fish also are located nowadays by echo-sounding.

## Acoustics of Rooms

When a musician is playing in a hall the sound reaches the ears of the audience (a) directly, and (b) by reflection from the walls and ceilings a short time later. This repetition of sound makes the music appear indistinct to the audience, and the hall will be unsuitable for concerts (i.e. acoustically bad) if no steps are taken to minimize the reflection of the sound. Thick curtains round the walls assist reception because they are good absorbers of sound; so do the specially designed seats at the Royal Festival Hall, London. The same acoustical problem was present in the early days of broadcasting. When people in a play, for example, spoke a few feet away from the microphone the sound reached it by reflection from the walls and ceiling as well as directly. British Broadcasting Corporation engineers have spent many years in research into the *Acoustics of Rooms*, as the subject is called. The walls and ceilings are now faced with a special type of sound-absorbent wood. Other rooms are converted into studios by means of heavy curtains or other suitable materials on the walls and ceiling.

## Refraction of Sound

Just like light waves, sound waves can be refracted. TYNDALL performed an experiment in which a large soap bubble was filled with carbon dioxide. and a high-pitched whistle was sounded in front of it. A sensitive flame, which reacted sharply to a high-pitched sound, was moved about on the other side of the soap bubble, and was found to be affected at one position. The carbon dioxide gas in the bubble had thus acted towards the sound waves in the same way as a convex lens to light, and had refracted the

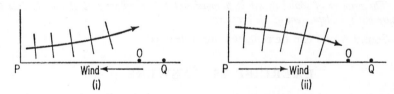

Fig. 25.10 Refraction of sound waves

waves spreading out from the whistle to a focus on the other side of the bubble.

It is a commonplace experience that a person shouting can be heard more easily some distance away if the wind is blowing towards the observer; if the wind is blowing in the opposite direction to the observer the latter experiences more difficulty in hearing the sound. These observations are explained by the refraction of the sound by the wind. Fig. 25.10(i) illustrates the case of the wind blowing in the opposite direction to a source of sound at P. The velocity of the air in contact with the ground is little influenced by the wind, while the velocity of the higher layers of air is diminished. The waves, which are represented by the straight lines in Fig. 25.10(i), are thus refracted (bent) upwards, and an observer at O on the ground has difficulty in hearing the sound. A reverse effect is obtained when the wind blows in the same direction PQ as the sound from the source P (Fig. 25.10(ii)). The velocity of the layer of air in contact with the ground is little affected, but the velocity of the upper layers is increased.

The waves are thus refracted downwards, and the sound is then more easily heard by an observer at O.

Refraction of sound also occurs as a result of temperature variation between layers of the air. In daytime, for example, the layers of air near the earth are warmer than those higher up. Now the velocity of sound in air increases when the temperature rises (p. 461). Consequently, the sound wavefront due to a person shouting travels faster at the ground than higher up. The wavefront therefore veers away from the earth. An observer some distance away does not hear the sound so easily because there is a loss of sound intensity reaching him. Conversely, at night the layers of air near the earth are cooler than higher up. Sound wavefronts veer towards the earth in this case, so that a person shouting will be heard over longer distances than in daytime.

## SUMMARY

1. Sound cannot pass through a vacuum; it requires a medium (such as air or water) to pass from one place to another.

2. Sound waves in air produce compressions (pressure above normal) and rarefactions (pressure below normal) of the air.

3. Sound is produced by a vibrating object. The loudness of a note depends on the amplitude of vibration; the pitch depends on the frequency (number of vibrations per second); the timbre or quality depends on the overtones present.

4. *The velocity of sound in air is proportional to the square root of the absolute temperature but is independent of the pressure.*

5. Sound waves can be reflected and refracted.

## EXERCISE 25 · ANSWERS, p. 480

**1.** (*a*) Explain why the thunder accompanying a flash of lightning is usually heard some seconds after the flash is seen.

(*b*) How can the properties of sound echoes be used to determine the depth of the sea? (*N.*)

**2.** Describe experiments, one in each instance, to show that: (*a*) a source of sound is vibrating; (*b*) a material medium is needed to transmit sound; and (*c*) the pitch of a musical note depends upon the frequency of the vibrations producing it.

Describe and explain a method of comparing the frequencies of two tuning forks. (*L.*)

**3.** What properties of a musical note depend on: (*a*) the amplitude of vibration of the source of sound; (*b*) the frequency of vibration of the source of sound? (*N.*)

**4.** Define *frequency* and *wavelength*. Deduce an expression for the velocity of sound in terms of the frequency and wavelength. What is the evidence that the velocity of a sound in air is independent of its wavelength?

Describe three different methods of producing a musical note, the source of vibration in each instance being different. In what ways will these notes be likely to differ from one another even though they are of the same pitch? (*L.*)

**5.** What is an echo? When a stationary fog-bound ship sounds a short blast, an echo from the coast is heard 5 s later. If the velocity of sound in air is 330 m/s, how far is the ship from the coast?

How does the velocity of sound in air depend upon the temperature of the air? (*N.*)

**PLATE 11**
Sound Recording – The Telephone

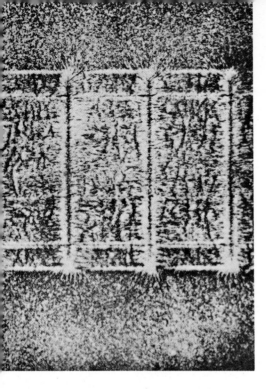

◀ 11(a) Magnetic field round tape used in Tape-recorders.

11(b) Gramophone record manufacture. Separation of stamper from mother record (p. 477).

◀ 11(c) Telephone earpiece and carbon microphone of Post Office telephone.

**PLATE 12**
Organ Pipes –
The Loudspeaker

12(**a**) Organ pipes. High notes are produced by short pipes and low notes by long pipes. By designing the shapes, each set of pipes produces overtones which make them sound like a flute or a clarinet or a trombone (latter corresponds to notes from pipes doubled over at base (p. 477)).

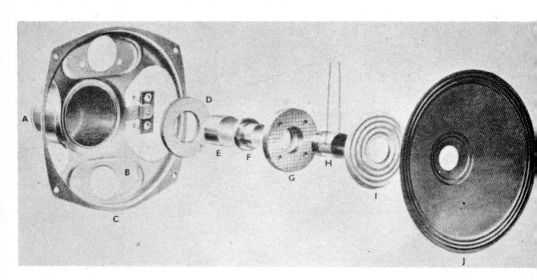

12(**b**) Loudspeaker parts. A, steel casing; B, transformer support; C, chassis; D, magnet ring; E, perman magnet; F, G, pole-pieces for radial field; H, speech coil on former; I, centring plate; J, cone; K, dust c

**6.** Explain the following: (*a*) the sound of an electric bell in a sealed vessel grows fainter as air is pumped out of the vessel; (*b*) a paper rider on a sonometer wire may jump off when the stem of a vibrating tuning fork is placed on the sonometer box; (*c*) sound waves may be used to determine the depth of the sea.

**7.** What is the nature of a sound wave? How is it propagated?

Describe experiments, one in each case, to show: (*a*) that the source of a sound is a vibrating body; (*b*) that a material medium is necessary to transmit sound. (*N.*)

**8.** An observer carrying a metronome which makes a clicking sound at $\frac{1}{2}$-s intervals notices that the echoes of the clicks from a wall 42 m away come midway between the clicks. Given that the velocity of sound lies between 300 and 400 m/s, calculate it. At what other greater distance from the wall could he hear the same effect?

Why cannot a bad radio receiver make an orchestra appear to play out of tune, though it can distort the loudness and quality of the individual instruments? (*O.* and *C.*)

**9.** Explain the terms *frequency, wavelength,* as used in sound.

Distinguish between transverse and longitudinal waves and give reasons for regarding sound as a longitudinal wave.

A given tuning fork produces waves in air of wavelength 60 cm. If the velocity of the waves in air is 330 m/s, calculate the frequency of the tuning fork. Prove any formula you use. (*O.* and *C.*)

## 26

# VIBRATIONS IN STRINGS AND PIPES

### Stationary Waves

When a violin, cello or guitar is played, a transverse wave travels along the bowed or plucked string. The wave is reflected at the fixed ends of the string, and as we saw on p. 335, interference takes place between two waves travelling in opposite directions. The resultant wave formed is

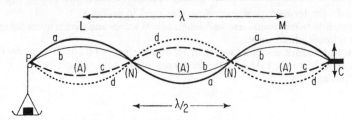

Fig. 26.1 Demonstration of stationary wave

called a *stationary wave*. It can be demonstrated by attaching a thread to a vibrator C, and adding suitable weights to a scale-pan at the other end after passing the thread round a pulley P (Fig. 26.1). Likewise, when an organ-pipe or flute is sounded a wave travels along the air inside the instrument and is reflected at the ends, so that a stationary wave is set up in the air.

As stationary waves play an important part in the theory of musical instruments, some of their properties are repeated here:

1. Points called *nodes* N in the stationary wave are permanently at rest.
2. Points called *antinodes* A in the stationary wave, midway between the nodes, are vibrating with a maximum amplitude.

Other points in the stationary wave between N and A are also vibrating, but with a smaller amplitude of vibration than at A (Fig. 26.1).

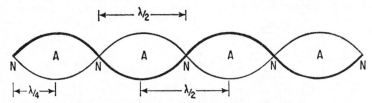

Fig. 26.2 Wavelength and stationary wave

It should now be carefully noted that, if $\lambda$ is the wavelength of the wave, the distance between successive antinodes AA or successive nodes NN is $\lambda/2$, and the distance between a node and a neighbouring antinode is $\lambda/4$ (Fig. 26.2). In Fig. 26.1, LM is $\lambda$. As we show later, these relations are

466

used in studying the frequency of the notes obtained from musical instruments.

# VIBRATIONS IN STRINGS

### Sonometer

The frequency of the notes obtained from a stringed instrument can be studied with the aid of a *sonometer*. This consists of a hollow box S, with a wire FDP attached to it at F (Fig. 26.3). The wire passes over movable bridges at C and B and then over a fixed pulley P. A weight $W$, such as 5 kg, is attached at the end to keep the wire at constant tension. With C kept fixed, the length of the string BC is altered simply by moving the position of the bridge B.

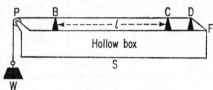

Fig. 26.3 Sonometer

If the length BC is made 20 cm the note obtained by plucking the wire gently in the middle is a high one. If the bridge is moved so that BC is increased to 30 cm, for example, a lower note is produced. At a length of 40 cm a note is obtained which is recognized as one octave lower in pitch than at 20 cm, or exactly half the latter frequency. Increasing the length of a vibrating wire at constant tension thus lowers the frequency.

### Variation of Frequency with Length

The exact relation between the frequency $f$ and the length $l$ of the string can be investigated by using six tuning forks of known frequency, between 256 and 512 c/s, for example.

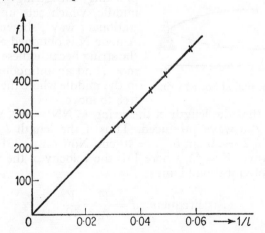

Fig. 26.4 Frequency and 1/length (tension constant)

### EXPERIMENT

Keep the tension in the string constant by using a fixed weight $W$. Sound the tuning fork of 256 Hz, and with C kept fixed, move the bridge B until the note obtained by plucking the midpoint of BC is exactly the same as that of the tuning fork. Record the length of BC. Repeat with the other tuning forks, and enter the values of $f$ and the length $l$ in a table of measurements, as below.

*Measurements.* The following results were obtained in an experiment:

| $f$ (Hz) | $l$ (cm) | $f \times l$ | $1/l$ |
|---|---|---|---|
| 256 | 35·0 | 8960 | 0·029 |
| 288 | 31·5 | 9070 | 0·032 |
| 320 | 28·0 | 8960 | 0·036 |
| 384 | 23·6 | 9060 | 0·042 |
| 426 | 21·2 | 9030 | 0·047 |
| 512 | 17·5 | 8960 | 0·057 |

*Calculations* (i) Find the value in each case of the product $f \times l$–these are practically constant. (ii) Calculate $1/l$ and plot a graph of $f$ v. $1/l$–a straight line passing through the origin is obtained (Fig. 26.4).

*Conclusion.* With a given string and tension, the frequency $f$ of the note obtained is inversely proportional to the length $l$, or $f \propto 1/l$.

## Fundamental Frequency Formula

These experimental results can be explained by considering the stationary wave set up along a string. The lowest possible note or *fundamental frequency* is produced by bowing or plucking the string gently in the middle, which sets up the simplest stationary wave, shown in Fig. 26.5. A node N is obtained at each end of the string because these points cannot move, and an antinode A is obtained in the middle where the string is most free to move.

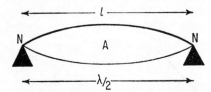

Fig. 26.5 Fundamental of bowed string

We now see that the length of the string $= NN = \lambda/2$, where $\lambda$ is the wavelength of the wave produced. Thus if the length $l = 20$ cm, for example, then $\lambda/2 = 20$ cm or $\lambda = 40$ cm. Now on p. 331 it was shown that, for any wave, $V = f\lambda$, where $V$ is the velocity of the wave and $f$ is the frequency of vibration. Thus

$$\text{Frequency} f = \frac{V}{\lambda} = \frac{V}{40}.$$

When the length of the string is increased to 30 cm this time the wavelength $\lambda = 60$ cm. The frequency thus changes to $V/60$. When the length of the string is increased to 40 cm the new frequency is $V/80$, which is exactly half the frequency $V/40$ obtained with a length of 20 cm. Consequently, it is one octave lower.

Generally it can be seen that, for a string of length $l$,

$$f = \frac{V}{\lambda} = \frac{V}{2l},$$

or

$$f \propto \frac{1}{l}.$$

This explains the experimental result obtained with the sonometer.

## How the Sonometer Can Be 'Tuned'

If a person has not a musical ear, the length of wire on the sonometer can be tuned to the note from a sounding tuning fork by one of two methods.

### (1) *Paper Rider*

A small piece of paper, a paper rider, is bent in the form of an inverted V, and placed in the middle of the wire whenever the length of wire is altered. The tuning fork is then sounded, and placed upright with its end firmly on the sonometer box so that its prongs are free; whereupon the vibrations of the fork are communicated to the wood of the sonometer box and then to the wire attached. The wire itself vibrates slightly when its natural frequency is not the same as that of the sounding fork, but when the natural frequency of the wire is exactly the same as that of the fork the wire vibrates through the largest amplitude (see p. 346). At this stage the paper rider is observed to tremble and to fall off the middle of the wire, which is then an antinode of the wave in the wire. The length of wire is now tuned to the same frequency as that of the fork.

### (2) *Beats*

The sonometer can also be tuned using the phenomenon of beats, described on p. 344. The length of the wire is altered by moving the bridge until a throb or beat note is heard when the fork and the wire are both sounded together. In this case we know that the frequency of the wire is near to that of the fork (p. 345). The length of the wire is then adjusted until the beats are as slow as possible, and the wire can now be considered practically 'tuned' to the fork.

## Wave Velocity · Fundamental Frequency

Using a constant length BC of wire on the sonometer, the tension $T$ in the wire can be increased by adding more weight $W$ (see Fig. 26.3). If BC is now plucked, a higher-pitched note is obtained. The frequency $f$ has thus increased. Now the velocity $V$ of the wave along the wire $= f\lambda$, and $\lambda = 2l =$ constant. Consequently, the velocity $V$ increases when the tension $T$ increases. If the tension is increased 4 times a note one octave higher is observed, so that the frequency has *doubled*. This suggests that $V \propto \sqrt{T}$. Tightening a violin string thus increases the velocity of the wave, and hence the frequency of the note produced.

When the length and tension are kept constant, and the wire itself is altered, either by using a different material or a different thickness, a new note is obtained. The velocity $V$ of the wave along a string thus also depends on its diameter and nature. Theory beyond the scope of this book shows that

$$V = \sqrt{\frac{T}{m}},$$

where $T$ is the tension and $m$ is the mass per unit length. The velocity $V$ is in m/s when $T$ is in newton and $m$ is in kg per m length of wire.

From the formula for frequency, $f = V/\lambda = V/2l$ (p. 469), it can now be seen that

$$f = \frac{1}{2l} \sqrt{\frac{T}{m}}.$$

Thus for a given length and string, the frequency $f \propto \sqrt{T}$. This means that if the tension $T$ is changed from 2 to 8 N, an increase of 4 times, the frequency increases twice as much. Further, suppose a note of 300 Hz is produced when the tension is 4 N. A note of 600 Hz is then produced by changing the tension to a value $T$ given by

$$\frac{600}{300} = \sqrt{\frac{T}{4\,\text{N}}},$$

assuming the length is constant. Thus

$$T = 2^2 \times 4\,\text{N} = 16\,\text{N}.$$

### Overtones in a Plucked String

Notes of higher frequency can be obtained by plucking the middle of a violin string with increasing force. In each case a stationary wave is set up along the string, and Fig. 26.6(i) illustrates the stationary wave set up along the string of length $l$ when the frequency is $f_1$. The middle of the stationary wave is an antinode A and the fixed ends of the string are nodes N; and it can be seen that the distance from a node to the next node, which is half a wavelength, is $\frac{1}{3}l$. The wavelength, $\lambda$, is thus given by $\frac{1}{2}\lambda_1 = \frac{1}{3}l$, from which $\lambda_1 = \frac{2}{3}l$.

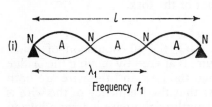

Frequency $f_1$

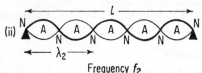

Frequency $f_2$

Fig. 26.6 Overtones of string

The frequency $f_1$ of the note obtained from the string can be calculated from the formula $V = f_1\lambda_1$, where $V$ is the velocity of the transverse wave along the string (p. 331). Thus $f_1 = \dfrac{V}{\lambda_1}$. But we have just seen that $\lambda_1 = \frac{2}{3}l$. Hence $f_1 = \dfrac{3V}{2l}$. Now the fundamental frequency $f_0$ can be obtained when the string is plucked in the middle, as illustrated in Fig. 26.5, and $f_0 = \dfrac{V}{2l}$, as explained on p. 469. Consequently $f_1 = 3f_0$, so that if $f_0 = 200$ Hz, $f_1 = 600$ Hz.

Yet another way in which the string can vibrate when plucked in the middle is illustrated in Fig. 26.6(ii). The middle is again an antinode A and the fixed ends of the string are nodes N; but the distance between two consecutive nodes is $\frac{1}{5}l$ in this case. Thus the new wavelength $\lambda_2$ of the wave set up along the string is given by $\frac{1}{2}\lambda_2 = \frac{1}{5}l$, and hence $\lambda_2 = \frac{2}{5}l$. The frequency $f_2$ of the note obtained is given by $f_2 = \dfrac{V}{\lambda_2} = \dfrac{5V}{2l}$, and since $f_0 = \dfrac{V}{2l}$, then $f_2 = 5f_0$.

It can now be seen that higher notes of frequency $3f_0$, $5f_0$, $7f_0$, etc., may accompany a note of fundamental frequency $f_0$ obtained when the string is plucked in the middle. These higher notes are the *overtones* of the fundamental notes and, as explained on p. 459, they give the quality or timbre of the note.

*Even harmonics* are obtained by plucking the string one-quarter and one-eighth of the way along. Fig. 26.7 illustrates the stationary wave in the former case, as the antinode is then $\frac{1}{4}$ of the way along the string. Thus the length $l$ of the string is one wavelength $\lambda$. The frequency $f$ of the note is

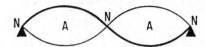

Fig. 26.7 Overtone with damping in middle

given by $f = \dfrac{V}{\lambda} = \dfrac{V}{l}$; and since the fundamental frequency $f_0 = \dfrac{V}{2l}$, from above, $f = 2f_0$. The reader should verify that the frequency is $4f_0$ when the string is plucked one-eighth of the way along from one end.

## VIBRATIONS OF AIR IN PIPES

Organ pipes, flutes and trombones all produce their characteristic notes when the air inside them is made to vibrate by blowing.

### Experiment

Take a number of test-tubes and fill them with different levels of water so that the air columns inside have different lengths (Fig. 26.8). Blow gently across the top of the tubes in turn, starting with A and finishing with E. Observe the pitch of the notes produced.

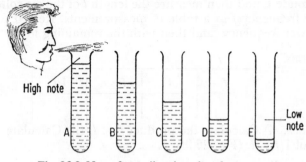

Fig. 26.8 Notes from vibrating air-column

*Conclusion.* The pitch is higher for a short column of air, so that the frequency increases as the length of the air column decreases.

## Resonance Tube Experiment

An investigation into the variation of the frequency $f$ of the note obtained with the length $l$ of the air column can be carried out with the apparatus shown in Fig. 26.9(i). It is called a *resonance tube* for a reason seen later. Here, by moving a side reservoir R up or down, the water level in T is varied, thus altering the length $l$ of the air column in T. Four or five tuning forks from 512 Hz downwards are required.

### EXPERIMENT

Raise R so that the water level is near the top of T. Start with the fork F of highest frequency, 512 Hz. Sound it gently, and hold the vibrating prongs

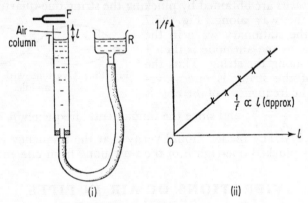

(i)                              (ii)

Fig. 26.9 Resonance tube experiment

over the air column as shown. Lower the reservoir R slowly, so that the water level in T drops and the length of air column increases. At first a low sound is heard, but as the level falls a louder sound is heard, rising to a maximum loudness at one stage. The air in the pipe has now been set into resonance by the sounding tuning-fork prongs, that is, the frequency of the fork is exactly equal to the natural frequency of the air column (see p. 345). By moving R slightly up or down, find the position when the sound is loudest, and then measure the length $l$ of the air column. Record $l$ and the frequency $f$ in a table of measurements. Repeat with the next fork of lower frequency, and then with the remaining forks.

*Measurements.*

| $f$ | $l$ | $f \times l$ | $1/f$ |
|---|---|---|---|
|  |  |  |  |

*Calculation.* (i) Calculate the product $f \times l$. (ii) Calculate $1/f$, and plot a graph of $1/f$ v. $l$. (Fig. 26.9(ii)).

*Conclusion.* Since, from the graph, $1/f$ is approximately proportional to

$l$, the frequency $f$ of the note from an air column of length $l$ is approximately proportional to $1/l$.

## Frequency of Closed Pipe

We can explain this experimental result by considering the simplest stationary wave set up in the air column when it is blown gently. The layer of air at the water cannot move, and this is therefore a node of the wave (Fig. 26.10). The air at the open end of the tube is free to move, and this is therefore an antinode. Thus the length of the air column $= \lambda/4$, where $\lambda$ is the wavelength (p. 466). The antinode is actually formed a small distance $c$ outside the end of the tube, as shown in Fig. 26.10 (i), and $c$ is called the *end-correction*. If we ignore $c$ for the present, then if $l$ is the length of the pipe,

$$\frac{\lambda}{4} = l, \quad \text{or } \lambda = 4l.$$

Now $V = f\lambda$, where $f$ is the frequency and $V$ is the velocity of sound in the air.

$$\therefore f = \frac{V}{\lambda} = \frac{V}{4l}.$$

Since $V$ is constant, it follows that $f \propto 1/l$, as we found in the experiment with the resonance tube. As stated above, this is not an exact relation, because the antinode is formed beyond the end of the pipe. In contrast, the relation $f \propto 1/l$ in the case of the sonometer (p. 469) is an exact one.

## Velocity of Sound in Air

The velocity of sound in air can be measured by using a fork of known frequency $f$ and a resonance tube.

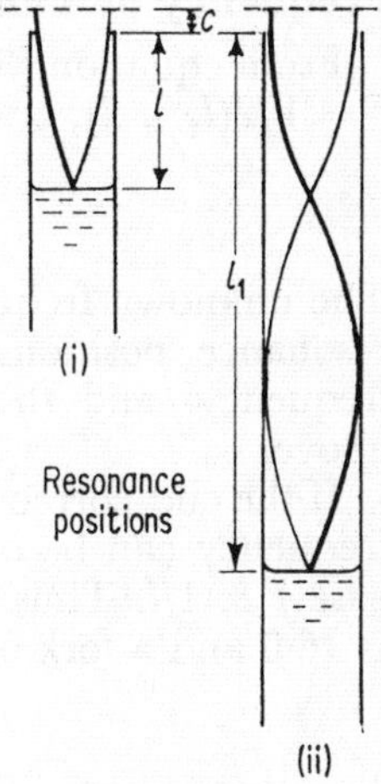

### Experiment

Using the fork, say 512 Hz, as described on p. 472, find the first position of resonance as the water level is slowly lowered from the top. Measure the length $l$ from the water level to the top of the tube. Now lower the water level again slowly by lowering the reservoir farther. Sounding the tuning fork, obtain a second position of resonance (Fig. 26.10(ii)). Measure the increased length $l_1$ from the new water level to the top of the tube. Record the temperature of the air in the laboratory.

*Calculation.* Work out the velocity of sound at the particular air temperature from the relation

$$V = f \times 2(l_1 - l)$$

Resonance
positions

(i)

(ii)

Fig. 26.10 Resonance positions

## Theory · Eliminating End-correction

If the end-correction is $c$, then, from Fig. 26.10(i),

$$\frac{\lambda}{4} = l + c \quad . \quad . \quad . \quad . \quad . \quad . \quad (1)$$

The second position of resonance occurs when a stationary wave is again formed in the tube with a node at the water level and an antinode at the top, but this time it extends a distance $3\lambda/4$, as shown in Fig. 26.10(ii).        Thus

$$\frac{3\lambda}{4} = l_1 + c \quad . \quad . \quad . \quad . \quad . \quad . \quad . \quad . \quad (2)$$

Subtracting (1) from (2) to eliminate $c$,

$$\therefore \frac{\lambda}{2} = l_1 - l$$
$$\therefore \lambda = 2(l_1 - l)$$
$$\therefore V = f\lambda = f \times 2(l_1 - l). \quad . \quad . \quad . \quad (3)$$

The end correction $c$ increases when a wider tube is used, but measurements using a second position of resonance enable it to be eliminated in this way, and thus helps in getting a more accurate result for $V$.

In one experiment with a fork of 512 Hz and an air temperature of 15°C, $l$ was found to be 16·2 cm and $l_1$ to be 50·1 cm. Thus $\lambda/2 = 50·1 - 16·2 = 33·9$ cm, and hence $\lambda = 2 \times 33·9 = 67·8$ cm. Hence $V = 512 \times 67·8 = 34\ 700$ cm/s $= 347$ m/s at 15°C.

The velocity of sound measured by the resonance-tube experiment is usually less than in free air. This is due to the damping effect of the sides of the tube on the motion of the air particles inside.

**Frequency of Tuning Fork**

From equation (3) above, the frequency $f$ of the tuning fork is given by $f = V/2(l_1 - l)$, or

$$f \propto \frac{1}{l_1 - l}.$$

The unknown frequency of a fork can thus be found by finding the two resonance positions, repeating the experiment with a fork of known frequency, and then using the relation between $f$ and $(l_1 - l)$ stated above.

If the end-correction is ignored, an approximate value for the unknown frequency can be obtained by using the first resonance position. In this case $f \propto 1/l$. Thus suppose the frequency $f$ produces resonance when $l$ is 16·0 and a fork of frequency 480 Hz when $l$ is 17·0 cm. Then

$$\frac{f}{480} = \frac{17}{16}, \quad \text{or } f = 510 \text{ Hz.}$$

**Closed Pipe**

A closed organ pipe consists basically of a pipe closed at one end N and having an opening at the other end near A (Fig. 26.11(i)). When air is blown into the pipe at X a stationary wave is formed in the air inside, as already explained. The simplest stationary wave in the pipe is one which has a node at the closed end N of the pipe, where the air cannot move, and an antinode A near the open end, where the air is free to move. Thus, roughly, the length $l$ of the pipe = distance from node N to nearest antinode A.

$$\therefore l = \text{NA} = \frac{\lambda}{4},$$

where $\lambda$ is the wavelength of the note obtained. See p. 473.

$$\therefore \lambda = 4l$$
But
$$V = f\lambda \text{ (p. 474)}$$

where $V$ is the velocity of sound in air and $f$ is the frequency of the note.

$$\therefore f = \frac{V}{\lambda} = \frac{V}{4l} \quad . \quad . \quad . \quad . \quad . \quad . \quad . \quad (1)$$

## Overtones

The frequency $V/4l$ is that of the lowest note obtained by blowing down the pipe, and is known as the *fundamental* frequency; its symbol is usually $f_0$. If air is blown slightly harder into the pipe a note of higher frequency is heard, with a corresponding stationary wave-form shown in Fig. 26.11(ii). In this case the length of the pipe $l = \frac{3\lambda_1}{4}$, where $\lambda_1$ is the wavelength, and hence $\lambda_1 = 4l/3$. Thus the higher frequency $f_1$ is given by

$$f_1 = \frac{V}{\lambda_1} = \frac{3V}{4l}.$$

$$\therefore f_1 = 3f_0, \text{ from (1).}$$

Thus the higher frequency $f_1$ is three times the fundamental frequency $f_0$.

By blowing harder, a note of higher frequency is obtained corresponding to the stationary wave shown in Fig. 26.11(iii). The frequency $f_2 = \frac{V}{\lambda_2} = \frac{5V}{4l}$, since $l = \frac{5\lambda_2}{4}$; thus $f_2 = 5f_0$. We now deduce that a closed pipe emits a fundamental note $f_0$ accompanied by notes of higher frequency $3f_0$, $5f_0$, $7f_0$, etc. The latter notes are the *over-*

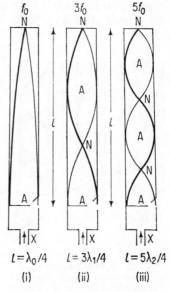

$l = \lambda_0/4$  $l = 3\lambda_1/4$  $l = 5\lambda_2/4$
(i)        (ii)        (iii)

Fig. 26.11 Fundamental and Overtones – closed pipe

*tones*, and accompany the fundamental note of frequency $f_0$; the overtones provide the 'quality' or 'timbre' of the fundamental note obtained from the pipe (see p. 459).

As an illustration, suppose the length of a closed pipe is 25 cm. Then, when the fundamental note of frequency $f_0$ is obtained, the wavelength $\lambda_0$ of the sound in air is given by $\lambda_0/4 = 25$ cm, from Fig. 26.11. Thus $\lambda_0 = 100$ cm. Now the velocity of sound $V$ in air is about 33 000 cm per sec.

Hence, as Frequency $= \dfrac{\text{Velocity}}{\text{Wavelength}}$,

$$f_0 = \frac{V}{\lambda_0} = \frac{33\ 000}{100} = 330 \text{ Hz}$$

Thus $f_1 = 990$ Hz. Similarly, the second overtone is given by $f_2 = 5 \times 330$ Hz, since the wavelength is $\frac{1}{5}$th that of the fundamental frequency, i.e. $f_2 = 1650$ Hz.

$$f_1 = \frac{33\ 000}{\frac{100}{3}} = 990.$$

Thus $f_1 = 990$ Hz. Similarly, the second overtone is given by $f_2 = 5 \times 330$ Hz, since the wavelength is $\frac{1}{5}$th that of the fundamental frequency, i.e. $f_2 = 1650$ Hz.

### Open Pipe

Fig. 26.12(i) illustrates an organ pipe open at both ends. When the lowest (fundamental) note of frequency $f_0$ is obtained by blowing air into the pipe, a stationary wave is set up by reflection of the wave at the open ends of the pipe. An antinode is obtained at both ends of the pipe, where the air is free to move; hence, since the distance between successive antinodes $= \dfrac{\lambda}{2}$ (p. 466), the length $l$ of the pipe $= \dfrac{\lambda_0}{2}$, where $\lambda_0$ is the wavelength of the sound wave.

Thus

$$\lambda_0 = 2l$$

$$\therefore f_0 = \frac{V}{\lambda_0} = \frac{V}{2l} \quad . \quad . \quad (1)$$

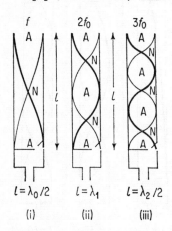

$l = \lambda_0/2$     $l = \lambda_1$     $l = \lambda_2/2$

(i)      (ii)      (iii)

Fig. 26.12 Fundamental and Overtones – open pipe

By blowing harder into the pipe, a note of higher frequency $f_1$ is obtained, corresponding to a stationary wave shown in Fig. 26.12(ii). Again, it should be noted, an antinode is obtained at both ends of the pipe. The length $l = \lambda_1$, the new wavelength.

Hence the new frequency, $f_1 = \dfrac{V}{\lambda_1} = \dfrac{V}{l}$. Consequently, $f_1 = 2f_0$, from (1). By blowing harder into the pipe a note of higher frequency $f_2$ is obtained, corresponding to Fig. 26.12(iii). It can be seen that $l = 3\lambda_2/2$, where $\lambda_2$ is the wavelength, and hence $f_2 = V/\lambda_2 = 3V/2l$. Consequently, $f_2 = 3f_0$ from (1). Thus, accompanying the fundamental note $f_0$ are *overtones* of frequency $2f_0$, $3f_0$, $4f_0$, etc. This is different from the case of the closed pipe, in which the overtones are odd multiples of $f_0$ (see p. 475).

## Cinema Organ

This has sets of pipes of different shapes and sizes. Each set is designed to produce special overtones to the fundamental, so that one set emits notes sounding like those from a flute, another set emits notes sounding like those from a clarinet, another produces notes like those from a trombone, and so on.

## Gramophone Record

At a British Association meeting in 1859 it was demonstrated that the wave-forms of sound can be faithfully reproduced. A bristle, attached to a parchment, pressed against the lamp-blackened surface of a cylinder which could be rotated. When a sound was spoken in front of the parchment the latter vibrated, and a trace of the sound wave was marked on the cylinder's surface as it was rotated; just as a trace was obtained on the blackened glass plate shown in Fig. 25.1(ii), p. 455, as it dropped past the vibrating bristle.

Nearly 20 years later the great American inventor EDISON succeeded in reproducing the sound from the wave-form produced. He used a rigid point instead of a bristle and covered the cylinder with tin-foil, so that a continuous mark of varying depth was made in the foil when the point vibrated and the cylinder was turned. On running the point over the trace again, the sound was produced. Edison called his apparatus the *Phonograph*, and cylindrical records were used on it.

In 1887 another American, BERLINER, succeeded in making a record of the sound on a flat disc, instead of a cylinder. Essentially, the method consisted of allowing a vertical needle to vibrate sideways while a horizontal lamp-blacked glass disc rotated, thus cutting grooves into a wavy shape into the lamp-black. By etching the traces into the glass and running a needle round them, the sound was produced, and Berliner called his apparatus a *Gramophone*. The invention of sound recording and the gramophone thus followed from the method of showing waveforms devised in 1859; and when we think of the pleasure of hearing the records made by great musicians we are reminded how the pioneer work of the physicist in the laboratory has contributed significantly to the cultural life of the community.

## Making Gramophone Records

Today, recordings of artists are first made on tapes (see p. 566). They are then run through the playback head, and the induced currents are passed into electromagnet coils which operate a magnetic stylus or pen. This presses on a revolving thick disc of nitrocellulose with an aluminium backing, so that wavy grooves are cut into the disc as the pen spirals towards the centre. The recorded disc is then sprayed with silver to make it a conducting surface and is placed in an electroplating bath so that a fine nickel deposit is obtained. By means of another electroplating bath, a backing of copper is now produced. The metal coating is stripped very carefully, and this becomes the *master* negative copy of the cellulose disc. On it there are ridges protruding from the surface, which have exactly the same wave-shape as the grooves in the cellulose disc.

The master negative is now placed in the electroplating bath, and a positive metal copy, called a *mother* disc, is obtained. This has a sound track identical to that on the original celullose disc, and is played to check for quality. A nickel-backed negative is then produced from the mother disc. It is called a 'stamper' or 'matrix'. Other stampers are made in the same way. Two stampers, one for each side of a record, are fitted into the mould plates of a powerful hydraulic press. On closing the press the stampers compress and mould warm plastic material, thus forming the actual record.

When the plastic material cools to room temperature it becomes hard and black and assumes the familiar appearance of the gramophone record after shaping and polishing. The sound is reproduced by a needle or stylus moving along the groove as the record rotates. The vibrations of the stylus produce electric signals, which are amplified and reproduced as sound through a loudspeaker.

## SUMMARY

1. Stationary waves set up along strings, and along the air in pipes, are due to two waves of the same frequency and amplitude travelling in opposite directions.

2. The 'nodes' of a stationary wave are the points permanently at rest–the 'antinodes' are the points midway between the nodes, and have maximum amplitude of vibration.

3. The distance between a node and the next antinode is $\lambda/4$, where $\lambda$ is the wavelength–the distance between successive antinodes (or nodes) is $\lambda/2$.

4. The fundamental frequency $f$ of a plucked string is given by $f = \dfrac{1}{2l}\sqrt{\dfrac{T}{m}}$,

where $l$ is the length, $T$ is the tension, $m$ is the mass per unit length of the string. This formula can be verified by the sonometer.

5. The fundamental frequency of a closed pipe $= V/4l$, where $l$ is the length and $V$ is the velocity of sound in air; the fundamental frequency of an open pipe of length $l = V/2l$.

6. The velocity of sound in air, and the unknown frequency of a tuning fork, can be measured by a resonance-tube experiment.

## EXERCISE 26 · ANSWERS, p. 480

1. Draw a simple diagram of a sonometer. Mark the positions of a node (N) and an antinode (A) when the wire is vibrating at its lowest frequency.

2. State the effect of a rise of temperature on; (a) the velocity of sound in air; (b) the note produced by a vibrating air column; (c) the frequency of the note emitted by a vibrating string. (N.)

3. A pipe closed at one end can be made to resonate with a vibrating tuning fork when the length of the air column is 16·5 cm. The next position of resonance occurs when the air column is 50·5 cm. If the velocity of sound in air during the experiment is 34 000

cm/s, find the end-correction and frequency of the fork. Draw sketches of the stationary wave at the resonance positions.

**4.** Define *frequency*, *amplitude* as applied to sound waves. What characteristics of a musical note are altered when these quantities are varied?

Describe an experiment to determine the frequency of a tuning fork, being provided with any apparatus you require, including a fork of known frequency.

Calculate the wavelength in air of: (*a*) a note of 400 Hz, and (*b*) a note an octave higher, when the velocity of sound in air is 33 000 cm/s. (*L.*)

**5.** State the relation between the frequency of the note obtained by plucking a stretched string and the tension in the string.

Describe an experiment to verify your statement.

A note is obtained by plucking a wire 50 cm long stretched by a load of 40 N. What load must be used to give a note which is an octave above the first when: (*a*) the length of the wire remains at 50 cm; (*b*) the length is reduced to 30 cm? (*L.*)

**6.** A pipe closed at one end can be made to resonate with a vibrating tuning fork when the length of the air column is 25 cm. The next position of resonance occurs when the length of the air column is 77 cm. If the velocity of sound in air during the experiment is 33 800 cm/s, calculate: (*a*) the wavelength of the note emitted; (*b*) the frequency of the fork. (*N.*)

. **7.** How would you find by experiment the frequency of vibration of .a tuning fork?

The frequency of the note emitted by a sonometer wire vibrating transversely is 300 Hz. What will be the frequency of the note when: (*a*) the length of the wire is reduced by one-third without changing the tension; (*b*) the tension is increased by one-third in the original length of wire? (*N.*)

**8.** Describe how you would measure the frequency of a steady note produced by an organ pipe.

What is there about the sound produced which makes us describe sound as a 'wave motion'?

If the frequency of the note is 400 Hz, what is its wavelength, and what is the wavelength of the note an octave lower?

(The velocity of sound in air is 340 m/s.) (*O. and C.*)

**9.** What is meant by *frequency*, *musical interval*?

How does the frequency of the note obtained by plucking a stretched wire depend on: (*a*) the length of the wire; (*b*) its tension?

Describe an experiment to verify the statement given in *either* (*a*) *or* (*b*).

Why is the intensity of the sound emitted by a plucked wire increased when it is mounted on a board? (*L.*)

**10.** How is sound produced using: (*a*) a violin; (*b*) an organ pipe; (*c*) a drum? State, in each instance, how the pitch of the note can be raised. Why do the qualities of sounds of the same pitch differ when emitted by different instruments?

Describe, in a general way, how the sound is transmitted from an instrument to the listener. (*L.*)

**11.** On the diagram of an open organ pipe, mark the position of any nodes and antinodes which are present when the pipe is sounding its fundamental note.

If the length of the pipe is 105 cm, what is the frequency of the note on a day when the velocity of sound is 336 m/s? (*N.*)

**12.** Explain: (*a*) what is meant by resonance in sound; (*b*) the production of an echo and why several echoes may be heared of a sound made inside a large hall.

Describe an experiment to determine the speed of sound in air. (*N.*)

**13.** A stretched wire adjusted to a length of 48 cm produces the same note when plucked as a fork of frequency 256 Hz. If the wire is then adjusted to 32 cm and the tension kept constant, what is the frequency of the fork which would then be in tune with the wire? (*N*.)

**14.** How does the frequency of transverse vibration of a stretched string depend on: (*a*) tension; (*b*) the length; (*c*) the diameter; (*d*) the density of the material of the string?

Describe an experiment you would carry out to determine the frequency of a tuning fork. (*N*.)

# ANSWERS TO NUMERICAL EXERCISES

## EXERCISE 25 (p. 464)

**5.** 825 m    **8.** 336 m/s, 126 m    **9.** 550 Hz

## EXERCISE 26 (p. 478)

**2.** (*a*) rises (*b*) rises (*c*) falls    **3.** 0·5 cm, 500 Hz    **4.** (*a*) 82·5 (*b*) 41·25 cm
**5.** (*a*) 160 N (*b*) 57·6 N    **6.** (*a*) 104 cm (*b*) 325 Hz    **7.** (*a*) 450 (*b*) 346 Hz    **8.** 85, 170 cm
**11.** 160 Hz    **13.** 384 Hz

# Current Electricity

# 27

# PRINCIPLES OF CURRENT, P.D., RESISTANCE

In this modern age, the science of Electricity holds one of the key positions. Electrical devices of one kind or another are needed in television, computers, satellites, telephony and commercial lighting, to name a few examples.

### The Battery · Electric Current

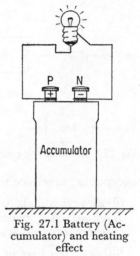

Electrical phenomena had been known for thousands of years, but few useful effects were obtained, because it was static or stationary electricity (see *Electrostatics*, p. 643). About 1800, however, an Italian scientist called VOLTA discovered that suitable chemicals were able to form a simple *cell*, which could keep electricity moving for some time in wires. A group of cells is called a *battery*. Later, the *accumulator* was invented. This is a very efficient cell; it keeps an electric current flowing for a long time. To distinguish between the poles or terminals of a cell, battery or accumulator, one is called the *positive* (+) *pole* P and the other the *negative* (−) *pole* N (Fig. 27.1).

Fig. 27.1 Battery (Accumulator) and heating effect

### Effects of Current

A small bulb glows when an accumulator is connected to it (Fig. 27.1). An electric current thus produces a *heating effect*. When a current

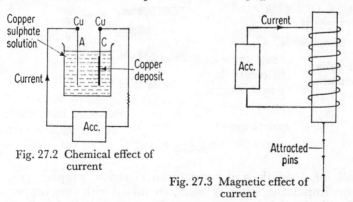

Fig. 27.2 Chemical effect of current

Fig. 27.3 Magnetic effect of current

flows through copper sulphate solution with two copper plates A and C dipping into it, some copper is seen deposited on one plate C after a time (Fig. 27.2). An electric current thus produces a *chemical effect*. When a current flows through a long coil of cotton-covered wire, called a 'solenoid', which is wound round a bar of soft iron, steel pins are attracted to the bar as long as the current flows (Fig. 27.3). An electric current thus produces a *magnetic effect*.

Summarizing, an electric current or flow of electricity can produce a heating, chemical or magnetic effect.

### Conductors, Insulators

As we have just seen, a small bulb lights up when an accumulator is connected to it (Fig. 27.1). An electric current is now flowing in the circuit. If the circuit is broken at A and B the light disappears (Fig. 27.4). Air, between A and B, does not therefore allow electricity to flow along it, and is called an *insulator*. When the wires at A and B are connected by any metal, or dipped into a salt solution, the bulb lights up again. These substances are called *conductors*. In this way, experiment shows that under normal conditions materials divide into one or other of these classes, with notable exceptions, as we see shortly. Thus

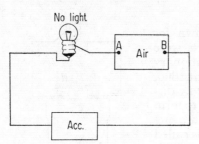

Fig. 27.4 Insulator and conductor test

*Conductors*: Metals, salt and inorganic acid solutions.

*Insulators*: Air, plastic materials, rubber, wood, paper, ebonite, pure water, organic acid solutions.

Silver is the best of the metallic conductors, copper is the next best. Pure copper is widely used for connecting wire·inside flex and plastic

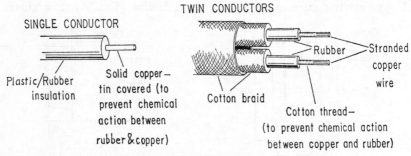

Fig. 27.5 Cables

materials for domestic and commercial electrical supplies (Fig. 27.5). When any impurities, however small, are mixed with pure copper an alloy is produced of much lower conductivity or, to put it the other way, of much higher electrical resistance. Nichrome and manganin are two copper alloys used in circuits as *resistance wire* (see p. 505). It should be noted that the insulating properties of materials, air or paper, for example, break

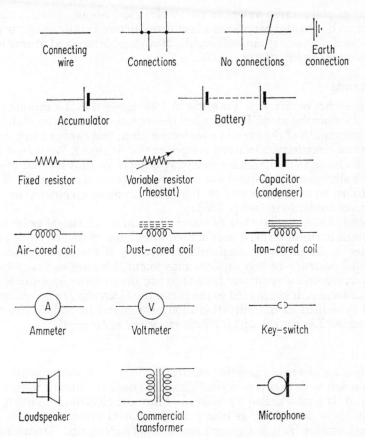

Fig. 27.6 Electric circuit symbols

down under certain conditions, and they become conductors. Flashes of lightning show the conducting path of the air during electrical storms.

Some of the symbols used in electric circuit diagrams are shown in Fig. 27.6.

## Semiconductors

There is another class of materials which come between the good conductor and the insulator. They conduct very slightly compared with copper, for example, and are known as *semiconductors*. Germanium and silicon are examples of semiconductors.

Semiconductors were neglected as useful materials for a long time. In 1946, however, SHOCKLEY, BARDEEN and BRATTAIN in America began an important study of germanium and silicon in an attempt to produce a solid substitute for a glass radio valve, which is a costly item to manufacture and easy to break. Using specially treated germanium, they invented the *transistor* in 1948. This solid material can act like a radio valve, and has many advantages over the valve (p. 711). In due course the transistor should displace radio valves in many electrical applications. A vast field of research into the properties of semiconductors is now under way, and many useful applications will result. Semiconductors and transistors are discussed more fully later, p. 706.

### Electrons

Researches by Sir J. J. THOMSON in 1896 showed that a minute particle existed within the atom. It was called the *electron*. The electron is about one-two-thousandth of the mass of a hydrogen atom and carries a tiny quantity of *negative* electricity. Electrons are present in all atoms. We do not get any electrical shocks by handling everyday objects, so that normal atoms are electrically neutral. Normal atoms therefore contain a quantity of *positive* electricity equal in amount to the total negative electricity on all the electrons inside them (see p. 738).

Metals are a special class of materials from an electrical point of view. Electrons in the outermost part of their atoms are relatively 'free', and they wander haphazardly through the framework of the metal. There is no external evidence of this random movement, because as many electrons are moving one way at any instant as are moving the opposite way. But when a battery is connected to the metal the electrons are driven one way. They now drift in one direction along the metal in addition to moving randomly. *The electron drift is the electric current in the metal.*

### Ions

When an electron (negative electricity) leaves an atom completely, the atom is left with excess positive electricity equal in amount to that on the electron. It is now called a *positive ion*. Since the electron is so light, the ion is a particle practically as heavy as the neutral atom. A *negative ion* is a neutral atom which has gained one or more electrons. Groups of atoms may lose or gain electrons, in which case the ion formed has about the same mass as all the individual atoms together.

In salt and acid solutions, called 'electrolytes', the current is carried by positive and negative ions. In gases the current is carried by electrons and ions. At very high temperatures, such as those existing in stars, electrons have been stripped from the atoms so that only electrons and ions are present. This state of matter is now called *plasma*.

### Holes

When semiconductors were investigated it was found that the current in these solid crystals was carried by electrons drifting in one direction and by a drift of positive electricity, equal in amount to that on the electrons, in the opposite direction. On account of the nature of the actual movement which creates the positive electricity effect, the latter is said to be due to *hole* movement. We thus speak of electrons (negative electricity) and holes (positive electricity) inside semiconductors (see p. 707).

The table on p. 487 is a summary of the carriers which carry electric current through the materials discussed previously. See also pages 542, 707.

| Material | Carriers |
|----------|----------|
| Metals | Electrons $(-)$ |
| Semiconductors | Electrons $(-)$<br>Holes $(+)$ |
| Liquids–electrolytes (salt and acid solutions) | Ions $(+$ and $-)$ |
| Gases | Electrons $(-)$<br>Ions $(+)$ |

## Electric Current

We now consider how an electric current is measured. The magnitude of an electric current is determined in a similar way to a water current. A water current can be measured by collecting the water running out at one end of a pipe in 10 s, for example, and measuring the volume. Suppose it is 60 cm³. The water current is then a flow of 60 cm³ per 10 s or 6 cm³/s. Note that if the current is steady everywhere along the pipe, then 6 cm³/s is the rate at which the volume passes every section of the pipe. If the water flows faster, then, if 90 cm³ passes every section in 10 s, the current is now 90 cm³/10 s or 9 cm³/s.

In a similar way, we think of a steady electric current as a constant quantity of electricity per second which passes every section of the wire. Quantity of electricity is measured in units called *coulombs*. Thus if 4 coulombs pass a section of the wire in 2 second, then

$$\text{Current} = \frac{\text{Quantity}}{\text{Time}} = 2 \text{ coulomb per second}$$

If the electric current is increased, then 6 coulomb may pass a section in 2 s, in which case the current is 3 coulomb per second. The carriers of the current in a wire are electrons (p. 486), each of which carries the very minute but definite amount of electricity, $1\cdot6 \times 10^{-19}$ coulomb approximately. Thus a current of 1 coulomb per second in a wire is due to a flow of about $6 \times 10^{18}$ electrons per second past any section of it.

All electrical quantities are denoted by an agreed set of symbols. Quantity of electricity is denoted by $Q$ and electric current by $I$. Thus if $Q$ coulomb flow past a section of a wire in $t$ second, then

$$I = \frac{Q}{t}$$

Hence                                $Q = I \cdot t,$

which means that if the current $I$ is 4 coulombs per s and the time $t$ is 10 s, the quantity $Q$ passing in that time $= 4 \times 10 = 40$ coulombs.

## Ampere · Current Instruments

The unit of electric current is called the *ampere* $(A)$. Nowadays, very accurate instruments at the National Physical Laboratory in England and other countries measure current directly in amperes (see p. 611). The ampere is taken as a standard unit, and 1 coulomb is then defined as the quantity of electricity passing a section of a wire in 1 second when the current is 1 ampere.

It can hence be seen that a current of '2 coulomb per second' is a current of '2 amperes (2 A)', and 4 coulomb per second = 4 A. The current in a 100-watt electric-lamp filament in our mains system is about 0·4 A. A 1-kilowatt electric-fire element carries a higher current of about 4 A. Currents vary in magnitude from the very small in transistor sets or radio sets, such as several millionths or thousandths ampere, to the very large in powerful electric motors, such as 500 A. The following smaller units are used:

$$1 \text{ microamp } (\mu A) = \text{one-millionth amp} = \frac{1}{10^6} A.$$

$$1 \text{ milliamp } (mA) = \text{one-thousandth amp} = \frac{1}{1000} A.$$

Electrical current-measuring instruments may be *ammeters*, *milliammeters* or *microammeters*, depending on the magnitude of the current. A more sensitive instrument than the milliammeter is the *galvanometer*, which may measure or detect currents thousands of times smaller than the microammeter.

### Position of Ammeters

The construction of instruments is discussed later (p. 603). Here we should note that an ammeter must always be placed in a circuit so that

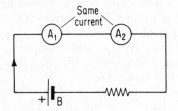

Fig. 27.7 Position of ammeter

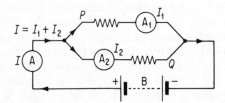

Fig. 27.8 Division of current – parallel resistors

the current to be measured flows directly through it. In Fig. 27.7, for example, the current flows through the ammeter $A_1$ and through the ammeter $A_2$. Both measure the current in the circuit; and as the current does not flow faster at different points, the readings on the instruments are the same. An ammeter is said to be placed *in series* in the circuit. In Fig. 27.8 the ammeter A measures the current through the battery B or 'main' current. The ammeter $A_1$ measures the current in the wire P. The ammeter $A_2$ measures the current in the wire Q. There is no reason why the current should be the same in P as in Q after it divides at X, since they have different resistances. Experimentally, we find

Reading on A = Reading on $A_1$ + Reading on $A_2$,

or $I = I_1 + I_2$,

where $I$ is the main current, or current flowing towards the junction X, and $I_1$ and $I_2$ are the respective currents in P and Q.

## Potential Difference

If one accumulator is connected to a suitable small bulb with resistance wire in series, a dim light will be seen. (Fig. 27.9). When two accumulators are used the light becomes more bright, so that the current has increased (Fig. 27.10). Now heat is a form of energy, and the energy must have its origin in the movement of electricity through the bulb filament. The carriers of the current, the electrons (p. 486), must hence release energy when they move from one end of the filament to the other.

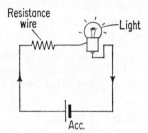

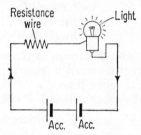

Fig. 27.9 One accumulator – dim light

Fig. 27.10 Two accumu- lators – brighter light

We have already met a case of movement from one position to another where change of position produces energy. This is the case of movement under gravity. An object held at a point A above the ground is said to have *potential energy* at A, and when it is released and moves to ground level its energy is converted mainly to heat at the ground. Here we recognize the existence of the earth's 'gravitational field', and at A and other points in the field, objects have an amount of potential energy, or, briefly, a potential, whose magnitude depends on the level above the ground. The *potential difference* between two points may be defined as the energy released per unit mass when an object falls from one point to the other.

An exactly analogous situation occurs in electricity, except that we must imagine movement of quantity of electricity, carried by electrons, for example, instead of movement of masses. Due to electrical exchanges between the atoms inside its chemicals, with which we are not concerned, the terminals of a battery or accumulator have an *electrical potential difference* (p.d.). One terminal, called the positive (+) terminal, is at a higher electrical potential than the other terminal, called the negative (−) terminal. If it could move, positive electricity would move along a conductor joined to the terminals from the positive to the negative terminal. Or, what amounts to the same thing, negative electricity would move along the conductor from the negative to the positive terminal.

When, therefore, a wire such as a bulb filament is joined to the terminals electrons move along the metal from one end to the other. Energy is then released in the wire in the form of heat.

## The Volt · Electrical Energy

The *volt* is the practical unit of potential difference. It is defined by reference to the joule, the practical unit of energy, as follows:

*1 volt is the potential difference between two points when 1 joule of energy is obtained when 1 coulomb moves between the points.*

A lead–acid accumulator has a p.d. at its terminals of about 2 volts (2 V). Batteries used in motor-cars usually have a p.d. of 12 V. The mains voltage is about 240 V in Great Britain. At power stations the p.d. at the generators is about 11 000 V. X-ray tubes may be operated at 40 000 V. A generator built specially for nuclear-energy experiments has the enormously high p.d. of 7000 million V. Units of p.d. other than volts are:

1 microvolt ($\mu$V) = one-millionth volt
1 millivolt (mV) = one-thousandth volt
1 kilovolt (kV) = 1000 volts

From the definition of the volt, it follows that when 5 coulombs move between two points having a p.d. of 1 V, 5 J of energy are released. If 5 coulombs move through a p.d. of 2 V, then 10 J are released. Thus if $Q$ coulombs move through $V$ volts,

Energy released $W = QV$ J.

Since $Q = I \cdot t$, where $I$ is the current flowing, then

$$W = IVt \text{ J.}$$

We shall consider later in more detail the topic of energy in electricity (p. 525).

## Voltmeter

A *voltmeter* is an instrument for measuring potential difference. Since it has to measure the p.d. between two points, its terminals must be connected to those two points. In Fig. 27.11, for example, the voltmeter $V_1$ measures the p.d. between the ends A and B of the wire P and the voltmeter $V_2$ measures the p.d. between the ends C and D of the wire Q. Note that, in distinction from an ammeter, *a voltmeter is always connected 'outside', or in parallel with, a circuit.* Further, since some of the current may be diverted from P, for example, through the voltmeter itself, the voltmeter chosen must have a large electrical resistance compared with that of P. In this case only a very small amount of current flows through the voltmeter, leaving P practically unaffected when the voltmeter is joined across it.

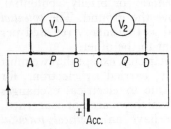

Fig. 27.11 P.D. across series resistors

## P.d. Across Resistances

A voltmeter provides useful information on the distribution of p.d. in electrical circuits. In Fig. 27.12, for example, the voltmeter $V_1$ may read 2 V and $V_2$ may read 4 V. A voltmeter connected to X and Y across *both* resistances then reads 6 V. If the current is altered by the rheostat S,

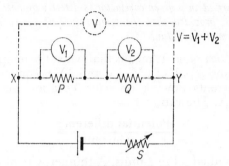

Fig. 27.12 P.D. in series circuit

$V_1$ may now read 0·5 V and $V_2$ read 1·0 V, and the p.d. across X and Y is then 1·5 V. Generally, then, for *resistance in series*, such as P and Q,

*Total p.d. across resistances = Sum of individual p.d.s.*

This result is explained from our energy definition of potential difference, because the total energy released when electricity moves across P and then across Q must be equal to the sum of the separate amounts of energy released in P and Q individually.

## Hydrostatic Analogy

Another popular (but not accurate) way of describing potential difference is by means of an analogy with flow of water through pipes. A water current can be obtained only if a pressure difference exists between

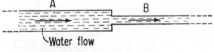

Fig. 27.13 Water flow

the ends of the pipe concerned. In the same way, an electric current can flow in a wire only if a potential difference exists between its ends. If two pipes A and B of different internal diameter are joined together, and a steady water current flows along them, the total difference of pressure between the extreme ends must be the sum of the pressure differences across A and B separately (Fig. 27.13). Further, if B is narrower than A a greater pressure difference is required to drive the same current through B as A. We say that the 'resistance' of B is greater than that of A to water flow.

In an analogous way, if in the electrical circuit of Fig. 27.12 the p.d. across Q is 2 V and that across P is 1 V, then, since the same electric current flows in each wire, the electrical resistance of Q is more than that of P. The electrons moving in Q are impeded more than in P.

## Resistance · Ohm's Law

In 1826 Ohm found by experiment that:

*The electric current in a given conductor is directly proportional to the potential difference applied, provided that the temperature of the conductor and its other physical factors remain constant.*

This is known as *Ohm's law*. It means that when the temperature of a given wire is constant, and a p.d. of 1 V, 2 V and 3 V is connected in turn across the wire, the currents flowing on each occasion may be respectively $\frac{1}{2}$ A, 1 A and $1\frac{1}{2}$ A. The ratio

$$\frac{\text{Potential difference}}{\text{Current}},$$

is then constant and equal to 2 units. A thinner wire of the same material and length may carry currents of $\frac{1}{8}$, $\frac{1}{4}$ and $\frac{3}{8}$ A respectively, when a p.d. of 1 V, 2 V and 3 V is connected in turn. The constant ratio for this wire is then 8 units. Although good conductors such as metals and alloys obey Ohm's law, many circuit components widely used in industry do not obey the law. Metal rectifiers, diode radio valves, and voltage-dependent resistors are some examples.

Ohm defined the *resistance* of a wire by the ratio

$$\frac{\text{P.d. applied}}{\text{Current}},$$

because the higher the p.d. required to maintain a given current, or the smaller the current flowing under a given p.d., the larger is the resistance. The symbol used for resistance is $R$ and its unit is the *ohm* ($\Omega$, Greek letter 'omega' = ohm). *One ohm is therefore the resistance of a wire when a p.d. of 1 volt across it maintains a current of 1 ampere.*

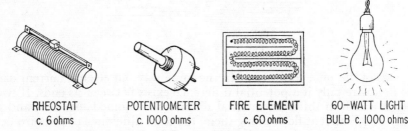

RHEOSTAT
c. 6 ohms

POTENTIOMETER
c. 1000 ohms

FIRE ELEMENT
c. 60 ohms

60—WATT LIGHT
BULB c. 1000 ohms

Fig. 27.14 Some resistors

The magnitude of resistance ranges from very small values, such as $\frac{1}{100}$ $\Omega$ for a piece of copper wire, to the very large values used in radio and television sets, such as a million ohms. A small rheostat in the laboratory may have a total resistance of a few ohms. A rotary rheostat or 'potentiometer' used in radio may have a total resistance of a thousand ohms. The 1-kW element of an electric fire made of Nichrome has a resistance of about 60 Ω. The tungsten filament of a 60-W electric lamp has a resistance of nearly 1000 Ω (Fig. 27.14).

Units of resistance other than the ohm are:

$$1 \text{ microhm } (\mu\Omega) = \text{one-millionth ohm}$$
$$1 \text{ kilo-ohm } (k\Omega) = 1000 \text{ ohms}$$
$$1 \text{ megohm } (M\Omega) = \text{one million } (10^6) \text{ ohms.}$$

## Fundamental Formulae

In physics, engineering and similar subjects measurements are essential. LORD KELVIN, one of the greatest scientists of the 19th century, renowned for many inventions as well as contributions to pure science, once said in reference to his subject: 'When you can measure what you are speaking about and express it in numbers, you know something about it, and when you cannot measure it, when you cannot express it in numbers, your knowledge is of a meagre and unsatisfactory kind.'

Electrical and radio engineers have particular need of formulae for calculating current, potential difference and resistance. We have seen that, by definition, resistance = p.d./current. Thus if $R$, $V$ and $I$ represent the magnitude of the three quantities,

$$R = \frac{V}{I}.$$

Hence

$$I = \frac{V}{R}$$

and

$$V = I \times R.$$

The current formula $I = V/R$ might be required, for example, in estimating the right fuse wire to use in an electrical circuit when $V$ and $R$ are known. The p.d. formula $V = IR$ might be required, for example, in estimating the right voltmeter range to use in testing a radio component with a voltmeter.

In these formulae $R$ must be in *ohms*, $V$ in *volts* and $I$ in *amperes*.

## Examples on Ohm's Law Formulae

1. Suppose a 2-volt accumulator is connected to a 10-ohm wire. Then

$$I = \frac{V}{R} = \frac{2}{10} = 0 \cdot 2 \text{ A.}$$

2. A current of 6 mA flows in a radio resistance of 20 kΩ. The p.d. $V$ across the latter is given by

$$V = IR = \frac{6}{1000} \text{ (A)} \times 20\ 000 \text{ (Ω)} = 120 \text{ V.}$$

3. The p.d. across a resistance wire is 12 mV, and the current flowing is 4 $\mu$A. Since 1 $\mu$A = 1 millionth A

$$R = \frac{V}{I} = \frac{\frac{12}{1000} \text{ (V)}}{\frac{4}{1\ 000\ 000}\text{(A)}} = 3000 \text{ Ω.}$$

## Measurement of Resistance · Voltmeter–Ammeter Method

The resistance $R$ of a coil of wire can be found by direct application of the definition $R = V/I$. An accumulator B is connected to the wire, with a rheostat and an ammeter A in series (Fig. 27.15). A voltmeter V is con-

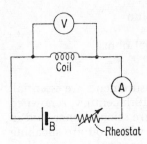

nected across the wire. When the current is switched on and the rheostat adjusted for convenient current and p.d. values, the readings on the ammeter and voltmeter are observed.

Suppose $V = 1.2$ V, $I = 0.3$ A. Then

$$R = \frac{V}{I} = \frac{1.2}{0.3} = 4 \ \Omega.$$

Fig. 27.15 Voltmeter–ammeter measurement of resistance

By varying the rheostat, other values of $V$ and $I$ may be obtained. The average value of the resistance can then be found either by calculation, or from the gradient of the straight-line graph obtained by plotting $V$ v. $I$. High currents are unsuitable, as the wire may get too hot and its temperature vary, in which case its resistance may change appreciably.

In this method of measuring $R$, some current is always diverted through the voltmeter when it is connected. The p.d. across $R$ is hence reduced, leading to error in the value for $R$ calculated from the readings. If the voltmeter resistance is very high compared with $R$ – at least ten times as large – only a very small current flows through the voltmeter and the error is low. The Wheatstone bridge method (p. 502) is a much more accurate method of measuring resistance, as it does not depend on reading values of p.d. and current from instruments.

### Resistance by Substitution

An AVO-meter is a commercial instrument widely used for measuring directly current ('amps' A), p.d. ('volts' V) and resistance ('ohms' Ω). As explained later on p. 495, switches on the instrument enable different ranges of current and p.d. to be obtained.

The principle of measuring resistance by the AVO-meter can be illustrated by using a box of known resistances R in series with a current measuring instrument M such as a milliammeter, a rheostat S and an accumulator B (Fig. 27.16).

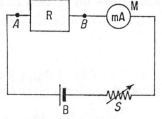

Firstly, R is made zero and S is then varied until the maximum reading, say 15 mA, is obtained on the instrument. The figure '0' is then marked on the scale beneath '15'. This is the zero of the *ohms scale*, since $R = 0$. A

Fig. 27.16 Making an ohm-meter

resistance of 10 Ω, for example, is then used from R. The current diminishes and the figure '10' is marked on the ohms scale at the new pointer position, since $R = 10$ Ω. In this way, taking more resistance from R, the scale can be calibrated in ohms. An infinitely large resistance corresponds to zero current, and the resistance scale thus goes the opposite way to the current scale, reading from right to left. Finally, having calibrated the scale in ohms, the box R is removed and leads are connected to the terminals at A and B. The rest of the circuit is left undisturbed.

*To measure the unknown resistance of a coil*, for example, the leads from A and B are connected across the coil. The resistance is then read directly from the ohms scale. This is called a method of measuring resistance by 'substitution', because in it we have substituted an unknown resistance by a known resistance which gives the same current reading in the same circuit.

## AVO-meter

When an AVO-meter is used to measure resistance the switch is first turned to 'resistance'. This connects a battery at the back of the instrument, so that when the metal ends of the leads are placed to touch each other a current flows and the needle is deflected. A rheostat inside the meter is then turned until the needle reaches the end of the current scale, corresponding exactly to 'zero' on the ohms scale. The connection between the leads is now broken, and they are placed across the ends of an unknown resistance R. The magnitude of R is read from the ohms scale directly. In using the AVO as an ohm-meter, always be sure first to connect the ends of the leads together and set the reading on the ohms scale to zero by means of the rheostat. Otherwise the reading on the ohms scale for an unknown resistance will be in error.

## Potential Divider · Resistance of Lamp

When a smooth variation of p.d. from zero to a particular value is required, a rheostat can be used with accumulators to form what is called a *potential divider* circuit. Fig. 27.17(i) shows a potential divider which produces a varying p.d. from 0 to 12 V, and this is applied to a lamp filament L. The 12-V battery P is joined to the two ends C and B of the rheostat resistance wire. The slider terminal D and the terminal C are joined to L. A voltmeter joined to C and D enables the p.d. across the lamp to be measured.

When the slider is moved to D so that contact with B is made the whole p.d. 12 V is connected to L, which then glows brightly. When the slider is moved to the position shown in Fig. 27.17(i) so that contact is made with X on the wire only the p.d. across the length CX is connected to L this time, and this is less than before. As S is moved farther to C the p.d. therefore diminishes, and the filament glow is weaker. When S makes contact with C no p.d. is applied to L.

EXPERIMENT

Connect the circuit shown in Fig. 27.17(i), using a 12-V accumulator or battery, a suitable rheostat which can carry a current of say 3 A, a lamp such as 12-V, 24-W used in a ray-box, an ammeter A reading to about 2 A and a voltmeter V reading to 12 V. Begin the wiring by joining the battery P to the two ends C and B of the rheostat—make sure you are not using the slider terminal here, otherwise you may damage the battery.

Vary the p.d. $V$ from zero in steps of 2 V up to a maximum of 12 V. Each time observe the reading on the ammeter A. Note the increased glow of the filament, showing that its temperature has risen.

*Measurements*

| $V$ (V) | $I$ (A) | $R = V/I$ ($\Omega$) |
|---------|---------|----------------------|
|         |         |                      |
|         |         |                      |

*Calculation.* At each value of $V$ work out the resistance $R$ of the filament. Enter the result in the table of measurements.

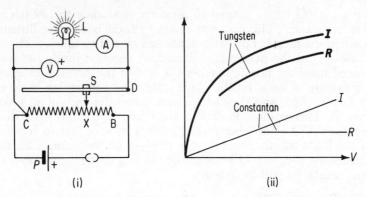

(i)                                        (ii)

Fig. 27.17 Variation of current and resistance

*Graphs.* Plot a graph of: (i) current $I$ v. p.d. $V$, and (ii) resistance $R$ v. p.d. $V$ (Fig. 27.17(ii)).

*Conclusions.* (i) The current is not proportional to the p.d. (ii) From the graph it follows that the resistance $R$ of the filament increases with the p.d. $V$, and hence with temperature rise. The filament is made of tungsten, a material which alters in resistance when its temperature changes (p. 505). 

For comparison, Fig. 27.17(ii) also shows the directly proportional variation of $I$ with $V$ by a coil of wire made of the alloy Constantan (p. 505). In contrast to the tungsten coil, the resistance $R$ depends very slightly on the p.d. $V$ or temperature, as shown.

## Thermistors

Pure metals such as copper and tungsten increase in resistance when their temperature rises. A class of materials called *thermistors*, made mainly from the semiconductor oxides of iron, cobalt and nickel, decrease in resistance when their temperature rises. The fall in resistance of a thermistor can be observed directly when its resistance is measured on an AVO-meter (p. 495). This has a battery which automatically passes current through the thermistor, thus causing its temperature to rise. Fig. 27.18(i) shows a typical resistance–current curve.

Thermistors are used to compensate for temperature changes in circuits where such changes are harmful. One case is a transistor receiver circuit, shown simplified in Fig. 27.18(ii). Here the transistor S is in series with a thermistor T of resistance $R$ and a battery B. To make it function correctly, the p.d. across the transistor S must be constant. Now part of the battery p.d. appears across S and the remainder across $R$. If the current $I$ in

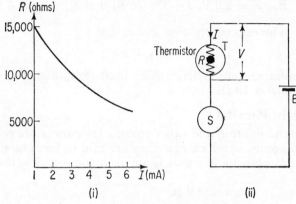

Fig. 27.18 Thermistor action

S rises the thermistor temperature rises. This lowers its resistance $R$. Consequently, the p.d. $V$ across the thermistor, which is given by $IR$, is automatically returned to its original value. This keeps the p.d. across S constant at its original value. The thermistor thus stabilizes the circuit.

## Resistances in Series

In electrical circuits such as those in radio receivers several resistances, such as $R_1$ and $R_2$ in Fig. 27.19, may be in *series*.

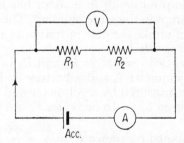

Fig. 27.19 Measurement of combined resistance

The combined or total or resultant resistance $R$ of $R_1$ and $R_2$ can be measured by connecting a voltmeter V across both, as shown, and observing the readings on V and the ammeter A. In one case,

$$V = 2 \cdot 4 \text{ V}, \quad I = 0 \cdot 2 \text{ A}.$$

$\therefore$ Combined resistance $R = \dfrac{V}{I} = \dfrac{2 \cdot 4}{0 \cdot 2} = 12 \ \Omega.$

$R_1$ and $R_2$ can each be measured separately, as explained on p. 494. Results were obtained as follows:

$$R_1: V = 3 \cdot 2 \text{ V}, I = 0 \cdot 4 \text{ A, so that } R_1 = \frac{V}{I} = 8 \, \Omega$$

$$R_2: V = 2 \cdot 0 \text{ V}, I = 0 \cdot 5 \text{ A, so that } R_2 = \frac{V}{I} = 4 \, \Omega$$

The experiment thus shows that

$$R = R_1 + R_2$$

when two resistances are in series. The total resistance of 4, 6 and 8 $\Omega$ in series is similarly 18 $\Omega$.

### Resistances in Parallel

In commercial lighting and other circuits resistances are connected to the same two points, in which case they are said to be *in parallel*. In Fig. 27.20(i), for example, some lamps L are each connected to the mains, so

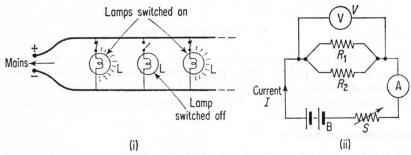

Fig. 27.20 Resistors in parallel

that switching one lamp on or off in a room has no effect on the p.d. applied to the other lamps in the same building. The lamps are in parallel across the mains. Thus unlike the case of resistance in series, *the p.d. is the same across conductors in parallel.*

Fig. 27.20 (ii) shows two resistances $R_1$ and $R_2$ in parallel in a circuit, with a rheostat S, an ammeter A and a battery B. The p.d. $V$ across the two resistances can be measured by a voltmeter and the current $I$ flowing towards them is measured by A. In one case, $V = 4 \cdot 0$ V, $I = 1 \cdot 5$ A.

$$\therefore \text{ Combined resistance } R = \frac{V}{I} = \frac{4 \cdot 0}{1 \cdot 5} = 2 \cdot 7 \, \Omega. \quad . \quad (1)$$

The resistances of $R_1$ and $R_2$ separately can be found from the circuit shown in Fig. 27.15. The results, using the same wires as in the series experiment, are $R_1 = 8 \cdot 0 \, \Omega$, $R_2 = 4 \cdot 0 \, \Omega$. The combined resistance $R$ of the two wires is certainly *not* given by the same relation as the series one, $R = R_1 + R_2$. Using a reciprocal relation for $R$, however, say

$$\frac{1}{R} = \frac{1}{R_1} + \frac{1}{R_2},$$

then
$$\frac{1}{R} = \frac{1}{8\cdot0} + \frac{1}{4\cdot0} = \frac{3}{8}$$

$$\therefore R = \frac{8}{3} = 2\cdot7 \ \Omega.$$

This calculation for $R$ agrees with the experimental result obtained previously, in (1). Hence for two parallel resistances $R_1$ and $R_2$, the combined resistance $R$ is given by

$$\frac{1}{R} = \frac{1}{R_1} + \frac{1}{R_2}.$$

Thus the combined resistances of 3 and 6 $\Omega$ in parallel is given by

$$\frac{1}{R} = \frac{1}{3} + \frac{1}{6} = \frac{3}{6}, \text{ or } R = \frac{6}{3} = 2 \ \Omega.$$

## Proof of Series and Parallel Formulae

*Series*

Suppose $I$ is the current in $R_1$ and $R_2$, and $V_1$ and $V_2$ are their respective p.d.s. (Fig. 27.21(i)). Then if $V$ is the p.d. across both resistances,

$$V = V_1 + V_2.$$

But $V_1 = IR_1$, $V_2 = IR_2$ and $V = IR$, where $R$ is the combined resistance of $R_1$ and $R_2$.

$$\therefore \ IR = IR_1 + IR_2$$
$$\therefore \ R = R_1 + R_2$$

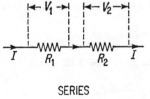
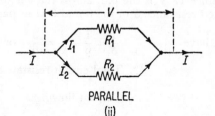

SERIES
(i)

PARALLEL
(ii)

Fig. 27.21 Series and parallel resistance

*Parallel*

Suppose $I_1$ and $I_2$ are the currents in $R_1$ and $R_2$ respectively and $I$ is the main current (Fig. 27.21(ii)). Then

$$I = I_1 + I_2.$$

Now the p.d. $V$ is the same across $R_1$ and $R_2$, and also across the single wire of resistance $R$, which we imagine can replace the two wires without altering the current $I$.

$$\therefore \ \frac{V}{R} = \frac{V}{R_1} + \frac{V}{R_2}.$$

Dividing by $V$,
$$\therefore \ \frac{1}{R} = \frac{1}{R_1} + \frac{1}{R_2}.$$

## Examples on Series and Parallel Circuits

The following examples illustrate how currents and p.d.s. are calculated in series and parallel circuits, and should be studied carefully.

**1.** A resistance of 6 Ω is in series with one of 4 Ω and a p.d. of 20 V is applied across the whole arrangement. Calculate the current in each wire and the p.d. across each.

20 V is the p.d. across *both* wires together.

The combined resistance = 6 + 4 = 10 Ω.

$$\therefore \text{ current } I = \frac{V}{R} = \frac{20}{10} = 2 \text{ A.}$$

Across the 6-Ω wire alone $V_1 = IR_1 = 2 \times 6 = 12$ V.
Across the 4-Ω wire alone $V_2 = IR_2 = 2 \times 4 = \phantom{0}8$ V.

### NOTE

A common error is to state that the current in the 6-Ω wire is $\frac{20}{6}$ A. This is *not* the case, because 20 V is not the p.d. across the 6-Ω wire but across both wires.

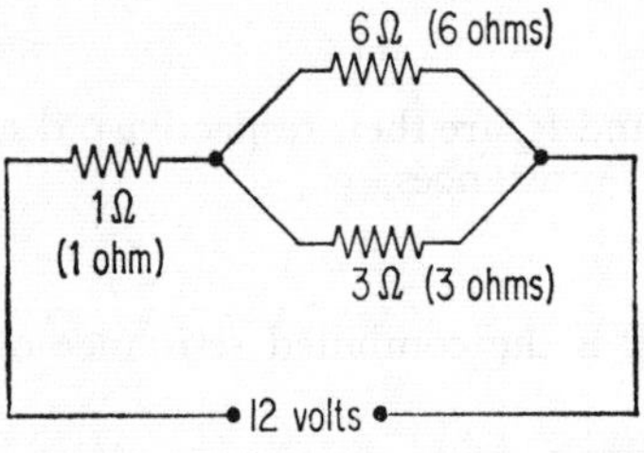

**2.** A circuit consists of a 1-Ω wire in series with a parallel arrangement of 6 and 3 Ω, and a p.d. of 12 V is connected across the whole circuit (Fig. 27.22). Calculate the currents in each of the three wires and the p.d. across each.

6Ω (6 ohms)

1Ω
(1 ohm)

3Ω (3 ohms)

12 volts

Fig. 27.22 Circuit calculation

Since 12 V is the p.d. across the whole circuit, we must first find the combined resistance $S$ of the 6 and 3 Ω in parallel and add the result to 1 Ω.

We have
$$\frac{1}{S} = \frac{1}{6} + \frac{1}{3} = \frac{3}{6}, \text{ or } S = \frac{6}{3} = 2 \text{ Ω.}$$

$$\therefore \text{ Total circuit resistance } R = 2 + 1 = 3 \text{ Ω.}$$

$$\therefore \text{ Current flowing } I = \frac{V}{R} = \frac{12}{3} = 4 \text{ A.}$$

This is the current in the 3-Ω wire.

To find the current $I_1$ in the 6-Ω wire, we need to know the p.d. $V_1$ across the 6 and 3 Ω in parallel. Now the current $I$ of 4 A in the 1-Ω wire, is also the current flowing through the single wire which could replace the parallel wires in the circuit. This is the wire of resistance $S$ of 2 Ω already calculated.

$$\therefore V_1 = I \times 2 = 4 \times 2 = 8 \text{ V}$$

$$\therefore \text{ Current in 6-Ω wire } I_1 = \frac{V_1}{6} = \frac{8}{6} = 1\tfrac{1}{3} \text{ A.}$$

Similarly, Current in 3-Ω wire $I_2 = \dfrac{V_1}{3} = \dfrac{8}{3} = 2\tfrac{2}{3}$ A.

As a check, note that $1\tfrac{1}{3} + 2\tfrac{2}{3} = 4$ A = Current in 1-Ω wire. Thus:

(*a*) the currents are respectively 4 A, $1\tfrac{1}{3}$ A, $2\tfrac{2}{3}$ A;
(*b*) the p.d.s. are respectively 4 V, 8 V, 8 V.

NOTE

A quick method of calculating the division of current between two parallel wires is to note that, since $I = V/R$ and the p.d. $V$ is the same across both wires, $I \propto 1/R$. Thus the current is shared *inversely* as the resistance. The two parallel wires in Fig. 27.22 have resistances of 6 and 3 Ω. Hence the current of 4 A is divided into $\frac{3}{9}$ through the 6-Ω wire, and the remaining $\frac{6}{9}$ through the 3-Ω wire. Thus $I_1 = (\frac{3}{9}) \times 4 \text{ A} = 1\frac{1}{3} \text{ A}$ and $I_2 = (\frac{6}{9}) \times 4 \text{ A} = 2\frac{2}{3} \text{ A}$.

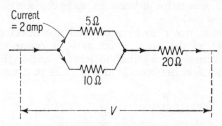

Fig. 27.23 Circuit calculation

**3.** Two wires of 10 and 5 Ω are in parallel, and the arrangement is in series with a 20-Ω wire (Fig. 27.23). If the current in the 5-Ω wire is 2 A, calculate the current in the 10-Ω wire and the total p.d. $V$ across the whole circuit.

The p.d. $V_1$ across the 5-Ω wire $= IR = 2 \times 5 = 10$ V.

$$\therefore \text{ Current in 10-}\Omega \text{ wire} = \frac{V_1}{10} = \frac{10}{10} = 1 \text{ A.}$$

$$\therefore \text{ Current in 20-}\Omega \text{ wire} = 1 + 2 = 3 \text{ A.}$$

$$\therefore \text{ p.d. } V_2 \text{ across 20-}\Omega \text{ wire} = I \times R = 3 \times 20 = 60 \text{ V.}$$

$$\therefore \text{ Total p.d. } V = \text{p.d. across parallel arrangement} + 60 \text{ V}$$
$$= 10 + 60 = 70 \text{ V.}$$

## Wheatstone-bridge Circuit

The resistance of a wire can be measured within a few per cent of its value by using a voltmeter and an ammeter, as explained on p. 494. A much more accurate method is attributed to CHARLES WHEATSTONE, the first Professor of Physics at King's College, London, who utilized a circuit now famous as the Wheatstone bridge.

The bridge is illustrated in Fig. 27.24. It consists of four resistances, $P$, $Q$, $R$ and $S$, joined to form a complete circuit, with a battery A across one pair $a$ and $c$ of opposite junctions, and a sensitive galvanometer G across the other junctions $b$ and $d$. A current then usually flows in G, but if the resistances are suitably adjusted the current becomes zero, and a 'balance' is said to be obtained. In this

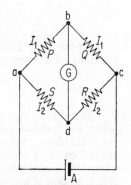

Fig. 27.24 Wheatstone bridge circuit

case, as we show on p. 503, there is a simple ratio relation between $P$, $Q$, $R$ and $S$. This is:

$$\frac{P}{Q} = \frac{S}{R}$$

*Metre Bridge*

The *metre bridge* was one of the earliest applications of the Wheatstone-bridge circuit. It consisted of a 100-cm length of uniform resistance wire soldered to terminals at $a$ and $c$, and strips of brass $am$, $np$ and $qc$ are connected so that gaps are left between $m$, $n$ and $p$, $q$.

The unknown resistance $P$ to be measured is connected in one gap, and a known resistance $Q$, of the same order as $P$, is connected in the other gap (Fig. 27.25). An accumulator is joined to $a$ and $c$, and a galvanometer G is connected to $b$, the junction between $P$ and $Q$, and to a tapping key which can

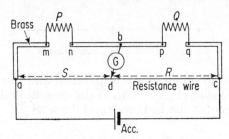

Fig. 27.25 Metre bridge

make contact with any point on the wire $ac$. A protective resistance (not shown) should be included with the galvanometer. When the key is at some point $d$ no current flows in G, and a balance is then obtained; the lengths $ad$, $dc$ are now measured.

Fig. 27.25 illustrates four resistances forming a complete circuit; they are $P$ and $Q$, the resistance $R$ of the wire $dc$ and the resistance $S$ of the wire $da$. As G is connected across two opposite junctions, and a battery is connected across the other pair of junctions, Fig. 27.25 is a Wheatstone-bridge circuit. Hence, from the relation above,

$$\frac{P}{Q} = \frac{S}{R}. \quad \text{But } \frac{S}{R} = \frac{\text{Length of } ad}{\text{Length of } cd}.$$

$$\therefore \frac{P}{Q} = \frac{\text{Length of } ad}{\text{Length of } cd}.$$

Suppose $Q = 10\cdot0$ $\Omega$, $ad = 45\cdot2$ cm and $cd = 54\cdot8$ cm. The unknown resistance $P$ is then given by

$$\frac{P}{10\ \Omega} = \frac{45\cdot2}{54\cdot8}, \text{ so that } P = \frac{45\cdot2}{54\cdot8} \times 10 = 8\cdot2\ \Omega.$$

The metre bridge thus enables resistance to be measured. It is most accurate when the balance-point is near the middle of the wire. The known resistance must be changed if this does not happen, after which the lengths of wire are measured. The resistances $R$ and $Q$ can be interchanged and the experiment repeated, and the average of the lengths then found. The Post Office designed a special box of known resistances for measuring an unknown resistance rapidly and accurately by the Wheatstone-bridge principle, as resistance measurement is frequently required in telecommunications.

## Proof of Wheatstone-bridge Formula

Suppose $I_1$ is the current along $ab$ when the bridge is balanced (Fig. 27.24). Then, since no current is diverted at $b$ through the galvanometer G, the current along $bc$ is also $I_1$. If the current along $ad$ is $I_2$, the current along $dc$ is also $I_2$ as no current is diverted at $d$. Further, since no current flows along $bd$, the potential at $b$ must be equal to that at $d$, and hence

$$\text{p.d. between } a, b = \text{p.d. between } a, d \quad . \quad . \quad . \quad . \quad (1)$$
and
$$\text{p.d. between } b, c = \text{p.d. between } d, c \quad . \quad . \quad . \quad . \quad (2)$$

Since Potential difference = Current × Resistance (Ohm's law), it follows from (1) and (2) that

$$I_1P = I_2S \quad . \quad . \quad . \quad . \quad . \quad . \quad . \quad (3)$$
and
$$I_1Q = I_2R \quad . \quad . \quad . \quad . \quad . \quad . \quad . \quad (4)$$

Dividing to cancel the currents,

$$\therefore \frac{P}{Q} = \frac{S}{R}.$$

## The Electric Strain Gauge

By means of the Wheatstone-bridge circuit, the changes of the resistance of a wire can be measured to a high degree of accuracy. In recent years a new method of examining the strains in aircraft surfaces and the surfaces of roads has been developed, which utilizes the change of resistance of a wire with strain. The apparatus is called an *electric strain gauge*, and consists simply of a fine eureka or Nichrome wire of several hundred ohms resistance, which is firmly attached by a special glue to the surface of the aeroplane, for example, under test. As the surface is strained, the metal wire is extended or compressed, and the wire undergoes a change of resistance bearing a definite relationship to the change of its dimensions, which in turn is related to the strain of the surface. By employing various wires attached to different parts of the surface the strains at many parts of the surface can be measured simultaneously when the aircraft is tested.

## Resistivity

By means of the voltmeter–ammeter method or the metre bridge, different lengths of a wire such as manganin can be cut from a reel of it and the resistance measured. Results obtained are as follows:

| $l$ (cm) | $R$ (ohms) |
| --- | --- |
| 50·0 | 1·2 |
| 100·0 | 2·4 |
| 150·0 | 3·6 |
| 200·0 | 4·8 |

We therefore conclude that, for a given diameter of wire, the resistance is directly proportional to the length, or $R \propto l$.

To find the effect of the cross-sectional area on resistance, the same length is cut from reels of constantan (eureka) or manganin wire which have different diameter. Results show that if $a$ is the area of cross-section, $R \propto l/a$.

The area is $\pi r^2$, where $r$ is the radius; or $\pi (d/2)^2$, which is $\pi d^2/4$, where $d$ is the diameter. Thus the thicker the wire, the lower is the resistance of a given length. If the diameter is doubled the resistance decreases to *one-quarter* of its original value for the same length of wire.

The two results can now be combined into one. If $l$ is the length of a given material and $a$ is the cross-sectional area, then, if $R$ is the resistance,

$$R \propto \frac{l}{a}$$

Thus
$$R = \rho \frac{l}{a}, \quad \ldots \quad \ldots \quad (1)$$

where $\rho$ is a constant known as the *resistivity* of the material.

Since $R = \rho$, from (1), when $l = 1$ and $a = 1$, resistivity can be defined as the resistance of unit length of material of unit cross-sectional area. Though wires have circular cross-section, a cube of side 1 cm is a simple shape of material of length 1 cm and cross-sectional area 1 cm². Thus resistivity is sometimes said to be the resistance between opposite faces of 'a unit cube' of material:

From (1)
$$\rho = \frac{R \cdot a}{l} \quad \ldots \quad \ldots \quad \ldots \quad (2)$$

Now the unit of $R$ is the 'ohm', the unit of $a$ is 'cm²' and the unit of $l$ is 'cm'. The unit of $\rho$ from (2) is thus *ohm cm*, cancelling the 'cm' of $l$ into the 'cm²' of $a$. If $R$ is measured in smaller units than the ohm such as the *microhm* (one-millionth ohm), the unit of resistivity is '*microhm cm*'. In SI units, $l$ is in metre, $a$ in metre², and $R$ in ohm in (1) above, and the unit of resistivity is then 'ohm metre', symbol $\Omega$ m.

## Measurement of Resistivity

The resistivity of a material such as manganin or copper can be found by cutting a suitable length of the wire from a reel and measuring

(1) the length $l$ with a metre rule;
(2) the diameter $d$ by a micrometer gauge;
(3) the resistance $R$ by a voltmeter–ammeter method (p. 494) or a metre-bridge method (p. 502).

As an illustration, suppose $l = 98 \cdot 5$ cm, $d = 0.64$ mm $= 0 \cdot 64 \times 10^{-3}$ m, $R = 3 \cdot 6 \ \Omega$. Now

$$\rho = \frac{R \cdot a}{l} = \frac{R \cdot \pi d^2}{4l}$$

$$\therefore \rho = \frac{3 \cdot 6 \times \pi \times (0 \cdot 64 \times 10^{-3})^2}{4 \times 0 \cdot 985}$$

$$= 0 \cdot 000 \ 001 \ 2 \ \Omega \ \text{m}$$
$$= 1 \cdot 2 \times 10^{-6} \ \Omega \ \text{m}$$

## Resistivity Values

Tables of the resistivity of different materials of wire have been compiled for use by manufacturers, and reels of wire of different diameter (gauge) are made. Copper has the low resistivity of $1 \cdot 72 \ \mu\Omega$-cm (1 $\mu\Omega$ is one-millionth of an ohm), and Nichrome has a resistivity of about 100

$\mu\Omega$-cm (or 0·0001 $\Omega$-cm). Nichrome is used as the material for *electric fires*, because it does not oxidize when white-hot. Constantan or manganin is used for making coils of *standard resistances*, whose resistances are known to a high degree of accuracy, because these materials show very little change of resistance when their temperature rises.

Once the resistivity of a material is known, a length of it can always be cut off to make a particular resistance for an appliance. As an illustration, suppose that a coil of 10 $\Omega$ is required to be made from a reel of manganin wire of diameter 0·40 mm, the resistivity of manganin being 44 $\mu\Omega$ cm, or 0·000 044 $\Omega$ cm. Since the area $a$ of the circular cross-section $= \pi r^2$, where $r$ is the radius of the circle, and $r = 0·02$ cm, it follows that $a = \pi \times 0·02^2$ cm$^2$. The length $l$ of wire required is given by $R = \dfrac{\rho l}{a}$ and since $R = 10$ ohms and $\rho = 0·000044$ ohm-cm, we have

$$10 = \frac{0·000\,044 \times 1}{3·14 \times 0·02^2} = \frac{0·000\,044 \times l}{3·14 \times 0·0004}$$

$$\therefore\ l = \frac{10 \times 3·14 \times 0·0004}{0·000\,044} = 285 \text{ cm.}$$

Thus a length of 285 cm of the wire will form a coil of 10 ohms resistance.

The variation of resistance with temperature is an important factor taken into account in the design of electric appliances. The resistivity of copper, used in cables and in transformers, increases when it becomes hot. The resistivity of tungsten, used in electric-lamp filaments, also increases when hot. Conversely, at very low temperatures near the absolute zero the electrical resistance of metals vanishes, and we have a condition of *super-conductivity*. If a current is established it can circulate for days on its own. Carbon and non-metals decrease in resistance with temperature rise.

## Example

2 metre of resistance wire, area of cross-section 0·50 mm$^2$, has a resistance of 2·20 $\Omega$. Calculate: (*a*) the resistivity of the metal; (*b*) the length of the wire which, connected in parallel with the 2 m length, will give a resistance of 2·00 $\Omega$.

(*a*) $$R = \frac{\rho l}{a}, \text{ or } \rho = \frac{R \cdot a}{l}$$

Now $a = 0·50$ mm$^2 = 0·50 \times 10^{-6}$ m$^2$.

$$\therefore\ \rho = \frac{2·20 \times 0·50 \times 10^{-6}}{2}\ \Omega \text{ m}$$

$$= 5·5 \times 10^{-7}\ \Omega \text{ m.}$$

(*b*) Suppose $R$ is the resistance of the wire required.

Then $$\frac{1}{2·00} = \frac{1}{R} + \frac{1}{2·20}$$

$$\therefore\ \frac{1}{R} = \frac{1}{2·00} - \frac{1}{2·20} = \frac{0·1}{2·2}$$

$$\therefore\ R = 22\ \Omega$$

But                                    $R \propto l$

$$\therefore \quad l = \frac{22}{2 \cdot 2} \times 2 \, \text{m}$$

$$= 20 \, \text{m}.$$

## SUMMARY

1. In metals the carriers of electricity are *electrons* (−ve particles, extremely light); in salt and acid solutions or electrolytes, the carriers of electricity are *positive or negative ions* (atoms which have lost or gained one or more electrons); in gases the carriers are *ions and electrons*; in semiconductors the carriers are *electrons and holes* (a hole drift is equivalent to a drift of +ve electricity equal numerically to that on an electron).

2. Connecting wire is made of pure copper, which has a very low resistance. Resistance wire is made of alloys, such as Nichrome or manganin.

3. Ohm's law states: The ratio *p.d./current* is a constant for a given conductor, provided the physical conditions of the conductor such as temperature remain constant.

4. Fundamental formulae are: $I = V/R$, $V = IR$, $R = V/I$.

5. The ammeter, measuring current, is placed in series in a circuit and has a low resistance. The voltmeter, measuring potential difference, is placed in parallel with the circuit, and should always have a high resistance compared with the component across which it is connected.

6. In a circuit containing resistances *in series*: (i) the current is the same in each resistance; (ii) the total p.d. = the sum of the individual p.d.s.; (iii) combined resistance = sum of individual resistances.

7. In a circuit containing resistances *in parallel*: (i) the p.d. across each resistance is the same; (ii) the combined resistance is given by $1/R = 1/R_1 + 1/R_2 + 1/R_3$ if there are three wires of resistances $R_1$, $R_2$, $R_3$ respectively.

8. Resistance can be measured by a voltmeter–ammeter method or by a Wheatstone-bridge method.

9. Resistivity is the resistance of a unit length with unit cross-sectional area of the material concerned. Units of resistivity: *ohm centimetre* or ohm metre (SI).

10. The resistance of a wire is given by $R = \rho l/a$.

11. The resistance of a pure metal increases with temperature rise. 'Standard resistances' are made with alloys which have a very low resistance change with temperature rise.

## EXERCISE 27 · ANSWERS, p. 640

1. What is: (*a*) an electric current; (*b*) a coulomb? (*N*.)

**2.** A resistor of 500 Ω, a voltmeter and a 2·00-V accumulator of negligible resistance are all connected in series. If the voltmeter reads 1·20 V, calculate its resistance. (*C.*)

**3.** Fig 27A shows an electric circuit in which some of the quantities are represented by numbers, others by letters. Calculate the values of *x* and *r*. (*C.*)

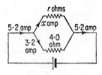

Fig. 27A

**4.** A cell of electromotive force 2·00 V and negligible internal resistance is connected in series with a resistance of 3·50 Ω and an ammeter of resistance 0·500 Ω. Calculate the current in the circuit. When a resistance of 0·100 Ω is connected in parallel with the ammeter, the current changes. Calculate the new current: (i) in the cell; (ii) in the ammeter. (*C.*)

**5.** State *Ohm's law* and show how it leads to a definition of resistance.

Describe an experiment to determine how the resistance of a uniform wire depends on its length. Give a diagram of the arrangement of the apparatus and state the result expected.

A resistance of 6 Ω is connected between two points A and B and resistances of 3 Ω and 4Ω in parallel between B and a point C. If a potential difference of 9 V is applied across A and C, what is: (*a*) the current in each resistance; (*b*) the potential difference between B and C? (*L.*)

**6.** State Ohm's law and define *electrical resistance*.

Distinguish between *resistance* and *resistivity*.

Show on a diagram and explain how you would connect up a voltmeter and an ammeter to an electric lamp in order to measure the resistance of the lamp.

A milliammeter reading up to 50 mA has a resistance of 0·2 Ω. How would you convert the instrument into a voltmeter reading up to 10 V? (*O. and C.*) (See also p. 605.)

**7.** State *Ohm's law*. Deduce an expression for the equivalent resistance of two resistances $R_1$ and $R_2$ in parallel and describe how you would check the result by experiment. (*L.*)

**8.** State Ohm's Law and mention ONE current-carrying device which does not obey it.

A wire is 200 cm long; 0·50 mm in diameter and has a resistance of 3·0 Ω. Another wire of the same material is 300 cm long and 0·60 mm diameter. What is the total resistance of the two wires when they are connected in series?

If wires of resistance 3·0 Ω and 4·0 Ω are put in parallel and a current divides between them so that 8·0 A pass through the first, how many amp will be flowing in the second wire?

What e.m.f. will be required to cause these currents?

From the total current passing through both, and the e.m.f. required, deduce their combined resistance. (*O. and C.*)

**9.** State Ohm's law, and define *resistance*.

How does the resistance of a piece of wire depend on its length, its diameter and the material of which it is made?

In Fig 27B, E is a 2-V cell of negligible internal resistance, and V a voltmeter of resistance 500 Ω that can be connected across XY by means of the switch S. Find the current given by the cell, and the p.d. between X and Y (a) with S open, (b) with S closed so that the voltmeter is in circuit. (O.)

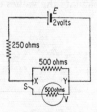

Fig. 27B

**10.** Find the resistance of a piece of wire 1 metre long and area of cross-section 0·25 mm². Resistivity of the material of the wire is $10^{-6}$ Ω m.

**11.** Draw a labelled diagram of a metre-bridge (Wheatstone) circuit set up to compare two resistances.

Describe how you would use it to determine the resistance of a length of wire.

State two precautions you would take to obtain an accurate result.

A wire 1·5 m long and diameter 1·0 mm is made of an alloy of resistivity $44 \times 10^{-8}$ Ω m. What is the resistance of the wire? (N.)

**12.** Define *electrical resistance, resistivity*.

How would you determine the resistance of an electric lamp while it is in use?

Calculate the resistance of a 5·46 m length of lead strip, 0·450 cm wide and 1·25 mm thick. What length of similar strip, but made of copper, would have to be put in parallel with the lead strip for the effective resistance of the combination to be 0·100 Ω?

(Resistivity of lead = $20·6 \times 10^{-8}$ Ω m; resistivity of copper = $1·72 \times 10^{-8}$ Ω m.) (C.)

**13.** Define specific resistance (or resistivity). In what unit is it measured?

How would you use a Wheatstone (metre) bridge to determine the resistivity of the material of a wire?

A wire of length 1 m and uniform diameter has a resistance of 1·05 Ω. What length of wire of the same material, but having half the diameter, would be needed to make a 5-Ω coil? (N.)

**14.** State the factors which determine the *resistance* of a metal wire and indicate how they do so.

Describe the metre-bridge method to determine the resistance of a coil of wire.

A resistance of 10 Ω is placed in the left-hand gap of a metre bridge and resistances of 15 Ω and 10 Ω joined in parallel in the other gap. What is the position of the balance point from the left hand end of the bridge wire? (L.)

# 28

# BATTERIES AND CIRCUITS

### Voltaic Cell

Millions of *batteries* are manufactured annually for use in transistor radio sets, bicycle lamps and torches. They maintain an electric current in a circuit (p. 513). *Accumulators* are made differently from batteries and have advantages over them, as we shall see later. They are used in cars for starting the engine and for lighting, in light trolleys to provide electric power for motors and in ships as a spare electrical supply if the main generator fails.

Batteries owe their origin to Volta's invention at the end of the 18th century. Galvani, a scientific friend, had told him that a dead frog's leg twitched when a knife was lightly placed in contact with it while the frog lay on a zinc plate. Volta came to the conclusion that this was an electrical phenomenon, due to the presence of two unlike metals with the chemical material in the frog's leg between them.

After some experiments he placed cloth soaked in brine alternately between a number of discs of zinc and copper (or silver) (Fig. 28.1). One unit, zinc and copper with brine between them, is called a *cell* (Fig. 28.1 (i)).

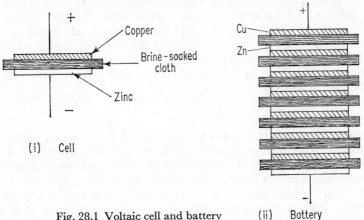

Fig. 28.1 Voltaic cell and battery    (ii)    Battery

The whole arrangement, consisting of a series arrangement of cells to produce a greater effect, is called a pile or *battery* of cells (Fig. 28.1(ii)). It was soon found that a wire became warm when connected to the ends of the battery, so that an electric current flowed in the wire.

Basically, the cell or battery converts chemical energy to electrical energy, that is, the chemical materials are gradually used up when the cell is in action. Usually, the cell has two unlike metal plates or *poles*, with chemicals between them. One pole, called the *positive* (+ve) *pole*, is at a higher electrical potential than the other, called the *negative* (−ve) *pole*, so that a current flows when they are joined by a wire. The terms 'positive' and 'negative' have no mathematical significance.

## Simple Cell

One of the earliest cells to be constructed, known as the *simple cell*, consists of a copper and zinc plate in dilute sulphuric acid (Fig. 28.2).

When a voltmeter is connected between the plates, the instrument shows

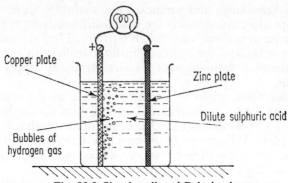

Copper plate

Zinc plate

Dilute sulphuric acid

Bubbles of hydrogen gas

Fig. 28.2 Simple cell and Polarization

that copper is at a higher potential than zinc, that is, copper is the positive and zinc is the negative pole. A suitable small electric bulb joined to the plates lights up, showing the flow of current from one pole to the other. At the same time observation of the plates inside the cell, made by projection through a plane-glass container on to a white screen, shows that a gas collects on the *copper* plate after movement from the zinc plate. It proves to be hydrogen. The light in the bulb goes out after a short time, indicating that the cell is no longer working.

## Electromotive Force (e.m.f.)

When a simple cell is first made, a potential difference of about 1 V is obtained when a voltmeter is connected between the plates. This is due to electrical actions between the metal surface atoms and the acid in contact, which result in a *contact potential difference* between the copper and the acid and between the zinc and the acid. The overall or net effect is a potential difference between the copper and zinc, with the copper at a higher potential than zinc by about 1 V.

The name *electromotive force, e.m.f.*, was given years ago to the p.d. between the terminals of a cell when no resistance is connected to it. It was considered that the e.m.f. was a 'force' inside the cell which maintains the electric current in a wire, for example, connected to the terminals. Now, however, we define e.m.f. in terms of energy, because the cell is a source of energy (see p. 526).

Note that the term 'e.m.f.' can be applied only to a *generator* of electrical energy, that is, a device which changes energy of one kind to electrical energy. Thus we can refer to the *e.m.f.* of a cell, which changes chemical to electrical energy, or to the *e.m.f.* of a dynamo, which changes mechanical to electrical energy. But the term is never used in connection with a wire which carries a current, because the wire, by itself, does not 'generate' energy.

### Defects of Simple Cell · Polarization

As already stated, bubbles of hydrogen gas collect on the copper plate when the simple cell is working. The gas is originally obtained when zinc reacts with sulphuric acid, and moves through the solution to the copper as explained on p. 544, where it collects. As a gas has a very high electrical resistance, the current in the circuit is therefore considerably reduced.

Hydrogen and zinc in the cell also have an e.m.f. between them, as they are unlike elements. This e.m.f. acts in the opposite direction to that between the copper and zinc, and is therefore known as a 'back e.m.f.'. The back e.m.f. due to hydrogen also reduces the current in the circuit in a short time, and when the cell stops working it is said to be *polarized*. The liberation of hydrogen at the copper plate thus produces *polarization*.

Another defect of the simple cell is that commercial zinc has impurities such as iron in it, which form tiny cells with the zinc. Bubbles of gas are therefore obtained from zinc even when the cell is not working, and the zinc is used up unnecessarily. This is called *local action*. To prevent it, mercury is rubbed over the zinc surface with a cloth. A film of mercury dissolves the zinc, which is then protected from the particles of iron present.

### Daniell Cell

Polarization of cells can be prevented by the addition of a suitable chemical, which is known as the *depolarizer*. This should not mix with the

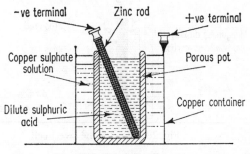

Fig. 28.3  Daniell cell

main chemical of the cell, such as the acid, called the *electrolyte*. However, the tiny electrical particles (ions) which carry the current inside the cell must move from one pole to another through both chemicals. Consequently, a porous pot, one with very tiny holes in it, is used to keep them separated, as shown in Fig. 28.3.

Many different types of cells were invented. The Daniell cell, Fig. 28.3, has:

(1) a +ve pole of copper – a copper container;
(2) a −ve pole of zinc – a zinc rod;
(3) an electrolyte of dilute sulphuric acid ($H_2SO_4$) – inside the porous pot,
(4) a depolarizer of copper sulphate ($CuSO_4$) solution – in the copper container.

When the cell is used: (i) zinc is gradually used up ($Zn + H_2SO_4 \rightarrow ZnSO_4 + H_2$); (ii) the hydrogen is eliminated by the copper sulphate solution ($H_2 + CuSO_4 \rightarrow H_2SO_4 + Cu$) so that hydrogen does not reach the copper pole.

The e.m.f. of the cell is initially about 1·1 V. It has an advantage for some electrical experiments of producing a small current for a long time without an appreciable drop in its e.m.f. The sulphuric acid and copper sulphate solutions have a tendency to mix through the porous pot, so the copper sulphate solution is poured out when the cell is not used.

## Leclanché Cell, Wet Type

Leclanché designed a cell still in use today. Fig. 28.4. It has:

(1) a +ve pole of carbon – inside the porous pot;
(2) a −ve pole of zinc – inside the glass vessel;
(3) an electrolyte of ammonium chloride ($NH_4Cl$) solution – inside the glass vessel;
(4) a depolarizer of manganese dioxide ($MnO_2$), mixed with powdered carbon – round the carbon pole.

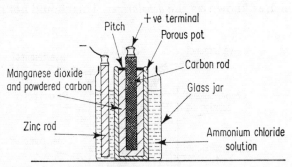

Fig. 28.4 Leclanché (wet) cell

When the cell is used: (1) zinc is gradually used up ($Zn + 2NH_4Cl \rightarrow ZnCl_2 + 2NH_3 + H_2$); (ii) hydrogen reaches the carbon pole and is then oxidized to water by the manganese dioxide ($H_2 + 2\,MnO_2 \rightarrow H_2O + Mn_2O_3$).

The e.m.f. of the cell is about 1·5 V. Its advantage is the cheapness of its chemicals and the relatively high e.m.f. The disadvantage is that it polarizes if used continuously, owing to the hydrogen liberated at the carbon pole. Manganese dioxide is a slow depolarizer and more hydrogen is produced than it can react with. The cell recovers if disconnected from the circuit concerned and left alone. It is therefore used in bell circuits, for example, where a current is required only at intervals. Unlike the Daniell cell, the depolarizer, manganese dioxide, is a solid and a poor conductor, and it is therefore mixed with powdered carbon, which is a good conductor.

## Leclanché Dry Cell

As the Leclanché cell described above is inconvenient to carry about owing to the liquid inside it, 'dry' cells are used. The electrolyte is now made into a paste or jelly of ammonium chloride with flour and gum, so that it is relatively 'dry'.

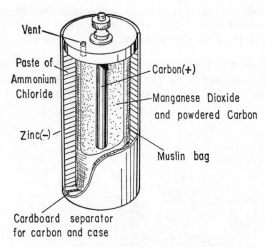

Fig. 28.5 Dry (Leclanché) cell

Fig. 28.5 shows a section of the dry cell. Note that: (i) zinc is now the container; (ii) muslin is used to contain the depolarizer; (iii) a cardboard disc is used at the bottom to prevent the carbon +ve pole touching the zinc −ve container, which would short-circuit the cell. As the e.m.f. of one Leclanché cell is 1·5 V, a 6-V dry battery has four cells inside it in series. A high-tension battery of 120 V has eighty cells in series.

When a battery is used in a torch, one end of the bulb filament, connected to the end of the bulb, makes contact with the carbon pole of the battery. The other end of the filament is connected to the metal casing round the bulb. When the switch of the torch is pressed the metal casing is connected to the zinc pole of the battery, which touches the bottom of the torch. The filament then lights up.

## Accumulators

The Leclanché and Daniell cells are known as *primary cells*. As current is drawn from them their chemicals are gradually used up, and eventually they have to be thrown away. Accumulators, called *secondary cells*, can be recharged by passing a current through them from an outside supply, that is, their chemicals are restored on the plates. They can therefore last for a long time. Further, unlike a primary cell, an accumulator can maintain a large current for a long time without polarization. Accumulators are used in motor-cars to provide the energy for the spark igniting the petrol in the engine (p. 625), and they are stored at Post-Office exchanges to provide the current required in telephone cables when a call is made.

### Lead–Acid Accumulator

There are now a number of different types of accumulators. One of the most widely used is the *lead–acid accumulator* (Fig. 28.6(i)). Another is the nickel–alkaline or 'Nife' cell. We shall deal only with the former. This has:

(1) a +ve pole of lead peroxide – chocolate brown in colour;
(2) a —ve pole of lead – slate-grey in colour;
(3) an electrolyte of dilute sulphuric acid.

The lead peroxide and lead are placed inside several grid plates made of a lead–antimony compound. All the positive plates are connected together and are interleaved with, but separated from, the negative plates.

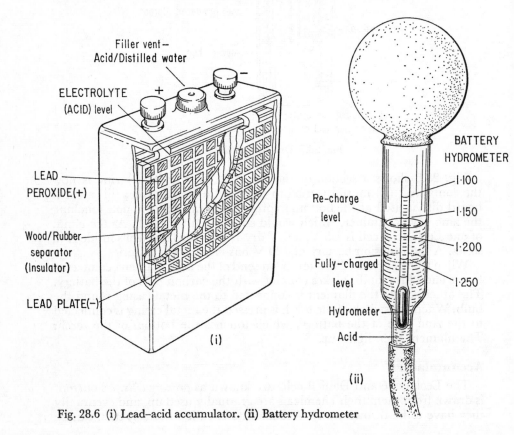

Fig. 28.6 (i) Lead–acid accumulator. (ii) Battery hydrometer

When the accumulator is freshly-made:

(i) it has an e.m.f. of about 2 V;
(ii) the acid has a specific gravity of about 1·25, depending on the
manufacturer's specification.

The *capacity* of an accumulator is classified in 'ampere-hours'. An
accumulator of 30 amp-hours could maintain a current of 1 A for 30 hr,
or a current of 2 A for 15 hr, before recharging is necessary. As higher
currents are drawn, however, the time before recharging becomes less
than the calculated theoretical figure.

## Using (Discharging) the Accumulator

When the accumulator is used a little lead sulphate is formed at both
plates (p. 545) and the e.m.f. and specific gravity of the acid slowly diminishes. The e.m.f. and specific gravity should not be allowed to fall below
the figures recommended by the manufacturer, for example, 1·9 V may be
the minimum for the e.m.f. and 1·18 may be the minimum for the specific
gravity. If this is not observed, excessive lead sulphate, a hard white
substance, is formed on the plates, and the accumulator may be permanently damaged, as it is then difficult to recharge.·

Accumulators thus require care and maintenance if they are to last for
some time. They should be 'topped-up' regularly with distilled water, for
example. The accumulator is said to be *discharging* when it is used, and after
a time recharging is necessary.

## Recharging Accumulators

To restore the materials of a discharged accumulator to their original
condition, a current is passed through the accumulator in the *opposite*
direction to that obtained when using it (Fig. 28.7). A current in one

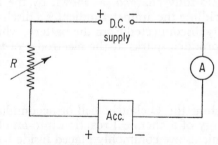

Fig. 28.7  Charging circuit

direction, or direct current (D.C.), is needed. If the mains supply is used
it is changed from an alternating (A.C.) to a direct voltage (D.C.) supply
by means of a rectifier incorporated in the battery-charger.

Care must always be taken to connect the positive terminal of the
charging supply to the *positive* (red terminal) side of the accumulator, and
the negative mains terminal to the negative (black or blue terminal) side

of the accumulator (Fig. 28.7). The current then passes into the accumulator in the opposite way to when the latter was used and recharges it. If a mistake is made and the positive mains terminal is joined to the negative pole side of the accumulator, the accumulator would help the mains to supply current and be used up even more, which is harmful to the accumulator. A suitable rheostat R is essential in the charging circuit, as the sulphuric acid has a very low resistance such as $\frac{1}{50}\,\Omega$, and an ammeter A is included so that the charging current is always known. The manufacturer's recommended charging current should be used.

As the accumulator is charged and restored to a healthy condition, the ammeter reading is observed to fall. This is due to the rise in e.m.f. of the accumulator. It occurs when the lead sulphate, which was formed on the plates while discharging, is changed back to lead peroxide and lead respectively. The acid specific gravity also rises. The charging is near the end of its process when 'gassing' occurs inside the accumulator, as the current begins also to decompose the water present. When the specific gravity reaches the recommended value such as 1·25, which is checked by a hydrometer, Fig. 28.6(ii), the accumulator is· fully recharged. The e.m.f. is then above 2 V, but soon falls to this figure after short use.

**Car Battery or Accumulator**

Some cars have an ammeter as one of the panel instruments, which is in series with the car battery, usually 12 V. The ammeter has a maximum range of 30 A on either side of the zero. When a car is started, the starter motor draws a very large current from the battery for a short time. Frequent short journeys by car thus tend to 'drain' the battery, because each time on starting a very large current flows. A home battery-charger is therefore advisable if a car is used in this way.

When the car is in motion a dynamo, driven by the fan-belt, recharges the battery. On a long journey in daylight, when lights are not used. the battery thus becomes fully charged, as shown by the ammeter reading. When the engine is idling the ammeter indicates a discharge. The dynamo is now automatically disconnected from the battery, which is then providing the energy to produce sparks at the electrodes of the sparking plugs (p. 277).

**Fuel Cells**

Electric cells such as the Leclanché cell or accumulator are devices for converting the energy of a chemical reaction into an electric current. As we have seen, chemicals are commonly placed inside batteries. Here they are continually consumed and eventually have to be replaced by further supplies. A much more efficient process is used in the *fuel cell*. Here the chemicals are stored outside the cell, and are supplied only when the electrical energy is needed.

Most fuel cells employ a reaction between hydrogen and oxygen gases (or air). Hydrogen and oxygen may combine at high temperatures. An explosion then occurs, water is formed and there is a considerable evolution of energy as heat. In the fuel cell, however, the combination of hydro-

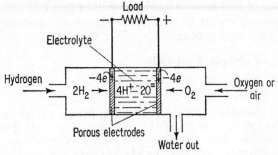

Fig. 28.8 Fuel cell

gen and oxygen takes place at room temperature. Moreover, the energy liberated is chiefly in the required form of electricity.

In one type of cell, shown diagrammatically in Fig. 28.8, hydrogen is supplied on one side of an electrode such as carbon or nickel, which is the negative pole of the cell. Oxygen or air is supplied on the other side of another porous carbon or nickel electrode, which is the positive pole of the cell. The electrodes contain catalysts, such as finely divided platinum, and have an electrolyte, such as potassium hydroxide, between them.

The cell uses no materials other than the hydrogen and oxygen (or air) supplied to it. It has a high efficiency, such as 60%, is compact and rugged, and is small. The fuel cell thus meets the need of artificial satellites, which require small, light batteries, and they have been used recently in rocket space experiments. Intensive research is now being conducted to exploit

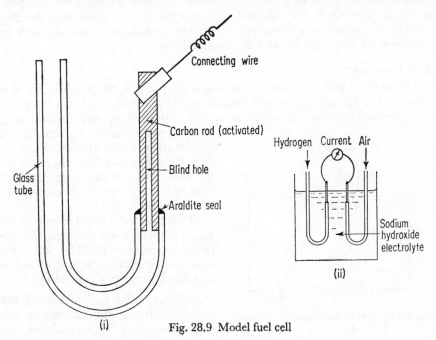

Fig. 28.9 Model fuel cell

the advantages of fuel cells. The Bacon cell, a fuel cell which operates at high temperatures, can deliver about 15 times the amount of power delivered by a lead–acid accumulator of the same size.

### Construction of Simple Fuel Cell

A simple fuel cell can be made as follows:

*Oxygen (or Air) Electrode*

Drill a blind hole down a carbon rod taken from a Leclanché cell and 'activate' the rod by heating it to dull-red heat, followed by cooling in water. Do this several times. Seal the rod into a J-shaped piece of glass tubing. Set up the cell (Fig. 28.9) and use pure oxygen as the gas (alternatively, use air, provided by an aspirator).

*Hydrogen Electrode*

This is made in a manner similar to the oxygen or air electrode. Hydrogen is fed into this electrode at sufficient pressure to cause bubbling of the gas from its interface with the electrolyte.

It is unnecessary to separate the two electrodes by a diaphragm. Concentrated sodium hydroxide can be used as electrolyte.

# E.M.F. AND CIRCUITS

### Electromotive Force (E.m.f.)

Later we shall define and explain how electromotive force (e.m.f.) is related to energy. Here we investigate how e.m.f. is related to the p.d. across different parts of a circuit to which a battery is connected. Suppose a cell or battery X is connected to a circuit of total resistance $R$ (Fig. 28.10). A voltmeter V is connected to the terminals A and B, and an ammeter M and keys $K_1$ and $K_2$, are included.

EXPERIMENT

(1) At first, with no current flowing from the cell, the reading on V is about 1·1 V for a freshly made Daniell cell (Fig. (28.10). This is the *e.m.f.* $E$ of the cell; it is the p.d. at the terminals of the cell on *open circuit*, that is, when it maintains no current (see p. 526 for an energy definition of e.m.f.). Experiment shows that the e.m.f. of the cell remains fairly constant as long as a small current is taken from it when used.

(2) When the circuit is closed by $K_2$, a current flows through the external resistance $R$. The ammeter may now indicate current of 0·3 A and the voltmeter a p.d. of 0·9 V.

*Deductions.* The p.d. of 0·9 V is the p.d. to maintain the current in the *external resistance*. This can be shown

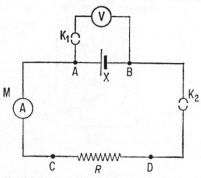

Fig. 28.10 E.m.f., terminal p.d., internal resistance

by placing the voltmeter across C and D, which is directly across $R$. The reading is again 0·9 V. It can also be seen from the circuit diagram that although the voltmeter is joined to the terminals A and B of the cell, *it is effectively measuring the p.d. across $R$.* Thus

Terminal p.d. = P.d. across *external* resistance.

When $R$ is made zero the voltmeter reading becomes zero. No p.d. exists across zero resistance.

### Internal Resistance

Since the e.m.f. of the cell is 1·1 V, this is the p.d. to maintain the current *everywhere* in the circuit while it flows. Now the p.d. across the external resistance is 0·9 V. It therefore follows that (1·1 − 0·9) or 0·2 volt is the p.d. across the *internal resistance* of the cell. The latter is the electrical resistance due to the chemicals between the poles of the cell.

The magnitude of the internal resistance $r$ can be found from Ohm's law. Since the current of 0·3 A measured by the ammeter also flows through the internal resistance, then

$$r = \frac{\text{P.d.}}{\text{Current}} = \frac{0 \cdot 2}{0 \cdot 3} = 0 \cdot 7 \ \Omega \ \text{(approx.)}$$

If the external resistance $R$ is made smaller the current may increase to 0·4 A and the p.d. $V$ at the terminals may decrease to 0·8 V. The p.d. across the internal resistance $r$ is now 1·1 − 0·8 or 0·3 V. Thus:

$$r = \frac{\text{P.d.}}{\text{Current}} = \frac{0 \cdot 3}{0 \cdot 4} = 0 \cdot 8 \ \Omega \ \text{(approx)}.$$

The internal resistance $r$ of the cell is thus fairly constant at a value of about 0·7—0·8 $\Omega$.

By using a voltmeter, one can always find whether a cell has a high or a low internal resistance. Suppose a voltmeter is placed across the terminals of a lead–acid accumulator, for example, on open circuit. The reading is then 2·0 V. If a resistance $R$ of a few ohms is connected to the accumulator so that a current flows, the voltmeter reading is only slightly below 2·0 V. Now the latter is the p.d. across the external resistance $R$. It therefore follows that the p.d. across the internal resistance, the difference between the e.m.f. and the terminal p.d., is extremely small. Hence the internal resistance of the accumulator is very low. This is due to the fact that the acid inside it is a very good conductor.

### E.m.f. and Internal Resistance

The conclusions can be summarized as follows:

(1) *E.m.f.* This is the p.d. across the terminals of the cell on open circuit. It is a constant value for a cell in good condition. In a closed circuit, the e.m.f. is the p.d. across the whole circuit, that is, across both the external resistance $R$ and internal resistance $r$. Hence the current $I$ is calculated by

$$I = \frac{E}{R + r}, \text{ from Ohm's law.}$$

(2) *Internal resistance.* When a current is maintained by a cell the current flows through its internal resistance $r$ as well as the external resistance $R$.

(3) *Terminal p.d.* This is the p.d. $V$ across the external resistance $R$. Hence $I = V/R$. If the e.m.f. $E$ of a cell is 1·5 V and the terminal p.d. $V$ is 0·8 V, the p.d. across the internal resistance is (1·5 − 0·8) or 0·7 V.

## Circuit Calculations

**1.** A cell of unknown e.m.f. $E$ and internal resistance 2 Ω is connected to a 5-Ω resistance. If the terminal p.d. $V$ is 1·0 V, calculate $E$.

Terminal p.d. $V = 1\cdot0$ V = p.d. across 5 Ω

$$\therefore \text{ current in 5-Ω wire } I = \frac{V}{R} = \frac{1\cdot0}{5} = 0\cdot2 \text{ A}$$

$\therefore$ P.d. across internal resistance $r$ of 2 Ω $= Ir = 0\cdot2 \times 2 = 0\cdot4$ V

$\therefore$ E.m.f. $E$ = p.d. across $R$ + p.d. across $r$
$$= 1\cdot0 + 0\cdot4 = 1\cdot4 \text{ V.}$$

**2.** A battery of e.m.f. 40 V and internal resistance 5 Ω is connected to a resistance of 15 Ω. Calculate the terminal p.d.

Since 40 volts $= E$ = p.d. across the whole circuit

$$\therefore I = \frac{E}{R + r} = \frac{40}{15 + 5} = 2 \text{ A}$$

$\therefore$ Terminal p.d. = p.d. across $R = IR = 2 \times 15 = 30$ V.

**3.** The terminal p.d. of a battery is 12 V when an external resistance of 20 Ω is connected and 13·5 V when an external resistance of 45 Ω is connected. Calculate the e.m.f. and the internal resistance of the battery.

For the whole circuit, $\qquad\qquad I = \dfrac{E}{R + r}$

For the terminal p.d., or p.d. across the external resistance $R$,

$$I = \frac{V}{R} = \frac{12}{20} = \frac{E}{20 + r} \quad . \quad . \quad . \quad . \quad . \quad (1)$$

and
$$I = \frac{V}{R} = \frac{13\cdot5}{45} = \frac{E}{45 + r} \quad . \quad . \quad . \quad . \quad . \quad (2)$$

Dividing (1) by (2) to eliminate $E$, and simplifying the left side,

$$\therefore 2 = \frac{45 + r}{20 + r}$$

$$\therefore 2(20 + r) = 45 + r, \text{ or } r = 5 \text{ Ω.}$$

From (1)
$$\frac{12}{20} = \frac{E}{20 + 5}$$

$$\therefore E = \frac{12}{20} \times 25 = 15 \text{ V.}$$

## Cells in Series

To obtain a larger e.m.f. in a circuit, and hence to increase the current, cells can be arranged in series to assist each other (Fig. 28.11). In this case:

(1) Total e.m.f. between terminals A and B = Sum of individual e.m.f.s;

(2) Total internal resistance = Sum of individual internal resistances. Thus if the three cells shown in Fig. 28.11 are identical, and each has an e.m.f. of 2 V and internal resistance 6 Ω, then

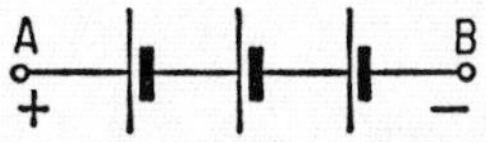

Fig. 28.11  Cells in series

$$\text{Total e.m.f. } E = 3 \times 2 = 6 \text{ V,}$$
$$\text{Total internal resistance } r = 3 \times 6 = 18 \, \Omega.$$

Hence if the external resistance $R = 12 \, \Omega$ the current flowing is:

$$I = \frac{E}{R+r} = \frac{6}{18+12} = \frac{1}{5} = 0 \cdot 2 \text{ A}$$

$$\therefore \text{ P.d. across terminals A and B} = \text{P.d. across } R$$
$$= IR = 0 \cdot 2 \times 12 = 2 \cdot 4 \text{ V.}$$

The p.d. across the total internal resistance $= 6 - 2 \cdot 4 = 3 \cdot 6$ V.

## Cells in Parallel

In charging accumulators it is sometimes an advantage to group a number of them in parallel.

Suppose three *identical* cells, each of e.m.f. 2 V and internal resistance 6 Ω, are arranged in parallel (Fig. 28.12). A voltmeter connected to the terminals A and B shows that the total e.m.f. is only 2 V, the e.m.f. of *one*

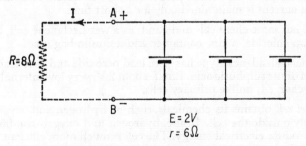

Fig. 28.12  Cells in parallel

cell. This is explained by the fact that as all the positive poles are joined to A and all the negative poles to B, in effect a new cell is obtained whose materials are the sum of those in the individual cells. Now the *e.m.f.* of a cell depends only on the nature of the chemicals used and not on their amount. Hence the e.m.f. is 2 V, the e.m.f. of *one* of the cells.

The internal resistance of the battery is different from that of a single cell. From Fig. 28.12 it can be seen that the internal resistances of the three cells are in parallel. Hence the total internal resistance $r$ is given by:

$$\frac{1}{r} = \frac{1}{6} + \frac{1}{6} + \frac{1}{6} = \frac{3}{6}$$

$$\therefore \; r = \frac{6}{3} = 2 \; \Omega.$$

The battery has thus an e.m.f. $E$ of 2 V and an internal resistance of 2 $\Omega$. If the external resistance is 8 $\Omega$,

$$\therefore \; I = \frac{E}{R + r} = \frac{2}{8 + 2} = 0 \cdot 2 \; \text{A}$$

This is the current in the 8-$\Omega$ wire. The current through each cell is one-third of 0·2 A or 0·07 A (approx.).

## SUMMARY

### Cells

1. *Simple cell* has a +ve copper and −ve zinc pole, and an electrolyte of dilute sulphuric acid. Defects: Polarization due to hydrogen at the copper pole–produces back e.m.f. and high internal resistance; local action–impurities in zinc produce tiny cells on the zinc, surface, which wastes zinc, and is avoided by amalgamating zinc.

2. *Daniell cell* has a +ve copper and −ve zinc pole, electrolyte of dilute sulphuric acid and depolarizer of copper sulphate. No hydrogen now produced. The e.m.f. is about 1 V, and a small current can be maintained for some time.

3. *Leclanché cell*, has a +ve carbon and −ve zinc pole, an electrolyte of ammonium chloride solution and a solid depolarizer of manganese dioxide. The hydrogen produced at the carbon pole is oxidized to water. The e.m.f. is about 1·5 V, and a current is maintained only for a short time.

4. *Dry cell* has same chemicals and e.m.f. as a wet Leclanché cell, but a paste of ammonium chloride, a zinc container and a muslin bag used.

5. *Accumulator*–lead–acid type has +ve lead peroxide and −ve lead poles and electrolyte of dilute sulphuric acid. E.m.f. about 2 V, very low internal resistance. It can be recharged, unlike primary cells.

6. The *fuel cell* obtains its chemicals, such as hydrogen and oxygen (or air), from a supply outside the cell. Inside, hydrogen and oxygen combine to form water and produce electrical energy. The cell is much more efficient than other types of chemical cells.

### E.m.f.

7. The e.m.f. of a cell is the total p.d. across the external and internal resistances (see p. 526 for energy definition of e.m.f.).

8. When a resistance $R$ is joined to the cell, current $I = E/(R + r)$ and p.d. at terminals = p.d. across $R = IR$.

9. Internal resistance can be investigated by using a voltmeter and ammeter.

10. Cells in series: total e.m.f. is sum of individual e.m.f.s., and total internal resistance is sum of individual internal resistances.

Cells in parallel: with *similar* cells, total e.m.f. = e.m.f. of one cell, and total internal resistance is given by the parallel resistance formula.

## EXERCISE 28 · ANSWERS, p. 640

1. What is the value of the electromotive force of: (*a*) a lead accumulator; (*b*) a simple (copper–zinc) voltaic cell; (*c*) a Leclanché cell? (*N.*)

2. Name TWO advantages which a lead accumulator has over a dry cell. (*N.*)

3. Give a diagram showing the structure of a Daniell cell, naming its components.
State one advantage and one disadvantage of the cell and explain on what factors its internal resistance depends. (*L.*)

4. Draw a diagram of a dry (Leclanché) cell and name the parts. (*C.*)

5. Draw a circuit diagram for the re-charging of an accumulator from a direct-current supply, at a definite charging-current; show clearly the polarity of the accumulator and of the supply terminals. If the polarity of the supply terminals were not marked, how would you determine the positive terminal by a magnetic method? (*C.*)

6. Describe a primary cell in common use and explain how the defects of a simple cell have been overcome.
A Daniell cell, of e.m.f. 1·1 V and internal resistance 2 Ω is connected in series with a switch and a resistance of 3 Ω. What is the reading of a high-resistance voltmeter connected across the poles of the cell when the switch is: (*a*) open; (*b*) closed? Explain the difference in the readings. (*L.*)

7. What is the nature of the positive plate of a lead accumulator when the accumulator is: (*a*) charged; (*b*) discharged?
State TWO precautions which must be observed in the use of an accumulator. (*N.*)

8. A battery consists of two cells joined in parallel. If each cell has an e.m.f. of 1·5 V and internal resistance 5 Ω, what current will flow through an external resistance of 5 Ω? (*N.*)

9. Describe one form (either wet or dry) of the Leclanché cell.
Explain what is meant by polarization. Show how depolarization is achieved in the form of cell you have described.
A 120-V dry battery sends a current of 0·24 A through an external circuit of resistance 440 Ω. What is the internal resistance of the battery? What current would pass if the external resistance were reduced to 240 Ω? (*O.*)

10. You are supplied with an accumulator, ammeter, voltmeter, rheostat, key, copper wire and an unknown resistance X. Describe, with the aid of a circuit diagram, how you could use this apparatus to determine the resistance X.
Explain why the method does not lead to an accurate result.
A cell of e.m.f. 1·5 V and internal resistance 2·0 Ω is connected in series with an ammeter of resistance 0·5 Ω and a 5·0-Ω resistance. What will be: (i) the ammeter reading, and (ii) the potential difference across the terminals of the cell? (*N.*)

11. Define *the ampere* and *the volt.* How would you investigate the way in which the current through a wire depends on the potential difference between its ends?
Three cells, each of e.m.f. 1·5 V and internal resistance 0·6 Ω are joined in series to form a battery. What current passes when the battery is connected across a 5-Ω resistance, and what is the p.d. between the terminals of the battery? (*O.*)

12. What is meant by the internal resistance of a cell?
You are supplied with a cell, an ammeter of negligible resistance, two standard resistances, a switch and connecting wire and no other instruments. How would you use this apparatus to find the internal resistance of the cell?

A cell of e.m.f. 1·5 V is connected in series with a coil of resistance $R$ and an ammeter of resistance 0·20 Ω. The difference of potential between the terminals of the cell is 1·35 V and the ammeter reading is then 0·3 A. Find the value of $R$ and the internal resistance of the cell. (N.)

**13.** Give an account of a simple form of voltaic cell (the 'simple cell'), and explain its mode of action. What is meant by the terms *polarization* and *local action*? Discuss briefly how these defects are minimized in EITHER the Daniell OR the Lechanché cell.

A battery is made up of three dry cells in series, each cell having e.m.f. 1·5 V and internal resistance 5 Ω. It is used to light a 75-Ω lamp. Find: (a) the current taken; (b) the p.d. across the terminals of the lamp. (O.)

**14.** Draw a labelled diagram of a common dry cell and describe what takes place inside it when it is producing a current.

Two identical cells, each of e.m.f. $E$ V and internal resistance $r$ Ω are joined: (a) in series; (b) in parallel. What is the e.m.f. and the internal resistance of the arrangement in each case?

A cell is joined in series with a resistance of 2 Ω, and a current of 0·25 A flows through it. When a second resistance of 2 Ω is connected in parallel with the first the current through the cell increases to 0·3 A. What is: (a) the e.m.f.; (b) the internal resistance of the cell? (N.)

**15.** State Ohm's law and describe the experiment you would perform in order to confirm it.

In the circuit shown in the diagram the e.m.f. of the cell is 1·5 V and its internal resistance is 2·5 Ω. Calculate the current through each resistance, and also the p.d. between the terminals of the cell. (O.) (Fig. 28A).

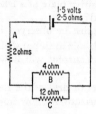

Fig. 28A

# 29

# ELECTRICAL ENERGY · HEATING EFFECT
# OF CURRENT

## Electrical Energy

The heating effect of an electric current serves a most useful purpose in the home and in industry. For example, it is used to light electric-lamp filaments, to make electric cookers function, electric fires radiate and electric blankets warm up.

The heating effect is obtained from a transformation of electrical energy to heat energy. Basically, it is due to the electrons 'colliding' with the atoms when moving through the metal structure. Electrical energy can also be transformed to many other useful forms of energy. In electric motors, for example, most of the electrical energy supplied is changed to mechanical energy of rotation. In the telephone earpiece most of the electrical energy is transformed into sound energy.

## Magnitude of Electrical Energy

As explained on p. 490, electrical energy is obtained when a quantity of electricity moves between two points which have a potential difference. Thus, from the definition of the volt, 1 J (joule) of energy is obtained when 1 coulomb moves between two points at a p.d. of 1 V. Thus if 10 coulombs move between two points having a p.d. of 2 V, 20 J of energy are obtained. Generally, if $Q$ coulombs move through two points having a p.d. of $V$ volts, then

$$\text{Energy } W = QV \text{ joules.} \quad . \quad . \quad . \quad . \quad (1)$$

A more useful formula in current electricity is one which utilizes the current $I$ which flows. Since $Q$ coulombs pass a section of a wire in $t$ sec when the current is $I$ amp, then $Q = I \cdot t$ (p. 487).

$$\therefore \quad \text{Energy } W = IVt \text{ joules, from above.} \quad . \quad . \quad (2)$$

## Electrical Machines

Suppose that a battery is connected to an electric motor, as represented in Fig. 29.1(i). The electrical energy liberated inside the motor is $IVt$ joules, from (2), where $I$ is the current in amps flowing, $V$ is the p.d. between the terminals in volts and $t$ is the time in seconds. Now a motor is a device which converts electrical energy to mechanical energy of rotation, and hence most of the $IVt$ joules is converted into mechanical energy. The rest is converted into heat in the wire. A radio transmitter converts some of the electrical energy from a battery into electrical energy in the form of radio waves, which travels out from the transmitter, and the rest of the energy is converted into heat in the circuit.

Suppose, however, that a battery is connected to an ordinary coil of wire. This time, unlike the case of an electric motor, *all* the electrical energy is converted into heat energy in the wire. This occurs in the coil of an electric iron or electric fire (Fig. 29.1(ii)). In this case, therefore,

$$\text{Heat energy } H = IVt \text{ joules}$$

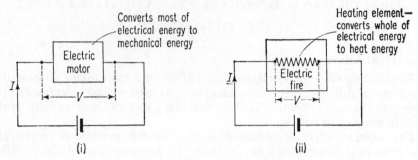

Fig. 29.1 Conversion of electrical energy

## Electrical Power

The *power* of any machine is defined as the rate at which it delivers energy or does work (see p. 126). Hence

$$\text{Power } P = \frac{\text{Energy delivered}}{\text{Time taken}}.$$

Thus the electrical power delivered to a motor or heating coil, for example, is given by

$$P = \frac{IVt}{t} = IV \text{ joules per second}$$

$$\therefore P = IV \text{ watts}$$

since 1 W is defined as the rate of working at 1 J per s (p. 126). Engineers sometimes remember this formula by:

$$Watts = Amps \times Volts$$

Any electrical appliance, which takes a current $I$ amp when the p.d. across its terminals is $V$ volts, uses power of $IV$ watts. Thus an electric motor, which is connected to a 50-V supply and takes a current of 5 A, uses power of $50 \times 5$ or 250 W. An electric fire carrying a current of 4 A when the mains p.d. is 240 V uses a power of 960 W. An electric-lamp filament carrying a current of $\frac{1}{4}$ A when the mains p.d. is 240 V uses a power of 60 W. The total power of the B.B.C. transmitters is more than a million watts.

## E.m.f. of Battery

The e.m.f. $E$ of a battery, a machine for producing electrical energy, can now be defined as *the total power produced per unit current*. Thus, from the definition, if a resistance $R$ is connected to the battery terminals, the total power produced is $IE$ watts, assuming $E$ is in volts and the current $I$ is in amperes. Part of the power is delivered to the resistance $R$ and the rest is absorbed by the internal resistance $r$. Hence

$$EI = \text{Power in } R + \text{Power in } r$$
$$= IV \qquad\qquad + Iv$$

where $V$ is the p.d. across $R$ and $v$ is the p.d. across $r$. Dividing by $I$,

$$\therefore E = V + v$$

or $E = $ P.d. across external resistance $+$ p.d. across internal resistance.

### Example

Define electromotive force. Calculate the total energy provided by a battery of e.m.f. $2 \cdot 50$ V when it causes a steady current of $0 \cdot 40$ A to flow for 15 min through an electric bulb.

If the battery had an internal resistance of $2 \cdot 00\ \Omega$, calculate the number of calories of heat dissipated in the electric bulb in that time. $(L)$.

$$\text{Total energy provided} = IEt = 2 \cdot 5 \times 0 \cdot 4 \times (15 \times 60)$$
$$= 900 \text{ J}$$

$$\text{Energy dissipated in internal resistance} = Ivt$$
$$= 0 \cdot 4 \times (0 \cdot 4 \times 2) \times (15 \times 60) \text{ J}$$
$$= 288 \text{ J},$$

since $v = Ir = (0 \cdot 4 \times 2)$ volts.

$$\therefore \text{Energy dissipated in electric bulb} = 900 - 288 = 612 \text{ J}$$

### Electric filament Lamp

The electric lamp changes electrical energy into heat and light. The great majority of present-day electric lamps contain a tungsten filament

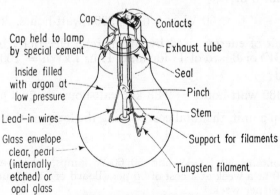

Fig. 29.2 Electric filament lamp

or wire, with a small amount of an inert gas, such as argon, to prevent oxidation and evaporation of the metal (Fig. 29.2). The lamps are connected in parallel across the mains (see p. 498).

## Fluorescent Lamps

A gas at low pressure inside a long glass tube glows when a high voltage is connected across metal terminals at each end of the tube. Some gases, such as mercury vapour, produce invisible ultra-violet rays when they glow, and this is the basis of the *fluorescent lamp*, now used in the home and in industry.

Most of the light coming from the lamp is obtained from fluorescent powders coating the inside of the tube, which are exposed to the ultra-violet light. Different coloured light is obtained by using different powders. There is no glare from such lamps, and little shadow. They have the advantage of being more efficient than the filament type of lamp, as a higher percentage of the electrical energy supplied is converted to light energy. The cost of running them is therefore much less than the filament type.

## Commercial Units of Electrical Energy and Power

In electrical engineering a larger unit of power than the watt is frequently used. This is the kilowatt, kW, which is 1000 W, so that 1500 W is 1·5 kW.

The *megawatt* is 1 million watts. One horse-power (h.p.), defined as the rate of working at 550 ft lbf per sec, is about 746 W or $\frac{3}{4}$ kW. The small electric motor used in one type of vacuum cleaner is $\frac{1}{6}$ h.p. or about 125 W.

When commercial electricity companies present their accounts for the cost of lighting and heating premises they are concerned with the amount of electrical *energy* consumed. Energy = Power × Time, from the definition of power, and hence a unit of energy could be a unit of power multiplied by a unit of time. In industry a unit of energy known as the *watt-hour* is used. It is the energy consumed when a power of 1 W is used for 1 hr. Thus the energy consumed by six 60 W lamps in a room, burning continuously for 8 hr,

$$= 6 \times 60 \text{ W} \times 8 \text{ hr} = 2880 \text{ watt-hours.}$$

A larger unit of energy, used by commercial electricity boards, is the *kilowatt-hour kWh* or Board of Trade unit. Thus 1 kWh = 1000 watt-hours = $3.6 \times 10^6$ J

Hence     2880 watt-hours = 2·88 kilowatt-hours or 2·88 units.

If the cost is 4p a unit, the total cost of using the lamps is therefore 11·52p.

*Example*

A factory contains 30 100-W and 50 60-W lamps, which are used continuously for 20 hr a week at a cost of 2p per (Board of Trade) unit. Calculate the total cost per per week.

$$\text{Total power used} = 30 \times 100 + 50 \times 60 = 6000 \text{ W.}$$

$$\therefore \text{ Number of kilowatt-hours} = \frac{6000}{1000} \times 20 = 120$$

$$\therefore \text{ Cost} = 120 \times 2p = \text{\textsterling}2\cdot40$$

## Power in Resistance

Generally, the power consumed in any electrical machine or appliance is given by $P = IV$. In the case of an electric lamp or electric fire the p.d. $V = IR$, where $R$ is the resistance of the element in the lamp or fire.

$$\therefore P = IV = I \times IR = I^2R.$$

Also
$$P = IV = \frac{V \times V}{R} = \frac{V^2}{R}.$$

Thus in addition to $P = IV$, the power dissipated in resistance wires can be calculated from the relations $P = I^2R$ or $V^2/R$.

Consider the case of an electric lamp rated as '60 W, 240 V'. This means that when the p.d. applied is 240 V the power consumed is 60 W. (It should be noted that when the p.d. applied is not 240 V, the filament resistance is altered because it depends on its temperature (p. 496). The power consumed is then not 60 W.) From $P = IV$ it follows that the current $I$ flowing in the lamp when used normally is given by

$$60 = I \times 240$$

$$\therefore I = \frac{60}{240} = \tfrac{1}{4}\text{ A.}$$

$$\therefore \text{ Resistance of filament } R = \frac{V}{I} = \frac{240}{\tfrac{1}{4}} = 960\ \Omega.$$

This result can also be obtained directly by using the relation $P = V^2/R$. In this case,

$$60 = \frac{240^2}{R}$$

$$\therefore R = \frac{240^2}{60} = 960\ \Omega.$$

The relation $P = V^2/R$ shows that, since $V$ is the same for lamps in parallel across the mains, a 100-W lamp filament has a *smaller* resistance than a 60-W filament.

## Heating Effect of Current

We have already seen that electrical energy can be converted entirely into heat in a wire.

An investigation into the heat produced can be carried out in the laboratory by placing a suitable resistance coil C of manganin or Nichrome inside a lagged calorimeter, and adding water sufficient to cover the coil (Fig. 29.3). Accumulators B, a small rheostat S and an ammeter A are connected in series with C, and a thermometer T is used to measure the temperature of the water.

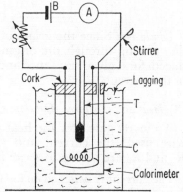

Fig. 29.3 Heat energy from electrical energy

## (1) *Heat and Time (Constant Current and Resistance)*

To investigate how the heat produced depends on the time, the current is kept constant by means of the rheostat at a suitable value such as 2·0 A. As the current is switched on, temperature and time ($t$) readings are taken at convenient intervals such as 1 min.

When the temperature rise $\theta$ is plotted against the time, a straight-line graph passing through the origin is obtained (Fig. 29.4). Since the temperature rise is proportional to the heat produced, it follows that *the heat is directly proportional to the time* for a given current and resistance.

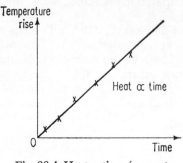

Fig. 29.4 Heat *v.* time (current constant)

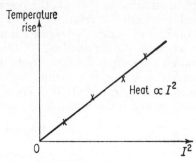

Fig. 29.5 Heat *v.* $I^2$ (time constant)

## (2) *Heat and Current (Constant Time and Resistance)*

To investigate how the heat produced depends on the current, currents of 1·0, 1·5, 2·0, 2·5 and 3·0 A respectively are passed for 10 min each time. The temperature rise $\theta$ is noted on each occasion, as below.

*Results*

| $I$ (amp) | $\theta$ (deg C) | $I^2$ |
|---|---|---|
| 1·0 | 2·4 | 1 |
| 1·5 | 5·6 | 2·25 |
| 2·0 | 9·8 | 4·0 |
| 2·5 | 15·8 | 6·25 |
| 3·0 | 21·2 | 9·0 |

*Conclusion.* Since the graph of $\theta$ v. $I^2$ (*not* v. $I$) is found to be a straight line passing through the origin (Fig. 29.5), it follows that the heat produced, for a given resistance and time, is directly proportional to the *square* of the current.

## (3) *Heat and resistance (Constant Current and Time)*

To investigate how the heat produced varies with the resistance of the wire used, the current and time are kept constant. Two resistances $R_1$ and $R_2$ are placed in series in a circuit, as shown in Fig. 29.6.

The current is switched on for 15 min, for example, and the temperature rise $\theta_1$ and $\theta_2$ in each calorimeter A and B is observed. Now the heat produced in calorimeter A, $H_1 = (m_1c + C_1)\,\theta_1$, where $m_1$ is the mass of water inside of specific heat capacity $c$ and $C_1$ is the heat capacity of the calorimeter. The heat produced in calorimeter B, $H_2 = (m_2c + C_2)\theta_2$, where $m_2$ is the mass of water and $C_2$ is the heat capacity of this calorimeter.

*Calculation*
$$\frac{H_1}{H_2} = \frac{(m_1c + C_1)\theta_1}{(m_2c + C_2)\theta_2},$$

and the ratio $H_1/H_2$ can hence be calculated.

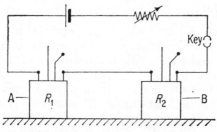

Fig. 29.6 Resistances in series.

*Conclusion.* Experiment shows that if the resistances $R_1$ and $R_2$ are known, or their ratio determined before the experiment by a voltmeter–ammeter method, then
$$\frac{H_1}{H_2} = \frac{R_1}{R_2}$$

**(4) *Heat and Resistance (Constant P.d. and Time)***

To investigate how the heat produced varies with the resistances $R_1$ and $R_2$ when their *potential difference* and the time are constant, the coils are

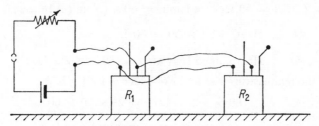

Fig. 29.7 Resistances in parallel

placed in parallel in a circuit (Fig. 29.7). The circuit is switched on for 15 min, for example, and the temperature rise $\theta_1$ and $\theta_2$ in each calorimeter is observed.

*Calculation*
$$\frac{H_1}{H_2} = \frac{(m_1c + C_1)\theta_1}{(m_2c + C_2)\theta_2},$$

and hence the ratio $H_1/H_2$ can be calculated.

*Conclusion.* Experiment shows that
$$\frac{H_1}{H_2} = \frac{R_2}{R_1},$$

or that on this occasion, unlike the case of constant current as in experiment (3), the heat produced is *inversely* proportional to the resistance.

### Joule's Laws of Electrical Heating

Joule, in 1843, was the first person to investigate the heating effect of an electric current. He used a method similar to that outlined in the first three cases described. His conclusions, known as *Joule's laws of electrical heating*, may be stated as follows:

The heat developed in a wire is proportional to: (i) the time (for a given resistance and current); (ii) the square of the current (for a given resistance and time); (iii) the resistance of the wire (for a given current and time).

These laws are easily derived from our previous work. On p. 525 it was shown that the heat energy $W$ in a wire is given by $W = IVt$ joules. Now $V = IR$, from Ohm's law, where $R$ is the resistance of the wire. Hence $W = IVt = I \times IR \times t = I^2Rt$ joules, which is a concise statement of Joule's laws. Since $I = V/R$, then the heat $= IVt = V^2t/R$ joules. Thus if $V$ and $t$ is constant, heat $\propto 1/R$. This is the conclusion of experiment (4) on p. 531. Note that, if the *current* is constant, heat $\propto R$.

### Examples

**1.** A current of 2 A is passed into a wire of resistance 50 $\Omega$ for 1 min. If the wire is totally immersed inside 80 g of water in a can of thermal capacity 42 J/K, calculate the temperature rise of the water.

$$\text{Heat supplied} = I^2Rt \text{ J},$$

where $I = 2$ A, $R = 50$ $\Omega$, $t = 1$ min $= 60$ s. (Note $t$ is in *seconds*.)

$$\therefore \text{ Heat} = 2^2 \times 50 \times 60 \text{ J}.$$

Now $$\text{Heat gained by water} = (80 \times 4\cdot2)x \text{ J},$$

where $x$ is the temperature rise and $c$ of water is $4\cdot2$ J/g K

$$\therefore (80 \times 4\cdot2 + 42)\, x = 2^2 \times 50 \times 60.$$

Solving, $$\therefore x = 32 \,°\text{C}.$$

**2.** Define the *watt* and the *joule*. If an electric kettle contains a 1000 watt heating unit, what current does it take from 230-V mains? How long will the kettle take to raise 1000 g of water at 15°C to the boiling point, if 90 per cent of the heat produced is used in raising the temperature of the water? How much would this cost if the charge is 1p for a kilowatt-hour?

The *watt* is the rate of working at 1 J/s. 1 J $= 1$ newton $\times$ 1 second.

(i) From $P = IV$, $\qquad 1000 = I \times 230$

$$\therefore I = \frac{1000}{230} = 4\cdot3 \text{ A}.$$

(ii) Heat to raise water to boiling point
$$= 1000 \times 4\cdot2 \times (100 - 15) = 357\,000 \text{ J}$$

Now      Energy supplied in $t$ second $= 1000t$ J

$\therefore$ Heat to raise water to boiling point $= 90\%$ of $1000t$

$$\therefore 900t = 357\,000$$

$$\therefore t = \frac{357\,000}{900} = 397 \text{ sec}$$

(iii) Number of kilowatt-hours $= \dfrac{1000\ (W)}{1000} \times \dfrac{397}{3600} \text{ (hr)} = 0\cdot11$

$$\therefore \text{Cost} = 0\cdot11\text{p.}$$

## Experiment on Heating Effect of Current

Electrical energy is expressed in joules, the same units as mechanical energy, and joules are now units of heat (p. 267). An experiment to compare the heat produced by a current with the energy supplied is shown in Fig. 29.8.

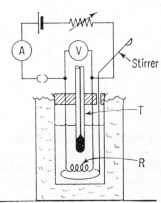

Fig. 29.8 Heat energy from electrical energy

EXPERIMENT

Weigh a calorimeter empty, and then with sufficient water to cover a heating coil R. Connect the circuit as shown, with a suitable ammeter A and voltmeter V. Switch on the current, start a stop-clock at the same time and observe the temperature rise $\theta$ after a measured time $t$ by means of the thermometer T. Keep the current $I$ amp or the p.d. $V$ volts constant during the time by suitably adjusting the rheostat, and note their values.

*Calculation.* Suppose $m$ is the mass of water used, $c$ is its specific heat capacity, and $C$ is the heat capacity of the calorimeter. Then

$$\text{Heat developed} = (mc + C)\theta \text{ J} \quad . \quad . \quad . \quad (1)$$

$$\text{Electrical energy supplied} = IVt \text{ joules} . \quad . \quad . \quad . \quad (2)$$

Compare your results in (1) and (2) assuming $c$ is $4\cdot2$ J/g K. Note that this experiment can be used to find $c$ for water, from (1) = (2).

## Hot-wire Ammeter

The heating effect of a current is utilised in the construction of an instrument known as a 'hot-wire' ammeter. The essential details of the instrument are shown in Fig. 29.9. PQ is a length of resistance wire connected to fixed terminals at P and Q, and the current to be measured flows through the wire, which then becomes hot. The length of the wire now increases, so that it begins to sag slightly; the point M on the wire is thus lowered. A silk thread passing round a grooved wheel A has one end connected to T, a point on a wire MN, and the other end to a spring

S. As the wire sags, A turns, and a pointer X connected to the axle moves over a scale. This is calibrated in amperes by passing known currents into PQ. The heating effect is proportional to the square of the current for a given resistance, so that the scale is not a uniform one; that is, equal steps

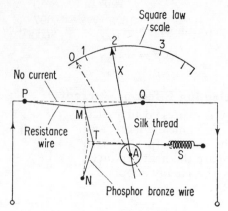

Fig. 29.9 Hot-wire ammeter

along the divisions of the scale do not represent equal increases in current (see p. 604). Heat is produced whichever direction the current flows. Hence the hot-wire ammeter can measure alternating current (A.C.) as well as direct current (D.C.).

## High tension (H.T.) Transmission

Since power $P = IV$, a given amount of power can be delivered either at high current $I$ and low voltage $V$, or at low current and high voltage; for example, a power of 2000 W is obtained if the p.d. (voltage) is 1000 V and the current supplied is 2 A. or if the p.d. is 100 V and the current supplied is 20 A. If the current flows along wires of fixed resistance the power converted to heat is proportional to the square of the current, since $P = I^2R$ (p. 529). Consequently, less power is wasted in the wires as heat if the given amount of power is delivered at low current and high voltage. This is the way commercial power is distributed throughout the country (see p. 633). The distribution is known as 'high tension (H.T.)' transmission, because the voltage is high.

## Fuses

The heating effect of an electric current is used in *fuses* to safeguard circuits against excessive currents. The fuse is simply a short piece of wire made of a tin–lead alloy, for example, with a low melting point. In modern electrical installations in the home and elsewhere *cartridge fuses* are used. These are small cylinders with a fuse wire inside between the metal ends. If the current is excessive the wire melts and breaks. After the fault is remedied a new cartridge fuse is easily replaceable inside the 13-A plug widely used today (see p. 535). 2-A, 5-A and 13-A cartridge fuses can be obtained.

Older types of installation have rows of porcelain fuse holders, each of which has a fuse wire fitted across two brass terminals. Fuse wire is rated as 5-A, 10-A or 15-A respectively, which means that the wire will melt if the current exceeds the particular value. The 15-A fuse wire is thicker than 10- or 5-A wire so that it can carry larger currents.

Suppose that a lighting circuit in a house has fuses of 5 A, and the mains is 240 V. A $\frac{1}{2}$-kW appliance connected in the circuit has then a current $I$ flowing in it given by $500 = I \times 240$, since power $P = IV$ (p. 526); thus $I = 2.1$ A. The appliance can hence be safely used, as the current is much below the value of the fusing current. If a 2-kW electric fire were connected across a 240-V mains, however, the current flowing would be given by $2000 = I \times 240$, so that $I$ would be about 8 A in this case. A 2-kW fire cannot therefore be used in the lighting circuit. It must be used in the power circuit, where the fuses are 13-A in modern wiring.

## Earthing . 13-A Plug

Electrical appliances such as an electric fire or an electric iron, used in the home or elsewhere, are highly dangerous if, for some reason, the insulation of the wire or element inside has broken down. If the wire, which is

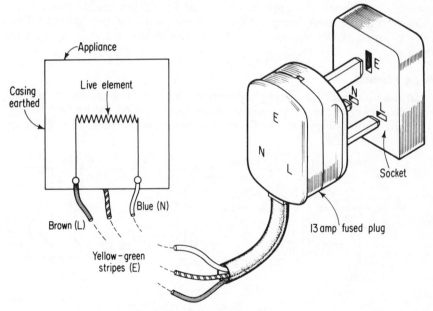

Fig. 29.10 Power socket – 13 amp

at a high potential relative to earth, then makes contact with the metal casing of an electric iron, for example, any person touching the casing will receive a dangerously high current which may prove fatal, as the human body is a good conductor of electricity.

To avoid this danger all electrical installations must have a good 'earth', that is, a low-resistance connection must be made directly to the earth, by means of an underground water pipe, for example. Two other connections, live and neutral, are made to the mains electrical supply coming from insulated underground cables. The domestic supply thus has in effect three terminals, 240 V A.C. between live L and neutral N and an earth E. Cables are then connected to sockets on walls which have three corresponding outlets L, N and E. The old type of two-pin sockets should be replaced by modern three-pin earthed sockets.

Fig. 29.10 illustrates diagrammatically an appliance with three leads inside the flex connected to it. One lead, coloured brown, is connected to the live or L-terminal inside a three-pin plug, another coloured blue is joined to the neutral or N-terminal, and the third, yellow-green stripes, is connected to the earth or E-terminal. The plug is then fitted into a socket, so that the appliance is connected to the mains terminals L, N and E when the plug is switched on. If the insulation of the wire in the appliance breaks down, this time the heavy current passes immediately to earth instead of passing through a person in contact with the apparatus. The fuse is then blown, and the supply is disconnected and a repair is made.

Extra care is needed in the bathroom, as water is a good conductor, and condensed steam may break down insulation in exposed sockets. No socket outlets, except a special safety shaving outlet, may be installed in the bathroom. Electric radiators with special insulation round the heating element may be used (see p. 317), but they should be permanently fixed by a competent electrician direct to the mains and a nylon cord switch used.

### Ring circuit

In the old system of wiring separate cables were taken from a fuse box to each socket outlet, and separate fuses were needed in the fuse box (p. 535). In the modern system of wiring a *ring circuit* is used (Fig. 29.11(i)).

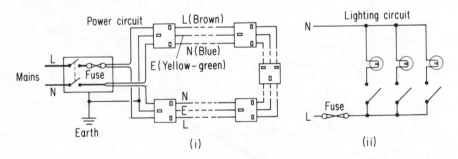

Fig. 29.11 Ring main and lighting circuit

Universal or standard 13-A socket outlets are now connected to a single ring or loop of cable from the mains. The sockets have shutters which slide over the live and neutral plug holes when the plug is removed, so that nothing can be pushed into the socket outlets by children.

The standard 13-A plug has a built-in cartridge fuse, as explained on p. 534, so that each appliance has its own fuse when connected in the circuit. The plug has flat pins, which make it easier to insert and remove. Fig. 29.11(ii) shows how lamps, which require only a 5A fuse, are connected across the mains.

## SUMMARY

1. Energy $W = IVt$ joules, where $I$ is in amperes, $V$ in volts, $t$ in seconds.

2. Power $P = IV$ watts, where $I$ is in amperes, $V$ is in volts. A Board of Trade unit $= 1$ kWh $= 1$ kW $\times 1$ h $= 3\cdot6 \times 10^6$ J.

3. Joule's laws of heating state: The heat produced is proportional to the square of the current, to the resistance and to the time. The heat $= I^2Rt$ joules $= V^2t/R$ joules $= IVt$ joules.

4. The specific heat capacity of water can be measured by passing a current through a coil immersed in water in a calorimeter; the heat supplied $= IVt$ joules.

5. A *fuse* is a wire with a low melting point which safeguards a circuit from excessive current. An *earth* connection to the casing is required to safeguard an appliance in case the insulation of the live element inside breaks down.

6. The hot-wire ammeter measures A.C. and D.C.; it has a square-law (non-uniform) scale.

## EXERCISE 29 · ANSWERS, p. 640

**1.** An electric lamp is marked 12 V, 36 W. Calculate: (a) the resistance of the lamp when in use; (b) the energy in joules expended each minute. (N.)

**2.** An electric lamp is marked 100 W, 250 V. If the lamp is connected to a 250-V mains, calculate: (a) the current taken; (b) the cost of using the lamp for 100 hr at 1p per kilowatt hour (Board of Trade unit). (N.)

**3.** Define *watt, kilowatt-hour*.

A washing machine for use on 240-V mains has a $\frac{1}{3}$-h.p. motor and a heating element rated at 2 kW. What current does it take when in use, and what is the cost of using it for 40 min each week for a period of 12 weeks if the cost of electricity is 2p per unit? (Assume 1 h.p. $= 0\cdot75$ kW.) (L.)

**4.** Give an expression for the rate at which heat is produced in a conductor carrying an electric current, stating the physical quantities involved and the units in which they are measured.

Water in an electric kettle connected to a 240-V supply took 6 min to reach its boiling point. How long would it have taken if the supply had been one of 210 V? (L.)

**5.** A current of 0·5 A flowing through a wire produces 21 J of heat in $\frac{1}{2}$ min. Find the resistance of the wire. (N.)

**6.** An electric kettle is rated at 1000 W, 250 V. (a) What is the resistance of the heating element when in use, (b) if electricity is charged at $1\frac{1}{2}$p per kilowatt hour and the kettle is used for 20 min each day, what does it cost per week? (N.)

**7.** Describe the structure and give the action of a hot-wire ammeter. Why is it suitable for measuring both alternating and direct current? (L.)

**8.** Describe briefly experiments, ONE in each case, to illustrate the chemical, magnetic and thermal effects by means of which a current of electricity may be detected.

A battery sends a current through two wires of resistance 10 and 20 Ω. Compare the rate of production of heat in the 10-Ω wire with that in the 20-Ω wire when they are connected (a) in series, (b) in parallel, across the battery. (N.)

**9.** (a) How would you determine by experiment the rate of production of heat by a current in a given resistance?

(b) Describe, with a diagram, the action of an ammeter suitable for the measurement of an alternating current. (N.)

**10.** How would you determine by experiment the relation between the electrical energy expended in a resistance coil and the heat produced?

A current of 2 A flows through a 5-Ω coil for $3\frac{1}{2}$ min. If all the heat generated is absorbed by 100 g of water, through how many degrees will the temperature of the water be raised? Assume sp. ht. capacity of water = 4·2 J per g per deg C (N.)

**11.** Describe an experiment to show that when the same current is passed through different wires the rates of production of heat are directly proportional to the resistance of the wires.

Two wires X and Y of the same material and the same length are joined in parallel to the terminals of a battery. The diameter of X is double that of Y. Find the ratio of the heat developed in X to that developed in Y during the same time. (N.)

**12.** It is required to raise the temperature of 20 litres of water from 15° to 75°C in 30 min by means of an electric heater. What must be the power of the heater and what current will it take if it is designed for use on 240-V mains? Explain why, in practice, the power must be greater than that calculated above. (L.)

**13.** State the factors which determine the rate of heat production in a resistance due to the passage through it of a steady electric current.

The heating element of an electric kettle, containing 1 litre of water, initially at a temperature of 20°C, is connected to a 250-V D.C. supply and the water commences to boil in 9 min. If the current through the heating element is 4 A, calculate: (a) the thermal capacity of the kettle; (b) the cost involved when the price of electrical energy is 6p per kilowatt hour. Neglect heat losses. (O. and C.)

**14.** Define *the volt, the watt* and *the kilowatt-hour*.

Explain how a fuse wire inserted in an electrical circuit protects the rest of the circuit against excessive current.

A small house with a mains supply at 250 V has two 2-kW electric heaters and six 100-W lamps. The power and the lighting circuit are entirely separate, and each has its own main fuse. What current passes through each of the fuses when both heaters and all the lamps are in use? (O.)

**15.** Define *volt, ampere* and *kilowatt-hour*.

How would you test the statement that the rate of production of heat in a given resistor is proportional to the square of the current passing through it?

A resistor is immersed in liquid in a calorimeter, the thermal capacity of the whole being 1260 J per deg C. When a steady current of 7 A passes for 5 min, the temperature rises by 16 deg C. Find: (a) the p.d. across the resistor; (b) its resistance in ohms. (O.)

**16.** Describe how you would investigate the factors which determine the heat generated in a wire by the passing of an electric current. What conclusions would you expect to reach?

An electric iron designed for use on 100-V D.C. mains takes a current of 3·2 A. What is the resistance of the heater and how much heat is evolved per second when it is in use?

Describe two alternative ways of using the same electric iron on a 200-V A.C. supply and briefly mention their relative advantages and disadvantages. (O. and C.)

**17.** Name and define the unit of power used in electricity.

A resistor of $R$ ohms has a potential difference of $V$ volts applied across it. Write down expressions for: (i) the power, and (ii) the heat developed in it per second.

Describe an experimental method for measuring the power of an electrical appliance.

Find the power of an electric kettle which will bring to the boil 1·25 litres of water initially at 16·0°C, in exactly 4 min, neglecting heat losses. The kettle is made of copper and weighs 800 g.

(Specific heat capacity of copper = 0·42 J/g K or 420 J/kg K.) (*C.*)

**18.** Define the *joule, watt* and *kilowatt-hour*.

An electric lamp has a metal filament. State, with reasons, whether at the moment of switching on it will produce heat at a greater or lesser rate than when in normal use.

A stream of water of 10·0 g per sec passes over a heating coil and its temperature rises 4·0°C as a result. The coil is supplied from a 12·0-V battery. What is the resistance of the coil?

What temperature rise would occur if a second 12·0-V battery were connected in series with the first? Ignore internal resistances of the batteries and any change in resistance in the coil. (*O.* and *C.*)

# 30

# ELECTROLYSIS

So far we have discussed the flow of electricity through metals, or solid conductors. In 1834 FARADAY began to investigate the behaviour of liquid conductors, such as acid and salt solutions; the study of the flow of electricity through liquids is called *electrolysis*, and the liquids are called *electrolytes*. By using a small light bulb in series with the liquid and a suitable battery, as shown on p. 484, it can be demonstrated, for example, that dilute sulphuric acid solution or a dilute solution of copper sulphate or common salt are good conductors or electrolytes. Pure water is a very poor conductor of electricity. Tap water conducts better owing to the dissolved salts in it.

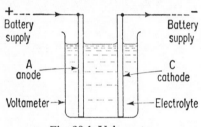

Fig. 30.1 Voltameter

The two plates or materials which lead the current into or out of the liquid are known as the *electrodes*. The *anode* A is the electrode on the positive side of the supply, so that the conventional current enters the solution by A. The *cathode* C is the electrode on the negative side of the supply, so that the current leaves the solution by C (Fig. 30.1). The whole vessel, containing electrolytes and electrodes, is known as a *voltameter*.

## Typical Examples of Electrolysis

There are many examples of electrolysis, but to understand the principles concerned we shall consider two typical cases.

### (1) *Electrolysis of Copper Sulphate Solution with Copper Electrodes*

If an electric current of about $\frac{1}{2}$ A is passed through a dilute solution of copper sulphate with copper electrodes for about 30 min, removal of the cathode shows that the part of it dipping into the solution is now covered with a bright fresh deposit of copper. When it is weighed the mass of the cathode is found to have increased by about 0·3 g. The anode, on the other hand, looks dull after removal from the solution, and when it is weighed it is found to have decreased by about 0·3 g. The density of the copper sulphate solution, determined by a hydrometer, is found to be unaffected by the electrolysis which has taken place.

### (2) *Electrolysis of Acidulated Water*

Pure water is a poor conductor of electricity. If a little sulphuric acid is added it conducts much better. To investigate the electrolysis of water, the acidulated water can be poured into a *Hoffmann voltameter*. This is a glass apparatus consisting of inverted graduated burettes with a platinum electrode at the bottom, as shown in Fig. 30.2. When a battery is connected and the circuit switched on, gases are formed at the anode A and cathode C and are collected above the electrodes. On testing, oxygen is found above A and hydrogen above C, the ratio of the volumes being exactly 1:2. The solution is found to be slightly more concentrated, so that water has been lost by electrolysis and hydrogen and oxygen formed.

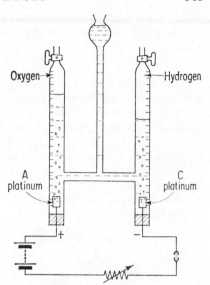

Fig. 30.2 Electrolysis of water – Hoffmann voltameter

### Conclusion

The electrolysis of copper sulphate solution and acidulated water both show that

*the metal or hydrogen is always deposited at the cathode.*

This general rule applies to all cases of electrolysis and should be memorized. Thus if silver nitrate solution is electrolysed silver appears at the cathode. If dilute hydrochloric acid is electrolysed hydrogen appears at the cathode.

*The nature of the electrode materials* play an important part in determining the final products at the electrodes. Thus if copper sulphate solution is electrolysed using platinum electrodes copper is again deposited at the cathode, but oxygen is collected at the anode.

## Ions and Ionic Crystals

An explanation of electrolysis was suggested by a Swedish chemist ARRHENIUS about 1890. He considered that a dilute solution of sodium chloride, for example, contained particles carrying quantities of electricity or charges, which he called *ions*. As stated on p. 486, an ion is an atom or group of atoms which have lost or gained one or more electrons, so that ions carry positive (+ve) or negative (−ve) charges.

Later researches into the structure of solids such as common salt show they have a crystal structure, with ions (not atoms) at the corners of a regular three-dimensional pattern or lattice (Fig. 30.3(i)). In sodium chloride (NaCl), for example +ve sodium ions ($Na^+$) and −ve chloride ions ($Cl^-$) are at the corners of a cubic structure. They are held together strongly to form a solid by the attractive force between positive and negative charges (p. 644). The sodium atom 'donates' an electron to the chlorine atom when both come together to form sodium chloride, so that a chloride ion has a charge $-e$ and a sodium ion has a charge $+e$, where $e$ is the *numerical* value of the electronic charge.

Thus

$$NaCl \to Na^+ + Cl^-.$$

Solid copper sulphate ($CuSO_4$) crystals likewise have $+$ve copper ions and $-$ve sulphate ions at the corners of a three-dimensional pattern, held strongly together to form a solid. This time, however, two electrons are donated by the copper atom to the sulphate group of atoms. Thus the $+$ve copper ion has a charge of $+2e$, and is hence written $Cu^{2+}$. The sulphate ion has a charge $-2e$ and is written $SO_4^{2-}$. Sulphuric acid ($H_2SO_4$) has two $+$ve hydrogen ions ($2H^+$) and a sulphate ion ($SO_4^{2-}$). Water is a poor conductor. One molecule in ten million is dissociated into a $+$ve hydrogen ion ($H^+$) and a $-$ve hydroxyl ion ($OH^-$). Thus $H_2O = H^+ + OH^-$.

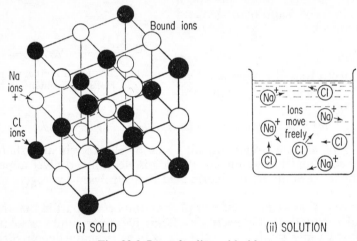

(i) SOLID                    (ii) SOLUTION

Fig. 30.3  Ions of sodium chloride

When water is added to solid common salt or copper sulphate the water now present between the ions diminishes the force of attraction between them to about $\frac{1}{80}$ of that when air was present. The ions thus drift apart (Fig. 30.3(ii)). The solid structure hence disappears.

### Movement of Ions

The ions in an electrolyte move freely and haphazardly in the solution, making frequent collisions; on the whole, there is no general movement of the ions in any one direction. When, however, a battery is connected to the electrodes a potential difference is set up across the electrolyte; and, as we have seen with electrons in the case of metals, the ions with a negative charge begin to drift across the liquid towards the anode A; the higher potential plate. At the same time the ions with a positive charge drift in the opposite direction towards the cathode C. If the battery is disconnected the drift of electricity between the plates ceases, and the ions once again have a completely random motion.

*Experiment.* Place a crystal of potassium permanganate in the middle of a wet filter paper on a glass plate. By clips attached to the paper on either side of the crystal, apply a high direct p.d. of about 200 V from a labpack or other source. Observe the movement of colour across the paper towards the anode. Reverse the p.d. and observe the colour movement again. Repeat using an *alternating* p.d. Explain the results. Try other coloured crystals.

## Explanation of Electrolysis

We are now in a position to explain the electrolysis of copper sulphate solution.

### (1) *With Copper Electrodes*

The copper sulphate solution contains positive copper ions ($Cu^{2+}$), positive hydrogen ions ($H^+$), negative hydroxyl ions ($OH^-$) and negative sulphate ions ($SO_4^{2-}$). When the battery is connected to the electrodes the positive ions drift towards the cathode. Now some ions are more easily discharged than others; in particular, the copper ions are more easily discharged than the hydrogen ions, and hence copper is deposited at the cathode. The discharge can be represented by

$$Cu^{2+} + 2e^- = Cu,$$

where $e^-$ represents an electron.

While copper is being deposited at the cathode the copper atoms of the anode electrode go into solution as ions. This is a process which occurs more easily than the discharge of hydroxyl and sulphate ions, which move to the cathode while the current flows, and it can be represented by

$$Cu = Cu^{2+} + 2e^-.$$

The electrons move along the wire joined to the battery.

Consequently, copper is lost from the anode, an equal amount is deposited on the cathode and the density of the copper sulphate solution is unaltered.

### (2) *With Platinum Electrodes*

When platinum electrodes are used instead of copper electrodes, copper is again deposited on the cathode, as explained above. At the anode, however, hydroxyl ions ($OH^-$) are discharged, a process which takes place more easily than the discharge of the sulphate ions ($SO_4^{2-}$), or the solution of platinum atoms from the anode as ions. Oxygen is formed at the anode from the discharge of the hydroxyl ions, a process which can be represented by

$$4OH^- = 2H_2O + O_2 + 4e^-.$$

The electrons liberated circulate in the wire joined to the poles.

Consequently, oxygen is obtained at the anode, copper is deposited on the cathode and the strength of the copper sulphate solution decreases.

## Electrolysis of Water

When a battery is connected to platinum electrodes in acidulated water, the positive hydrogen ions ($H^+$) drift towards the cathode, where they are discharged

$$2H^+ + 2e^- = H_2.$$

The negative hydroxyl ions ($OH^-$) and sulphate ions ($SO_4^{2-}$) drift towards the anode, where the hydroxyl ions are discharged and oxygen is formed

$$4OH^- = 2H_2O + O_2 + 4e^-.$$

It can be seen that the amount of sulphuric acid in solution remains constant as the current flows, and the water in the solution is decomposed into hydrogen and oxygen.

## Movement of Ions in Cells · Simple Cell

The action of a simple cell in changing chemical energy to electrical energy, described on p. 510, can now be more fully discussed. When a zinc rod is placed in dilute sulphuric acid it is considered that some zinc ions go into solution. This leaves the rod with a net negative charge or surplus of two electrons per atom. The copper atoms on the copper plate of the simple cell, however, remain unaffected.

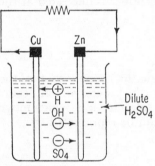

Fig. 30.4 Movement of ions – simple cell

When the copper and zinc are joined by a wire the electrons flow through the wire to the copper from the zinc. Inside the cell the electrical flow is completed by the movement of ions. Hydrogen ($H^+$), hydroxyl ($OH^-$) and sulphate ($SO_4^{2-}$) ions then drift as shown in Fig. 30.4. The positive charge on each hydrogen ion is neutralized by the electrons flowing towards them, so that atoms and then molecules of hydrogen are given off. From the zinc plate, zinc atoms pass into solution as zinc ions. The net effect is: (i) hydrogen is given off at the copper plate; (ii) zinc is used up and zinc ions are formed in solution.

## Daniell Cell

A similar explanation accounts for the action of a Daniell cell. Fig. 30.5 shows the ions drifting in the solution. The reader should explain why: (i) copper (not hydrogen) is deposited; (ii) zinc is used up and zinc sulphate is formed. See also p. 512.

## Leclanché Cell

In this cell, described on p. 512, carbon and zinc are used as poles. When a wire is connected to the poles electrons move in the wire from zinc

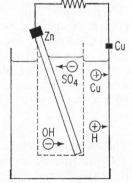

Fig. 30.5 Movement of ions – Daniell cell

to carbon, and ammonium ($NH_4^+$), hydrogen ($H^+$), chloride ($Cl^-$) and hydroxyl ($OH^-$) ions move in the cell between the electrodes. See also p. 512.

## Lead–Acid Accumulator

In the lead–acid accumulator (p. 514) electrons in a wire move from the lead (−ve) pole to the lead peroxide (+ve) pole when the wire is joined to the two poles. Inside the electrolyte, hydrogen ($H^+$), sulphate ($SO_4^{2-}$) and hydroxyl ($OH^-$) ions move between the poles.

## Fuel Cell

When a supply of hydrogen gas enters the negative side of the fuel cell (see p. 517), the molecules diffuse through the pores of the electrodes and are dissociated into atoms by the catalyst. The atoms are then ionized by interaction with the electrolyte. Four hydrogen ions ($H^+$) and four electrons ($-4e$) are obtained from two molecules of hydrogen ($2H_2$), (p. 517). The electrons move through the wire joined to the poles to the positive electrode on the other side of the cell. Oxygen gas or air is supplied here, and this also diffuses through the electrode on this side and forms atoms. The four electrons which arrive at the oxygen electrode reduce two oxygen atoms, forming oxygen ions ($O^{2-}$). These combine with the four hydrogen ions which flow across to this electrode and complete the electrical circulation of charges. Two molecules of water are hence formed. The net effect is the chemical formation of water at low temperature, with the production of electrical energy.

## Industrial Applications

Electrolysis is used in a wide variety of practical applications.

Pure copper, which must be used in the manufacture of cables, is produced commercially by electrolysis of copper compounds, which are contained in large tanks. Cathode plates, dipping into the electrolyte, are raised after a time, and the pure copper deposited is stripped from the cathode. Sodium and aluminium are manufactured commercially by electrolysis.

Silver plating of objects such as spoons is carried out by making them the cathode of a voltameter containing silver compounds and passing a current through the solution. Steel objects, such as the bumpers of cars or the parts of a bicycle, are often protected by chromium-plating, which presents a bright appearance and prevents rusting. The handlebars of a bicycle, for example, are made the cathode of a voltameter containing a solution of acid and other additives, and a current is passed through the solution until a suitable deposit of chromium is obtained on the frame.

An important case of electrolysis occurs in the manufacture of *electrolytic capacitors*, used extensively in radio receivers. The capacitor is made by the electrolysis of ammonium borate, for example, between aluminium electrodes (Fig. 30.6). A very thin film of aluminium oxide is formed on the anode plate. The oxide is an insulating medium or dielectric, and is so thin that the capacitance of the capacitor is extremely high. High capacitance in a small volume is a valuable feature of the electrolytic capacitor, which is therefore widely used. When the capacitor is used in a transistor circuit, for example, care must be taken to connect the terminals from the anode aluminium plate to the *positive* side of the circuit; otherwise the oxide film on the cathode will break down by a reversal of the electrolysis action. The positive terminal of the capacitor is clearly marked on it by the manufacturers.

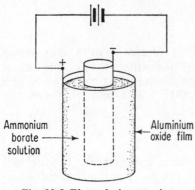

Ammonium borate solution

Aluminium oxide film

Fig. 30.6 Electrolytic capacitor

## Investigation of Faraday's First Law of Electrolysis

We have already seen that a metal or hydrogen is deposited on the cathode during electrolysis. Faraday performed many experiments to find how the mass of the element deposited varied with the magnitude of the current and the time.

### EXPERIMENT

A quick though approximate investigation can be made with a Hoffmann voltameter filled with acidulated water (Fig. 30.2). A rheostat, ammeter and a suitable battery are included in the circuit. The element collected at the cathode is hydrogen, whose density varies with pressure. The pressure change during the experiment, shown by the difference in water levels, is small, and we shall therefore assume that, approximately, the density of hydrogen is constant. In this case the *mass* of hydrogen collected is proportional to the volume, which we can read off.

### 1. *Constant Current*

The current is kept constant at a convenient value such as 0·5 A by means of the rheostat and ammeter. The volume $v$ of hydrogen produced is read at five consecutive equal intervals of time $t$ with the aid of a stop-watch. Typical results are shown in the table on p. 547. The ratio volume/time is found to be practically constant, or if a graph of volume v . time is plotted it is found to be a straight line which passes through the origin.

*Conclusion.* For a constant current in a voltameter,

$$\text{Mass deposited} \propto \text{Time.}$$

| Volume of hydrogen $(v)$ $(cm^3)$ | Time $(t)$ $(sec)$ | Volume/Time $(cm^3$ per sec$)$ |
|---|---|---|
| 4·4 | 80 | 0·055 |
| 5·0 | 90 | 0·055 |
| 5·8 | 100 | 0·058 |
| 6·2 | 110 | 0·056 |
| 7·0 | 120 | 0·058 |
| 7·4 | 130 | 0·057 |
| 8·0 | 140 | 0·057 |

Current Constant

| Volume of hydrogen $(cm^3)$ | Current $(A)$ | Volume/ Current $(cm^3$ per amp$)$ |
|---|---|---|
| 8·0 | 0·6 | 13·3 |
| 9·4 | 0·7 | 13·4 |
| 12·1 | 0·9 | 13·4 |

Time Constant

## 2. Constant Time

This time several currents are passed into the voltameter for the same time on each occasion, such as 2 min. The results for three currents are shown in the table. The ratio volume/current is practically a constant.

*Conclusion.* Assuming the volume of hydrogen is practically proportional to its mass, then for a constant time,

$$\text{Mass} \propto \text{Current.}$$

## Faraday's First Law · Electrochemical Equivalent

The two results of the experiment outlined can now be combined into one general law. It is known as *Faraday's first law of electrolysis*, after the discoverer, who performed many careful experiments:

*The mass of an element deposited by electrolysis in a given voltameter is directly proportional to the current and to the time.*

Thus if $m$ is the mass, $I$ the current and $t$ the time, then

$$m \propto It,$$
or
$$m = zIt \quad . \quad . \quad . \quad . \quad . \quad . \quad (1)$$

where $z$ is a constant called the *electrochemical equivalent* of the element deposited.

The explanation of Faraday's law follows directly from our knowledge that ions carry the current through electrolytes (see p. 542). The product '$I \times t$' or 'current $\times$ time' is the *quantity* of electricity $Q$ which has passed through the voltameter, since 1 *coulomb* is the quantity which has passed in 1 second when the current is 1 ampere (see p. 487). The expression in (1) is thus better stated as:

$$m = zQ \quad . \quad . \quad . \quad . \quad . \quad . \quad (2)$$

Now the ions which carry the current have both charge and mass, and the mass of an ion is practically the same as the mass of its atom (p. 487). Thus if 50 ions, for example, reach the cathode at the end of a given time, 50 atoms of copper are deposited and 50 units of charge have passed through the electrolyte. If 80 ions reach the cathode, then 80 atoms are deposited and 80 units of charge have passed through the electrolytes. It can thus be seen that Faraday's first law is a consequence of the ionic structure of crystals.

## Measurement of Electrochemical Equivalent

From $m = zQ$, the electrochemical equivalent $z = m/Q$, or $z$ is
the mass of an element deposited by 1 coulomb,
or the mass deposited by 1 ampere in 1 second.

The units of electrochemical equivalent are *gramme per coulomb*, g/C, or, in SI units, kg/C.

*Method*. The electrochemical equivalent (e.c.e.) of an element such as copper can be measured by using two copper plates as electrodes with copper sulphate solution in a beaker as the electrolyte (Fig. 30.7). An ammeter M, a rheostat, accumulators and a key K are in series in the circuit.

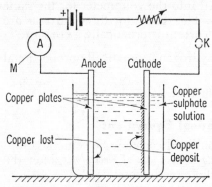

Fig. 30.7 Electrolysis – copper voltameter

The current must not be too high, otherwise copper will not be deposited firmly on the cathode and will flake off, nor too low, otherwise the mass of copper deposited in a reasonable time is too small to be measured accurately. Another plate, a 'dummy' cathode, is therefore placed in the beaker first and the rheostat then varied until a suitable current, 0·8 A, for example, is obtained.

The actual copper cathode is cleaned with emery paper, washed and dried, and weighed. It is then placed in the beaker and the current is simultaneously switched on. A constant current is now passed for a suitable time such as a half-hour. The cathode is removed, washed gently with distilled water and carefully dried by holding it above a low flame. The plate must not be placed too near the flame, otherwise the copper will become oxidized. The cathode is now reweighed.

*Measurements.* Mass of cathode initially $= 46\cdot342$ g
Mass of cathode + copper deposit $= 46\cdot822$ g
Current $= 0\cdot80$ A
Time $= 30$ min $= 1800$ s.

*Calculation.* Mass of copper deposited $m = 0\cdot480$ g $= 0\cdot480 \times 10^{-3}$ kg

From $$m = zIt,$$

$$z = \frac{m}{I \times t}$$

$$= \frac{0\cdot480 \times 10^{-3}}{0\cdot8 \times 1800}$$

$$= 3\cdot3 \times 10^{-7} \text{ kg/C}$$

Similar experiments show that the e.c.e. of hydrogen, the lightest element, is 0·000 010 4 g per coulomb, and that the e.c.e. of silver, a heavy element, is 0·001 118 g per coulomb. The hydrogen ion carries the same quantity of electricity or charge as the silver ion, but it is very much lighter. Consequently, the mass per coulomb, or e.c.e., of hydrogen is much smaller than that of silver (see p. 551).

## Measurement of Current

Silver is an element which can be obtained in a high degree of purity, and many careful and elaborate experiments have been performed to find its electrochemical equivalent very ac-curately. The result is quoted above.

The procedure used to measure the e.c.e. can also be utilized to check a particular reading on an ammeter A (Fig. 30.8). The in-strument is used in place of M in the circuit of Fig. 30.7. The beaker is filled with a suitable silver compound, and using a dummy cathode the reading on M is first set to the required current value by means of the rheostat. After a suitable measured time the mass deposited is determined.

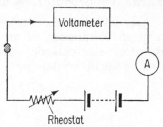

Fig. 30.8 Calibration of ammeter

As an illustration, suppose the instrument reading was kept constant at 1·60 A, and 2·173 g of silver was deposited in 20 min. The true current $I$ is then given by

$$m = zIt.$$

$$\therefore I = \frac{m}{z \times t} = \frac{2 \cdot 173}{0 \cdot 001\ 118 \times 20 \times 60}$$

$$= 1 \cdot 62 \text{ A.}$$

The correction to be applied to the instrument reading is thus 1·62 − 1·60 or 0·02 A. In a school laboratory, calibration experiments on am-meters can be carried out by using copper sulphate solution, and assum-ing the e.c.e. of copper is 0·000 33 g per coulomb.

It is interesting to note that years ago scientists agreed internationally to define 1 ampere as that current which deposits a weight of 0·001118 g of silver in 1 second. Nowadays current is measured accurately by an 'ampere-balance', which measures the force between current-carrying wires (p. 611).

## Example

A current of 2 A is passed through a copper voltameter. Given that the copper is deposited evenly on an electrode of area 66 cm², find the thickness of the layer deposited after the current has been flowing for 30 min. (Electrochemical equivalent of copper = 0·000 33 g/C. Density of copper 9·0 g/cm³.)

The mass $m$ of copper deposited is given by

$$m = zIt$$
$$= 0.000\,33 \times 2 \times (30 \times 60) = 1.188 \text{ g}$$

Since

$$\text{Volume} = \frac{\text{Mass}}{\text{Density}}, \text{ the volume of copper deposited} = \frac{1.188}{9.0} \text{ cm}^3$$
$$= 0.132 \text{ cm}^3.$$

$$\therefore \text{ Thickness} = \frac{\text{Volume}}{\text{Area}} = \frac{0.132}{66} \text{ cm} = 0.002 \text{ cm}.$$

### Faraday's Second Law of Electrolysis

Faraday's law of electrolysis refers to the mass of an element deposited in a given voltameter. If an investigation is required into the masses of different elements deposited in different voltameters, then a circuit can be set up containing a copper voltameter, a silver voltameter and a

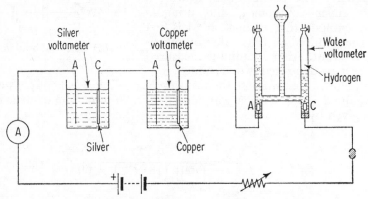

Fig. 30.9 Faraday's second law

water voltameter, a rheostat and an ammeter in series with a battery (Fig. 30.9). A suitable current can now be passed for a convenient time, after which the masses deposited can be determined.

*Results.*      The mass $m_1$ of hydrogen collected   $= 0.020$ g.
        The mass $m_2$ of copper deposited      $= 0.64$ g.
        The mass $m_3$ of silver deposited      $= 2.18$ g.

*Calculations.* $m_1 : m_2 : m_3 = 0.02 : 0.64 : 2.18 = 1 : 32 : 109.$

But the ratio of the *chemical equivalents* of hydrogen, copper and silver $= 1 : 31.5 : 108.$

*Conclusion.* The masses deposited are proportional to the chemical equivalents.

It should be carefully noted that because the same current flowed through each voltameter for the same time, the same quantity of electricity passed each voltameter. Faraday's *second law of electrolysis* can therefore be expressed as follows:

*When the same quantity of electricity passes through different electrolytes the masses of the elements deposited are proportional to their chemical equivalents.*

## Explanation

This law, like the first one, is also explained by the fact that the carriers of the current are ions. Each ion carries a charge proportional to its *valency*, or to the number of atoms of hydrogen it will combine with or replace in a compound. Thus if $e$ is the numerical value of the charge on an electron the sodium ion carries a charge $+e$. A copper ion from copper sulphate solution carries a charge $+2e$. A hydrogen ion carries a charge $+e$. A silver ion carries a charge $+e$.

Suppose a charge of $4e$ flows through a water and copper sulphate voltameter in series. Then 4 hydrogen ions and 2 copper ions are deposited at the respective cathodes, and form 4 hydrogen and 2 copper atoms. The mass of 4 hydrogen atoms is 4 units say, and the mass of 2 copper atoms is then $2 \times 63$ units, since the atomic weight of copper is 63, or $2 \times 2 \times 31 \cdot 5$ units, where $31 \cdot 5$ is the equivalent weight of copper. Hence

Mass of hydrogen deposited : Mass of copper deposited $= 1 : 31 \cdot 5$
    $=$ Equivalent weight of hydrogen : Equivalent weight of copper.

## Example

A copper and water voltameter are in series, and at the end of a period of time $3 \cdot 0$ g of copper is deposited. Calculate the mass of oxygen deposited.

Since the same quantity of electricity passes through each voltameter,

$$\frac{m}{3 \cdot 0} = \frac{\text{Chemical equivalent of oxygen}}{\text{Chemical equivalent of copper}},$$

where $m$ is the mass of oxygen.

$$\therefore \frac{m}{3 \cdot 0} = \frac{8}{31 \cdot 5}$$

$$\therefore m = \frac{8 \times 3}{31 \cdot 5} = 0 \cdot 8 \text{ g (approx)}.$$

## Faraday Constant

From Faraday's second law, it follows that the same quantity of electricity passing through different voltameters liberates the equivalent weight in grams or *gram equivalent* of all the elements concerned. Now on p. 549 we saw that $0 \cdot 001\ 118$ g of silver are deposited by 1 coulomb. The gram equivalent of silver is 108 g. Hence this is liberated by a quantity of electricity equal to $108/0 \cdot 001\ 118$ coulombs or 96 500 coulombs approximately. Also, the electrochemical equivalents of copper and hydrogen are respectively about $33 \times 10^{-5}$ and $1 \cdot 05 \times 10^{-5}$ g per coulomb. The gram equivalents of copper and hydrogen are thus liberated by, respectively, $32 \cdot 5/(33 \times 10^{-5})$ and $1/(1 \cdot 05 \times 10^{-5})$ coulomb, which is about 96 500 coulombs in each case. The quantity of electricity which liberates a gram equivalent of any element in electrolysis is called the *faraday constant* and given the symbol $F$.

$$1 F = 96\ 500\ \text{coulomb (approx.)}$$

From the constant we can easily calculate, for example, the mass of silver deposited when a constant current of 2 A flows for 30 min through a silver voltameter. The quantity·of electricity which flows $= I \times t = 2$ (A) $\times 30 \times 60$ (s) $= 3600$ coulomb. Now 96 500 coulomb liberates 108 g of silver. Hence the mass of silver deposited

$$= \frac{3600}{96\ 500} \times 108 = 4\ \text{g (approx.)}.$$

## SUMMARY

1. Electrolytes are liquids which conduct an electric current. They contain positive (metal or hydrogen) ions and negative ions which carry the current through the liquid.

2. When copper sulphate solution is electrolysed with copper electrodes, copper is deposited at the cathode, an equal mass of copper is lost from the anode and the density of the solution is unaltered. With platinum electrodes, copper is deposited at the cathode, oxygen is liberated at the anode and the density of the solution decreases.

3. *Faraday's laws of electrolysis* state: (*a*) The mass of an element deposited or liberated in electrolysis is proportional to the current and to the time. (*b*) When the same current passes through different electrolytes for the same time the mass of any element deposited is proportional to its chemical equivalent.

4. The 'electrochemical equivalent' $z$ of an element is the mass per coulomb deposited or liberated in electrolysis. The mass $m$ g of an element deposited in $t$ s by a current of $I$ A $= zIt$ g, where $z$ is in 'g per coulomb'.

## EXERCISE 30 · ANSWERS, p. 640

**1.** Which of the following are electrolytes: petrol, mercury, common salt solution, molten copper, dilute sulphuric acid? (*N.*)

**2.** Explain the terms *anode, cathode*.

Describe an experiment to check the accuracy of the 1 A reading of an ammeter, assuming that a value for the electro-chemical equivalent of copper is available.

What is the cost of depositing a layer of silver 0·2 mm thick on an object of total surface area 150 cm$^2$ if a current of 1 A used for an hour costs 2p? (Density of silver is 10·5 g/cm$^3$; electrochemical equivalent of silver is 0·001 12 g/C.) (*L.*)

**3.** A Daniell cell of e.m.f. 1·1 V, maintained a constant current of 0·35 A in a circuit for 20 min. Find: (i) the energy supplied by the cell in this time; (ii) the change in mass of its positive pole, indicating whether this was an increase or decrease. (E.c.e. of copper $= 0·000\ 33$ g/C.) (*L.*)

**4.** (*a*) Describe an experiment to determine the electrochemical equivalent of copper.

(*b*) Describe a lead accumulator and state the changes which occur as a result of its being discharged. State, with a reason, two precautions which should be taken to maintain accumulators in good condition. (*N.*)

**5.** State the *laws of electrolysis* and define the terms *anode* and *cathode*.

Describe an experiment to determine how the mass of copper deposited in a copper voltameter varies with the quantity of electricity passing through it. Give a circuit diagram and full experimental details.

An electric current passes through two voltameters in series, containing copper sulphate and silver nitrate respectively. What is the mass of silver deposited in a given time if the mass of copper deposited in that time is 1 g? (At. wt. of copper is 63, at. wt. of silver is 108, valency of copper is 2, valency of silver is 1.) (*L.*)

**6.** Explain how a current can pass through an electrolyte and discuss how the process resembles and differs from current flow in a metal.

A 12-V 90-A-hr accumulator suffers a drop in voltage of 2 V when supplying a current of 20 A. What electrical energy would be taken from the supply when a charging current of 10 A is passed through the accumulator for 9 hr?

If the current is continued after the battery is fully charged, what becomes of the energy supplied? (*O. and C.*)

**7.** State Faraday's laws of electrolysis.

Describe an experiment to decompose water by electrolysis.

Calculate the current required to liberate 10 cm³ of hydrogen per min in electrolysis. (1 litre of hydrogen weighs 0·09 g at the temperature and pressure at which the volume is measured in this case. The electrochemical equivalent of hydrogen is 0·000 010 5 g/C.) (*N.*)

**8.** Explain the terms *electrolysis, ionization, electrochemical equivalent*.

State Faraday's laws of electrolysis, and describe how you would attempt to test the second law experimentally.

Calculate the time for which a steady current of 0·2 A must pass through a water voltameter in order to liberate a quantity of hydrogen which would occupy a volume of 50 cm³ at s.t.p.

(Take the electrochemical equivalent of hydrogen to be 0·000 010 45 g/C and its density at s.t.p. to be 0·000 09 g/cm³.) (*O.*)

**9.** Explain how the ionic theory accounts for Faraday's laws of electrolysis.

Give an account of the processes occurring when dilute sulphuric acid is electrolysed between platinum electrodes.

A cell with copper electrodes and containing copper sulphate solution is in series with one containing dilute sulphuric acid and platinum electrodes. When 3·175 g of copper have been liberated, what volume of hydrogen, measured at s.t.p. is released?

(Atomic weights: copper, 63·5; hydrogen, 1·008. Valency of copper, 2. Density of hydrogen at s.t.p., 0·0899 g/litre.) (*O. and C.*)

**10.** Explain what is meant by the statement that the electrochemical equivalent of copper is 0·000 33 g per colomb and describe an experiment to verify it.

Copper is electrolytically refined by depositing it on the cathode of a suitable voltameter. If the current is adjusted to be $\frac{1}{50}$ A for each cm² of the cathode surface, find the time required to deposit a layer of copper 2 mm thick.

(Density of copper = 8·9 g/cm³.) (*L.*)

**11.** How would you use a copper voltameter to check the accuracy of an ammeter when it shows a reading of 1·0 A.?

A current of 0·25 A passes for 20 min through a silver voltameter. Find the change in weight of the cathode.

(Electrochemical equivalent of copper = 0·000 33 g/C. Chemical equivalent of copper = 31·5, chemical equivalent of silver = 108·0.) (*N.*)

**12.** State Faraday's laws of electrolysis and explain the meaning of electrochemical equivalent. Describe some form of water voltameter, draw a diagram and explain how it works.

A copper voltameter and a 12-Ω coil are connected in series and an electric current passed through them. Find the weight of copper deposited in the voltameter in 30 min when a p.d. across the terminals of the 12-Ω coil is maintained at 15 V.

The electrochemical equivalent of copper = 0·000 33 g/C. (*O.* and *C.*)

**13.** State the laws of electrolysis and describe TWO experiments to illustrate these laws.

What are the chief differences between the passage of electricity through metals and through ionized solutions?

A steady current passes through a silver voltameter and a 10-Ω resistance arranged in series. A high-resistance voltmeter placed across the terminals of the 10-Ω coil reads 6 V. Given that 0·8048 g of silver is deposited on the cathode of the voltameter in 20 min, calculate the electrochemical equivalent of silver. (*O.* and *C.*)

**14.** Describe how you would use the electrolysis of a solution containing copper ions to calibrate an ammeter. Draw a diagram of a suitable arrangement to show the circuit details.

If the ammeter records a steady value of 1·5 A for 30 min, while the mass of copper deposited is 0·99 g, calculate the error in the ammeter. Assume that errors in the clock and the balance are negligible and that the E.C.E. of copper is 0·000 33 g/C. (*O.* and *C.*)

# 31

# MAGNETISM

Modern civilization relies to a great extent on efficient magnets, some permanent and others temporary, and on magnetic materials. They play an important part in the generation of electrical power, in communications and in computers, for example. The telephone earpiece, the microphones used in broadcasting stations and the loudspeaker in a radio receiver all depend on magnets and magnetic materials to function efficiently. A television cathode-ray tube has a shield made of magnetic material. Tape-recorders utilize a tape of magnetic material.

## Lodestone

The first observations on magnetism began over 2000 years ago in China. It was known that a mineral called *lodestone*, which is a magnetic oxide ($Fe_3O_4$), always pointed approximately north and south when it was suspended. It was used for navigation across country and sea; the name 'lodestone' was given because it acted as a 'leading' stone. It was also found that iron and steel were attracted to the ends of the lodestone, which was a natural magnet. Little progress, however, was made in the study of the subject until 1603, when Dr Gilbert became interested in the phenomenon. He was a physician to Queen Elizabeth, and his extensive researches, culminating in a learned volume called *De Magnete*, have led him to be regarded as the founder of the science of magnetism.

## Fundamental Observations

Experiment shows that:

(i) iron filings cling mainly round the ends of a bar magnet (Fig. 31.1 (i));
(ii) the bar magnet, suspended so as to swing freely in a horizontal plane, always comes to rest with its axis pointing roughly north–south (Fig. 31.1(ii)).

The ends of the magnet, where the attracting power is greatest, are called its *poles*. The pole which points towards the north is called the 'north-seeking' or north (N) pole of the magnet, and the other is called the south (S) pole.

(iii) If a N-pole of a magnet is brought near to the N-pole of a suspended magnet repulsion occurs. If the N-pole is brought near to the S-pole attraction occurs (Fig. 31.1(iii)). Thus

*like or similar poles repel* (i.e. two N- or two S-poles), and *unlike poles attract* (i.e. a N- and S-pole).

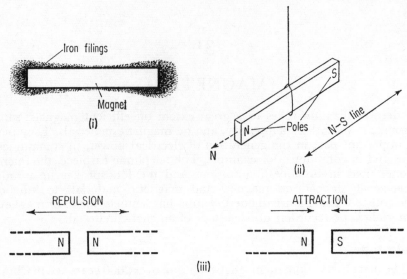

Fig. 31.1  Properties of magnets

## Methods of Making Magnets · Electrical Method

The best and quickest method, or industrial method, of making a magnet is to use the magnetic effect of an electric current (p. 580). A coil of insulated wire of many turns, called a *solenoid*, is used to carry a direct or steady current, obtained from a battery, for example. If a bar magnet is required the bar X is placed inside the solenoid, the current is then switched on for a few seconds and then switched off (Fig. 31.2(i)). On testing, X is found to be a magnet with N and S poles as shown. We shall see later that a rule for polarity can be stated as follows:

If the current flows *clockwise* in the coil when one end of it is viewed, then X has a *south* (S) pole at this end. If the current flows anti-clockwise the end of X is north (N).

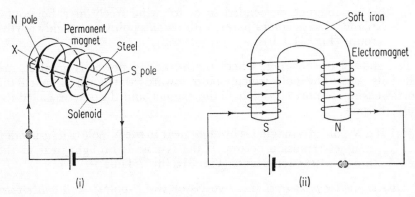

Fig. 31.2  Permanent and temporary magnets

If a *ring* or *horse-shoe* magnet is required, an insulated solenoid is wound round the material and a steady current is switched on and then off (Fig. 31.2(ii)).

## Other Methods of Making Magnets

### (1) *Single Touch*

A simple method of making a steel knitting needle, or a piece of clock spring, into a magnet consists of laying the specimen X on a table, and stroking it repeatedly in one direction with a pole of a magnet Y (Fig. 31.3(i)). This is known as the method of single touch. Y must be raised clear of X each time the end of the latter is reached, otherwise the magnetism induced in X in one movement of Y would be annulled on the opposite movement if Y always moved along the surface of X. Experiment shows that the end of X last touched has an *opposite* polarity to the stroking pole.

### (2) *Divided Touch*

Fig. 31.3(ii) illustrates the method of magnetizing by divided touch, which makes a strong magnet more quickly than the method of single

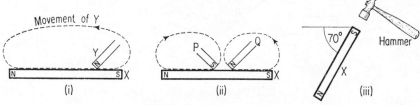

Fig. 31.3 Making magnets

touch. Beginning at the middle of the specimen X, two magnets P and Q are used to magnetize one half of X; the procedure for each half is the same as in the method of single touch. The end of X last touched is again opposite in polarity to the magnetizing pole, so that the poles of P and Q used must be opposite in polarity to make X a normal magnet.

If two south poles of P and Q are used a N-pole appears at each end of X and a S-pole in the middle. These are called *consequent poles*. X is not then a 'magnet' in the normal sense of the term. If it is suspended it will not settle in any particular direction, as the resultant magnetism is zero.

### (3) *Hammering in the Earth's Field*

In the 17th century a weak magnet was made by hammering one end of the specimen X, which pointed north; the best position to hold X is about 70° inclined to the horizontal (Fig. 31.3(iii)). Instead of a magnet, such as Y in Fig. 31.3(i), the magnetic influence of the earth is utilized in this case (see p. 573), and in Great Britain the lower end of the specimen is found to possess a north polarity. This is the case for countries such as the United States and Russia, which are also in the northern hemisphere. In Australia or South America, which are in the southern hemisphere, the lower end has south magnetic polarity.

## Demagnetization

Magnets can be partially demagnetized by hammering them hard when they are pointing east–west, that is, about 90° to the earth's magnetic field direction, or by heating them strongly.

The most effective method of demagnetization is to use an alternating current. A low mains voltage, such as 12 V A.C., is connected in series with a suitable rheostat R and a solenoid S (Fig. 31.4). *The solenoid is placed with its axis pointing east–west.* The object to be demagnetized, a bar magnet or wrist-watch, for example, is then placed inside S and the alternating current is switched on. After a few seconds the object is slowly withdrawn from the solenoid to a long distance away from it, and will then be

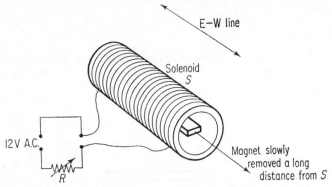

Fig. 31.4 Demagnetization by A.C.

found to be demagnetized. The alternating current is one which reverses every $\frac{1}{100}$ sec, and hence reverses the magnetism in the material one hundred times per second. This has the effect of making the material magnetically softer, and the magnetism soon shrinks to a very small value and disappears.

## Magnetic and Non-magnetic Materials

Iron, nickel, cobalt and certain alloys of these metals can be made into strong magnets. They are hence strongly magnetic and are known as *ferromagnetic* materials.

If a bar of pure or soft iron is placed inside a solenoid and the current is switched on the iron becomes a powerful magnet and can pick up pieces of iron. When the current is switched off, however, the pieces fall to the ground. *Temporary magnets are therefore made of soft iron.* Electromagnets, which are temporary magnets, hence use iron (p. 556).

Steel is an alloy of iron, made by adding a small percentage of carbon to pure iron. It is generally a much harder metal than pure or soft iron. If a bar of steel is placed inside a solenoid and the current is switched on the steel becomes a magnet and can pick up pieces of iron. When the current is switched off the iron remains attracted to the steel. Thus *permanent magnets are made of steel*. The magnetic properties of steel hence form a complete contrast to that of iron. Steel is more difficult to magnetize than iron, but loses its magnetism much less easily than iron.

A vast amount of research has been expended in finding alloys which can be made into powerful magnets. Special alloys such as mumetal have been developed for the electromagnet and the transformer, in which temporary magnets are used. Alni, alcomax and ticonal are alloys of nickel and cobalt which are used for making powerful permanent magnets.

Experiments show that elements other than iron, nickel and cobalt can be made only into very feeble magnets, over a million times weaker than those made from ferromagnetic materials. We shall call them *non-magnetic* substances. Copper, brass and wood, for example, are non-magnetic.

### Ferrites

A special class of strongly magnetic materials are those known as *ferrites*. They are made by chemical combination of a metal oxide such as zinc oxide with ferric oxide ($Fe_2O_3$). Ferrites can be magnetized strongly and, even more important, they are not demagnetized easily once they are magnetized. Ferrite rods, which are black and hard, are used in aerials of transistor receivers. They are particularly useful for reception of microwaves, which are radio waves of very high frequency (p. 333). Usually there are considerable power losses in magnetic materials due to circulating electric currents inside them (see p. 625), but ferrites have a high electrical resistance, and hence the current and power loss are extremely small. A fine iron oxide power is used to coat the tapes used in magnetic tape-recorders (p. 566).

### The Magnetic Filter

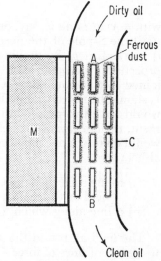

Fig. 31.5 Magnetic filter

The attraction of a magnet for iron is utilized in industry for cleaning oil and for separating iron from non-ferrous materials mixed with it. Fig. 31.5 illustrates the essential features of a magnetic filter system, which is used for extracting small iron particles contaminating oil or any other liquid. M is a powerful magnet made of ticonal, and C is an iron filter cage which is magnetized by M. The dirty oil is poured into the filter at A, and the iron particles collect in the iron cage under the powerful attractive force on them, as shown. The clean oil issues through B, the bottom of the filter cage.

### Magnetic Fields · Lines of Force

If a small compass needle is placed at a point *a* near to the north pole N of a bar magnet, the needle or light magnet immediately swings round on its support and settles in a definite direction (Fig. 31.6). At other points near the magnet, such as $b$, $c$ and $d$, the needle settles in a different direction. In the area or region round the magnet, then, a magnetic force is obtained. We call this region a *magnetic field*, and a compass needle is a sensitive method of finding the direction of the magnetic force at a point in the field.

The direction of the magnetic forces in an area of the field can be mapped out by marking the north and south ends of a compass needle

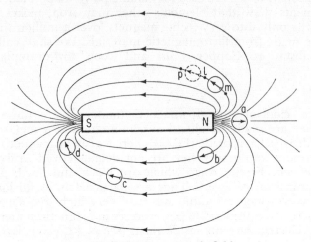

Fig. 31.6 Plotting magnetic field

with dots at $l$ and $m$, say, on a sheet of paper. The compass is then moved until the S-pole of the needle is at $l$ and the new position of its N-pole is marked at $p$. By moving the compass in this way and repeating the procedure, the paper can be covered by a series of dots, which are then joined by a smooth curve. Other lines of force are drawn in the same way. It should be noted that no lines of force cross, otherwise the magnetic field would have two possible directions at the point of intersection. By agreement, the *direction* of the field at a point is the direction which a *north* pole would tend to move if placed at that point. A line of force can be defined as either:

   (i) a line such that the tangent to it at any point is the direction of the field at this point; or
(ii) a line along which a free N-pole would tend to move.

The field all round the magnet, in three dimensions, can be imagined to be full of lines of force. This way of picturing the forces in a magnetic field is due originally to Faraday, and it was placed on a sound mathematical basis by a later scientist, Maxwell. Today, magnetic lines of force in magnetic fields are used extensively in their work by electrical engineers and designers of nuclear fusion devices, and by astronomers in their theories of the origin of the universe.

**PLATE 13**
Computer Memory Store –
Van de Graaff Generator

3(a) Memory store of computer. Ferrite rings,
magnified, linked by conductors carrying information (p. 571).

(b) Memory store undergoing final test before
e.

13(c) Van de Graaff electrostatic generator
used at Aldermaston, producing 6 million volts
for nuclear investigations (p. 656).

PLATE 14
Electrical Instruments

14(a) Moving-coil instrument. Cut-away view, showing the magnet, the curved poles to provide a radial field, and the coil; the latter is wound on an aluminium former whose outline is just visible and which helps to make the pointer come to rest quickly. One end of the upper spring can be moved by the lever shown, thus adjusting the zero position of the pointer (p. 603).

14(b) Demonstration Hot-Wire ammeter. The larger range, 0–5 amp, can be used by connecting the shunt, the open coil of wire on the left, across the resistance wire by means of the link at the bottom. The rod at the top can act as a zero adjuster by moving one end of the wire, which is mounted at the top of a flexible steel strip (p. 534).

14(c) Section of a Moving-Iron repulsion instrument, showing the fixed and moveable curved iron pieces removed from the solenoid below, the damping vane, and the dashpot in which the vane moves (p. 593).

## Some Magnetic Fields

A vivid demonstration of magnetic fields can be obtained by sprinkling iron filings on a sheet of paper in the field round magnets, and then tapping the paper lightly. The filings become tiny magnets by the process of induction (see p. 564), and tapping the paper gives them the necessary freedom to settle in the direction of the magnetic field, like miniature compass needles.

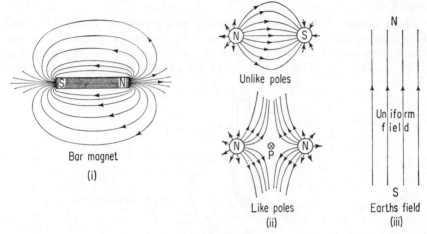

Fig. 31.7  Magnetic fields

Fig. 31.7 illustrates a few typical lines of force in some magnetic fields obtained with a compass needle. The lines exist in space, so that the sketches represent a section in a horizontal plane. Points worth noting are:

(i) *Bar Magnet.* The lines are oval in shape and come out of the N-pole into the air and enter the S-pole. The lines are continuous; one should imagine them inside the magnet passing from the south pole S to the north pole N in the magnetic material (Fig. 31.7(i)).

(ii) *Unlike and like poles.* The field between the N- and S-poles has many lines and is therefore strong. In contrast, there are no lines round P between the N- and N-poles facing each other, and hence there is no magnetic force or field here (Fig. 31.7(ii)).

(iii) *Earth's field.* Locally, the field consists of parallel lines pointing northwards (Fig. 31.7(iii)). The density of the lines at any part of the field is hence constant. The density of the lines round a *single pole*, however, decreases the farther we go from the pole (Fig. 31.7(ii)). The earth's field is an example of a 'uniform field', one whose direction and strength is constant, whereas the field round a single pole is non-uniform.

## Neutral Points in Fields

So far we have met the effect produced by one magnetic field. The result of combining two magnetic fields can be investigated by plotting with a compass needle the field round: (i) a single N-pole in the earth's field, or

(ii) a bar magnet in the earth's field pointing in the magnetic north–south direction. In either case the combined or resultant field has a special feature. At a point P the compass needle refuses to point in any particular direction, that is, there is no net magnetic field here. Thus:

*A neutral point is a point where the combined or resultant magnetic field is zero.*

### Single Pole

There is one neutral point P round a single pole N. It lies due south (Fig. 31.8(i). If an *n*-pole is imagined at P, the earth's field would urge it northwards and the field due to N would repel it southwards (Fig. 31.8(ii)). The directions of the forces are exactly opposite to each other

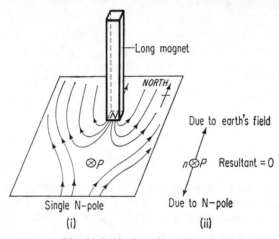

Fig. 31.8 Single pole and earth

and cancel at P. A neutral point would not appear due north of N, because this pole and the earth would both urge an *n*-pole placed here due northwards.

### Bar Magnet, N-pole Pointing South

There are two neutral points (Fig. 31.9(i)). If an *n*-pole is imagined at $P_1$ the net attraction of the bar magnet is southwards because its S-pole is nearer $P_1$ than its N-pole. The earth's field, however, urges the *n*-pole northwards and completely neutralizes the magnet's force, as shown. At $P_2$, the other neutral point, the magnet's attraction for an *n*-pole is still southwards, and again completely opposes the force due to the earth's field.

### Bar Magnet, N-pole Pointing North

Again there are two neutral points, X and Y, but this time they lie on an east–west line, as shown (Fig. 31.9(ii)). An *n*-pole at X is repelled by N along the line NX, and attracted by S along the line SX. The resultant

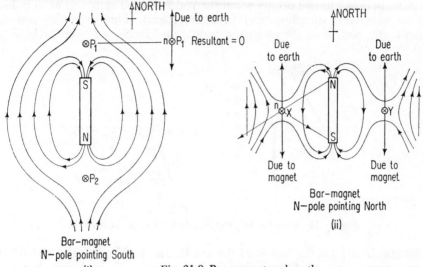

Fig. 31.9 Bar magnet and earth

force due to the magnet is southwards. This is opposite in direction to the force due to the earth. A similar state of affairs exists at Y.

### Parallel Bar Magnets

When two bar magnets are placed parallel and close to each other, then, neglecting the earth's field this time, neutral points are obtained at the positions marked X in Fig. 31.10(i), (ii). At these points the magnetic

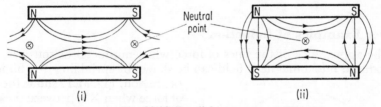

Fig. 31.10 Parallel magnets

force due to one magnet is equal and opposite to that due to the second magnet.

## Magnetic Induction

*Mu-metal* is a magnetic alloy which can be magnetized strongly by very weak magnetic fields (p. 559).

### EXPERIMENT

If a long rod of mu-metal is held pointing downwards at an angle of about 60° to the horizontal the lower end A is observed to repel the N-pole of a suspended magnetic needle (Fig. 31.11(i)). The other end B is observed to repel the S-pole of the needle. The rod AB is therefore a magnet.

A surprising change occurs when the rod is turned round so as to point in an east–west direction and held horizontal. This time the end A *attracts* the N-pole of the magnetic needle and its S-pole (Fig. 31.11(ii)).

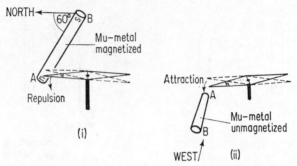

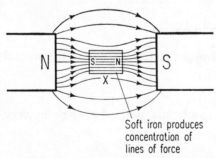

Fig. 31.11  (i) Induction. (ii) Non-induction in earth's field

The end B attracts the S-pole of the needle and its N-pole. The rod AB is hence unmagnetized. On turning the rod round to lie in the north–south direction, the rod becomes a magnet again.

*Explanation.* When it is placed in the direction of the earth's magnetic field, that is, north–south, the magnetic force magnetizes the rod. Since the rod is not in contact with a magnet, as in the method of single touch, the magnetism in the rod is called *induced magnetism*. The phenomenon is called *magnetic induction*, and this is an example of magnetic induction by the earth's field. If the rod is placed east–west, at right angles to the earth's field, the induced magnetism due to the earth along the length of the rod is zero.

## Other Examples of Induction

The appearance of the lines of force when mu-metal or soft iron X is introduced into a magnetic field can be shown by sprinkling iron filings in the region. Fig. 31.12 shows the lines of force when X is between the poles N and S of two strong magnets. Compared with the appearance when air is between the poles, the lines of force now crowd into the iron. The concentration of the lines may be increased many hundreds of times with soft iron alloys such as mu-metal. This is due to the strong induced magnetism in the metal.

Fig. 31.12  Soft iron in magnetic field

Observe that the lines of force in the field *enter* the S-pole of the induced magnet and leave at the N-pole. This provides a simple rule for polarity for the case of an induced magnet.

When a magnet is placed near an unmagnetized steel pin, the pin first becomes magnetized by induction and is then attracted to the magnet (Fig. 31.13(i), (ii), (iii)). Induction thus precedes attraction. This explains why unmagnetized iron and steel are attracted to magnets. Steel objects,

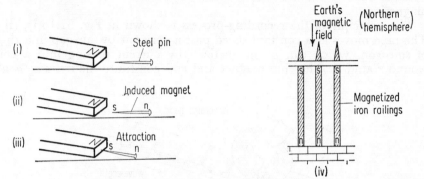

Fig. 31.13 Induction by magnet and earth's field

such as pliers or railings or girders in buildings, are often found to be magnetized (Fig. 31.13(iv)). This is due to induction by the earth's field.

## Magnetic Shields

The concentration of lines of force in soft iron can be used to shield objects from unwanted magnetic fields. Fig. 31.14(i) shows an object X with a soft iron ring round it and a powerful magnet Y outside. All the lines of force pass through the ring. None passes into the air inside the ring.

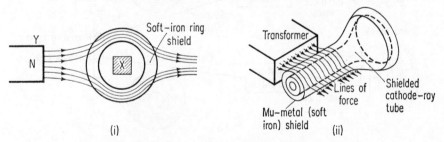

Fig. 31.14 Magnetic screening

Thus X is shielded from the magnet. The electron beam inside cathode-ray tubes in television receivers is shielded from the undesirable influence of external magnetic fields by placing a soft iron or mu-metal cylinder round the neck of the tube (Fig. 31.14(ii)).

## Principle of Magnetic Tape recorder

In 1898 Poulsen, a Danish scientist, invented an instrument called a *Telegraphone,* in which he succeeded in recording sound magnetically. The instrument used spools of steel wire to record the sound, but the invention lapsed as the machine was cumbersome and not easy to operate.

The modern magnetic tape-recorder developed from about 1940, with the production of a new magnetic oxide which is called gamma-ferric oxide. The oxide is made in finely powdered form, mixed with an adhesive or binder and then coated on to a very thin plastic-base smooth tape, which forms a backing for the oxide. Once magnetized, the oxide retains its magnetism.

The principle of the recording-process is shown in Fig. 31.15(i), (ii). The tape moves at a constant speed past a very narrow air-gap in a ring of soft-iron, which can be magnetized by current passing into a coil wound round it. The ring magnet and coil constitute the *recording head*.

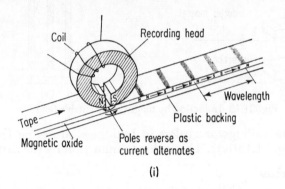

(i)

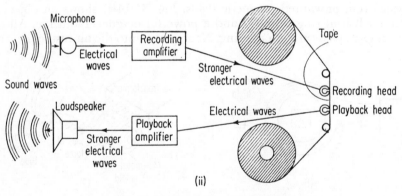

(ii)

Fig. 31.15 Magnetic tape recording

Speech or music is converted by a microphone into a varying or alternating current of exactly the same frequency, and this flows in the coil. On one half of a current cycle the oxide arriving at the gap becomes a small magnet by induction. On the other half of the same cycle the current reverses, and as the tape has moved on, the next small section of oxide arriving at the gap is now magnetized in an opposite direction. As the tape moves along, therefore, a series of small magnets are obtained, pointing opposite ways. The strength of the magnets depends on the current strength, which in turn varies with the loudness of the sound. The number of pairs of magnets obtained per second on the tape is equal to the frequency of the sound. Thus the oxide records the sound magnetically.

In play-back the magnetized tape is run at the same constant speed past the same or another soft iron ring, each magnet closing the gap between the poles as before. This time, however, the introduction of lines of force into the gap by the magnets produces an e.m.f. and current in a coil round the ring (see *Electromagnetic Induction*, p. 616). The current is amplified and passed to a loudspeaker, thus reproducing the sound.

## Domain Theory of Magnetism

The search for a theory to explain all the observed phenomena in magnetism has occupied the attention of many famous scientists. Even today the problem cannot be said to have been fully solved. It is, however, usually agreed that the magnetism in a specimen is due to the electrons surrounding the nucleus of the atom, which carry a quantity of electricity and move and spin inside the atom (see p. 738). These moving and spinning electrons constitute tiny circulating electric currents, which together produce the magnetism in a bar magnet, for example. See p. 584.

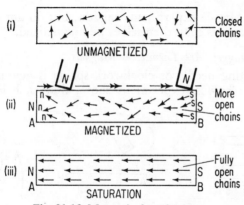

Fig. 31.16 Magnetic domain theory

Ferromagnetic materials have a crystal structure. When they are closely examined numerous small regions of the tiny crystals appear to have very strong resultant magnetism, although the material as a whole shows no magnetism. These tiny regions are called *domains*. In each domain the resultant magnetism points in different and random directions.

The magnetism in a domain can be represented like a vector (see p. 82). We therefore draw an arrow pointing in the direction concerned and mark the ends *n*, *s* to show the magnetism. In an unmagnetized material the resultant magnetism of all the domains is zero. Fig. 31.16(i) shows how this occurs in simple cases. The domain magnetism forms 'closed chains' along the material. The resultant *n*- and *s*-poles in neighbouring domains then cancel each other's magnetic effect, and the material appears to be unmagnetized.

## Magnetization

When the pole of a magnet is used repeatedly for stroking the material (Fig. 31.16(ii)), the domains move round in stages under the magnetic influence. Their resultant south poles, *s*, for example, turn to face a *N*-stroking pole as it moves from A towards B. The closed chains then become open, as illustrated roughly in Fig. 31.16(ii), and as the *s*-poles at the end B now have no north poles to neutralize them, unlike the case of the closed chains in Fig. 31.16(i), the end B acts like a south magnetic pole, S. Similarly, the 'free' N-poles *n* of the domains at the end A makes this end act like a N-pole N. Experiment shows that the strength of a magnet cannot be increased beyond a certain limit, when the magnet is said to be 'saturated'. On the domain theory this occurs when the magnetism in all the domains points in one direction (Fig. 31.16(iii)).

## Experimental Evidence

Positive evidence of the existence of domains was first obtained by Bitter in 1930, who sprinkled a fine magnetic powder on the smooth surface of a ferromagnetic substance. The powder settled into patterns, now called *Bitter patterns*, which clearly showed the boundaries of the domains.

The following simple experiments suggest, but do not prove, that tiny regions inside magnets may themselves be magnetized.

### Small Pieces of Magnets Are Themselves Magnets

If a steel knitting needle or clockwork spring is magnetized and then broken into two halves, experiment shows that each half is a magnet

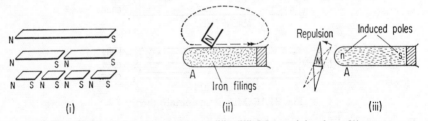

Fig. 31.17 (i) Breaking magnet. (ii), (iii) Magnetizing iron filings

(Fig. 31.17(i)). If each half is now broken again four magnets are obtained. No matter how small the broken parts may be, each of them is a magnet. This suggests that magnets may contain small regions which are themselves magnetized.

### Magnetizing Iron Filings in a Test-tube

A test-tube is filled with iron filings and laid on the table. The end A of the tube has then no magnetism, since it attracts each end of a compass needle on testing, but when the tube is stroked by the N-pole of a strong magnet (Fig. 31.17(ii)), as in magnetizing by single touch, the filings at the top of the test-tube can be seen to be aligning themselves horizontally. The end A is now found to exhibit a north polarity, on testing with a compass (Fig. 31.17(iii)). On shaking the test-tube vigorously so that the filings are again mixed, no magnetism is shown by the ends of the tube. This experiment suggests that small regions in an unmagnetized specimen, equivalent in size to iron filings, may align themselves when the specimen is magnetized.

### Closed and Open Chains

A group of compass needles can be arranged to form a closed chain of small magnets, as illustrated in Fig. 31.18(i). When a magnet is brought near them the needles turn round and form an open chain (Fig. 31.18(ii)).

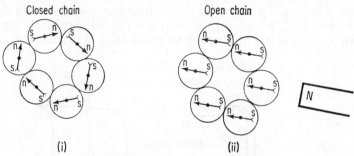

Fig. 31.18 Open and closed chains

When the magnet is taken away the needles return to the closed chain formation, which is a stable pattern. This experiment suggests that a closed chain of domains may be made into an open chain by means of a magnet.

## Keepers of Magnets

If magnets are stored in a box in a haphazard manner their open chains tend to become partly closed owing to: (i) the effect of other magnets, and (ii) the vibration which occurs in buildings due to outside traffic and inside machinery. To prevent demagnetization, magnets are stored in boxes in pairs, with opposite poles facing each other, and with pieces of soft iron known as keepers, K, placed across both ends (Fig. 31.19).

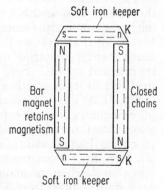

Fig. 31.19 Action of keepers

An open chain of domains will return to a closed chain formation if it is disturbed, and in general, a closed chain is much more stable than an open chain. When the soft-iron keepers K are placed across the two bar magnets, they become magnetized by induction with poles as shown in Fig. 31.19. The domains of the two bar magnets and of the two keepers now form a closed chain throughout the whole of the four pieces of metal, and when stored in this manner, the magnets retain their magnetism for a very long time. The keepers are made of soft iron and not of steel because the former metal is easier to magnetize by induction.

### Hysteresis Curves or Loops

As we have seen, magnetic materials are needed for many different purposes. They are usually judged as a suitable material or otherwise from their *hysteresis curve or loop*. This is a curve which shows how the magnitude of its flux density (see p. 632), or roughly the strength of its magnetization, which is a measure of its magnetic state, varies when it is magnetized by a magnetic field.

Fig. 31.20(i) shows the kind of curve or loop obtained with pure or soft iron. The applied magnetic field varies one way $(+)$ and then the opposite way $(-)$, so a complete loop is obtained. Magnetic saturation is shown at the point S and strong magnetization is produced by a small

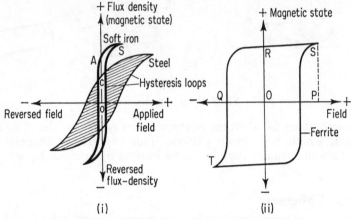

(i)      (ii)

Fig. 31.20 Variation of flux density (magnetic state)

magnetic field. When the field is removed, for example, by switching off the current in a solenoid (p. 556), the flux density or magnetism remaining in the iron is represented by OA, which is called the *remanence* of the iron. Fig. 31.20(i) also shows the hysteresis loop for a sample of steel. This is a much harder material magnetically than soft iron; its remanence is represented by OC. Soft iron, however, easily demagnetizes, whereas steel retains its magnetism.

Fig. 31.20(ii) is a typical loop of a *ferrite* material, used in making computer memory cores as explained shortly. This has much more of a square shape than those of soft iron or steel, and the remanence OR is high and not much less than the saturation value PS. When the magnetic field is switched off, then, little change occurs in the magnetic state of the ferrite. On the other hand, when the magnetic field is reversed from OP to OQ, the magnetic state changes from S to T, which is a very large change.

### Ferrite Memory Cores in Computers

For recording all its information, such as statistics of sales in industry, a digital computer uses the *binary system* of numbers. In this system any number can be represented by a row of only two numbers, '0' and '1'. (See

*Topics in Modern Mathematics*, by J. G. Thomas (Blackie), for a clear account of Binary Numbers.)

The binary numbers are stored by means of ferrite cores, tiny rings, F, which may have holes only 0·3 in wide, for example (Fig. 31.21(i)). The binary numbers or 'information' is written in by a 'write wire' W and later read or printed by a 'read wire' R. The wires are insulated and thread the core. See Plate 13, p. 556.

When a current suddenly flows along W the ring F is magnetized in a clockwise direction, as shown by the circular line in Fig. 31.21(i). If the current in W flows in the opposite direction the ring is magnetized in an anti-clockwise direction (Fig. 31.21(ii) (see also p. 584)). The core can thus exist in either of two magnetic states. One of them, say that corresponding to the clockwise direction in Fig. 31.21 (i), corresponds to a '0' state or number, and the other to a '1' state (Fig. 31.21(ii)).

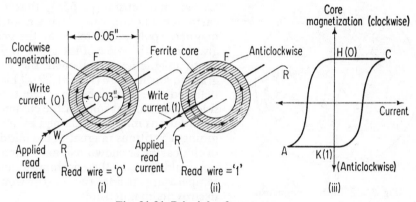

Fig. 31.21 Principle of memory core

The variation of the magnetic intensity of the ferrite ring due to a current is shown in Fig. 31.21(iii). With a clockwise current the magnetization first reaches a value represented by C in Fig. 31.21(iii), and then, when the current is switched off, it falls very slightly to H. This is the magnetic state corresponding to '0', say. If the current is anti-clockwise the magnetization first reaches a value corresponding to A in Fig. 31.21(iii) and then returns to K when the current is switched off, which is the magnetic state corresponding to '1'.

### Reading Information

To read the information stored in the form of '0' or '1', a current is now passed along the wire W in the same direction as that shown in Fig. 31.21(i). This makes the magnetization of the core change very slightly from H to C in Fig. 31.21(iii). Another wire R, the read wire, is then hardly affected by the passage of the current in W.

Suppose, however, that a '1' was stored in the core, as in Fig. 31.21(ii), corresponding to a magnetic state represented by K in Fig. 31.21 (iii). When a current is passed along W in the same direction as before, the magnetism changes from K to C in Fig. 31.21(iii). This is a large change in magnetization. As explained later (p. 626), a large induced voltage is then obtained in the read wire R. In this way the binary numbers are stored by the ferrite core, and then read by R and printed on a tape.

In practice, many cores and complex circuits are needed in memory cores to make them work efficiently. Each core, for example, is threaded by five wires. The interested reader is referred to *Electronic Computers*, by Hollingdale and Tootill (Penguin), for further information. See Plate 13.

## THE EARTH'S MAGNETISM

Centuries ago it was known that a pivoted magnetic needle came to rest pointing approximately north and south, and the magnetic compass

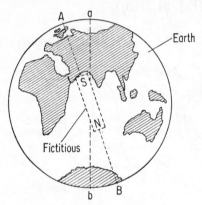

Fig. 31.22 Earth's magnetism

needle is still used for navigation purposes. It follows, from the fundamental law of magnetism (p. 555), that the earth acts like a magnet, with a south magnetic pole A located near the geographic north pole *a* of the globe and a north magnetic pole B near the geographic south pole *b* (Fig. 31.22). This is believed due to circulating currents inside the earth, but some of the common observations may be explained by imagining a fictitious magnet at the middle of the earth, as shown. No actual magnet could exist, as the enormously high temperature inside the earth would destroy its magnetism.*

### Angle of Dip

If a non-magnetic uniform piece of metal, such as brass, is suspended at its centre of gravity M, the brass remains horizontally in equilibrium (Fig. 31.23(i)). If, however, a magnetic needle is suspended at its centre of gravity G so that it is free to take up any position in a magnetic north–

* Scientists have still not solved the mystery of the origin of the earth's magnetism.

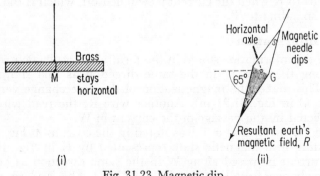

Fig. 31.23 Magnetic dip

south plane, it 'dips' at an angle of about 65° to the horizontal in this country, with its N-pole pointing downwards (Fig. 31.23(ii)). Now, in general, a magnetic needle points in the direction of the total (resultant) strength or intensity of the magnetic field in which it is situated. It therefore follows that the total or resultant intensity $R$ of the earth's magnetic field acts at an angle of about 65° to the horizontal in this country.

## Horizontal and Vertical Components

An ordinary compass needle is pivoted so that it can only move in a horizontal plane. We now see, however, that the total earth's field $R$ acts not only in a north–south direction, but also downwards; the compass needle does not show the latter effect because it cannot move in a vertical plane. The strength or intensity of the earth's field in a horizontal plane is less than $R$ and is called the *horizontal component* intensity. We shall denote this by $H_0$. If $R$ is drawn to scale in magnitude and direction in a vector diagram of fields then it will be represented by $bd$ in

Fig. 31.24 acting at an angle $\theta$ to the horizontal, which is the angle of dip, say 65°. The component $H_0$ of $R$ is then represented by the side $ab$ of the rectangle $abcd$, as explained on p. 166. It can be seen that

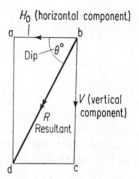

Fig. 31.24 Components of earth's field

$$\frac{H_0}{R} = \cos \theta,$$

or $H_0 = R \cos \theta.$

Likewise, the earth's magnetic field has a *vertical component V* represented by $bc$ in Fig. 31.24. It can be seen that $V = R \sin \theta$. Also $V$ is greater than $H_0$ in Great Britain since $\theta$ is about 65°; $bc$ or $V$ is then longer than $ab$ or $H_0$. The vertical plane such as $abcd$ which contains the earth's total intensity is called the *magnetic meridian*, and points in a magnetic south and north direction.

A magnetic needle which can settle in the direction of the resultant intensity of the earth's field is said to be 'freely suspended' in the field. The *angle of dip* at a place on the earth's surface is the angle made with the horizontal by a freely suspended magnetic needle there. Otherwise defined, the angle of dip is the angle made with the horizontal by the resultant intensity of the earth's field at the place concerned.

## Variation of Angle of Dip over Earth's Surface

If a freely suspended magnetic needle is taken to various points on the earth's surface, the needle 'dips' at various angles to the horizontal. Near the geographic N- and S-poles, for example, the needle dips vertically, and the angle of dip is then 90°. Near the equator the freely suspended needle is horizontal, and the angle of dip is then zero. In Great Britain the angle of dip was about 67° in 1966, but it should be noted that the angle at a given place on the earth varies very slightly every hour and every day, and that at the end of each year there is a change of a small fraction of a degree.

Fig. 31.25 illustrates how a fictitious magnet in the middle of the earth can account for the variation of the angle of dip all over the world. Some of the lines of force due to the magnet are shown. The tangent to the

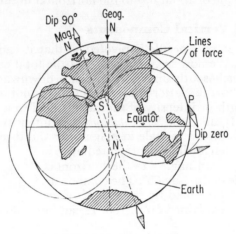

Fig. 31.25 Variation of dip

line of force passing through a place P is the direction of the magnetic field there (p. 560). Now at this place the tangent is parallel to the earth's surface, since the latter is the tangent to the circle representing the globe. It follows that the resultant intensity of the earth's field is horizontal at P, and the angle of dip is therefore zero. Near the geographic N-pole, a tangent to the line of force is vertical, and hence the angle of dip is 90° here. This region contains the magnetic N-pole of the earth, and in a magnetic survey an aeroplane equipped with a specially designed magnetic dip-needle has located this pole about 1000 miles from the geographic N-pole. At T on the globe the tangent to the line of force is less than 90° to the earth's surface, and hence the angle of dip is less than 90°.

### Measurement of Angle of Dip or Inclination

At meteorological stations all over the world, measurements of the angle of dip are made daily. A dip circle is an instrument capable of measuring the angle of dip directly, and is illustrated in Fig. 31.26. It consists of a vertical circle A graduated in

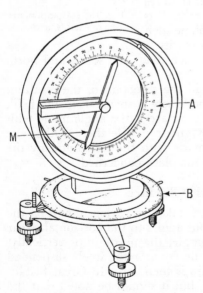

Fig. 31.26 Dip circle

degrees, which can rotate about a vertical axis passing through the centre of a horizontal circle B, also graduated in degrees. A magnetic needle M is placed at the middle of A so that it can rotate about a horizontal axis through its centre of gravity.

To measure the angle of dip:

1. The vertical circle A is first turned *until the needle is vertical*. The needle is now only affected by the vertical component because it sets vertically. The horizontal component $H_0$, which acts in the magnetic meridian, thus has no effect in the plane of A, and consequently, the magnetic meridian must be perpendicular to the plane of A.

2. The reading on the lower graduated circle B is now noted, and *the plane of A is turned through an angle of 90°* so that the magnetic needle M now lies in the magnetic meridian. The inclination of M to the horizontal is read on A, and this reading is the angle of dip. See also Fig. 31.24.

### Angle of Declination (or Variation)

As stated previously, a vertical plane containing the earth's total intensity is known as a magnetic meridian. The direction of the magnetic meridian depends on the particular place of the earth where such a plane is drawn, and is easily found by drawing the vertical plane through a compass needle when it settles down, as the needle points in the magnetic north and south directions. The geographic meridian is the vertical plane passing through the geographic N- and S-poles, and at a given place on the earth this meridian G makes an angle $x$ with the magnetic meridian M (Fig. 31.27). This angle is known as the *angle of declination* or the *variation*. Seamen need to know the declination to navigate by a magnetic compass, so that the true (geographic) north is obtained. For navigation by the compass, therefore, maps are prepared showing the declination at different parts of the world. This is done by drawing *isogonic lines* on the maps, which are lines connecting all places having the same declination.

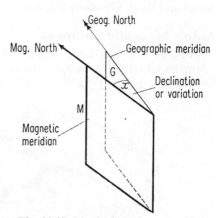

Fig. 31.27 Magnetic and geographic meridians

As the poles of a magnet are not at the ends (p. 555), the magnetic north and south direction is not the line joining the ends of a magnetic needle. The true direction can be found by suspending the needle in a stirrup so that it can oscillate in a horizontal plane just above a sheet of paper (Fig. 31.28(i)). When it comes to rest, the positions P and Q of its ends are noted on the paper. The magnet is then turned over and allowed to come to rest again, and the new positions P′ and Q′ are again noted on the paper. Now the line joining the two ends is a fixed line in the magnet, and hence it always makes the same angle with the magnetic north and south when the magnet comes to rest. Thus PQ and P′Q′ are inclined at equal angles to the magnetic north and south, and thus the latter direction is given by the bisector of the angle POP′. The magnetic meridian, of course, is the vertical plane passing through the bisector.

The geographic north and south direction is the direction of the shadow of a vertical pole X at noon, on a day when the sun is shining. Alternatively, the direction of the shadow is noted two hours before noon, for example, and two hours after noon, and the geographic north and south is that direction corresponding to the bisector of the angle between the two shadows (Fig. 31.28 (ii)).

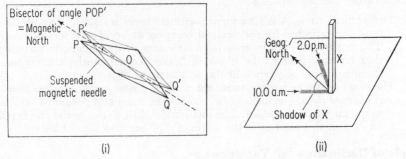

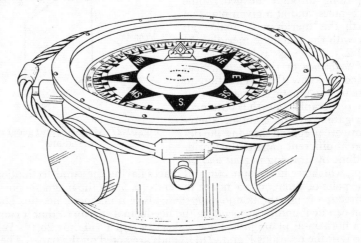

Fig. 31.28 Determination of magnetic and geographic meridians

This method was first given by Dr. Gilbert (p. 555). Knowing the magnetic north and south direction, the angle of declination is easily obtained.

### Ship's Magnetic Compass

Although it has been displaced in large ships by another form of compass which is gyroscopic, the magnetic compass is still used extensively in ships for navigation. Fig. 31.29 illustrates one form of compass, which

Fig. 31.29 Ship's compass

consists of a few magnets fixed side by side, underneath a compass card contained in a non-magnetic bowl. The card is pivoted at its centre by means of a pointed support, one end of which rests on the bottom of the bowl. The latter is mounted on an axle (not shown), about which it is free to swing. The axle itself is fixed to an outer ring, which in turn is free to swing about a perpendicular axis. This is known as *gimbal* mounting. It has the advantage that the card now always remains in a horizontal position, even if the ship pitches or rolls.

The iron and steel used in making a ship causes the compass needle to deviate from the true magnetic north–south direction. The deviation is found by experiment before the ship starts on its journey, and a correction is applied to the reading of the card.

## SUMMARY

1. Like magnetic poles repel. Unlike poles attract.

2. Magnets can be made: (i) by the electrical method ('clockwise current–south pole'); (ii) by single or double touch ('end last touched has opposite polarity to stroking pole'); (iii) by hammering or induction in the earth's field. (In Great Britain, the end of the bar pointing downwards is a N-pole.)

3. Soft (pure) iron is more easily magnetized than steel (iron alloy with carbon), and loses its magnetism more easily. Magnetic iron oxide powder is used in tape-recorders, as it is difficult to demagnetize. Ferrites are compounds of iron oxide and a metal oxide which are magnetic and have a high electrical resistance, thus reducing energy losses in radio reception, for example.

4. According to the domain theory, tiny regions or domains in an un-magnetized bar have strong magnetism, but the resultant or total magnetism is zero, since the directions of magnetism are all different. Keepers are pieces of soft iron used in storing magnets; they help to form stable closed 'chains' and prevent demagnetization.

5. A *magnetic field* is a region in which a magnetic force may be detected. A *line of force* is a line such that the tangent to it at any point is the direction of the magnetic field there; alternatively, it is a line along which a N-pole would tend to move if it were free.

6. Magnetic fields can be plotted with the aid of a compass needle. The appearance of the whole field can be seen by using iron filings. The field round a single pole has one neutral point; the field round a bar magnet has two neutral points (N-pole pointing southwards, neutral points are on *axis* of magnet). At a neutral point the field due to the single pole or the magnet is exactly equal and opposite in direction to the earth's magnetic field, so the resultant field is zero.

### Earth's Magnetism

7. Angle of dip = angle made with horizontal by total or resultant earth's magnetic field. It is 90° at the magnetic poles and zero at the magnetic equator.

8. Angle of declination = angle between magnetic and geographic meridians at the place concerned.

9. Magnetic meridian = vertical plane in the magnetic north–south direction. Geographic meridian = vertical plane in the geographic north–south direction.

## EXERCISE 31

**1.** Describe *three* different methods of magnetizing similar steel bars indicating, in each instance, the polarity of the magnet formed.

How would you expect the three magnets to compare in strength, and what test would you make to determine which is the strongest? (*L.*)

**2.** Explain the following:

(*a*) Two steel needles hang from the lower end of a vertical bar magnet, but do not hang vertically.

(*b*) A horse-shoe magnet is often supplied with a 'keeper'.

(*c*) When a bar magnet is placed on a table there are two points near it when there is no horizontal magnetic field. (*N.*)

**3.** What is meant in magnetism by *line of force* and by *neutral point*?

Why can iron filings be used to plot the lines of force in a strong magnetic field, but are unsuitable if the field is weak?

A bar magnet stands vertically on a table with its N-pole downwards. Describe how you would plot the lines of force round the end of the magnet. Give a diagram of the field and mark the position of the neutral point. (*L.*)

**4.** Give an account of a theory which explains the difference between a piece of unmagnetized steel and a steel magnet and describe two experiments in support of this theory.

Give *three* tests to determine whether or not a short steel needle is magnetized.

Explain why iron filings can be used to plot the lines of magnetic force round a bar magnet. (*L.*)

**5.** Explain: (*a*) what is meant by magnetic induction; (*b*) why one end of a bar of unmagnetized soft iron may attract either end of a compass needle.

Draw a diagram of the magnetic field in a horizontal plane when: (i) a bar magnet is placed horizontally with its axis in the magnetic meridian and its N-pole pointing north; (ii) a bar magnet and a bar of soft iron are placed horizontally in line with a gap between them (neglect the influence of the earth's magnetic field). (*N.*)

**6.** Describe and explain a method of magnetizing a steel needle AB so that the end A shall have north-seeking polarity.

What do you know about the magnetic strength of the poles formed at A and B?

Describe an experiment to verify your statement. (*N.*)

**7.** (*a*) Describe TWO different ways in which you could find whether a given steel bar is magnetized or not. (*b*) What is meant by a neutral point in a magnetic field? Show the positions of the neutral points in the case of a bar magnet placed horizontally in the earth's magnetic field with its N-pole pointing north. A force diagram is NOT required. (*c*) Explain what is meant by the statement that the horizontal intensity of the earth's magnetic field at a certain point is 0·18 oersted. (*N.*)

**8.** Draw the magnetic lines of force which pass through the iron on the diagram shown. (Neglect any effect of the earth's magnetic field.) (*C.*)

**9.** Draw a diagram showing the magnetic field in a horizontal plane containing the magnetic axis of a bar magnet if this axis also lies in the magnetic meridian, and the N-seeking pole of the magnet points to magnetic north.

Explain how your diagram can give information about the strength and direction of this field; define, and indicate the position of each neutral point. (*O. and C.*)

**10.** What do you understand by a *line of magnetic force*? Describe how you would plot the pattern of the lines of force in the neighbourhood of a bar magnet.

Sketch the pattern you would expect to obtain in the case of a long bar magnet which is lying on the bench in the magnetic meridian, with its S-seeking pole pointing northwards.

Explain, with the help of a diagram, how you would expect the pattern to be modified if a piece of soft iron were placed close to one pole of the magnet. (*O.*)

## Terrestrial Magnetism

**11.** What is the angle of dip at: (*a*) the magnetic equator; (*b*) the magnetic poles? (*N.*)

**12.** Explain what is meant by: (*a*) the magnetic meridian at a point; (*b*) the angle of declination (variation). (*N.*)

**13.** Define *declination* (magnetic variation), *dip*.

Describe how you would obtain the value at a given place of *either* the declination *or* the dip.

Explain why a vertical iron girder is often found to be magnetized. State the polarity of the girder. In what position must an unmagnetized iron girder be placed in order: (i) to become as strongly magnetized as possible by the earth's field; (ii) to remain unmagnetized along its length? (*L.*)

**14.** What is meant by the total intensity of the earth's magnetic field?

State the relation between the angle of dip and the horizontal and vertical intensities.

Describe an experiment to find the angle of variation (declination) in your town.

Describe how the angle of variation varies with position over the surface of the earth and how it changes with time at any one position. (*N.*)

**15.** Define *angle of dip*. Describe any one simple experiment which illustrates the effect of magnetic dip. State clearly the result of the experiment when it is conducted in: (i) the northern hemisphere; (ii) the southern hemisphere. (*C.*)

**16.** Explain the terms *declination, dip* and *horizontal component of the earth's magnetic field*. Describe how you would determine the value of the dip at a given place.

How does the value of the dip vary from place to place over the earth's surface? (*O.*)

# 32

# MAGNETIC EFFECT OF CURRENT

## Oersted's Discovery

In previous chapters the chemical effect (electrolysis) and the heating effect of an electric current were discussed. The most important effect of an electric current, however, is its magnetic effect, which was discovered by Oersted in 1820 while he was lecturing to students.

Experiment

Oersted's experiment can be repeated by connecting a long piece of insulated wire AD to an accumulator B, with a small rheostat R and a key K in the circuit (Fig. 32.1). When AD is held parallel to and over a

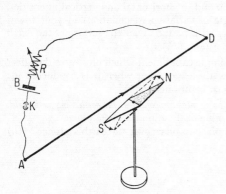

Fig. 32.1 Magnetic effect of current in straight wire

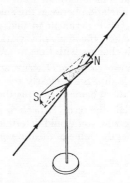

Wire below: needle deflection reverses

Fig. 32.2 Current below needle

pivoted horizontal light magnetic needle NS (i.e. in a north–south direction) the needle is deflected. As the current is increased by means of R the deflection increases. For large currents the deflection is nearly 90°, that is, nearly perpendicular to the current direction. If a current-carrying wire is placed under a magnetic needle instead of above it and in line with its axis, the needle deflects the opposite way (Fig. 32.2).

From the way the needle turns when the current-carrying wire is held parallel to it we know that the forces on the poles are *perpendicular* to the current direction. This is illustrated by Figs. 32.1 and 2. We therefore conclude that:

(1) a current has a magnetic field all round it;
(2) the magnetic field is in a direction perpendicular to the current.

## Forces Due to Magnetic Field

The reader should keep in mind that the direction of a magnetic field is taken, by convention, to be the direction a *north* pole would tend to move if placed in the field (p. 560). Fig. 32.3(i) shows the direction of the magnetic field when the current is parallel to the magnetic needle. The N-pole is then urged in the same direction by the force $F_1$ and the S-pole in the opposite direction by the force $F_2$.

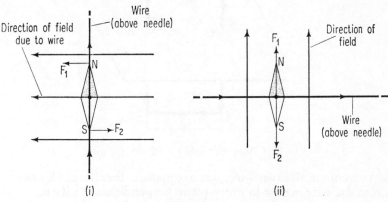

Fig. 32.3 Forces due to current

If the wire is placed perpendicular to the compass needle NS (i.e. in an east–west direction) and the current is switched on, the needle does not move this time. Two forces $F_1$ and $F_2$ still act on the poles of the needle (Fig. 32.3(ii)). But since the magnetic field is perpendicular to the current, $F_1$ and $F_2$ act in the same straight line in opposite directions. $F_1$ and $F_2$ are practically equal in magnitude, and hence their resultant is zero; the needle does not therefore move. In Fig. 32.3(i) the forces $F_1$ and $F_2$ are also perpendicular to the wire, but they do not act in the same straight line. Here their effect is to turn or deflect the needle about its pivot.

## Magnetic Field Round Straight Wire

The magnetic field round a straight current-carrying wire can be investigated by using a large rectangle or square ABCD of about 50 cm side made with 10 turns of insulated copper wire (Fig. 32.4). The plane of the coil is placed vertically and a horizontal board X is positioned through its centre.

A current of a few amperes is then passed into the coil, some iron filings are lightly sprinkled round and between the vertical sides AD and BC, and the board is tapped gently. Since AD and BC are reasonably far apart, each can be considered as an isolated straight wire. AD carries a downward current, while BC carries an upward current.

The field pattern round a straight wire is seen to consist of *circles* with the wire as centre. This means that the field is symmetrical round the wire, as we may expect when there are no other magnetic influences. The same effect is shown by placing compass needles all round and near to the wire. At first, with no current flowing, they all point north–south. When the current is switched on the needles swing round and settle in a circle. The direction of the field round AD is opposite to that round BC,

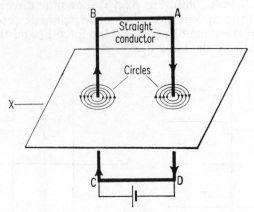

Fig. 32.4 Magnetic field round straight wire

as the currents in the two wires are in opposite directions. Of course, the field round a wire occurs in every plane perpendicular to itself.

## Rules for Magnetic Field Direction

Several useful rules have been given for the direction of the magnetic field (the direction of movement of a N-pole) in the magnetic field of a *straight wire* carrying a current.

### (1) *Maxwell's Corkscrew Rule*

If a right-handed corkscrew is turned so that its point travels along the current direction, the direction of rotation of the corkscrew gives the direction of the magnetic field.

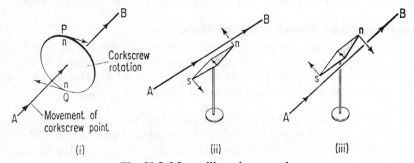

Fig. 32.5 Maxwell's corkscrew rule

Fig. 32.5(i) shows a straight current-carrying wire AB. At a point P above it a N-pole *n* moves from left to right, as shown, according to Maxwell's rule. At a point Q below it a *n*-pole moves from right to left in the opposite direction. If the corkscrew rule is applied to the wire AB (Fig.

32.5(ii)) the *n*-pole of the magnetic needle below it moves to the left as at Q in Fig. 32.5(i). The *s*-pole of the needle moves in the opposite direction. If the magnetic needle is placed above the wire AB (Fig. 32.5(iii)) the *n*-pole moves to the right, as at P in Fig. 32.5(i), and hence the *s*-pole moves to the left.

### (2) *Clenched Fist Rule*

Grasp the wire with the right hand so that the thumb points in the current direction and clench the fist (Fig. 32.6). The direction of the

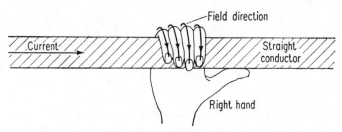

Fig. 32.6 Right-hand rule

curled fingers gives the direction of the field. It should be shown by the reader that this rule applies to the cases already considered.

### Neutral Point

When the field round a downward straight current-carrying wire is carefully plotted a neutral point P is obtained on one side of it owing to the presence of the earth's magnetic field (Fig. 32.7). The field direction here due to the current is due south, since this is the direction of the tangent to the line of force at P. The direction of the earth's magnetic field at P is due north. The two fields neutralize each other completely at P.

On the other side of the wire, due west, the field of the current acts due north. This is in the same direction as the field of the earth. No neutral point is therefore possible here. At points due north or due south of the wire the current

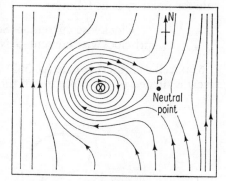

Fig1 32.7 Combined field of straight wire and earth

field is perpendicular to that of the earth, and hence no neutral points are obtained. The only place where a neutral point can be obtained is due east of the wire, with the direction of the current as shown.

### Circular Current

The magnetic effect due to a wire can be increased by winding it into a narrow circular coil of many turns. All parts of the wire are now close to the area surrounded by the coil. The total magnetic effect in this region is thus increased.

A simple narrow circular coil can be made by winding thin insulated wire round the finger. When a current is passed into the wire and one face of the coil AB is brought near to a suspended magnetic needle the

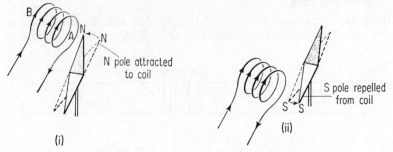

Fig. 32.8 Magnetic effect of circular coil

*n*-pole may be attracted (Fig. 32.8(i)). When the same face is brought near to the *s*-pole this is repelled (Fig. 32.8(ii)). We conclude that this face of the circular coil acts like a south magnetic pole, as indicated in Fig. 32.9(i).

### Rules for Polarity

The other face B of the circular coil can be investigated in the same way. It is found to have north polarity. The circular coil thus acts like a *very short magnet*, with one face north and the other face south. A simple 'clock' rule for polarity is as follows:

Clockwise current → South pole (Fig. 32.9(i)).
Anti-clockwise current → North pole (Fig. 32.9(ii)).

Maxwell's corkscrew rule (p. 582) also provides the polarity. If a corkscrew is turned in the direction of the circular current, then the direction

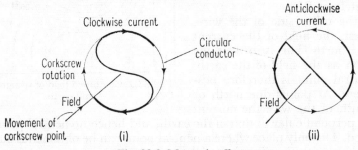

Fig. 32.9 Magnetic effects

of movement of the point is the direction of the field. Fig. 32.9 also shows the field direction, deduced from Maxwell's rule, when one face of the coil in Fig. 32.8 is viewed from A and when the other face is viewed from

B. The current flows clockwise when viewed from A and anti-clockwise when viewed from B. The field direction agrees with the fact that lines of force come out from a N-pole and enter the S-pole of a magnet (see Fig. 31.7(i), p. 561).

## Demonstration of Polarity

The magnetic polarity of the faces of a narrow circular coil can be demonstrated by joining a small coil to the terminals of a piece of copper and zinc contained in a small vessel V of dilute sulphuric acid (Fig. 32.10). The vessel is floated on water in a trough T by means of cork C, and a current flows in the coil as a result of the action of the simple cell (p. 510). When the S-pole of a magnet is brought near to the face A of the coil the whole vessel drifts across the water away from the magnet. If the magnet is turned round and the N-pole is presented to the face A the vessel drifts across the water towards the magnet. This simple experiment shows that the current in the coil has a magnetic effect which causes the face A to behave like a south magnetic pole.

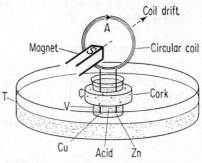

Fig. 32.10 Repulsion of circular coil

## Magnetic Field Round Narrow Circular Coil

The magnetic field all round a narrow circular coil can be investigated by means of a vertical coil C of, say, thirty turns of thin insulated wire and

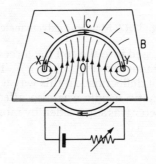

Fig. 32.11 Field of narrow circular coil

a horizontal board B passing through its centre (Fig. 32.11). The board could be placed perpendicular to C above or below the centre, as the magnetic field round C occurs in all perpendicular planes.

The coil is connected to a battery and a rheostat, and a current of a few amperes is passed into it. Iron filings are now lightly sprinkled on B and then lightly tapped. The field pattern is then seen to consist of:

(1) circles round the edges at X and Y; and
(2) straight lines near the centre at O.

The same field pattern is observed by placing a number of plotting compass needles round X and Y and near O. Circles are obtained round X and Y because the wires here are practically straight (see p. 582). The circles increase in radius from X towards O, or from Y towards O. Round a small region at O the lines of force are straight and parallel, and this is therefore a *uniform field* (see p. 561).

### Early Galvanometer

One of the earliest galvanometers or current-measuring instruments consisted of a narrow circular coil, with a small pivoted magnetic needle at its centre which could move over a horizontal circular graduated scale (Fig. 32.12(i)). The plane of the coil was first turned until it was in the magnetic meridian, that is, parallel to the magnet.

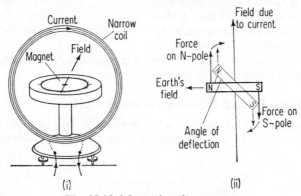

Fig. 32.12  Magnetic galvanometer

When the current is passed into the coil the forces on the poles due to the magnetic field rotate the magnet (Fig. 32.12(ii)). The greater the current, the greater is the angle of deflection. In this way a measure of the current was obtained. The magnetic needle was also affected by stray magnetic fields, and the galvanometer was superseded by more reliable and accurate instruments (p. 603).

### Solenoid

The most useful form of current-carrying wire is a *solenoid*. This is the name given to a long circular coil of wire whose turns are usually close together (Fig. 32.13). Unlike the narrow circular coil, the solenoid has a powerful magnetic field along practically the whole length of its axis, as we shall see shortly. It often has an iron core and is widely used in many practical applications in industry (p. 589).

The magnetic effect of a solenoid can be demonstrated by connecting a battery to a long cylindrical rheostat AB, which is an insulated solenoid, or to a long circular coil of copper wire (Fig. 32.13). When a suspended magnetic needle M is placed near the end B the S-pole of the needle is attracted towards it. When the needle is taken to the other end A the N-pole is attracted towards it.

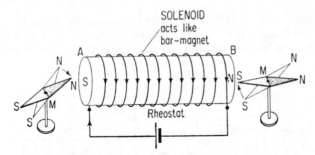

Fig. 32.13  Magnetic effect of solenoid

The solenoid thus acts like a *bar magnet*. The rule for magnetic polarity is the same as for the narrow circular coil, that is, if the current flows clockwise when one end is viewed that end acts like a S-pole, and if it flows anti-clockwise that end acts like a N-pole.

Two simple demonstrations of the magnetic effect of a solenoid are

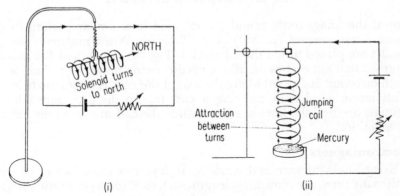

Fig. 32.14  Magnetic effects

illustrated in Fig. 32.14. If it is suspended so that it can turn, a current-carrying solenoid will always settle in a north–south direction, thus showing that it acts like a bar magnet (Fig. 32.14(i)). In Fig. 32.14(ii) a light flexible coil, with one end dipping into a pool of mercury, will jump up and down repeatedly. This is due to the attraction between neighbouring turns of the coil, which are north and south respectively, thus contracting the coil. After the circuit is broken at the mercury surface the coil falls down again, and the action is repeated.

### Magnetic Field Round Solenoid

The magnetic field round a solenoid can be obtained by using a long cylindrical rheostat as a solenoid (Fig. 32.15(i)), or by passing insulated wire through many holes in a perspex base and forming a coil, which has the advantage of allowing the field inside the solenoid to be seen (Fig. 32.15(ii)).

As the turns of the rheostat solenoid are usually vertical, a board B is placed horizontal, that is, in a plane perpendicular to the current-carrying turns. When iron filings are spread on B and the board is lightly

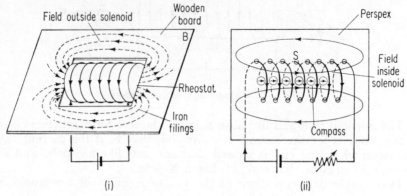

Fig. 32.15  Magnetic field of solenoid

tapped the filings settle round the solenoid in a similar pattern to that round a bar magnet (p. 560) (Fig. 32.15(i)). When magnetic compass needles are placed inside the solenoid S (Fig. 32.15(ii)), and the axis of S is turned to point east–west, all the needles swing round from their north–south direction and point along the axis of the coil. A magnetic field thus exists inside a solenoid carrying a current. Further, as shown by the compass needles, the lines of force outside the coil are a continuation of those inside.

### Electromagnets

Using insulated wire and windings in layers, a solenoid can be made with many turns per centimetre length, such as 50 turns per centimetre, and

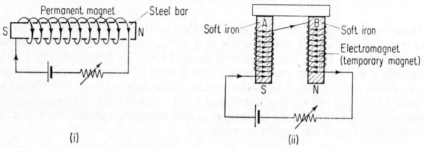

Fig. 32.16  Permanent magnet and electromagnet

if a large current of several amperes is passed into the solenoid a powerful magnetic field is obtained practically everywhere inside. A long solenoid can therefore be used for making *permanent magnets*, such as bar magnets (Fig. 32.16(i)).

An *electromagnet* or temporary magnet can be made by winding solenoids in opposite directions round two soft iron cores A and B (Fig. 32.16(ii)). The lower end of A then acts as a S-pole and the lower end of B as a N-pole.

Electromagnetic devices are used in a wide variety of applications in industry. In communications, for example, numerous magnetic relays and telephone receivers are employed which contain electromagnets.

### Magnetic Relay

The principle of a magnetic relay is illustrated in Fig. 32.17. The relay consists of a solenoid with a projecting soft-iron core A inside it, and a bent bar of soft iron B helps to complete an iron 'circuit'. At C a bent piece of

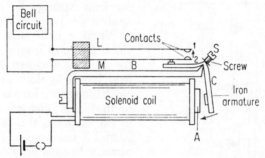

Fig. 32.17 Magnetic relay

soft iron called the armature (or moving part) can rock about an edge of B. A screw S controls the degree of armature movement.

When a telephone subscriber lifts the receiver a current flows along telephone wires to the solenoid of a relay. The magnetized iron core then

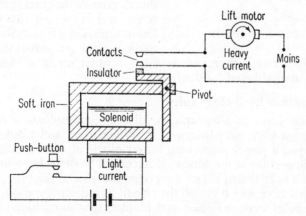

Fig. 32.18 Application of magnetic relay

attracts the armature and the contacts 1 and 2 close. The circuit connected to the light springs L and M is now made, and this in turn may operate a warning bell on a switchboard at a distant exchange. The complex system of telephone circuits throughout the country requires millions of relays for automatic and efficient operation.

A relay is an electromagnetic switch, and may require a current of only a few milliamperes to operate. It can control, however, circuits carrying much higher currents. When the button in a lift, for example, is pressed a relay closes the electric-motor circuit (Fig. 32.18). The motor requires a large current to operate, and as direct contact with the circuit would be extremely dangerous if the insulation broke down, the relay helps to protect the operator.

### Electric-bell Circuit

The electric bell is another example of a useful electromagnetic device. In the circuit shown in Fig. 32.19 a soft-iron spring X presses lightly against the point of a screw N.

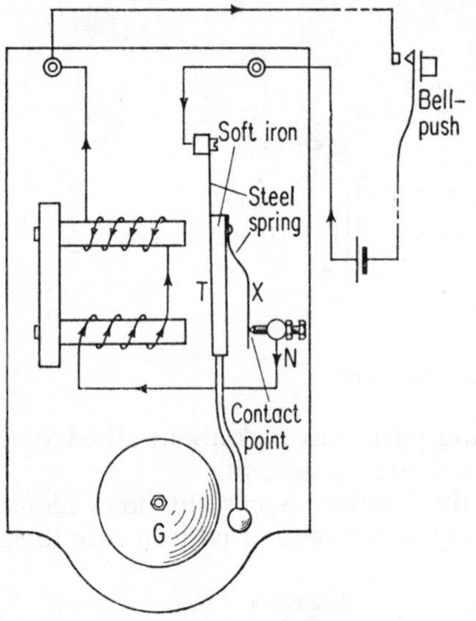

Fig. 32.19 Electric bell

When a bell-push is pressed the electric circuit is completed through the contact between X and N. The iron bar T is then attracted to the electromagnet, breaking the circuit, so that T is released and falls back again to make the contact between X and N once more. Thus the iron is attracted repeatedly by means of the 'make-and-break' contact. A clapper is attached to the iron, and this strikes a gong G repeatedly. The rate at which the gong is struck depends partly on the elasticity of the spring to which the iron is attached and on the degree of contact between the screw and X, which can be varied. Each time the circuit is broken a spark is obtained across the contact.

The tip of the screw is therefore tungsten or other metal of high melting-point which is unaffected by the heat.

### Communication by Telegraphy

The earliest form of telegraph, or direct communication by electric current along a wire, was invented by Wheatstone and Cooke in 1837. They employed a panel containing letters of the alphabet, together with five magnetic needles at the centre which could be deflected when current flowed in coils near them. When a message was passed the operator at one station pressed keys which closed the circuit at a distant station. Here two magnetic needles were deflected and pointed to a particular letter.

The telegraph was the sensation of the day, and many people flocked to send messages *via* the new system to friends some miles away. It was soon superseded by less tedious and more efficient methods. Today, *teleprinters* can be used to pass messages directly from one place to another. The teleprinter is like a typewriter, with letters on its keyboard which are operated electromagnetically when a message is sent.

## Communication by Telephone

Communication by telephone requires a *microphone*, an apparatus which converts sound energy to electrical energy. There are several different types of microphone, but the earliest was the *carbon microphone*, which was invented about 1878 and is still used today.

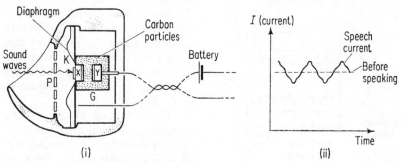

Fig. 32.20 Carbon microphone

This microphone contains carbon particles G between two carbon blocks X and Y (Fig. 32.20(i)). When a caller lifts the telephone to make a call, a battery from a central exchange is automatically connected to X and Y so that a steady current $I$ flows in the circuit, (Fig. 32.20(ii)). The magnitude of $I$, from Ohm's law, is given by $I = E/R$, where $E$ is the battery e.m.f. and $R$ is the total resistance of the carbon between X and Y, the connecting wire in the circuit, and the internal resistance of the battery.

K is a thin conical diaphragm; P is a perforated plate protecting it from damage. When a person speaks into the microphone, sound waves impinge on K and make it vibrate. K thus moves very slightly to and fro in accordance with the varying pressure of the sound wave. When it moves to the right the carbon particles are compressed; when K moves to the left the particles are loosened. Since the contact between the particles is worse in the latter case, the resistance between X and Y is increased; and, conversely, the resistance between X and Y is decreased when the particles are compressed. From Ohm's law it follows that a *varying current flows in the circuit whose magnitude alters at exactly the same frequency as the pressure variation of the sound waves*. This is illustrated in Fig. 32.20(ii). Sound energy is thus converted to electrical energy in the microphone circuit. The varying electric current travels along to a telephone earpiece at the receiver's end of the telephone line. Here, as we now explain, the electrical energy is changed back to sound energy.

### Telephone Earpiece

After years of continued efforts, Alexander Graham Bell designed the first telephone in 1876. Nowadays we are so used to the existence of the telephone that we are apt to overlook Bell's service to society in enabling people to communicate easily with each other; but if we think of the ease with which we can get in touch with the doctor today we obtain some idea of the practical value of Bell's researches.

The telephone earpiece has the opposite function to that of the microphone. It converts electrical energy back into sound energy. In one form a permanent magnet X is bolted to pieces of soft iron, and coils of insulated wire, represented by B and D, are wound round the other part of the soft iron. A flexible soft-iron circular plate A, made of suitable magnetic material such as Stalloy, is near to the latter, and when the ear-piece is used, A is close to the ear, (Fig. 32.21).

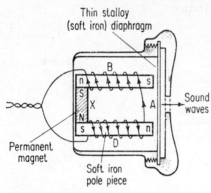

Fig. 32.21 Telephone earpiece

The presence of the magnet X results in the ends of the soft-iron pieces close to A becoming a north $n$ and a south pole $s$ respectively. When a person speaks into the microphone at the other end of the telephone line the sound energy is converted into electrical energy, as already explained, and electric currents, varying in magnitude at the frequency of the sound, travel along and pass into the coils B and D, which are connected together. Since the current varies, the strength of the magnetic effect due to the current also varies. Thus the total attractive force on the soft-iron plate A now varies, and A begins to vibrate to and fro. This movement gives rise to sound waves in the air of the same frequency, and the notes produced are a close replica of those spoken into the microphone. In modern telephones the earpiece and carbon microphone are at either end of the same unit (see Plate 12).

### Moving-iron Ammeter

The magnetic effect of a current is used in one form of electrical ammeter, an instrument used to measure current. Basically, the ammeter consists of a solenoid D, with a fixed piece of soft iron F and a moving piece of soft iron M inside the solenoid (Fig. 32.22(i)). A simplified section of the instrument is shown in Fig. 32.22(ii).

When a current is passed into D, F and M both become magnetized with similar or like poles facing each other at their respective ends. The movable iron M is then repelled. The larger the current, the more strongly are F and M magnetized, and hence the greater is the repulsion of M. The moving iron comes to rest when the turning effect produced by repulsion is equal to the opposing turning effect of the phosphor-bronze

**PLATE 15**
**Electrons and Photo-sensitive Devices**

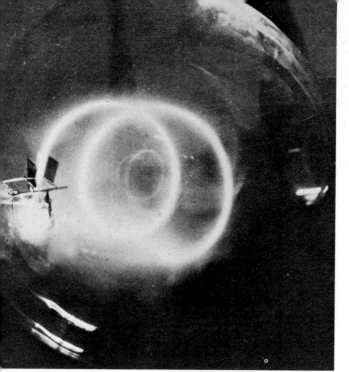

(a) Spiral electron path in a LEYBOLD fine-beam tube, due to a magnetic field at a small angle to the electron beam. The beam passes initially through a conical anode and then through two plates which may be used to apply an electric field (p. 686).

**15(b)** ▶

(b) Photoelectric cell. The large metal surface is a caesium-antimony cathode, which is particularly sensitive to daylight and to light predominantly blue. The small disc shown is the anode, and the two metals are contained in a vacuum (p. 697).

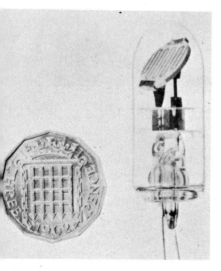

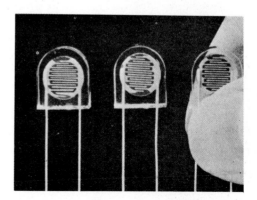

(c) Light-sensitive resistor, semiconductor, designed for detecting flame failure.

**15(d)** Light-sensitive resistors, semiconductors, falling from 10 megohms (dark) to 100 ohms (bright) – used to trigger electric circuits in computers.

**PLATE 16**
**Photo-electric Applications**

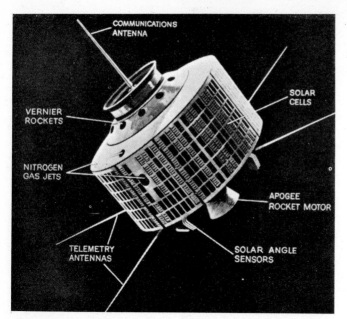

**16(a)** Communications satellite SYNCOM, for beaming television from one part of the world to another.

**16(c)** Sound track film, extreme left, showing sound variations of several frames (p. 697).

◀ **16(b)** Television camera tube – image orthicon type (p. 698).

spring which controls the deflection. The movement of the attached pointer thus gives a measure of the current. When the current is switched off the iron is returned to zero by the spring.

If an alternating current, one which reverses in direction continuously, is passed into the coil the magnetism in the iron pieces F and M reverses continuously. Whether the ends of F and M are both north or both south, however, M is always repelled by F. Consequently, the moving-iron

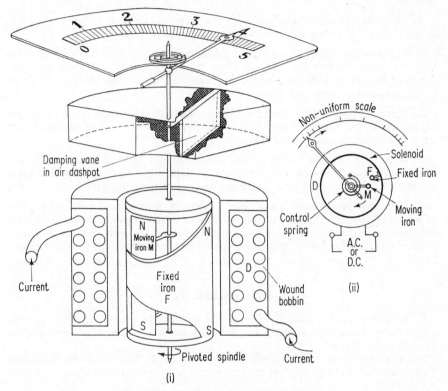

Fig. 32.22 Moving-iron repulsion ammeter

repulsion-type instrument, as it is known, can be used to measure alternating current as well as direct (steady) current. This is an advantage of the instrument. The disadvantage is that the scale is not a regular or uniform one, that is, equal intervals along the scale do not represent equal changes of current. By special shaping of the iron pieces, part of the scale can be made uniform.

## SUMMARY

1. When a conductor carries an electric current a magnetic field exists at right angles to the plane of the conductor.

2. Maxwell's corkscrew rule states: If a right-handed corkscrew is turned so as to move in the direction of the current in a conductor the direction of

rotation of the corkscrew is the direction of the lines of force. In a coil a clockwise current makes that face act like a S-pole; anti-clockwise current makes the face act like a N-pole.

3. The lines of force round a straight current-carrying wire are circular. For a flat circular coil they are straight in the middle of the coil in a plane through its centre perpendicular to the coil, but circular near the edges. The lines of force pass through and round a solenoid, similarly to the lines round a bar magnet.

4. *Electromagnets* are usually solenoids wound round soft iron; when the current is switched off the magnetism in the iron practically disappears.

5. The *magnetic relay* has a solenoid and soft-iron core. When a current flows a pivoted piece of soft iron (armature) is attracted and two contacts are then made or broken in another circuit. An *electric-bell circuit* has an electromagnet and a make-and-break device.

6. The *carbon microphone* contains carbon granules whose electrical resistance varies when the pressure varies. Sound-wave pressure is thus changed into current variations of the same frequency.

7. When the varying current from a microphone flows through the speech coils of a *telephone receiver*, a thin iron plate or diaphragm is attracted with a varying force of the same frequency. Sound waves are then produced.

8. The *moving-iron repulsion meter* has a solenoid, a fixed and a movable piece of iron, and a spring to control the movement of the pointer. When a current, alternating or direct (steady), flows in the solenoid the irons are always magnetized with similar poles facing, and therefore repel. The scale is non-uniform.

## EXERCISE 32

**1.** (*a*) With the aid of a labelled circuit diagram describe how you would magnetize a steel rod so that a marked end would be a N-pole. State two ways of increasing the strength of the magnet.

(*b*) A wire is supported horizontally in a north–south direction above a compass needle. Describe and explain what happens when a current flows along the wire towards the north. (*N*.)

**2.** Give a diagram of an electric bell and explain its action. State, with reasons for their use, the materials of which the various parts are made. (*L*.)

**3.** Describe how an electric current may be used: (*a*) to magnetize a steel bar; (*b*) to demagnetize it.

Draw a diagram of the magnetic field produced by a current flowing in a long straight wire in a plane at right angles to the wire. State a rule which gives the relation between the direction of the current and that of the field.

A long vertical wire carrying a current passes through a horizontal bench. Give a diagram of the resultant magnetic field on the surface of the bench around the wire due to the current and the earth. Mark the positions of any neutral points formed in this field. (*L*.)

**4.** (*a*) Describe the structure of a carbon microphone and explain its action.

(*b*) Give an account of the structure and mode of action of a telephone earpiece. (*L*.)

**5.** Explain what is meant by *magnetic field* and by *magnetic induction*.

Describe an experiment to plot the magnetic field inside and outside of a solenoid carrying a direct current, in a plane containing its axis. Draw a diagram of the field marking the directions of the lines of force and the direction of the current.

If you were provided with bars of iron and steel of the same dimensions, and any other necessary apparatus, how would you demonstrate the main differences in the magnetic properties of iron and steel? State these differences and indicate, giving reasons, which of these metals you would use for: (a) a compass needle; (b) the core of an electro-magnet. (L.)

6. Explain carefully TWO of the following phenomena: (a) a suitably-placed bar of soft iron is drawn into a solenoid that is carrying an electric current; (b) each of two parallel wires carrying currents in the same direction experiences a force tending to draw it towards the other; (c) when a current is passed through a vertical plane circular coil set in the magnetic meridian, a small magnet placed at the centre of the coil is deflected from the meridian. (O.)

7. Draw a labelled diagram of a telephone earpiece. Describe and explain the sounds heard in the earpiece: (i) when it is connected to a dry cell for a few seconds and then the contact is broken; (ii) when it is connected to a suitable low-voltage alternating-current supply of frequency 50 Hz. (C.)

8. Draw diagrams to show the magnetic field due to: (a) a current in a long, straight wire; (b) a current in a single circular turn. Indicate clearly the direction of current and magnetic field.

An insulated vertical wire passes through a tank of water. A single turn of wire with its plane vertical and carrying a steady current floats on a light raft on the water. Describe and account for the behaviour of the raft when a steady current is passed upwards through the straight wire. Neglect the earth's magnetic field. (O. and C.)

9. Draw diagrams, with the direction of current and field clearly indicated, to show the pattern of the magnetic lines of force due to an electric current flowing in: (a) a long straight wire; (b) a plane circular coil; (c) a solenoid.

Describe and explain what happens to a piece of soft iron when it is placed inside a solenoid in which a current is flowing.

Draw a labelled diagram of an electric bell, and explain its action briefly. (O.)

# 33

# FORCE ON CONDUCTOR · THE MOTOR PRINCIPLE

Soon after Oersted's discovery of the magnetic effect of a current, Ampère showed in 1821 that, under certain conditions, a force was exerted on a current-carrying conductor situated in a magnetic field. From Ampère's fundamental work in the laboratory useful instruments such as the electric motor, the moving-coil ammeter and voltmeter, and the moving-coil loudspeaker were developed.

## Demonstration of Force

### Experiment

The force on a current-carrying conductor in a magnetic field can be demonstrated with the apparatus shown in Fig. 33.1(i). A smooth brass cylindrical rod AL is placed across two horizontal brass rails PQ and RT, and a battery C and small rheostat D are connected in series to the ends of the rail. A powerful horse-shoe magnet is then positioned so that the rod

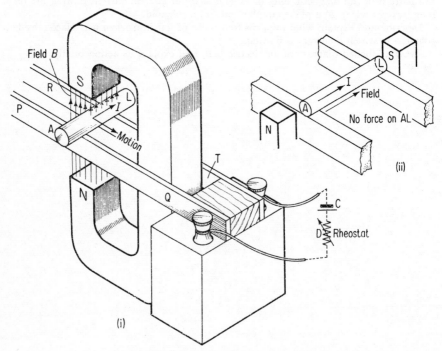

Fig. 33.1 Conductor in magnetic field
596

AL is between its poles and the magnetic field is vertical. The magnetic field, denoted by $B$, is then *perpendicular* to the length of the rod.

1. When the circuit is made and a current $I$ flows along AL the rod is observed to roll along the rails. A force thus acts *at right angles* to the field, and to the length AL of the conductor.
2. When the current is increased by altering the rheostat, the rod moves faster. The force on the rod thus increases when the current $I$ increases.
3. When the magnet is moved slightly sideways so that the strength of the magnetic field at the rod AL is decreased, the rod moves more slowly. The force thus decreases when the magnetic field strength decreases.
4. When the magnet is turned round so that the field is *parallel* to the length of the rod, and the circuit is switched on, the current-carrying rod remains practically still (Fig. 33.1(ii)). There is no force now on the rod.

*Conclusions.* A force acts on a current-carrying conductor when it is at right angles to a magnetic field. The direction of the field is perpendicular both to the current and to the field direction. The magnitude of the force increases when the current increases and when the field strength increases.

When the angle between the conductor and field decreases from 90° the force decreases, and becomes zero when the conductor is parallel to the field.

## Reason for Force

The force on a current-carrying conductor in a magnetic field was considered by Ampère to be due to the general law in mechanics that *action and reaction are equal and opposite* (see p. 113). A current in a conductor such as the rod AL in Fig. 33.1 sets up a magnetic field all round it. The magnet NS, being in this field, is therefore acted upon by a force. By the law of action and reaction, an equal and opposite force is exerted *on the conductor*. The magnet does not show any movement, because it is heavy and not in a position to move readily, but the rod AL is relatively light and can easily roll. The force on it is opposite to that on the magnet, and can be considered as the 'reaction'.

## Interaction Between Fields

Another useful way of explaining the force is to consider the interaction between two magnetic fields. Fig. 33.2(i) shows the circular lines of force round the end A of the current-carrying rod AL in Fig. 33.1. The current flows downwards into the paper. Fig. 33.2(ii) shows the straight lines near A due to the magnetic field between the poles N and S. Fig. 33.2(iii) shows the resultant field. On the left of A the field is strengthened and the lines are denser. On the right of A the field is weakened and the lines are relatively less dense.

The conductor moves from the strong to the weaker part of the field, just as if there was a greater stress in the field due to dense lines of force which makes it move. When the current direction is reversed, or the field direction is reversed, the direction of the force reverses for the same reason. If the field is *parallel* to the current, as shown in Fig. 33.1(ii), the density of lines of force on both sides of AL in a direction perpendicular to the length of the rod is now practically unaffected by the lines due to the magnetic

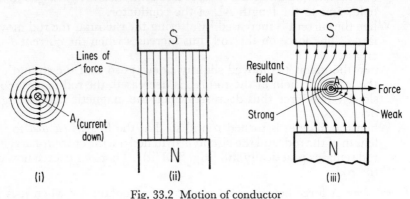

Fig. 33.2 Motion of conductor

field. Fig. 33.2(ii), in fact, is roughly the field appearance. Thus no force acts on a current-carrying conductor when it is parallel to a magnetic field.

## Direction of Force

The direction of the force when a current-carrying wire is situated in a perpendicular magnetic field is given by *Fleming's left-hand rule*. It states:

*If the thumb and first two fingers of the left hand are held at right angles to each other, with the forefinger pointing in the direction of the magnetic field B and the middle finger pointing in the direction of the current I, then the thumb points in the direction of Motion of the conductor, i.e. in the direction of the force acting on it.*

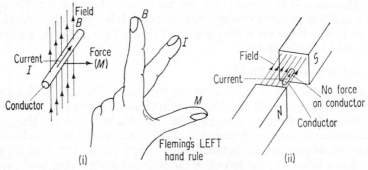

Fig. 33.3 Force on conductor

The rule is illustrated in Fig. 33.3. Note carefully that the rule applies only if the magnetic field and current are perpendicular, or inclined, to each other.

## Faraday's Motor Experiment

With the apparatus shown in Fig. 33.4, due originally to Faraday, the force on a conductor can be made to rotate it continuously.

A wire AB, pivoted at A, is suspended so that its lower end B just touches the surface of mercury in a dish. The north pole N of a strong magnet projects through the middle of the dish, and a battery, rheostat and a key are in series with AB.

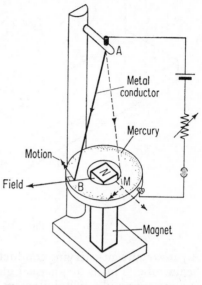

Fig. 33.4 Faraday's motor experiment

When the circuit is made the current-carrying wire AB is seen to rotate continuously round the N-pole. This was the first electric 'motor' invented, which is a device for changing electrical energy continuously to mechanical energy of rotation. As the current is increased by the rheostat the wire is observed to rotate faster, showing that the force on it has increased.

*Explanation.* The lower part of the wire such as B is situated in the strong magnetic field due to the N-pole. This field radiates outwards from N as shown. From Fleming's left-hand rule, a force now moves the wire as shown. When the wire moves to another position such as AM the field is still perpendicular to the wire, and the force on it again acts in the same direction. The wire thus keeps on rotating about N because of the radial magnetic field in which it is situated.

## Barlow's Wheel

Another simple motor demonstration, known as Barlow's wheel, consists of a metal disc with spokes which just touch the surface of mercury in a small dish (Fig. 33.5).

A battery and rheostat are connected so that a current flows down through a radius OA. When a horse-shoe magnet is placed round the wheel so that the magnetic field *B* is perpendicular to OA the wheel is observed to rotate. When the current is increased, or the field is increased, the wheel rotates faster.

Here again Fleming's left-hand rule shows that the force rotates the wheel about its axis at O. Other radii, such as OD, which reach the mercury surface are immediately acted upon by a force, since they then carry a current.

## Electrons in Magnetic Fields

So far we have seen that a current-carrying copper wire is deflected when it is perpendicular to a magnetic field. The metal is a solid conductor.

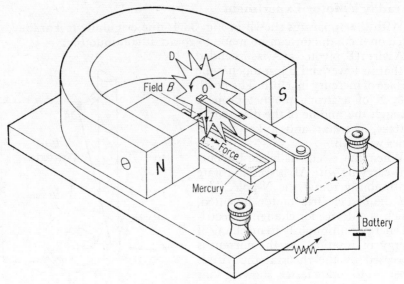

Fig. 33.5 Barlow's wheel

A *gaseous* current-carrying conductor can be obtained with a LEYBOLD fine beam tube. This is a spherical glass bulb containing hydrogen at a very low pressure and a metal filament F which emits electrons when heated to a high temperature (Fig. 33.6(i)). The electrons move from F towards a gap in a conical electrode A which is kept at a high positive potential

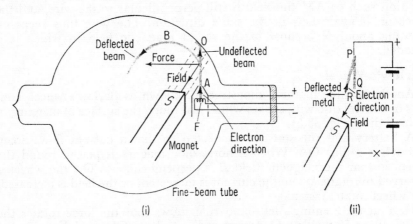

Fig. 33.6 Deflection of electrons in gas and metal

relative to F, and they move through the gap at a high speed. On account of the gas pressure and other factors the fast-moving electrons are focused in a fine straight beam AO which is luminous.

When the S-pole of a powerful Ticonal magnet is brought up horizontally close to the bulb the electron beam AO is deflected sideways to B

(Fig. 33.6(i)). A force thus acts on the beam when it is perpendicular to a magnetic field. A similar experiment can be carried out with a light metal conductor PQ of copper, loosely suspended at its ends P and Q (Fig. 33.6(ii)). When the S-pole S of a Ticonal magnet is in position and the circuit is completed so that a current flows along the wire, PQ is seen to be deflected sideways to PR. This is exactly the same effect produced by the magnet on the electron beam AO. The identical behaviour of the electron beam and the metal is due to the fact that in a copper atom the outermost electron is shielded from the attraction of the nucleus and wanders about relatively freely through the metal (p. 486). The electrons are deflected in the magnetic field, carrying the metal with it, whereas in the case of the fine beam the electrons are much more free to move.

## Forces on Rectangular Coil

So far we have investigated the force on a straight wire carrying a current. If the wire is bent so that it forms a *rectangular coil* it can be suspended between two points by flexible strips of foil in the magnetic field between two poles N and S (Fig. 33.7).

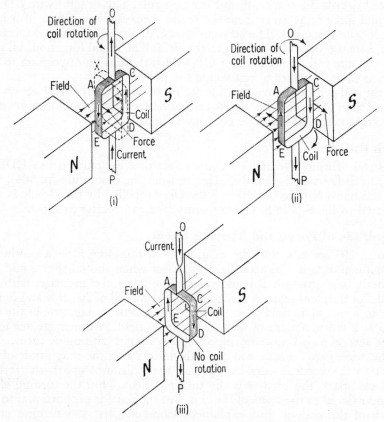

Fig. 33.7 Rectangular coil in magnetic field

EXPERIMENT

1. First, the plane of the coil can be arranged to lie in the plane of the magnetic field, with the long sides AE and CD vertical and the short sides AC and ED horizontal. When a current is now passed into the coil, ACDE *twists* about OP and settles in a position X practically facing the poles (Fig. 33.7(i)).

2. If the circuit is broken and the battery terminals reversed the current flows in the opposite direction in the coil. The rotation of the coil now reverses (Fig. 33.7(ii)).

3. If the coil is turned so that it faces the N- and S-poles of the magnet, that is, the plane of the coil is perpendicular to the field, and a current is now passed into the coil, there is no movement (Fig. 33.7 (iii)).

The movement of the coil can be explained *either* by Fleming's rule *or* by a clock rule.

### Fleming's Rule

Suppose the current flows in the coil when its plane is initially parallel to the magnetic field (Fig. 33.7(i)). The straight wire AE is then acted on by a force towards the reader, from Fleming's rule. The straight wire CD has an equal force on it away from the reader because the current in it is in the opposite direction. The straight wires AC and ED have no force on them, because they are parallel to the field. The equal forces on AE and CD *rotate* the coil about the axis OP, and if there is no opposition to the rotation the coil comes to rest facing the poles.

If the coil is initially in a position facing the poles, forces still act on the vertical sides of the coil. But unlike the previous case, the two forces are now *directly* opposite, and hence do not rotate the coil.

### Clock Rule

Instead of using Fleming's rule, we note that the current in ACDE in Fig. 33.7(i) flows anti-clockwise. Hence, from the clock rule (p. 584), this face acts like a N-pole. This face is therefore repelled by the N-pole N and attracted to the S-pole S. Consequently, the coil rotates as shown.

### Magnitude of Force and Turning effect

On p. 597 we saw that the magnitude of the force $F$ on a conductor perpendicular to a magnetic field increases when the current $I$ and the magnetic field strength $B$ is increased. The force also increases with the length $l$ of the conductor. A narrow rectangular coil of 30 turns and height 2 cm in a magnetic field is equivalent to a length on one side of the coil of $30 \times 2$ cm or 60 cm of wire in the same field. A much greater force hence acts on a current-carrying coil if it is wound with many turns.

The *turning effect* on the coil depends not only on the magnitude of the forces on the vertical sides but also on their distance apart—the farther they are apart, the greater is the turning effect. Thus the turning effect is proportional to the *width* of the coil. But the force is proportional to the length of the coil, as just explained. Consequently, the turning effect depends on the magnitude of *length* × *breadth* or *area* of the rectangular coil.

Summarizing, the turning effect due to current in the coil is proportional to the magnitude of the current, the magnetic field strength, to the number of turns, and to the area of the coil.

## The Moving-coil Instrument

The most accurate commercial instrument for measuring current and potential difference uses the action of the forces on a rectangular-shaped conductor in a magnetic field. The essential features are: (i) an insulated

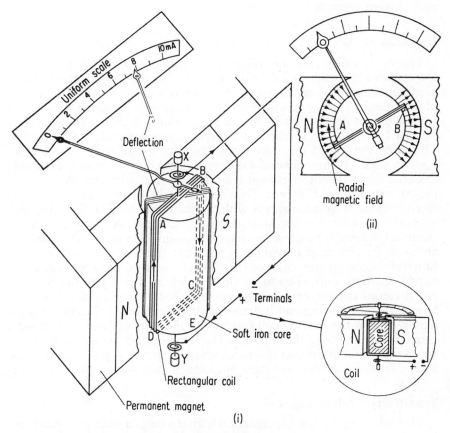

Fig. 33.8 Moving-coil instrument

rectangular coil ABCD; (ii) a powerful magnetic field produced by the curved poles N and S of a horse-shoe magnet; (iii) a soft-iron cylinder E to concentrate the lines of force towards its centre; (iv) two springs, oppositely wound, to control the rotation of the coil about a spindle moving in jewelled bearings at X and Y. Fig. 33.8(i) illustrates a side view of the apparatus, while Fig. 33.8(ii) is a view from above. When the terminals of the instrument are connected in an electrical circuit a current flows through the coil via the springs, each spring being connected to one terminal.

*Action of Instrument*

Using Fleming's rule or the clock rule, the forces on the sides AD and BC rotate the coil about XY in a clockwise direction. The springs oppose the rotation, and when the coil is in equilibrium the turning effect due to the forces on the coil is equal to the opposing turning effect due to the elastic forces in the spring. The spring thus controls the angle of deflection of the coil. If it is a strong spring the coil will rotate through a small angle. If the spring is weak the angle of rotation will be bigger. When the current is switched off the spring untwists and restores the coil to the initial or zero current position.

## Uniform Scale

The forces which turn the coil always act perpendicular to the plane of the coil owing to the radial field (compare *Faraday's experiment*, p. 599). Their turning effect or moment about the axis of the coil is thus always proportional to the magnitude of the force. Since the force on a given conductor in a fixed magnetic field is proportional to the magnitude of the current (p. 597), the rotation of the coil is proportional to the current. Thus a current of 1·5 mA deflects the coil through an angle from its zero position which is three times the deflection caused by a current of 0·5 mA. On account of this proportional relation, equal divisions on the scale represent equal steps of current. This is a great advantage, because the scale can be subdivided geometrically with precision into smaller divisions, such as tenths of current units, and read with accuracy of this order. Some electrical instruments, however, have a non-uniform scale. The hot-wire ammeter (p. 534) and moving-iron ammeter (p. 593) are cases in point. Equal divisions on the scales of these instruments do not represent equal steps in current, and when the pointer is deflected to a position between two divisions the current cannot be read to such accuracy as on the uniform scale of a moving-coil instrument. If the soft-iron cylindrical core is removed from the moving-coil instrument the lines of force pass straight across from the N- to the S-pole of the magnet and the radial field disappears. Although the coil still turns when a current is passed into it, the angle of rotation is now no longer proportional to the current.

## Sensitivity · Advantages

The 'sensitivity' of the instrument is high if a large deflection is produced by a given current such as 1 mA. This depends to a large extent on the type of spring used. If it is a weak spring the sensitivity is high; if it is a strong spring the sensitivity is less. The sensitivity also increases if the strength of the magnetic field increases and the number of turns and area of the coil increase.

The moving-coil instrument has such a powerful magnetic field inside it that stray magnetic fields outside have no effect on the deflection of the coil. The instrument can also be easily adapted to measure a wide range of current and of potential difference, as we shall now see. On p. 495 we showed how it could be used also to measure resistance.

## Extension of Current Range

Many moving-coil instruments with a spring are made as *milliammeters*, which read milliamperes on their scales. One such instrument, for example, reads 0–10 mA and its coil has 5 Ω resistance exactly.

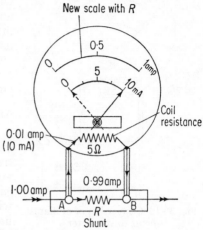

To make the scale divisions read a higher current range such as 0–1 A, a suitable resistance $R$ is placed *in parallel* with the coil across the terminals (Fig. 33.9). $R$ is often called a 'shunt' as it diverts or shunts current through itself, as we see shortly.

Suppose the maximum current of 1·00 A flows towards A with the shunt fitted. The pointer must then be deflected full scale to the 10-mA mark. This means that a current of 10 mA flows through the coil of resistance 5 Ω. Consequently, the rest of the current, which is $(1 - 0·01)$ or 0·99 A, flows through $R$, since 10 mA = 0·01 A.

Fig. 33.9 Conversion of milliammeter to ammeter

Using the coil resistance $R_1$ of 5 Ω, p.d. $V$ across AB $= IR_1 = 0·01 \times 5 = 0·05$ V.

This is also the p.d. across $R$, the shunt resistance.

$$\therefore R = \frac{\text{P.d.}}{\text{Current}} = \frac{0·05}{0·99} = 0·05050 \ \Omega.$$

A shunt of this resistance thus converts the range to 0–1 A.

To convert the current range to 0–5 A, a smaller shunt resistance is needed because it must now divert a larger current equal to $(5 - 0·01)$ or 4·99 A. The current through the coil must again be 10 mA, so the p.d. across AB is still 0·05 V. The new resistance is hence given by 0·05 V/4·99 A which is 0·010 02 Ω.

## Conversion to Voltmeter

To convert the milliammeter of 0–10 mA, 5Ω resistance, to a voltmeter, a high resistance $S$ is needed *in series* with the coil (Fig. 33.10). The new terminals are then A and C instead of A and B.

To illustrate how $S$ is calculated, suppose the instrument is converted to act as a voltmeter reading 0–3 V. Then, when a p.d. of 3 V is joined to A and C the reading is full scale as shown. This means that a current of 10 mA flows in the coil AB, and hence along ABC.

$$\therefore \text{Total resistance between A and C} = \frac{\text{P.d.}}{\text{Current}}$$

$$= \frac{3 \text{ V}}{0·01 \text{ A}} = 300 \ \Omega.$$

$$\therefore S = 300 \ \Omega - \text{coil resistance} = 300 - 5 = 295 \ \Omega.$$

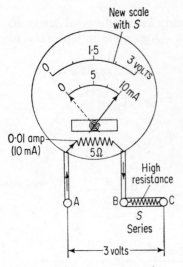

Fig. 33.10 Conversion of milliammeter to voltmeter

A series resistance of 295 Ω is thus needed to obtain a voltmeter reading 0–3 V. Higher ranges of p.d. are obtained by adding suitable larger resistances in series, whose values are calculated in the same way as that just described. The AVO-meter, a commercial instrument widely used, can provide many different ranges of current and p.d. simply by turning a switch, which connects suitable resistors either in parallel or in series.

## Moving-coil Galvanometer

A very sensitive current-measuring or detecting instrument, measuring extremely small fractions of millionths of amperes (microamperes), is the *mirror galvanometer*. It is made: (1) by dispensing with the coiled springs used in the milliammeter and using fine phosphor-bronze wire to control the rotation of the coil; (2) by dispensing with the pointer and using a beam of light to measure the deflection. When a very small current flows in the coil, the coiled spring is too powerful to allow much deflection of the coil. Phosphor-bronze wire, however, has a very weak control, and

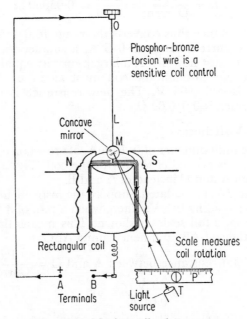

Fig. 33.11 Moving-coil galvanometer

the deflection is then considerable. A beam of light is much superior for measuring small deflections than a pointer.

A mirror galvanometer is shown in Fig. 33.11. The phosphor-bronze wire OL is coated with a conducting powder and the current passes into the coil through terminals A and B. The deflection of the coil is measured by means of the mirror M rigidly attached to it. M is a concave mirror of large radius of curvature, usually 1 m, and focuses an illuminated vertical crosswire in a tube T on to a translucent scale P graduated in millimetres. The circular spot of light moves across P when a current flows in the galvanometer, and its deflection is a measure of the magnitude of the current. In one type of galvanometer 1 $\mu$A (one-millionth of an ampere) may deflect the spot 20 cm along the scale.

### Example

Describe the structure and explain the action of an instrument used for measuring or indicating an electric current. Illustrate with a clearly labelled diagram.

An accumulator of e.m.f. 2·0 V and negligible internal resistance is connected in series with a resistor of resistance 50 $\Omega$ and an ammeter of resistance 5 $\Omega$. A voltmeter of resistance 450 $\Omega$ is connected in parallel with the 50-$\Omega$ resistor. Calculate the readings of: (a) the ammeter; (b) the voltmeter. (L.)

The combined resistance R of voltmeter and resistor of 50 $\Omega$ is given by

$$\frac{1}{R} = \frac{1}{450} + \frac{1}{50} = \frac{10}{450} = \frac{1}{45}$$

$$\therefore R = 45 \ \Omega.$$

$$\therefore \text{Total circuit resistance} = 45 + 5 = 50 \ \Omega.$$

$$\therefore \text{Current} = \frac{E}{50} = \frac{2}{50} = 0\cdot04 \text{ A} = \text{ammeter reading.}$$

Also, Voltmeter reading = P.d. across combined resistance R
$$= IR = 0\cdot04 \times 45 = 1\cdot8 \text{ V.}$$

### Electric Motor

The electric motor is widely used for driving electric trains and many machines in industry. In its simplest or basic form it consists of:

(1) a *coil of wire*, *abcd*, which can turn about a fixed axis O;
(2) a powerful *magnetic field* in which the coil turns;
(3) a *commutator*, which, in its simplest form, is a split copper ring whose two halves A and B are insulated from each other (Fig. 33.12).

The ends of the coil are each soldered to one half the commutator, which rotates with the coil. As it turns, fixed brushes L and M press against the commutator, so that a current from a connected battery PQ always flows through the commutator metal into the coil.

### Principle of Action

Suppose the coil *abcd* is horizontal and is situated between the poles N and S of a permanent magnet (Fig. 33.12(i)). When a current flows in the direction *dcba* this face of the coil appears to flow clockwise to an observer looking at it from above. Hence, from the clock rule on p. 584, this face of the coil acts like a south magnetic pole (see Fig. 32.9). Now like poles repel and unlike attract. The coil thus begins to rotate about the horizontal axis O in an anti-clockwise direction, as shown by the arrow X.

While the coil turns, the two halves of the commutator A and B rotate with it. A keeps in contact with the brush L and B with the brush M. When the coil just passes the vertical, however, A now makes contact with the brush M and B with the brush L (Fig. 33.12(ii)). The battery connections *to the coil* are thus reversed. The current therefore reverses in the coil. The side of the coil facing the N-pole of the magnet thus acts as a north-magnetic pole, and hence, by the action between the poles, the coil continues to spin round in the same direction about the axis O.

The direction of rotation of the coil in Fig. 33.12 can also be found by applying Fleming's left-hand rule (p. 598) to the current-carrying conductors, *ab* and *cd*. It will then be found that the force on *ab* is upwards in Fig. 33.12(i) and the force on *cd* is downwards, so that the coil spins round

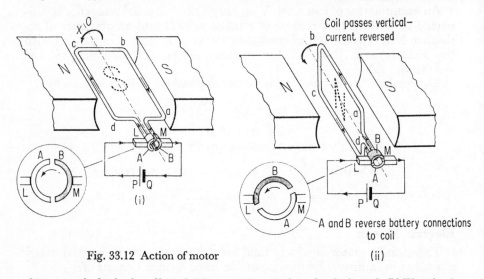

Fig. 33.12  Action of motor                    (ii)

in an anti-clockwise direction, as we have already deduced. If Fleming's rule is applied to Fig. 33.12(ii) when the coil has just passed the vertical it will be found that the forces on *ab*, *cd* have reversed, so that the coil continues to move in the same direction. Fig. 33.13 shows a model motor.

### The Commutator

By tracing the direction of the current in the coil while it rotates, it can be seen that the current is reversed by the commutator every time the coil passes the vertical, or twice in one complete revolution. The coil thus keeps on rotating in the same direction. A more powerful motor is obtained if the commutator is arranged into a large number of separated equal sections, and the coil is wound into an equal number of parts in different planes

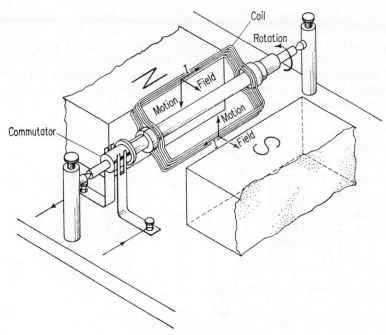

Fig. 33.13 Simple motor

whose ends are connected to the commutator sections. A larger turning effect is also obtained in a simple or commercial type of motor if a greater current or a more powerful magnetic field are employed (p. 602). Further discussion of the motor is outside the scope of this book.

## The Moving-coil Loudspeaker

The moving-coil loudspeaker is used in radio receiver sets and in gramophones and tape-recorders. Like the telephone earpiece, it converts electrical energy to sound energy. It operates on a different principle from the telephone earpiece, because it utilizes the force on a current-carrying conductor in a magnetic field.

The loudspeaker contains a permanent magnet B, with circular concentric north N and south S poles (Fig. 33.14(i)). These poles create a radial magnetic field, so called because the lines of force between the poles spread out like the radii of a circle (see Fig. 33.14(ii)). A coil of wire A, known as the speech-coil, is wound round a small cylindrical 'former', and is placed between the two circular poles of the magnet. A large paper cone C is rigidly attached to the former, and is loosely connected to a circular board F, which surrounds the cone, known as a baffle board.

When a radio receiver set functions, electric currents varying at sound frequencies flow through A, the speech-coil. Suppose the current $I$ flows in the direction shown in Fig. 33.14(ii) at some instant. Then, applying Fleming's left-hand rule (p. 602) to the small portion X of the speech-coil, whose circular section D is shown, the force on X is upwards towards the reader;

this is true for all parts of the speech-coil, as can easily be verified. If the current reverses later, the force on the coil also reverses. Further, the magnitude of the force is proportional to the magnitude of the current (p. 597); so that as the speech current varies, the force on the coil increases and decreases. Since the coil is wound on the former to which the cone C is attached, it follows that the cone vibrates at the same frequency as the varying current in the coil.

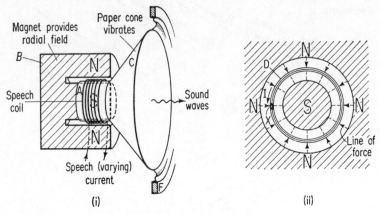

Fig. 33.14 Moving-coil loudspeaker

The vibratory movement of the large mass of air in contact with the cone gives rise to a loud sound, and hence the loudspeaker converts electrical energy to sound energy.

## Forces Between Currents

So far we have considered what happens when a current-carrying conductor is situated in a magnetic field due to a permanent magnet. A demonstration of what happens when a current-carrying conductor A is

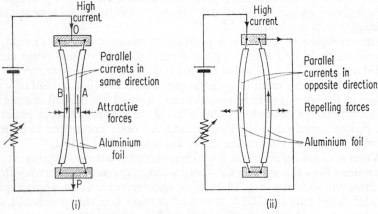

Fig. 33.15 Forces between parallel currents

situated near another current-carrying conductor B can be shown by passing a large current *in the same direction* through two parallel strips of aluminium foil suspended between O and P (Fig. 33.15(i)). The two strips, A and B, are then observed to attract each other. When the currents are arranged to flow in the *opposite direction* the two strips are now observed to repel each other (Fig. 33.15(ii)).

*Explanation.* The magnetic field round a straight conductor consists of circles round it. The magnetic field at A due to the current in B is therefore perpendicular to the conductor A, as shown in Fig. 33.16(i). Applying Fleming's left-hand rule, it can be seen that A then moves towards B.

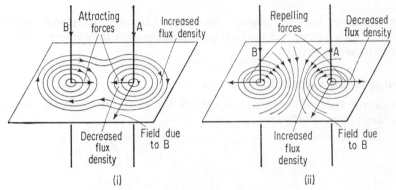

Fig. 33.16 Attraction and repulsion

The combined field round A due to both currents shows that the density of the lines is increased on the left side of A and decreased on the right side. Thus, as explained on p. 597, A moves towards B. A similar explanation holds for the repulsion of A when the currents are in the opposite direction (Fig. 33.16(ii)).

## Ampere Balance

Since 1948 the ampere has been defined by reference to the force between parallel current-carrying conductors, as follows:

*The ampere is that current which produces a force of $2 \times 10^{-7}$ newton per metre in a vacuum between two parallel infinitely long conductors of negligible cross-sectional area 1 metre apart when flowing in each of the conductors.*

This definition is accepted for international use. It was adopted because the force between parallel conductors can be measured with high accuracy. Fig. 33.17 shows the principle of an *ampere balance,* used for measuring current at the National Physical Laboratory. The coils A and B are fixed and C is attached to the balance. The coils are all in series, and when a current flows the coil C is repelled by A and attracted by B. C thus moves downward with a force depending on the magnitude of the current. The force is found by counterbalancing with a weight on the other side of the balance, and the current is then calculated. Electrical-instrument

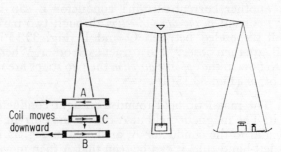

Coil moves
downward

Fig. 33.17 Ampere balance

manufacturers have a 'master' ammeter, obtained from the National
Physical Laboratory, for calibrating the ammeters they make, which are
placed in series with it.

### A Simple Current Balance

A simple current balance, to illustrate the main principle, can be made
as follows (Fig. 33.18):

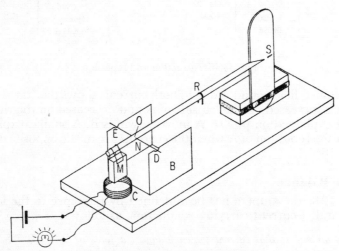

Fig. 33.18 Simple current balance

1. Attach a small $\frac{1}{2}$-in alcomax magnet M near the end of a plastic
   straw O with a small piece of Sellotape.
2. Balance the straw on the needle N to find the centre of gravity. Then
   push the needle through the straw at O, about a millimetre away
   from the centre of gravity on the opposite side of the axis of the
   straw to the magnet.
3. Balance the straw and magnet by placing the needle in grooves D
   and E cut in the aluminium support B, and then put a rider R,
   made of a few centimetres of 26 s.w.g. bare copper wire, on the

straw, as shown. The straw should balance when R is about 5 cm from O, and the rider may be moved with a piece of wire.

4. When the straw is horizontal, mark a reference line S on the wooden spatula near the end of the straw, as shown.

EXPERIMENT

Make sure there are no draughts to upset the equilibrium of the straw. Connect a 2·5-V bulb and a 1·5-V dry cell, for example, in series with the coil C so that M is attracted by the current flowing in C. If necessary, reverse the battery terminals (Fig. 33.18). Move the rider R along the straw until the straw returns to the reference mark S. Mark the position of the rider on the straw or measure its distance from O. This is now a measure of the current, which may be called 'one unit' of current.

Add a second similar lamp in parallel with the first lamp, move R to restore equilibrium as before, observe the new position of R and estimate the current in terms of the 'unit current'. Repeat by placing three lamps in parallel. Make a table of your observations and your deductions.

## SUMMARY

1. A force acts on a current-carrying conductor when it is perpendicular to a magnetic field or inclined to the field. The direction is perpendicular to both the current and the field direction. No force acts on the conductor if it is parallel to the field.

2. Fleming's left-hand rule states: If the thumb and first two fingers of the left hand are held at right angles to each other, with the forefinger pointing in the direction of the field and the middle finger in the current direction the thumb points in the direction of motion of the conductor.

3. Electrons such as those in a fine-beam tube are deflected by a perpendicular magnetic field. Metals have electrons which carry the current in them.

4. The moving-coil instrument has a rectangular coil, a permanent magnet, a soft-iron core to provide a radial field and two springs. The current is proportional to the deflection, that is, the scale is uniform – this is due to the radial field. The sensitivity increases if the spring is weak or the magnetic field is strong.

5. The mirror galvanometer has a long phosphor-bronze wire suspension, which exerts a very weak control over the deflection, and a mirror, a beam of light and a scale to measure the deflection.

6. The simple electric motor has: (i) a coil of insulated wire; (ii) a magnetic field; (iii) a commutator which reverses the current in the coil after half a revolution so that the coil keeps turning round.

7. Parallel wires attract each other when carrying currents in the same direction, and repel each other when the currents are in opposite directions. An ampere balance measures current by means of the force between parallel currents.

## EXERCISE 33 · ANSWERS, p. 640

**1.** The diagram represents the section of a wire between the opposite poles of two bar magnets. Draw the lines of force between the poles when a current flows down the wire (into the paper). (*N.*)

**2.** Describe how a compass needle can be used to determine the direction of the current in a straight wire when the wire is: (*a*) vertical; (*b*) horizontal and lying in the meridian. In each instance, give a diagram showing how the direction of the current is obtained.

A rectangular coil of wire is suspended by a torsion fibre between the poles of a horse-shoe magnet with the plane of the coil parallel to the field. Explain why the coil is deflected when a current passes through it. State the factors which determine the magnitude of the deflection of the coil and indicate how they do so. Give a rule by which its direction of motion can be obtained and illustrate with a diagram. (*L.*)

**3.** (*a*) Draw a labelled diagram of the pointer type of moving-coil galvanometer and explain its action.

(*b*) What single resistance will have the same effect in a circuit as three resistances of 2, 4, 5 Ω connected in parallel? If the total current flowing in the circuit is 1·52 A what will be: (i) the current in the 5-Ω resistor; (ii) the electrical energy converted into heat in the 5-Ω resistor in 2 min? (*N.*)

**4.** Describe an experiment to demonstrate that a mechanical force acts on a current-carrying conductor when it is suitably placed in a magnetic field. On a diagram indicate the directions of the current, the field and the motion of the conductor. (*L.*)

**5.** Describe the structure and mode of action of a moving-coil galvanometer. (*L.*)

**6.** Two wires hang side by side but not touching. State what happens when electric currents flow through the wires in the same direction. Sketch the lines of magnetic force surrounding the wires. (*N.*)

**7.** Describe an experiment to show that a wire carrying an electric current experiences a force when suitably situated in a magnetic field. In what direction does the force act?

If such a wire experiences no force, what can you conclude about the magnetic field?

Show, with a simplified diagram, how such forces are used to produce continuous rotation in a direct current motor. (*O.* and *C.*)

**8.** A moving-coil galvanometer has a resistance of 40 Ω and gives a full-scale deflection of 2 mA.

(*a*) What is the potential difference across its terminals when this current is flowing?

(*b*) How can the galvanometer be converted into a voltmeter? (*N.*)

**9.** Describe, with the aid of a diagram, the essential parts of an electric motor and explain its action.

Why is it not possible to calculate the current taken by a motor by dividing the applied volts by the resistance in ohms of the motor?

What is the efficiency of a 230-V motor which takes a current of 0·5 A when its output is 0·1 h.p.? (1 h.p. = 746 W.) (*N.*)

**10.** Describe the structure of a moving-coil loudspeaker and explain its action. (*L.*)

**11.** Describe the structure of a voltmeter and explain how it works.

A 12·0-V battery of negligible internal resistance and resistances of 30 and 70 Ω respectively are connected in series. Calculate the potential difference between the ends of the 70-Ω resistance. Draw a circuit diagram showing how a voltmeter could be used to record this potential difference.

If a resistance of 60 Ω is now connected in parallel with the one of 30 Ω, find the new value of the potential difference between the ends of the 70-Ω resistance. (*L.*)

**12.** Draw a diagram to show the magnetic field set up by a current flowing down a vertical wire.

Describe an experiment to show that a straight conductor carrying a current and placed in a magnetic field experiences a mechanical force. State a rule to determine the direction of the force.

Name TWO practical applications of this principle. (*N.*)

**13.** A rectangular coil, carrying an electric current, is pivoted about an axis AB which is parallel to one pair of its sides. The coil is in a uniform magnetic field whose lines of force are perpendicular to AB, and is held so that its plane is inclined to the field direction. Draw a diagram showing the relative directions of the current, the magnetic field, and the mechanical forces experienced, as a result of the interaction of the current and the magnetic field, by the two sides of the coil which are parallel to AB.

Explain how use is made of this effect in the design of the moving-coil galvanometer.

A galvanometer of resistance 5 Ω gives a full-scale deflection with a current of 15 mA. How would you convert it into a voltmeter reading up to 1·5 V? (*O.*)

**14.** Describe an experiment to demonstrate the existence and direction of the magnetic effect of a current in a long straight wire. Explain how the same experiment shows that there is a force exerted by a magnet on a current-carrying conductor.

Sketch some of the magnetic lines of force due to a current flowing upwards in a long vertical straight wire: (*a*) if the earth's magnetic field can be neglected; (*b*) if the earth's field cannot be neglected. (*O. and C.*)

**15.** Describe an experiment which shows that a wire inclined at an angle to a magnetic field and carrying an electric current experiences a force. Draw a diagram showing how the directions of current, field and force are related.

Explain in detail how this effect is applied to measure current in the moving-coil galvanometer and discuss the features which determine: (*a*) the sensitivity of the instrument; (*b*) the uniformity of its scale.

Illustrate your answer with a diagram. (*O. and C.*)

# 34

# ELECTROMAGNETIC INDUCTION ·
# THE DYNAMO

The year 1791 saw the birth of one of the greatest of English experimental scientists. MICHAEL FARADAY was the son of poor parents, who sent him to work at a bookshop at the age of 15; there, undeterred by his lack of education, he began to take an interest in the books around him, especially on the scientific side. When he was 20 he attended the lectures at the Royal Institution given by SIR HUMPHRY DAVY, and at the end of the course sent him his notes, beautifully written and illustrated, and asked if there was a vacancy at the Institution. Davy was impressed by the work, and offered him the post of laboratory assistant. From this position he rose to be Davy's personal assistant and, at the age of 33, when his pre-eminence in science was already established, he was made Director of the Royal Institution on Davy's retirement. By his pioneer work on the physical principles of electromagnetism, he led the way to the dynamo and the generator, without which commercial electrical lighting and heating could never have been obtained.

## Using Magnetism to Produce Electricity

It had been known since 1820 that an electric current gave rise to magnetic effects (p. 580). Faraday, like many experimental physicists at that time, began experiments to obtain the reverse effect. In 1832 he succeeded in 'using magnetism to obtain electricity'. Fig. 34.1 shows simple apparatus to investigate this phenomenon. C is a solenoid or coil with many turns connected to a centre-scale zero galvanometer G, so that a current in either direction in C produces an observable deflection.

EXPERIMENT 1. *Magnet Moving, Coil Still*

(i) When the N-pole of a powerful magnet such as a Ticonal one is pushed towards one end of C, the needle in G deflects to the right (Fig. 34.1(i)). An electric current is therefore obtained in the circuit. It is called an *induced current*, since there is no cell or battery in the circuit, and this is an example of *electromagnetic induction*.

(ii) When two and then three similar magnets are pushed into the coil greater induced effects are produced.

(iii) When a magnet is at rest, even though it may then be right inside the coil C, the induced current is zero (Fig. 34.1(ii)).

(iv) When the N-pole is removed quickly the needle is deflected to the left (Fig. 34.1(iii)). A current in the opposite direction to that in (i) is therefore obtained.

(v) When the magnet is moved faster towards or away from the end of the coil the induced current is increased.

(vi) When soft iron such as mu-metal is placed inside the coil as a core, and the experiment is repeated, a much larger induced current is obtained.

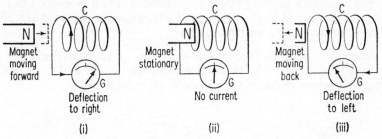

Fig. 34.1 Induced current – moving magnet

*Conclusions.* An induced current flows only while the magnet is moving. The magnitude of the induced current increases when the strength of the magnet and the speed of the magnet are increased, and when a soft-iron core is inside the coil.

EXPERIMENT 2. *Coil Moving, Magnet Still*

(i) Keeping the N-pole stationary, move the coil C quickly towards N. An induced current is obtained in C (Fig. 34.2(i)).

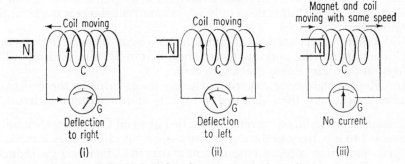

Fig. 34.2 Induced current – moving coil

(ii) Repeat, moving the coil C quickly back from N. An induced current is now obtained in the opposite direction (Fig. 34.2(ii)).

(iii) Move the coil and magnet together at the same speed. No induced current is obtained (Fig. 34.2(iii)).

You should also see if induced currents are obtained by holding an insulated length of straight wire joined to a sensitive galvanometer, and moving it in different directions in the field between the poles of a horse-shoe magnet.

*Conclusion.* Together with the conclusion in Experiment 1, it can now be stated that an induced current flows when there is *relative motion* between the coil and magnet.

## Primary and Secondary Coils

EXPERIMENT

Instead of using a moving magnet, a solenoid A can be placed in series with an accumulator L and a tapping key K, since the solenoid acts like a magnet when carrying a current (p. 587).

(i) When A is close to C and K is depressed, a slight deflection is obtained in G (Fig. 34.3(i)). With a soft iron inside both A and C, as shown, a very much larger deflection is obtained (Fig. 34.3(ii)).

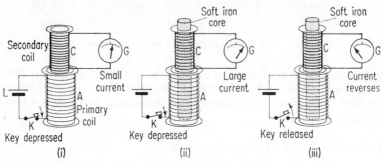

Fig. 34.3  Primary and secondary coils

(ii) With the key K held down firmly, no deflection is observed on G. The induced current thus flows only while a current *change* is made in the solenoid A.

(iii) When K is released so that the solenoid circuit is broken a deflection in G is obtained in the opposite direction to that in (i) (Fig. 34.3(iii)). The deflection then dies down quickly to zero.

(iv) Repeat the above experiment using fewer turns in the coil C. A smaller deflection is then obtained in making and breaking the circuit.

*Conclusion.* An induced current is obtained in a coil C near another coil A connected to a battery in the circuit only when the circuit of A is made or broken. A soft-iron core, and more turns of wire in C, increases the induced current. Coil A is known as the 'primary' coil and coil B as the 'secondary' coil.

## Origin of Induced E.m.f.

An electric current flowing in a circuit is due to the presence of an electromotive force (e.m.f.). The e.m.f. produced by electromagnetic induction is called an induced e.m.f.

If a battery is connected to a coil it can literally be seen as the source of e.m.f. The source or origin of an induced e.m.f. is not as obvious. In a battery a movement of ions between the electrolyte and plates produces an e.m.f. at the electrode terminals. An induced e.m.f. in a coil, however, has its origin in an electron drift inside the metal in one direction, produced, for example, when a magnet is pushed towards the coil. One end of the coil, even though it is not connected to a galvanometer, then has an excess negative charge due to the excess electrons at this end. The other end of the coil then has an equal excess positive charge.

The two ends of the coil, like the two poles of a battery, now have an electrical potential difference. We call it an *electromotive force* and not a potential difference because it is produced by an interchange of energy. In the case of a battery, chemical energy is converted to electrical energy. In this case, the mechanical energy of the moving magnet is converted to electrical energy.

### Faraday's Law of Electromagnetic Induction

Faraday considered that the medium in a magnetic field was in a state of stress or strain, and he pictured lines of force or flux everywhere in the medium (p. 560).

When a magnet is placed near a coil, a number of lines of flux links or 'threads' the turns of the coil (Fig. 34.4(i)). If the magnet is moved nearer, more lines link the turns. An induced e.m.f. is therefore always obtained

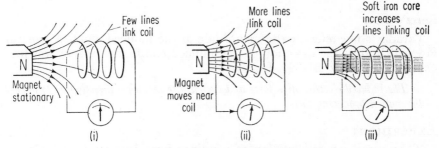

Fig. 34.4 Flux change and induced e.m.f.

when a *change* is made in the number of lines linking the coil. A slight magnet movement produces a small change, and the induced e.m.f. is therefore small. A large magnet movement produces a greater change and a greater induced e.m.f.

As we saw from the experiments in p. 616, the induced current and hence e.m.f. also increases when the speed of movement of the magnet increases. The magnitude of the e.m.f. thus depends not only on the change in the number of lines linking the coil but on the *time* taken to make the change. The two effects can be combined into one statement, that the induced e.m.f. depends on the *rate of change* of the lines linking the coil.

Experiments performed with a coil (or a dynamo on a bicycle wheel) rotating at a different measured number of revolutions per second shows that the induced e.m.f. produced, which is proportional to the measured induced current, is *directly* proportional to the rate of change of the lines. Faraday stated a general law:

*The induced e.m.f. in a circuit is directly proportional to the rate of change of the number of lines of flux linking the circuit.*

## Using Faraday's Law

Scientists such as electrical engineers perform calculations with numbers of lines of flux. As an illustration, suppose the pole in Fig. 34.4(i) is moved from one position A, where 200 lines link the turns of the coil, to another position B, as in Fig. 34.4(ii), where the number of lines passing through the coil increases to 2000. If the magnet is moved from A to B in $\frac{1}{2}$ sec the rate of change of the lines linking the coil $= (2000 - 200)/\frac{1}{2} = 3600$ lines per sec. If the magnet is now moved quickly away from the coil so that the number of lines through the coil diminish from 2000 to 200 in $\frac{1}{100}$ sec, the rate of change of the lines is now $1800 \div \frac{1}{100}$ sec $= 180\,000$ lines per sec. The induced e.m.f., from Faraday's law, is now 50 times as great as before. Thus if the induced e.m.f. in the first case was 20 millionths of a volt the induced e.m.f. increases to 1000 millionths of a volt.

This is the order of magnitude of induced e.m.f. when a small coil is used. By using an extremely rapid change, a large number of turns of wire in a wide coil, and a soft-iron core inside it, the induced e.m.f. may reach several thousand volts, as explained on p. 624.

## Lenz's Law

Faraday's law refers to the magnitude of the induced e.m.f. A general law which gives the direction of the induced e.m.f. and current was first expressed by Lenz in 1834. It states:

*The induced electromotive force and current are in such a direction as to oppose the motion producing them.*

### Experiment

Take a coil whose direction of windings from its terminals is clearly visible. Connect a battery B, a suitable large resistance R and a centre-zero sensitive galvanometer G in series with the coil (Fig. 34.00(i)). Switch on the current and suppose the deflection in G is to the right, as shown.

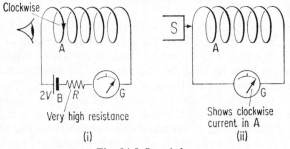

Fig. 34.5 Lenz's law

The current now flows *clockwise* in the end A of the coil when viewed from the left side.

Now completely remove the battery B and resistance R and connect the ends of the coil to G (Fig. 34.5(ii)). Plunge the S-pole of a magnet into the end A and observe the deflection in G. This is to the right, as shown.

From our previous experiment, it follows that the induced current flows clockwise in A.

*Conclusion.* The induced current acts to repel the S-pole, that is, to oppose its movement.

Similar experiments with a N-pole show the truth of Lenz's law. We now know, from the principle of the conservation of energy, that work or energy must always be spent to produce energy. Consequently, a force of opposition acts on a magnet as it moves towards a coil. The work done in moving the magnet against the force is converted into electrical energy.

### Examples of Lenz's Law

Fig. 34.6 shows examples of Lenz's law.

(i) In Fig. 34.6(i) a N-pole moves towards the face of a coil whose ends are joined. The coil opposes the movement of N. Since like poles repel, a N-pole is induced and hence a current flows in A which is anti-clockwise.

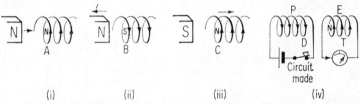

(i)          (ii)          (iii)          (iv)

Fig. 34.6  Direction of induced current

(ii) In Fig. 34.6(ii) a N-pole is withdrawn from the coil. A S-pole is therefore induced in the face B. The induced current hence flows clockwise.

(iii) In Fig. 34.6(iii) a coil is moved away from a stationary S-pole. A N-pole is therefore induced at the face C to oppose the movement. The current therefore flows anti-clockwise.

(iv) In Fig. 34.6(iv) the circuit is made in the primary coil P, so that a N-pole is suddenly induced at the end D. This is similar to moving the N-pole of a magnet quickly to the end E of the secondary coil T. An induced N-pole is therefore obtained at E to oppose the motion, that is, the current flows anti-clockwise. When the current in P is switched off the current in E flows clockwise to oppose the withdrawal of the N-pole at D.

### Demonstrations of Induced Currents

So far a coil of wire has been used for demonstrating induced currents. Provided the materials are conducting, however, induced currents can flow in a solid block of a substance or in a liquid or a gas. In a solid metal the induced currents whirl round like the eddies in water and are hence called *eddy currents*.

Demonstrations of such induced currents can be made by using a coil with a large number of turns of wire wound round a long soft-iron core, extended well beyond the end of the coil (Fig. 34.7). Parts of a commercial transformer (p. 631) are suitable. The coil is connected to the mains, which supplies it with alternating current reversing in direction one hundred times every second.

### (i) *Making Water Boil*

Some tap water is placed inside a metal ring R which can be held (Fig. 34.7(i)). When R is placed round L and the alternating current switched on, steam is observed coming from the water after a time and the metal ring is hot.

*Explanation.* The alternating current in C produced a continuous and rapid change of lines of force through the ring. This is made of metal and hence an induced current circulates round it continuously. The metal therefore becomes hot and the water evaporates after a time.

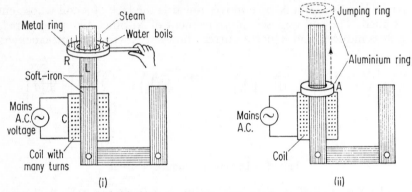

Fig. 34.7 Demonstration of induced currents

### (ii) *Jumping Ring*

When an aluminium ring is placed over the soft iron core at A, and the mains supply is switched on, the ring is seen to fly straight up into the air. Fig. 34.7.

*Explanation.* This is a consequence of Lenz's law. When the alternating current flows in the coil an induced current is obtained in the ring. The induced pole always opposes the inducing pole near it. In Fig. 34.6(iv), for example, an induced N-pole is obtained when the inducing pole brought near it is north. Consequently, repulsion occurs and the ring flies off.

### Arago's disc · Speedometer

Following an early experiment on induction by Arago, the poles of a horse-shoe magnet can be spun at constant speed above a horizontal metal disc pivoted at its centre. The disc then begins to turn and follow the motion of the poles, and soon spins at nearly the same speed as the magnet.

The rotation of the disc is an example of Lenz's law. An induced current flows in the disc which opposes the motion producing it. Consequently the disc rotates in the same direction as the magnet to make their relative velocity zero, and eventually it rotates at nearly the same speed. When the magnet stops, the disc soon stops; their relative velocity again becomes zero.

The car speedometer utilizes this phenomenon. A magnet is geared to the car wheels so that it spins at a rate proportional to the speed of the car when the car is in motion. An induced current is then set up in a pivoted aluminium disc in front of the magnet. The rotation of the disc about the axis is controlled by a hairspring, and a pointer attached to the axis is then deflected through an angle which is proportional to the speed. This is read from a speedometer dial, previously calibrated at known speeds.

## Induction Furnace

In industry, eddy (induction) currents are used to produce high temperatures necessary to melt metals. A type of furnace known as an induction furnace is used, as illustrated in Fig. 34.8. The metal S to be melted is contained in a crucible A lined with a special heat-resistant material, and coils in the form of a copper tube C are wound round the crucible. An alternating voltage V, which varies at the rate of several thousand cycles per second (p. 636), is connected to the coils, and induced (eddy) currents circulate in the metal S as the result of the continuous change of lines of force in it. A continuous supply of heat is then obtained in S, and as no heat is lost from A, the temperature of the metal rises until it finally melts. In this way very pure steel can be made. The copper tube C also becomes hot by eddy currents circulating in itself, so the tube is cooled by passing water through it while the induction furnace is working.

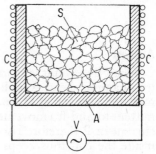

Fig. 34.8 Induction furnace

As extremely pure metal is needed to manufacture transistors and other semiconductor devices, a special form of induction furnace (shown in Plate 17) is used to refine germanium and silicon.

## Induction Coil

The *induction coil* is an apparatus which supplies a p.d. of several thousand volts using a battery of only 6 V, for example (Fig. 34.9). It consists of:

1. A primary coil P of a few hundred turns of thick insulated copper wire, to which a battery B with a switch K is connected.
2. A secondary coil S of several thousand turns of thin insulated copper wire wound round P, with terminals at X and Y.
3. A soft-iron core in the form of a bundle of soft iron wires (see p. 625).
4. A make-and-break arrangement, as in the electric bell, consisting of a soft-iron head A at the end of a springy steel strip of metal D which makes light contact with the tungsten tip of a screw B. A capacitor C is placed across D and B to 'quench' the spark obtained at the tip of the screw when the machine is working.

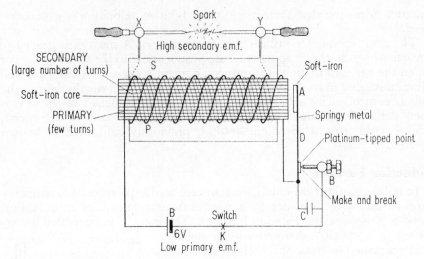

Fig. 34.9 Induction coil

## Action

When the circuit is made by K a current is obtained in the coil P. The iron hammer-head A is then attracted towards P. The circuit is now immediately broken at D, and A falls back to renew contact at D. This intermittent to-and-fro movement of A goes on at a definite frequency, and the current in P therefore rises and falls rapidly. The number of lines of flux inside the coil thus keeps changing rapidly. Now the lines also link the secondary turns. Consequently, an induced e.m.f. is obtained at the terminals X and Y of the secondary coil.

The e.m.f. at X and Y when the primary circuit is made is in the opposite direction to the e.m.f. obtained when the primary circuit is broken. The effect is similar to bringing a magnet towards and then away from the secondary coil, as explained on p. 621. Now the time taken by the current to fall to zero at the break is at least ten times smaller than the time taken to rise to its full value at the make. From Faraday's law, the induced e.m.f. at the break is very much greater than at the make. It may be 5000 V, for example, in the former case and only 200 V at the make.

On the average, then, the e.m.f. at the secondary terminals X and Y is high and in one direction. When the battery is connected, a spark, which may be several centimetres long, passes between X and Y. The high voltage between X and Y is thus sufficient to break down the insulation of the air and to make the gas conduct. In the laboratory an induction coil is a useful supply where an e.m.f. of several thousand volts is required, as in producing discharges, or luminous columns, of gases From the law of conservation of energy, the power developed in the secondary is equal to that in the primary, neglecting losses. Thus a much *smaller* current flows in the secondary than in the primary.

## Eddy Current Losses

When the induction coil is functioning, a varying number of lines of force pass through the soft-iron inside the coils (Fig. 34.10), and as iron is a conductor, induced currents are obtained in it. These currents are called

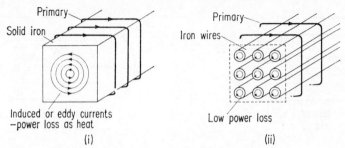

Fig. 34.10 Eddy current and power loss

*eddy currents*, and they produce heat in the iron. If the iron in Fig. 34.10(i) were solid the amount of heat produced would be a wasteful high percentage of the power expended by the primary battery. To diminish the loss of power due to eddy currents, the core is made of a bundle of iron wires, which are insulated from each other. This prevents eddy currents circulating across the wires. (Fig. 34.10(ii)).

## Ignition Coil

A small induction coil is used as the ignition coil in cars to provide a spark to fire the petrol–air mixture in the cylinder of the engine.

One end of the secondary coil is connected to the central metal electrode A in the sparking plug, which is insulated by porcelain from the metal casing round the plug (Fig. 34.11). The other end of the coil is connected to an electrode B joined to the casing, which has a small clearance from A. In the primary circuit of the induction coil, contact

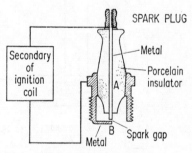

Fig. 34.11 Spark plug

points are opened at instants corresponding to the piston power stroke, when a spark is obtained at the electrodes of the plug.

## Induced E.m.f. in Straight Conductor

So far we have met the induced e.m.f. and current in circular coils such as solenoids. *Straight conductors* are used in commercial generators producing high voltages for the electrical grid system used throughout the country (p. 634).

EXPERIMENTS

1. Place a cylindrical metal rod AB across two brass horizontal rails P and Q, and complete the circuit by joining a sensitive galvanometer G to

the ends of P and Q (Fig. 34.12(i)). Arrange a powerful horse-shoe magnet round AB so that the field is *perpendicular* to AB. Now roll AB along the rails. An induced current is observed in G as shown.

Note that the direction of the field, the length AB and the movement of AB are all perpendicular to each other.

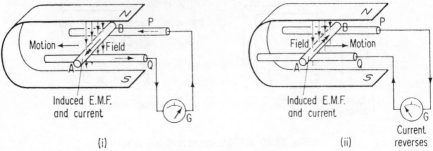

Fig. 34.12 Straight conductor

2. Roll AB in the opposite direction (Fig. 34.12(ii)). An induced current is observed in G in the opposite direction to that previously obtained.

3. Turn the magnet round so that its field is now *parallel* to the rod AB (Fig. 34.13(i)). Roll AB along the rails in either direction. No induced current is observed in G.

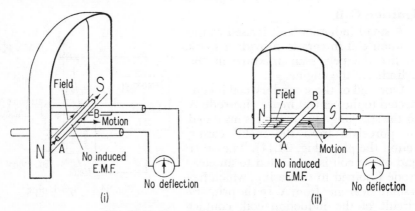

Fig. 34.13 No e.m.f. in moving conductor

4. Finally, turn the magnet round so that the field is parallel to the direction of movement of AB as it rolls (Fig. 34.13(ii)). No induced current is observed in G.

*Conclusion.* An induced current flows in a straight wire when it is perpendicular to a magnetic field and is moved perpendicular to its length and to the field.

### Cutting Lines of Flux · Fleming's Rule

When a magnet is pushed into a solenoid, the e.m.f. is considered due to lines of flux 'linking' the coil. The induced e.m.f. in a straight conductor, however, is considered due to the 'cutting' of lines as it moves through the field. The magnitude of the induced e.m.f. is directly proportional to the rate of cutting of lines of force. Consequently, the faster the wire is moved perpendicular to the field, the greater is the induced e.m.f. When the wire is moved parallel to the field, however, no lines are cut. This is also the case when the wire is parallel to the field and moved perpendicular to the field.

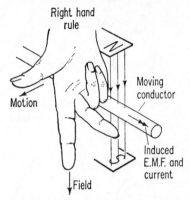

Fleming gave a rule for deducing simply the direction of the induced current in a straight conductor: 'If the thumb and the first two fingers of the RIGHT hand are held at right angles to each other, with the Fore-finger held in the direction of the Field and the thuMb in the direction of Motion, the induced current ($I$) flows in the direction of the mIddle finger.'

Fig. 34.14 Fleming's right-hand rule

The rule is illustrated in Fig. 34.14. It should be carefully distinguished from Fleming's *left*-hand rule, which gives the motion of a current-carrying conductor in a magnetic field (p. 598).

## Principle of Simple Dynamo

The *dynamo* is a source of e.m.f. which utilizes electromagnetic induction. A simple dynamo consists of a coil rotating at a steady speed in a magnetic field about a fixed axis.

Fig. 34.15 illustrates various positions of the coil during its rotation, from which it can be seen that a change occurs continuously in the number of lines linking the coil. Thus when the coil is horizontal the lines pass over its face but none pass *into* it (Fig. 34.15(i)). When the coil rotates farther, more and more lines pass through or link the coil. When the coil reaches the vertical the number of lines linking it is then a maximum (Fig. 34.15(iii)).

## Magnitude of E.m.f.

The coil can be considered made of four straight wires AB, BC, CD and DA. AB and DC rotate in the plane of the field and do not cut any lines. No induced e.m.f. is therefore obtained in these wires. The other two wires BC and AD cut lines of force as they rotate. When the plane of the coil reaches the horizontal, for example, as in Fig. 34.15(i), the wires BC and AD cut the lines perpendicularly. The induced e.m.f. is then a maximum. As the coil rotates towards the vertical, as in Fig. 34.15(ii), the wires BC and AD sweep across the lines at an angle less than 90°, and hence the induced e.m.f. decreases. When the coil reaches the vertical BC and AD are both then moving horizontally or parallel to the field (Fig 34.15(iii)). No lines are cut at this instant, and hence the induced e.m.f. is then zero.

Summarizing, the induced e.m.f. is a maximum when the plane of the

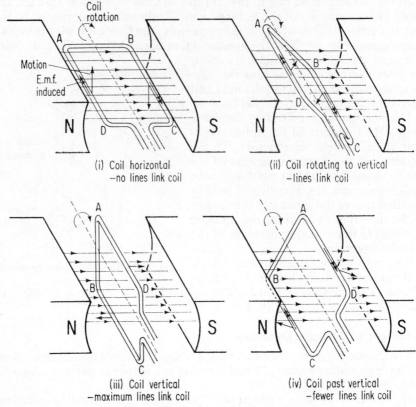

(i) Coil horizontal
—no lines link coil

(ii) Coil rotating to vertical
—lines link coil

(iii) Coil vertical
—maximum lines link coil

(iv) Coil past vertical
—fewer lines link coil

Fig. 34.15 Action of simple dynamo

coil is horizontal, and it diminishes to zero when the plane of the coil is vertical.

### Direction of Current and E.m.f.

The direction of the induced current in the coil, if closed, may be found by applying Fleming's right-hand rule. Thus, since AD is a straight conductor moving vertically upwards in the magnetic field between the poles N, S (Fig. 34.15(i)), the direction of the induced current is DA. Since BC is a straight conductor moving vertically downwards, the induced current in it is in the direction BC. Thus considering the whole coil, a current flows in the direction DABC. As the coil rotates towards the vertical the conductors AD, BC are still moving respectively up and down in the same direction, and hence the induced current remains in the same direction (Fig. 34.15(ii)). When the vertical is reached (Fig. 34.15(iii)) the conductors AD, BC are moving parallel to the magnetic field, and hence at this instant the induced current is zero (see p. 626).

When the vertical is passed, AD and BC move respectively downwards and upwards, and the current in the coil reverses (Fig. 34.15(iv)). As the coil rotates farther, the direction of the current remains the same, but

when the coil again reaches the vertical the direction of the current reverses. In general: (i) the current reverses when the coil passes the vertical,
(ii) the current is a maximum at the instant when the coil is horizontal
and zero when it is vertical. Fig. 34.16 shows the variation of e.m.f.

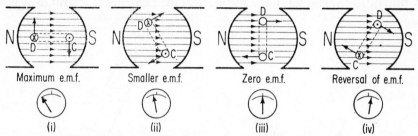

Maximum e.m.f.    Smaller e.m.f.    Zero e.m.f.    Reversal of e.m.f.

(i)               (ii)              (iii)            (iv)

Fig. 34.16 Direction and magnitude of induced e.m.f.

## A.C. Voltage

Fig. 34.17 illustrates the variation with time of the current $I$ obtained in
the coil; as is usual in graphs, the reversal of current is indicated by
plotting the values below the time-axis. This current is known as the
simplest alternating current (A.C.), and the e.m.f. in the coil is known as

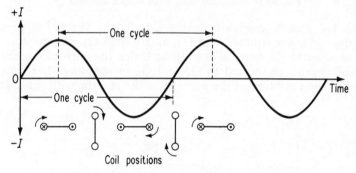

Fig. 34.17 Variation of induced current

an alternating e.m.f. (or, popularly, as an A.C. voltage). The electrical
supply in most areas of the country is 240 V A.C. It has a frequency of
50 Hz, that is, it takes $\frac{1}{50}$ sec to go once through a complete cycle (see Fig.
34.17). In the United States of America, the mains frequency is 60 Hz.

## Obtaining the Induced Current

### (1) Alternating Current

The alternating current in the rotating coil can be obtained in a lamp R
by using two copper rings P and Q, to which each end of the coil is permanently connected (Fig. 34.18(i)). These slip-rings press against metal
brushes $G_1$ and $G_2$ as they rotate with the coil, and the lamp is connected
to the brushes.

## (2) *Direct Current*

A dynamo can supply a current in one direction by using a commutator. The latter consists of a split copper ring (p. 608); the ends of the coil are connected permanently to each half X and Y of the ring (Fig. 34.18(ii)). As the coil rotates to the vertical the half-ring X gradually leaves

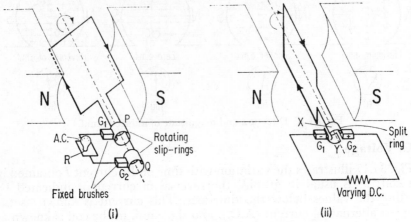

Fig. 34.18 A.C. and D.C. from simple dynamo

the brush $G_1$, and Y gradually leaves $G_2$. As the vertical is passed X changes over to make contact with $G_2$, and Y with $G_1$. But the current in the coil reverses as the vertical is passed; hence the current continues to flow in the resistor connected to $G_1$ and $G_2$ in the same direction. Fig. 34.19 shows the variation of the current in the resistor with time. It may

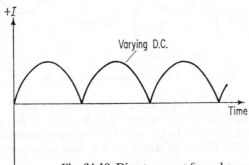

Fig. 34.19 Direct current from dynamo

be termed a varying direct current, since it has an average value in one direction.

The current obtained, however, is not a very steady one. A very steady direct current is obtained in practice by having many insulated straight conductors equally spaced round a soft-iron drum which rotates in the magnetic field. The commutator is divided into a corresponding number of equal segments, to which each conductor is connected.

## The Transformer

As we shall see later, alternating e.m.f.s or A.C. voltages require to be changed from low to high values in order to transmit A.C. power economically over long distances. The voltage also needs to be changed from a high to a lower, and much less dangerous, value when brought into the home, for example.

A *transformer* changes the magnitude of an A.C. voltage. It consists simply of two insulated coils P and S, called primary and secondary coils, wound round a soft-iron core (Fig. 34.20). The core is made of laminae or

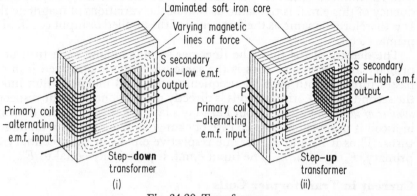

Fig. 34.20 Transformers

sheets of soft iron insulated by varnish from each other. This reduces the heat losses of energy inside the iron, which would be considerable if a solid block of metal were used and induced or eddy currents circulated inside it (see p. 625).

One type of demonstration mains transformer has a large number of primary turns and about 60 times fewer secondary turns of thicker wire (Fig. 34.20(i)). When the 240 V A.C. mains is connected to the primary a low voltage of about 4 V is recorded on a high-resistance A.C. voltmeter such as the AVO-meter connected across the secondary.

The transformer thus has stepped down the A.C. voltage in the ratio 4 : 240 or 1 : 60. This is the same as the ratio of the number of secondary turns, $n_s$, to the number of primary turns, $n_p$. A ray-box transformer which produces 12 V A.C. from a 240 V A.C. mains supply is a step-down transformer, with 20 times fewer turns in the secondary compared with the primary.

When the number of secondary turns is greater than that of the primary turns a correspondingly greater alternating e.m.f. is produced at the secondary terminals (Fig. 34.20(ii)). Generally, a commercial transformer, which has a soft-iron core, obeys the following law for e.m.f.s:

$$\frac{E_s}{E_p} = \frac{n_s}{n_p},$$

where $E_s$ and $E_p$ are the e.m.f.s in the secondary and primary respectively and $n_s$ and $n_p$ are the corresponding number of turns .The ratio $n_s/n_p$ is called the *turns-ratio* of the transformer. A step-up transformer has a turns-ratio greater than 1; a step-down transformer has a turns-ratio less than 1.

## Explanation of Transformer Action

The law $E_s/E_p = n_s/n_p$ can be explained by considering the flux changes in the secondary and primary coils.

When an alternating e.m.f. is connected to the primary an alternating current flows, that is, a current which reverses continually. This sets up variations of magnetic flux in the iron core, which also link the secondary coil. An induced e.m.f. is therefore set up in the secondary coil. The frequency of this e.m.f. is the same as that of the variations of magnetic flux. It is therefore the same as the frequency of the applied or input e.m.f. in the primary coil.

The density of the flux (the lines per sq cm) linking each turn of the primary and secondary coils is the same, since the magnetized iron core passes through both coils. Consequently, the total magnetic flux linking the secondary and primary coils is directly proportional to their respective *number of turns*, $n_s$ and $n_p$. From Faraday's law, the magnitude of the e.m.f. in a coil is proportional to the rate of change of magnetic flux linking the turns. Thus if $E_s$ and $E_p$ are the respective e.m.f.s in the secondary and primary, $E_s/E_p = n_s/n_p$. The input e.m.f. is practically equal to $E_p$.

## Current in Transformer Coils

The transformer, like any other machine, obeys the Principle of the Conservation of Energy. Thus the energy per second or power obtained in the secondary at any instant is equal to the power supplied to the primary, neglecting losses of power. Hence if $I_s$ and $I_p$ are the secondary and primary currents when the e.m.f.s are $E_s$ and $E_p$ respectively,

$$I_s \times E_s = I_p \times E_p$$
$$\therefore \frac{I_s}{I_p} = \frac{E_p}{E_s} = \frac{n_p}{n_s}$$

Thus if the transformer is a step-up one with turns-ratio 60 : 1 the currents are *stepped-down* in the ratio 1 : 60. Suppose the primary e.m.f. is 4 V A.C. and the secondary e.m.f. is 40 V A.C. Then if the primary current at an instant is 1 A, the power in the primary at this instant is 4 W. The current in the secondary is approximately $\frac{1}{10}$ A and the power in the secondary is then approximately $\frac{1}{10} \times 40$ or 4 W, the same as in the primary coil.

In the same way, the current in the secondary of a step-down transformer is *greater* than the current in the primary. The large current flowing in the 4-V secondary of a 240-V mains step-down transformer when a 5-A fuse wire is connected across the secondary terminals will melt the wire.

In practice, there is a loss of energy or power in a transformer; for example, some energy is wasted as heat in the coils, and in the iron core owing to eddy currents. Thus the efficiency of the transformer is less than 100%. As an illustration, consider a 240-V step-down mains transformer, used for lighting 8 12-V, 24-W ray-box lamps, which in practice draws a current of 1 A in the primary. Then

$$\text{power supplied to primary} = 240 \text{ V} \times 1 \text{ A} = 240 \text{ W},$$

and $\qquad$ power developed in secondary $= \quad 8 \times 24 \text{ W} = 192 \text{ W}.$

In this case, $\qquad\qquad$ efficiency $= \dfrac{192}{240} \times 100\% = 80\%.$

## High-tension Transmission

Since power is the product of current and voltage (p. 526), a given amount of A.C. power can be transmitted either at high voltage and low current or at low voltage and high current. Now cables of copper wire are needed to connect the output power from a commercial power station to distant towns. These cables carry the current, and hence power is lost as heat in the metal. Now the smaller the current, the less is the power loss. Consequently, a low current, and therefore a high voltage, is the most economic way of transmitting power over long distances. As the cables needed to carry a low current can be relatively thin, the cost of the copper used is also reduced.

## Demonstration of A.C. Power Transmission

Power transmission can be demonstrated with the apparatus shown in Fig. 34.21.

*Without Transformer*

An A.C. supply of 12 V is connected to the ends A and B of two resistance wires M and N, each 1 m long of 28 s.w.g. eureka wire (Fig. 34.12(i)). The circuit is completed at the other ends Y and Z by connecting a 12-V, 24-W ray-box lamp D across them, and a similar lamp L

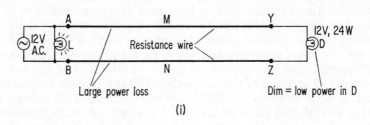

(i)

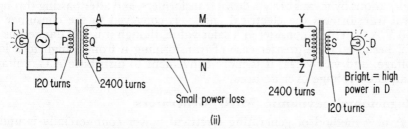

(ii)

Fig. 34.21 Power loss and high-tension transmission

is placed directly across the supply. It is then observed that L glows brightly, but D hardly glows at all.

In this case, then, very little power reaches D at the end of the wires; most of the power supplied at A and B is lost as heat in the two wires. Only a small p.d. is now obtained across the lamp D, and the current in the lamp is small.

*With Transformer*

Transformers are now used to transmit the power from the supply, and are arranged as shown in Fig. 34.21(ii). P and Q are the primary and secondary of a step-up transformer at the supply end, so that a high voltage is obtained at the ends A and B of the wires or 'transmission lines'. At the far ends, at Y and Z, the A.C. voltage is stepped-*down* by a similar transformer with primary R and secondary S, and the secondary is connected to the lamp D. It is now observed that, unlike the previous case where no transformer was used, D glows brightly.

In this case, then, practically all the A.C. power from the supply is produced at D; very little power is now lost as heat in the wires. The A.C. voltage at the ends of the lamp D is practically 12 V this time. Thus efficient transmission of A.C. power can be carried out by first stepping-up the voltage from the supply, so that a high voltage is obtained at the beginning of the transmission lines and low current flows.

In engineering language high voltage is called 'high tension'. Thus high-tension transmission is used to distribute commercial electrical power. It has the additional advantage that, with low current, the copper cable need be made only with thin wire, thus economizing in the amount of copper needed.

## The Grid · Use of Transformers

The A.C. generators in main power stations, like the great station at Battersea, London, are driven by steam turbines, which may require thousands of tons of coal and thousands of gallons of water daily. On this account power stations should be near a river, so that coal may be brought by ships or barges and water be readily available. The e.m.f. of the dynamos can be built up to about 11 000 volts at the main power stations. On account of the advantages of high-tension transmission, the e.m.f. is stepped up to 132 000 V by transformers. This voltage is transmitted to different parts of the country along cables supported by steel pylons (the grid system). Voltages of the order of 30 000 V are obtained at sub-stations by means of step-down transformers, and after passing through local transformers the electrical energy is conveyed to the consumer at 240 V A.C. This is considered a safer value, though it is still dangerous to life. Because of the greater ease of transforming it from a high to a low voltage, and vice-versa, it is more convenient to distribute A.C. voltage than D.C. voltage over an area.

## Magnetohydrodynamic (MHD) Generators

A new method of generating electrical power commercially is under active development in Great Britain. It is called 'magnetohydrodynamic

(MHD)' generation. Conventional generators at power stations in this country have rotating wires or armatures driven by steam engines, together with slip-rings. Frictional losses occur at all the bearings. In contrast, MHD generators have no moving parts. Power losses due to friction are therefore eliminated, leading to greater efficiency.

The basic principle of a MHD generator consists of passing a conducting gas at high velocity through a strong magnetic field (Fig. 34.22). From Fleming's right-hand rule for moving straight conductors (p. 627), an e.m.f. is induced at right angles to the field, as shown. The gas is produced by an apparatus similar to that used in rocket engines, which expel

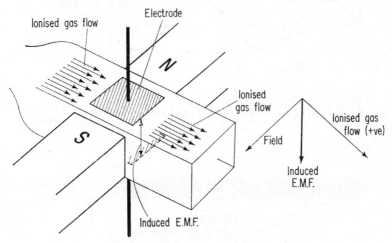

Fig. 34.22 Magnetohydrodynamic principle

hot gases at high velocities. The high temperature, about 2000°C, makes the gas more conducting as many ions are produced under these conditions, and to increase the conductance further it is 'seeded' with small quantities of the metal potassium.

The hot gases emerging from the MHD generator will be utilized for useful purposes, such as making steam which powers other plant. Although problems connected with MHD generators are under active investigation, their total efficiency is expected to be at least 10% better than conventional generators.

### Alternating-current Frequencies

The simplest type of alternating current (A.C.) has a waveform similar to that obtained from the dynamo discussed on p. 628, and is illustrated in Fig. 34.23(i); it is known as a sinusoidal A.C. This is a current which increases from zero at an instant O to a maximum value at an instant A; it then decreases to zero at an instant B and reverses. It increases to a maximum value in the opposite direction (instant P), and finally decreases to zero at an instant Q and reverses again. The variation of current now begins all over again, and hence the variation from the instant O to the instant Q is known as a complete *cycle*. From the time P to the time T the current also goes through one complete cycle. The frequency of the A.C. mains in this country is standardized at 50 Hz; in America the frequency is 60 Hz. An A.C. mains receiver set sometimes gives rise to a low hum when it is first switched on, indicating a note of double the mains frequency or 100 Hz in this country (1 Hz = 1 cycle per sec, p. 331).

If a person speaks into a microphone, alternating currents of audio-frequency are obtained. In speech and music the range of audio-frequency is from about 20 to 15 000 Hz (Fig. 34.23(ii)). The electrical circuits of

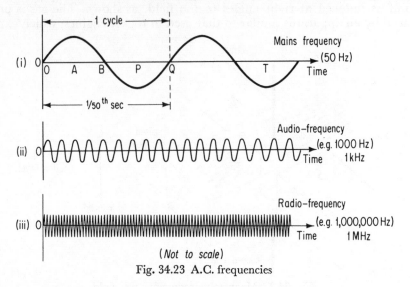

Fig. 34.23 A.C. frequencies

radio transmitters usually generate alternating currents of much higher frequency than 15 000 Hz. At a station broadcasting on a wavelength of 300 m, the current has a frequency of a million cycles (megacycle) per second, 1 MHz, and some of the electrical energy is radiated into space in the form of radio waves of this frequency (Fig. 34.23(iii)). These high frequencies are termed radio-frequencies.

### Effects of Alternating Current

When a ray-box light bulb is connected to the 12-V A.C. supply from a transformer the filament lights up. An alternating current in a wire therefore produces a heating effect. It does so because the drifting electrons in the wire encounter a resistance to their motion by 'collision' with the stationary ions in the crystal lattice whichever direction they flow along the wire. An alternating current can thus be measured by a hot-wire ammeter (p. 534).

On the other hand, when an alternating current flows through a copper sulphate solution with copper electrodes no deposit is obtained on the cathode. Alternating current does not therefore produce electrolysis. The ions, which carry the current through the solution, drift first in one direction and then in the opposite direction, so that the net deposit on the cathode is zero.

When an alternating current of mains frequency is passed into a moving-coil ammeter no deflection is obtained. The frequent reversals of current are too rapid to be followed by the coil and the needle, as they are too sluggish and heavy. On the other hand, a moving-iron ammeter produces a deflection when an alternating current is passed into it (see p. 593).

## Peak Values · R.m.s. Value

The maximum value of an alternating current during its cycle is known as its *peak value*. It is *not* the value recorded by a hot-wire ammeter because the current reaches the peak value only at an instant and changes thereafter.

The pointer deflection or instrument reading depends on the average heat produced in the wire of the hot-wire ammeter (p. 534). This depends on the average value of the *square* of the current or $I^2$ (p. 534). A practical or commercial value of current called the *root-mean-square* value or r.m.s. value is therefore used in connection with alternating current. It may be defined as *the square root of all the values of $I^2$ taken over a cycle*, or as the *steady or direct current producing the same heating effect per second as the alternating current*. Thus if a direct current of 2·0 A produces a temperature rise of 1·5°C per minute in a coil of wire, and an alternating current produces the same temperature rise per minute in the same coil, then the r.m.s. value of the alternating current is 2·0 A. The peak value is higher than the r.m.s. value, for example, it may be 2·8 A in this case.

Likewise, commercial values of alternating voltages are expressed in r.m.s. values, not peak values. The '240 V A.C.' mains voltage common in Great Britain means that 240 V is the r.m.s. voltage of the mains. The peak voltage is about 100 V higher than 240 V, or nearly 340 V. An electrical appliance used on the 240-V mains must therefore be strong enough to withstand a momentary voltage of nearly 340 V.

## SUMMARY

1. An induced e.m.f. and current are obtained in a coil in a magnetic field whenever there is a change in the number of lines of force or magnetic flux linking the coil.

2. The magnitude of the induced e.m.f. is proportional to the rate of change of the lines of force; the direction is such as to oppose the motion producing the induced e.m.f., which is known as Lenz's law.

3. The induction coil produces an average high p.d. in the secondary in one direction when a low p.d. source, supplying direct current, is connected to the primary. A make-and-break device creates a very rapid change in the lines of force at the break.

4. An induced e.m.f. is obtained in a *straight conductor* when it is perpendicular to a magnetic field and moves perpendicular to both the direction of the field and the length of the conductor.

5. The direction of the induced e.m.f. and current in a straight conductor is given by Fleming's *right*-hand rule. This states: If the first three fingers of the right hand are held at right angles to each other, with the Forefinger pointing

in the direction of the field and the thuMb in the direction of motion, then the mIddle finger points in the direction of the induced current $I$ and e.m.f.

6. The simple dynamo has a coil rotating steadily in a magnetic field. An alternating current (A.C.) is obtained by connecting the respective ends of the coil to two slip-rings; a direct current (D.C.) is obtained by connecting the respective ends to the two halves of a split-ring commutator.

7. The commercial transformer consists of a primary and a secondary coil wound round soft-iron sheets. The voltage to be changed is connected to the primary. $E_s/E_p = n_s/n_p$.

8. An alternating current produces no electrolysis, and does not make the needle of a moving-coil instrument move. It can be measured by a hot-wire ammeter or by a moving-iron ammeter. The *peak value* is the maximum value of current or voltage during a complete cycle; the *root-mean-square* (*r.m.s.*) value is the square root of the mean (average) of all the squares of current (or voltage) values in a complete cycle, or the direct current (or voltage) which produces the same heating effect per sec in a resistor.

9. Magnetohydrodynamic (MHD) generation is the generation of an induced e.m.f. by using ionized gas flowing at high speed across a powerful magnetic field.

## EXERCISE 34 · ANSWERS, p. 640

1. State two laws of electromagnetic induction. (*N.*)

2. Describe an experiment to show the production of an induced current in a coil of wire. What factors determine the magnitude of the current? With the aid of a diagram, show how the direction of the current in the coil can be predicted.

Describe the structure and explain the action of an appliance which depends on electromagnetic induction for its operation. (*L.*)

3. State *Lenz's law*.

The ends of a solenoid are connected together by a wire. Give diagrams showing the direction of the current (if any) induced in the solenoid when the N-pole of a magnet is: (i) thrust into the solenoid; (ii) at rest in the solenoid; (iii) rapidly withdrawn.

A copper disc is rotated by an electric motor. Explain why the disc comes to rest more quickly when the current is cut off if the disc rotates between the poles of a horse-shoe magnet. (*L.*)

4. (a) Describe, with a diagram, an experiment to induce a current in a coil of wire and indicate the direction of the current you would expect to obtain.

(b) With the aid of a labelled diagram describe and explain the action of a transformer. (*N.*)

5. A rectangular coil of wire, attached to a commutator, rotates between the poles of a permanent magnet to form a dynamo. (a) What is the position of the coil when the e.m.f. is greatest? (b) What is the function of the commutator? (*N.*)

6. Describe how you would demonstrate the production of an induced current using (a) a bar magnet, (b) an accumulator, together with other necessary apparatus. Indicate clearly the direction of flow of the currents and give a reason to justify your answers.

How could the strength of the induced currents be increased in each of these experiments? (*N.*)

7. With the aid of labelled diagrams explain the action of any TWO of the following: (a) an induction coil; (b) a transformer suitable for supplying about 3 A at 12 V from a 240-V A.C. mains; (c) a diode valve used as a rectifier. (*N.*)

**8.** State the essential condition for the production of an induced e.m.f. in a conductor. What determines its magnitude?

Give an account, with a diagram, of the structure and mode of action of a transformer which supplies 12 V when connected to 240-V mains. If this transformer takes 0·55 A from the mains when used to light five 12-V, 24-W lamps in parallel, find: (a) its efficiency; (b) the cost of using it for 10 h, at 6p per kWh. (L.)

**9.** Explain what is meant by *electromagnetic induction* and state the law which indicates the magnitude of the induced e.m.f. Describe an experiment in support of your answer and indicate on a diagram how the direction of the induced current is determined by the cause producing it.

Describe EITHER a simple A.C. dynamo OR a transformer, and explain how it works. (L.)

**10.** Describe the structure and mode of action of a simple D.C. generator.

What alterations would you make in the simple generator in order to convert it into one of higher e.m.f.?

Explain the source of the electrical energy supplied by the generator. (N.)

**11.** Draw a labelled diagram of a simple step-up transformer.

What would be the effect of passing alternating current through acidulated water in a water voltameter? (N.)

**12.** Draw a labelled diagram of an ammeter which could be used to measure alternating current. (N.)

**13.** State the two laws of electromagnetic induction.

A transformer is designed to work from 240-V A.C. mains and to give a supply at 8 V to ring house bells. The primary coil has 4800 turns.

(a) Would you expect the secondary coil to be of thicker or of thinner wire than the primary?

(b) About how many turns would you expect it to have?

(c) Why is the iron core made of laminations (or sheets) of iron instead of being in one solid piece?

(d) What would happen if the transformer were connected to 240-V D.C. mains?

(e) Do you think the primary current will increase or decrease when a bell is being rung?

Give reasons for your answers. (O. and C.)

**14.** State the laws governing the production of electric current in a circuit which is cutting lines of magnetic force. Describe TWO experiments to illustrate your statement.

Fig. 34A shows two coils A and B of insulated wire wound on a U-shaped soft iron core. Coil A is in circuit with a battery and a tapping key C, while the terminals of coil B are joined by a wire.

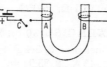

Fig. 34A

Indicate on a diagram: (a) the direction of the magnetic field through the iron core when the current is established in coil A; (b) the direction of the induced current set up in coil B at the instant the tapping key C is closed. Give a reason.

What would be the effect on the magnitude of the induced current of resting a soft iron bar across the ends of the iron core? (O. and C.)

**15.** Describe and interpret ONE experiment which demonstrates the laws of electromagnetic induction. State the laws.

Suggest explanations for the following: (a) An A.C. transformer has a core of thin soft-iron sheets ('stampings'). (b) The starting current of a D.C. motor is usually much greater than the current after the motor has reached its running speed. (O. and C.)

**16.** State the laws which determine the magnitude and direction of a current produced by electromagnetic induction. Describe a simple experiment to verify ONE of the laws.

A plane coil of wire is rotated about a diameter, which is perpendicular to a uniform magnetic field. Sketch a graph of the induced e.m.f. against the angle between the coil and the field. Indicate clearly the values of the angle, on the x-axis. What is the effect on the e.m.f. of using a coil of: (a) greater number of turns; (b) greater area? (O. and C.)

**17.** Draw and describe a simple form of D.C. dynamo pointing out the principles underlying its design.

How would you expect the e.m.f. to depend on: (a) the speed of rotation; (b) the direction, or sense, of rotation? Give reasons for your answer.

If a headlamp of a bicycle is disconnected from the wheel-driven dynamo the lower-wattage tail lamp frequently 'burns out'. Suggest a reason for this. (O. and C.)

## ANSWERS TO NUMERICAL EXERCISES

### EXERCISE 27 (p. 506)

**2.** $750\Omega$  **3.** $x = 2\,\text{A}, r = 6\cdot4\,\Omega$  **4.** $0\cdot5\,\text{A}$. (i) $0\cdot56$ (ii) $0\cdot093\,\text{A}$
**5.** (a) $1\frac{1}{6}, \frac{1}{2}, \frac{2}{3}\,\text{A}$ (b) $2\,\text{V}$  **6.** series $199\cdot8\Omega$  **8.** $6\frac{1}{8}\Omega, 6\,\text{A}. 24\,\text{V}, 1\frac{5}{7}\Omega$
**9.** (a) $0\cdot002\,67\,\text{A}, 1\cdot33\,\text{V}$ (b) $0\cdot004\,\text{A}, 1\,\text{V}$  **10.** $4\Omega$  **11.** $0\cdot84\Omega$  **12.** $0\cdot20\Omega, 6540\,\text{cm}$
**13.** $1\cdot19\,\text{m}$  **14.** $62\cdot5\,\text{cm}$

### EXERCISE 28 (p. 523)

**6.** (a) $1\cdot1$ (b) $0\cdot66\,\text{V}$  **8.** $0\cdot2\,\text{A}$  **9.** $60\Omega, 0\cdot40\,\text{A}$  **10.** (i) $0\cdot2\,\text{A}$ (ii) $1\cdot1\,\text{V}$
**11.** $0\cdot66\,\text{A}, 3\cdot3\,\text{V}$  **12.** $R = 4\cdot3\Omega, 0\cdot5\Omega$  **13.** (a) $0\cdot05\,\text{A}$ (b) $3\cdot75\,\text{V}$
**14.** $2E, 2r; E, r/2$ (a) $1\cdot5\,\text{V}$ (b) $4\Omega$  **15.** $\text{A} - 0\cdot2\,\text{A}, \text{B} - 0\cdot15\,\text{A}, \text{C} - 0\cdot05\,\text{A}; 1\cdot0\,\text{V}$

### EXERCISE 29 (p. 537)

**1.** (a) $4\Omega$ (b) $2160\,\text{J}$  **2.** (a) $0\cdot4\,\text{A}$ (b) $10\text{p}$  **3.** $9\cdot4\,\text{A}, 36\text{p}$  **4.** $7\cdot8\,\text{min}$  **5.** $2\cdot8\Omega$
**6.** (a) $62\cdot5\Omega$ (b) $3\frac{1}{2}\text{p}$  **8.** (a) $1/2$ (b) $2/1$  **10.** $10°\text{C}$  **11.** $4/1$  **12.** $2\cdot8\,\text{kW}, 11\cdot7\,\text{A}$
**13.** (a) $607\,\text{g}$ (b) $0\cdot9\text{p}$  **14.** $16\,\text{A}, 2\cdot4\,\text{A}$  **15.** (a) $9\cdot6\,\text{V}$ (b) $1\cdot4\Omega$  **16.** $31\cdot25\Omega, 320\,\text{J/s}$
**17.** $1955\,\text{W}$  **18.** greater rate. $0\cdot9\,(6/7)\Omega, 16\cdot0°\text{C}$

### EXERCISE 30 (p. 552)

**2.** $16\text{p}$  **3.** (i) $462\,\text{J}$ (ii) $0\cdot14\,\text{g}$ increase  **5.** $3\cdot4\,\text{g}$
**6.** $1170\,\text{Wh or } 4\cdot21 \times 10^6\,\text{J}$  **7.** $1\cdot43\,\text{A}$  **8.** $35\cdot9\,\text{min}$  **9.** $1\cdot12\,1$  **10.** $74\cdot9\,\text{h}$  **11.** $0\cdot34\,\text{g}$
**12.** $0\cdot74\,\text{g}$  **13.** $0\cdot001\,118\,\text{g/C}$  **15.** $0\cdot17\,\text{A}$

### EXERCISE 33 (p. 614)

**3.** (b) $1\frac{1}{19}\Omega$ (i) $0\cdot32\,\text{A}$ (ii) $61\cdot4\,\text{J}$  **8.** (a) $80\,\text{mV}$ (b) series resistor  **9.** $65\%$
**11.** $8\cdot4, 9\cdot3\,\text{V}$  **13.** series $95\Omega$

### EXERCISE 34 (p. 638)

**8.** (a) $91\%$ (b) $7\cdot9\text{p}$
**13.** (a) thicker (b) $160$ (c) eddy current losses (d) burn out (e) increase

# Electrostatics

# 35

## STATIC ELECTRICITY

### Origin of Electricity

About 300 B.C. the Greeks knew that particles of wheat were attracted to amber necklaces. They called this effect 'electricity' after the word 'electron', which meant amber. Subsequently, many substances were found to possess the ability to attract light particles once they had been rubbed, and the particles were then said to be *electrified* or *charged*, or to have a *charge*. Today you can easily see this power of attraction if you rub a plastic biro or pen holder, or a rubber balloon, vigorously on the sleeve of your jacket, and bring it near to small pieces of paper. The paper is immediately attracted towards the plastic pen or rubber balloon.

### Positive and Negative Charges

EXPERIMENT

Charge a small ebonite rod A by rubbing it with fur, and suspend it by thread in a wire stirrup (Fig. 35.1(i)). Charge a similar ebonite rod B by

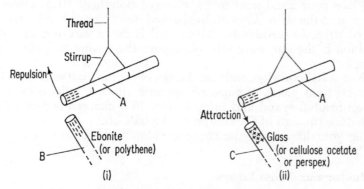

Fig. 35.1 Repulsion and attraction

rubbing it with fur and bring it near to the suspended rod A. Notice that A is repelled by B.

Now charge a glass rod C by rubbing it with silk and bring it near to the suspended ebonite rod A (Fig. 35.1(ii)). Notice that, this time, A is attracted by C.

*Conclusion.* The charge on the ebonite rod B is different from that on the glass rod C.

Benjamin Franklin called the charge on ebonite rubbed with fur *negative electricity*, and that on glass rubbed with silk *positive electricity*. All charges fall into one of these two classes. Observation shows that the kind of charge on a rubbed substance depends on the rubbing material. Glass rubbed with material other than silk may have a negative charge.

## Fundamental Rule

From the above experiment a general rule can be stated:

*Like (similar) charges repel, unlike (opposite) charges attract.*

Thus two positive charges repel each other and two negative charges repel each other. But a positive and a negative charge attract each other.

## Testing Charges

EXPERIMENT

1. Take a long polythene strip and rub it vigorously with a duster or cloth. Hang two rubbed strips beside each other. Observe the repulsion between them, showing they are charged. Bring an ebonite rod rubbed with fur near to the strips. They are repelled. Hence the polythene strip is *negatively* charged.

2. Now rub a rod of cellulose acetate vigorously with silk. Observe the attraction of pieces of paper or cotton thread, so that the rod is charged. Bring it close to a charged polythene strip and notice the attraction of the strip. The acetate rod has thus a *positive* charge.

3. Place your hand near to the charged polythene strip. Observe the attraction of the strip. Thus an uncharged object, your hand, attracts the charged strip. Consequently, 'attraction' is not a sure test of a charge. Repulsion is the only sure test. (Compare the testing of 'poles' in magnetism.)

4. Rub a balloon vigorously on the sleeve of your coat or frock. Observe it attracts light pieces of paper. Suspend a rubbed balloon by a thread from an insulating rod and bring it near to a charged rod of polythene and deduce the sign of the charge on the balloon.

Bring your hand near the charged balloon and observe the effect. Is your hand charged?

## Conductors and Insulators

In 1729 Gray found that a charged conductor lost its charge when suspended by a metal wire, but retained the charge when suspended by a dry silk thread. It follows that metal wire must conduct electricity, whereas dry silk does not.

Substances can be classed as *conductors* and *insulators* (non-conductors). Examples of conductors are all metals, the earth and the human body. Examples of insulators are plastic, bakelite, ebonite, paper, glass and sulphur. A metal rod held in a person's hand cannot be charged permanently by rubbing it, since the electrical charge will leak away along the metal through the body to earth. If, however, an ebonite rod is attached to the metal rod, the latter becomes negatively charged when rubbed with fur. (How would you show this?) The ebonite, an insulator, prevents the charge leaking away to the earth.

## Electrons and the Atom

When you have read some of the remaining chapters of this book you will realize that the atoms of all substances contain particles called *electrons*, which are negatively charged, and an equal number of particles called *protons*, which are positively charged. The total amount of negative electricity is equal to the total amount of positive electricity, so that a normal atom is electrically neutral. The protons are buried deep inside the central core or nucleus of the atom, whereas the electrons are outside the nucleus, and one or more may be at the surface of the atom.

When an ebonite rod is rubbed with fur some of the electrons on the surface of the atoms of the fur move to the ebonite. Since electrons carry negative electricity, the ebonite rod now has a negative charge. The fur, having lost negative electricity, now has a surplus of positive electricity, since all the atoms to begin with were electrically neutral, having equal numbers of protons and electrons. Consequently, the rubbing material has a charge as well as the substance being rubbed. Thus rubbing causes electricity to be *transferred* from one substance to another. *Rubbing does not create electricity.*

In a similar manner, we can explain the results of our other experiments. It is supposed that electrons are rubbed off the atoms of the cellulose acetate when rubbed with silk, so this becomes positively charged because it has a deficiency of electrons. Also, electrons are rubbed off the duster on to the polythene strip when rubbed, giving the latter a negative charge since it then has a surplus of electrons.

## Metal Conductors

Atomic theory can explain why a metal is usually a good conductor of electricity. After studying the sections of the book on atomic structure you will become familiar with the theory that atoms consist of electrons orbiting round a central nucleus which is positively charged. It is thought that the silver atom has one electron in its outermost part, which is weakly bound to the atom because it is shielded from the attraction of the nucleus by other electrons which have an opposite effect. The lone electron can thus wander from one atom to the next. Hence electricity (electrons) can move along the metal. Similar theory applies to other metals. Conversely, the electrons in the atoms of insulators are so firmly 'bound' to their particular atoms that they do not normally move.

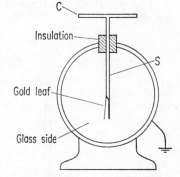

Fig. 35.2 Gold-leaf electroscope

## Gold-leaf Electroscope

An electroscope is a device which can detect electric charges. The simplest instrument is the *gold-leaf electroscope*. For specialized experiments a more sensitive instrument is required, for example, the *Wulf electroscope*, which is used in experiments on radioactivity (p. 728).

The gold-leaf electroscope consists of a metal cap C and stem S, with a gold leaf (or leaves) attached to the end of the stem (Fig. 35.2). The leaf is surrounded by a draught-proof metal case with two plain glass sides so that it is clearly visible, and is carefully insulated from the case by surrounding the stem by plastic material. The cap C is one terminal of the electroscope; the case, usually earthed, screens the leaf from outside influences and is the other terminal, and the leaf is the sensitive or moving part of the instrument.

## Charging Electroscope · Testing Conductors and Insulators

### EXPERIMENT

1. Stroke the cap C of an electroscope with a charged strip of polythene or a charged rod such as ebonite. Observe that the leaf opens and stays open. The metal cap and stem conduct some of the charge to the bottom, where the leaf is repelled from the metal as both carry similar charges. The electroscope is now negatively charged.

2. Touch the cap of the electroscope with your finger. The leaf now collapses. The charge on the leaf, stem and cap passes away through the body to the earth.

3. Test the conducting and insulating properties of a number of materials, such as rods or strips of copper, brass and other metals and wood, glass, paper, polythene, mica and perspex. Hold one end of the material and touch the cap of a charged electroscope with the other end. Observe how fast the leaf falls. Rapid collapse means that the charge escapes easily, and hence the material concerned is a good conductor. A slow collapse means that it is a poor conductor. No collapse means that the material is an insulator.

Examine the effect of some of the substances classed as insulators when they are slightly damp. Why, then, should rods such as ebonite and glass be warmed before charging them by rubbing?

## Testing Charges with Electroscope

The gold-leaf electroscope can be used to test the sign of the charge on an object.

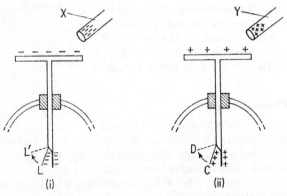

Fig. 35.3 Testing negative and positive charges

The charge on the leaf of the electroscope is first made negative by touching the cap with a rubbed polythene strip or an ebonite rod rubbed with fur, so that the leaf diverges. The unknown charge on a rod X, for example, is then brought near the cap (Fig. 35.3(i)). If the leaf diverges farther from L to L', the charge on X must be negative.

Observe that when a positively charged rod Y is brought near to the cap the divergence of the leaf is diminished. But this also occurs when an uncharged object such as the hand is brought near the cap. Consequently, the only sure way of testing a positive charge is to charge the electroscope positively, and then to bring the charge near to the cap, when the leaf diverges farther, from C to D (Fig. 35.3(ii)).

## Charging from Mains

EXPERIMENT

Join the live lead of the 240-V A.C. mains (care!) to the cap of a sensitive gold-leaf electroscope and the other (neutral) lead to the case. Switch on the voltage supply. The leaf diverges. Remove the live lead with an insulating rod (care!). Test the sign of the charge. Repeat the experiment several times. Observe that on some occasions the charge is positive and on others it is negative.

*Conclusions.* The electricity from the mains is the same kind of electricity as that on a charged rod. The charge on the leaf may be positive or negative, depending on whether, at the moment of disconnection, the voltage was on the positive or negative half of the cycle.

When the lead is disconnected from the cap the leaf may be observed on occasion to rise or fall slightly from its original deflection. This is due to the fact that at the moment of disconnection the A.C. voltage may be higher or lower than its average or steady value (p. 629).

## Electrostatic Induction

On p. 563 we saw that magnetism could be induced in soft iron without actually touching the metal. In the same way electric charges can be obtained on an object without touching it, and this process is called *electrostatic induction*.

EXPERIMENTS

1. Place two metal cans A and B on polythene insulating tiles as shown in Fig. 35.4(i)), so that the cans touch one another. Place a positively charged rod R near to one can, then separate the conductors by touching only the polythene tiles while keeping the rod in position (Fig. 35.4(ii), and finally remove the rod R (Fig. 35.4(iii)).

Using the polythene tiles, move each of the two cans in turn to a positively charged electroscope. In one case the leaf diverges less and in the other case the leaf diverges farther. The cans have thus become charged with opposite charges, without contact with the charged rod R originally used. Verify with a negatively charged electroscope. The charges produced are called *induced charges*; the charge on R is the *inducing charge*.

2. Next, repeat the charging of the cans by induction. Before bringing the cans near the charged electroscope, however, allow the cans to touch. Neither now affects the electroscope, showing that the induced charges were equal in magnitude and opposite in sign.

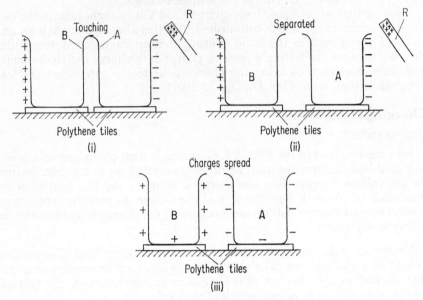

Fig. 35.4 Induced charges

## Explanation of Induction

*Stage* 1

When the positively charged rod R is placed near to the cans A and B while they are touching, the electrons (negative charges) in the metal cans are attracted. They therefore move from B to A, where they remain as close to R as possible (Fig. 35.4(i)). Since B has lost negative charge from one end, it is left with an equal positive charge, as shown.

*Stage* 2

The cans are now separated, keeping the rod R in position (Fig. 35.4(ii)). The positive charge on B and the negative charge on A are now separated. Observe carefully that if R were *not* kept in position at this stage the electrons would move back from A to B.

*Stage* 3

With R removed, the negative and positive charges now spread all over the respective metal cans (Fig. 35.4(iii)).

## Charging Electroscope by Induction

Fig. 35.5 illustrates the four stages showing how a gold-leaf electroscope can be charged by induction.

Suppose that a negatively charged rod R is brought near to the cap of an electroscope (Fig. 35.5(i)). Since unlike charges attract, electrons in the metal are repelled downwards, leaving a positive charge on the cap. The leaf thus diverges. When the cap is touched electrons pass through the body to earth and the leaf collapses; but the positive charge remains (Fig. 35.5(ii). When the finger is removed the positive charge remains at the same place (Fig. 35.5(iii), but the leaf remains uncharged. However, when the rod is removed, in effect some of the positive charge spreads downwards, and the leaf therefore diverges (Fig. 35.5(iv)). The electroscope has thus become charged without being touched by the rod R; i.e. it has acquired an induced charge. On bringing a positively charged glass rod

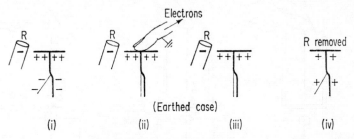

Fig. 35.5  Charging electroscope by induction

near to the cap, the divergence of the leaf increases, showing that the induced charge is positive, i.e. opposite to the negative inducing charge, the charge on R.

It should be noted that if the rod is taken away before the finger is removed in Fig. 35.5(ii), electrons pass through the body from the earth and neutralize the positive charge. In this case no charge appears on the leaf, and it does not diverge.

## Ice-pail Experiment

Faraday investigated the phenomenon of induction further by using a deep conductor in the form of an ice-pail to surround a charge completely. In the laboratory a deep small can, such as a tall metal calorimeter, can be used.

### EXPERIMENT

Take a deep metal can C and place it on the cap of a gold-leaf electroscope (Fig. 35.6(i)). Charge a small sphere A, suspended by insulating thread, with a positive charge for example.

(i) Lower the charge A in C, taking care not to touch the sides. When A is deep inside C observe the divergence of the leaf–this is appreciable. Move A about near the bottom of C and observe the divergence of the leaf –this remains constant. Now remove A. Observe that the leaf does not diverge now.

*Conclusion.* When A is lowered deeply into C an induced negative charge appears on the inside of C and an induced positive charge on the outside of C and on the cap and leaf of the electroscope. Since the divergence is unaltered when A is moved about, the induced positive charge is constant. Further, as the leaf closes when A is removed from C, the induced negative and positive charges are equal.

(ii) Repeat the experiment, that is, lower the charge A deeply into C and note the leaf divergence. This time, allow A to *touch* the bottom of the can (Fig. 35.6(ii)). Observe the divergence of the leaf at the instant of touching—there is no change in the divergence.

Remove A. Take it to an uncharged electroscope. Observe the leaf—there is no divergence, showing that A has now no charge (Fig. 35.6(iii)).

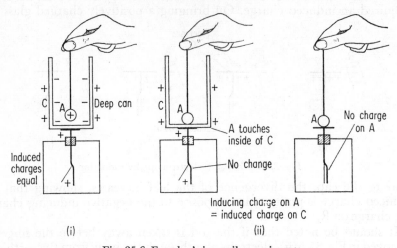

Fig. 35.6 Faraday's ice-pail experiment

*Conclusions.* 1. When A touches the inside of the can the negative induced charge on the inside exactly neutralizes the positive charge on A. Thus the induced charge is equal and opposite to the 'inducing charge', as the charge on A is called.

2. As A has no charge on it after touching the can, the induced positive charge on the can must exist entirely on its outside surface—otherwise A would have had some positive charge by contact with the can.

It should be carefully noted that a metal conductor, such as the can employed in the experiments, must *completely surround* a charged sphere before the conclusions stated are obtained. On this account: (*a*) a deep can must be used; (*b*) the charged sphere must touch the bottom of the can. Ideally, the conclusions apply to a hollow 'closed' conductor.

## Measuring Magnitude of Charge

The ice-pail experiment described leads us to further experiments.

EXPERIMENTS

1. Place a deep can D on a polythene insulating tile, and charge the can (Fig. 35.7(i)). Now touch the inside of D with a small metal sphere S held by an insulating thread. Remove S, and test it by taking it to an un-

charged electroscope. Observe that the leaf does not diverge, showing there is no charge inside D.

Touch the outside of D with S, take S to the uncharged electroscope and observe the divergence of the leaf (Fig. 35.7(ii)).

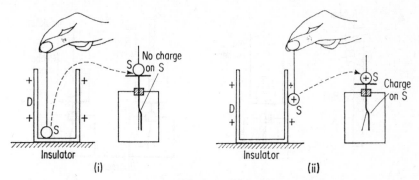

Fig. 35.7 Charge inside and outside hollow conductor

*Conclusion.* The charge on a conductor resides on its outside surface only, as previously found.

2. Charge a sphere M. Touch the cap C of a gold-leaf electroscope with M, and observe that the leaf diverges (Fig. 35.8(i)). Remove M, discharge the electroscope and bring M back to touch C (Fig. 35.8(ii)). Observe that the leaf diverges. Repeat the experiment.

*Conclusion.* The charge on the sphere M is *not* given up completely to the electroscope by touching the cap—some charge still remains on M.

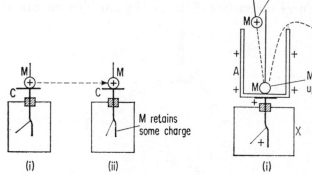

Fig. 35.8 Part charge to electroscope

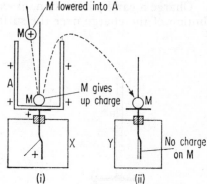

Fig. 35.9 Whole charge to electroscope

3. Place a deep can A on the cap of a gold-leaf electroscope X (Fig. 35.9(i)). Charge a sphere M, lower it inside A until it touches the bottom (Fig. 35.9(ii)). Remove M and test if it has any charge by taking it to an uncharged electroscope Y. The leaf does *not* open now, showing M has given up all its charge to the can A and the electroscope X.

*Conclusion.* The magnitude of the charge on a metal sphere can be compared with other charges by transferring them in turn to the inside of a deep metal can connected to an electroscope. *The divergence of the leaf is proportional to the magnitude of the charge in this case, as all the charge is given up.*

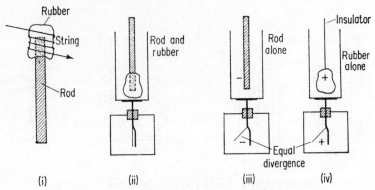

Fig. 35.10 Equal charges on rod and rubber

Fig. 35.10 shows how Faraday demonstrated by a deep can that the charges produced by friction on a rod and rubber were equal and opposite. An ebonite rod was fitted with a fur cap, and this was rotated by means of a silk thread or string (Fig. 35.10(i)). By means of a deep can and an electroscope he showed that the rod and rubber together then produced no deflection (Fig. 35.10(ii)), but that the rubber alone and the rod alone each produced exactly the same deflection (Fig. 35.10(iii), (iv)).

## Distribution of Charge on Conductor

Charge a can by induction, as explained on p. 648. Examine the distribution of the charge over the surface of the can by touching the can at

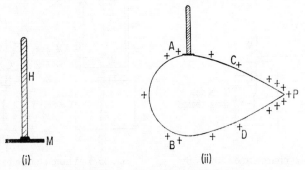

Fig. 35.11 Surface density of charge

different places with a small metal disc M with an insulating handle H, called a *proof-plane* (Fig. 35.11(i)). The size of the charge removed by the proof-plane depends on the surface-density of the charge at the place concerned. Using the results of Faraday's ice-pail experiment, a rough estimate of the charge removed can be obtained each time by lowering the proof-plane inside a deep can placed on top of an electroscope cap, and noting how much the leaf diverges. The divergence is a measure of the charge.

*Inside* the can the proof-plane collects no charge. Consequently, there is no charge inside the can. The charge on the can is therefore only on the *outside* surface (p. 650).

*Outside* the can the proof-plane collects a charge which is larger at the edges or lips of the can and smaller at other parts. The density of the charge on a part of a surface is therefore greater if that part has a greater curvature, that is, if it is more sharply pointed. Fig. 35.11(ii) shows the approximate variation of charge-density on the surface of a pear-shaped conductor.

## Action at Points

### EXPERIMENTS

1. A demonstration of the high density of charge at a pointed conductor is illustrated in Fig. 35.12(i). A metal 'windmill' W, with sharp points, is balanced on a metal pivot O on an insulating stand. A connecting wire

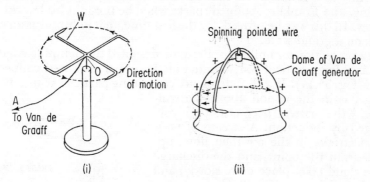

Fig. 35.12 Reaction of pointed conductor

OA is then joined to the terminal of a Wimshurst or van de Graaff high-voltage generator (p. 656). When the generator is started up the windmill becomes charged and spins about the pivot in a direction opposite to the points, as shown. If the experiment is carried out in the dark room a circular glow can be observed as the points rotate, showing the presence of a discharge in the air near the points. The spinning effect due to action at points can also be demonstrated by pivoting a thick metal with sharp points at each end on the top of the high-voltage terminal of a van de Graaff machine, as shown in Fig. 35.12(ii). The metal spins when the machine is working.

2. The discharge or electrical flow can also be shown by pressing down on the centre of the windmill with an ebonite or other insulating rod A so that the windmill does not move (Fig. 35.13), and then bringing a candle-

flame close to one point B. The flame veers to one side, showing a 'wind' or air movement away from the point.

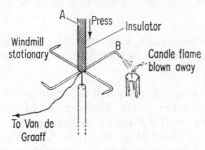

Fig. 35.13 Electric discharge

*Explanation.* Suppose the windmill is positively charged. The very high charge-density at a point attracts the outer electrons from the neutral molecules of air round it, leaving the molecules with a positive charge. These are repelled rapidly from the point because it has a similar charge. The air movement produces the 'wind' which blows the candle-flame. By Newton's law of action and re-action, the metal windmill turns in the opposite direction if free to move, as shown in Fig. 35.12.

### The Lightning Conductor

In a thunderstorm in 1749 Benjamin Franklin, at the risk of his life, flew a kite with a metal point at the top and a metal key at the lower end, using a silk (insulating) thread. For a short time there was no effect. Then suddenly, as a result of the rain, the silk thread became completely conducting, and Franklin obtained sparks when he brought his knuckles near the key. In this way he proved, for the first time, that lightning is an electric spark on a grand scale.

As a protection for buildings, Franklin designed the lightning conductor. This is a long vertical rod, pointed at the upper end, and connected to earth (Fig. 35.14). If a thundercloud, containing particles with a negative charge, is in the region above the conductor the latter becomes positively charged by induction. A stream of positive electricity in the air then flows upwards from the point, and the discharge of the cloud takes place more slowly, and with less violence, than if the conductor were absent.

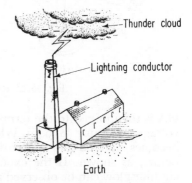

Fig. 35.14 Lightning conductor

### Electrostatic Generators

In the 19th century machines were invented to produce a large quantity of electricity by electrostatic methods. Lord Kelvin designed several 'generators' of static electricity, but their development came to a halt as a result of the greater efficiency of the dynamo, which provides a large and continuous flow of electricity.

### The Electrophorus

The first, and simplest, electrostatic generator was the 'electrophorus', or 'charge carrier', which consisted of an ebonite or perspex disc. A, and a metal plate B with an insulating handle H. A is rubbed and may become negatively charged (Fig. 35.15(i)). B is then placed on it and momentarily touched with the finger (Fig. 35.15(ii)). Finally, B is removed by means of H (Fig. 35.15(iii). On testing, B is found to possess a positive charge, which is opposite to that on A.

When B is placed on A, the two make very little contact as their surfaces are not perfectly plane, and hence electrons in B are repelled to the upper surface, leaving a positive charge on B near A. The electrons pass through the body to earth when the finger is placed on B (Fig. 35.15(ii)), and the positive charge spreads over B when it is lifted by the handle H (Fig. 35.15(iii). Thus B has acquired a charge by induction, leaving the negative charge on A unaltered. The charge on B can be transferred by contact to another conductor Y, and B can then be brought back to A and recharged

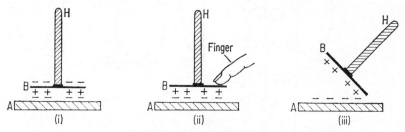

Fig. 35.15 Action of electrophorus

in the same way. In this manner a large charge can be built up on Y from a small charge. It should be noted that, in charging an electrophorus, the finger must be removed before B is taken away. The energy in a spark drawn from B is derived from the work done in moving it away from A against the force of attraction (Fig. 35.15(iii)).

## Model van de Graaff Generator

In 1930, van de Graaff, an American physicist, designed a new type of electrostatic generator, following an idea gained from an account of Kelvin's work on the subject.

A small van de Graaff generator, used in school laboratories, is illustrated in Fig. 35.16. It has a belt B of rubber moving round two rollers X and Y. X is made of perspex. Y has a metal surface on perspex and is driven by a motor. A set of metal points, represented by L, faces the roller Y. Another set, represented by U, faces the upper roller X and is connected to the inside of a large metal dome C at the top. The dome is the high-voltage terminal of the generator when it is in action.

The generator is self-exciting. When the motor is switched on the rubber belt at the top becomes negatively charged by friction with the moving roller, so that X now acquires a positive charge. When the belt descends and reaches the points L it induces an opposite charge on L, which then discharges on to the belt. The belt thus loses its charge. As it leaves the metal surface of the roller Y, the belt becomes charged again, but this time it has a positive charge, as shown. The belt now carries a positive charge up to the roller X. As it passes the points U it induces a negative charge on the points and an equal positive charge on the metal dome C. The points discharge the belt in the way already explained, leaving the charge on C, and when the belt leaves X the whole action is repeated.

The net result is that an increasing positive charge is collected by C. The amount of charge stored depends on the size of the dome, the efficiency

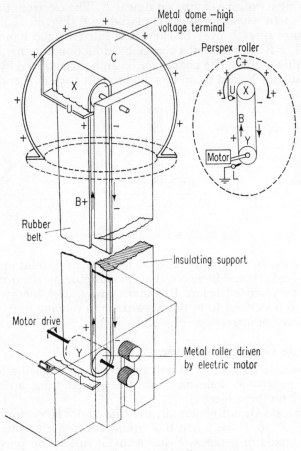

Fig. 35.16 van de Graaff generator

of the insulation and the air humidity. The apparatus works much better on a dry day or when warmed gently before use. High voltages of many tens of thousands of volts can be generated, and a violent spark can be obtained by discharging the dome by a neighbouring earthed metal sphere (see Plate 16). It may be noted that the energy of the spark comes from the energy of the motor, which raises the positive charge on the belt against the force of repulsion due to the positive charge on the dome C.

Large van de Graaff generators are used nowadays in the study of atomic structure and atomic energy. A high potential difference of several million volts can be obtained from it to accelerate electrical particles to high energies and then to disrupt atoms.

## Demonstrations with van de Graaff generator

Numerous demonstrations can be made with the van de Graaff machine. The following are additional to those already described:

EXPERIMENTS

1. Place an earthed sphere close to the high-voltage terminal and observe the spark which passes in the air as a result of the high voltage.

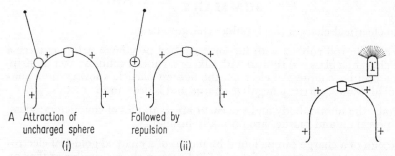

A  Attraction of          Followed by
   uncharged sphere       repulsion
        (i)                    (ii)

Fig. 35.17 Attraction and repulsion          Fig. 35.18 Repulsion of charged hair

2. Bring a light polystyrene sphere, at the end of a long insulating nylon thread, to the high-voltage terminal. Observe the attraction of the sphere (Fig. 35.17(i)) and then its subsequent repulsion (Fig. 35.17(ii)).

3. Attach a 'head of hair' to the high-voltage terminal. Observe the effect when the machine is switched on (Fig. 35.18). Explain the effect.

4. *Induction*. Stand on an insulating stool. Stretch your arms sideways

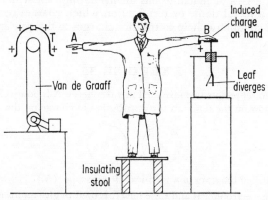

Fig. 35.19 Induction by human conductor

and place the fingers of one hand, A, near (but not too close to) the high-voltage terminals (Fig. 35.19). Touch the cap of a gold-leaf electroscope with the other hand B. Observe that the leaf diverges, showing that an induced charge is obtained at B. Test the sign of the charge on the electro-scope with the aid of a charged rod.

Now compare the sign of the charge on the high-voltage terminal with that on the charged rod. To do this bring the electroscope near to the terminal and notice if further or less divergence occurs when the charged rod is brought to the cap.

*Conclusion.* The charge on B has the same sign as the high-voltage terminal. Explain the result.

## SUMMARY

1. Like electrical charges repel, unlike charges attract.

2. An ebonite rod rubbed with fur, or a rubbed polythene strip, acquires a negative charge; a glass rod rubbed with silk, or a rubbed cellulose acetate strip, acquires a positive charge. An electron, the lightest particle known, and a con-stituent of all atoms, carries a negative charge and is 'free' inside metals.

3. Metals, the human body and the earth are examples of conductors; glass, ebonite, porcelain and plastics are normally insulators.

4. The sign of a charge can be found by means of a charged gold-leaf electro-scope; an increased divergence of the leaf indicates a charge similar to that on the electroscope.

5. An induced charge can be obtained on an insulated conductor by: (i) bringing a charge near; (ii) touching the conductor; (iii) removing the finger; (iv) taking the charge away.

6. A pointed charged conductor has a high density of charge at the point. This is shown by the rotation of a metal windmill. The lightning conductor has a pointed end.

7. The van de Graaff machine has an insulating belt which is driven con-tinuously, and a metal dome or terminal on which charges continue to be in-duced. Pointed conductors help to transfer charges to the belt and to the dome.

## EXERCISE 35

1. The diagram represents two insulated metal spheres. A is positively charged and B is uncharged. Show on the diagram the distribution of charge on the spheres. (*N.*)

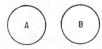

2. Draw a labelled diagram of a gold-leaf electroscope. (*N.*)

3. Describe an electrophorus and state how it is used to give an insulated conductor a positive charge. Give diagrams to illustrate the distribution and nature of the charge in each stage of the operation.

Why is it possible to obtain a series of approximately equal charges from this instru-ment without recharging it and what is the source of energy of these charges? (*L.*)

4. Describe experiments, one in each instance, to show that: (i) equal and opposite

charges are produced when ebonite is rubbed on fur; (ii) the inside surface of a charged hollow can has the same potential as the outside surface but has no charge; (iii) silk thread is a better insulator than cotton thread. (*L.*)

**5.** (*a*) Describe and explain what happens when a charged rod is brought near a pith ball suspended by a dry silk thread.

(*b*) Why has one end of a lightning conductor a sharp point? Describe a laboratory experiment to support your explanation. (*N.*)

**6.** Describe the structure and explain the action of an electrophorus.

(i) Describe the distribution of charge over a charged pear-shaped conductor supported on an insulated stand. (ii) Explain what would happen if a steel needle projects from the surface of a charged conductor. (*N.*)

**7.** (i) How would you show by experiment that equal and opposite electrostatic charges are produced by friction?

(ii) Describe and explain how, with a negatively charged ebonite rod and two insulated uncharged metal spheres, you could obtain a positive charge on one sphere and a negative charge on the other. (*N.*)

**8.** Describe, with the aid of diagrams, how a gold-leaf electroscope can be charged positively by induction and explain what happens at each stage of the process.

A tall hollow metal can stands on the cap of a gold-leaf electroscope. A charged insulated metal ball is made to touch the outside of the can. Why is the divergence of the leaves different from that which would be observed if the ball had been lowered into the can to touch the bottom?

Describe an experiment to show that the total charge induced by a conductor equals the inducing charge. (*L.*)

**9.** A hollow metal can stands on a slab of paraffin wax. Explain, with the aid of diagrams, how the can may be given a negative charge by induction.

State and explain how you would arrange for the induced charge on the can to be equal in magnitude to the inducing charge.

How would you show that charges produced by friction are equal and opposite? (*L.*)

**10.** How would you demonstrate: (*a*) that the charge on a charged hollow conductor resides entirely on its outer surface; (*b*) that a charged conductor gives up all its charge on making contact with the inside of a hollow metal can?

Describe the van de Graaff generator, and explain its mode of action. (*O.*)

**11.** Draw a diagram of a gold-leaf electroscope and label the important parts.

An electroscope is charged negatively. When a large positive charge is slowly brought near, the leaf first collapses and then rises again. Explain this.

If a charged ball is lowered into a deep can placed on the cap of an uncharged electroscope the leaves diverge until the ball is inside the can. Explain why this happens and why the divergence is unaltered when the ball touches the inside of the can. Why must the can be tall and narrow rather than shallow and wide-mouthed? (*O. and C.*)

**12.** Describe and explain what happens when a charged ebonite rod is brought near to a pith ball suspended by a silk thread. You should consider the possibility that the pith ball may touch the rod.

In what way, and why, does the behaviour of the pith ball differ from that of a small piece of soft iron similarly suspended when one end of a long bar magnet approaches? (*O. and C.*)

**13.** Explain what is meant by *electrostatic induction*. Why does this take place only on conductors?

If you were given two insulated metal spheres A and B, an ebonite rod and a piece of fur, how would you charge A positively and B negatively by induction?

How would you use a gold-leaf electroscope to show that A and B did indeed bear charges of these signs? (*O.*)

**14.** Describe a gold-leaf electroscope, illustrating your answer with a diagram. Explain how you would: (*a*) charge it positively, given a negatively charged insulator; (*b*) use it to demonstrate that there are two kinds of electric charge; (*c*) use it, with additions, to demonstrate that the charge on a hollow conductor lies entirely on the outside. (*O.* and *C.*)

**15.** (*a*) A rubber balloon filled with air can be made to stick to the ceiling after it has been rubbed. Explain this and explain why the trick works only if the balloon and the rubbing material are dry. Does it matter if the ceiling is dry?

(*b*) You are given two charged metal objects supported by insulating strings. How would you compare the size and settle the sign of the charges without destroying the charge? You may use a gold-leaf electroscope and simple additional apparatus. (*O.* and *C.*)

**16.** Draw a gold-leaf electroscope and label the essential parts. How would you use it with other apparatus: (*a*) to find the larger of two charges, one on a large brass ball and the other on a small brass ball; (*b*) to demonstrate that charges produced by friction are equal in magnitude and opposite in sign? (*O.* and *C.*)

# 36

# ELECTRIC FIELDS · CAPACITORS

## ELECTRIC FIELDS

### Meaning of Electric Field

A small positive charge on a light metal foil sphere S, suspended by a nylon thread, is repelled when placed close to the positively charged terminal X of a van de Graaff machine (Fig. 36.1). When it is placed at other points such as A farther away the force on the charge is less than before. The region round X is called an *electric field*; it is a region where a force is exerted on an electric charge placed in it. If the force on a given charge is high, the electric field there is said to have a high *intensity* or

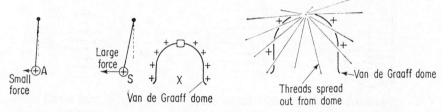

Fig. 36.1  Electric field round van de Graaff          Fig. 36.2  Repulsion effect

*strength.* Since the force on a charge diminishes when it is placed farther away from the high-voltage terminal, the electric intensity decreases.

If light nylon threads are stuck all round the high-voltage terminal or dome, the threads become electrified and stand out from the dome when the generator is started (Fig. 36.2). Like iron filings in a magnetic field, which take up positions showing the directions of the field, the threads show the direction of the electric field round the charged dome. When an earthed metal sphere is brought near the threads they curl towards the sphere and meet it at right angles. The electric field round the dome is thus altered by the presence of the metal sphere.

### Electric Fields

Demonstrations of electric fields may be made by supporting separated metal strips A and B in a Petri dish with the aid of a perspex sheet, gently pouring a light oil between them to form a shallow pool and then sprinkling semolina particles on to the oil (Fig. 36.3). When a van de Graaff high-voltage terminal is joined to A, and B is earthed, an electric field can be produced between A and B. The particles then show a pattern of the lines of electric force. This is similar to using iron filings to demonstrate magnetic lines of force.

Other fields can be demonstrated in a similar way. Fig. 36.4 shows the pattern of some of the electric lines of force in typical fields. Arrows are drawn on lines of force to show the direction of movement of a *positive* charge.

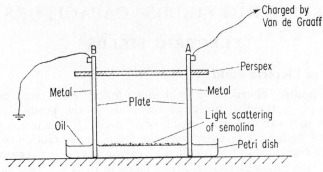

Fig. 36.3 Electric field demonstration

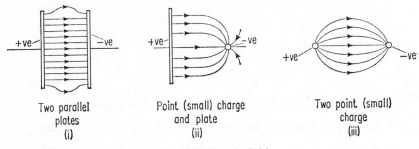

Two parallel plates (i)    Point (small) charge and plate (ii)    Two point (small) charge (iii)

Fig. 36.4 Electric fields

## Movement of Charges in Electric Fields

EXPERIMENTS

1. Connect a metal plate A to the high-voltage terminal of a van de Graaff machine (Fig. 36.5(i)). Place another plate B parallel and close to the plate A, and connect B to the earth side of the machine. By induction or contact, now charge a light metal foil ball C, suspended by a nylon thread, with a positive charge. Position the ball C between the plates, and then start the van de Graaff machine. Observe the movement of C from one plate to the other. Check that C moves from A to B if A is positively charged.

Repeat the experiment, charging C negatively. Observe the movement of C in the opposite direction.

*Conclusion.* A positive charge moves in an electric field from the positive to the negative terminals. A negative charge moves the opposite way.

2. Place two separated metal bars P, Q on an insulating surface such as a plastic lid, and connect one of them to the high-voltage terminal of a van de Graaff machine and the other to earth (Fig. 36.5(ii)). Alterna-

tively, connect the two terminals of a Wimshurst machine to P and Q respectively

Now sprinkle a number of small silvered balls, used in cake decorations, between P and Q.

When the generator is started up the balls make contact with P or Q and are charged by contact. They are then repelled and fly across to the other metal bar. Here they are charged again by contact and are then repelled

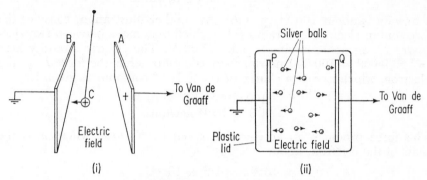

Fig. 36.5 Forces due to electric fields

once more. The charged balls move to and fro rapidly between P and Q, showing the presence of a strong field.

### Electric Intensity and P.d.

The force on a small light charged sphere, suspended between two parallel metal plates connected to a source of p.d. as in Fig. 36.5(i), is observed to increase when the p.d. $V$ between the plates is increased. If the p.d. is kept constant and the distance $d$ apart of the parallel plates is diminished the force is observed to increase once more. In this way it is found experimentally that the force on a given charge depends on the *potential gradient* between the plates, which is defined as the ratio $V/d$. Doubling the p.d. between the plates, and halving their distance apart at the same time, increases the force on the charge four times.

The *electric intensity E* in a field, such as that between the plates, may be defined as the *force per coulomb* there. If a charge equal to a coulomb is moved from one plate to the other, then, if $E$ remains constant,

$$\text{Work done} = \text{Force} \times \text{Distance} = E \times d.$$

But the work done in moving a coulomb between two points is defined as the potential difference between the points (p. 490). Thus

$$E \times d = V,$$

or

$$E = \frac{V}{d}.$$

The electric intensity is thus numerically equal to the potential gradient in the field. When $V$ is in volts and $d$ is in metres as in the SI system of units (p. 116), then $E$ is in volts per metre or newtons per coulomb.

As an illustration of what this means, suppose a charge of $\frac{1}{10000}$ coulomb is placed between two plates 2 cm apart having a p.d. of 240 V between them. Then, changing 2 cm to metres,

$$E = \frac{V}{d} = \frac{240}{0 \cdot 02} = 12\ 000 \text{ newtons per coulomb}$$

$$\therefore \text{ Force on } \tfrac{1}{10000} \text{ coulomb} = 12\ 000 \times \tfrac{1}{10000}$$
$$= 1 \cdot 2 \text{ newtons.}$$

1 newton is about 100 gf (p. 116). As another illustration, consider the force on an electron in a radio valve which may move between two electrodes 2 mm apart having a p.d. of 200 V. The electric intensity here $= 200/0 \cdot 002 = 100\ 000$ newtons per coulomb. Thus the force $F$ on the electron, which carries a charge of $1 \cdot 6 \times 10^{-19}$ coulomb, is given by

$$F = 100\ 000 \times 1 \cdot 6 \times 10^{-19}$$
$$= 1 \cdot 6 \times 10^{-14} \text{ newtons}$$

This force produces an acceleration given by $F = ma$, where $m$ is the mass of the electron, about $9 \times 10^{-31}$ kg.

$$\therefore 9 \times 10^{-31} a = 1 \cdot 6 \times 10^{-14}$$
$$\therefore a = 2 \times 10^{16} \text{ m per sec}^2 \text{ (approx)}$$

This enormous acceleration is due to the small mass of the electron, which is accelerated to very high speeds even though the actual force on it is tiny by everyday standards. Protons, which are electrical particles about 2000 times heavier than the electron (see p. 733), are accelerated to very high speeds in *proton accelerators* for nuclear energy experiments. Here potential differences of millions of volts are used.

## Electrical Screening

When a gold-leaf electroscope L is held near the high-voltage terminal D of a van de Graaff machine the leaf diverges (Fig. 36.6(i)). We can see

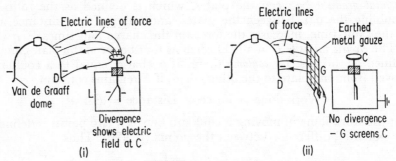

Fig. 36.6 Electrostatic screening

why this occurs if we suppose D has a negative charge. The metal cap C of the electroscope, which contains electric charges (electrons), is in the electric field of D. Hence the electrons in C are repelled. They are driven to the lower end of the stem and to the leaf, which therefore diverges.

When an earthed metal gauze G is interposed between D and the electroscope the leaf closes (Fig. 36.6(ii)). The electroscope is *screened* from the influence of D. The electric lines of force from D now terminate on G— no lines of force reach C.

In radio circuits the metal electrodes inside radio valves are screened from outside electrical influences by coating the glass with metal paint which is earthed, thus shielding the electrodes. Faraday was the first person to demonstrate screening by sitting inside an earthed box or cage with an electroscope. He recorded that violent discharges from a Wimshurst machine had no effect on the electroscope leaf. Today, engineers test high voltages with electrical instruments situated inside metal cages called 'Faraday cages'. The cages screen the instruments from the intense electric field due to the presence of the high-voltage cables outside, so that accurate readings are obtained.

### Electric Potential

In the chapter on energy and work in Mechanics we discussed the energy possessed by an object raised to a point above the earth's surface. This energy is called 'potential energy', and the amount $W$ is given by $W = wh$, where $w$ is the weight of the object and $h$ is the height above the earth (p. 119).

Since the effect of gravity is due to the earth, we speak of the earth's 'gravitational field'. Objects at different points in the gravitational field have different amounts of potential energy. Similar ideas occur in the case of the electric field. As we have seen, forces act on charges placed in these fields, and if they are taken to any point work will therefore be done. Electric charges thus possess potential energy in the field, and they are said to have an 'electric potential', or, briefly, a 'potential', in the field.

### Movement of Electricity · Earth Potential

An object in the earth's gravitational field falls if it is allowed to drop. Thus it moves naturally from one point to another where its potential

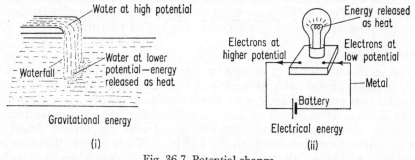

Fig. 36.7 Potential change

energy is less (Fig. 36.7(i)). In the case of an electric field an electron (negative electricity) moves from one point to another at a higher potential (Fig. 36.7(ii)). A positive charge would move from one point to another at a lower potential, the opposite way to an electron. This movement was demonstrated on p. 663 in a simple experiment. The idea of 'potential' in electricity is often considered analogous to 'temperature' in heat. Thus the temperatures of two bodies decide which way heat flows from one to the other when they are joined by a conductor. Similarly, the potentials of two points decide which way electricity will flow between them when they are joined by a metal–the electrons in the metal would flow from the point of lower to the point of higher potential. If there is no flow the potentials of the two points are the same.

If the actual potential at a point is required, then one must agree on a 'zero' of potential. In temperature measurement the 'zero' of the Celsius scale is chosen as the temperature of melting ice, which is a constant value. In potential measurement the electric potential of the earth is chosen as 'zero' potential. The earth, a conductor, is so large that its electric potential is practically unaffected by any gain or loss of electricity daily, and can be

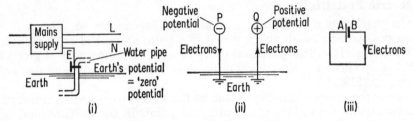

Fig. 36.8 Potential and potential difference

regarded as constant. A point in an electrical circuit joined to the earth thus has a 'zero' potential, as shown in Fig. 36.8(i) (see p. 535). An isolated negatively charged rod or sphere has a potential lower than that of the earth or a negative potential (Fig. 36.8(ii)). Points in the electric field due to the negative charge have likewise a negative potential. An isolated positively charged rod or sphere has a positive potential, and points in the electric field due to the positive charge have a positive potential. *Electrons* flow from a point of low potential to one of higher potential. In the case of a battery, (Fig. 36.8(iii)), electrons thus flow from the negative terminal B to the positive terminal A, which has the higher potential.

### Energy Gain

Just as masses can gain energy by moving between two points in the earth's gravitational field, so electric charges can gain energy by moving between two points in an electric field. One case occurs when electrons move in a television tube from one end to the other in a vacuum and gain sufficient energy to produce light on impact with the screen. Another occurs when electrons move from one end of an X-ray tube to the other in a vacuum and gain sufficient energy to produce X-rays when they strike a metal (p. 701).

To calculate the energy gained, we remind ourselves that 1 J of energy is gained when 1 coulomb of electricity moves between two points having a p.d. of 1 V (p. 490). Suppose one million ($10^6$) electrons, each carrying a charge of $1.6 \times 10^{-19}$ coulomb, moves from one point to another in a television tube where the p.d. is 5000 V. Then energy gained = quantity in coulombs × p.d. in volts

$$= 10^6 \times 1.6 \times 10^{-19} \times 5000$$
$$= 8 \times 10^{-10} \text{ J.}$$

### Gold-leaf Electroscope Measures P.d.

We can now return to a special feature of the gold-leaf electroscope which had to be omitted until the topics of potential and potential difference had been discussed.

Fig. 36.9(i) illustrates a negatively charged rod R placed near to the

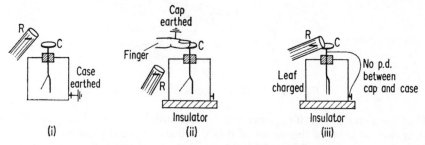

Fig. 36.9 Electroscope and potential difference

cap C of a gold-leaf electroscope whose case is earthed, causing the leaf to diverge (p. 646). C has a negative potential, since it is insulated and in the neighbourhood of R, and as the case is earthed, there is a p.d. between the cap and the case. If the case is insulated and the cap is earthed the leaf diverges when the charged rod is brought near to the case (Fig. 36.9(ii)). The case has now a negative potential and C is at earth (zero) potential, so that a p.d. exists between the cap and the case. However, if the case is insulated, and the cap and case are connected by a metal wire, the leaf does not diverge when the charged rod is brought near (Fig. 36.9(iii)). Even when the rod now touches the case or cap and gives up some of its charge to the electroscope, the leaf does not diverge. In this special case there is no p.d. between the cap and the case since they are connected.

It follows from these experiments that the leaf of an electroscope diverges only when there is a potential difference between the cap and case; further experiments show that the divergence is a measure of the p.d. between the cap and the case.

### Demonstration of Potential in Electric Field

EXPERIMENTS

1. As we have just seen, the divergence of the leaf of an electroscope is a measure of the potential of the cap when the case is earthed. If a gold-

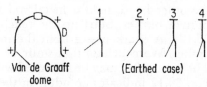

Fig. 36.10 Testing potential round charge

leaf electroscope case is held near to the high-voltage terminal D of a van de Graaff generator, the leaf diverges to a position such as 1 in Fig. 36.10.

The place where the cap is situated has thus a potential. When the electroscope is moved farther away from D the divergence of the leaf diminishes, as shown by 2, 3, 4 in Fig. 36.10. Thus the charge on the metal D produces a potential in the field round it which diminishes as we go farther away from D.

2. The existence of a potential can also be seen by using a small neon bulb N, as in a mains tester which has one terminal connected to the end of a thick pointed metal rod A (Fig. 36.11). When the bulb is held by the other terminal, and the point of A is placed near the high-voltage terminal

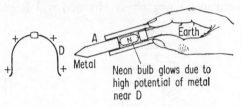

Fig. 36.11 Testing potential with neon bulb

D of the van de Graaff generator the neon bulb lights up. The potential of A is thus at least of the order of 100 V relative to earth potential.

3. To investigate the potential variation over the surface of a charged conductor, wind a wire round an insulating rod and connect one end to an electroscope cap. Holding the rod, touch the surface with the other end of the wire. The leaf opens. Move the end of the wire all over the surface and observe the divergence. This experiment shows that *the potential is the same* at all parts of the surface. This is the case even though the conductor has a variation of charge-density on its surface, as for a pear-shaped conductor (p. 652).

## Effects of Static Electricity

Current electricity, or electricity which is moving, produces a heating, chemical and magnetic effect (p. 483). Effects produced by using static electricity can be shown by the following experiments:

### Experiments

### 1. *Heating Effect*

Stand on a stool with insulating legs and touch the metal dome or high voltage terminal of a small van de Graaff generator with one hand. With the other hand, hold a metal rod with a sharp point near the top of a Bunsen burner. Start up the machine and turn on the gas slowly (Fig. 36.12). A spark passes from the metal point to the burner and the gas ignites. We conclude that static electricity can produce a heating effect. The same effect may be produced with an electrophorus (p. 654). Although Fig. 36.12 indicates an adult in the demonstration (*Care is essential*), it is safer to attach the metal rod to an insulating support and not to hold it.

Note that the heat produced by the spark is derived from the electrical energy obtained from the generator.

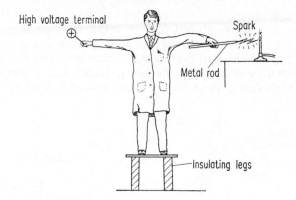

Fig. 36.12 Charging and discharging–heating effect

## 2. *Chemical Effect*

Take a piece of blue litmus paper and dip it into sodium sulphate solution. Place the paper across the metal dome and a neighbouring earthed metal sphere quite close to the van de Graaff machine and operate the machine for some time. Note that a colour change to red occurs at one terminal A, say, showing an acid reaction. Repeat the experiment using red litmus paper dipped into sodium sulphate solution. Note that a colour change to blue occurs at the other terminal B. We conclude that static electricity can produce a chemical effect.

## 3. *Magnetic Effect*

Wind a wire from one terminal of a sensitive galvanometer round one end of an insulated perspex rod, and earth the other terminal. Stand on a stool with insulating legs and touch the dome of a van de Graaff machine with one hand. Operate the machine slowly for a very short time so that you acquire a charge. Now grasp the wire on the perspex rod. A galvanometer deflection occurs. Since the galvanometer coil rotates between the poles of the magnet inside the instrument, the static electricity has produced a magnetic effect. (*Care is essential*–see Experiment 1.)

### Static and Current Electricity

From these and other demonstrations, we conclude that static electricity is the same kind of electricity as current electricity. It may be noted that in static electricity the potential relative to earth of a charged object is high, but the quantity of electricity obtained by discharging the object is very small. For example, when the knuckle is brought near to a charged ebonite rod a small spark passes to the knuckle from the rod (Fig. 36.13(i)). The spark indicates the breakdown of the insulation of the air (p. 654), and this shows that the potential or 'voltage' of the rod relative to the earth may be a few hundred volts. This is not a lethal effect, however, because only a very small quantity of electricity, or a very small current of the order of millionths of an ampere, passes through the body.

On the other hand, current electricity may be produced by cells with low potential differences or 'voltages' such as a few volts, but the amount of electricity or current produced may be relatively high (Fig. 36.13(ii)). On this account it is dangerous to touch the mains live terminals. The mains p.d. is 240 V, which at first may be considered not dangerous. But the current discharged through the body, which has a low resistance, may be very high and of the order of many amperes, leading to fatal injury.

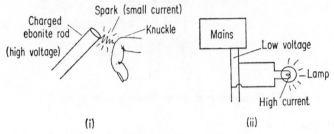

Fig. 36.13 Static and current electricity

## CAPACITORS

### Charging a Capacitor

A *capacitor*, also called a *condenser* in early days, is an apparatus for storing electric charge, that is, a quantity of electricity. It is an essential component in all television and radio receivers and transmitters, besides being used in the Post Office telephone and telegraph services. The simplest practical form of a capacitor consists of two parallel metal plates P and R separated by an insulating medium such as air, mica or paper. When a battery B is connected to the plates they each acquire a charge in a very short time, and in this condition the capacitor is said to be 'charged' (Fig. 36.14).

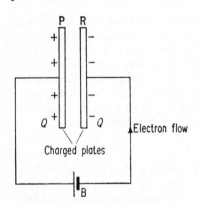

Fig. 36.14 Charging capacitor

The electrical action occurring is due to the fact that a battery is a device whose terminals are always maintained at a difference of potential; it thus acts like an 'electric pump', driving electrons round any metal circuit connected to it.

Electrons thus begin to move in the direction PBR when the battery B is connected, and hence R gains a negative charge and P an equal positive charge. Since like charges repel, the electrons already on R begin to oppose the movement of the electrons in the wires, and the electron flow quickly ceases. The two plates P and R then have equal but opposite charges of magnitude $Q$ units, for example, and the charge stored in the capacitor is said to be $Q$ units, the charge on either plate.

## Discharging a Capacitor

If a wire is connected between the two plates P and R of a charged capacitor a spark passes as the connection is made, and no charge remains on either plate after a very short time (Fig. 36.15). Suppose the plate R was originally negatively charged and that P had a positive charge (see Fig. 36.14). When the wire A is connected to the plates, electrons from R flow to P and completely neutralize the positive charge there; the charges on R and P, it will be remembered, were originally equal (p. 670). Thus R and P have no charges left on them, and the capacitor is now said to be discharged.

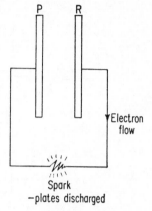

Fig. 36.15 Discharging capacitor

## Capacitance C

Suppose a capacitor such as the pair of plates P and R store a charge $Q$ of 2 units when a p.d. $V$ of 100 V is connected to them. Experiment then shows that the charge stored is 4 units when the p.d. is increased to 200 V and 6 units when the p.d. is increased to 300 V. We can see that more charge should be stored on the capacitor when the p.d. is increased, since more electrons are driven to the plate R when the battery p.d. is increased. It then follows that, for a given capacitor,

$$\frac{Q}{V} = C, \text{ a constant.}$$

where $Q$ is the charge stored when a p.d. $V$ is applied. The constant $C$ is called the *capacitance*. Using the same p.d., a large pair of plates stores more charge than a small pair of plates, other factors being equal. It has therefore a larger capacitance.

## Values of Capacitance

The practical unit of capacitance is the *microfarad* ($\mu$F). A capacitor has a capacitance of 1 $\mu$F if it stores a charge of 1 micro-coulomb (one-millionth coulomb) when a p.d. of 1 V is applied. In radio receivers capacitors of the order of 1 $\mu$F are considered relatively large. Very large capacitors, such as 64 $\mu$F or more, are needed in some parts of radio circuits. A large capacitor is needed in the induction coil, p. 624.

At the other end of the scale, the capacitor used for tuning to different stations in radio receivers is small. Its maximum capacitance is 0·0005 $\mu$F, or 2000 times smaller than 1 $\mu$F. Still smaller capacitors are used in very fine tuning to produce a change of pitch. Thus a small variation of 'treble' or 'bass' in a good-quality record-player is obtained by turning a small capacitor of the order of micro-microfarads. One micro-microfarad, one-millionth of a microfarad, is known as a *picofarad* (pF).

## Factors Affecting Capacitance

The capacitance of a capacitor depends on the area of its plates, their distance apart and the nature of the insulating medium (air, glass, paper

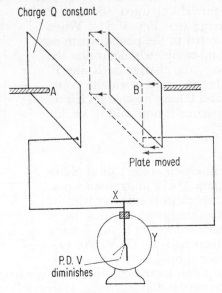

Fig. 36.16 Effect of distance apart

or mica, for example) between them. The factors can be investigated in a

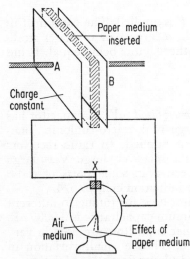

Fig. 36.17 Effect of dielectric constant

simple way by giving a charge $Q$ to the plates and then measuring the p.d. $V$ with the aid of a gold-leaf electroscope. On p. 667 it was shown that the amount of divergence of the electroscope leaf is a measure of the p.d. between the cap and case. The capacitance $C$ can then be deduced from the ratio $Q/V$.

### 1. *Distance Between Plates*

To investigate this factor, connect two large insulated plates A and B to the cap X and case Y of a gold-leaf electroscope (Fig. 36.16). Then, by using a rubbed polythene or ebonite rod, give A a charge $Q$. The leaf diverges. Now move the plate B nearer to A. The divergence of the leaf *diminishes*. Since $Q$ is unchanged and $V$ is diminished, it follows, from $C = Q/V$, that the capacitance $C$ is increased when the distance between the plates is decreased.

## 2. *Medium Between Plates*

As before, charge the plate A and observe the divergence of the leaf (Fig. 36.17). Now carefully place a thick book between the plates, thus introducing a paper medium in place of air. The divergence of the leaf *diminishes*. Observe if the same effect is produced when glass is used in place of paper. Since $V$ diminishes and $Q$ is unchanged, then, from $C = Q/V$, the capacitance is increased when paper or glass is used between the plates in place of air.

### *Dielectric Constant*

The material or medium between the plates of a capacitor is called the *dielectric*. When a dielectric such as waxed-paper completely fills the space between the plates, the capacitance is increased about 3 times compared with air as the dielectric. The *dielectric constant* of waxed-paper is said to be 3. Mica has a higher dielectric constant of 7.

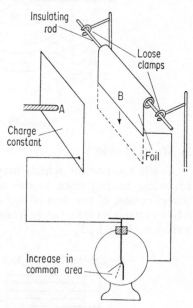

### 3. *Common Area of Plates*

Replace the plate B by a roll of tin-foil or aluminium foil (Fig. 36.18). With an appreciable area of foil facing the plate A, charge A and observe the divergence of the leaf. Now wind up the foil. The divergence of the leaf increases. Thus, from $C = Q/V$, the capacitance decreases when the common area between the capacitor plates decreases. When the foil is wound back so that the common area increases, the divergence of the leaf (and hence $V$) decreases. The capacitance $C$ thus increases when the common area between the plates increases.

36.18 Effect of common area

### *Summarizing*

The capacitance of a capacitor increases when the common area of its plates increases, when the distance apart of the plates decreases and when the insulating medium between the plates is a material such as paper or glass in place of air. These facts help the manufacturer to design suitable practical capacitors.

## **Practical Capacitors**

### 1. *Variable Air Capacitor*

This capacitor consists of a set of metal plates which can be rotated by a spindle so that they interleave with another set of fixed metal plates (Fig. 36.19). The two sets of plates are insulated from each other. The common area, and hence the capacitance, can thus be varied easily (Fig. 36.19(i), (ii)).

This type of capacitor is used in radio receivers for 'tuning' to different broadcasting stations. When a different station is required a knob on the radio set is rotated, and this turns the movable set of plates, thus altering the capacitance in the 'tuning circuit'.

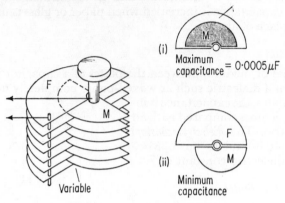

Fig. 36.19 Variable air capacitor

## 2. *Mica Capacitor*

Another capacitor which may be used in a radio receiver is a fixed capacitor having mica as the medium between the plates (Fig. 36.20). These consist of two sets of metal plates each connected to a terminal T. Mica capacitors may range from small values such as 0.0001 $\mu$F to higher values such as 1 $\mu$F.

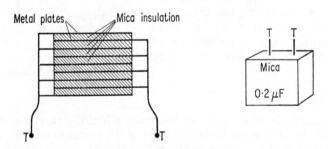

Fig. 36.20 Mica capacitor

## 3. *Paper Capacitor*

Paraffin-waxed paper is a much cheaper dielectric than mica, and has the further advantage of saving space, because, unlike mica, it can be rolled. The paper capacitor is made by placing two sheets of the paper between two sheets of tin-foil and then rolling the arrangement (Fig. 36.21). The terminals of the capacitor are connected to each tin-foil.

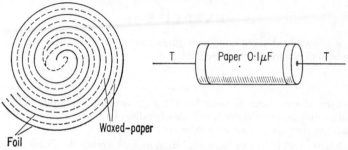

Fig. 36.21 Paper capacitor

### 4. Electrolytic Capacitor

In radio circuits, capacitors of very high capacitance such as $32\,\mu F$ may be required. Such a capacitor would have to be economical in size to be of practical use. The problem of its design is solved by passing a current between two aluminium plates immersed in ammonium borate solution (Fig. 36.22). A very thin film of aluminium oxide is formed on the anode plate by electrolytic action, and the oxide constitutes a very thin dielectric (insulating medium) between the anode and the solution when the action stops. Now it was pointed out on p. 672 that the capacitance is greater if the distance between the two plates is smaller. The very thin dielectric consequently makes the capacitance between the anode plate and the solution exceptionally high. A capacitor of several hundred

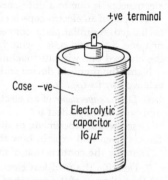

Fig. 36.22 Electrolytic capacitor

microfarads can be made by this method. The electrolytic capacitor, as it is called, has the anode plate as one terminal; the cathode plate can be used as the other, because the solution has a low resistance (see also p. 545).

### SUMMARY

1. Electric fields exist all round metals which carry electric charges. Objects can be 'screened' by earthed metal plates or cages.

2. Electrons move from one point to another point at a higher potential when they are joined by a wire. Earth potential is chosen as 'zero' potential.

3. The volt is the p.d. between two points if 1 J of energy is needed to move 1 coulomb of charge between them. The energy change $W$ when $Q$ coulombs move between two points at p.d. of $V$ volts is $QV$ joules.

4. The capacitance $C$ of a capacitor in *farads* $= Q/V$, where $Q$ is the charge in coulombs, and $V$ is the p.d. in volts. 1 $\mu F$ = one-millionth farad.

5. The capacitance of a capacitor increases: (i) the smaller the distance between the plates; (ii) the greater the common area of the plates; (iii) the higher the dielectric constant of the medium between the plates.

## EXERCISE 36

**1.** A is a positively charged conductor placed above a bench BC which is earthed. Draw the lines of electric force of the field.

Ⓐ

B _____ C

**2.** Draw the electric lines of force between: (i) two small positive charges; (ii) a small positive and negative charge; (iii) two parallel-plane plates with unlike charges; (iv) a small positive charge and an earthed plate near it.

**3.** A voltmeter requires to be 'screened' from outside electrical influences. How is this done? Explain, with diagrams, why your method results in screening.

**4.** How does the capacitance of a parallel-plate capacitor alter, if at all, when: (a) the distance between the plates is increased; (b) the area of one plate is reduced; (c) a sheet of glass is placed between the plates. (N.)

**5.** Describe and explain an experiment to show that the divergence of the leaves of an electroscope is due to a difference of potential between the cap and the case.

The cap of an electroscope is connected to a large insulated metal plate and the system is charged. A similar plate, connected to earth, is placed opposite to and parallel with the first one. State and *explain* what happens when: (a) the earthed plate is moved nearer to the charged one without touching it; (b) a sheet of glass is placed between the plates.

Name the practical device which is based on the above experiment and state what is achieved by its use. (L.)

**6.** Define *capacitance* of an electrical capacitor. On what factors does its value depend, and how do they affect it?

Describe an experiment to illustrate how the capacitance depends upon *one* of the factors given and explain how the experiment does so.

Describe the construction of one type of capacitor in common use. (L.)

**7.** Describe the gold-leaf electroscope. How would you use this instrument to show: (a) that all parts of the surface of an insulated charged conductor are at the same potential; (b) that the size of this potential is reduced by bringing up an earthed metal plate close to (but not touching) the conductor?

Describe a simple form of capacitor. What is meant by the *capacitance* of a capacitor? On what factors does the capacitance depend? (O.)

**8.** What do you understand by the term *potential* in electrostatics? How would you show experimentally that the potential of a charged conductor is the same all over its surface? Explain how the potential of such a conductor is altered when (a) an insulated metal plate, (b) an earthed metal plate is brought close to it.

Describe a simple form of variable capacitor. State what is meant by the *capacity* (or *capacitance*) of such a capacitor, and specify the factors on which its value depends. (O.)

# Atomic Physics

# 37

# ELECTRONS · ELECTRONICS · X-RAYS

## Conduction in Gases

At normal pressures air is a poor conductor of electricity. When two pointed electrodes are connected to a source of high voltage such as an induction coil, however, a spark passes between them in air when they are a short distance apart (see p. 624). This is called an electrical 'discharge'. A minute electrical current then flows in the air, and the spark shows the path of the current. Gases at low pressure hence conduct electricity.

When an electrician wishes to test whether a high voltage is present at some point in a circuit, he presses the metal end of a screwdriver fitted with a small neon lamp against the point. If the voltage is above about 150 volts, the neon gas glows. When the metal end of the screwdriver is placed near the live terminal or dome of a van de Graaff generator, the gas glows, owing to the presence of a high voltage or potential round the dome (see p. 668). A similar result is obtained by placing the neon lamp between two large metal plates and connecting a high-voltage supply to the plates.

## Discharge Tube

Gases, then, conduct electricity when the voltage or potential difference across them is large enough. In order to study conduction in gases, the gas

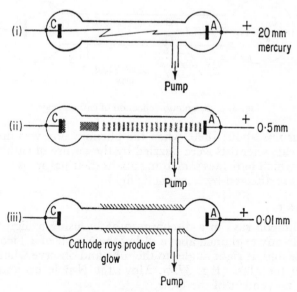

Fig. 37.1 Discharge tube effects

679

concerned is contained in a glass tube called a *discharge tube*, with metal electrodes C and A at either end (Fig. 37.1). A side tube enables the gas to be pumped out while the voltage is applied, as a smaller amount of gas will be a better conductor. (*Warning*. SAFETY regulations must be followed when using the discharge tube.)

EXPERIMENT

Use air as the gas and set up the discharge tube in the dark-room. Connect the induction-coil terminals across C and A, and then switch on the voltage. Pump the air slowly out of the tube and observe what happens.

At first nothing is seen. As most of the air is pumped out, long blue streamers or sparks flash between A and C–the air pressure is now reduced to about 20 mm mercury and the air conducts (Fig. 37.1(i)). When the pressure is lowered further, the discharge changes to pink and then widens and forms a long column. The column next breaks up into 'discs' and moves back from C to A, leaving a glow round the cathode and a dark space between the cathode and the shortening column (Fig. 37.1(ii)). As the pumping proceeds the dark column increases in length, the cathode glow becomes fainter and at a very low pressure of about 0·01 mm mercury, when very little air is left in the tube, the dark column fills the tube. The glass is now seen to be glowing with a green light (Fig. 37.1(iii)).

## Cathode Rays · The Electron

The green glow is due to fluorescence of the glass. It is due to invisible rays coming from the cathode which strike the glass. They were therefore

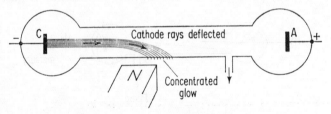

Fig. 37.2 Magnetic deflection of cathode rays

called *cathode rays* about 1890, when the phenomenon was first observed. For many years scientists were puzzled by the nature of cathode rays. An experiment which indicates something about their nature is to see whether or not they are affected by a magnetic field.

EXPERIMENT

Obtain cathode rays in a discharge tube by using a high voltage and pump, as already explained above. Bring the N-pole of a Ticonal (strong) magnet close and at right angles to the tube and observe what happens to the glow on the glass (Fig. 37.2). Move the N-pole up and down and observe the movement of the glow.

Reverse the magnet so that the S-pole is near the tube and repeat.

*Conclusions*. 1. The glow moves up and down as the magnet moves. The cathode rays are therefore easily affected by a magnetic field. They are hence not electromagnetic waves and are probably extremely light particles.

2. From the deflection of the glow in Fig. 37.2 it follows that the particles act like an electric current—we have seen on p. 599 that a metal conductor carrying an electric current is deflected when it is perpendicular to a magnetic field. Applying Fleming's left-hand rule to the deflection of the glow, we find the current flows from A to C; but the current direction in Fleming's left-hand rule represents the movement of positive electricity, which is equivalent to *negative* electricity moving from C to A. The particles which create the glow hence carry negative electricity, and they were called *electrons* (see also *Perrin tube*, p. 684).

## Thomson's Investigation · Electrons and Ions

In 1896 Sir J. J. Thomson carried out an experiment to compare the mass of the electron with the mass of the hydrogen atom. Until this time the hydrogen atom was believed to be the lightest particle in existence. Thomson found that the electron had a mass far less than that of the hydrogen atom, and experiments show that:

*Mass of electron $= \frac{1}{2000}$ of mass of hydrogen atom (approximately).*

When an electron is torn off a neutral atom of air the particle left is practically as heavy as before, because the electron is extremely light. It now carries a positive charge equal numerically to that in the electron, and it is called a positive *ion*. If an electron attaches itself to a neutral atom of air the latter becomes a negative ion. Positive and negative ions in liquids have already been met in electrolysis (p. 541).

In the discharge tube a powerful electric field is set up between the metal electrodes C and A by the high voltage applied (p. 680).

There are always a few electrons and positive ions in a gas, and as the air is pumped out the ions are accelerated to higher speeds and bombard the metal cathode violently. Some electrons are then obtained from the metal atoms. The electrons are accelerated by the electric field and produce more ions and electrons by collison with the gas molecules, and so the process is repeated. Many electrons are thus obtained in the discharge tube. This is described as the production of electrons using a *cold cathode*.

## Hot Cathodes

In 1902 Richardson found that electrons are readily obtained from hot metals. Metals contain many free electrons (p. 486), and if a fine tungsten wire, for example, is heated to a high temperature by connecting a battery of a few volts, the extra energy given to the electrons enables them to break through the surface of the metal and exist outside as an electron 'cloud'. This is called *thermionic emission*. The effect is analogous to vaporization near the boiling point, except that molecules break through a liquid surface in this case and form a vapour outside.

Later it was found that the oxides of barium and strontium emit electrons at much lower temperatures than tungsten wire. A tungsten wire glows white hot when it emits electrons. In contrast, the oxides of barium and strontium appear dull when they emit electrons, as the temperature required for emission is much lower.

The emitter of electrons is generally called the *cathode*. In early radio valves (p. 689) the cathode was a hot fine tungsten wire (Fig. 37.3(i)). This is a *directly heated* valve. Modern valves, however, also contain a tungsten wire, but it is used only as a source of heat. It warms a cylinder near it whose surface is coated with a mixture of barium and strontium

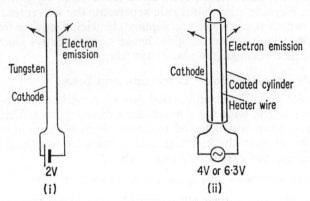

Fig. 37.3 Directly and indirectly heated cathodes

oxides, which readily emits electrons (Fig. 37.3(ii)). The cylinder is therefore the cathode in this case and the wire is the heater. This is consequently an *indirectly heated* cathode. The voltage required for the heater is low, such as 6·3 V or less, and it can be obtained from a D.C. supply such as accumulators, or an A.C. supply. In the latter case it is conveniently obtained from a step-down mains transformer (p. 631).

### Fine Beam and Evacuated Tubes

Since electrons are easily produced by using a hot cathode, experiments on electrons can be very conveniently carried out using this source.

In a *fine beam tube* a conical metal anode A with a hole at the top is placed over the cathode C, and maintained at a high positive potential of a few thousand volts relative to the cathode (Fig 37.4). The tube containing the cathode and anode has a very small amount of hydrogen inside, in which case the fast-moving electrons which pass through the hole in the anode produce a fine beam of light as they ionize the gas molecules (compare p. 679). The beam thus shows the path of the electrons. A cylinder W round the cathode, called a Wehnelt cylinder, is kept at a few volts positive relative to the cathode and helps to focus the electrons. Any movement of electrons inside the tube is clearly shown by the movement of the fine beam.

When a tube with a hot cathode is *evacuated* no gas molecules impede the moving electrons. As shown on p. 684, the anode is now made as a cylinder, and after passing through, the high-speed electrons produce a bright spot of light on fluorescent material painted on the glass. Any movement of the electrons is shown by a movement of the spot of light on the fluorescent screen.

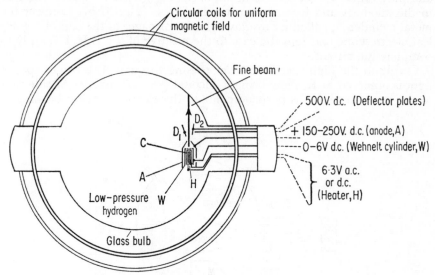

Fig. 37.4 Fine beam tube

### Energy and Velocity of Electron Beams

The glow of light in the fine beam and evacuated tubes is derived from the energy of the electrons passing through the anode. As we saw on p. 525, a quantity of electricity $Q$ coulombs gains energy in moving through a p.d. of $V$ volts equal to $QV$ joules. In this case electrons starting from the cathode are attracted to the anode where the potential is higher by 400 volts, say. Suppose a total mass of electrons equal to $0.1$ g and carrying a total charge of $18 \times 10^5$ coulomb moves to the anode. Then

$$\text{Energy gained} = QV = 18 \times 10^5 \times 400 = 72 \times 10^7 \text{ J.}$$

The fast-moving beam at the anode has kinetic energy given by $\frac{1}{2}mv^2$ joules, if $m$ is in kilogramme and the velocity $v$ is in metre per second (p. 122). Thus the velocity is given, if the beam is unimpeded as in a vacuum, by

$$\frac{1}{2} \times \frac{0.1}{1000} \times v^2 = 72 \times 10^7$$

$$\therefore v = \sqrt{\frac{2 \times 72 \times 10^7 \times 1000}{0.1}}$$

$$= 4 \times 10^6 \text{ m per s (approx).}$$

### Charge on Electron · Perrin Tube

The charge carried by electrons can be investigated by using the evacuated TELTRON Perrin tube, following an apparatus originally due to Perrin, a French scientist, about 1895 (Fig. 37.5).

EXPERIMENT

Connect the low-voltage supply, such as 6 or 6·3 V, to the cathode F and the +3000 V (+3 kV) terminal of the high-tension (E.H.T.) supply to the anode A and its negative terminal to F (Fig. 37.5). Connect the metal cylinder P, called a Faraday cylinder or cage, to the cap of a gold-leaf electroscope and join the case to the positive of the E.H.T. supply to complete the circuit.

Owing to the high-speed electrons passing through A, a bright spot of light is observed at R. Now bring a Ticonal (strong) magnet M sideways to the tube, so that it is perpendicular to the electron stream beyond A.

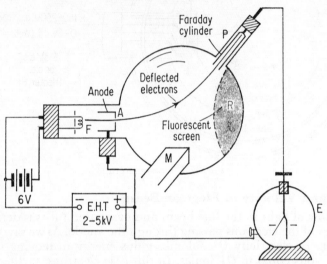

Fig. 37.5 Perrin tube – sign of charge on electrons

The spot is then deflected. If necessary, reverse M so that the spot is deflected towards the metal cylinder P. As M is brought nearer to the tube the spot moves towards P, that is, the electron beam is deflected towards P, and at one stage the beam enters *completely inside* P. The electroscope leaf then diverges, showing that electrons carry a charge. Disconnect the wire to the electroscope with an insulating rod, to prevent the electroscope discharging through the glass of the apparatus or through the power supply.

Switch off the voltage supplies. Bring a rubbed polythene or other suitable rod which has a negative charge near the electroscope cap. The divergence of the leaf increases.

*Conclusion.* Electrons carry negative charges.

You should also check the sign of the charge from the direction of deflection of the electron beam by using Fleming's left-hand rule, as explained on p. 681.

*Note.* 1. Secondary electrons are emitted when the cathode electrons strike the metal cylinder P. If the latter strike the outside of P the secondary electrons may leave P, and the electroscope then has a positive charge.

This is overcome by making the deflected cathode electrons pass straight down the middle of P, in which case the secondary electrons are captured by P.

2. If the charge on the electroscope leaks away repeatedly, try earthing the positive terminal of the high-tension (+2 kV) supply, *insulating the battery* connected to the cathode, and joining the electroscope case to earth. Take care not to touch the battery in this case.

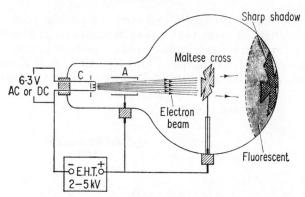

Fig. 37.6 Maltese cross shadow

Other properties of electrons can be shown by using hot cathodes in TELTRON tubes:

1. They travel in straight lines—a sharp image of a metal obstacle such as a Maltese cross is obtained on the glass of the tube (Fig. 37.6).
2. They have momentum and energy—light metal foil vanes turn round when the rays are incident on them and the metal glows, showing heat is produced.
3. They produce fluorescence in certain materials.

## Effect of Magnetic and Electric Fields

The LEYBOLD fine beam tube enables the movement of an electron beam to be studied when magnetic and electric fields are applied. A uniform *magnetic field* is produced by passing a current through two large parallel circular coils on either side of the tube. The coils act respectively like a north and a south magnetic pole which face each other (p. 584), so that the whole of the beam is situated in the field between the coils. The magnetic field strength can be altered if required by varying the strength of the current.

### EXPERIMENT

Connect the required voltages to the tube, for example, 6·3 V D.C. or A.C. to the heater; a few volts positive with respect to the cathode to the Wehnelt cylinder; about 250 V positive with respect to the cathode to the anode; and a 6-V accumulator with a rheostat and switch in series to the two coils (see Fig. 37.4).

Switch on the cathode heater first, then the Wehnelt cylinder and the anode voltage. A fine vertical luminous beam should be seen in the tube. Switch on the current in the coils, and examine the effect of increasing it from 0 to 2 A. Observe the *circular path* of the electrons and the change in radius as the magnetic field varies in strength (Fig. 37.7). Turn the tube round so that the magnetic field is not perpendicular to the beam and observe now the *spiral path* of the luminous beam. (See also p. 600.)

The beam (a gaseous current) is deflected in a circular path because the force per unit length at every part is perpendicular to its direction, owing to the uniform magnetic field applied, and is constant in magnitude. The force is an example of a centripetal force (p. 109); it is always at right angles to the beam and acts towards the centre of a circle of which the path

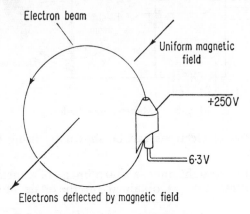

Fig. 37.7 Magnetic deflection

of the beam is an arc. If the field is not uniform and perpendicular everywhere along the beam, the path is not circular.

The spiralling motion of electrons and ions due to the earth's magnetic field in outer regions of space produce natural phenomena. The glow in the sky at the north pole of the earth, called the Northern Lights or *aurora borealis*, is attributed to the 'bunching' or focusing of electrons when they spiral on entering the earth's field. The concentrated electron beam produces a glow on impact with atoms of air. *Van Allen radiation belts* are layers of spiralling ions and electrons which are trapped high in the atmosphere by forces due to the earth's magnetic field. Their existence was discovered by Van Allen after analysing the measurements from instruments in satellites.

### Electric Field

The fine beam tube also enables the effect on electrons of an *electric field* to be observed. For this purpose two plates are used. The beam passes centrally between the plates P and Q after emerging from the anode (Fig. 37.8).

Experiment

In this experiment the two circular coils are no longer required. Connect a battery B of a few hundred volts between P and Q, and join the anode to the centre of the battery supply.

Observe the effect of making P positive with respect to the anode, and Q negative. The beam is deflected towards P, as shown. Reverse the outside battery terminals, so that P is negative and Q is positive. The beam is deflected towards Q. An alternating p.d. can also be applied between P and Q with a suitable transformer, in which case the electrons fan out and trace a luminous line on the glass.

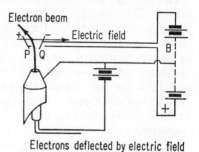

Electron beam

Electric field

Electrons deflected by electric field

Fig. 37.8 Electric deflection

## Cathode-ray Tube and Oscillograph

The discovery of cathode rays or electrons led to a vast field of practical applications grouped under the heading *Electronics*. Electrons are so light that they are highly sensitive to electrical or magnetic variations, as previous experiments show.

The glow produced by fast-moving electrons on a fluorescent screen led to the important invention of the *cathode-ray tube*. This is used for television and in radar, and in the form of the *cathode-ray oscillograph* (*C.R.O.*) it is used for measurements and demonstrations connected with radio, television and telecommunications.

Basically, the C.R.O has: (1) a cathode C which emits electrons; (2) a plate G which is kept at a negative potential, so that the number of electrons per second reaching the screen, or *brightness* of the picture, can be

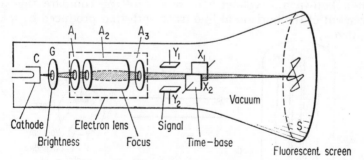

Fig. 37.9 Cathode-ray tube

controlled; (3) an *electron lens*, consisting of metal plates or cylinders, $A_1$, $A_2$ and $A_3$, all kept at a high potential relative to C of the order of a thousand volts (Fig. 37.9). The potential of $A_2$ varies the focal length of the lens so that the electron beam can be *focused* on to the screen S, and the high potentials of the metals, which are anodes, produce a fast-moving electron beam whose energy can be converted into light energy of fluorescence.

The arrangement of cathode and anodes is sometimes called an *electron gun*.

In radio or television, varying or alternating voltages are usually obtained. They are always connected for examination to two metal plates, $Y_1$, $Y_2$, called *Y-plates*, through which the electron beam passes before striking the fluorescent screen. The beam moves up and down in response to the changing voltage, thus tracing a line on the screen. The waveform or 'shape' of the voltage can be seen when a *time-base* circuit is switched on. This special circuit provides a voltage connected to metal plates $X_1$, $X_2$ called *X-plates*, through which the beam also passes, so that the beam is deflected horizontally. Thus as the beam moves up and down in response to the voltage in the Y-plates it is also swept horizontally, and a waveform is produced with a horizontal *time-axis*. In this way, for example, a faulty waveform can be traced in a radio receiver, and a radar operator can check from its waveform that a radio signal reflected back from an aeroplane is the same as that sent out.

## Using the C.R.O.

Make sure the time-base is connected to the X-plates, and *switch on* the instrument. Allow it to warm up, when a horizontal line appears on the screen. Adjust the *brightness* to a minimum, and then focus the line. Now connect the signal or other voltage to be examined to the Y-plates, when a rapidly moving waveform is obtained. When the frequency of the voltage on the Y-plates is the same as the frequency of the time-base voltage, or a whole number of times the frequency, the waveform becomes stationary on the screen. The *time-base frequency* is therefore varied by switches until a near-stationary picture is obtained, and finally the *synchronization* control is turned slightly, which makes the picture stationary.

To measure (i) an *a.c. voltage*, compare the height of its waveform with that of a known a.c. voltage, (ii) a *d.c. voltage*, compare the vertical displacement of the horizontal line trace with that produced by a known d.c. voltage.

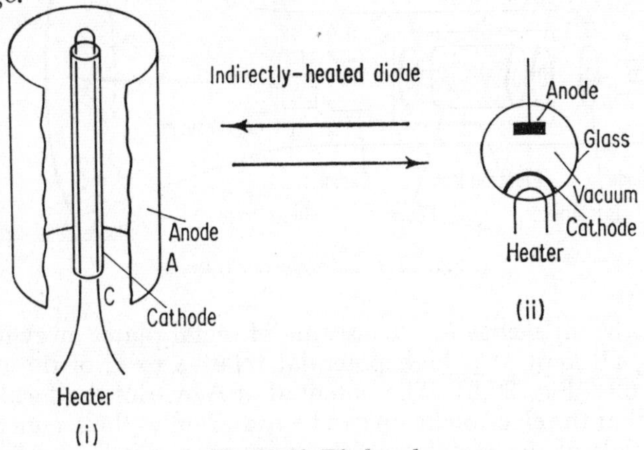

Fig. 37.10 Diode valve

**PLATE 17**
Semiconductors – Transistors –
Microelectronics

7(a) Manufacture of transistors and junction
iodes begins by reducing germanium oxide
owder to germanium and then fusing it into
bar as shown.

17(b) Purifying germanium to 1 part in $10^8$.
High-frequency currents in the coils melt the
impure germanium inside the tubes by induction, and the impurities move to one end as the
molten germanium is drawn along the tubes.

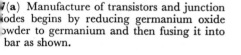

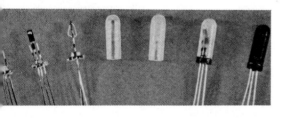

17(c) Stages in completing transistors.
*from left to right*
(i) Three leads for emitter, base, collector
(ii) Transistor attached to base (central) lead
iii) Emitter and collector leads attached
iv) Glass capsule
(v) Capsule filled with protective grease
vi) Transistor housed in sealed capsule
ii) Capsule coated with black lacquer to exclude light,
    to which transistor is sensitive

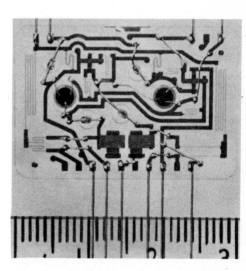

17(d) Microelectronics. A thin film circuit,
used in binary counters. It has an overall width
of 3 cm and consists of two transistors (centre),
capacitors (such as lower centre), and resistors
(such as thin lines on extreme left and right).

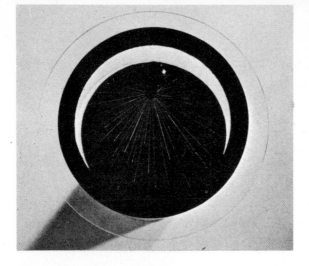

**PLATE 18**
**Nuclear Investigations**

**18(a)** Alpha-particle streaks in a cloud chamber, showing long tracks (high energy particles) and short streaks (smaller energy particles) (p. 726).

**18(b)** ▶
**18(b)** Synchro-cyclotron atomsmashing machine at Berkeley, California, U.S.A. Protons are released in the middle of the central box seen and then whirled round with ever-increasing speed by the giant magnet whose poles are above and below the box. After reaching a very high energy the protons are made to bombard an element and the products of the nuclear explosion are photographed (p. 739).

◀ **18(c)** Nuclear collision by α-particle, which transmutes a nitrogen to an oxygen atom. The streaks show α-particles moving through nitrogen gas. One α-particle (*right, centre*) has collided and merged with a nitrogen nucleus, and this has broken into an oxygen nucleus (*short track*) and a proton (*long thin track crossing others*). This historic photograph was taken by P. M. S. Blackett, Esq., in 1925 using a Wilson cloud chamber (p. 735).

## Diode Valve

The first radio valve was invented by Sir J. A. Fleming in 1902. By that time Marconi, the gifted Italian inventor, had made radio communication between England and America, and the need arose for a reliable device which could detect radio waves. This is discussed later (p. 692).

Fleming's valve contained: (1) an emitter of electrons, a filament or cathode C; (2) a nickel plate or *anode*, A, which is a collector of electrons (Fig. 37.10(i), (ii)). It is called a *diode* (two-electrode) valve. In order that the electrons emitted from C should not be impeded from reaching A, the electrodes are contained inside a glass envelope from which all the air is pumped out.

## Diode Characteristic

The characteristic action of a diode can be investigated with the TELTRON type diode and the circuit shown in Fig. 37.11(i). The per-

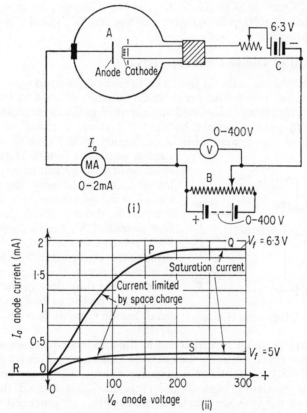

Fig. 37.11 TELTRON diode valve and characteristics

formance of the diode is assisted by a circular disc parallel to the anode A, attached to one side of the filament or cathode. A variable high-tension source B, 0–400 V, is joined with its positive pole or side to A and its negative pole to C, with a 0–2-mA meter (MA) in series. By suitable accumulators or other sources, connect the required heater voltage, 6·3 V, including a small rheostat to alter the heater current if required.

Start with a heater voltage of 6·3 V. Increase the anode voltage $V_a$ in suitable steps from zero to about 300 V or more. Each time observe the anode current $I_a$ on the meter. Observe that the current rises and then remains constant (Fig. 37.11(ii)). Adjust the heater rheostat so that the heater current is diminished slightly, and repeat the experiment.

Now *reverse* the battery terminals of B, so that the negative terminal is joined to A and the positive to C, the cathode. Increase the anode voltage from B, but this time observe that no current flows.

*Graph.* Plot the graphs of $I_a$ v. $V_a$ for each of the three cases. Fig. 37.11(ii) shows a typical result. The graph is called the *characteristic* of the diode, and it consists of QOR or SOR. Note that the part OR, which shows no current when the anode voltage is negative relative to the cathode, is part of the diode characteristic.

## Explanation of Diode Characteristic

In 1902 Richardson found that the number of electrons emitted per unit area per second by a particular cathode depends only on its temperature. When the temperature increased the number emitted increased.

When electrons are emitted by the hot cathode a 'cloud' of electrons is formed round its surface. This is analogous to the case of a saturated

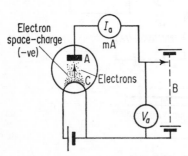

Fig. 37.12 Diode valve circuit

vapour outside a liquid. Here the molecules return to the liquid as fast as they leave. In the same way, the number of electrons per sec returning to the cathode is equal to those emitted. A small positive potential $V_a$ on the anode A, say 20 V relative to the cathode C, attracts some of the electrons in the cloud, which then move across the empty space to the anode. Here they flow in the anode circuit round to the cathode, and hence a small anode current $I_a$ is obtained (Fig. 37.12).

The electrons which swarm across to the anode always create a negative charge in the space between the cathode and anode. This is called a *space charge* (Fig. 37.12). As it has a negative charge, the space charge repels some of the electrons back to the cathode. Thus not all the electrons emitted from the cathode reach the anode when its potential is small. As the anode potential is increased, the attractive effect on electrons emitted from the cathode increases. The repulsion effect of the space charge is then diminished, and hence the anode current $I_a$ increases. At one stage, when the anode potential reaches a high value corresponding to P, *all* the electrons emitted by the cathode at its particular temperature are collected by the anode; and as further increase in anode potential does not now increase the current, as shown by PQ in Fig. 37.11(ii), it is called the saturation current. If the saturation current is measured and the charge on one electron is known (p. 487), the number of electrons per unit area per second emitted by the cathode at its particular temperature can be calculated. When the cathode temperature is decreased by lowering the heater current, the saturation current increases along a lower level S because fewer electrons per sec are then liberated (Fig. 37.11(ii)).

No anode current flows when the anode potential is negative relative to the cathode. The electrons, which carry negative charges, are all repelled back to the cathode in this case. The diode is called a 'valve' because current (electrons) flows only in one direction through it, from cathode to anode, and only when the anode is positive in potential relative to the cathode.

### Resistance of Diode

We can think of the diode as a special kind of electrical resistance component. A metal wire such as copper or tungsten at constant temperature has the same electrical resistance whichever way the terminals of a battery

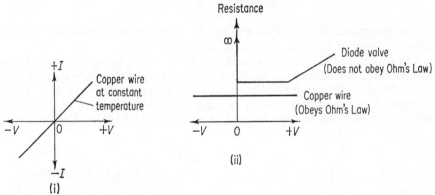

Fig. 37.13 Metal and Diode characteristics

are connected to it, because the current $I$ has the same value when the p.d. $V$ connected to it is reversed (Fig. 37.13(i)). In contrast, the diode conducts relatively easily when the anode potential is positive relative to the cathode, so that its 'resistance' is then low. But when the anode potential is negative relative to the cathode no current flows, and so its resistance is now infinitely high. The metal wire obeys Ohm's law, that is, the current is always proportional to the p.d. applied (Fig. 37.13(i)). The diode valve does not obey Ohm's law (Fig. 37.13(ii)).

### Diode as Rectifier

Suppose an alternating voltage, one which reverses continually in direction (see p. 629), is connected to a diode valve with a resistance $R$ of a few thousand ohms in series (Fig. 37.14(i)). For convenience one may use an A.C. voltage $V$ of 12 V from a ray-box transformer.

A cathode-ray oscillograph (C.R.O.) can be used to investigate what happens in the circuit. With a double-beam oscillograph, the variation of the applied voltage $V$ can be observed on the screen by connection to the $Y_1$-plate (Fig. 37.14(ii)). The voltage $V_1$ across $R$ can be compared with $V$ by connection to the $Y_2$-plate. $V_1$ is seen to be the same wave as $V$, but with one half of its voltage waveform missing, so that $V_1$ is a voltage which varies

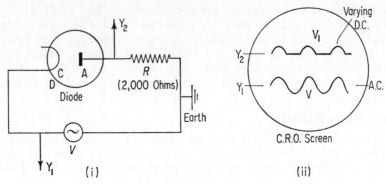

Fig. 37.14 Diode valve as rectifier

in *one* direction. We say that the applied voltage $V$ is 'rectified' by the diode valve and that the valve is a 'rectifier'.

On account of its rectifier action, the diode valve is used in radio communication. Radio waves arriving at a receiver aerial induce an alternating voltage of the same frequency, but the average current is zero if this is applied to a resistance wire, because the wire conducts equally well whichever way the voltage is applied to it (Fig. 37.15(i)). On the other

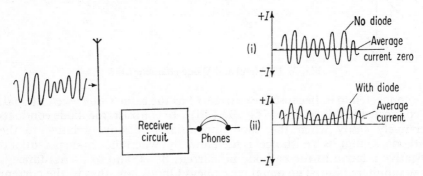

Fig. 37.15 Diode valve detection

hand, if the alternating voltage is connected to a diode valve a current flows in one direction only (Fig. 37.15(ii)). This time an average current (*not* zero) is obtained, and eventually the current can be detected by a telephone earpiece. The radio wave is thus detected as a result of the rectifier action of a radio valve.

One of the earliest rectifiers was a carborundum crystal with a steel point pressing against it. The crystal-steel junction had a low resistance to p.d. applied in one direction and a high resistance in the opposite direction, and is therefore a rectifier. A cuprous oxide–copper junction, made by oxidizing one face of a copper disc, acts as a rectifier. Copper-oxide rectifiers, as they are known (Fig. 37.16), are used in *battery-chargers*, which produce current in one direction for charging batteries connected in the circuit (see p. 515).

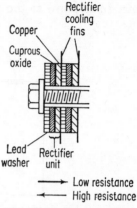

Fig. 37.16 Copper oxide (metal) rectifier

## Triode Valve

In 1907 Lee de Forest invented the *triode valve*, which had three electrodes. As we shall soon see, it is a much more active electrical component than the diode because it can amplify alternating voltages. The triode has an open coil of wire called the *grid* G between the cathode C and anode A (Fig. 37.17). Since G is close to C, the electron flow towards A is very sensitive to changes in the potential of G relative to C. If G is made only a few volts positive relative to C, for example, the attractive force on the electrons in the cathode is so strong that the cathode disrupts. Consequently, G is usually *negative* in potential relative to C. In this case G has a repulsive force on the electrons, but as the wire is open the electrons reaching G flow through it towards the anode and are collected. An anode current is thus obtained. When the potential of G is made less negative the current increases; when it is made more negative the current decreases. In this way, like a policeman controlling the entry of a large crowd through gates, the potential of the grid

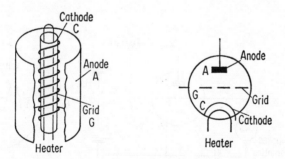

Fig. 37.17 Triode valve

can control smoothly the number of electrons which pass it and reach the anode and form the anode current.

Fig. 37.18 shows a TELTRON triode connected in a circuit for investigating the action of the grid when its potential relative to the cathode is varied negatively. The circuit can be used to investigate the triode characteristics, including the effect of altering the anode voltage.

Fig. 37.19(i) shows a small alternating voltage $V$ connected between the grid G and cathode C. The battery, G.B., is called a 'grid-bias' battery, and its purpose is to make the grid potential stay negative while $V$ is varying. In response to the changes in grid potential, which are due to $V$, the

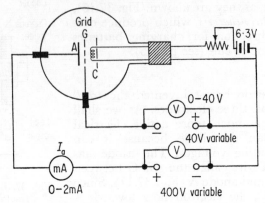

Fig. 37.18 Triode valve (TELTRON) circuit

anode current $I_a$ varies at the same frequency. A resistance $R$ of several thousand ohms is included in the anode circuit. This produces an alternating *voltage*, and measurement with a cathode-ray oscillograph, for example, shows that it may be many times greater than the applied alternating voltage $V$ (Fig. 37.19(ii)). As the latter may be due to a radio signal received, the triode valve can thus act as a voltage amplifier. The

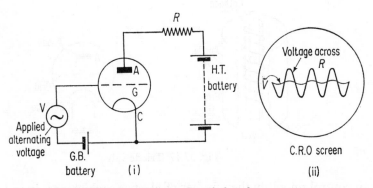

Fig. 37.19 Amplification by triode valve

triode is a much more active electrical device than the diode valve, which cannot amplify. Other valves with four electrodes (*tetrodes*) and five electrodes (*pentodes*) can produce much greater amplification than the triode.

## Electron Emission by Photoelectric Effect

We have seen that electrons can be emitted from metals when they are given energy in the form of heat. Let us investigate whether electrons can be emitted from the surface of metals if they are given energy from high-frequency *light* or radiation, such as ultra-violet light.

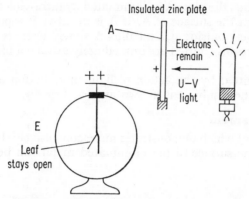

Fig. 37.20 Illumination of positively charged plate

### Experiment

Use a zinc plate as the metal. First prepare the surface by placing a small droplet of mercury on the plate and wiping it repeatedly over the zinc with cotton-wool moistened with dilute sulphuric acid. Wipe the plate dry with clean tissue. It is ready for the experiment when mirror-bright.

Insulate the zinc plate A and connect it to the cap of a gold-leaf electroscope E (Fig. 37.20). By means of a suitable rubbed rod, or by induction, charge A positively, so that the leaf diverges. Now place an ultra-violet lamp, such as a Philips TUV 6-W lamp, close to A. Switch on the lamp

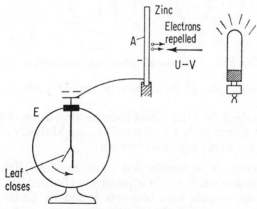

Fig. 37.21 Photoelectric effect

(*make sure you do not look at the lamp, as ultra-violet light is harmful to the eyes*).
Observe that the leaf stays open.

Now switch the lamp off, discharge A by touching it and then charge A
*negatively* so that the leaf diverges. Switch on the lamp. The leaf now closes
(Fig. 37.21). Repeat the experiment and verify that the plate is discharged
as soon as it is illuminated by ultra-violet light.

*Conclusion.* When the plate A is illuminated by ultra-violet light, electrons
are emitted from the surface of A. If A is negative it repels the electrons
(negative charges) outside it, and thus A slowly loses its charge. If A is
positive the electrons emitted are immediately attracted back, so A retains
its charge.

Electrons emitted by the action of light or radiation are called *photoelectrons*. The whole subject is known as *photoelectricity*.

### Electron Movement

An experiment which demonstrates more conclusively that electrons are
emitted from the surface of the illuminated metal can be carried out as
follows:

EXPERIMENT

Attach an insulated zinc plate A and an insulated metal gauze D to the
respective caps of gold-leaf electroscopes (Fig. 37.22). Place A close to D

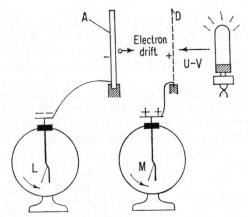

Fig. 37.22 Photoelectric effect – electron drift in field

but not touching. Charge A negatively and D positively, so that both
leaves diverge.

Now illuminate A by ultra-violet light passing through the open spaces
of the gauze D. Observe that both leaves L and M slowly close. Repeat the
experiment to confirm your observations.

*Conclusion.* When A is illuminated by ultra-violet light electrons are
emitted from its surface. Since D is positively charged, the electrons move
towards D in the electric field between A and D. Thus: (i) the positive
charge on D diminishes, (ii) the negative charge on A diminishes.

Summarizing, electrons can be liberated from metals by using energy in the form of heat, called thermionic emission, or by using energy in the form of light–photoelectric effect. In the former case the electrons come from atoms inside the metal, so that many more electrons are emitted.

## Sound Track

The discovery of the photoelectric effect has led to many useful devices for converting light to electrical energy. Years ago, before the invention of the *sound track*, sub-titles were used on films shown in cinemas, and an orchestra played suitable music. The sound track is a strip of film adjacent to the frames which has light and dark bands across it, so that a beam of light incident on it is interrupted periodically (Fig. 37.23). The

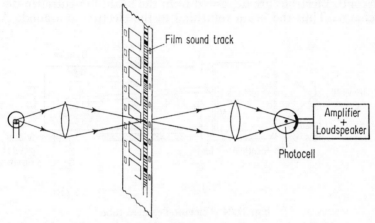

Fig. 37.23 Action of film sound track

transmitted light falls on a photo-sensitive metal surface in a vacuum cell, which emits a number of electrons in proportion to the intensity of the light. Thus a varying electric current is produced when the electrons are collected by another metal or anode, at a positive potential relative to the emitter. The current is then amplified and passed through a loud-speaker, so that sound corresponding to the film track is heard.

## Television Camera Tube

The principle of a modern television camera tube, called an *image orthicon* tube, is shown in Fig. 37.24. Light from a particular scene is focused by the camera lens on to a *photocathode* P, a very thin layer of photosensitive material made from caesium, silver and bismuth, for example. P emits electrons from parts of its area in proportion to the light intensity falling there. A bright part of the scene focused on P thus produces a high number of emitted electrons, and a dark part a small number. The photoelectrons are then accelerated to a target T through a mesh M, which is kept at a few volts positive relative to P. T is a very thin sheet or membrane of special glass which is photoactive, and secondary electrons are produced at T by the arrival of the photoelectrons. The secondary electrons are captured by M, leaving T with a positive charge. In effect, then, the whole surface of T has a positive charge density proportional to the light intensity of the scene in front of the camera.

An *electron-gun* arrangement, comprising a metal G at a negative potential relative to the cathode C and an anode $A_1$ at a high positive potential, provides an electron beam focused on T with the aid of anodes such as $A_2$, which is a few hundred volts positive in potential. A wire mesh N at a suitable potential decelerates the arriving electrons, so that their velocity is very low on reaching T. By using two pairs of coils whose axes are perpendicular to each other and which carry suitably varying current, the electron beam is made to scan the target area, moving sideways repeatedly across it, as in reading a page in a book, at the rate of 625 lines per second. Electrons are deposited from the beam to neutralize the positive charge. Thus the beam returning to the electron-gun anode $A_1$ has

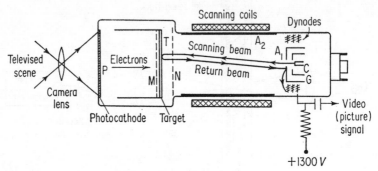

Fig. 37.24 Television camera tube

fewer electrons where the scene was bright, but is practically unchanged where the scene was dark. In this way the returning electron beam is modulated by the picture. From the anode $A_1$ the returning electrons are multiplied in number by secondary emission from metals called *dynodes*, shown shaded in Fig. 37.24. An electric picture or video signal is obtained from the last dynode, and this is passed to an aerial for distant transmission and reception.

Summarizing, the image orthicon camera tube has a *target* at one end where the picture is 'changed' into electric charges, a *scanning beam* of electrons to neutralize the charges so that the returning beam is modulated by the picture, and an *electron multiplier* at the electron-gun end of the tube which increases the picture or video signal.

### Charge on Electron

We conclude the chapter with a brief account of the charge on an electron, whose numerical value is denoted by the symbol $e$.

The first clue that there was a basic unit of electricity was due to Helmholtz many years ago from a study of electrolysis. Here it is known that particles of matter (ions) have associated with them a definite amount of electricity. As explained on p. 551, the gram equivalent of any element is deposited by the same quantity of electricity, about 96 500 coulombs, called the 'faraday'.

From Avogadro's hypothesis, that equal volumes of gases contain an equal number of molecules at the same temperature and pressure, it can be shown that the same number of atoms are present in the gram-equivalent of any monovalent element. The number is about $6 \times 10^{23}$ and is called *Avogadro's number*. It therefore follows, since all the atoms carry 96 500 coulombs, that each has a charge of 96 500/($6 \times 10^{23}$) or $1\cdot6 \times 10^{-19}$ coulomb. This is the charge carried by all monovalent ions in electrolysis, no matter what their chemical nature may be. A divalent ion, which has half as many atoms in its gram-equivalent, carries twice as much charge. Consequently, the charge $1\cdot6 \times 10^{-19}$ coulomb is a basic unit of electricity. After Sir J. J. Thomson discovered the electron (p. 681), it was found that this particle carried a charge of $1\cdot6 \times 10^{-19}$ coulomb.

## Millikan's Determination of *e*

Millikan, a famous American physicist, carried out many measurements from 1909 to determine accurately the charge *e* on an electron. Here we can only deal with the principle of the method he used.

Two parallel metal plates A and B are inside a constant-temperature

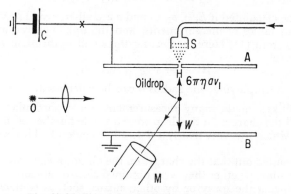

Fig. 37.25 Principle of Millikan's experiment

chamber (not shown) (Fig. 37.25). The top plate A has a fine hole H at its centre, and a battery C of a few thousand volts can be connected across A and B. A fine oil spray S produces tiny droplets of oil, which have charges on them owing to frictional forces as they are produced. Some of the droplets fall through H. The region between the plates is illuminated by O and a particular drop can be observed by the reflected light on looking through a low-power microscope M. The drop falls slowly with a constant or terminal velocity (p. 104), and by means of two fine lines in the eye-piece and a stop-watch, the velocity of fall can be measured.

Millikan's basic idea to measure the charge on the drop was to observe its velocity change on switching on the battery. This immediately changes the velocity of fall because a force proportional to the charge now acts on the drop. The velocity change is a measure of the charge:

*Method.* Observe a particular oil drop as it falls between A and B with its terminal velocity. Measure its velocity $v_1$ under gravity.

Now switch on the battery to reduce the velocity–if necessary, reverse the battery terminals. Observe the new velocity $v_2$. By using a radioactive substance to ionize the air, or an X-ray tube as Millikan did originally, the *same* drop gains more charge. The new velocity is measured when the battery is switched on again, and this can be repeated for many different charges. For a particular drop Millikan carried out over a hundred measurements of its velocity, frequently holding it under observation for several hours.

*Theory.* Stokes showed that the frictional or drag force $F$ on a sphere of radius $a$ moving with a velocity $v$ through a fluid of viscosity $\eta$ was $6\pi\eta av$. Thus, for free fall under gravity when the terminal velocity is reached,

$$W = 6\pi\eta\, av_1 \quad . \quad . \quad . \quad . \quad . \quad . \quad (1)$$

where $W$ is the weight of the oil drop less the upthrust in air (see p. 104).

The *electric field intensity*, $E$, between the plates A and B is the potential gradient between them, or $V/d$, where $V$ is the p.d. applied by the battery and $d$ is the distance apart of the plates (p. 663). As $E$ is the force per *unit* charge, the force on a charge $ne$, where $n$ is a whole number, is $E \cdot ne$. Thus when the field is switched on to oppose the force due to gravity, then, if $v_2$ is the new velocity of the drop,

$$W - E \cdot ne = 6\pi\eta\, av_2 \quad . \quad . \quad . \quad . \quad . \quad . \quad (2)$$

Hence, using (1), $\qquad\qquad E \cdot ne = 6\pi\eta\, a(v_1 - v_2).$

Thus $ne$ can be calculated if $E$, $\eta$, $v_1$, $v_2$ and $a$ are all known. The viscosity of air is known from tables of measurements, and the unknown radius $a$ of the oil drop is found from (1). Here, neglecting the small upthrust in air, if $d$ is the density of the oil,

$$W = \tfrac{4}{3}\pi\, a^3 dg = 6\pi\eta\, av_1,$$

from which, on simplifying, the radius $a$ can be calculated.

*Results.* Millikan made many measurements on a particular oil drop. He found that all the charges on the drop were a whole number of times the lowest value found, which was very nearly $1 \cdot 6 \times 10^{-19}$ coulomb. This is the electronic charge, $e$.

Millikan pointed out that the charges on the oil drops were always a multiple of the same value $e$, whether they were obtained electrostatically in the frictional process of blowing the spray or by other means, such as contact with ions of air produced by X-rays. Further, the results show that, generally, an electric charge is 'granular' or particulate, that is, it is always a whole number of times $e$, never fractional.

*Note.* As an alternative to Millikan's method, the voltage $V$ applied to the plates can be varied until the drop is 'balanced'. In this case,

$$\text{Force due to electric field} = E \cdot ne = \frac{V}{d} \cdot ne = W.$$

The weight $W$ of the drop can be found by timing its fall under gravity and measuring its radius $a$, as already explained, so that the particular charge $ne$ can be calculated. When the charge on the oil drop is changed, a new voltage $V$ is needed to balance the drop. In this way many values of $ne$ can be found, and $e$, the lowest value, obtained.

Details of Millikan's oil-drop apparatus, suitable for a school laboratory, can be obtained from manufacturers listed in the front of the book.

## X-RAYS

In 1895 a German physicist, Roentgen, found that a penetrating radiation was coming from a discharge tube when cathode rays were produced inside. He covered the tube with black paper, but found that a screen nearby, coated with a barium salt, fluoresced every time the tube was operated. As Roentgen himself said when reporting this discovery: 'The most striking feature of this pheno-menon is the fact that an active agent here passes through a black card-board envelope, which is opaque to the visible and the ultra-violet rays of the sun ... an agent which has the power of producing active fluor-escence.'

He called the unknown radiation *X-rays*. It is now known that X-rays are produced when fast-moving elec-

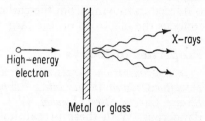

Fig. 37.26 Production of X-rays

trons strike heavy atoms such as those of metals or even atoms of glass (Fig. 37.26). On collision, some of the kinetic energy of the incident elec-trons is converted into this form of radiation energy. This is the reverse of the photoelectric effect discussed on p. 695, where some of the incident light energy on a metal plate is given to electrons.

### X-ray Tubes

X-ray tubes are widely used in medical treatment and in scientific laboratories. Fig. 37.27 shows the essential features of the interior of a modern X-ray tube.

A filament or hot cathode emits electrons. It is usually heated by the

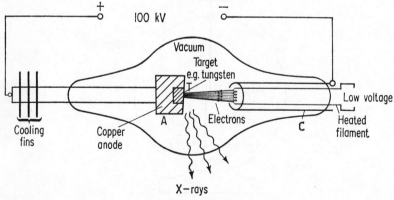

Fig. 37.27 X-ray tube

current supplied from a small mains transformer (not shown). T is a metal 'target', made of copper or tungsten, for example, which emits X-rays when it is bombarded by the electrons. The target is embedded in a copper block A at the end of a metal rod. A is the anode, and it is maintained at a high voltage, such as 50 000 V or many hundreds of thousands of volts, depending on the use of the tube, so that electrons reach A with high energy. Only a very small part of this energy is converted into X-radiation. The rest is converted into heat, so that A becomes very hot. Radiating fins may therefore be used to dissipate the heat conducted along the rod into the air, as shown. In large X-ray tubes used in hospitals the anode is cooled by water flowing behind the anode continuously. The X-ray tube is evacuated of air so that the electrons are not hindered in moving from the cathode to A, and a metal cylinder C is used to·focus the electrons on the target area.

## Properties of X-rays

*In experiments on X-rays it is absolutely essential to adhere strictly to the safety precautions issued in the latest pamphlets of the Department of Education and Science, Curzon Street, London, W.1.* Exposure of the skin to X-rays is very dangerous. With these precautions obeyed, experiments may be carried out to investigate some properties of X-rays. The GRIFFIN & GEORGE and TELTRON X-ray tubes are designed for school laboratory use.

### 1. *Penetration Properties*

Using a fluorescent screen or photographic film, place various materials of different densities and thicknesses between the tube and the screen. Observe whether the images are clear or diffuse. Suitable materials are the paper of a book, wood, perspex, aluminium, lead and metallic objects such as coins and keys. POLAROID film produces quick results.

The results show that less-dense materials of the same thickness allow X-rays to pass easily through them. Denser metals partially stop X-rays, but only thick lead blocks appear to stop them completely.

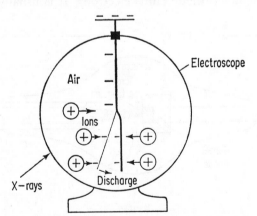

Fig. 37.28 Ionization of air by X-rays

### 2. *Ionization Property*

Take a gold-leaf electroscope and charge it. Make sure that the leaf is open wide and stays open. Now place an X-ray tube near it and switch it on, so that the air all round the electroscope is exposed to the radiation.

Observe that the leaf collapses. Repeat with an opposite charge on the leaf.

*Explanation.* The energy of the X-rays is able to strip some electrons from the molecules of air, which then become positive ions (Fig 37.28). If the electrons attach themselves to neutral air molecules the latter become negative ions. Thus X-rays ionize air. Positive ions drift towards the cap and leaf and discharge the leaf; negative ions drift to the case.

## Some Applications of X-rays

Since X-rays penetrate the flesh but are stopped by harder substances such as bone, Roentgen immediately realized their importance in medicine. Nowadays, most hospitals are equipped with X-ray machines. A fracture of a bone can easily be detected from a radiograph of the limb

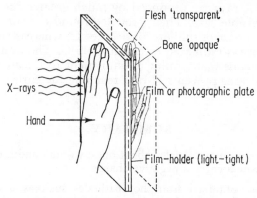

Fig. 37.29 X-ray photograph of hand

concerned, the radiograph being the 'picture' obtained when X-rays are incident through the limb on to a photographic plate (Fig. 37.29).

X-ray machines are also used industrially for detecting flaws and defects in steel plates that are invisible to the eye; X-rays pass more easily through the flaws than through the rest of the material.

## Nature of X-rays

The nature of X-rays was not definitely known for many years after their discovery. It was thought they were of the same nature as light rays. If so, X-rays should be diffracted at very narrow openings like light rays, as explained in *Waves*, p. 334. The diffraction gratings used for light rays, however, produced no diffraction of X-rays. In 1911, however, Laue realized that if X-rays had a much shorter wavelength than light rays a grating was required which had openings spaced very much closer than those used for light rays, which were mechanically ruled. He then made the brilliant suggestion to two research students, Friedrich and Knipping, that the atoms of a crystal could form a natural diffraction grating for very short wavelengths, as they were regularly spaced about $10^{-8}$ cm apart. The experiment was tried, and a diffraction pattern was produced on a photographic plate after the X-rays were transmitted by a crystal.

It is now known, therefore, that X-rays are electromagnetic waves like light waves, but their wavelengths are about a thousand times shorter and of the order of $10^{-8}$ cm. See p. 331.

After Laue's discovery was announced the two eminent British scientists Sir William Bragg and his son Sir Lawrence Bragg began researches into X-rays, for which they were awarded the Nobel Prize in 1915. They measured the wavelengths of X-rays and then began a study of the structure of crystals using X-rays. This has led to a knowledge of the structure of molecules of fibres, for example, and the manufacture of new materials, and, more recently, to the discovery of the structure of the molecule of DNA, a vital constituent in the living cell, for which British scientists were awarded the Nobel Prize in 1965.

Recent investigations with Skylark rockets containing X-ray film in a small tube have shown that X-rays are present in regions round the visible disc of the sun. X-rays are produced by much greater energy changes in the atom than visible rays. Their presence round the sun indicates temperatures of the order of millions of degrees C, whereas the visible sun's outer temperature is estimated at about 6000°C. The atmosphere of the earth, which extends many miles, is equivalent to a lead shield about a metre thick, and thus prevents X-rays reaching the earth.

## SUMMARY

1. Electrons are minute particles about one-two-thousandth of the mass of a hydrogen atom which carry a negative charge.

2. Electrons are obtained from hot cathodes such as a directly heated tungsten wire or an indirectly heated surface coated with a mixture of barium and strontium oxides.

3. Electrons: (i) travel in straight lines; (ii) are deflected in a circular path by a uniform perpendicular magnetic field; (iii) are deflected by a perpendicular electric field applied between two plates.

4. A cathode-ray tube has a hot cathode, a brightness control, an electron lens, Y-plates for connecting the voltage to be examined or received and X-plates to which a time-base voltage is usually applied.

5. A diode valve has an anode A and a cathode C inside an evacuated glass bulb. It conducts when A is positive in potential relative to C and has an infinitely high resistance when A is negative in potential relative to C. The current is space-charge limited at first, and reaches a saturation value when all the electrons per second emitted from the cathode are collected by the anode as the anode potential is increased.

6. The triode valve has a grid between the cathode and anode which controls the anode current in a more sensitive way than the anode alone, leading to voltage amplification.

7. Electrons are emitted when ultra-violet light falls on a zinc surface. This is utilized in the photoelectric cell and the television camera to change light energy into electrical energy.

X-rays are produced when high-speed electrons bombard a metal target such as tungsten. A hot cathode produces electrons, and a high voltage, such as 50 kV, gives the electrons high energy.

8. X-rays are electromagnetic waves of very short wavelength, such as $10^{-8}$ cm or less.

## EXERCISE 37 · ANSWERS, p. 745

1. Electrons can be liberated by a 'cold cathode' or a 'hot cathode'. Explain what this means.

2. Write down four properties of electrons (cathode rays). Briefly describe how each property can be demonstrated.

3. 'Electrons are: (i) extremely light; (ii) deflected by a magnetic field; (iii) deflected by an electric field.' Describe, and illustrate with diagrams, the experiments you would carry out to demonstrate (i), (ii) and (iii).

4. Draw a diagram of a *Perrin tube*. Explain, with a complete circuit diagram, how it is used to determine the sign of the charge on an electron.

5. What are the properties of electrons which make them so useful in a cathode-ray tube?
Draw a labelled diagram of a cathode-ray tube and explain how it works.

6. Describe, with sketches, the difference between a directly and indirectly heated cathode. Draw a complete circuit diagram showing the respective battery supplies in a diode valve with: (i) a directly heated cathode; (ii) an indirectly heated cathode.

7. A zinc plate connected to the cap of a gold-leaf electroscope is given, firstly, a negative charge and then a positive charge. In each case the plate is illuminated by ultra-violet light. Describe and explain what happens.

8. Describe briefly how a narrow beam of electrons may be caused to flow along an evacuated tube. How may the beam be deflected? How is the deflection made visible and the sign of the charge on the particles determined?
How fast will the electrons in a beam be moving after falling through a potential of 20 V if every g of electrons carries a charge of $1.8 \times 10^7$ coulomb? (*O. and C.*)

9. Draw a labelled diagram of a diode valve. (*N.*)

10. With the aid of labelled diagrams explain the action of any TWO of the following: (*a*) an electrophorus; (*b*) a transformer; (*c*) a diode valve used as a rectifier. (*N.*)

11. Describe the diode valve and give details of the cathode circuit. State the essential condition, in each instance, to produce an anode current which is (i) zero, (ii) a maximum, assuming that the cathode is at a constant temperature suitable for the normal working of the valve. (*L.*)

12. What do you understand by the 'photoelectric effect'? Describe one application of photoelectricity.

13. Draw a labelled diagram of an X-ray tube. Explain how the X-rays are produced. How would you obtain X-rays of greater penetrating power from the tube?

14. 'X-rays are particles, cathode-rays are waves.' Is this statement true? If not, correct it, and explain how the nature of X-rays was demonstrated. Explain the connection between X-rays and visible rays.

# *38*

## PRINCIPLES OF SEMICONDUCTORS·
## JUNCTION DIODE·TRANSISTORS

### Semiconductors

From the point of view of conductors of electricity, solids are divided into good conductors such as metals, poor conductors such as silicon and germanium called *semiconductors* and insulators such as polythene or perspex. In 1945 Shockley and fellow-workers in America began to investigate the properties of semiconductors, with discoveries which led to some of the greatest commercial developments in electricity since Faraday's discovery of electromagnetic induction in 1834.

### Crystal Structure of Semiconductor

The semiconductor element germanium has a crystal structure, that is, it has a large number of ions or atoms distributed throughout its volume at various sites which have a geometrically recurring pattern. The outer part or shell of the atom has four electrons, called 'valence electrons' because they are available to take part in phenomena such as chemical changes (Fig. 38.1). The atoms or ions are kept in position in the metal by chemical bonds between their valence electrons.

At the lowest possible temperature, absolute zero, a semiconductor is an insulator. No electrons move. At room temperature, however, the in-

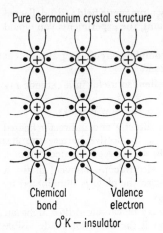

Pure Germanium crystal structure

Chemical bond — Valence electron

0°K – insulator

Fig. 38.1 Pure (intrinsic) semiconductor at 0°K

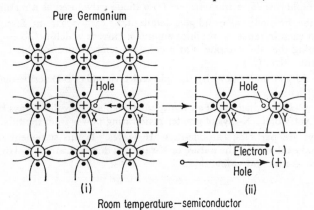

Pure Germanium

Room temperature—semiconductor
Fig. 38.2 Pure (intrinsic) semiconductor at room temperature

706

creased thermal energy of the electrons is now sufficient for some of them to escape from the attraction of its atomic nucleus. Having left the atom, X say, the electron is now a 'conduction electron'. It leaves a vacancy or *hole* in X (Fig. 38.2(i)), and since the net charge on the atom is now positive, a valence electron from a neighbouring atom, Y say, is then attracted to fill the hole in X. But this leaves a hole in Y (Fig. 38.2(ii)). Consequently, another electron moves to Y, creating yet another hole. Now the movement of an electron (a negative charge) to the left, for example, is equivalent to a movement of a positive charge equal to that on an electron to the right. *We therefore regard the hole as a positive charge which can move through the crystal.* The current in a semiconductor is carried by both electrons and holes, just as, in an electrolyte such as copper sulphate solution, the current is carried by both negatively and positively charged ions.

## N-type and P-type Semiconductors

As we have seen, pure germanium has four valence electrons. If the crystal is 'doped' with a tiny amount of phosphorus, such as one part in a million, the phosphorus atoms are absorbed into the crystal structure and take the place of germanium atoms at various places. Now phosphorus

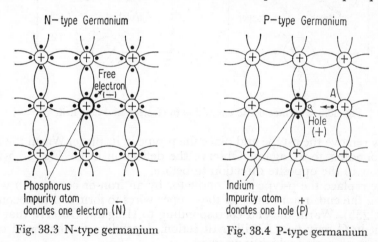

Fig. 38.3 N-type germanium          Fig. 38.4 P-type germanium

has *five* valence electrons. Only four are needed to keep the phosphorus atom in the crystal structure. Consequently, there is a surplus electron (Fig. 38.3). Thus ten million atoms of phosphorus impurity, for example, in germanium will create a surplus of ten million electrons or negative charges. The doped germanium is therefore called n-type (n for negative) germanium, or simply 'n-germanium'. The n-germanium is a better conductor than pure germanium, because it has more carriers of electricity. Notice that there are still holes, which are positive (p) charges in the crystal but they are 'minority' carriers in n-germanium.

Pure germanium can also be doped by adding a tiny amount of indium. The indium atoms are then absorbed into the crystal structure. But indium has *three* valence electrons. One electron from a neighbouring germanium atom A is therefore attracted towards the indium atom, thus completing four valence bonds. This leaves a hole in A (Fig. 38.4). Thus if 10 million atoms of indium are added, 10 million holes are created. The doped germanium is now called 'p-germanium', because it now contains a great majority of positive (p) charges. The minority carriers present are negative charges or electrons.

## Carriers in P- and N-type Semiconductors

A simple experiment shows the difference between the majority carriers in n- and p-type germanium.

EXPERIMENT

Connect copper wires to a microammeter G and attach the free ends to an n-type semiconductor (Fig. 38.5). Warm one end H gently with a lighted taper. Observe the deflection in G (see Fig. 38.5(i)).

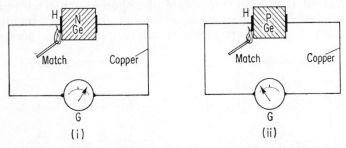

Fig. 38.5 Thermoelectric e.m.f. due to electron and hole drift

Now remove the n-type and place the p-type in its place. Warm the end corresponding to H again. Observe the deflection in G (Fig. 38.5(ii)). It is now in the opposite direction to before.

Now replace the p-type semiconductor by an iron or constantan wire, twisting the ends round those of the copper wires to form a thermocouple (see p. 235). Warm the end corresponding to H again. Observe that the current in G may flow in either direction, depending on the metal used and on the junction which is warmed.

*Conclusion.* 1. In each case a thermoelectric e.m.f. is developed (see p. 235).
2. Inside n-germanium, the movement of carriers which produces the e.m.f. is opposite to that inside p-germanium.

## Rectification with P- and N-Semiconductors

EXPERIMENT

Clamp firmly together a sample of p- and n-type germanium in a wooden clamp, with two leads A and B pressing against the outer surfaces, and join a centre-zero milliammeter G in series (Fig. 38.6(i)).

Connect the positive pole of a 1·5-V battery D to A and the negative pole to B, as shown (Fig. 38.6(i)). An appreciable deflection is then obtained in G. Now reverse the poles of the battery (Fig. 38.6(ii)). A very small current is obtained.

*Conclusion.* The p- and n-type together conduct well when p is positive in potential relative to n. It conducts poorly when p is negative in potential relative to n. Thus the p–n materials together act like a *rectifier*.

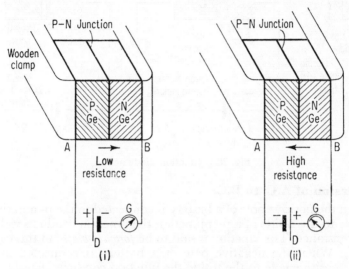

Fig. 38.6 P–N junction diode

## Junction Diode

The reason for the p–n rectification is due to a movement of charges at their junction. The high concentration of positive charges or holes in the p-germanium makes some of them diffuse into the n-germanium across the junction, just as the molecules of a perfume diffuse out into the air when the bottle is unstoppered. Likewise, the high concentration of negative charges or electrons in the n-germanium diffuse into the p-germanium. The drift of charges ceases in a short time because it leaves the materials oppositely charged, which opposes the drift. A 'back e.m.f.', a *potential barrier* to further drift of charges, now exists across the junction (Fig. 38.7(i)). Thus p- and n-germanium in contact have a junction deficient in positive and negative charges. In a commercially made *junction diode* the thickness of the p–n junction has been estimated at about $\frac{1}{1000}$ cm.

Suppose a battery B of a few volts, which is more than the potential barrier, is connected with its positive pole to p and its negative pole to n (Fig. 38.7(ii)). The battery e.m.f. urges positive charges from p to n and negative charges from n to p. The charges therefore now begin to drift once more across the junction. The p–n junction thus conducts. Suppose the battery and galvanometer terminals are reversed (Fig. 38.7(iii)). This time it can be seen that the battery e.m.f. is added to the potential barrier, instead of opposing it, so the positive and negative charges in the junction do *not* move on this occasion. The minority carriers (the negative charges in the p-germanium and the positive charges in the n-germanium), however, are urged across the junction, and hence a small current flows.

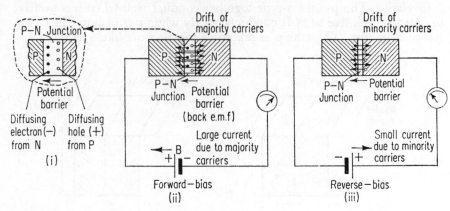

Fig. 38.7 Junction-diode action

## Conversion of A.C. to D.C.

When the positive pole of a battery is connected to the p- and the negative pole to the n-side of a p–n junction the junction conducts well or has a low resistance. The junction is said to be *forward-biased* in this case (Fig. 38.8(i)). When the negative pole of a battery is connected to the p- and the positive pole to the n-side the junction conducts slightly or has a high resistance. The junction is now said to be *reverse-biased* (Fig. 38.8(i)).

The p–n junction can hence be used as a rectifier, that is, it can change

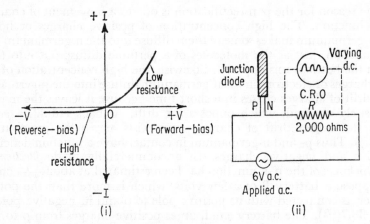

Fig. 38.8 Junction diode as rectifier

A.C. to D.C. voltage. The A.C. voltage is placed in series with the *junction diode*, as we now call the p–n semiconductor, together with a suitable resistance R (Fig. 38.8(ii)). When the Y-terminals of an oscillograph are connected across R the rectified waveform is seen on the screen. The diode radio valve is also a rectifier (p. 691). Unlike the junction diode, however, which is a solid rectifier, the current in this case is always carried by electrons.

## Transistor

In 1948 the American physicists Bardeen and Brattain found that a current could be amplified by sandwiching a thin layer of n-germanium between two thicker layers of p-germanium (Fig. 38.9(i)). This three-

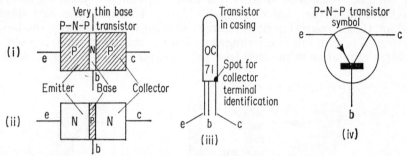

Fig. 38.9 P–N–P Transistor

element doped semiconductor is called a p–n–p *transistor*. An n–p–n transistor is shown in Fig. 38.9(ii).

Though the actions inside them are different, it is useful to compare the transistor to a triode valve, which also acts as an amplifier (p. 693). One of the p-germanium materials is called the *emitter e* because, as we see later, it can emit or send positive charges for collection by the other p-germanium, which is therefore called the *collector c*. The emitter is therefore like the cathode in the triode valve, which emits electrons, and the collector is like the anode in the valve, where the electrons arrive and flow out-

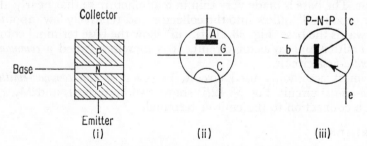

Fig. 38.10 Transistor compared to triode valve

side in the anode circuit, creating a current. The collector current in a transistor is thus analogous to anode current in a triode valve. Fig. 38.9 (iii), (iv) shows the external appearance of a transistor and the circuit symbol used.

The flow of electrons in the triode is controlled by the grid potential, the grid being positioned between the cathode and anode (p. 693). In the same way the collector current in a transistor can be controlled by the thin n-germanium sandwiched between the emitter and collector, which is called the *base* of the transistor. Thus the terminals connected to *e*, *b* and *c* in the transistor are analogous to the terminals joined to the cathode, grid and anode in the triode valve (Fig. 38.10).

## Circuit Connections

We can think of the emitter-base of the p–n–p transistor as equivalent to a p–n diode backed by another p–n diode, the collector-base, the central n-germanium being shared (Fig. 38.11(i)). To make the first diode conduct, that is, to send positive charges from the emitter towards the base, a battery X or p.d. must be arranged so that the emitter (p-type) is positive in potential relative to the base (n-type), as explained on p. 709. Thus the

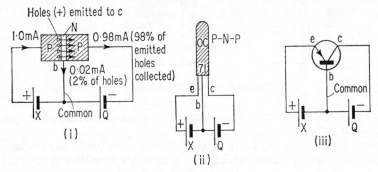

Fig. 38.11 Action of transistor

emitter-base is forward-biased. If the neighbouring p–n diode, the collector-base is *reverse*-biased by a battery Q, that is if the base (n-type) is positive relative to the collector (p-type) the positive charges are urged towards the collector. The base is made very thin in manufacture so that nearly all the charges (about 98%) flow into the collector and relatively few (about 2%) drift towards the base. Fig. 38.11(ii), (iii) show the base terminal common to the emitter and collector circuit. It is therefore called a *common-base* or *grounded-base* circuit.

The most commonly used circuit, however, is the *common-emitter* or *grounded-emitter* circuit. Fig. 38.12(i) shows two batteries L and M, with a common connection to the emitter terminal.

EXPERIMENT

Connect the circuit shown in Fig. 38.12(ii) or 12(iii), using a p–n–p audio transistor such as OC71, a variable rheostat R of 50 000 Ω, a micro-ammeter H and a milliammeter K. Make sure the batteries are the right way round, or else you will ruin the transistor.

Observe that the collector current $I_c$ is much greater than the base current $I_b$. Alter the base current by 50 $\mu$A, for example, and observe the much greater change in the collector current.

*Conclusion.* A current change in the base circuit is amplified in the collector circuit.

The base current thus has a sensitive control over the collector current,

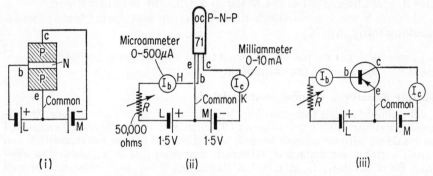

Fig. 38.12  Common-emitter circuit

that is, it controls the flow of charges from the emitter to the collector. A base-current change may be amplified some thirty times or more, depending on the circuit components used.

## Transistor as Amplifier

Fig. 38.13 shows a simple common-emitter circuit which employs only one battery. The emitter can be considered as an 'earthed' or common terminal for the collector and for the base circuits. The emitter-base is forward-biased because E is positive in potential relative to B. The current which flows through the resistor R makes B higher in potential than C. Hence C is negative in potential relative to B and so the collector-base is reverse-biased, as required.

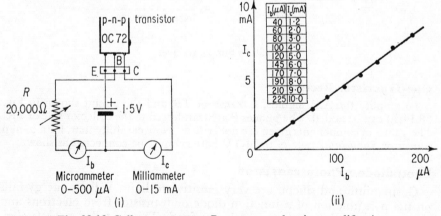

Fig. 38.13  Collector current *v.* Base current, showing amplification

## EXPERIMENT

Set up the circuit shown. Vary the rheostat R and observe the change in the collector current $I_c$ as a result of a change in base current $I_b$.

Then replace the microammeter in the base circuit by a high-resistance telephone earpiece X and the milliammeter G by another telephone earpiece Y, preferably of low resistance. Using X as a microphone, speak softly into it, and observe the effect heard in Y with the help of a friend.

Connect Y to an oscillograph, and observe the waveform obtained when speaking softly into X.

*Conclusion.* This simple circuit shows that the common-emitter arrangement can act as a current amplifier.

### Simple Receivers · Point-contact Diode

The simple circuit shown in Fig. 38.14(i) can receive home or local broadcasting stations. AB is a coil of about 100 turns of thin (e.g. 36 s.w.g.) enamelled or insulated wire on a small former (a suitable coil can be purchased); *C* is a small variable air capacitor, $0.0005\mu F$ maximum; D is a *point-contact diode* (OA 60 or similar); $C_1$ is a $0.01\mu F$ capacitor; P are high resistance phones (e.g. 2000 ohms per earpiece). The aerial can be joined to A, or to T, about one-third from B along the coil, for better reception. A good earth at B, such as a water-pipe, will increase the sensitivity. The aerial wire should be placed as high as possible. Fig. 38.14(i) is a semiconductor version of the old 'crystal' receiver.

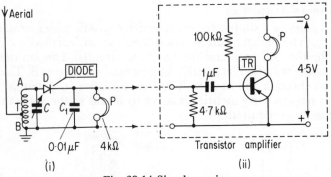

Fig. 38.14 Simple receiver

### One-Transistor Receiver

To amplify the A.F. signals, a transistor TR and the circuit shown in Fig. 38.14(ii) can be added. The phones P are transferred to the collector side of TR. The values of components given are not critical. Note carefully that, for a p–n–p transistor, the *positive* pole of the 4·5 V battery must be connected as shown.

### Photodiode · Phototransistor

Germanium and silicon are very sensitive to light. If light energy falls on the p–n junction of a junction diode or transistor fresh electrons and holes are created and a current is obtained. This is called a *photovoltaic* effect. The ordinary junction diode and transistor has therefore to exclude light. As the materials are protected in a glass envelope, this is coated with an opaque paint.

*Photodiodes* and *phototransistors*, however, exploit the light sensitivity of the p–n junction. Photodiodes can be used, for example, to start or stop an electric circuit on exposure to light. The Mullard OCP71 phototransistor produces a current change of a few milliamps from dark to light. The current change can easily be observed on a milliammeter when a battery of a few volts is connected between the collector and emitter with the base left out of the circuit. In practice, the phototransistor employs all three terminals with the benefit of the amplification action. The OCP71 is sensitive to the infra-red and can be used to demonstrate the existence of this invisible radiation (p. 441).

Other light-sensitive devices are the cadmium sulphide cell, whose resistance drops when exposed to light, selenium cells, which generate a current when exposed to light, and silicon solar cells, which produce a photovoltaic effect on exposure to light.

## Microelectronics

The components in radio circuits consist mainly of diodes, transistors, capacitors and resistors, suitably connected together. As we know when we look at a domestic receiver set, a fair amount of space is taken up when one component is joined to another by solder. Circuits are therefore 'printed' with thin metal film and the components are joined at appropriate places. This reduces the size of the circuit, and the connections are more reliable.

The thin-film technique has now been combined with the manufacture of transistors so that a complete circuit, having many transistors and other components, can be made with a piece of silicon about the area of one small letter on this page and half a millimetre thick! It is called a *micro-integrated circuit*, and the whole subject is known as *microelectronics*. The circuits are built into a tiny wafer of silicon, parts of whose area are masked and the others made into transistors by diffusing suitable impurities into them at successive stages. The reduced image of the whole circuit, drawn extremely accurately by competent draughtsmen, is projected on to the wafer by precision-type lenses. The wafer is initially coated with photosensitive material, so that an image of the circuit appears on the wafer and transistors, capacitors and resistors can be accurately positioned. All the components are suitably connected with the aid of a thin metal film deposited on the face of the wafer by using an appropriate mask. A film called *Microelectronics*, which shows stages in a similar manufacturing process, may be borrowed from Mullard Ltd, London.

Microintegrated circuits are extremely fast in operation, as the carriers have to travel only very short distances, and they are very reliable. They are built up under controlled manufacturing conditions, and the absence of wire connections makes them withstand successfully vibrations and high accelerations in rockets. Consequently, they are used in rockets and satellites for space research. A microintegrated circuit with 16 transistors and many other components may occupy an area only about $\frac{1}{4}$ cm square and a half a millimetre thick, and this has led to a remarkable reduction in size of large computers, where hundreds of thousands of components are required.

For low-frequency amplification and rectification in radio receivers the radio valve is likely to be superseded by the transistor and by the junction diode. Special fast-acting semiconductor diodes called *tunnel diodes* are used in computers. Besides their radio applications, semiconductors are used as efficient voltage generators and in refrigeration, and can be made into sensitive detectors of light and of radioactive particles.

## SUMMARY

1. A pure semiconductor at room temperature has an equal number of electrons and holes.

2. Hole movement is equivalent to movement of a positive charge equal numerically to the charge on an electron.

3. A p-type semiconductor is one which has a majority of holes and a minority of electrons. An n-type semiconductor is one which has a majority of electrons and a minority of holes.

4. A p–n *junction diode* has p- and n-type semiconductors in contact. It is a rectifier, conducting well when the potential of the p-type is positive relative to the n-type semiconductor, and slightly when the reverse is the case.

5. The p–n–p *transistor* has a p-type emitter, a thin n-type base and a p-type collector. It acts as an amplifier of current when the emitter-base is forward-biased and the collector-base is reverse-biased.

## EXERCISE 38 · ANSWERS, p. 745

Read the following accounts and complete the missing words to which clues are given:

1. *Semiconductors.* A semiconductor has a resistance between that of a good c..[1]..r and an i..[2]... Two examples of semiconductors are g..[3].. and ..[4]..n. The carriers of electric current in a semiconductor are e..[5]..s and h..[6]..s. Electrons carry a ..[7].. charge; holes move as if they had a ..[8].. charge. Metals have only ..[9]..s as carriers of current. In a pure semiconductor there are an ..[10]..l number of electrons and ..[11]..s. Temperature rise in a semiconductor must never be excessive, otherwise many ..[12]..s and ..[13]..s are liberated and the semiconductor is ruined.

2. *Junction Diode.* An n-type semiconductor has an excess of ..[14]..s, which are ..[15]..e charges. This is produced by adding a tiny amount of i..[16]..y to a pure semiconductor. A p-type s..[17].. has an excess of ..[18]..s, which are p..[19].. changes. A 'p–n junction' is the narrow junction between ..[20]..-type and ..[21]..-type semiconductors. The junction conducts well when the ..[22].. pole of a battey is joined to the ..[23]..-type and the ..[24].. pole of the n-type semiconductor. This is called 'f..[25]..d-bias'. The junction conducts very slightly when the battery ..[26].. pole is joined to the p-type and the positive pole to the ..[27]..-type semiconductor. The p–n junction thus acts like a ..[28].. radio valve, and is used as a ..[29]..r to change alternating to ..[30].. voltage.

3. *Transistor.* A p–n–p transistor has ....[31].... semiconductor regions. The emitter of the transistor is ..[32]..-type, the b..[33].. is n-type and the c..[34].. is p-type. The emitter and base are f..[35]..-biased, that is, the ..[36].. pole of a battery is joined to the ..[37].. and the ..[38].. pole to the ..[39]... The collector and base are r..[40]..-biased, that is, the negative pole of a battery is joined to the ..[41].. and the positive pole to the ..[42]... A transistor amplifies c..[43].., whereas the triode radio valve amplifies v..[44]...

In a *common-emitter* circuit used as an..[45].. circuit, the small current to be amplified is in the ..[46]..-emitter circuit and the amplified ..[47].. is obtained in the ..[48]..-emitter circuit.

# 39

## RADIOACTIVITY · ATOMIC STRUCTURE · NUCLEAR ENERGY

### RADIOACTIVITY

#### Becquerel · Rutherford

In the previous chapter we saw that electrons can be liberated from atoms in metals. Since these particles are obtained fairly easily, the electrons are likely to be in the outermost parts of the atoms.

The first clues on the heart or nucleus of the atom, and of atomic structure, were provided by a phenomenon discovered by Becquerel in 1896. He placed a uranium compound on a photographic plate covered with lightproof paper, and found to his surprise that, on developing, the plate was fogged. He traced this effect to some unknown radiation coming from the uranium compound. The phenomenon, called *radioactivity*, was investigated in 1897 by a New Zealand scientist, then Ernest Rutherford and later Lord Rutherford, in whose honour the Rutherford High Energy Laboratory at Didcot, England, is named. This laboratory is part of the National Institute for Research in Nuclear Science, which was opened in 1957.

Though it takes more than two weeks to complete, an experiment on radioactivity can be performed as follows:

EXPERIMENT

*Radiation from Uranium Salts*

Sprinkle a few crystals of uranyl acetate on the sensitive surface of an unexposed photographic plate in a dark room. An ILFORD Ordinary Plate is suitable. Carefully wrap up the plate in light-proof paper and leave it undisturbed for a few weeks. Then process the plate and observe the effect produced by the radiation from the crystals. Repeat the experiment, this time placing the crystals directly on to the light-proof wrapping of the plate. POLAROID film gives much quicker results.

#### Geiger–Müller (G–M) Tube

Geiger was a student of Lord Rutherford at Cambridge, and he carried out many important experiments in radioactivity (see p. 732). Working with Müller, he also designed a detector of radiation, which is widely used today. It is called a Geiger–Müller (G–M) tube (Fig. 39.1).

Basically, the G–M tube is a small closed glass tube T with a thin mica-end window W. It may contain a gas such as argon at very low pressure, together with a small amount of halogen gas such as bromine vapour. A central wire A, insulated from the tube, passes down the middle and is one electrode. The inside of the tube is coated with a conductor and forms the second electrode B. A high voltage, such as 450 V for a Mullard MX 168 G–M tube, is connected with its positive terminal to the wire A through a high resistance $R$, and its negative terminal is joined to the tube B.

Energetic particles or radiation can produce ions in a gas by collision with its atoms, stripping electrons from the atoms. They are hence called 'ionizing particles or radiation'. If an ionizing particle or radiation enters a G–M tube some argon atoms are ionized. The electrons produced then drift to the wire A and the positive ions to the tube B, and a small current

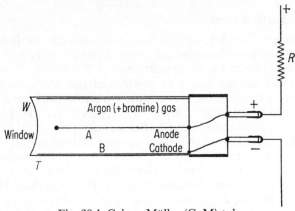

Fig. 39.1 Geiger–Müller (G–M) tube

is obtained for a short time. It is called a 'pulse current'. A corresponding pulse voltage is then obtained in the high resistance $R$, and this can be amplified and passed to an electronic unit such as a *scaler* or *ratemeter*.

### Scaler · Ratemeter

The scaler counts the number of pulse currents produced in the G–M tube by an ionizing radiation, and the number per second or *count rate is proportional to the intensity of the radiation*. The scaler uses a number of 'magic eye' tubes called *dekatron tubes* together with an *electromechanical register* (Fig. 39.2). In a small scaler, scales of units and tens are marked round the respective faces of two tubes, and after every 100 pulses are obtained the number is automatically passed to the register, which records hundreds, thousands and so on. After a suitable interval of time measured by a stop-clock the number or count is read from the register and the scales round the two tubes. The count rate, the number per second, can then be calculated.

Scalers count the actual number of pulses in the G–M tube. If the average rate or counts per second is required directly, the tube is connected to a *ratemeter*. This is an electronic circuit with a microammeter, which records on it a current proportional to the count rate.

A G–M tube, then, detects ionizing particles or radiation. Used in conjunction with a scaler or ratemeter, the number of the ionizing particles per minute or the intensity of the radiation can be measured.

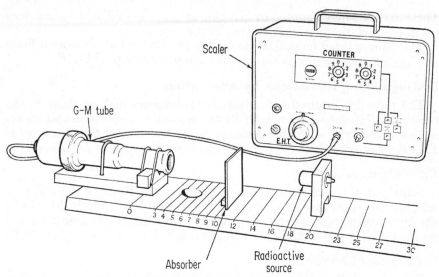

Fig. 39.2 Geiger–Müller (G–M) tube and Scaler

## Audible Effects

The pulse currents obtained from a G–M tube can also be passed to a loudspeaker. They are then heard as a succession of 'clicks'. Such an arrangement is often used in radiological laboratories to give audible warning when the radioactivity reaches a dangerous level.

'Clicks' can also be heard in a small transistor set placed near the G–M tube when the set is switched on and not tuned to any station. The discharges in the G–M tube produce radio waves which are picked up by the transistor set.

## Sources of Radiation

In the PANAX demonstration kit and others used for radioactivity experiments, sealed radioactive sources are used. Though the radiation hazard is negligible, the correct handling procedure must always be used. The sources are mounted in metal containers, *which must be handled only with forceps, and the open end of the sources must not be pointed towards the body*. With the aid of the forceps, the containers are fixed into perspex source holders.

| Source No. | Source strength (approx) | Material | Radiation |
|:---:|:---:|:---|:---:|
| S1 | 0·125 μc | Americium-241 | alpha, α |
| S2 | 0·125 μc | Strontium-90 | beta, β |
| S3 | 5 μc | Cobalt-60 | gamma, γ |
| S4 | 9 μc | Strontium-90 | |
| S5 | | Thorium carbonate | |
| S6 | | Uranium oxide–natural | |
| S7 | | Thorium oxide–natural | |

These may be safely handled by gripping at the sides so that the hand does not come near the small active area in the container.

The sources listed on p. 719 are used in the PANAX kit shown on Plate 24 – the microcurie ($\mu$c) is a unit of radioactivity (see p. 731).

## Distinguishing Radiations by Absorption

The radiation emitted by radioactive substances such as those in the table on p. 719 can be investigated using a suitable G–M tube and scaler. *In these and other radioactivity experiments safety regulations of the Department of Education and Science, and age regulations for experimenters, must be followed.*

### EXPERIMENTS

Switch on the scaler and allow it to warm up. Remove the end-cap, which protects the thin window, from the G–M tube. Connect the tube to the high-voltage supply inside the scaler marked E.H.T., and turn the dial on the E.H.T. until it reads 450 V, a suitable p.d. for this tube (Fig. 39.2).

### 1. Background Radiation

Even though the radioactive sources are all sealed in the containing box and this is placed well away from the tube, observe that the scaler indicates a reading which varies irregularly. This random radiation is due to cosmic rays which pass into the G–M tube and produce ionization. If it is a small count rate it may be ignored, but in accurate experiments it must be taken into account.

### 2. Natural Radioactivity

Place the natural sources S5, S6, S7 in turn near the G–M window using forceps always, and observe the increased count rate. This demonstrates the natural radioactivity of the sources. Observe the 'clicks' in a transistor set near the G–M tube (p. 719).

### 3. Alpha Particles

Now place the source S1, Am-241, in the perspex source holder (Fig. 39.2). Move the G–M window very close to the source and observe the fast count rate. Now place a sheet of thin paper between the source and the tube. Observe the greatly decreased count rate. Remove the paper and then bring it back.

*Conclusion.* The radiation emitted by Am-241 is absorbed by thin paper. The radiation actually consists of particles called $\alpha$ *(alpha)-particles*. If the tube window is placed more than about 4 cm from the source the count rate is small, showing that the range of the $\alpha$-particles is about 4 cm in air at atmospheric pressure. The window of the G–M tube used is made of specially thin perspex so that the $\alpha$-particles are not absorbed when the source is placed close to it.

### 4. Beta Particles

Return the $\alpha$-source to the box and place the source S2 – Strontium-90 – in position in front of the tube. Observe the fast count rate on the scaler. Place a sheet of thin paper between the source and the tube. Observe that

**PLATE 19**
Nuclear Reactions —
Man-made and Natural

19(a) British Experimental Pile O, BEPO, at Harwell, a natural uranium-fuelled graphite-moderated reactor. For research purposes, scientists are loading a large tube containing materials into the reactor core (p. 741).

19(b) ▶

19(b) High-altitude Research of nuclear reactions. A balloon is used to fly stacks of special photographic film to high altitudes, such as 20 miles, for detection of cosmic rays and interactions.

## PLATE 20
### Radioactive Materials

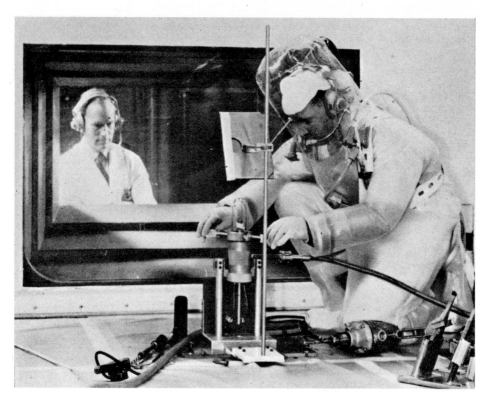

**20(a)** High-activity Handling at Harwell. A maintenance mechanic in protective clothing adjusting equipment in a highly active cell, which has walls over 1·5 m thick. The work is directed from behind the glass shield, which is part of a tank also more than 1·5 m thick containing zinc bromide solution (p. 736).

**20(b)** Gearbox inspection by gamma-rays from a radio-isotope, which is mixed with the oil before the car run (p. 736).

the count rate is only slightly reduced. Now place a sheet of aluminium a few millimetres thick between the source and the tube. The count rate is now considerably reduced.

*Conclusion.* The radiation from Strontium-90 is not absorbed by thin paper. It does not, therefore, consist of α-particles. The radiation, which is absorbed by a sheet of aluminium, consists of particles called β (*beta*)-*particles* (see p. 722).

### 5. *Gamma Rays*

Return the β-source to the box and place the source S3–Cobalt 60– in front of the tube. Observe the high count rate. Place paper and then an aluminium plate between the source and tube. Observe that the paper has no effect on the count rate and that the aluminium has a slight effect.

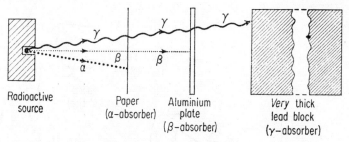

Fig. 39.3 Absorbers of α-, β-particles and γ-rays

Now use increasing thicknesses of lead block. For a thickness of about 2 cm the count diminishes considerably.

*Conclusion.* The radiation from Cobalt-60 is not absorbed by paper or aluminium, and is therefore not α- or β-particles. It consists of electromagnetic waves of extremely short wavelength (p. 331), and is called γ (*gamma*)–radiation, or γ-rays. The radiation is penetrating and stopped only by thick lead blocks (Fig. 39.3).

## Alpha-particles

We have already mentioned that α-particles are easily absorbed by paper and by air. Their range in air at atmospheric pressure is several cm. The mica window of a G–M tube, although very thin, would also normally absorb α-particles, and thus a specially thin perspex window is used for detecting α-particles by G–M tubes, as stated previously.

A *solid state detector* can detect α-particles more easily than a G–M tube. It consists of a thin film of semiconductor of small area, and when it is placed a short distance from the source, α-particles penetrate the film and liberate more electrons and holes (p. 707). An increased current then flows in the semiconductor, and it is amplified and passed to the scaler.

α-particles are deflected slightly by powerful magnetic fields. Consequently, they are heavy particles which carry an electric charge. The direction of the deflection, using Fleming's left-hand rule, shows that α-particles have a positive charge. They were identified by Lord Rutherford as atoms of helium (the next heavier gas to hydrogen) which have lost two electrons, or as helium nuclei, which have about four times the mass of hydrogen atoms (p. 733).

## Beta particles

An experiment which provides information about $\beta$-particles is illustrated in Fig. 39.4.

### Experiment

Place a collimated source of $\beta$-particles, that is, one with a narrow slit to define the direction of the $\beta$-particles, directly in front of the G–M tube shown. Observe the count rate. Now place a strong magnet in front

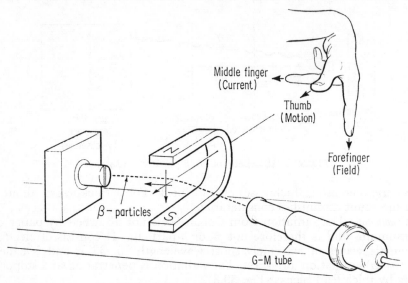

Fig. 39.4 Magnetic deflection of $\beta$-particles

of the source so that the magnetic field is perpendicular to the direction of the $\beta$-particles. Observe that the count rate diminishes. The $\beta$-particles are thus deflected by the magnetic field.

To identify the charge, move the G–M tube round on either side of the undeflected direction of the $\beta$-particles. On one side, as illustrated in Fig. 39.4, the count rate is observed to increase again. The $\beta$-particles have been deflected therefore in this direction.

*Conclusion.* Applying Fleming's left-hand rule, the $\beta$-particles are found to carry a negative charge. Further experiments show, in fact, that they are *electrons*.

## Gamma rays

Powerful magnets are unable to deflect $\gamma$-rays. An experiment which provides information about their nature can be carried out as follows:

EXPERIMENT

Place a small γ-source S (such as S3) in front of a G–M tube. Observe the count rate at a suitable short distance from the window of the tube. Now move the source back to twice the distance, and observe the diminished count rate. Remove the source to three times the distance and again observe the count rate.

From your measurements, see if the intensity of the γ-radiation, which is proportional to the count rate, has halved or gone down by a quarter, approximately, when the distance is doubled. Then see if the intensity has decreased to one-third or to one-ninth, approximately, when the distance is increased three times.

*Conclusion.* Measurements show an 'inverse-square law', that is, the intensity *I* is proportional to $1/d^2$, where *d* is the distance from the source. This is the same law obtained for the intensity variation with distance for radiation from light sources (see p. 448). γ-rays, in fact, consist of electro-magnetic waves like light and travel with the speed of light, but their wavelengths are about ten thousand times shorter than visible light and shorter than X-rays (p. 331).

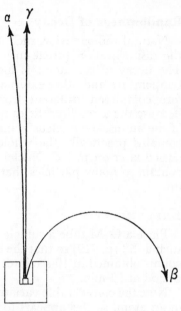

Field into paper

Fig. 39.5 Effect of magnetic field – diagrammatic

*Summary.* The following table summarizes some main points about α- and β-particles and γ-radiation.

| | Nature | Electrical charge | Mass | Velocity | Relative penetration | Absorber |
|---|---|---|---|---|---|---|
| Alpha, α-particle | Helium nuclei | $+2\,e$ | 4 units | c. $\frac{1}{20}$ velocity of light | 1 | Thin paper |
| Beta, β-particle | Electrons | $-1\,e$ | $\frac{1}{1840}$ unit | 3–99% velocity of light | 100 | Metal plate |
| Gamma, γ-ray | Electromagnetic radiation | No charge | Negligible | Velocity of light | 10 000 | Large lead blocks |

*e* = numerical value of charge on electron

Fig. 39.5 illustrates diagrammatically the effect of a perpendicular magnetic field, into the paper, on α- and β-particles and γ-rays. The

$\beta$-particles, which are very light and carry a negative charge, are deflected very considerably. The $\alpha$-particles, which are relatively very heavy and carry a positive charge, are deflected slightly in the opposite direction. The $\gamma$-rays are undeflected because they carry no charge.

## Randomness of Decay

Natural radioactivity, such as that from the salts of uranium, is due to the disintegration (break-up) or *decay* of some of the uranium atoms. The decay of the atoms is independent of chemical combination or of temperature and other external conditions such as pressure, and is therefore completely different from any known chemical change. This is because the $\alpha$- or $\beta$-particles or $\gamma$-radiation are emitted from the *nucleus* of the atoms, the nucleus being the small central core of an atom which contains practically the whole mass of the atom. The nucleus is discussed later on p. 732. Nuclei of many heavy elements such as uranium contain so many particles that they tend to become unstable and break up.

### Experiment

Place a G–M tube a sufficient distance from a weak radioactive source such as S2 (p. 719) so that the counts on a scaler occur slowly. Record the counts obtained in 10-sec intervals, one after the other, over a continuous period of 15 min.

Note the considerable variation in the counts. Each represents the decay of an atom, so this appears to be a random process. Plot a graph of the number of intervals $n$ in which a definite number $p$ of decays are obtained against this number $p$. Comment on the graph obtained.

These and other experiments show that radioactive decay is an entirely random process, that is, we can never identify or predict at any instant which atoms in uranium are about to become unstable. All we can say is that the number per second which decay at any instant is proportional to the actual number of atoms present. In a similar way, the number of car accidents per day tends to be proportional to the number of cars on the road. At a later time fewer atoms are present because many have disintegrated. Consequently, the number per second decaying at this instant is less than before. But the constant of proportionality remains the same, that is, if one-hundred-thousandth of all the atoms decay per second to start with, then one-hundred-thousandth of the remaining atoms decay per second at any later time.

## Half-life

The time taken for half the atoms initially present to decay is called the *half-life* of the radioactive element. Different radioactive elements have different half-lives, which are thus characteristic of the element. Radium has a half-life of about 1600 years, radon has a half-life of about four days and polonium a half-life of 3 min. Thoron–the radioactive gaseous element formed by the decay of thorium–has a half-life of about 8 min.

*Half-life of Radon*

Place a G–M tube T close to a small cell C, which is connected by rubber tubing to a polythene bottle G containing thorium-228 (Fig. 39.6). Connect T to the scaler. The thorium decays into thoron, a radioactive gas, some of which is transferred to C by squeezing the bottle G smartly.

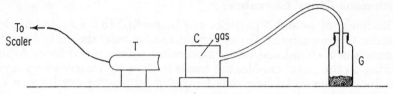

Fig. 39.6 Half-life of Thoron by scaler

Now obtain the number of counts in 10 s, and then, as the thoron decays, repeat this observation after every 30 s for about 5 min.

*Measurements*

| Time | Counts | Count rate |
|------|--------|------------|
|      |        |            |

*Graph*. Draw a graph of count rate v time. From it, find the time for half the thoron atoms initially present to decay.

*Conclusion*. The half-life of thoron is.......

As an illustration of the meaning of half-life, suppose a radioactive element has a half-life of 3 min. Then if the count rate obtained at a particular time is 512, in 3 min time the rate diminishes to 256. In another 3 min the rate diminishes to half of 256 or to 128. In another 3 min it again diminishes to half of what it was to start with, or to half of 128, so the count rate is 64. Thus 9 min from a count rate of 512, the rate diminishes to 64. If the count-rate is 400 for another element whose half life is $x$ min, and this rate diminishes to 100 in 16 min, then $(x + x)$ represents a time in which the rate diminishes to 100. Hence $2x = 16$ min; the half life $x$ is thus 8 min.

## Ionization and Penetration Power

Alpha-particles are relatively heavy and have a considerable amount of energy, as they are emitted with high velocity. When they pass into a gas such as air, for example, they are therefore able to strip electrons from the atoms, producing ions. An alpha-particle creates a considerable number of ions and electrons in its path. Correspondingly, since the energy of the alpha-particle is used up quickly, it has a relatively short range in air (p. 720).

Beta-particles are very light. Although their velocities are high, it follows that their energy is relatively low. Consequently, a small degree of ionization is produced in air by a beta-particle. A gamma-ray produces a much lower ionization than a beta-particle. In a given distance in air the ratio of the ionization produced is approximately

$$\alpha : \beta : \gamma = 10\,000 : 100 : 1,$$

and the penetration power or range is approximately $1 : 100 : 10\,000$.

### Continuous Cloud Chamber

The tracks of α- and β-particles can be studied in a continuous cloud chamber. One form is the Taylor *diffusion cloud chamber* shown in Fig. 39.7. It consists of a cylindrical perspex chamber with a metal base. A black covering on the metal enables the tracks to show up clearly when viewed from above. The metal base may be cooled by placing dry ice in a container below it.

EXPERIMENT

Using a dropper, place some methylated spirit inside the top of the chamber by moistening the felt or sponge ring inside the lid. A drop or two may also be placed on the black base and allowed to spread over it. The base of the container is then unscrewed and 'dry ice' is pressed in contact with the metal plate. The foam or sponge inside is replaced to keep the

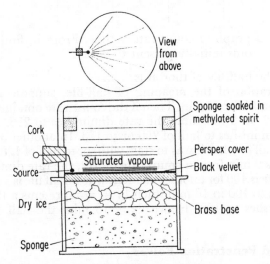

Fig. 39.7 Diffusion cloud chamber

'dry ice' in contact with the plate, and the base is screwed on again. It is important that the chamber should be levelled, as convection currents are then reduced. Rubbing the perspex top provides an adequate electric field for removal of ions already present.

The metal base of the chamber is cooled to about −60°C by the dry ice. Alcohol evaporates continuously from the lid, which is at room temperature, and the vapour sinks or diffuses continuously to the bottom of the chamber over the whole of its area. Just above the cold metal base the vapour is supersaturated, that is, its vapour pressure is above the saturated vapour pressure at that temperature.

The radioactive sample, such as radium, is introduced into the chamber by a wire holder located in cork in the side of the chamber. Looking down through the top of the glass vessel, observe the tracks or trails formed by particles emitted by the radium from the tip of the metal holder. They show the passage of α-particles, which ionize the atoms of air (p. 718). The vapour condenses round the ions, forming droplets which scatter light so that the tracks are visible. Vapour trails are sometimes seen at the rear of high-flying aircraft. Note that the tracks of the α-particles appear straight. Some tracks are shorter than others, thus indicating that the particles have different energy.

## Wilson Cloud Chamber

The first cloud chamber was designed by C. T. R. Wilson in 1912. It provided invaluable information on radioactivity and atomic structure

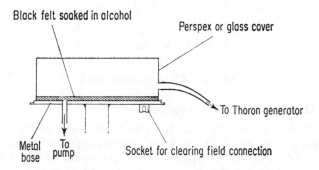

Fig. 39.8 Wilson cloud chamber

because, for the first time, it became possible to photograph the tracks of ionizing particles (see Fig. 39.9).

Basically, the cloud chamber consists of an expansion chamber in which supersaturated water vapour condenses on ions produced by the passage of an ionizing particle. Simultaneously the chamber is illuminated and the water droplets reflect light and are photographed, so that the particle track is seen.

In a simple laboratory form of Wilson cloud chamber, a few cm³ of 50% alcohol are placed on the felt pad at the bottom of the chamber (Fig. 39.8). By means of a hand vacuum pump, air is rapidly drawn out of the chamber. Thoron gas is then transferred from a bottle by squeezing so that it passes into the chamber. Thoron emits α-particles, and the tracks can then be seen. If required, the ions can be cleared by applying a p.d. across the chamber.

γ-rays are poor ionizing agents compared with α- or β-particles. If a powerful source of γ-rays were available, 'spidery' tracks would be seen. Not only can the tracks of α- and β-particles be seen but their ranges and energies can be measured. An α-particle, being relatively heavy, normally travels in a straight line. β-particles are more readily deviated by collisions with the air molecules. Fig. 39.9.

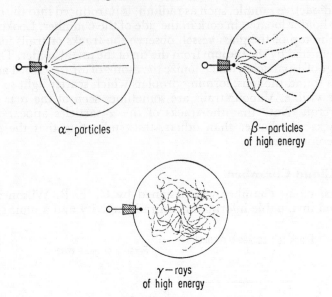

α–particles

β–particles
of high energy

γ–rays
of high energy

Fig. 39.9 Cloud chamber tracks

## Particle Detection by Electroscope

In electrostatics we saw that high voltages were concerned but the charges were very small. A gold-leaf electroscope indicates such small charges (p. 646). It has a capacitance of the order of $10^{-12}$ farad, so if voltages of a few hundred volts, such as on a rubbed ebonite rod, produce a deflection of the leaf, this means that the instrument can detect a charge of the order $Q$ given by $Q = CV = 10^{-12} \times 100 = 10^{-10}$ coulomb. If the charge flows away from the leaf in a time $t$ of 10–100 sec, the current detected would be $Q/t$ or $10^{-11}$–$10^{-12}$ A. Electroscopes, then, can measure currents smaller than moving-coil galvanometers, and they were therefore used in early investigations on radioactivity.

Ions in air, produced by α-particles, carry charges of the order $e$, the electronic charge, about $10^{-19}$ coulomb. Suppose a million ions are produced by an α-particle. The total charge on them is about $10^{-19} \times 10^6$ or $10^{-13}$ coulomb, and if they are collected by an electrode in 1 sec the current flowing is $10^{-13}$ A. A sensitive electroscope can thus detect the ions produced by an α-particle.

## Pulse (Wulf) Electroscope

An electroscope suitable for radioactivity investigations is shown in Fig. 38.10. It is called the *pulse electroscope*, and consists of an adjustable side electrode S inside a case C which can be earthed (Fig. 39.10). The moving part of the instrument is a light metal flag, mounted on a taut phosphor-bronze suspension. A top electrode T is connected to the flag, and T is used to test the existence and nature of electrical particles, as we shall see. A support rod R is connected to the case.

The side electrode S is kept at a potential of about 2000 V (2 kV), and it is moved towards the flag until it is about 1–2 mm away. In this position the flag is attracted towards S, and after it touches S and is charged to the same potential the flag breaks away from S. If the flag is discharged at regular intervals, for example by opposite charges moving towards the electrode T connected to the flag, the latter is attracted towards S again and flies back again, at a rate depending on the number of charges per second reaching T. The flag thus 'beats' at a definite rate. *Ions* produced between R and T hence can cause the flag to 'beat'. When the distance between S and the flag is about 1 mm the rate of beating is about one per second.

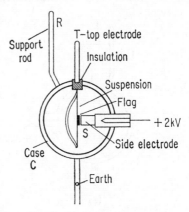

Fig. 39.10 Pulse electroscope

## Demonstrations of Ions Produced by Flame

The heat energy of a flame is sufficient to remove electrons from some gas atoms in the flame, thus creating ions. These ions can be detected by the pulse electroscope.

The side electrode S is maintained at a potential of +2 kV (2000 V), and the case C is earthed (Fig. 39.10). S is moved near to the flag until it beats once, that is, the potential of the flag and the electrode T is the same as S. A lighted match is now held between R and T, and the flag is observed to beat at a regular rate. This shows the existence of negative ions, which drift towards T and cause the flag to discharge at a rate depending on the number of ions per second reaching the electrode. When the side potential is changed negatively to −2 kV the flame again causes the flag to beat faster, showing the presence of positive ions as well as negative ions.

## Demonstration of α-particles

The α-particles emitted by radioactive substances produce ionization of air or gas molecules. Radioactive sources can therefore be detected by using a pulse electroscope. To demonstrate the emission of α-particles, a weak radium source is taken out of its container using a lifting tool, and clamped on R by the handle (Fig. 39.10). The potential of the side electrode is made +2 kV, and then −2 kV, and the leaf is observed to beat at a faster rate on each occasion; thus showing that the α-particles emitted by the radium source produce both positive and negative ions in air.

## Measurement of Range of α-particles in Air

To measure the range of α-particles emitted by a radium source, an ionization chamber is used (Fig. 39.11(i)). This chamber is clipped on to the support R, with its gauze window uppermost, and is slid down R as far as possible. The weak radioactive source S is now raised by the lifting tool, and fastened to the support rod R so that S faces downward, directly above the gauze window.

The rate of beating of the leaf is observed at different distances $d$ of the source S from the window of the ionization chamber. As $d$ increases, the rate of beating of the flag diminishes, showing that less ions per second are produced in the ionization chamber. A graph similar to that shown in Fig. 39.11(ii) is obtained, from which it follows that the range of the α-particles is about 7 cm.

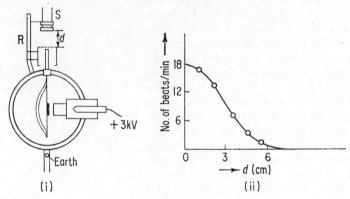

Fig. 39.11  Range of α-particles

### Measurement of Half-life Period of Thorium Emanation

The half-life period (p. 724) of thoron gas can be measured with the aid of a chamber attached by a tube to a plastic bottle containing thorium-228. See p. 725. The chamber is clipped on to the support rod R and slid down the rod so that the top electrode of the electroscope enters the centre of the chamber as far as possible. The side electrode is kept at a potential of $+3$ kV, and the gap between the side electrode and the flag is made 1–2 mm.

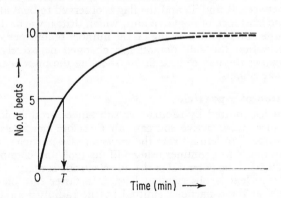

Fig. 39.12  Half-life determination

The bottle is shaken to release any thoron gas trapped, and then squeezed once or twice, keeping it upright. The thoron gas is transferred to the chamber, where it decays with the emission of α-particles and produces an ionization current. The flag should begin to beat at the rate of 1 beat or more per 10 sec. The time of successive beats is noted, and this time lengthens as the radioactive gas decays.

The actual number of the beats is plotted against the time (Fig. 39.12). A smooth curve is drawn through the points and extended to level off horizontally. If the number of the beats corresponding to the horizontal part of the curve is ten, as shown, the half-life period $T$ is the time to reach the fifth beat. This time is read from the graph.

## Health Hazards

$\alpha$-particles are stopped by the outer layers of skin as their penetration power is small (p. 720). If they penetrate into the body, however, they are dangerous and damage organs.

$\beta$-particles are more penetrating than $\alpha$-particles. Inside the body they are dangerous.

$\gamma$-rays and energetic $\beta$-particles destroy cells in tissue when the body is exposed to them, and upset the natural chemical reactions. This may lead to injury or death. X-rays have the same effect (p. 702).

The body is exposed to radiation originating from outer space, but the amount per year is small and harmless. Exposure to cosmic rays high above the earth increases the amount considerably. Luminous paint used for the dials of watches is radioactive. When an atomic bomb explodes, apart from the blast hazard, dangerous products are released which are swept into the atmosphere. Strontium-90, for example, may settle thousands of miles away from the centre of an explosion on grass, or soil, where eventually it may find its way into the body through the animals which consume the grass or the food produced from the soil. Strontium-90 is absorbed by bone, and as it emits $\beta$-particles it upsets the formation of the red corpuscles in the body. Other dangerous products are also emitted in an atomic explosion.

The best shields against $\gamma$-radiation in the laboratory are thick blocks of lead, iron and high-density concrete. Workers in radiological laboratories are required to wear badges containing photographic film sensitive to radiations, and these are checked at least once a month to find out whether the wearer has been exposed to a dangerous dose of radiation.

In view of the increase in radioactive man-made products since 1945, it is interesting that part of the German fleet scuttled at the end of the 1914–18 war in Scapa Flow will be raised to provide steel which is practically non-radioactive.

## Some Units of Radioactivity

The *curie* is a unit of radioactivity. It is defined as $3 \cdot 7 \times 10^{10}$ disintegrations per second; this is about the number of disintegrations per second occurring in 1 g of radium. The *microcurie* ($\mu$c) is a millionth of a curie, or $3 \cdot 7 \times 10^4$ disintegrations per second. The radioactivity of some weak sources used in school experiments is given on p. 719.

The ionization produced by a radioactive substance is expressed in *roentgens*. The roentgen is defined in terms of the electric charge produced when ions are formed in a unit volume of air at standard temperature and pressure. The *milliroentgen*, denoted by *mr*, is one-thousandth roentgen.

From all natural sources a person receives about 140 mr per year; cosmic rays provide about 35 mr and potassium about 25 mr.

## ATOMIC STRUCTURE

### Geiger and Marsden's Experiment

As already seen, the α-particle is a heavy particle moving at high speed. It therefore has considerable energy, and Rutherford decided to use it to disrupt the atom.

At Rutherford's suggestion, Geiger and Marsden in 1909 investigated the effect of α-particles incident on thin films of metal. A radioactive source was contained in an evacuated vessel V, and emitted α-particles in a narrow beam on to a very thin metal foil F of gold or other metal of high atomic weight (Fig. 39.13). The path of α-particles could be traced by the pin-points of light (scintillations) produced as they struck a fluorescent screen S, in the focal plane of a microscope M. Geiger and Marsden investigated the scattering of α-particles by the atoms of the metal. They observed scintillations not only in the straight-through position, $\theta = 0°$, but also in directions which were at large angles such as 45° to the incident beam. A few scintillations were even observed in a direction FB, which was opposite to the incident beam. Some α-particles were thus deflected through very large angles, as if they had reached, inside the metal atom, a centre which repelled them violently.

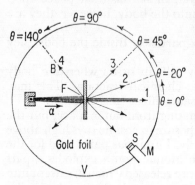

Fig. 39.13 Scattering of α-particles by gold foil (Geiger and Marsden)

### The Nucleus · Atomic Structure

Rutherford, with characteristic genius, saw immediately the significance of these observations. He said at the time that the repulsion of α-particles in an opposite direction was as surprising as if a shell fired from a gun had been repelled by tissue paper! Sir J. J. Thomson had thought that an atom might be pictured as a cloud of positive charge, with negative electrons spread uniformly through the cloud like currants in a pudding. Rutherford now proposed, however, that there was a *concentration* of positive electricity inside the atom which repelled the positively charged α-particles. On this basis he calculated how many α-particles would be scattered through any given angle (Fig. 39.14), and the relationship was verified by Geiger and Marsden in subsequent experiments.

In 1911 Rutherford suggested the first adequate structure of the atom. The atom, he said, consists of a concentration of positive electricity in

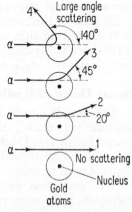

Fig. 39.14 Scattering by nuclei

a very tiny volume in the heart or centre of the atom, called the **nucleus,** with electrons moving round the nucleus. Most of the atom is empty. If we imagine the nucleus magnified to the size of a cricket ball the farthest part of the atom would be a few miles away, and the space between the ball and the farthest part would be empty except for electrons. The nucleus of the hydrogen is called a **proton.** Since there is one electron round the hydrogen nucleus, the proton carries a charge of $+e$, where $e$ is the numerical value of the charge on the electron. See Fig. 39.15(a).

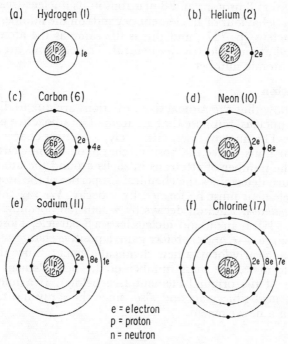

Fig. 39.15 Atomic structure and atomic numbers of some atoms

## Protons, Neutrons . Atomic Nuclei

By using α-particles as 'bullets', Rutherford and collaborators found that *protons* were ejected from the nuclei of many elements when collisions occurred. Atomic nuclei thus contain protons. In 1932 Sir James Chadwick carried out experiments which showed that a particle can be obtained from a nucleus *which carries no charge* and has a mass about the same as a proton. It is called a *neutron.*

We now believe that the nucleus of an atom is built up of protons and neutrons. (Other particles have been discovered in the nucleus, but this will not concern us.) The particles inside a nucleus are called **nucleons.** The hydrogen nucleus is the proton, p, which has a positive charge $+e$ numerically equal to the charge $e$ on an electron, but it is nearly 2000 times as heavy as the electron. The next heavier element is helium, whose atomic number is 2; this has therefore a charge of $+2e$ on its nucleus, that is, the nucleus contains 2 protons. The mass of the helium nucleus is 4, and hence the nucleus must also contain 2 neutrons. The symbol for the helium nucleus is $^{4}_{2}$He, the upper number representing the atomic mass or *mass number* and the lower number representing the *atomic number* or the number of protons in the nucleus. The proton itself is represented by $^{1}_{1}$H, the neutron by $^{1}_{0}$n. Each has a mass 1, but the neutron has no charge. An electron is represented by $_{-1}^{0}$e, since it has negligible mass and charge$-1$e. The nitrogen nucleus is written $^{14}_{7}$N; it has an atomic mass of 14 and atomic number 7, that is, there are 7 protons and 7 neutrons in the nucleus. Oxygen, $^{16}_{8}$O, has 8 protons and 8 neutrons in its nucleus (see Fig. 39.16). The chemical behaviour of an element depends on the number of electrons round the nucleus (p. 737), and this is the same as the atomic number, since a normal atom is electrically neutral. The elements are thus identified by their atomic number.

**Transmutation**

We know now that the radioactive radiations are due to disintegration of an atomic nucleus. Suppose that an atom of a radioactive element loses an α-particle. The nucleus loses, effectively, a positive charge of $+2e$ and a mass of four units. Since the positive charge on the nucleus of an atom determines the number of electrons in shells around the atomic nucleus, and this in turn determines the chemical properties of the atom, it follows that a new element must be formed by α-decay. We say that there has been a *transmutation*. Radium decays by α-emission, forming the rare gas radon (see p. 734). If an atomic nucleus loses a β-particle, it loses effectively a negative charge, $-e$, and therefore gains a positive charge, $+e$. This type of decay again clearly forms a new chemical element. Since the β-particle has negligible mass, the mass number of the new element will be the same as that of the original element (see below). If a nucleus loses a γ-ray it loses negligible mass and also, since the γ-ray has no charge, it does not form a new element.

**Examples**

1. Radium (Ra)-226 decays to radon (Rn) by α-emission (helium nucleus):

$$^{226}_{88}\text{Ra} \rightarrow {}^{222}_{86}\text{Rn} + {}^{4}_{2}\text{He}$$

The mass number, 226, and the atomic number 88, are both changed.

2. Radon decays to radium-222 by β-emission (electron):

$$^{222}_{86}\text{Rn} \rightarrow {}^{222}_{88}\text{Ra} + 2{}^{0}_{-1}\text{e}$$

The mass number, 222, is unchanged; the atomic number is changed.

3. A uranium nucleus, U-238, atomic number 92, emits two α-particles and two β-particles (electrons) and forms a thorium (Th) nucleus. What is the symbol of this nucleus?

Change in mass number $= 2 \times 4 = 8$ (α-particle has mass number 4; β-particle has negligible mass)

$\therefore$ mass number of Th nucleus $= 238 - 8 = 230$

Change in atomic number $= + 2 \times 2$ ($\alpha$-particles) $- 2 \times 1$ ($\beta$-particles)
$$= 2.$$
$\therefore$ atomic number of Th nucleus $= 92 - 2 = 90$
$$\therefore \text{ Symbol} = {}^{230}_{90}\text{Th}.$$

The nuclear decay may be expressed by:

$$ {}^{238}_{92}\text{U} \rightarrow {}^{230}_{90}\text{Th} + 2\,{}^{4}_{2}\text{He} + 2\,{}^{0}_{-1}\text{e} $$

An *isobar* is a nucleus with the same mass number as another but with a different atomic number. Isobars are usually produced in $\beta$-emission, as in Example 2. Here radon and radium are isobars, each having the same mass number 222.

In the nuclear equations given in the examples, note carefully that the mass number on one side (at the top) is equal to the total mass number of the products on the other side. Similarly, the charge on one side (at the bottom) is equal to the total charge on the other side.

### Artificial Transmutation

Artificial disintegration was first achieved when energetic $\alpha$-particles were used by Rutherford to disrupt the nitrogen nucleus, and he found that protons were emitted after the collision. The $\alpha$-particle is a helium nucleus with 2 protons, the nitrogen nucleus has 7 protons, and hence the total charge concerned is equal to $+9e$. After the disintegration the products are one proton together with another atom X (Fig. 39.16). The latter must have 8 protons because the total charge before collision equals the total charge after collision, and hence X is an *oxygen* atom, because the atomic number of this element is 8. Thus a nitrogen atom has been transformed into an oxygen atom, a process called *transmutation* (Fig. 39.16). See also Plate 18.

Also, mass of $\alpha$-particle + nitrogen nucleus $= 4 + 14$
$$= 18,$$
and mass of proton produced
$$= 1.$$
$\therefore$ Mass of oxygen atom produced $= 18 - 1 = 17$

We can therefore write the nuclear reaction as:

$$ {}^{4}_{2}\text{He} + {}^{14}_{7}\text{N} \rightarrow {}^{17}_{8}\text{O} + {}^{1}_{1}\text{H}. $$

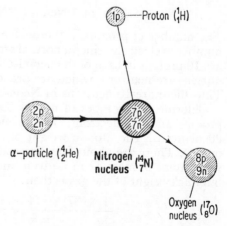

Fig. 39.16 Disintegration of nitrogen nucleus

### Masses of Nuclei · Mass Spectrograph

Atoms which have lost one or two electrons become *positive ions*, which are practically as heavy as atoms because electrons are so light. Density measurement of an element, such as that discussed on p. 17, gives the average mass of *all* the atoms in the element if we divide it by the number of atoms per unit volume. This is called the 'chemical atomic weight or mass'. It does not provide the mass of individual atoms.

In 1911, however, Sir J. J. Thomson deflected individual ions by using an electric and magnetic field. If the ions of a gas have the same charge the heavier ions are deflected less than the lighter ones, and so the curvature of the paths of the two ions are different. The *mass spectrograph*, as the instrument is called, enabled the masses of the ions to be measured. In Thomson's experiment ions of different masses produced different curves on a photographic plate, and measurements of their dimensions enabled the masses to be calculated.

### Isotopes

Thomson found that a sample of neon gas produced two curves on the photographic plate, one fainter than the other. When the masses were calculated it was found that the stronger curve was due to atoms of mass 20 and the fainter one to atoms of mass 22. The atomic weight of neon is 20·2. Neon gas actually has about 90% of atoms of mass 20 and 9% of mass 22. The 'atomic weight' as measured by chemists gives the average mass of the atoms in the gas; the individual atoms cannot be distinguished. In Thomson's mass spectrograph, however, the masses of individual atoms (stripped of one or two very light electrons) can be measured.

Atoms of the same element which have the same chemical properties but different masses are called *isotopes*. The chemical properties depend on the number of electrons round the nucleus, because chemical combination is due to an exchange of outer electrons between elements. The *charge* on the nuclei of the two isotopes of neon is hence the same – it is $+ 10e$ in this case. The symbols for the isotopes are written as:

$^{20}_{10}$Ne and $^{22}_{10}$Ne, or $^{20}$Ne and $^{22}$Ne, or Neon-20 and Neon-22.

The number of protons in the nucleus of an atom is always equal to the number of electrons, since a normal atom is electrically neutral. Thus there are 10 protons in each of the neon isotopes. The remaining particles in the nucleus are neutrons, which are practically as heavy as protons (p. 733). Thus there are 10 neutrons in Neon-20, but 12 neutrons in Neon-22.

Chlorine has isotopes of masses 35 and 37 respectively. Each has 17 protons and 17 electrons. The former has 18 neutrons and the latter 20 neutrons. The atomic weight of chlorine is ordinarily 35·5 approximately. There are about four times as many atoms of mass 35 compared with atoms of mass 37 in a given sample of gas, because the average or 'atomic' weight of the gas is then

$$\frac{(1 \times 37) + (4 \times 35)}{1 + 4} = \frac{177}{5} = 35 \cdot 4.$$

Hydrogen has isotopes which are important in the study of nuclear energy (p. 742). Most atoms in ordinary hydrogen gas have masses 1, but a tiny percentage, 0·015%, have masses twice as heavy or mass 2. This is known as *heavy hydrogen* or *deuterium* and the symbol is $^2_1$H or D.

**PLATE 21**
Laboratory Apparatus · I

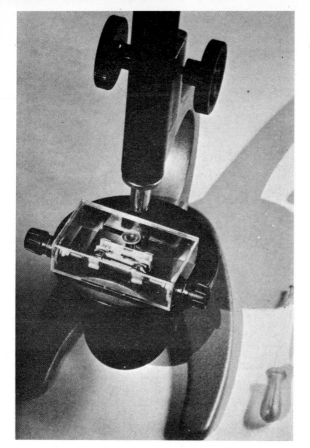

21(**a**) Brownian motion apparatus (p. 27).

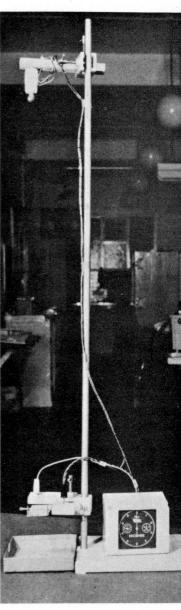

21(**b**) Free-fall experiment for $g$ (p. 75)

21(**d**) Van de Graaff model
electrostatic generator (p. 656).

21(**c**) Ticker-timer and tape, with trolley (p. 59).

PLATE 22
Laboratory Apparatus · II

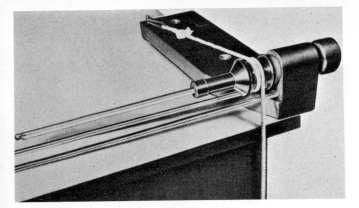

22(**a**) Mechanical equivalent of heat, showing solid metal calorimeter and thermometer (p. 274).

22(**b**) (Waves) Ripple tank, showing two vibrators and interference of water waves (p. 324).

22(**c**) ▶
22(**c**) Boyle's Law by Bourdon gauge (p. 197).

22(**d**) (Waves) Microwave apparatus, showing refraction of microwaves by prism containing liquid paraffin (p. 398).

# PLATE 23
## Laboratory Apparatus · III

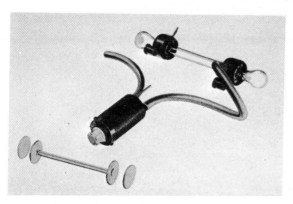

23(a) Discharge tube for conduction in gases at low pressure (p. 679).

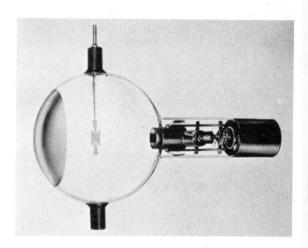

23(b) Maltese cross (p. 685).

23(c) Perrin tube with coils (p. 684).

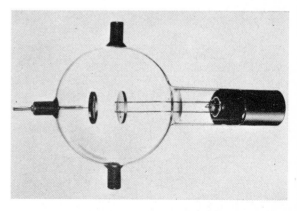

23(d) Diode valve (p. 689).

# PLATE 24
## Laboratory Apparatus · IV

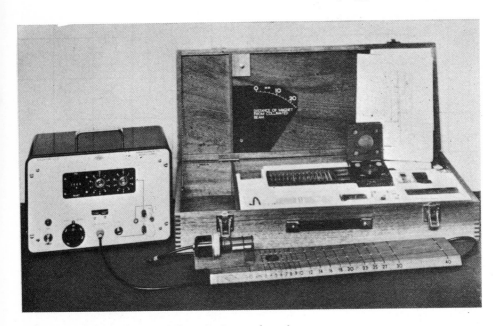

**24(a)** Scaler (left); Geiger–Müller tube (on graduated board); large case with long box containing aluminium and lead absorbers; wide open box containing weak sources of radiation; and a stronger source, in centre, covered with a lead disc (p. 719).

**24(c) ▶**
**24(c)** Diffusion cloud chamber, with source holder and dry ice container (p. 726).

**24(b)** LEYBOLD fine beam tube, showing circular deflection of electrons in uniform magnetic field (p. 683):
▼

Thus

Hydrogen, $_1^1H$, has one proton and one electron.

Deuterium, $_1^2H$, has one proton and one neutron in the nucleus, and one electron. A deuteron is a nucleus of deuterium.

A third isotope of hydrogen can be made, but it is unstable. It is called *tritium*, symbol $_1^3H$.

## Uses of Radioactive Isotopes

Isotopes can be made artificially by firing neutrons or protons or deuterons at elements. Thus the isotope sulphur-35 can be made by bombarding the normal atom sulphur-34 by neutrons:

$$_{10}^{34}S + _0^1n \rightarrow _{10}^{35}S.$$

Isotopes made in this way are unstable. Their nuclei decay and emit α- or β-particles and γ-rays. They are therefore called *radioisotopes*. The half-life of sulphur-35 is about 87 days. The half-life of iodine-131 is about 8 days. The half-life of cobalt-60 is about 5 years.

As they are radioactive, radioisotopes can be traced by instruments such as Geiger–Müller tubes. Radioisotopes are widely used in industry and medicine. Thus compounds of carbon-14 have been mixed with insecticide and used to trace their action on plants. The γ-rays from cobalt-60 are used to sterilize medical supplies as bacteria are destroyed on exposure to the rays. Water-marks in printing may be made by exposing the paper to the weak β-particles from sulphur-35. On Plate 19 a scientist can be seen examining a gear-box by using a radioisotope.

## The Periodic Table

The elements can be grouped into families, the members of each family having similar physical and chemical properties. When the elements are listed in ascending order of *atomic number* members of a particular family occur at intervals, or *periodically*, and this is known as the periodic law. It suggests that atoms have a structure which varies periodically with increasing atomic number.

*Examples*

| Hydrogen[1] | | | | | | | Helium[2] |
|---|---|---|---|---|---|---|---|
| Lithium[3] | Beryllium[4] | Boron[5] | Carbon[6] | Nitrogen[7] | Oxygen[8] | Fluorine[9] | Neon[10] |
| Sodium[11] | Magnesium[12] | Aluminium[13] | Silicon[14] | Phosphorus[15] | Sulphur[16] | Chlorine[17] | Argon[18] |
| Potassium[19] | Calcium[20] | ... | | | | | |

The families generally run vertically downwards. Thus, lithium, sodium and potassium belong to the *Alkali Metals*, fluorine and chlorine to the *Halogens* and so on. The table shows the first twenty elements and their atomic numbers.

Mendeléeff, who produced the first periodic table in 1869, classified the elements in order of atomic weight. Nickel has a smaller atomic weight than cobalt, as the great majority of its isotopes have a smaller mass number than the atoms of cobalt. The atomic number, however, is greater than that of cobalt. In order of atomic number or chemical properties, therefore, nickel is actually higher in the table.

### Electron Shell Structure

The electronic structure of an atom determines its chemical properties (see p. 733). The least chemically active atoms are those of the rare gases, helium (2 electrons), neon (10), argon (18), krypton (36), xenon (54) and radon (86), and it is reasonable to assume that these have a very stable electronic structure. Atoms having one electron more than these, lithium (3), sodium (11), potassium (19), rubidium (37) and caesium (55), are all very reactive chemically and have properties closely resembling each other. The same is true of atoms having one electron less, fluorine (9), chlorine (17), bromine (35) and iodine (53).

Atoms are said to be built up of *shells* of electrons. When these shells are completely filled the element is inert; and when it has one electron too many or too few the element is particularly reactive chemically. Successive shells are complete when they contain $2n^2$ electrons, where $n =$ 1, 2, 3, 4, etc., the number of the shell from the nucleus. Thus shell 1 contains two electrons, shell 2 eight electrons, shell 3 eighteen electrons, shell 4 thirty-two electrons and so on. To explain the case of argon, it is necessary to assume that shell 3 is split into *sub-shells*, and this assumption is also needed when dealing with shells of higher number.

The structures of a few simple atoms and shells are shown simplified in Fig. 39.15. Further discussion of electron shell structures can be obtained from more advanced books on physics and chemistry.

# NUCLEAR (ATOMIC) ENERGY

### Einstein's Mass-energy Relation

In 1905 Einstein showed, from his *Theory of Relativity*, that when a decrease of mass $m$ occurs an amount of energy $E$ is produced which is given by

$$E = mc^2,$$

where $E$ is in joules when $m$ is in kg and $c$ is $3 \times 10^8$, the numerical value of the velocity of light in metres per second (see p. 332). A decrease in mass of 1 *milligramme* would thus produce an amount of energy $E$ given by

$$E = 1/10^6 \times (3 \times 10^8)^2 \, \mathrm{J}$$
$$= 9 \times 10^{10} \, \mathrm{J}$$
$$= 90\,000 \text{ million J.}$$

Now a 100-W lamp uses 100 J of energy per second, and would therefore use 0·36 million J of energy in 1 h. Hence 90 000 million J would keep

250 000 100-W lamps burning for 1 h, and this large amount of energy would be obtained from a decrease of mass of only 1 mg.

As we shall see shortly, a considerable mass change may occur in a nuclear reaction with a uranium atom. When such reactions take place in only a small mass of uranium they involve many millions of atoms, and a great amount of energy is released. This has led to the development of the *nuclear reactor* as a source of power in Great Britain, and many reactors are now operating efficiently throughout the country (p. 741).

Einstein's mass-energy law was shown to be true in experiments on radioactivity, where energetic particles such as α-particles are emitted when a nucleus disintegrates, and the decrease of mass was found to be numerically equal to the energy produced. The deep significance of Einstein's law is that *mass is a form of energy*. When we say that the masses of the substances in a reaction are conserved we also mean that the total energy is constant. The atomic mass of carbon-12, $^{12}C$, is taken as 12·00 atomic mass units (a.m.u.), and since mass and energy can be changed from one to the other by Einstein's relation, energy changes in nuclear reactions are often expressed in terms of a.m.u.

## Binding Energy

The particles of the nucleus, the protons and neutrons, are called *nucleons*. The nucleons are kept together in a very tiny volume by powerful forces whose nature is not yet known. To tear the nucleons apart requires energy, and this amount of energy is known as the *binding energy* of the nucleus. Plate 18 shows one of the atom-smashing machines used to break up the nucleus. The mass of the nucleus should be less than the total mass of the separated nucleons; the difference in mass, from Einstein's law, represents the amount of binding energy.

An an illustration, we know by measurements that:

$$\text{Mass of proton} = 1\cdot0073 \text{ a.m.u.,}$$
and $$\text{Mass of neutron} = 1\cdot0086 \text{ a.m.u.,}$$

Now the carbon nucleus $^{12}_{6}C$ has 6 protons and 6 neutrons, and measurement of the mass of the nucleus shows it is 12·0038 a.m.u.

But Mass of 6 protons = $6 \times 1\cdot0073 = 6\cdot0438$ a.m.u.,
and Mass of 6 neutrons = $6 \times 1\cdot0086 = 6\cdot0516$ a.m.u.
∴ total mass = 12·0954 a.m.u.
∴ Mass of nucleus is *less* than the mass of its nucleons by 12·0954 − 12·0038, or 0·0916 a.m.u.
∴ Binding energy of carbon nucleus = 0·0916 a.m.u.

The binding energy of a uranium nucleus, which has 92 protons and 146 neutrons, is much greater than that of a carbon nucleus. If a heavy nucleus can be partly disintegrated the masses of the particles produced are less than the masses of the particles when they were locked together in the nucleus, and the difference in mass is converted into energy.

### Nuclear Fission · Chain Reaction

As we saw on p. 735, energetic α-particles can disrupt an atomic nucleus. Neutrons, however, can penetrate to the nucleus much more easily than α-particles; they carry no charge and are therefore not repelled by the positive nuclear charge as α-particles (positive charges) are. From 1934, Fermi and others began to use neutrons as 'bullets' to fire at atomic nuclei. Usually only a small fragment was 'chipped' from the nucleus, and consequently, only a small amount of energy was released.

In 1938 a study was made of the results of firing neutrons at the heaviest nucleus, uranium. It was then found that the nucleus this time had split into two large nuclei, such as barium, Ba, and krypton, Kr. Several neutrons were also produced, a point of special significance as we see later (Fig. 39.17). The reaction can be written:

$$^{235}_{92}U + ^{1}_{0}n \rightarrow ^{143}_{56}Ba + ^{90}_{36}Kr + 3^{1}_{0}n$$

The break-up of the uranium nucleus into two large nuclei was called *nuclear fission*. The total mass of the uranium nucleus and incident neutron

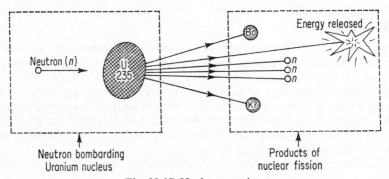

Neutron bombarding
Uranium nucleus

Products of
nuclear fission

Fig. 39.17 Nuclear reaction

was less than the total mass of the products obtained, barium, krypton and neutrons, and the difference in mass was converted into energy accompanying the reaction. Using's Einstein's mass-energy relation on p. 738, a calculation showed that if all the atoms in 1 kg of uranium undergo fission the energy released would be as much, theoretically, as that released by burning 3 million kg of coal!

The practical possibility of achieving this enormous energy was due to the release of several neutrons when nuclear fission of a uranium atom took place. Under control, these neutrons could each produce fission of another atom, with the release of more neutrons, which in turn would produce fission of more atoms, thus creating a *chain reaction* (Fig. 39.18). The fission of millions of uranium atoms would thus take place very rapidly throughout the uranium, and an enormous amount of energy would be produced in a short time. The speed of the neutrons is a critical factor in nuclear fission, and nuclear reactors such as those at Harwell in England contain material to modify their speed. If the neutrons are slowed down, the fission is more effective.

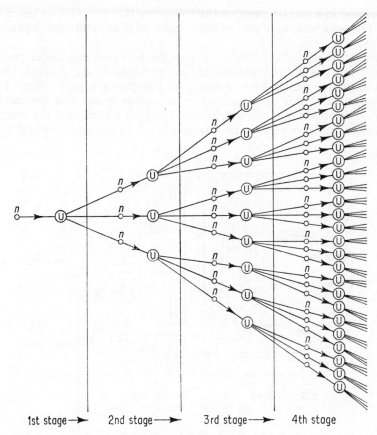

1st stage ➤   2nd stage ➤   3rd stage ➤   4th stage

Fig. 39.18 Chain reaction

## Nuclear Reactor Principle

Natural uranium contains about 1 part by weight of atoms of atomic weight 235 and 139 parts by weight of atoms of atomic weight 238. Nuclear fission occurs with the atoms of atomic weight 235 when uranium is used in nuclear reactors.

Fission takes place only if the bombarding neutron is slow. Graphite is therefore used round the uranium to moderate the speed of the neutron, so that the chain reaction is prevented from dying out. If there are too many neutrons the reaction will proceed too fast and get out of hand, and this is controlled by neutron-absorbing boron steel rods. These rods are moved in and out of the reactor from the control room by small electric motors, which raise them through the floor to a suitable position. In the event of an electrical failure the rods would fall and shut off the reactor automatically.

The reactor is started by a charge machine, which lowers uranium rods, each about 1 m long and 2 cm in diameter, into about 1700 fuel channels in the graphite core (Fig. 39.19). The boron steel rods are then raised slowly, and in a certain position the chain reaction proceeds at the desired rate. The reactor is now said to have 'gone critical', and heat is produced steadily.

Carbon dioxide gas is blown through the fuel elements under pressure, and the hot gas is led into heat exchangers outside the reactor. Here it heats up water which flows through an independent pipe system in the exchanger, and this is converted into high-pressure and low-pressure steam. The steam is used to drive the turbines which turn the electrical generators, and the electricity obtained is fed to the national grid system

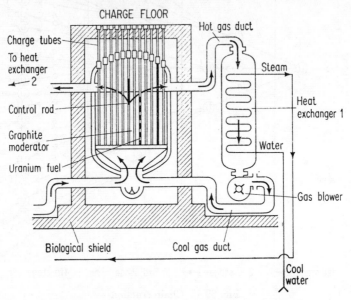

Fig. 39.19 Nuclear reactor, producing heat energy

to supplement the electrical energy produced by burning coal at conventional power stations. The nuclear reactor can operate day and night for a few years with the same uranium fuel. The reactor core is shielded by a welded-steel pressure vessel and a concrete biological shield, which protects operators outside from the intense radiation inside the core.

### Nuclear Fusion

We have seen that a considerable amount of energy is released when a heavy nucleus such as uranium disintegrates into two large parts; this is nuclear fission. Experiments show that a considerable release of energy is also obtained when the nuclei of the lightest elements are *fused* to form a heavier nucleus, and this is known as *nuclear fusion*.

One transformation of matter into energy is the fusion of deuterons, $^{2}_{1}H$, the nuclei of heavy hydrogen, into helium nuclei. As the sun contains a considerable amount of hydrogen, it was suggested in 1939 that the energy of the sun was basically due to nuclear fusion. It has been estimated that, theoretically, the energy released from the fusion of all the atoms in 1 kg of deuterium, heavy hydrogen, is equivalent to that released by burning nearly 3 million kg of coal.

In order to obtain nuclear fusion, the nuclei must approach near enough to each other to overcome the repulsion of their like charges. The lightest elements have the smallest nuclear charge, and hence the chance of nuclear fusion is greatest for such elements. The most practical way of achieving nuclear fusion is to raise the temperature of deuterium gas to millions of degrees, which is the temperature in the heart of the sun. At these very high temperatures deuterons fuse together, a process known as *thermonuclear fusion*. Very heavy electrical discharges are sent through the gas to heat it to such enormously high temperatures, and researches into methods of retaining the energy in the gas, and hence to promote fusion, are at present being made in Great Britain and abroad. If methods for nuclear fusion were successful a cheap source of power would be available. Heavy hydrogen forms about 1 part in 45 000 of water, and if heavy hydrogen were extracted from sea-water it could provide limitless power at a very economic price.

## CONCLUSION

We have now come to the end of our journey through the pathways of elementary physics. Throughout it the reader will have noted how experiments in the laboratory, performed by scientists in their efforts to find out how things work, have led to world-wide applications affecting our everyday lives. Sir Isaac Newton laid the foundations of the science of mechanics, with its practical applications in the design of bridges, buildings, ships and aeroplanes. Faraday discovered the laws of electromagnetic induction, and this made possible the dynamo and the distribution of electrical power for lighting and heating in our homes and factories. Oersted's discovery of the magnetic effect of a current led to the invention of the telephone and the telegraph; Ampère's experiments on the force on a conductor in a magnetic field led to the electric motor used for driving vehicles and machines; Snell's discovery of the law of refraction led to the design of efficient telescopes and microscopes; Appleton's experiments on the ionosphere led to the invention of radar, which permits safe navigation by sea and air; Sir J. J. Thomson's discovery of the electron led to the invention of the cathode-ray tube, used both in radar and in television; Shockley's researches in semiconductors led to the transistor; and Rutherford's discovery of the nucleus of the atom led to the development of nuclear reactors.

Today, thousands of scientists of different nationalities are engaged in laboratories all over the world in the quest for knowledge. Will that knowledge be used to increase human wealth, health and happiness? Or will it be used for purposes of destruction and war? The choice will not be made by scientists; every citizen of the world must decide to work for peace and plenty.

## SUMMARY

1. α-particles are positively charged nuclei of helium (mass about four times that of the hydrogen atom and carrying a charge of $+2e$). They are stopped by thin paper and ionize air considerably. Owing to their heavy mass, they are not deflected appreciably in a magnetic field.

2. β-particles are usually electrons and negatively charged. They are stopped by aluminium plate several millimetres thick, and ionize air less than α-particles. They are easily deflected in a powerful magnetic field, as they are so light.

3. γ-rays are electromagnetic waves. They are stopped by very thick lead blocks and ionize air slightly. They are not deflected by magnetic fields.

4. Detectors of ionizing particles and radiation are: (i) Geiger–Müller tube or solid-state detector and scaler (or ratemeter)–the count rate is proportional to the intensity; (ii) diffusion or Wilson cloud chamber–tracks can be seen; (iii) pulse electroscope–beating of flag depends on intensity. The G–M tube and solid-state detector detects all three radiations, the cloud chambers detect mainly α- and β-particles, and the pulse electroscope detects α-particles.

5. The atom consists of a small central nucleus containing protons and neutrons; a proton is a hydrogen nucleus. The number of protons is called the *atomic number*, the total number of protons and neutrons is called the *mass number*. The number of electrons round the nucleus in the rest of the atom is normally equal to the number of protons, and determines the chemical nature of the atom.

6. An *isotope* is an atom which has the same atomic number and hence the same chemical properties as another atom with a different mass. A *radioactive isotope* is one which emits ionizing radiations, and many are used in medicine and industry

7. *Nuclear fission* is the 'splitting' of a heavy atom, such as uranium, into two heavy parts; it is produced when suitably moderated neutrons are fired into a mass of uranium-235, with the release of nuclear energy. *Nuclear fusion* occurs in the sun – hydrogen nuclei fuse to form helium nuclei, with the release of nuclear energy.

## EXERCISE 39 · ANSWERS, p. 745

1. Write an account of the precautions you would take in the school laboratory in doing an experiment with radioactive materials. What warning system could be installed to monitor the radiation in the air in case it becomes excessive?

2. Describe briefly what happens inside a Geiger–Müller tube when an α- or β-particle or γ-ray enters. What is the difference between a *scaler* and a *ratemeter*?

3. Describe an experiment to investigate the relative absorbing powers of α- and β-particles and γ-rays. State the result of the experiment.

4. What is meant by the *half-life* of a radioactive element? A scaler records 1000 counts per min when a radioactive element is first used, and this falls to 125 counts per min in 6 min. Calculate the half-life. When will it fall to about 30 counts per min?

5. The emission of α- or β-particles from a radioactive element is said to be 'irregular'. What is the evidence for this, and what is the significance of the irregular emission?

6. A ratemeter instrument records: (i) 100 divisions and then 95 divisions when a radioactive element X is moved near and then twice the distance from the G–M tube; (ii) 120

divisions and then 30 divisions when another radioactive element Y is moved twice as far away. What radiation is probably emitted by X and by Y? Give reasons. How can the reading on the ratemeter be reduced to zero in the case of Y?

**7.** Describe, with a labelled diagram, an experiment to investigate the sign of the charge on β-particles. State two reasons why it is difficult to perform a similar experiment on α-particles. Draw three sketches showing the action of a perpendicular magnetic field on α- and β-particles and γ-rays respectively.

**8.** Describe a continuous cloud chamber of the diffusion type. Draw sketches showing the effects produced with α- and β-particles, and explain what happens.

**9.** Draw a simple diagram of a hydrogen atom and label the particles of which it is composed. (*N.*)

**10.** If a proton is considered to have mass $m$, what is the mass of: (*a*) a neutron; (*b*) an electron? (*N.*)

**11.** A uranium nucleus of mass number 238 and atomic number 92 can change to radon (Rn) by the emission of four α-particles and two β-particles (electrons). (i) What is the mass number and atomic number of the radon? (ii) Write down the nuclear change by an equation. (iii) What other emission may occur? Give a reason. (iv) If bismuth-214 is produced by the emission of one more β-particle and more α-particles, what is the number of α-particles emitted and the atomic number of bismuth?

**12.** The nitrogen nucleus is represented by $^{14}_{7}N$. What does this mean? An α-particle, represented by $^{4}_{2}He$, is fired at a nitrogen nucleus. A proton, $^{1}_{1}H$, is emitted as a result of the collision. Write down the nuclear reaction, and state what other nucleus is left.

**13.** Write an account of *the atom*, or of *nuclear fission* and its practical application.

**14.** Outline one way of detecting α-particles and describe, so far as you can, how it works.

The counting rate recorded by a detector fixed in front of an α-particle emitter is 256 per sec. This figure is an average rate worked out from a count lasting for several minutes. What is the average counting rate 20 days later, for the same arrangement, if the half-life of the emitter is 5 days?

If the number of α-particles were recorded at this time for 1 sec precisely, would you expect to find the number you have just calculated? Give a reason for your answer. (*O and C.*)

## ANSWERS TO NUMERICAL EXERCISES

### EXERCISE 37 (p. 705)

**8.** $8.5 \times 10^{5}$ m/s

### EXERCISE 38 (p. 716)

**1.** conductor **2.** insulator **3.** germanium **4.** silicon **5.** electron **6.** holes **7.** negative **8.** positive **9.** electron **10.** equal **11.** holes **12, 13.** holes, electrons **14.** electrons **15.** negative **16.** impurity **17.** semiconductor **18.** holes **19.** positive **20, 21.** p/n **22.** positive **23.** p **24.** negative **25.** forward **26.** negative **27.** n **28.** diode **29.** rectifier **30.** direct **31.** three **32.** p **33.** base **34.** collector **35.** forward **36.** positive **37.** emitter **38.** negative **39.** base **40.** reverse **41.** collector **42.** base **43.** current **44.** voltage **45.** amplifier **46.** base **47.** current **48.** collector

### EXERCISE 39 (p. 744)

**4.** 2 min. In 10 min **6.** X — β-particles, Y — γ-radiation. Lead block **10.** (*a*) $m$ (*b*) $m/2000$, approx **11.** (i) 222, 86 (iv) 2, 83 **12.** $^{17}_{8}O$ **14.** 16

*The questions in the following Revision Papers cover some of the main topics discussed in the book. Section A has a mixture of questions, Section B has questions on Mechanics and Fluids, Section C on Optics, Heat and Sound, and Section D on Electricity. [Where necessary assuming $g = 10$ m/s², weight of 1 kg mass = 10 N, specific heat capacity of water = 4·2 J/g K or 4200 J/kg K.]*

# Paper 1

## Section A

**1.** (a) A steel tape of correct length at 15·0°C is used to measure a distance on a day when the temperature is 10·0°C. Is the result too large or too small?

If the linear expansivity of steel is $11·0 \times 10^{-6}$/K, what is the error in measuring a distance of 20 m?

(b) 'Cool as a cucumber' is a common saying. Would a cucumber be cooler than its surroundings if it were kept in a sealed polythene bag?

Two cucumbers are left out on a plate, one is fresh and moist, the other is older and dried up. Which would you expect to be the cooler?

(c) Comment on the following statement and correct it where necessary: 'The strength of the steam is converted into electricity in the power station.'

(d) An empty test-tube containing a dry powder is pushed (base downwards) at an angle of about 45° to the horizontal into a beaker of water. When viewed from above the tube appears bright, as though full of mercury. Explain this using a simple ray diagram.

(e) A slide 7·5 cm × 7·5 cm is used in a projection lantern to produce a picture 1·8 m square on a screen in a lecture hall. The slide is 30 m from the lens. What kind of lens is used?

(f) Describe the Brownian movement, indicating what light it sheds on the nature of matter. Why is it only observed for very minute particles? (*O. and C.*)

## Section B

**2.** A bullet of mass 15 g has a speed of 400 metres per second. What is its kinetic energy? If the bullet strikes a thick target and is brought to rest in 2 cm, calculate the average net force acting on the bullet. What happens to the kinetic energy originally in the bullet?

**3.** Define *uniform acceleration*. How would you show that a body acted on by a constant force moves with uniform acceleration? Explain carefully how the conclusion is drawn from the observations made.

A trolley is released from rest so that it travels freely under gravity down a slope inclined at 5° to the horizontal. Find how long it takes to travel 3 m down the slope, and its velocity at the end of this time. (*O.*)

**4.** What are the conditions for a body to be in equilibrium under the action of three non-parallel coplanar forces?

Describe an experiment to find the resultant of forces of 30 N and 50 N acting at a point in directions inclined at an angle of 60°. Explain why your method gives the required result.

A uniform cylinder of mass 200 g rests with its axis horizontal in the right angle formed by two smooth planes inclined at 30° and 60° respectively to the horizontal so that its curved surface is in contact with the planes. What force does the cylinder exert on each plane? (*L.*)

**5.** State the principle of Archimedes and describe how it may be verified experimentally. A wooden block, of relative density 0·6 and of mass 100 g, floats in water with a block of metal, of relative density 8·0 and of mass 30 g, tiéd underneath it. What fraction of the volume of the wood is submerged? (*L.*)

**6.** Define *pressure* and use the definition to obtain an expression for the pressure at the base of a column of liquid of height *h* and density *ρ*. Describe and explain an experiment to compare the densities of two liquids by using this result. (*L.*)

## Section C

**7.** A luminous object is placed 15 cm from: (*a*) a plane mirror; (*b*) a concave mirror of focal length 25 cm. In each case, draw a ray diagram to illustrate the formation of the image, and give details of the nature, position and magnification of the image. How would you verify experimentally the value given for the focal length of the concave mirror: (i) approximately and quickly; (ii) more precisely?

**8.** A converging lens of focal length 5 cm is used to produce on a screen an image of a film with magnification 3. What are the distances of the screen and the film from the lens? Draw a ray diagram showing how the image is formed. (If the results are obtained by calculation the sign convention used must be stated.)

Give a labelled diagram showing the structure of a projection lantern. Draw the paths of suitably chosen rays to illustrate the action of the instrument. Give two reasons why the image produced is said to be real and account for any slight coloration which is sometimes seen although white light has been passed through a black-and-white slide. (*L.*)

**9.** Explain the distinction between *saturated* and *unsaturated* vapour, and between evaporation and boiling.

Heat is supplied at a rate of 500 watts to a pressure cooker containing water and fitted with a safety valve. Steam escapes at such a rate that the loss of water is 10·4 g/min. If heat is supplied at the rate of 700 watts 15·6 g of water is lost per minute.

Suggest an explanation of these figures and deduce (*a*) the latent heat of steam in joules per g at the temperature of the cooker, and (*b*) the rate of loss of heat from the cooker at this temperature by other processes than evaporation. (*O. and C.*).

**10.** (*a*) Define *melting point* of a substance. Why can hard snowballs usually be formed by pressing soft snow between the hands? State when this is impossible and give the reason.

(*b*) Describe an experiment to show how the volume of a fixed mass of gas at constant pressure varies with the temperature as recorded on a mercury-in-glass thermometer. State the result of the experiment. (*L.*)

**11.** Explain the following:

(*a*) the sound is louder when the stem of a vibrating tuning fork is held against the top of a table, (*b*) the pitch of the note from a sounding organ pipe changes as the temperature rises, (*c*) the same note has a different quality when produced by different musical instruments. (*N*).

## Section D

**12.** Explain: (*a*) what is meant by magnetic induction; (*b*) why one end of a bar of unmagnetized soft iron may attract either end of a compass needle.

Draw a diagram of the magnetic field in a horizontal plane when: (i) a bar magnet is placed horizontally with its axis in the magnetic meridian and its N-pole pointing North; (ii) a bar magnet and a bar of soft iron are placed horizontally in line with a gap between them (neglect the influence of the earth's magnetic field). (*N.*)

**13.** State Ohm's law and define *resistance*. Explain how you would measure the resistance of an electric filament lamp: (a) at its working temperature; (b) cold.

An electric lamp is marked 240 V 150 W. What would you expect its resistance to be when the lamp is working? When the lamp is connected across a 2-V cell it passes a current of 0·01 A. Is this contrary to Ohm's law? Give a reason for your answer. (*O.* and *C.*)

**14.** State the relation between the electric *current* flowing through a wire and the *potential difference* across its ends, and define *resistance* and *resistivity*. A voltaic cell of e.m.f. 1·5 V is connected in series with an ammeter of resistance 0·2 Ω and a resistance R. The reading of the ammeter is ⅓ A, and the p.d. between the terminals of the cell is 1·3 V. Calculate the value of R and the internal resistance of the cell. (*O.* and *C.*)

**15.** State Faraday's laws of electrolysis and describe an experiment to illustrate *one* of them.

Calculate the volume of mercury liberated in 40 min by a current of 2 amps. passing through a solution of mercury salt. (Electro-chemical equivalent of hydrogen = 0·000 010 44 g/C, chemical equivalent of mercury = 200·6, density of mercury = 13·6 g/cm³.) (*L.*)

**16.** (a) Describe the structure of a diode valve. How can it be shown experimentally that a current flows through the valve in normal use when the anode has a positive potential but does not do so when it has a negative one? Give a diagram of the circuit used.

(b) Describe a step-down transformer and explain how it works.

## Paper 2

### Section A

**1.** (a) One and one make two when we are adding grams, but not always when we are adding newtons. Explain why there is this difference and give' TWO more examples of quantities of each type.

(b) Explain what is meant by the *surface tension* of a liquid and describe briefly THREE common effects which are due to it.

(c) A gardening book says that the soil brought up in worm casts is 119 790 kg per acre. What might a non-gardening scientist say about this figure?

(d) 'Don't do anything suddenly' is the advice given to drivers of motor-cars on icy roads. Explain fully in terms of mechanics why this is good advice.

(e) Draw a rough qualitative graph of current (vertically) and voltage (horizontally) to show how these are related in the case of a pure metal for small currents in either direction of flow. Draw a similar graph for a metal rectifier and comment on the shape of both graphs.

(f) A uranium nucleus of mass number 238 and atomic number 92 can change to radium by the emission of three α-particles and two electrons. Write down the equation for this and state the nuclear mass number and charge of the radium. By emitting five more α-particles and some electrons it can change to lead of nuclear charge 82. How many electrons are emitted, and what is the mass number of the lead nucleus? (*O.* and *C.*)

### Section B

**2.** The following results were obtained in an experiment to investigate the stretching of an elastic cord:

| Stretching force in N. | 0 | ½ | 1 | 1½ | 2 | 2½ | 3 |
|---|---|---|---|---|---|---|---|
| Length of cord in cm | 340·0 | 340·6 | 341·2 | 341·8 | 342·4 | 343·1 | 344·1 |

On the squared paper provided, plot a graph to illustrate Hooke's law. State this law and point out its validity for the specimen used in the experiment.

Describe an experiment by which these results may have been obtained. (*L.*)

**3.** Define the terms energy and power.

Explain how the movement of a pendulum illustrates the principle of the conservation of energy. A body falls freely from rest and the distances $s$ travelled after times $t$ are given below:

| $t$ s. | 0 | 1 | 2 | 3 | 4 | 5 |
|--------|---|----|-----|------|------|-------|
| $s$ m | 0 | 4·9 | 19·6 | 44·1 | 78·4 | 122·5 |

Plot a graph of distance against time and use the graph to determine the velocity of the body 2·5 seconds after release. (*N.*)

**4.** Give the meaning of the terms *kinetic energy, potential energy, work*.

A car can be stopped by its brakes in a distance of 10 m from a speed of 48 km/h on a level road. Assuming that the brakes exert a constant retarding force, find the value of the retardation.

What does the stopping distance become if the car is travelling at 48 km/h (*a*) up, (*b*) down a hill of slope 1 in 9? Give your answers correct to two significant figures (*O. and C.*)

**5.** What do you understand by the *principle of moments*? Describe how you would verify the principle experimentally for parallel forces. A uniform rod of length 50 cm and weight 1 N is pivoted 8 cm from one end A. Loads are weighed by attaching them to A, and moving a weight of 5 N along the rod, on the opposite side to the pivot, till the rod balances horizontally. Find the position of the weight when a load of 24 N is attached to A and calculate the total upthrust at the pivot. (*N.*)

**6.** Give the meaning of each of the terms *mechanical advantage, velocity, ratio, efficiency*, of a simple machine, without mentioning the other two. Derive the relation between them.

Explain the principle of the hydraulic press or jack, illustrating your answer by a diagram.

In a form of hydraulic press the bore of the pump is 2·5 cm and that of the main cylinder 30 cm. The pump is operated by a lever of velocity ratio 5. What is the velocity ratio of the whole press? If the efficiency is 85%, what is the greatest force the press can exert when a force of 22 N is applied to the end of the lever? (*O. and C.*)

## Section C

**7.** (*a*) Describe how you would locate, by experiment, the image of a pin formed by a plane mirror. State the position and nature of the image and draw a ray diagram to show how it is formed.

(*b*) Find, graphically or otherwise, the deviation produced in a ray of light when it passes through a triangular glass prism of refracting angle 30° if it is incident normally on the first face and the refractive index of the glass is 1·5. (*L.*)

**8.** How would you set up two lenses as a telescope suitable for viewing an object at the far end of a long corridor?

Explain the purpose of each lens, suggesting values for their focal lengths. Draw a diagram showing the paths through the telescope of two rays from a non-axial point on the object. (*L.*)

**9.** Why are fixed points necessary in order to establish a thermometer scale? Define those usually employed and describe how you would determine whether they are marked correctly on a given mercury thermometer.

A thermometer is found to indicate −0·5° at the lower fixed point and 100·5° at the upper one. What does this thermometer read when the true temperature is 60·0°C?

Mention one advantage and one disadvantage of an alcohol thermometer as compared with a mercury thermometer. (*L.*)

**10.** (*a*) Explain the process of heat convection. Describe an experiment to show that convection takes place in air. (*b*) A sealed flask of capacity 500 cm³ contains gas at a pressure of one atmosphere. If the temperature of the gas is raised from 10°C to 35°C, without any change in volume, calculate (i) the new pressure, (ii) the heat required to raise the temperature of the gas. [Density of gas = 1·30 g/litre, specific heat capacity of gas = 0·71 J/g K or 710 J/kg K.]

How do you explain the increase in pressure of the gas? (N.)

**11.** What are the differences in the waveforms of: (a) a loud note and the same note played softly; (b) a high note and a low note?

Describe an experiment to determine an accurate value for the velocity of sound in air using a resonance tube and a tuning fork of known frequency.

How, if at all, would the result of the above experiment be affected if the experiment were carried out at: (i) a lower temperature; (ii) a lower atmospheric pressure? (L.)

## Section D

**12.** Describe a method of magnetizing a steel knitting needle XY so that the end X shall have a north-seeking polarity. Give a diagram and state reasons for the procedure adopted. How would you prove the poles to be equal in strength?

State the magnetic properties desirable in: (a) the magnet of a moving-coil ammeter; (b) a transformer core. Name the type of material to be used in each case. (N.)

**13.** Describe an experiment to determine how the electrical energy expended in a resistance wire when a current flows through it is related to the heat produced in the wire.

A hot water tank containing 40 000 g of water is heated by an electric immersion heater rated at 3 kilowatt, 240 volt. Calculate (i) the current in the heating element, (ii) the time required to raise the temperature of the water from 20°C to 60°C assuming that 80 per cent of the heat supplied is retained by the water (c = 4·2 J/g K or 4200 J/kg K for water). (N.)

**14.** Describe an experiment to show that a straight conducting wire carrying a current experiences a mechanical force when placed in, and perpendicular to, a magnetic field. State a rule to determine the direction of the force.

(b) A galvanometer has a resistance of 5 ohms and gives a full-scale deflection with a potential difference of 50 millivolts between its terminals. How can it be adapted to measure currents up to 5 amperes? (N.)

**15.** Give a brief account of the main features of the earth's magnetism and indicate how you would locate the magnetic south pole if you were in the vicinity and equipped with a dip circle.

Draw a careful diagram of the magnetic field around a bar magnet when lying in the magnetic meridian with its N-pole pointing North. Mark the position of any neutral points. Indicate by arrows the direction of the field. (O. and C.)

**16.** Describe an electrophorus and state how it is used to give an insulated conductor a positive charge. Give diagrams to illustrate the distribution and nature of the charge in each stage of the operation.

Why is it possible to obtain a series of approximately equal charges from this instrument without recharging it and what is the source of energy of these charges? (L.)

## Paper 3

## Section A

**1.** (a) A car moves round a circular track at constant speed. Indicate on a diagram the direction of its acceleration and velocity at any point, and the direction of the resultant force there. What name is given to the force which keeps the car moving in a circle?

(b) 'Force is proportional to change of momentum.' Is this a true statement? If not, rewrite it correctly. Using the molecular theory, explain why the pressure of a gas at constant temperature is doubled when its volume is halved.

(c) What is a *longitudinal wave*? Explain how you would produce a longitudinal wave travelling along a long loose coil. How would you set up a stationary wave in the coil? Draw a sketch of the wave in each case and point out any special features.

(d) Draw sketches showing how the current $I$ in (i) a Nichrome coil, (ii) a tungsten coil as in a lamp, (iii) a thermionic diode valve vary with the applied p.d. $V$. Explain your graphs.

(e) 'X-rays are unaffected by magnetic fields; electrons are deflected by magnetic fields.' What does this show about the nature of X-rays and electrons? Draw a sketch of the deflection of an electron beam due to a powerful magnet, showing clearly the direction of motion of the electrons and the field.

(f) What is meant by the *half-life* of a radioactive substance? The count rate of a radioactive substance diminishes from 200 to 50 in 120 s. What is the half-life?

## Section B

**2.** Describe an experiment to determine the value of the acceleration due to gravity. A ball is thrown vertically upwards from the ground with an initial velocity of 20 m/s. Determine: (a) the height to which it rises; (b) the potential and kinetic energies of the ball, if its mass is 225 g, (i) when at its maximum height, (ii) when half-way up, showing clearly how you obtain the values and stating the units in which they are expressed. (*L.*)

**3.** (a) The acceleration of a vertically moving rocket, burning fuel at a constant rate increases as the fuel burns away. Why is this?

(b) A man of mass 70 kg stands in a lift. What force does he exert on the floor when the lift is: (i) descending at constant speed; (ii) descending with the acceleration due to gravity (10 m/s²); (iii) descending with an acceleration of 2·5 m/s²; (iv) descending with a retardation of 2·5 m/s².

What would your answers to (iii) and (iv) be if the lift were ascending instead of descending? (*O. and C.*)

**4.** Describe how you would make a simple form of hydrometer with the help of a uniform test tube and some lead shot. What is the principle underlying its action? The stem of a hydrometer has a diameter of 4 mm, and it floats in water with the bulb and part of the stem immersed, the volume immersed being 5 cm³. How much more of the stem will be immersed in a liquid of relative density 0·80? (*L.*)

**5.** Explain what you understand by the phrase 'the resolution of a force into components at right angles'. Illustrate your answer by a diagram.

Describe an experiment to find the resultant of forces of 9 and 12 N acting at right angles to each other.

Two smooth planes are joined together along a horizontal line. One of the planes is vertical and the other is inclined at an angle of 30° to the vertical. A spherical ball of mass 4 kg rests between the planes. What are the reactions of the planes on the ball? (*N.*)

**6.** State Boyle's law and describe an experiment to verify it. A cylindrical iron tube, closed at one end and 30 cm long, is coated on the inside with a soluble pigment. It is carefully lowered into water, with the open end downwards and when withdrawn it is found that the pigment has been dissolved over a length of 5 cm from the bottom. What was the maximum pressure of the air in the cylinder and to what depth was the bottom of the cylinder sunk? (Barometric height = 75 cm of mercury.) (*L.*)

## Section C

**7.** Explain, by the aid of suitable diagrams, the meaning of *total internal reflection* and *critical angle*. A man stands on a ladder in a swimming bath, so that his head is just above the water, and observes the light from a lamp held at the bottom of the bath. As the lamp is moved horizontally away from him he observes that the lamp cannot be seen when it is more than 2·5 m away from him, measured horizontally. Explain this observation and calculate the depth of the water. Take the refractive index of the water as $\frac{4}{3}$. (*L*.)

**8.** Explain the terms *focal length* and *linear magnification* as applied to a converging lens.

How would you use a converging lens of focal length about 5 cm to project a highly magnified image of an illuminated slide on the wall of the laboratory? Give a ray diagram of the arrangement.

If you were also provided with a converging lens of focal length 50 cm, describe how you would arrange the two lenses as a telescope to view a lamp at the other end of the laboratory.

State the differences between the image of the slide formed by the single lens and that of the lamp seen in the telescope as regards: (*a*) their nature; (*b*) the type of magnification. (*L*.)

**9.** Define *cubic expansivity*. Distinguish between the *real* and *apparent* cubic expansivity of a liquid, and state a relation between them.

A glass bottle of volume 10·0 cm³ at 0°C contains 132·6 g of mercury of density 13·60 g/cm³ at 0°C. What is the volume of the mercury at 0·C? Assuming the cubic expansivities of mercury and glass to be $18 \times 10^{-5}$/K and $3 \times 10^{-5}$/K respectively, find the temperature at which the mercury just fills the bottle. (*L*.)

**10.** Describe an experiment to determine the specific latent heat of steam. State the sources of error and point out the precautions you would take to reduce them.

A 200-g brass weight is held suspended in liquid oxygen contained in an open vacuum (Thermos) flask. It is then quickly transferred to a beaker of water at 0°C. What weight of ice will form on the brass?

(Boiling point of oxygen is −183°C, specific heat capacity of brass = 0·38 J/g K or 380 J/kg K and specific latent heat of fusion of ice = 336 J/g or 336 000 J/kg.) (*N*.)

**11.** How does the frequency of the note emitted by a vibrating wire depend on: (*a*) the length; (*b*) the tension? Describe an experiment to verify the relation between frequency and length. A wire 1 m long gives the same note as a tuning fork of 288 Hz. What are the frequencies of the notes emitted by the two segments of the wire when a bridge is placed at a point 60 cm along the wire from one end? (*L*.)

## Section D

**12.** Define *resistance* and establish the formula for the combined resistance of two resistances connected in parallel.

A battery consisting of 6 cells, each of e.m.f. 2·0 V and internal resistance 0·1 Ω, is connected to a 9-Ω coil. Find the current in the coil. If now a 6-Ω coil is connected in parallel with the one of 9 Ω, find the new value of the current in the latter.

What would be the reading of a high-resistance voltmeter, connected to the terminals of the battery, in each of the above circuits? (*L*.)

**13.** Define *electrochemical equivalent* and explain the terms *anode* and *cathode*.

Describe an experiment to show the relation between the mass of copper deposited in the electrolysis of a solution of copper sulphate and the quantity of electricity which passes. Give a circuit diagram, full experimental details and state the result.

What time is required for a current of 1·2 A to deposit 0·18 g of copper in a copper voltameter?

(E.c.e. of copper = 0·000 33 gramme per coulomb.) (*L*.)

**14.** (*a*) The ends of a coil of wire are joined together to form a closed circuit. How could you produce in the coil without breaking the circuit: (i) a momentary direct current; (ii) an alternating current? What factors determine the magnitude and direction of the current in (i)?

(b) Describe a hot-wire ammeter and explain its action. Why is it suitable for measuring either alternating or direct current? (L.)

## Paper 4

### Section A

**1.** (a) The diameter of a molecule can be roughly estimated by allowing a tiny droplet of olive oil to fall on the surface of water. Explain: (i) why the droplet spreads; (ii) how the area it covers can be estimated; (iii) why the method gives only a rough estimate. State the order of magnitude of the diameter of a molecule.

(b) A passenger in a car moves forward when the brakes are suddenly applied, and a jet aeroplane moves forward when hot gases are expelled from the engine. Explain briefly the laws of mechanics concerned in the two cases.

(c) Two close vibrators dipping into water in a ripple tank produce an interference pattern on the water. Explain why this occurs. Explain why two close sodium burners, emitting light of the same frequency, do not produce an interference effect on a screen placed in front of them.

(d) A luminous beam is produced in a fine-beam tube by using a source of low p.d. and one of high p.d. Explain: (i) what produces the light; (ii) what is the source of energy of the light; (iii) how the beam can be deflected in a circular path; (iv) how the beam can be deflected by an electric field. Draw diagrams in illustration of your answers.

(e) Explain how 'evaporation' and 'saturated vapour' are accounted for by molecular theory.

(f) A Geiger–Müller tube is connected to a scaler and a count rate of 1256 per min is obtained when a source A of an unknown radiation is placed in front of the tube. A count rate of about 310 per min is obtained when A is placed twice as far away. What radiation does A probably emit? Explain your answer. What material could be used to reduce the count rate to practically zero if it is placed in front of A?

### Section B

**2.** In what circumstances is *work* done by a force? State and explain a unit in which it is measured.

A train, starting from rest, is uniformly accelerated at $\frac{1}{3}$ m/s² until it attains a constant velocity of 48 km/h. Construct a velocity–time graph for the first minute of its motion and find the distance it travels in this time.

If the motion of the train is opposed by a steady resistance of 2200 N, find the horse-power required to maintain the constant velocity of 48 km/h on a horizontal track. What extra power is required to maintain this velocity along a track which rises 1 m in every 200 m of track, if the train weighs $1 \cdot 2 \times 10^6$ N? (L.)

**3.** State *two* differences between friction and viscosity.

Describe how you would proceed in order to measure the force of friction between a horizontal wooden surface and a metal block sliding on it. How would the results be affected by placing a second metal block on top of the first?

A body of mass 2·5 kg is placed on a horizontal surface, a horizontal force of 5 N is applied to it and the body just moves. Calculate: (a) the work done against friction when the body is moved a distance of 1·2 m; (b) the velocity acquired by the body in 2 s if a horizontal force of 10 N is applied. (Assume that friction remains constant.) (L.)

**4.** Subtract a velocity of 4 km/h due East from a velocity of 3 km/h due North. Mark the size and direction of the answer on a diagram which can be either freehand or drawn to scale.

A metal ball of mass 1 kg hangs on a long thin wire. What horizontal force must be applied to the ball to hold the wire at an angle of 30° to the vertical? What is then the tension in the wire? What is the smallest force which, suitably applied, would produce this deflection of 30°? (*O. and C.*)

**5.** (*a*) Define *efficiency* of a machine

The radius of the wheel of a wheel and axle is 45 cm and that of the axle 7·5 cm. What is (i) the effort required, (ii) the work done by the effort, when the machine raises a bucket of water weighing 120 N from a well 5 m deep, if its efficiency is 70%?

(*b*) State the theorem of the triangle of forces.

A 120 N weight is suspended by two strings attached to it and inclined at 30° and 45° respectively to the horizontal. Find the pull in each string. How should the strings be arranged so that the pull in each is as small as possible? (*L.*)

**6.** State Boyle's law and describe an experiment to verify it.

A flask of capacity 1 litre is slowly exhausted of air by means of a pump. The cylinder of the pump has an internal area of 5 cm² and the length of the stroke is 20 cm. If the initial air pressure in the flask is 77 cm of mercury, what will be the pressure after: (*a*) one stroke; (*b*) two strokes? (*N.*)

## Section C

**7.** Draw ray-diagrams to explain the action of a lens used (*a*) as a 'burning glass', (*b*) as a magnifier (simple microscope) producing an image at infinity. What type of lens would be required?

A ball bearing of diameter 1 mm is clearly seen through a converging lens of 12 cm focal length held close to one eye. What must be the distance of the ball from the lens if the eye is focused for infinity? If the image of the ball appears the same size as the sun, deduce the size of the image of the sun produced by a converging lens of 50 cm focal length. (*O. and C.*)

**8.** When white light strikes a glass prism it gives rise to an *impure spectrum*. Explain what is meant by this, and show by a drawing how lenses may be used with the prism to obtain a pure spectrum. A flag has stripes of red, white and blue. How will it appear if viewed in: (*a*) red light; (*b*) green light? Give reasons. (*O. and C.*)

**9.** How does the volume of a fixed mass of an ideal gas vary with its temperature when its pressure is kept constant? Describe an experiment that you would perform to test the relationship for air.

A gas is contained at a pressure of 1 atmosphere and temperature 20°C inside a sealed glass globe which fractures when the pressure inside rises above 2 atmospheres. Determine the maximum temperature to which the globe may be raised before it breaks. You may neglect the expansion of the globe. (*L.*)

**10.** Describe *one* method of measuring each of the following: (*a*) specific heat, (*b*) melting-point, and in each case explain the principle of the method used.

A copper sphere of mass 100 gm is cooled to the temperature of liquid air ($-185$°C) and then completely immersed in water at a temperature of 0°C. Calculate the mass of ice formed. (Specific latent heat of fusion of ice = 336 J/g or 336 000 J/kg; mean specific heat capacity of copper = 0·3 J/g K or 300 J/kg K.) (*O. and C.*)

**11.** Explain carefully what is meant by *resonance*. Describe how you would arrange a state of resonance between a column of air and a given tuning fork, and explain how you would use the arrangement to enable you to determine the velocity of sound in the air, the frequency of the fork being known. (*N.*)

## Section D

**12.** Give an outline of the elementary theory of magnetism and describe TWO magnetic effects which are evidence in support of it.

Sketch the lines of magnetic force produced by: (*a*) a bar magnet lying along the magnetic meridian with its S-pole pointing north; (*b*) two parallel wires perpendicular to the paper and carrying currents in the same direction, into the paper; (*c*) a coil of wire wound round an iron ring.

Mark any neutral points and put arrows on the lines to show the direction of the field in cases (*a*) and (*b*). In cases (*b*) and (*c*) ignore the earth's magnetic field. (*O*. and *C*.)

**13.** Given a supply of resistance wire, how would you determine the value of its resistance per metre length? 1 metre of nichrome wire, area of cross-section 0·50 mm$^2$, has a resistance of 2·20 Ω. Calculate: (*a*) the resistivity of nichrome; (*b*) the length of the wire which, connected in parallel with the 1 metre length, will give a resistance of 2·00 Ω. (*N*.)

**14.** Give a diagram showing the lines of magnetic force due to a current in a straight conductor, in a plane perpendicular to the conductor.

Describe an experiment to show that a current-carrying conductor which is perpendicular to a magnetic field experiences a mechanical force. Show clearly the directions of the current, the magnetic field and the force.

Describe a simple electric motor and explain why a commutator is necessary to obtain continuous rotation. (*L*.)

**15.** Assuming that you have an insulated negatively charged conductor, how would you charge a gold-leaf electroscope: (*a*) positively; (*b*) negatively?

How would you use the electroscope to test the sign of the charge on a conductor without transferring any charge?

State and explain the effect of bringing an earthed conductor near the cap of a charged electroscope. (Give diagrams to illustrate your answers.) (*L*.)

**16.** Explain how the laws of electromagnetic induction are applied in the design of an a.c. transformer. Draw a diagram of such a transformer and give reasons for the shape and kind of material used in the core. What decides the ratio of the primary to the secondary turns?

Why may a transformer designed to transform from 240 to 4 volts be used to transform 4 volts to 240, but not 240 volts to 14 400? (*O*. and *C*.)

# ANSWERS TO REVISION PAPERS

## PAPER 1 (p. 746)

**1.** (a) too large. 0·11 cm (e) (i) 28·8 cm (ii) 7·5 m long at least   **2.** 1200 J, 60 000 N
**3.** 2·6 s, 2·3 m/s   **4.** 1·73, 1 N   **5.** 0·76
**7.** (a) 15 cm, virtual, $m = 1$ (b) 37·5 cm, virtual, $m = 2·5$   **8.** 20, 6⅔ cm
**9.** (a) 2308 J/g, 6000 J/min   **13.** 384Ω   **14.** $R = 3·7Ω, r = 0·6Ω$   **15.** 0·74 cm³

## PAPER 2 (p. 748)

**1.** (f) 226, $+88e$; 4 electrons, 206   **3.** 24·5 m/s   **4.** 8·9 m/s² (a) 9 (b) 11 m
**5.** 43 cm from A, 30 N   **6.** V.R. = 700; 13 500 N   **7.** (b) 18°   **9.** 60·1°C
**10.** (i) 1·09 At. (ii) 11.5 J   **13.** (i) 12·5 A (ii) 46⅔ min   **14.** parallel 5/499Ω

## PAPER 3 (p. 750)

**1.** (f) 60 s   **2.** (a) 20 m (b) (i) 45 J, 0 (ii) 22·5, 22·5 J
**3.** (b) (i) 700 (ii) 0 (iii) 525 (iv) 875 N; ascending: (iii) 875 (iv) 525 N
**4.** 9·9 cm   **5.** 8, 6·9 kg   **6.** 209 cm   **7.** 2·2 m   **9.** 9·75 cm³, 171°C   **10.** 38 g
**11.** 480 720 Hz   **12.** 1¼, 1⅐ A 11¼, 10⅔ V   **13.** 455 s.

## PAPER 4 (p. 753)

**1.** (f) $\gamma$-radiation. Lead   **2.** 600 m; 29⅓, 80 kW   **3.** (a) 6 J (b) 4 m/s
**4.** 5 km/h, 53° W of N; 6 N, 12 N, min = 5 N
**5.** (a) (i) 28 N (ii) 857 J (b) 88 N, 107 N; both vertical   **6.** (a) 70 (b) 63·6 cm mercury
**7.** (a) 12 cm (b) 4 mm (5/12 cm)   **9.** 313°C   **10.** 16·5 g
**13.** (a) $1·1 \times 10^{-6}$ Ω m (b) 10 m

# MULTIPLE-CHOICE QUESTIONS

The two model Papers each has seventy questions covering all branches of the subject. Read the directions in each section carefully before answering the questions.

**Paper 1**   (Time: 75 min)   (ANSWERS ON p. 770)

### Section I

**Directions.** The group of questions below consists of five-lettered headings followed by a list of numbered questions. For each numbered question select the heading which is most closely related to it. Each heading may be used once, more than once, or not at all.

*Questions 1–4*

The graphs A to E in Fig. 1 are velocity–time graphs. Which is the most suitable graph for the following cases:

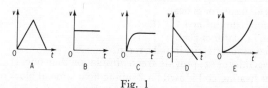

Fig. 1

**1.** A ball thrown vertically upwards and returning to the ground.
**2.** An ice puck pushed forward on a smooth horizontal surface.
**3.** An object released from an aeroplane at a great height.
**4.** A lift moving between two floors.

*Questions 5–8*

Five electrical instruments or machines are listed below:
   A carbon microphone,  B telephone receiver,  C moving-coil galvanometer,
   D simple motor,  E dynamo
Which of these
**5.** uses the attractive force on soft iron.
**6.** works by electromagnetic induction
**7.** uses a commutator.
**8.** does not rely on the magnetic effect of current for its working

*Questions 9–12*

Five examples of waves are listed below:
   A sound waves,  B light waves,  C infra-red waves,  D X-rays,  E γ-rays
Which of these are
**9.** electromagnetic waves about 1000 times shorter in wavelength than light waves.
**10.** used to take photographs of objects in the dark.
**11.** longitudinal waves.
**12.** produced by nuclear changes in the atom

### Section II

**Directions.** Each of the questions or incomplete statements in this section is followed by five suggested answers. Select the *best* answer in each case.

*Questions* 13–27

**13.** When released from a height a ball falls 5 metres in 1 second. In 4 seconds from release it falls

A 400 m,   B 100 m,   C 80 m,   D 40 m,   E 10 m.

**14.** An object of mass 2 kg falls from a height of 20 m above the ground. If $g = 10$ m/s², the loss of potential energy just before the mass strikes the ground is

A 20 J,   B 40 J,   C 200 J,   D 400 J,   E 800 J.

**15.** The 'newton metre' is a unit of

A moment of a force,   B force,   C energy,   D power,   E momentum.

**16.** A rectangular solid box has dimensions 2 m by 1 m by 0·5 m and weighs 100 N. When the box is placed on flat ground the maximum pressure exerted is

A 100 N,   B 200 N/m²,   C 100 N/m²,   D 50 N/m²,   E 200 N.

**17.** A rough measure of the size of a molecule can be obtained by

A measuring the height to which water rises in a narrow capillary tube.

B finding the speed with which bromine vapour spreads in air.

C observing Brownian motion of smoke particles.

D measuring the area of the circle into which a small oil drop spreads in water.

E measuring the friction force between the molecules of a solid.

**18.** Double glazing is used in domestic installations because

A glass is an insulator of heat.

B refraction is diminished.

C air is an insulator of heat.

D light passes more easily into the room.

E two glass thicknesses provide greater insulation than one.

**19.** A 50 watt heating coil is used for 2 min to heat a metal block of 500 g and specific heat capacity 1·0 J per g deg C. The temperature rise is

A 6°C,   B 12°C,   C 15°C,   D 24°C,   E 25°C.

**20.** A closed flask contains air at 1 atmosphere and 27°C. When the flask is heated to 127°C, the pressure becomes

A 4·7 atm,   B 4 atm,   C $3\frac{1}{3}$ atm,   D $1\frac{1}{3}$ atm,   E $1\frac{1}{6}$ atm.

**21.** A lens of focal length 50 cm is used as a magnifying glass and produces a magnification of 5. The object distance from the lens is then

A 45 cm,   B 40 cm,   C 10 cm,   D 8 cm,   E 4 cm.

**22.** Colours are produced when white light passes through a glass prism because

A the light waves interfere,   B the glass colours the light,   C different colours travel at different speeds in glass,   D the different colours are filtered   E diffraction of light occurs.

**23.** Light waves differ from sound waves because

A light is an electromagnetic wave

B light waves are long and sound waves are short.

C interference is obtained with light waves but not with sound waves.

D the speed of light is independent of the medium in which it travels.

E sound waves do not travel in water.

**24.** A wire X is half the diameter and half the length of a wire Y of similar material. The resistance of X to that of Y is

A 8:1,   B 4:1,   C 2:1,   D 1:1,   E 1:4.

**25.** A 120 W, 240 V lamp has a resistance when working of

A 480 ohms,   B 120 ohms,   C 60 ohms,   D 20 ohms,   E 2 ohms.

**26.** When a downward current flows in a straight vertical conductor, the direction of its magnetic field at a point due north of the wire is

A upward,   B north,   C south,   D west,   E east.

**27.** The mass of an atom depends on the number of

A neutrons,   B electrons,   C protons,   D neutrons plus protons,   E α-particles

## Section III

**Directions.** For each of the questions below, *one* or *more* of the responses given are correct. Decide which of the responses is (are) correct. Then choose

    A if 1 and 2 and 3 are correct.
    B if 1 and 2 only are correct.
    C if 2 and 3 only are correct.
    D if 1 only is correct.
    E if 3 only is correct.

*Summary of directions*

| A | B | C | D | E |
|---|---|---|---|---|
| 1, 2, 3, | 1, 2 | 2, 3 | 1 | 3 |
| only | only | only | only | only |

*Questions 28–43*

**28.** Mercury, and not water, is used in a barometer because
    1. Mercury has a high density.
    2. Mercury has a low vapour pressure.
    3. Water is a poor conductor.

**29.** Fig. 2 shows five successive positions of two balls falling under gravity, one dropping vertically and the other simultaneously projected forward. The diagram shows

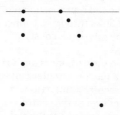

Fig. 2

    1. The vertical acceleration is non-uniform.
    2. The velocity in a horizontal direction is constant.
    3. The vertical motion is that of constant acceleration.

**30.** In any machine
    1. The velocity ratio is usually greater than the mechanical advantage.
    2. The efficiency = mechanical advantage/velocity ratio.
    3. The velocity ratio is independent of friction.

**31.** The following phenomena show that molecules have forces of attraction between them.
    1. Surface tension in liquids.
    2. Elasticity of metals.
    3. Most gases cool on expansion into a vacuum.

**32.** The following phenomena show that molecules are in rapid motion
    1. A small oil drop spreads into a large circular patch on water.
    2. Particles of smoke are seen to move haphazardly in air.
    3. Bromine gas spreads quickly into a vacuum.

**33.** An object is in stable equilibrium if
    1. its potential energy is a minimum.
    2. its potential energy is a maximum.
    3. the centre of gravity is lowered when the object is slightly displaced.

**34.** Wet washing can dry quickly when hung on a line outside the house
   1. if the barometric pressure is low.
   2. because maximum surface area is exposed to the air.
   3. on a windy day.

**35.** A pressure cooker
   1. depends on the rise in boiling point of water with decrease in pressure.
   2. would not be useful for people living at high altitudes.
   3. depends on the increase in the boiling point of water with increased pressure.

**36.** A concave-type mirror is used for
   1. a shaving mirror.
   2. an astronomical large telescope.
   3. a dentist's mirror.

**37.** The accommodation of the eye is
   1. due to a change in the size of the pupil.
   2. produced by movement of the retina.
   3. produced by ciliary muscles.

**38.** In connection with waves,
   1. X-rays and light rays are both electromagnetic waves.
   2. sound waves and X-rays have similar wavelengths.
   3. sound waves can not be diffracted.

**39.** In a moving coil galvanometer
   1. the turning effect on the coil is proportional to the current squared.
   2. the radial magnetic field produces a uniform scale.
   3. the turning effect on the coil is proportional to the current.

**40.** When an electric current flows through a copper voltameter with copper electrodes,
   1. the concentration of the solution is constant.
   2. the anode loses copper.
   3. the copper deposited on the cathode is due to copper ions.

**41.** When a current flows in a narrow circular coil with its plane pointing north–south,
   1. the lines of force at the centre point east–west.
   2. the magnetic field is uniform at the centre.
   3. the magnetic field is uniform outside the coil.

**42.** When a vertical electron beam passes through a uniform horizontal magnetic field
   1. the beam is deflected into a circular path.
   2. a horizontal force acts on the one beam.
   3. a vertical force acts on the beam.

**43.** When a uranium nucleus $^{235}_{92}U$ emits an $\alpha$-particle $^{4}_{2}He$, the nucleus remaining has
   1. an atomic number of 90.
   2. an atomic mass of 231.
   3. an atomic mass of 235.

## Section IV

**Directions.** Each question below consists of an *assertion* (statement) in the left hand column and a *reason* in the right hand column. Choose one answer from A to E according to the following.

   A  if both assertion and reason are true statements and the reason is a *correct explanation* of the assertion,

   B  if both assertion and reason are true statements but the reason is NOT a correct explanation of the assertion.

   C  if the assertion is true but the reason is a false statement,

   D  if the assertion is false but the reason is a true statement,

   E  if both assertion and reason are false statements.

*Summary*

|   | Assertion | Reason |   |
|---|-----------|--------|---|
| A | true | true | Reason is a correct explanation |
| B | true | true | Reason is NOT a correct explanation |
| C | true | false | |
| D | false | true | |
| E | false | false | |

*Questions* 44–58

ASSERTION

REASON

**44.** Right-angle 45° prisms are used in prism binoculars.

Clear images are then seen.

**45.** Red light and infra-red light may both produce interference phenomena.

Waves can interfere with each other.

**46.** A loud note has a high frequency.

Loudness is proportional to pitch.

**47.** Sea breezes are natural convection currents.

In daytime the air above the land rises and its place is taken by cooler air.

**48.** When an object is placed 20 cm from a converging lens of focal length 10 cm, an image is formed beside the object.

The radius of curvature of a lens is twice its focal length.

**49.** In conduction through a solid, the heat is carried by the molecules which move from one end to the other of the solid.

The molecules gain energy when heated.

**50.** The earth moves in an elliptical orbit round the sun.

The force on the earth due to the sun is constant while the earth moves.

**51.** A steel needle may be floated on water.

The upthrust on an object immersed in water = the weight of water displaced.

**52.** When the elastic limit is not exceeded the extension of a wire is proportional to the load.

In small displacements, the force of attraction between molecules is proportional to the displacement.

**53.** When an object rests on an inclined plane, the reaction is equal to the weight.

Action and reaction are equal and opposite.

**54.** Inside a spacecraft in orbit, objects appear 'weightless'.

The force on objects in circular motion is directed towards the centre.

**55.** The resistance of the tungsten filament of a lamp increases when the lamp is switched on.

Pure metals increase in resistance when their temperature is increased.

**56.** In electrolysis, the mass deposited is proportional to the quantity of electricity passed.

The current is carried by negative and positive ions.

**57.** Two similar batteries in parallel, each of e.m.f. $E$, provide an e.m.f. of $2E$.

The e.m.f. does not depend on the nature of the chemicals used.

**58.** The following nuclear change can occur when an α-particle is emitted from a nucleus X and another nucleus Y is formed:

$$^{202}_{86}X \rightarrow\ ^{198}_{84}Y + ^{4}_{2}He$$

The total mass is constant but the total charge is not constant.

## Section V

**Directions.** These groups of questions deal with practical situations. Each situation is followed by questions with five suggested answers. Select the *best* answer for each question.

*Questions* 59–62

A ball of mass 0·5 kg, released from a height of 20 m above the ground, rebounds to a height of 5 m. Assuming $g = 10$ m/s² and the ball is in contact with the ground for 0·1 second,

**59.** the velocity just before striking the ground is

     A 40 m/s,    B 20 m/s,    C 15 m/s,    D 10 m/s,    E 5 m/s.

**60.** the force on the ball as it falls is

     A 20 N,    B 10 N,    C 7·5 N,    D 5 N,    E 0·5 N.

**61.** the force acting on the ball on impact with the ground is

     A 250 N,    B 200 N,    C 150 N,    D 20 N,    E 10 N.

**62.** the loss of energy due to impact is

     A 5 J,    B 10 J,    C 25 J,    D 50 J,    E 75 J.

*Questions* 63–66

Fig. 3 shows two rays M and N from a *very distant* object incident on the objective lens O of an astronomical telescope. The image $I_1$ of the object is formed at Q. As shown, the rays which form the image $I_1$ go on to the eyepiece lens E and this lens forms the final image seen at $I_2$ by the eye near the eyepiece. If the focal length of the lens O is 100 cm and that of the lens E is 5 cm, then

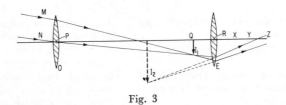

Fig. 3

**63.** the distance PQ is

     A 500 cm,    B 200 cm,    C 105 cm,    D 100 cm,    E 5 cm.

**64.** the final image $I_2$ is

     A erect,    B inverted,    C larger than the object viewed,    D real,    E 10 cm from the eyepiece.

**65.** the best position for the eye on using the telescope is at

     A Q,    B R,    C X,    D Y,    E Z.

**66.** If the final image $I_2$ is formed 25 cm from the eyepiece, focal length 5 cm, the image $I_1$ is formed at a distance from the eyepiece of

     A $4\frac{1}{6}$ cm,    B 5 cm,    C $6\frac{1}{4}$ cm,    D $12\frac{1}{2}$ cm,    E 20 cm.

*Questions* 67–70

In Fig. 4, a diode valve X is in parallel with a resistor Y. A voltage supply Z, with terminals H and K, is connected across the arrangement. Under suitable conditions, X and Y each conducts a *medium* current when a battery or an a.c. voltage is applied. If the two medium currents flow in the same direction in one part of the circuit, a *high* current is obtained there; if they flow in the opposite direction, *zero* current is obtained.

With the voltage supply that given in questions 67–70, write down the magnitudes of the currents each time at the places 1, 2 and 3 in the circuit of Fig. 4. Give your answer as *high* or *medium* or *zero*.

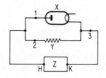

Fig. 4

**67.** A battery connected with its positive pole to H and its negative pole to K.

**68.** A battery connected with its negative pole to H and its positive pole to K.

**69.** An a.c. voltage connected between H and K.

**70.** A battery, in series with an a.c. voltage of lower maximum p.d. than the p.d. of the battery, connected between H and K; the negative pole of the battery is on the same side as H and the positive pole is on the same side as K.

## PAPER 2 (Time: 75 min)   (ANSWERS ON p. 770)

### Section I

**Directions.** The group of questions below consists of five-lettered headings followed by a list of numbered questions. For each numbered question select the heading which is most closely related to it. Each heading may be used once, more than once, or not at all.

*Questions 1–4*
Fig. 5 shows five graphs of quantities in heat, A, B, C, D, and E.

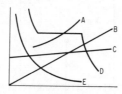

Fig. 5

Which of these best represents the variation of

**1.** the volume of a fixed mass of gas with temperature in °C, the pressure being constant.

**2.** the volume of a fixed mass of gas with pressure, the temperature being constant.

**3.** the boiling point of a liquid with external pressure.

**4.** the temperature of a pure liquid with time before and after freezing.

*Questions 5–8*
The following are five optical instruments:

    A astronomical telescope,  B projection lantern,  C simple microscope,  D prism binoculars,  E opera glasses.

Which of these

**5.** uses a diverging lens.

**6.** produces a magnified real image.

**7.** produces a magnified virtual image of a close object.

**8.** uses internal reflection.

*Questions 9–12*

The following are five electrical conductors or components:

A  copper wire,  B  Manganin wire,  C  Nichrome wire,  D  diode valve,  E  carbon lamp filament.

Which of these

**9.** has a fairly constant resistance as its temperature rises and does not oxidise at high temperatures.

**10.** decreases in resistance as its temperature rises.

**11.** increases in resistance as its temperature rises.

**12.** has a high or low resistance depending on the direction of the potential difference across it.

## Section II

**Directions.** Each of the questions or incomplete statements in this section is followed by five suggested answers. Select the *best* answer in each case.

*Questions 13–27*

**13.** The acceleration of a moving object may be found from

A  the area under its velocity–time graph.

B  the slope of the velocity–time graph.

C  the area under its distance–time graph.

D  the slope of the distance–time graph.

E  the slope at the peak of its distance–time graph.

**14.** If the solid cube X is just tilted about O by the force $F$ in Fig. 6, the weight of the cube is

A  1 N,  B  5 N,  C  10 N,  D  15 N,  E  20 N.

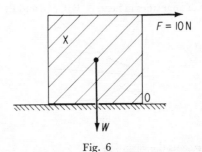

Fig. 6

**15.** A glass stopper weighs 40 gf in air, 30 gf in water, and 28 gf in brine. The density of the brine is

A  0·93 g/cm³,  B  0·7 g/cm³,  C  1·07 g/cm³,  D  1·20 g/cm³,  E  1·40 g/cm³.

**16.** A force of 10 N, acting continuously, increases the kinetic energy of an object from 20 J to 60 J. The distance moved by the object is then

A  400 m,  B  200 m,  C  60 m,  D  20 m,  E  4 m.

**17.** The bulb of a mercury thermometer contains 0·5 cm³ of mercury up to the 0°C graduation. Assuming the expansivity of mercury relative to the glass is 0·0002/K, the volume of the capillary tube between 0°C and 100°C is

A  0·25 cm,  B  0·05 cm³,  C  0·01 cm³,  D  0·001 cm³,  E  0·0005 cm³.

**18.** The specific latent heat of steam is 2300 J per g and the specific heat capacity of water is 4·2 J/g K. When 2 g of steam at 100°C condenses to water at 40°C, the heat given up is

A  5104 J,  B  4936 J,  C  4636 J,  D  4684 J,  E  4600 J

**19.** When a gas is compressed at constant temperature, the gas molecules
A move faster so that the pressure is increased.
B move at the same speed so that the pressure is unchanged.
C gain kinetic energy.
D increase slightly in size.
E make more impacts per second on the walls of the container.

**20.** If an object is 30 cm from a concave mirror of focal length 15 cm, the image will be
A erect, B the same size as the object, C diminished, D virtual, E twice the object size.

**21.** When total internal reflection is just about to occur at an air-water boundary as the incident angle is increased from zero, the refracted ray then
A travels in the water at 90° to the normal.
B travels in air along the normal.
C makes the critical angle with the normal in air.
D travels in the air at 90° to the normal.
E makes an angle of 90° with the incident ray.

**22.** A vibrating string has a tension of 40 N and produces a note of 200 Hz when plucked in the middle. When the length of string is unaltered and the tension is increased to 160 N, the frequency becomes
A 1600 Hz, B 800 Hz, C 400 Hz, D 200 Hz, E 25 Hz.

**23.** A parallel arrangement of 3 and 6 ohms is placed in series with an 8 ohm resistor. If a p.d. of 30 V is connected across the whole circuit, the current in the 3 ohm resistor is
A 15 A, B 10 A, C 3 A, D 2 A, E 1 A.

**24.** In the electrolysis of copper sulphate solution, 1·98 g of copper is deposited in 100 min by a steady current. If the electrochemical equivalent of copper is 0·000 33 g per coulomb, the current is
A 6 A, B 3 A, C 1 A, D 0·5 A, E 0·1 A.

**25.** In an iron-cored transformer, the number of secondary turns is 100 and the number of primary turns is 2000. When 240 V a.c. is connected from the mains, the output is
A 240 V, B 12 V, C 100 V, D 20 V, E 4800 V.

**26.** A horizontal electron beam passes between two parallel horizontal plates X and Y with X above Y. When a high tension battery is connected with its negative pole to X and its positive pole to Y, the beam is deflected
A upwards, B sideways, C downwards, D upwards at 45° to the horizontal, E downwards at 45° to the horizontal.

**27.** A nucleus of an element X which is $^{202}_{84}X$ emits an α-particle and then a β-particle. The final nucleus formed has an atomic number of
A 83, B 82, C 198, D 200, E 80.

### Section III

**Directions.** For each of the questions below, one or more of the responses given are correct. Decide which of the responses is (are) correct. Then choose
A if 1, 2 and 3 are correct.
B if 1 and 2 only are correct.
C if 2 and 3 only are correct.
D if 1 only is correct.
E if 3 only is correct.

*Directions summarised*

| A | B | C | D | E |
|---|---|---|---|---|
| 1, 2, 3 | 1, 2 | 2, 3 | 1 | 3 |
| only | only | only | only | only |

*Questions* 28–42

**28.** A vertical string, suspended from a fixed point, has a small mass swinging to-and-fro at the other end.

   1. The potential energy of the mass is a minimum in the middle of the swing.

   2. Its kinetic energy is a maximum in the middle of the swing.

   3. The sum of the potential and kinetic energies is constant throughout the swing.

**29.** Archimedes' Principle can be used to support or deduce the following statements:

   1. A vacuum is needed at the top of a mercury barometer.

   2. Hydrometers have narrow stems to make them more sensitive.

   3. The relative density of a solid is the ratio of the weight in air to the apparent loss in weight in water.

**30.** When a ladder rests in equilibrium with one end against a smooth wall and the other on rough ground,

   1. The force $F$ at the ground, the reaction $R$ at the wall, and the weight $W$ of the ladder all pass through one point.

   2. The weight $W$ is equal and opposite to the vertical component of $F$.

   3. The reaction $R$ is equal and opposite to the force $F$.

**31.** The pressure exerted at a place X below a liquid,

   1. does not depend on the magnitude of the area placed at X.

   2. depends on the liquid density.

   3. depends on the direction of the force at X.

**32.** A rough measure of the size of a molecule can be found by

   1. measuring the radius of the circle into which a known small volume of oil spreads on water.

   2. measuring the mass of hydrogen in a 1-litre container.

   3. using a microscope to measure the diameter of sand grains.

**33.** In an experiment to verify Charles' law for gases, the following are constant

   1. pressure.

   2. mass of gas.

   3. volume.

**34.** In an experiment to measure the specific latent heat of ice by mixtures

   1. the ice must first be dried before adding it to the water.

   2. The mass of ice used is weighed on filter paper on a balance.

   3. ice is added until no more is melted.

**35.** When a mass of liquid is heated in a flask

   1. its density normally decreases.

   2. the liquid level is usually seen to fall at first.

   3. the kinetic energy of its molecules increases.

**36.** When an object is placed 20 cm from a converging lens of focal length 15 cm,

   1. a virtual image is obtained.

   2. the magnification is 3.

   3. an inverted image is formed.

**37.** When water waves travel to a shallower region,

   1. the frequency changes.

   2. the wavelength is unchanged.

   3. the velocity decreases.

**38.** A 60W and a 120W domestic lamp are used on the 240V mains.

   1. The filament of the 60W lamp has a higher resistance than that of the 120W lamp.

   2. The total current through the mains is 1 ampere.

   3. The p.d. across each lamp is 120V.

**39.** A coil of wire, joined to a galvanometer, is moved quickly away from a magnet pointing along its axis.

1. There is a change in the flux (lines of force) linking the coil.
2. An induced e.m.f. is obtained.
3. The magnitude of the induced current is increased if the coil and magnet are moved together.

**40.** A horizontal straight wire is held above and at right angles to a magnetic compass needle. When a current is passed into the wire,
1. the magnetic field due to the current is perpendicular to the needle.
2. the needle is not deflected.
3. the magnetic field due to the current is parallel to the needle.

**41.** In radioactivity,
1. α-particles have relatively a very small mass compared to an electron.
2. γ-rays are not very penetrating.
3. β-particles have relatively a very small mass compared to an atom.

**42.** In a diode valve
1. the current is carried by electrons.
2. the current is carried by negative charges.
3. the valve conducts when the anode is positive relative to the cathode.

## Section IV

**Directions.** Each question below consists of an *assertion* (statement) in the left hand column and a *reason* in the right hand column. Choose one answer from A to E according to the following:

A   if both assertion and reason are true statements and the reason is a *correct explanation* of the assertion.

B   if both assertion and reason are true statements, but the reason is NOT a correct explanation of the assertion.

C   if the assertion is true, but the reason is a false statement.

D   if the assertion is false, but the reason is a true statement.

E   if both assertion and reason are false statements.

*Summary*

| | Assertion | Reason | |
|---|---|---|---|
| A | true | true | Reason is a correct explanation |
| B | true | true | Reason is NOT a correct explanation |
| C | true | false | |
| D | false | true | |
| E | false | false | |

*Questions 43–58*

| ASSERTION | REASON |
|---|---|
| **43.** A simple microscope consists of a converging lens of short focal length. | The image must be erect. |
| **44.** For a constant mass of gas at constant temperature, a graph of pressure against volume is a straight line passing through the origin. | Boyle's law is true. |
| **45.** Bright and dark bands are seen in a Young's experiment. | The bands are formed by interference of reflected rays. |
| **46.** A pool of water looks shallower than is actually the case. | Light is sometimes refracted towards the normal on entering air from water. |
| **47.** The latent heat of fusion of ice is greater than the latent heat of vaporisation of water. | The molecules are farther apart in a vapour than in a solid. |

**48.** The shorter the length of a pipe, the higher is the pitch of the note obtained.

A stationary wave is produced in the pipe.

**49.** A falling object has a constant momentum.

The velocity is proportional to the distance fallen.

**50.** A ball-bearing in the middle of a saucer is in a neutral equilibrium.

Neutral equilibrium occurs when the height of the C.G. remains constant on displacement.

**51.** The acceleration due to gravity above the earth increases with the height.

The force due to gravity is proportional to the height.

**52.** Neglecting frictional forces, the total potential and kinetic energy of the bob of a pendulum is constant throughout the swing.

The P.E. and K.E. always have the same value throughout the swing.

**53.** A smoke particle moves erratically in air.

The force due to the bombarding air molecules is an irregular one.

**54.** A simple electric motor works better when a low a.c. voltage is connected.

A current-carrying coil rotates in a magnetic field perpendicular to its axis of rotation.

**55.** An electron beam is unaffected by a magnetic field parallel to its direction.

The force on the beam is perpendicular to the field.

**56.** A 500 W lamp uses 1 kilowatt-hour in 2 hours.

Energy used $= IV/t$, where $I$ is the current, $V$ is the p.d., and $t$ is the time.

**57.** A simple dynamo produces a greater e.m.f. with a stronger magnet.

The induced current opposes the motion producing it.

**58.** The magnetic field lines inside a solenoid are circular.

The magnetic field inside a solenoid is non-uniform.

## Section V

**Directions.** These groups of questions deal with a practical situation. Each situation is followed by questions with five suggested answers. Select the *best* answer for each question.

*Questions 59–62*
In Fig. 7, a hydraulic press P is used to raise a load of 1000 kgf, which is 10 000 N. A force $F$ of 25 N is applied at the end of a lever pivoted at O to just raise the load.

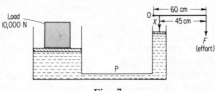

Fig. 7

**59.** The force $X$ applied to the press is then
    A 1500 N,   B 1125 N,   C 100 N,   D $33\frac{1}{3}$ N,   E $13\frac{1}{3}$ N.

**60.** If the load rises 0·1 cm when $X$ moves down 20 cm, the work obtained is
    A 20 000 J,   B 10 000 J,   C 2000 J,   D 10 J,   E 5 J.

**61.** The work done by the force $X$ in moving the load 0·1 cm is
    A 20 J,   B 25 J,   C 75 J,   D 80 J,   E 100 J.

**62.** Excluding the lever, the efficiency of the press is
    A 90%,   B 80%,   C 75%,   D 60%,   E 50%.

*Questions 63–66*

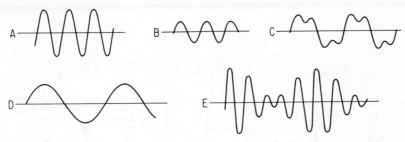

Fig. 8

Identify the following sound waves from the sketches in A to E (Fig. 8):

**63.** note from musical instrument.

**64.** soft note.

**65.** bass note.

**66.** beat note obtained on sounding two tuning forks of close frequency.

*Questions 67–70*

In Fig. 9, RT is a conductor loosely suspended from pivots at X and Y. X and Y are joined to terminals at L and M.

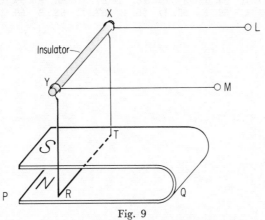

Fig. 9

**67.** When a battery is joined with its positive pole to L and its negative pole to M, then RT

   A remains stationary, B moves upwards towards YX, C moves towards Q, D moves towards P.

**68.** The magnet is moved to the right so that RT is between the poles N and S. When the battery is re-connected to L and M, the force on RT.

   A remains zero, B increases, C decreases, D reverses in direction.

**69.** The battery is now disconnected from L and M and a galvanometer is joined between the two terminals. If RT is now pushed towards Q, the galvanometer shows the induced current is

   A zero, B in the direction RT, C in the direction PQ, D in the direction TR.

**70.** When RT is made to swing to-and-fro between P and Q, the induced current is

   A alternating, B zero, C in the direction PQ, D in the direction QP.

# ANSWERS TO MULTIPLE-CHOICE QUESTIONS

### Paper 1 (p. 757)

**1.** D   **2.** B   **3.** C   **4.** A   **5.** B   **6.** E   **7.** D   **8.** A   **9.** D   **10.** C
**11.** A   **12.** E   **13.** C   **14.** D   **15.** A   **16.** B   **17.** D   **18.** C   **19.** B   **20.** D
**21.** B   **22.** C   **23.** A   **24.** C   **25.** A   **26.** E   **27.** D   **28.** B   **29.** C   **30.** A
**31.** A   **32.** C   **33.** D   **34.** C   **35.** E   **36.** A   **37.** E   **38.** D   **39.** C   **40.** A
**41.** B   **42.** D   **43.** B   **44.** B   **45.** A   **46.** E   **47.** A   **48.** E   **49.** D   **50.** C
**51.** B   **52.** C   **53.** D   **54.** B   **55.** A   **56.** B   **57.** E   **58.** C   **59.** B   **60.** D
**61.** C   **62.** E   **63.** D   **64.** B   **65.** E   **66.** A   **67.** (1)M (2)M (3)H
**68.** (1)0 (2)M (3)M     **69.** (1)M (2)M (3)H     **70.** (1)0 (2)M (3)M

### Paper 2 (p. 763)

**1.** C   **2.** E   **3.** A   **4.** D   **5.** E   **6.** B   **7.** C   **8.** D   **9.** C   **10.** E
**11.** A   **12.** D   **13.** B   **14.** E   **15.** D   **16.** E   **17.** D   **18.** A   **19.** E   **20.** B
**21.** D   **22.** C   **23.** D   **24.** C   **25.** B   **26.** C   **27.** A   **28.** A   **29.** C   **30.** B
**31.** B   **32.** D   **33.** B   **34.** D   **35.** A   **36.** C   **37.** E   **38.** D   **39.** B   **40.** C
**41.** E   **42.** A   **43.** B   **44.** D   **45.** C   **46.** C   **47.** D   **48.** B   **49.** E   **50.** D
**51.** E   **52.** C   **53.** A   **54.** D   **55.** C   **56.** C   **57.** B   **58.** E   **59.** C   **60.** D
**61.** A   **62.** E   **63.** C   **64.** B   **65.** D   **66.** E   **67.** D   **68.** B   **69.** D   **70.** A

# BASIC EXPERIMENTS

The following is a list of some of the basic experiments described in the book which may be carried out to illustrate the principles of the topic concerned. The list is not intended to be exhaustive, and teachers will no doubt have many alternative or additional experiments. The reader is recommended to Nuffield booklets for details of numerous experiments in their modern course.

Manufacturers who can supply the main parts of any apparatus not readily available in a school laboratory are:

Griffin & George Ltd, Alperton, Wembley, Middlesex–All apparatus.

M.L.I. (Sales) Ltd, Putney High Street, London, SW15–All apparatus.

White Electrical Instrument Co. Ltd, Spring Lane, Malvern Link, Worcestershire–Electrical instruments.

## MODERN PHYSICS

Teltron Ltd, 32–36 Telford Way, London, W3–Discharge tube, Perrin tube, Diode, Triode, X-ray tube.

Scientific Teaching Apparatus Ltd, Colquhoun House, 27–37 Broadwick Street, London, W1–Leybold fine beam tube.

Panax Equipment Ltd, Holmethorpe Estate, Redhill, Surrey–Radioactivity experiments.

Rainbow Radio (Blackburn) Ltd, Mincing Lane, Blackburn–Unilab Microwave apparatus.

Mullard Educational Service, Mullard House, Torrington Place, London, WC1–Information on Radio circuits and Transistors.

### Matter and Molecules
Density and relative density of solids and liquids, pp. 17–20, 209.
Density of air, p. 22.
Existence of molecules–Brownian motion in gases and liquids, pp. 27, 30.
Estimate of molecular size–oil-film, p. 24.
Motion of molecules–diffusion, p. 29.
Spaces between molecules, p. 24.
Forces between molecules in solids–elasticity experiments, pp. 32–5,
  friction, p. 36.
Forces between molecules in liquids–surface tension experiments, pp. 41–2,
  cohesion and adhesion, p. 44,
  viscosity, p. 45.

**Waves and Wave Effects**

Ripple tank and wave propagation, p. 324.
Measurement of wavelength and velocity, p. 330.
Microwaves, diffraction and interference, pp. 333, 341.
Light waves, diffraction and interference, pp. 334, 340.
Stationary wave, p. 335. Resonance p. 345.
Effect of medium on sound waves, p. 326.

**Optics**

Pin-hole camera, p. 354, shadows, p. 355.
Laws of reflection at plane surfaces, pp. 359, 360.
Image location and size, plane mirror, pp. 362–3.
Reflection at curved spherical surfaces, pp. 367–9.
Concave mirror images, p. 372.
Measurement of focal length, radius of curvature, p. 370.
Concave mirror magnification, p. 377.
Refraction demonstrations, pp. 381–5.
Laws of refraction, p. 385.
Refractive index of glass, p. 390. Apparent depth, p. 388.
Total internal reflection, critical angle, p. 393.
Lenses demonstrations, pp. 404–5.
Converging lens images, p. 408.
Focal length measurements, p. 407.
Magnification by converging lens, p. 410.
Model eye, defects of vision, p. 420.
Compound microscope, p. 425.
Telescope, p. 426.
Dispersion, pp. 432–3.
Recombination of colours, p. 434.
Pure spectrum, p. 436, spectra, p. 438.

**Sound**

Sound vibrations, p. 455. Velocity of sound, p. 460.
Seebeck's wheel and frequency, p. 457.
Sonometer experiments, p. 467.
Resonance tube, pp. 472–3.

**Current Electricity**

Measurement of resistance, pp. 494–6, p. 502.
Resistances in series and parallel, pp. 497–8.
Resistivity, p. 504.
E.m.f. and internal resistance, pp. 518–19.
Joule's laws of heating, p. 529, heating by current, p. 533.
Faraday's first law of electrolysis, p. 546.
Electrochemical equivalent, p. 548.
Calibration of ammeter, p. 549.
Making magnets, p. 556, demagnetization, p. 558.

# INDEX